ATOMIC ABSORPTION AND FLUORESCENCE SPECTROSCOPY

ATOMIC ABSORPTION AND FLUORESCENCE SPECTROSCOPY

G. F. KIRKBRIGHT
Department of Analytical Chemistry,
Imperial College,
London

M. SARGENT
Thornton Research Centre,
Shell Research Limited,
Cheshire, England

1974 ACADEMIC PRESS
LONDON · NEW YORK · SAN FRANCISCO

A Subsidiary of Harcourt Brace Jovanovich, Publishers

ACADEMIC PRESS INC. (LONDON) LTD
24-28 Oval Road
London NW1 7DX

U.S. Edition published by
ACADEMIC PRESS INC.
111 Fifth Avenue,
New York, New York 10003

Library of Congress Catalog Card Number: 74-5647

ISBN: 0-12-409750-2

Printed by William Clowes & Sons Limited
London, Colchester and Beccles

Preface

The decision to prepare this volume was taken at a time (1969) when few texts devoted exclusively to analytical atomic absorption spectroscopy (AAS) had been published, and none concerned specifically with atomic fluorescence spectroscopy (AFS) was available. In the past four years several additional texts on AAS, which also contain sections on AFS, have been written. Despite the appearance of these new volumes, however, we have not been discouraged from preparation of the present volume. It was felt that there was an unfilled need for a general textbook which gave more detailed, though not indigestible, theory of both techniques, and which presented a more exhaustive consideration of instrumentation and practical techniques for the working analytical chemist as well as for the student of analytical chemistry. We recognize that it is dangerous to attempt comprehensive coverage of the theory and practice of techniques at a time when they have been developing rapidly; although it is still possible that the present volume may be rendered rapidly out of date by startling new advances in AAS and AFS, it is the opinion of the authors that while applications of the techniques continue to increase, some stabilization in the fundamentals of the instrumentation and methodology has occurred. Two exceptions which might be mentioned specifically are in the instrumentation and methodology associated with the use of non-flame cells for AAS and AFS and, the application of laser sources in AFS. Atomic fluorescence spectroscopy is considered throughout as a technique which is complementary to AAS and which holds considerable promise for many applications.

The volume deals in detail with the individual components of atomic absorption and fluorescence spectrometers (sources, atom cells, dispersive devices etc.). Sufficient fundamental theory is presented to avoid a misleading and oversimplified approach but the emphasis in the text is placed on the practice of the techniques for analysis. Although the applications literature is not exhaustive the authors feel that sufficient data is given to the reader in search of

AAS or AFS methods for a particular application; he should be able to gain an estimate of the number of alternatives possible and obtain easy access to the pertinent original publications. As mentioned at several points in the text, we consider the advent of the publication 'Annual Reports on Analytical Atomic Spectroscopy', in which comprehensive world-wide coverage of both fundamental and applications literature in both AAS and AFS is presented regularly, has rendered unnecessary any attempt at up-to-date comprehensive coverage of the applications literature in a volume of this type.

During the past decade many workers have made significant contributions to the understanding of the fundamental principles of AAS and AFS and the analytical practice of these techniques. In particular, however, we wish to pay tribute to the work of C.S.I.R.O., Australia, where analytical AAS originated in the work of Alan Walsh and his colleagues, and to Prof. C. Th. J. Alkemade, University of Utrecht, who has so often pointed the way for developments in these techniques and has made many contributions to understanding of their fundamental principles. The extensive contributions of Winefordner and co-workers at the University of Florida, which have so assisted the development of the techniques, must also be acknowledged. Perhaps to a greater extent than for many other instrumental techniques of analysis a very considerable contribution to AAS theory and practice has been made by the instrument manufacturers. In the case of several companies these contributions have been so important and sustained that the work of their employees has received international recognition.

It is a pleasure to acknowledge the encouragement given to us during preparation of this book by Professor Tom West. We wish to express our thanks to the many authors and publishers who have given permission for the reproduction of figures and other material from the original literature. We thank close colleagues for helpful discussions and acknowledge the assistance given by Mike Adams, Dai Bevan, Arthur Ward and P. J. Wilson in collection of much of the applications data given in Chapter 13. We also wish to thank Mike Adams for the preparation of the Subject Index. We are indebted to the Technical Editing staff at Thornton Research Centre, particularly Bob Taylor and Tony Shaw, for help with the preparation of manuscript and diagrams. Many secretaries have been involved with the task of preparation of draft and corrected typescripts. We particularly wish to thank Mrs. Janet Broome and Miss Maureen Coleman for their efforts in this respect. We also should thank our families for their patience and forebearance during the preparation of this book.

G. F. Kirkbright.

M. Sargent.

November, 1973.

Contents

Chapter 5 Spectral Light Sources

Chapter 6 Flames

Chapter 7 Non-flame Absorption and Fluorescence Cells

Chapter 8 Introduction of Liquid Samples into Flame Atom Cells

Chapter 9 Wavelength Selection

Chapter 10 Atomic Absorption and Fluorescence Instrumentation

Chapter 11 Practical Techniques of Atomic Absorption and Fluorescence Spectroscopy

Chapter 12 Interferences

Chapter 13 Analytical AAS and AFS Characteristics of the Elements and Applications Data

Chapter 14 Special Techniques in AAS and AFS

CHAPTER 1

Introduction

1.1 HISTORICAL ASPECTS

The techniques to be described in this book are concerned with one small region of the electromagnetic spectrum: the wavelength range extending from approximately 180 nm to 800 nm and embracing the region of visible light. Within the restrictions imposed by these somewhat arbitrary limits to "the spectrum" it may be said that the history of spectroscopy dates back over two and a half centuries to the discovery of the visible spectrum by Newton.[1] Since that time, spectroscopists at any given period have concentrated their efforts on one of two fields: analytical chemistry or the interpretation of physical and atomic properties. At the present time atomic absorption spectroscopy in particular represents a major area of interest and growth in the former field.

Following Newton's study of the spectrum of the sun was a period of almost one hundred years entirely concerned with *emission* spectroscopy. However, in 1802 Wollaston[2] reported the presence of dark bands in the continuum emission spectrum of the sun and, after a more detailed study by Fraunhofer (1814), Brewster[3] (1820) was able to ascribe them to *absorption* of radiation within the sun's atmosphere. Another forty years passed before Kirchoff and Bunsen[4,5,6] showed that one of these dark bands in the emission spectrum of the sun corresponded exactly to the yellow emission band obtained when sodium vapour is heated in a flame. Their work enabled Kirchoff to enounce the fundamental relationship between emission and absorption spectra: any species that can be excited to emit radiation at a particular wavelength will also absorb radiation at that wavelength. Thus a new technique could be added to the already well-established one of emission spectroscopy and the work of Bunsen and Kirchoff was continued and placed on a sound theoretical basis by the physicists and astronomers of the early twentieth century. Their research, which has been summarized by Mitchell and Zemansky,[7] was mainly performed with atoms at low pressures in enclosed vessels and was primarily concerned with the interpretation of atomic structure and spectra.

Although there were occasional practical applications to the study of elements present in solar and stellar atmospheres, virtually no attempt was made to apply the discoveries to practical chemical analysis; only the determination of mercury vapour in air[8] was reported as a rather exceptional example. This striking contrast with emission spectroscopy, which by the 1920s was an important analytical tool in the form of arc and spark spectrography, must be viewed in the light of two important factors. Firstly, before about 1930 the emission technique was mainly used for qualitative analysis, and there was no advantage to be gained from using absorption spectra for this purpose (in fact the reverse is true, as is witnessed by the present-day applications of the absorption technique). Secondly, even after 1930 the difficulty of making absorption measurements for quantitative analysis was considerably greater than that experienced with emission. Not only was the photographic recording of the absorption spectra more tedious, but the theory seemed to indicate that they would only prove useful for quantitative analysis if observed under very high resolution. Thus, when atomic absorption spectrometry was introduced as an analytical tool by Walsh[9,10] in 1953 and Alkemade and Milatz[11,12] in 1955, emission methods of analysis were firmly established and difficult to replace. However, these workers were able to overcome two major problems which twenty years earlier had seemed almost insuperable. The technique proposed by Walsh of measuring only the peak absorbance at the centre of the spectral line (see Section 3.5.2) removed the practical difficulty of making quantitative absorption measurements with a simple monochromating device, and the availability of photoelectric light detectors of very high sensitivity meant that it was no longer any more troublesome to record absorption rather than emission signals. By presenting this new "simple" absorption technique in combination with modern instrumentation and a high-temperature flame as the atom reservoir, Walsh in particular was able to demonstrate the advantages of atomic absorption spectrometry over emission spectrometry and flame photometry for many elements. Today a wide range of instruments is available commercially, atomic absorption is widely used in a greater range of routine determinations than the arc, spark and simple flame emission techniques, and new publications on atomic absorption spectrometry greatly outnumber those on the emission methods.

The re-emission by an atom of radiation as *fluorescence* after the absorption of light from a suitable source was first reported in 1905 when Wood[13] succeeded in exciting atomic fluorescence of the D lines of sodium vapour. He used an evacuated test tube containing sodium vapour as the atom reservoir and a gas flame containing sodium chloride as the illuminating source. The result was predicted by the classical theory of light in the same way as the resonance of standing waves observed for sound, vibrating strings, etc. Wood, therefore, termed this fluorescence "resonance radiation" and the name is still encountered today for the atomic fluorescence of resonance lines such as the D lines. Similar work was carried out with other readily volatile elements, enclosed cells again being used to contain the atomic vapour, and was mainly concerned with fundamental studies of atomic spectra. This work has been described in some detail by Mitchell and Zemansky[7] and by Pringsheim.[14]

The fluorescence of atoms in flames was first reported in 1923 by Nichols and Howes[15] for calcium, strontium, barium, lithium and sodium in a Bunsen flame. Similar observations were also reported by Badger[16] and by Mannkopff[17] for a number of other volatile elements. Further interest in the subject then appears to have lapsed until the use of atomic fluorescence spectrometry in 1956 by Boers, Alkemade and Smit[18] to study quenching processes in flames. Following this, a further report was given by Alkemade[19] in 1962 on sodium fluorescence in flames together with the suggestion that it might have analytical possibilities. This suggestion was followed up by Winefordner, Vickers and Staab[20,21] who reported the first successful analytical application of atomic fluorescence spectrometry with the determination of zinc, cadmium, and mercury. Since that time the number of publications on the technique has increased rapidly.

1.2 FUNDAMENTALS

It is worthwhile initially to outline the instrumental arrangement most commonly used for analytical absorption and fluorescence measurements, and to define some of the more widely used terms. This should avoid confusion when these terms arise in subsequent chapters.

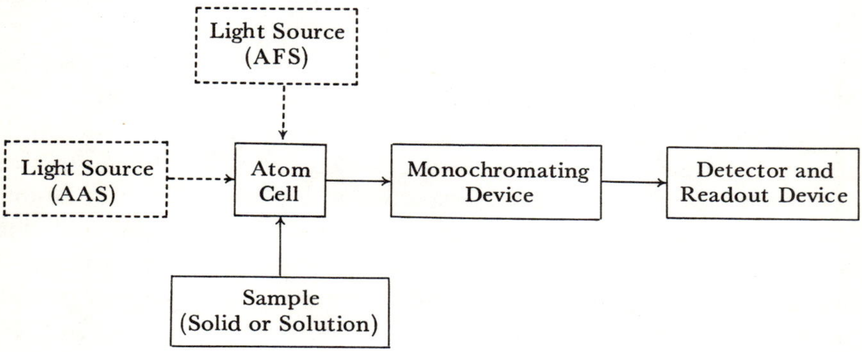

Fig. 1-1. Basic apparatus for analytical atomic spectroscopy.

The basic apparatus required is shown diagrammatically in Fig. 1-1, and a typical commercial atomic absorption spectrophotometer conforming to this arrangement is illustrated in Fig. 1-2. In order to observe atomic absorption or fluorescence signals it is necessary to have a population of free, neutral atoms of the element of interest. Thus, the most important component of the instrument is the *atom cell* or reservoir. This may take a wide variety of forms, ranging from a simple quartz cell at room temperature (for mercury) to a high-temperature flame or plasma with its ancillary equipment. If the atoms contained in this cell are to absorb radiation, and possibly to re-emit a fraction of this as fluorescence, they must be irradiated with light of the same wavelength as that of a strong absorption spectral line of the element that has been atomized. This radiation is obtained from a *light source* which may emit the spectrum of a single element, the spectra of several elements whose common analysis is frequently required or a continuum spectrum (see Section 5.1). As with the atom reservoir, the complexity of the light source and its ancillary equipment can vary considerably.

The light transmitted after the atomic absorption process has occurred, or the fluorescence emission stimulated from the atoms in the reservoir, will often be composed of radiation of both the required wavelength (i.e. the absorption line(s)) and a number of unwanted wavelengths. The latter may simply take the form of stray daylight, or may arise from other spectral lines emitted by the light source or the atom cell. It is usually necessary, therefore, to select the wavelength of the radiation from the atom cell before an attempt

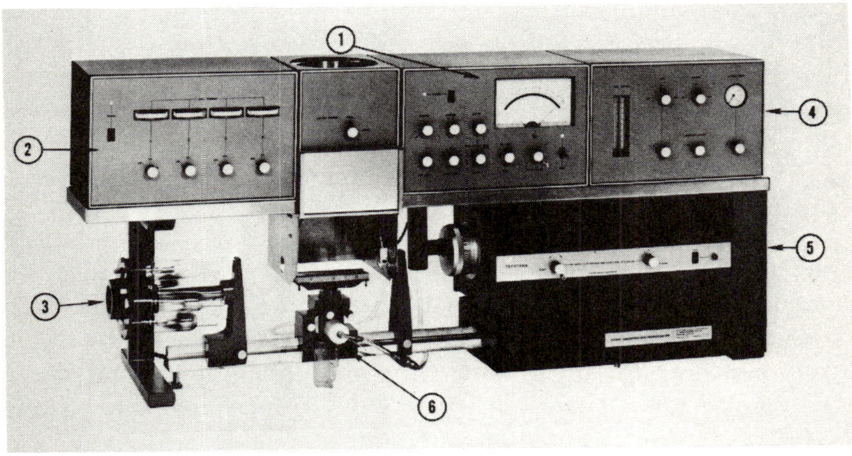

Fig. 1-2. Commercial atomic absorption spectrophotometer. 1, amplifier; 2, lamp power supply; 3, lamp turret; 4, gas control unit; 5, monochromator; 6, burner and sample nebulizer.

is made to measure the absorption or fluorescence signal. This is done with a *monochromating device*, the design of which will depend on the complexity of the instrument and the type of work to be carried out with it. For simple instruments intended only for measurements with one or two elements giving single strong spectral lines at wavelengths that can be detected a simple optical filter may suffice. On the other hand, most commercial instruments designed for maximum versatility utilize relatively complex diffraction grating monochromators (see Chapter 9).

Whatever monochromating device is employed, it must be placed in the correct position with respect to the light source and the atom cell. In Fig. 1-1 it is shown to lie in the direct optical path from the source to the cell for absorption measurements (so that the amount of radiation absorbed is easily determined) and at 90° to this optical path for fluorescence measurements (so that fluorescence emission may be detected without interference from any radiation received direct from the light source). Although these configurations are frequently adhered to in commercial instruments for the *optical path* between light source and monochromating device, it must be remembered that the arrangement of the individual components may not be as straightforward as in Fig. 1-1; in many cases the light beam may be deflected by several mirrors or prisms so that a practically convenient instrument design is obtained. Less frequently, the optical configuration itself may vary. This is particularly true with fluorescence measurements when the relative positions of the light source and monochromating device are not of great fundamental importance provided that radiation cannot pass directly between them. Thus, if it is necessary to irradiate the atom cell with several light sources, perhaps for the simultaneous analysis of several elements, the 90° optical configuration may be changed.

Once the absorption or fluorescence signal has been obtained it must be recorded and measured. The device used to convert the radiation constituting the signal into a form that can be easily measured (and recorded if necessary) is known as the *detector*. The traditional detector is, of course, the photographic plate or film; as mentioned earlier, however, this is not very convenient for quantitative measurements and quite impracticable for absorption measurements. Hence almost all modern instruments use photoelectric devices (see Chapter 10). These convert the optical signal to an electrical signal so that a *readout device* is required in order to measure the signal. Again, this may vary considerably in complexity to range from a simple meter to measure the voltage generated by a photoelectric cell to elaborate electronic assemblies using computing techniques.

In addition to the instrumental components discussed above other more general terminology is in use the early definition of which is worthwhile. The analytical techniques utilizing the measurement of atomic absorption and fluorescence are usually referred to as respectively *atomic absorption spectrometry, spectroscopy or spectrophotometry* (*AAS*) and *atomic fluorescence spectrometry, spectroscopy or spectrophotometry* (*AFS*). The instruments used for the measurements are referred to as *spectrophotometers or spectrometers*. These different derivatives of spectr- are often used quite interchangeably in the literature but there is usually little danger of confusion arising. The observation of the presence of an element by its absorption or fluorescence spectrum is referred to as its *detection* whereas the quantitative analysis is known as its *determination*. The actual element detected or determined is frequently termed the *analyte* and it is present in the *analytical sample*, the actual composition of which may be referred to as the *sample matrix*. The word *sample* is also generally used for the solid or solution added to the atom cell, even though this may have been obtained by complicated treatment of the original analytical sample. Thus, care is sometimes necessary to avoid confusion, especially when mention is made of the *analyte concentration* in the sample. As absorption measurements in particular are made as the difference between two signals it is usually necessary to have a sample giving zero signal, that is, a *blank* which contains no analyte element. Finally, other species present in the sample or the atom cell can sometimes cause an erroneous determination of the analyte; these are known as *interferences* and are revered by analytical chemists.

The discussion of AAS and AFS necessitates frequent mention of the emission technique, and much of the nomenclature given above applies to this also. The corresponding names *atomic emission spectrometry, spectroscopy or spectrophotometry* (*AES*) are widely used; it is also common to find the technique referred to as *flame emission spectrometry, spectroscopy or spectrophotometry* (*FES*), as the flame is a widely used atom cell (in the present discussion we shall exclude traditional arc or spark emission spectrography from our definition of AES).

1.3 COMPARISON OF TECHNIQUES

It is readily recognizable from the preceding pages that essentially the same equipment and techniques may be applied to both AAS and AFS. In fact, apart from the spectral light sources, this may also apply to AES. Thus, in an analytical laboratory where inorganic trace analysis is frequently carried out by atomic spectroscopy, sufficient equipment and expertise may be available for all three techniques. It is then possible to select one of the techniques on its own merits as the most suitable for a particular determination. In view of this, it is intended to make a brief comparison of techniques by discussion of their individual advantages and disadvantages, and then to make a general comparison of all three techniques with other instrumental methods of inorganic analysis.

1.3.1 Comparison of AES, AAS and AFS

It is important first to emphasize the opinion of the authors implied above that the three techniques are complementary and no one has outstanding advantages over the other two. This is true whether comparison is based on the criterion of "limit of detection" or that of ease and convenience of analysis. First, at the time of writing there exists for each technique a group of elements for which it may provide the most sensitive method of analysis. There also remain many elements whose determination may be undertaken with equal sensitivity by two or even all three techniques. Secondly, each technique has its own instrumental and experimental advantages and disadvantages. Many are common to more than one technique, and a direct comparison may be made by detailing favourable and unfavourable points for each technique. This approach should assist when a decision is to be made concerning the technique best suited to a particular analytical service.

Advantages of AES

1. No spectral light sources are required.
2. For many elements a wide choice of sensitive analytical lines exists; many of these lines may lie in the visible region of the spectrum where intensity measurements are less complicated than in the ultraviolet region.
3. The technique can provide flexibility in dealing with a wide range of sample concentrations.
4. Qualitative as well as quantitative analysis is readily carried out.
5. Simultaneous multi-element analysis is possible.

Disadvantages of AES

1. Large changes in the intensity of the analytical signal result from small temperature changes in the atom cell.
2. Background and other line or band spectra emitted from the atom cell may cause greater interference than in the other two techniques.
3. Considerable operator skill is required routinely to obtain reliable results.
4. A high quality monochromator is essential if the technique is to be reliable and versatile.

Advantages of AAS

1. The instrumentation is simple to set up and operate because of the useful wavelength and intensity reference provided by the background light source.

2. The technique is easily automated.
3. Double-beam operation is readily arranged to provide for reduction of the effect of variation in intensity of the light source.
4. Sharp-line light sources provide better spectral selectivity than is usually obtainable with AES.
5. The analytical signal in AAS for many elements is less sensitive to temperature changes in the atom cell than it is in AES.

Disadvantages of AAS

1. A large number of relatively expensive spectral light sources are usually required. High sensitivity cannot be obtained when continuum or broad-line light sources are used with low resolution commercial instrumentation.
2. Many of the most useful absorption lines are quite far into the ultraviolet region of the spectrum where measurement is more difficult; for most elements only a few analytical lines are available.
3. When spectral interferences do occur they are less apparent than in AES and can therefore cause a more serious chance of error.
4. It is relatively difficult to obtain a long range of concentration over which linear calibration is possible.

Advantages of AFS

1. Sensitivity attainable is controlled by the intensity of the light source employed.
2. The technique is less sensitive to temperature changes in the atom cell than AES is.
3. Much of the equipment required can be less complex than that needed for AES or AAS.
4. Continuum or fairly broad-line light sources can be used with a less serious loss of sensitivity than for AAS.
5. High sensitivity is available far into the ultraviolet region of the spectrum where AES is insensitive.

Disadvantages of AFS

1. A spectral light source is required, usually one for each element, as in AAS.
2. Quenching of the fluorescence signal is greatly affected by the gas species present in the atom cell.
3. The very hot flames needed as atom cells for the determination of elements such as the refractory metals and to avoid some chemical interferences favour AES rather than AFS.

4. Scattering of light from the source may be more difficult to deal with than in AAS.
5. Although AFS usually gives a wider linear concentration range than does AAS, the extension is only on the low side of the range.
6. AFS is much less sensitive than AES for elements giving their main absorption and fluorescence lines in the visible region.

To summarize, AAS is probably the technique best suited to rapid, routine operation for analysis by relatively unskilled operators. AES is the more versatile method for use by a skilled analytical chemist in a laboratory where small numbers of non-routine determinations are required, often at short notice. AFS can provide extremely high sensitivities for certain elements less easily determined by the other two techniques, but is often said to combine some of their disadvantages.

1.3.2 Comparison with Other Instrumental Methods of Inorganic Trace Analysis

In the comparison that follows we shall be more concerned with the advantages of AAS and AFS to an analytical laboratory carrying out a reasonable volume of routine work, rather than with minor points of interest to specialized research laboratories. Perhaps the major advantage of AAS and AFS over most of the comparable techniques is their relative freedom from interferences and the need for extensive sample pretreatment. For electrochemical methods such as polarography and coulometry most types of sample require dissolution followed by a preliminary separation of the analyte to obtain sufficient selectivity. Similar disadvantages apply to molecular absorption spectrophotometry in solution, which in addition also requires the availability of many different organic reagents when a laboratory undertakes a wide range of analyses. As AAS in particular can also offer greater speed and simplicity of measurements than can these techniques, one of its early successes has been in the field of clinical and biochemical analyses where they had formerly been widely applied.

Even with instrumental techniques such as radiochemistry, automated arc spectrometry and X-ray spectrometry, AAS is able to compete in terms of lower capital cost and greater speed of analysis. This is particularly true at the present time for the former applications of these techniques where a moderate number of analyses of varying materials was required. However, the automation of AAS on a commercially viable scale has already begun and it may soon rival the large, automatic optical and X-ray spectrometers used extensively in metallurgical and other industries. Although the use of solid samples for AAS on this scale is not feasible at present it should be remembered that the direct analysis of solid samples requires the use of metallic standards to obtain freedom from matrix effects. Thus, there may be no serious disadvantage in returning to dissolution of samples with AAS determination in conjunction with standards prepared from aqueous solutions of pure reagents, particularly if the dissolution process can also be automated.

AAS and AFS exhibit some disadvantages compared to other techniques. These include the difficulty of determining major sample constituents without a large degree of dilution (with its attendant errors) and of measurement of non-metallic constituents by direct methods. At the present time a number of specific metals and sample types also present difficulties for routine analysis. Although some of the less widely applied methods, such as mass spectrometry and activation analysis, do not suffer these problems to the same extent, it should be remembered that in general they require much wider use of skilled operators. This adds greatly to the cost of the techniques. Furthermore, the current intensification of research as a result of the commercial possibilities of AES, AAS and AFS seems to be removing these problems more quickly than is the case for those inherent in other instrumental methods of inorganic analysis. Thus, it seems probable that the current rate of expansion at the expense of other techniques more traditionally used in analytical service laboratories will continue for some time.

1.4 PREVIOUS BOOKS AND REVIEWS

Until quite recently the number of textbooks on atomic absorption spectrometry was very limited. However, at the time of writing a good selection of books exclusively devoted to AAS have been published, and many review articles concerned with AAS and AFS have appeared. We shall list the books devoted entirely to atomic absorption and fluorescence with a selection of the most useful books dealing with flame photometry generally. The most useful and the most recent of the reviews and bibliographies are also indicated.

Books dealing entirely with AAS (which in some cases contain a short chapter on AFS) have been written by Elwell and Gidley,[22] Ivanov,[23] L'vov,[24] Robinson,[25] Ramirez-Munoz,[26] Slavin[27] and

Rubeska and Moldan.[28] Recently, L'vov has produced a greatly enlarged English edition[29] of his earlier monograph and this probably forms the most thorough account to-date of Russian work, with particular reference to the use of the graphite cuvette technique. Books concerned with specialized applications of AAS have also been written by Rousselet[30] (biology), Angino and Billings[31] (geology) and Christian and Feldman[32] (agriculture, biology and medicine). The most useful of the books dealing with general flame spectrometry are those by Herrmann and Alkemade,[33] Mavrodineanu and Boiteux,[34] Pungor,[35] Mavrodineanu[36] and Dean and Rains.[37] The last listed are the first two parts of a three-volume work which contain chapters by individual authors on theoretical and fundamental topics. At the time of writing no books devoted entirely to AFS have been published. The most useful and generally applicable of the many reviews concerned with AAS are those by Gilbert,[38] Walsh,[39] Mavrodineanu,[40] Slavin,[41] Kahn,[42] Koirtyohann,[43] Price,[44] West,[45] Willis[46] and Fuwa.[47] Mention must also be made of the papers presented at the Atomic Absorption Symposium held at the 1968 meeting of the ASTM. These have been published in book form,[48] as have also the plenary lectures from the 1969 AAS Symposium at Sheffield, England[49] and those of the Paris AAS Symposium in 1971.[50] More specialized reviews of AAS are also available (mainly concerned with particular applications); two recent general ones worthy of mention are by Warn[51] (comparison of AAS with other instrumental methods) and Pickett and Koirtyohann[52] (comparison of AAS with AES). The lesser current popularity of atomic fluorescence spectrometry as an analytical technique is reflected in the much smaller number of reviews published to date. Mention must be made of those by Winefordner and Mansfield,[53,54] Demers,[55] Ellis and Demers,[56] Price,[57] West,[58] Winefordner and Smith,[59] Smith[60] and Winefordner and Elser.[61] Winefordner[62] has also reviewed the relative merits of AAS and AFS.

Although bibliographies are of less immediate use than reviews, it is worth recording some of the most extensive ones. Mavrodineanu[63] has assembled 353 references on AAS with a further 87 on hollow cathode lamps, covering the period from the earliest work on the subject up to 1965. This comprehensive bibliography contains principally references on flame emission spectroscopy, many of which contain material relevant to fundamental work in AAS and AFS. It has already been extended in similar form to cover the period ending December 1968, additional earlier references also being included.[64] Of recent bibliographies, the most useful is that published annually in Perkin-Elmer's *Atomic Absorption Newsletter* by Slavin[65] and available in complete form from the manufacturer.[66] Other manufacturers have produced similar bibliographies on AAS[67,68] and AFS.[69] Another up-to-date source of references on AES, AAS and AFS is the new journal entitled *Atomic Absorption and Flame Emission Spectroscopy Abstracts.*[70] Although this was initiated only in January 1969, early volumes contain many references to publications of several previous years. Finally, the reader is directed to the biannual reviews published in *Analytical Chemistry*. Those by Margoshes and Scribner on emission spectrometry published in 1964,[71] 1966[72] and 1968[73] contain a number of references on AAS and AFS while those on flame spectrometry by Winefordner and Vickers for 1970 and 1972[74] provide the most extensive bibliographical review to date on AFS, together with an excellent section on AAS. These articles provide comprehensive notes of recent developments in the field and their high standard will no doubt be maintained in future years.

To conclude this section we include details of important works which have come to our attention since the list of books and reviews given above was compiled. *Annual Reports in Analytical Atomic Spectroscopy*, published by the Society for Analytical Chemistry, offers a comprehensive and critical annual review of the whole topic. Reynolds and Aldous[75] have written a practical guide to the experimental techniques of AAS and Price[76] has written a useful text concerned exclusively with AAS. The second volume of the work by Dean and Rains on emission and absorption spectrometry is now available[77] and deals with instrumental components and practical techniques. Parsons and McElfresh[78] have prepared an atlas of spectral lines used in flame spectroscopy.

REFERENCES

1. I. Newton, "Optics, or a Treatise of the Reflections, Refractions, Inflections and Colours of Light". 1st Ed. 1704, reissued by Dover Publications, New York, 1952.
2. W. H. Wollaston, *Phil. Trans. Roy. Soc. London,* Ser A, 1802, **92**, 365.
3. D. Brewster, *Report of 2nd Meeting of British Association*, 1832, 320.
4. G. Kirchoff, *Phil. Mag.,* 1860, **20**, 1.
5. G. Kirchoff and R. Bunsen, *Phil. Mag.,* 1860, **20**, 89.
6. G. Kirchoff and R. Bunsen, *Phil. Mag.,* 1861, **22**, 329.
7. A. C. G. Mitchell and M. W. Zemansky, "Resonance Radiation and Excited Atoms", University Press, Cambridge, 1934. (Republished 1961.)

8. T. T. Woodson, *Rev. Sci. Instr.*, 1939, **10**, 308.
9. A. Walsh, Australian Patent 23041 (1953).
10. A. Walsh, *Spectrochimica Acta,* 1955, 7, 108.
11. C. T. J. Alkemade and J. M. W. Milatz, *Appl. Sci. Res.*, 1955, **4B**, 289.
12. C. T. J. Alkemade and J. M. W. Milatz, *J. Opt. Soc. Am.*, 1955, **45**, 583.
13. R. W. Wood, *Phil. Mag.*, 1905, **10**, 513.
14. P. Pringsheim, "Fluorescence and Phosphorescence", Interscience, New York, 1949.
15. E. L. Nichols and H. L. Howes, *Phys. Rev.*, 1923, **22**, 425 and 1924, **23**, 472.
16. R. M. Badger, *Z. Physik.*, 1929, **55**, 56.
17. R. Mannkopff, *Verhandl. d. Deutsch. Phys. Ges.*, 1933, **14**, 16.
18. A. L. Boers, C. T. J. Alkemade and J. A. Smit, *Physica,* 1956, **22**, 358.
19. C. T. J. Alkemade, *Proc Xth Colloq. Spectros. Inter., Maryland, U.S.A. 1962,* pp. 143-170. (Spartan Books, Washington, 1963.)
20. J. D. Winefordner and T. J. Vickers, *Anal. Chem.*, 1964, **36**, 161.
21. J. D. Winefordner and R. A. Staab, *Anal. Chem.*, 1964, **36**, 165.
22. W. T. Elwell and J. F. Gidley, "Atomic Absorption Spectrophotometry", 2nd ed., Pergamon Press, London, 1966.
23. N. P. Ivanov, "Atomno-absorbcionnyj Analiz", IREA, Moscow, 1966.
24. B. V. L'vov, "Atomno-absorbcionnyj Spektralnyj Analiz", Izdatelstvo "Nauka", Moscow, 1966. (Complete English translation published by U.S., A.E.C. and National Science Foundation, Report No. AEC-tr-6979, 1969.)
25. J. W. Robinson, "Atomic Absorption Spectroscopy", Dekker, New York, 1966.
26. J. Ramirez-Munoz, "Atomic Absorption Spectroscopy", Elsevier, Amsterdam, 1968.
27. W. Slavin, "Atomic Absorption Spectroscopy", Interscience, New York, 1968.
28. I. Rubeska and B. Moldan, "Atomic Absorption Spectrophotometry", Iliffe, London, 1969.
29. B. V. L'vov, "Atomic-Absorption Spectrochemical Analysis", Hilger, London, 1970.
30. F. Rousselet, "Spectrophotometry by Atomic Absorption Applied to Biology", C.D.V. and S.E.D.E.S., Paris, 1967. (In French.)
31. E. E. Angino and G. K. Billings, "Atomic Absorption Spectrophotometry in Geology", Elsevier, New York, 1968.
32. G. D. Christian and F. J. Feldman, "Atomic Absorption Spectroscopy: Applications in Agriculture, Biology and Medicine", Wiley-Interscience, New York, 1970.
33. R. Herrmann and C. T. J. Alkemade, "Flame Photometry", 2nd ed., Springer, Berlin, 1960 (in German). Translated into English as "Chemical Analysis by Flame Photometry", Interscience, New York, 1963.
34. R. Mavrodineanu and H. Boiteux, "Flame Spectroscopy", Wiley, New York, 1965.
35. E. Pungor, "Flame Photometry Theory", Van Nostrand, Princeton, 1966.
36. R. Mavrodineanu (ed.), "Analytical Flame Spectroscopy", Macmillan, London, 1970.
37. J. A. Dean and T. C. Rains (eds.), "Flame Emission and Atomic Absorption Spectrometry; Vol. 1, Theory", Dekker, New York, 1969.
38. P. T. Gilbert, *Anal. Chem.*, 1962, 34, 210R.
39. A. Walsh, *Proc. XIIth Colloq. Spectroscop. Inter.*, Hilger and Watts, London, 1965, p. 43.
40. R. Mavrodineanu, *Encycl. Ind. Chem. Anal.*, 1966, 1, 160.

41. W. Slavin, *Appl. Spectr.*, 1966, **20**, 281.
42. H. L. Kahn in "Trace Inorganics in Water", R. A. Baker, Advances in Chemistry Series, No. 73, American Chemical Society, Washington, 1968, p. 183.
43. S. R. Koirtyohann, in "Developments in Applied Spectroscopy", ed. by W. K. Baer, A. J. Perkins and E. L. Grove, Vol. 6, Plenum Press, New York, 1968, p. 67.
44. W. J. Price, "Atomic Absorption Spectroscopy", chapter of "Spectroscopy", ed. by D. R. Browning, McGraw-Hill, New York, 1969.
45. T. S. West, *Minerals Science and Engineering*, 1969 **1**, 3 (of June issue).
46. J. B. Willis, in R. Mavrodineanu (ed.), "Analytical Flame Spectroscopy", Ch. 10, Macmillan, London, 1970.
47. K. Fuwa, in J. D. Winefordner (ed.), "Spectrochemical Methods of Analysis", Ch. 3, Wiley-Interscience, New York, 1971.
48. American Society for Testing and Materials, Atomic Absorption Spectroscopy, 443, ASTM, Philadelphia, 1969.
49. R. M. Dagnall and G. F. Kirkbright (eds.), "Atomic Absorption Spectroscopy", Butterworth, London, 1970.
50. Proc. 3rd CISAFA, Paris, Sept. 1971. Plenary lectures published in "Methodes Physiques d'Analyse", GAMS, Paris, September 1971.
51. J. R. W. Warn, *Measur. Instr. Rev.*, 1968, **15**, 505, 576, 669.
52. E. E. Pickett and S. R. Koirtyohann, *Anal. Chem.*, 1969, **41**, 28A (of issue No. 14).
53. J. D. Winefordner and J. M. Mansfield, in "Fluorescence", ed. by G. G. Guilbault, Dekker, New York, 1967.
54. J. D. Winefordner and J. M. Mansfield, *Appl. Spectros. Reviews*, 1968, **1**, 1.
55. D. R. Demers, *Appl. Spectros.*, 1968, **22**, 797.
56. D. W. Ellis and D. R. Demers, in "Trace Inorganics in Water", R. A. Baker, Advances in Chemistry Series, No. 73, Washington, 1968.
57. W. J. Price, "Atomic fluorescence spectroscopy", chapter of "Spectroscopy", ed. by D. R. Browning, McGraw-Hill, New York, 1969.
58. T. S. West, *Minerals Science and Engineering*, 1970, **2**, 31 (of January issue).
59. J. D. Winefordner and R. Smith, in R. Mavrodineanu (ed.), "Analytical Flame Spectroscopy", Ch. 11, Macmillan, London, 1970.
60. R. Smith, in J. D. Winefordner (ed.), "Spectrochemical Methods of Analysis", Ch. 4., Wiley-Interscience, New York, 1971.
61. J. D. Winefordner and R. C. Elser, *Anal. Chem.*, 1971, **43**, (4) 24A.
62. J. D. Winefordner, *Rec. Chem. Progr.*, 1968, **29**, 1.
63. R. Mavrodineanu, "Bibliography on Flame Spectroscopy Analytical Applications 1800-1966", NBS Miscellaneous Publications 281, Washington, 1967.
64. R. Mavrodineanu, in R. Mavrodineanu (ed.), "Analytical Flame Spectroscopy", Ch. 13, Macmillan, London, 1970.
65. S. Slavin, *Atomic Absorption Newsletter*, 1971, **10**, 17 (preceding parts in the issues for 1962-1970).
66. Perkin-Elmer Corporation, "Analytical Methods for Atomic Absorption", Norwalk, Connecticut (revised regularly).
67. Fisher Scientific Co. "Bibliography for Atomic Absorption Spectroscopy", Pittsburg, 1968.
68. C. A. Lermond and L. Capacho-Delgado, "Atomic Absorption Bibliography", Bausch and Lomb, Inc., Rochester, New York, 1968.
69. R. Reiss, "Atomic Fluorescence Bibliography", Aztec Instruments, Inc., South Norwalk, Conn., 1968.
70. P. R. Masek and I. Sutherland (eds.), *Atomic Absorption and Flame*

Emission Spectroscopy Abstracts, Science and Technology Agency, London. (Issued bi-monthly with annual index.)
71. B. F. Scribner and M. Margoshes, *Anal. Chem.*, 1964, **36**, 329R.
72. M. Margoshes and B. F. Scribner, *Anal. Chem.*, 1966, **38**, 297R.
73. M. Margoshes and B. F. Scribner, *Anal. Chem.*, 1968, **40**, 223R.
74. J. D. Winefordner and T. J. Vickers, *Anal. Chem.*, 1970, **42**, 206R; 1972, **44**, 150R.
75. R. J. Reynolds and K. Aldous, "Atomic Absorption Spectroscopy: A Practical Guide", Griffin, London, 1970.
76. W. J. Price, "Analytical Atomic Absorption Spectrometry", Heyden and Son Ltd., London, 1972.
77. J. A. Dean and T. C. Rains (eds.), "Flame Emission and Atomic Absorption Spectrometry; Vol. 2, Components and Techniques", Dekker, New York, 1971.
78. M. L. Parsons and P. M. McElfresh, "Flame Spectroscopy: Atlas of Spectral Lines", Plenum Press, New York, 1971.

CHAPTER 2

Spectroscopic Theory

2.1 INTRODUCTION

A primary difficulty for the authors of a volume of this type is the decision whether to introduce the minimum of necessary theory, with the result that the book appears incomplete to many readers, or to give a fairly comprehensive theoretical treatment which other readers will ignore. We have attempted to compromise by dividing the theoretical part of this book into three separate sections. In this first theoretical chapter the most important points of general atomic spectroscopic theory are summarized or revised. This summary also provides an introduction to the techniques for the non-specialist reader and allows ease of reference to points of general theory in later sections. In the following two chapters more detailed quantitative discussions are presented of atomic absorption spectroscopy and atomic fluorescence spectroscopy respectively; it is hoped that this will prove sufficient for the research worker who is involved in a more fundamental approach to the techniques than are many analytical chemists. Frequent reference is made to sources of more detailed information, so that these chapters should provide a starting point for those interested in such problems as calculation of the effects of line width on absorbance, measurement of absolute concentrations of atoms, and the study of the shapes of growth curves etc.

As the notation *atomic* absorption and fluorescence spectrometry implies, we shall be mainly concerned with *atomic spectra.*[1] This type of spectrum is associated with quantized changes in atomic energy produced by variations in the orbital motions of the valence electrons. The wavelengths transporting the energy appear, in general, to be isolated; hence, the widely used term "line spectra". Although these are the most important from our point of view, it must be remembered that there are two other types of spectra that are also encountered in work concerned with AAS and AFS.

Molecular or band spectra[2] are associated with the energy changes of molecules and give rise to closely packed groups (bands) of spectral lines. They are much more complex than atomic spectra because the internal energy of a molecule comprises three independent parts—changes in vibrational and rotational energy occur in addition to electronic excitation. The main interest of these spectra in AAS and AFS lies in the interferences which they can cause in analysis; both molecular emission and absorption in the atom cell can cause inaccuracies in the measurements made at atomic lines.

The third type of spectra are *continuum spectra.** In these spectra the energy is distributed in an uninterrupted manner between all wavelengths within a given domain. A true continuum of this kind is often observed as the black-body radiation from a hot solid or liquid[3] or as the result of an atomic or molecular energy transition in which one of the energy states is unquantized.[4] Extensive, rather diffuse, band systems are also often referred to as "continua" but this should be treated as a description rather than a definition. Continua are mainly encountered in AAS and AFS as spectral light

* This type of spectrum is often referred to as continuous. The alternative definition of continuous spectral sources will be found in Section 5.1.1 and to avoid confusion it is better not to use continuous as an alternative to continuum.

sources which may be used over a range of wavelengths rather than only at isolated spectral lines.

2.2 ATOMIC SPECTRA AND ATOMIC ENERGY LEVELS

2.2.1 The Atom and Its Spectra

As mentioned earlier, experimental spectroscopy had achieved a considerable body of results well before the end of the last century. Thus it was recognized that the line spectra observed with spectroscopes and spectrographs were associated with atoms and that each line could be considered as "monochromatic" radiation. As light was considered to have a wave character (following the experiments of Young, Fresnel and others), each line of the spectrum was analogous to a periodic wave and its position in the spectrum could be defined in terms of its *wavelength*, λ, its *frequency*, $\nu = c/\lambda$, where c is the velocity of light, or its *wavenumber*, $\sigma = 1/\lambda$.* The velocity of light is taken as $2{\cdot}99793 \times 10^{10}$ cm/s and the frequency is measured in cycles/s, or Hertz (Hz). This gives the convenient unit of cm^{-1} for the wavenumber.† The centimetre can be used as the unit of wavelength but is inconveniently large for the *optical region* of the spectrum being discussed here. The traditional unit of wavelength is the Ångström unit (1 Å = 10^{-8} cm) but it is now being replaced by the nanometre (1 nm = 10^{-9} m = 10^{-7} cm; hence 1 nm = 10 Å). Other units which have been fairly widely used are the micron (1 μm = 10^{-6} m) and the millimicron (1 mμ = 1 nm = 10^{-9} m). The visible region of the spectrum extends from about $\lambda = 770$ nm (red) to $\lambda = 380$ nm (violet) although optical spectra may be studied for a considerable range on either side of this. We shall be mainly concerned with the region extending from about 190 nm in the ultraviolet to 850 nm in the infrared. Finally, it should be noted that the value of c quoted above is the velocity of light in a vacuum. Thus values of λ, ν and σ evaluated with it will refer to radiation travelling through a vacuum. Since these vacuum units are the quantities related to the properties of the atom, the wavenumber (which is directly applicable to the energy levels of the atom) is usually given in tables in vacuum units. For convenience, the wavelength is normally

* The symbols λ, ν and σ are those most commonly used for these terms, although variations are often found in the literature.

† The reciprocal centimetre is often referred to as the Kayser, which is abbreviated to the symbol K.

quoted as its (slightly smaller*) value measured in air without correction to the vacuum wavelength.

As the number of recorded spectra increased and instruments capable of better resolution were produced, it was found that the distribution of the spectral lines of each individual element was not as random as had once been supposed. Thus it was discovered empirically that the wavelengths of the lines of the simple atomic spectra, such as those of hydrogen and the alkali metals, could be fitted to simple series formulae with great accuracy. These series were even more straightforward when the wavelengths of the lines were replaced by their vacuum wavenumbers. Furthermore, it had already been noticed that many of the lines in these simple spectra occurred in small groups named *multiplets*, such as the doublets of the alkali metals or the triplets of the alkaline earths, and in 1883 Hartley observed that there was a constant difference between the wavenumbers of the two components of some doublets (or two of the three components of some triplets). This is illustrated in Table 2-1 for some of the sodium doublets; it can be seen that the two lines

TABLE 2-1
Some sodium doublets showing constant wavenumber differences

Wavelength, nm	Wavenumber, cm^{-1}
1140·35	8766
1138·12	8783
819·48	12200
813·33	12217
616·08	16227
615·42	16244
589·59	16956
589·00	16973
568·82	17577
568·26	17594
515·36	19399
514·91	19416
498·28	20064
497·85	20081

* Even this value must be quoted for "standard air" since the refractive index varies slightly with location, atmospheric conditions etc.

of each doublet are separated by 17 cm^{-1}. Some years later, after the phenomenon of constant differences had been shown to apply to many other spectra, Ritz found that it was a direct consequence of a more general rule which he named the *combination principle*. This states that for any atom (or molecule) there is a set of *spectral terms* which are such that the vacuum wavenumber of any line of the spectrum of that species is the difference between two of the terms. It was also observed that the number of spectral terms of an atom is always much smaller than its number of spectral lines; thus the same term must appear in more than one wavenumber. This is best illustrated on a diagram, such as that in Fig. 2-1 which again uses the sodium spectrum as an example. This type of diagram is known as a *term diagram*, a *Grotrian diagram* (after its originator) or a *level diagram*. The spectral terms are represented by horizontal lines with vertical lines between them to show the spectral line corresponding to the wavenumber difference of two terms. For convenience, the term with the smallest wavenumber is taken as zero on this particular diagram. Reference to the diagram shows for example, that the term 16973 cm^{-1} appears in no less than seven line wavenumbers.

A most important advance in the study of atomic spectra was made by Bohr shortly after the empirical discovery of the combination principle. He postulated that any atom is allowed only certain discrete and characteristic energy values, and that emission or absorption of radiation by an atom occurs when it undergoes a transition between these stationary states or *energy levels*. Furthermore, these energy levels were seen to be directly proportional to the empirical spectral terms of the atom; it had already been shown by Planck that the energy and frequency of a particular radiation are related such that

$$E = h\nu = hc\sigma \tag{2-1}$$

where E is the energy and h is a constant (6·6254 erg s) now known as *Planck's constant*. Hence the frequency of radiation corresponding to a transition of an atom between two levels (or states) of energy E_m and E_n respectively will be given by

$$\nu_{mn} = (E_m - E_n)/h$$

and the corresponding spectral terms are

$$\sigma_m = E_m/hc \quad \text{and} \quad \sigma_n = E_n/hc$$

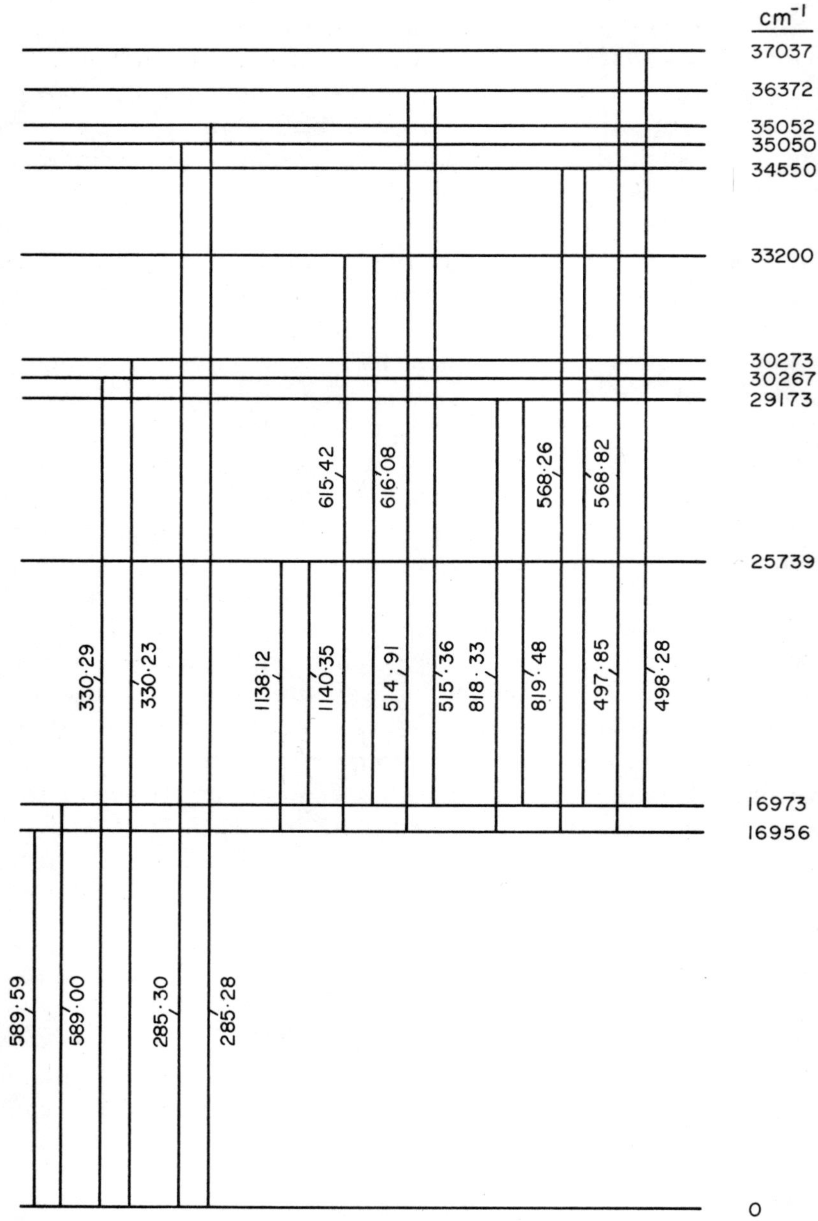

Fig. 2-1. Spectral term diagram for sodium (wavelengths of lines are given in nm).

Thus the experimental term diagrams become *energy level diagrams* of the atom simply by a change of scale.

These straightforward postulates immediately removed many of the problems existing in the interpretation of the atomic spectra that had been observed experimentally at that time. It was now obvious rather than surprising that there should be a simple relationship between the wavenumbers of many of the lines of each spectrum, and the vast number of lines in many of the spectra could be explained; even the smallest source emitting light contains a great number of atoms and these will exist in many of the different stationary states possible for that type of atom (i.e. that element). Thus, the different lines of the spectrum of that element are emitted by different atoms and not all by the same atom. This greatly simplified the task of correlating the structure of the atom and its observed spectrum.

In a similar way, the mechanism by which the different spectra could be produced was also explained by the postulates. The level of lowest energy corresponds to the spectral term with the smallest wavenumber; it is unique and is the state in which an atom will normally exist. It is referred to as the fundamental or *ground state*, whereas any level of higher energy is said to correspond to an *excited state*. A spectral line can only be emitted if the atom is in an excited state, so that energy will be released when it returns to the state of lower energy (either the ground state or an intermediate state). Thus the atom will emit radiation only after the energy required to raise it to an excited state has been provided by an external source, such as a flame, an arc, a spark, etc. On the other hand, the absorption spectrum can be obtained without such an external source as at room temperature most of the atoms are in the ground state. Hence when the atoms are irradiated with light of the correct wavelength (i.e. that corresponding to the energy difference between an excited state and the ground state) some of the energy will be absorbed to raise atoms to an excited state. This also explains the greater simplicity of an absorption spectrum compared to the corresponding emission spectrum; only a small fraction of the possible transitions for a given element will have the ground state as one of their two spectral terms.*

* It is, of course, possible to observe in absorption a transition which has an excited state as its lower spectral term, but only if an external source of energy is used to raise sufficient atoms to that excited state.

2.2.2 *Convention and Notation*

Bohr's postulates of 1913, two of which were quoted in the previous section, were based on the assumption that the energy states of an atom depended on the way in which its electrons moved about the nucleus and that this movement occurred in circular orbits; changes in angular momentum would occur in units of $h/2\pi$ when an electron moved from one orbit to another and hence the energy must also change in multiples of a fixed unit, now known as a *quantum* (or *photon*). Although this theory had proved most satisfactory in explaining the known experimental facts, the real test was whether it could be used to calculate theoretically the allowed energies of the atoms. In practice, it was found that satisfactory results were obtained only when the nucleus was accompanied by a single electron, such as with hydrogen. The revision necessary to apply the theory to all atoms and their spectra was carried out during the period 1920 to 1930. The *quantum theory* was devised to explain the structure of the atom and the way in which it gives rise to the optical spectra. At about the same time, all of the quantum theory was underpinned and explained in a logical fashion mathematically via the *quantum* or *wave mechanics*. The reader will find a useful presentation of the most important aspects of the wave mechanics in reference 1. It should be borne in mind that the quantum theory is the basis of much of the convention and notation used in any discussion of atomic spectra.

The optical spectra originate only from the positional and spin changes of electrons in the outer orbits, which with most elements are incompletely filled by electrons. These are known as the *optical electrons*. The different energy states possible for any atom are found by calculating the different orbital possibilities of its optical electrons. Each possibility corresponds to a different discrete energy of the atom as a whole and not just to energy of the electrons.* The orbits of the electrons may be readily described by a set of *quantum numbers*. These same numbers are used to calculate the number of spectral energy levels that may result from particular groups of electrons; derivations of them provide a useful notation by which the individual terms may be referenced for ease of identification. These reference numbers and symbols also indicate the order of increasing energy for groups of terms of each atom, so that the task of relating

* The atom will, of course, also have its translational kinetic energy but this need not be discussed in connection with the spectral energy levels on which it has no direct effect (see also Section 2.5).

experimentally measured spectral terms to their quantum number equivalents is considerably simplified.

It is usual to refer to an electron as being of a particular type depending on the orbit that it occupies. These types are indicated by the first two quantum numbers. To avoid confusion, the second quantum number is replaced by a letter* according to

$$\begin{matrix} 0 & 1 & 2 & 3 & 4 & \ldots \\ s & p & d & f & g & \ldots \end{matrix}$$

Thus the electrons (but more properly their orbits) are referred to as $3s$, $5d$, $2p$, etc. The combination of electrons giving rise to a particular energy of the atom is known as a *configuration*. For example, for an unexcited gallium atom it is $1s^2 2s^2 2p^6 3s^2 3p^6$-$3d^{10} 4s^2 4p$. The superscripts to the right of the letters indicate the number of electrons of each type. As we are only concerned with the optical electrons, this particular configuration would normally be abbreviated in spectral tables to $4p$.

The spectral term symbols are also based on the second quantum number, the same letters as above being used but in upper case; e.g. the symbol for the term corresponding to the configuration of gallium given above is P. If there is only one optical electron, as in this example, this letter must be the same as that used for the electron. When two or more optical electrons are present, their second quantum numbers are added vectorally, giving several resultant numbers. The same lettering scheme is used with these. The same type of term may thus arise from a variety of optical electron configurations and to avoid confusion the configuration should be indicated with the term, e.g. for the gallium example we should write $4p\ P$.

These terms are not directly analogous to the empirical spectral terms mentioned earlier, because the electron spin can result in each term being subdivided into further *states* (and it is these that correspond to the empirically determined spectral terms or energy levels). The individual states are indicated by a third quantum number, known as the J value, which is written as a subscript to the right of the term symbol. Thus the gallium configuration given above actually results in the states $4p\ P_{1/2}$ and $4p\ P_{3/2}$. Of these the former corresponds to the lower energy and is the ground state. The

* The odd choice of the first three letters is an historical result; they are related to the "sharp", "principal" and "diffuse" series of lines which may be distinguished in the sodium spectrum.

J values are useful when measurements of spectral line intensities are made, since the quantity $(2J + 1)$ is the *statistical weight*, g, of the state (see Sections 2.3 and 2.4).

The resultant electron spin of the term governs the number of possible values of J; it is used to calculate a quantity known as the *multiplicity* of the term. This usually* corresponds to the number of J values (i.e. states) for that term and is shown as a superscript to the left of the term symbol, e.g. $4p\ ^2P_{1/2}$ and $4p\ ^2P_{3/2}$ for the gallium example. In this case the term is referred to as a *doublet*; if the multiplicity were 3 it would be a *triplet*, etc. The same word *multiplet* is applied to a group of spectral lines corresponding to transitions between initial levels belonging to one term only and final levels belonging only to one other term; i.e. the upper (or lower) levels differ only in their value of J. A group of lines is designated a triplet, if the terms are triplets, but the actual number of lines will often be greater than this. For example, the calcium triplet observed between 442 and 446 nm ($4p\ ^3P$–$4d\ ^3D$) is actually composed of six lines.

2.3 THE EMISSION AND ABSORPTION OF ENERGY

2.3.1 General

As described in Section 2.2, for any atom there exists an extensive set of possible energy levels or states, and spectral emission or absorption results from a transition between particular pairs of these levels. The pairs between which transitions are possible are predicted by the quantum mechanics via the *selection rules*.† This results in a "picture" of the spectrum of the element with the positions of the lines indicated, but the quantum theory is also able to assist in predicting the relative *intensities* of these lines (i.e. the quantity of energy emitted or absorbed).

The intensity of a spectral line produced by a group of emitting or absorbing atoms depends on two main factors. First, given a fixed population for all energy levels (i.e. at any instant equal numbers of atoms exist in each of the possible states for a given element), the

* S terms obtained from a single s-type optical electron have a multiplicity of 2 but only one J value and corresponding state, e.g. the ground state of the alkali metals. This is simply a result of the combination of numbers involved (the second J value would be $-\frac{1}{2}$ if negative quantum numbers were allowed).

† The rules are not as precise as some other parts of the quantum mechanics and certain "forbidden" lines are observed for a number of elements.

probability of any particular transition occurring will depend on the intrinsic properties of the atom as expressed by the quantum mechanics. Thus, in a group of atoms, the greater the probability of a particular transition the more intense is the corresponding spectral line. This *transition probability* is a fundamental property of the atom and is discussed below.

Secondly, the number of times a particular transition occurs will depend on the number of atoms existing at any given time with the energy of its initial level. The population of each level will depend on the properties of the group of atoms as a whole, since it must depend on the amount of energy available to the atoms. Hence, environmental factors, such as temperature and pressure, will be important and the intensity of the spectral lines will vary with the means used to obtain a population of free atoms or to excite emission lines. This effect of the distribution of atomic population is discussed in Section 2.4.

2.3.2 The Einstein Transition Probabilities

On the assumption that the energy levels have equal populations (or, if this is not the case, if a correction is made as described in Section 2.4) it is often possible to calculate the relative intensities of the lines of a single multiplet of a particular element by a fairly simple process using the quantum numbers of the initial and final states. The calculation is further simplified by the availability[5] of tables of the results that can be generally applied. However, the method has seen wide practical use only for certain special applications (such as the calculation of intensity ratios of hyperfine structure components; see Section 2.5.2) mainly because the results are only reliable with the lighter elements and because the restriction to a multiplet of a particular element limits the usefulness anyway.

The more common method of classifying and explaining the relative line intensities is based on Einstein's theory of radiation which defines three transition probabilities; under given conditions, the intensity of a line corresponding to a particular transition will be directly proportional to its transition probability. The transition probabilities may be calculated using quantum mechanics from first principles, in the same way that the possible states, the selection rules, etc. are derived. The calculations are complex and only possible when simplifying assumptions are made. Considerable numbers of calculated values have been published[6] but the accuracy depends on the degree of approximation and the element, and most of the results have been for the lighter elements (atomic number, Z, in the range 1 to 30). In view of this many of the most useful results have been obtained experimentally from observed spectra.

Three Einstein coefficients are defined: for spontaneous emission, absorption, and stimulated emission (fluorescence) and are usually written A_{ij}, B_{ji} and B_{ij} respectively. The meaning of the subscripts should be clear from Fig. 2-2. A_{ij} is the easiest to define. It is simply the probability that an atom in state "i" will spontaneously emit a quantum ($h\nu$) and pass to the state "j". Hence A_{ij} has the unit s^{-1}. The other two coefficients are more difficult to define, as the probability of an absorption or fluorescence transition will obviously depend on the amount of incident radiation. Einstein was able to overcome this problem by defining the transition probability in these

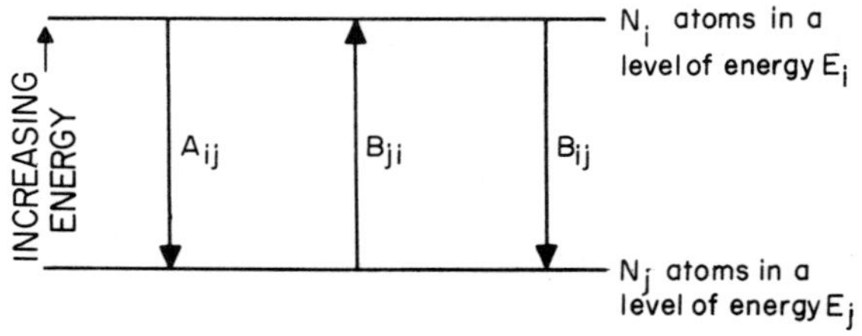

Fig. 2-2. The Einstein transition probabilities.

cases such that it is obtained by multiplication of the appropriate coefficient and the *radiation density* at the frequency corresponding to the particular transition (ρ_ν).* Hence $B_{ji}\rho_\nu$ and $B_{ij}\rho_\nu$ have the unit s^{-1}. Since the radiation density (ρ_ν) is energy per unit volume, it has the units erg cm^{-3} or g cm^{-1} s^{-1}. Thus B_{ji} and B_{ij} have the units cm g^{-1}. Note should be taken of this important difference between the A and B coefficients; strictly speaking only the former should be referred to as a transition probability.

Einstein was able to obtain several useful relationships between the three coefficients by considering the thermodynamic equilibrium between the radiation and the atoms and comparing it with the equilibrium of a black-body radiator at the same temperature:

$$A_{ij}/B_{ij} = 8\pi h\nu^3/c^3 \qquad (2\text{-}2)$$

$$g_iB_{ij} = g_jB_{ji} \qquad (2\text{-}3)$$

where g_i and g_j are the statistical weights of the two states. These

* In certain cases the transition probability is defined in terms of isotropic radiation *intensity*. The coefficients are then related to the two used here by B (intensity) = $(4\pi/c)B$(density). Care should be taken not to use simultaneously equations based on different definitions of the probabilities without applying the correction factor.

equations may also be used to relate all three transition probabilities, via A_{ij}, to the *oscillator strength* or *f-value* of the atom. This was regarded in the classical electron theory of dispersion (which preceded the quantum theory) as a measure of the degree to which the atom resembles an oscillating electron in its ability to emit and absorb a particular spectral line. The f-value is directly proportional to A_{ij}, as shown by Ladenburg's formula:

$$A_{ij} = 8\pi^2 e^2 f_{ij}/\lambda^2 mc \qquad (2\text{-}4)$$

where e and m are the electron charge and mass respectively. Substitution of the values of the constants gives

$$A_{ij} = 0{\cdot}6670 \times 10^{14} f_{ij}/\lambda^2 \qquad (2\text{-}5)$$

where A_{ij} is in s^{-1} and λ is in nm. The f-value is still widely used as it results in a convenient form for many formulae. This is particularly so since the absorption oscillator strength, f_{ji}, is related to f_{ij} by

$$g_j f_{ji} = g_i f_{ij} \qquad (2\text{-}6)$$

In view of this, tables of f-values are usually given as "gf" values which may be used for emission or absorption lines corresponding to the transition between two particular states of the atom.

2.3.3 *The Intensities of Spectral Lines*

The intensity of an emission, absorption or fluorescence line may now be expressed in terms of the number of atoms existing in its initial and/or final states. The simplest case is that for a non-self-absorbed emission line.* From the notation of Fig. 2-2, the number of transitions per second (i.e. the number of photons) corresponding to spontaneous emission will be $A_{ij} N_i$. (For convenience, N_i is usually taken as the number of atoms per unit volume, but this is not essential.) Hence, the emission intensity may be represented by

$$I_{\text{em}} = h\nu A_{ij} N_i \qquad (2\text{-}7)$$

In the case of absorption and fluorescence, the numbers of transitions are $B_{ji}\rho_\nu N_j$ and $B_{ij}\rho_\nu N_i$ respectively. The total reduction in intensity observed will depend on the difference between these two numbers.† Hence for absorption we may write

$$I_{\text{ab}} = h\nu(B_{ji}N_j - B_{ij}N_i)\rho_\nu \qquad (2\text{-}8)$$

* Self-absorption of spectral lines is discussed in Section 2.5.2.

† Spontaneous emission is independent of this and is not included; in practice, the two types of emission may be distinguished by suitable choice of experimental conditions (see Chapter 10).

whereas for fluorescence the intensity is simply

$$I_{fl} = h\nu B_{ij}\rho_\nu N_i \tag{2-9}$$

As it is generally more convenient to work in terms of incident intensity rather than radiation density, we may write $I = (c/4\pi)\rho_\nu$. This assumes, however, that the incident light of intensity I is isotropic in the same way as is the spontaneously emitted light of intensity I_{em}. If, as is usually the case, the incident light is in the form of a pencil of light of solid angle Ω, the equivalent intensity will be $I_0 = (\Omega/4\pi)I$. For a beam of unit solid angle this becomes $I/4\pi$. Hence we obtain

$$I_{ab} = (h\nu/c)(B_{ji}N_j - B_{ij}N_i)I_0 \tag{2-10}$$

$$I_{fl} = h\nu B_{ij}N_i I_0/c \tag{2-11}$$

where B_{ji} and B_{ij} are the coefficients defined in terms of radiation density, as explained earlier.

These three expressions for I_{em}, I_{ab} and I_{fl} form the basis of most of the practical formulae used in analytical atomic spectrometry, although they are usually more useful after substitution for N_i and N_j in terms of the total number of atoms present in all states, as explained in the next section. The expressions for I_{em} and I_{fl} are more directly applicable to practical situations than that for I_{ab}; it will be seen later that other absorption expressions allowing less difficult experimental measurements are commonly used for analytical atomic absorption spectroscopy.

2.4 DISTRIBUTION OF ATOMIC POPULATION

2.4.1 General

It has already been mentioned that the relative populations of the states of the atom play a major part in determining the relative intensities of the spectral lines corresponding to the various transitions. Furthermore, these populations depend not merely on intrinsic properties of the atom, but also on the properties of the emitting or absorbing gas as a whole. If the particular type of energy in question (i.e. electronic for an atom but possibly also vibrational or rotational for a molecule) has attained equilibrium the relative populations of the quantized levels may be easily found from the Maxwell-Boltzmann Law. For some of the commonly used atom cells we may assume that thermal equilibrium does indeed exist for the atoms whose absorption or fluorescence is being measured and apply the law accordingly. In some atom cells equilibrium may exist only in certain regions or may, strictly speaking, not apply to all species present. For example, flames are widely used as atom cells and whereas the law may produce quite satisfactory results in the regions normally used for analytical measurements, this may not be the case in the primary reaction zone. Similarly, the law may give unsatisfactory results for some of the flame gas molecules in a region where it is applicable to added metal atoms.

A quite different situation exists in the case of many *electric discharges*, which are frequently used as the spectral light source for atomic absorption or fluorescence measurements and sometimes as the atom cell. Excitation in the discharge results from collisions with electrons or ions of very high energy. The Boltzmann factor $\exp -E/kT$ (see Section 2.4.2) then approaches unity and the ratio of the populations of any two levels is simply the ratio of their statistical weights, i.e.

$$N_i/N_j = g_i/g_j$$

Under these conditions the transition probability plays an even greater part than usual in determining the relative intensities of the lines, so that lines corresponding to high excitation energies may be relatively many times more intense than in, say, a flame source. As shown elsewhere, the presence of such lines in a spectral light source can have an important effect on the choice of a suitable absorption line.

2.4.2 The Maxwell-Boltzmann Law

The law states that if at an absolute temperature T there are N_i particles with energy E_i and N_j particles with energy E_j, then

$$N_i/N_j = \exp -(E_i - E_j)/kT \qquad (2\text{-}12)$$

provided that both energy states are non-degenerate. (k is the Boltzmann constant.) In fact, many of the energy states that we have regarded up to now as single levels represented by a particular value of J are actually degenerate because they further split into $(2J + 1)$ sublevels on application of a magnetic field. When there is no magnetic field (as will usually be the case in examples of interest here), each of these sublevels has the same energy. Therefore, in the notation used above, the number of particles with energy E_i is actually $(2J_i + 1)n_i$ where n_i is the population of each of the non-degenerate sublevels. Hence $N_i = g_i n_i$ and this definition of the statistical weight now becomes more logical. Since the law must be applied to non-degenerate levels of energy E_i and E_j we obtain

$$N_i/N_j = (g_i/g_j)_j(n_i/n_j) = (g_i/g_j) \exp -(E_i - E_j)/kT \qquad (2\text{-}13)$$

This general form of the law extends to any state which is non-degenerate (e.g. the ground state $^1S_0$ of the alkaline earths), since its statistical weight will simply be unity.

The equation obtained above is extremely useful and is widely used for calculations involving relative line intensities. However, for atomic absorption and fluorescence is often more convenient if the population of any one level (e.g. the ground state) can be expressed in terms of the total number, N, of particles present. This is easily achieved by expressing N as the sum of the populations of all levels, i.e. $N = \sum_i N_i$ (or $\sum_j N_j$, etc.). As the equation may be rewritten as

$$N_i/N_j = (g_i \exp(-E_i/kT)/(g_j \exp(-E_j/kT))$$

we have

$$N_i/N = g_i \exp(-E_i/kT)/\sum_j g_j \exp(-E_j/kT) = g_i \exp(-E_i/kT)/F(T) \qquad (2\text{-}14)$$

where $\sum_j$ is known as the *partition function*, $F(T)$, for the type of energy in question, and the difficulty of calculating it varies considerably with specific cases. With this proviso the equation applies quite generally to both atoms and molecules and to all types of energy levels.

A number of expressions for the partition function under different circumstances, and also some approximations, are available (see, for example, Ref. 2, Ch. 20-4). However, with the examples likely to be encountered in AAS and AFS it is usually more convenient with atomic spectra simply to perform the summation manually using the published g_i and E_i values for the energy levels. This is especially true when the calculation is performed for relatively low-temperature atom cells, such as flames. It should be obvious from the equation that even for temperatures as high as 3000 K, the value of N_i will be very small if E_i is more than a few electron volts, for example, if $T = 3000$ K and $E_i = 2$ eV, $\exp(-E_i/kT) = 4 \times 10^{-4}$ so that N_i is negligible compared with N. It is, therefore, rarely necessary to consider levels more than about 1 eV above ground state when the summation is carried out.

2.4.3 Relevance of Atomic Population Distribution to AAS and AFS Measurements

The Maxwell-Boltzmann Law only relates the population of each energy level to the total number, N, of *free neutral atoms* present. Under analytical conditions the usefulness of this information is frequently limited by inadequate knowledge of the relation between N and the actual sample concentration. The fraction of the total number of analyte atoms in the sample which N represents varies substantially depending on the analyte element, the type of sample and the nature of the absorption or fluorescence cell. The form of this relation in a given case may be of considerable interest if, for example, an absolute analysis is required. The problem is very much an experimental one, however, as the relations between the contributing factors (such as degree of ionization; volatility and stability of the compounds formed by the element; and efficiency of sample utilization) are very complex. Further discussion will be found in Section 6.5.

In spite of this difficulty, a knowledge of the *relative* distribution of sample atoms amongst the possible energy levels may still prove to be most useful. It provides a simple basis by which methods of improving sensitivity in flame absorption cells may be understood as well as assisting in the selection of the optimum absorption line for a particular element. Unfortunately, this knowledge of the relative populations of the energy levels obtained in flame atom cells has also led to a widespread misunderstanding of the comparative sensitivities of the atomic absorption and flame emission techniques. As shown in Section 2.4.2, at normal flame temperatures the atomic population of the excited state responsible for analytically useful emission of radiation is usually negligible compared to that of the ground state. This has often been quoted as proof of the belief that AAS must be considerably superior in sensitivity when compared with flame emission spectrometry (FES). That this is not the case has been clearly demonstrated by Alkemade,[7] by simple considerations from basic principles. Briefly it has been demonstrated that if experimental variables are neglected, the relative ultimate sensitivities of the two techniques depend not on the distribution of atom population caused by the flame cell, but on the effective temperature of the background light source used for AAS.

Two other fallacies that were quite common in early AAS work also resulted from an incomplete understanding of the population distribution in the atom cell, particularly of the significance of the relatively large ground-state population. In view of this, it has sometimes been stated that, if a constant free-atom population is assumed, the number of absorbing atoms will *always* be independent of temperature. In fact, this is true only for atoms such as mercury that give simple spectra in which the only low-lying level is the ground state itself. When several levels lie below *ca.* 1 eV, their relative populations will vary with temperature to produce a similar variation in the absorption at a particular line. Although the variations observed are often quite small, some combinations of statistical weights and energy levels can give rise to large changes with temperature for certain elements (e.g. tin).

Related to this problem is another which concerns the actual population of the ground state; it is often thought that at normal flame temperatures the ground state invariably has a far greater population than any other level. In fact this may not be the case for a number of elements. For example, the level of tin at 0·210 eV has a higher population than the ground state at temperatures of about 2500 K and above. For aluminium, in which the "ground state" is a doublet with a separation of 0·0139 eV, the population of the higher level is almost double that of the true ground state even below 1000 K (see also Section 11.1.1).

2.5 WIDTH OF SPECTRAL LINES

2.5.1 General

AAS used by analytical chemists at the present time is based on an experimental technique that relies for its simplicity on measuring the "peak absorbance" of a spectral line by using a spectral light source emitting a "sharp" line. The fact that even the sharpest line that can be produced has a finite width has an extremely important bearing on the application of both AAS and AFS. It introduces a number of problems in the practice of the techniques, and has resulted in considerable recent effort to produce better light sources giving greater absorption and fluorescence sensitivity. From the theoretical point of view, the fact that a spectral line may never be represented by a unique frequency (or wavelength) introduces considerable complexity into the derivation of formulae that may be used to interpret practical observations. Furthermore, this problem is recognized as one of the major difficulties encountered in any attempt to carry out absolute analyses by AAS. In view of this situation, the factors contributing to the width of a spectral line will be considered in some detail. The effects on practical AAS measurements of variations in the source (and absorption) lines will be discussed in Chapter 3.

When it is recognized that no spectral line is infinitely narrow it is no longer adequate for the purposes of AAS to represent it by the usual vertical, straight "line" of photographic spectra. A common notation for the intensity distribution of an emission line is shown in Fig. 2-3. The intensity is written as I_ν, as it is a function of the frequency ν, with a maximum value, I_0, at the central frequency of the line, ν_0. The width of the line cannot easily be quoted as the distance $\nu_1 - \nu_2$ since the position of these points is ill-defined.

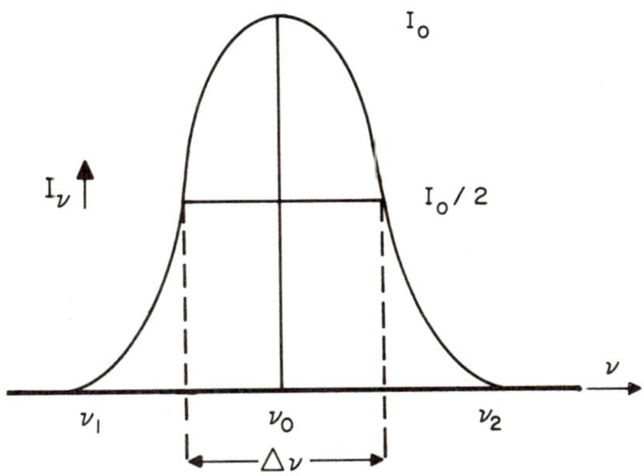

Fig. 2-3. The shape of a spectral line.

Instead, it is customary to define the breadth of any line in terms of its *half-width*, $\Delta\nu$. This is the distance between the two points on the profile where the intensity is equal to half the maximum intensity. Some authors define the half width as $\Delta\nu/2$, so that care must be taken to avoid confusion. The same nomenclature is used for absorption lines with I_ν replaced by the absorption coefficient, k_ν.

2.5.2 *Line Broadening Effects*

A spectral line from any source, or from any absorption or fluorescence cell, will be broadened by a number of factors. Some of these are always more important than others and the contributions of different factors will vary with the source. The most common broadening effects are listed below and an indication is given of their relative importance to AAS and AFS. It should be remembered that any numerical values quoted are merely examples and that wide variations of the different quantities may be observed in practice.

(*i*) *Natural broadening*

This is the result of the finite lifetime of any atom in any excited state. If the mean time the atom spends in a state E_i is Δt_i, then according to Heisenberg's Uncertainty Principle there will be an uncertainty ΔE_i in the value of E_i. This is defined by

$$\Delta E_i \Delta t_i = h/2\pi \tag{2-15}$$

where h is Planck's constant.

Thus, the frequency ν_0 for a transition between levels E_i and E_j given by $\nu_0 = (E_i - E_j)/h$ is no longer unique and the line has a finite width, $\Delta\nu_N$. It may be shown that

$$\Delta\nu_N = (1/\Delta t_i + 1/\Delta t_j)/2\pi \tag{2-16}$$

This equation may be simplified for a resonance line, as E_j will be the ground state with an infinitely large value of Δt_j. Furthermore, Δt_i will then be equal to the radiative lifetime* of the excited level, which for a resonance level is defined as $1/A_{ij}$. Hence

$$\Delta\nu_N = A_{ij}/2\pi \tag{2-17}$$

Thus $\Delta\nu_N$ is of the same order as A_{ij}, about $10^8\ \mathrm{s}^{-1}$. In terms of wavelengths this would represent about 10^{-5} nm at 300 nm, which is negligible in comparison with the other factors.

(*ii*) *Doppler broadening*

This is caused by the thermal motion of the emitting or absorbing atoms. If there is a relative velocity v between the observer and an atom emitting light of frequency ν, the light received by the observer would appear to be of frequency $(\nu + \mathrm{d}\nu)$ or $(\nu - \mathrm{d}\nu)$, depending on the direction of motion, where $\mathrm{d}\nu = \mathrm{v}\,\nu/c$. As the atoms of a gas move in all directions, the observer will receive light of many slightly different frequencies. Thus the line appears broadened. The observed frequencies will have the same distribution as the velocities of the

* This assumes such a low pressure that the number of radiationless transitions from E_i is negligible, which may not always be true (see Sections 2.5.2 (6) and 4.5).

atoms in the gas, i.e. a Gaussian curve will represent the shape of the line.

The Doppler half-width, $\Delta\nu_D$, may be calculated from the formula

$$\Delta\nu_D = [2\sqrt{2R \ln 2}\, \nu_0 \sqrt{T/M}]/c \qquad (2\text{-}18)$$

where R is the gas constant, c the velocity of light and M the atomic weight of the species concerned.

After evaluation of the constants this becomes

$$\Delta\nu_D = 7{\cdot}162 \times 10^{-7}\, \nu_0 \sqrt{T/M}$$

Thus the Doppler width depends on the atomic weight of the species, the temperature of the source and the wavelength of the observed line. In terms of wavelength units, it usually lies in the range 5×10^{-4} to 50×10^{-4} nm for absorption lines in the common flame cells and is somewhat smaller in the spectral light sources, which mostly operate at a rather lower temperature.

(*iii*) *Lorentz broadening*

Collision of an emitting or absorbing atom with other atoms or molecules of a foreign gas* results in a broadening, shift and asymmetry of the line profile. The resultant half-width, $\Delta\nu_L$, is known as the Lorentz, collisional or pressure-broadened width of the line. The interpretation of the collision processes in terms of the effect of the collisions on the energy levels of the atom, and hence the spectral line, is still not complete. According to the classical Lorentz theory, when an emitting or absorbing atom collides with a foreign particle there is an interruption of the vibration within the atom of an electron responsible for the spectral line. This results in a broadening of the spectral line that is directly proportional to the number of collisions per second per atom. Thus by the same reasoning that was applied to natural broadening, the value of $\Delta\nu_L$ is given by $1/\pi\tau$ (or $1/2\pi\tau$ for resonance lines) where τ is the *collisional lifetime* of the atom.† Hence, $\Delta\nu_L$ for a resonance line is given by

$$\Delta\nu_L = \sigma_L^2 N \sqrt{2\pi RT(1/M_1 + 1/M_2)}/\pi \qquad (2\text{-}19)$$

where σ_L^2 is the collision cross-section for Lorentz broadening of the atom concerned (and is usually an effective value determined experimentally); N is the number of foreign species per unit volume

* The reason for exclusion of collisions by like atoms will be found in the next section.
† See Section 4.5.2 for the evaluation of τ.

(and is calculated from $N = 9740 \times p \times 10^{15}/T$ where p is the foreign gas pressure in torr); R is the gas constant; and M_1 and M_2 are the atomic or molecular weights of the colliding species.

The equation is useful for our purposes in that it allows a reasonable approximation to the actual value of $\Delta\nu_L$ to be calculated. However, the simple Lorentz theory is unable to deal with the accompanying shift and asymmetry of the line and even the dependence of $\Delta\nu_L$ on $T^{-1/2}$* is not in complete agreement with experimental evidence (which, however, is still of insufficient accuracy to allow complete justification of any one theory). More modern theories, which may be applied to particular regions of the line profile and particular ranges of gas pressure, have been developed to account for both the line shift and asymmetry. They have been reviewed in some detail by Breene,[8] but a useful approximation is

$$\Delta\nu_s = 0{\cdot}36\ \Delta\nu_L$$

where $\Delta\nu_s$ is the frequency shift of the line towards the red. Hence the shift of the line peak is of the same order as the collisional half-width.

The relatively high concentration of foreign molecules in flame absorption and fluorescence cells results in a value of $\Delta\nu_L$ similar to that of $\Delta\nu_D$. Hence the shift of the absorption line may be considerable and lead to a reduction in the "peak" absorbance observed at the central frequency (ν_0) of the line when a narrow, unshifted source line is used. Values of $\Delta\nu_L$ observed in low-pressure spectral sources containing only inert gas atoms are much smaller, and in devices such as hollow-cathode lamps and electrodeless discharge tubes $\Delta\nu_L$ and $\Delta\nu_s$ will usually be negligible compared with $\Delta\nu_D$. For example, the collision cross-section for broadening of the calcium 422·673 nm line by neon atoms has been found[9] to be $7{\cdot}4 \times 10^{-15}$ cm^2. If the above equation is used, and converted to wavelength units, the Lorentz half-width for a neon pressure of 10 torr is 0·00001 to 0·00003 nm depending on the temperature, whereas the Doppler half-width under the same conditions may vary from 0·002 to 0·0008 nm.

(*iv*) *Holtzmark broadening*

The term Holtzmark or resonance broadening is applied to the broadening of an excited level of an atom by interaction with atoms

* $T^{-1/2}$ at constant foreign gas pressure because of the dependence of N on T^{-1} as noted above.

of the same kind in the ground state, provided that the two levels are linked by a strong dipole transition. Thus, the electron vibration mentioned earlier is affected not merely by the collision but by a much stronger electrostatic interaction. In fact, if the same equation is used to approximate values of the resonance half-width, $\Delta\nu_R$, (with $M_1 = M_2$ for like atoms) it is found that values of σ_R are often several hundred times greater than comparable values of σ_L. It is important to note that although Lorentz and Holtzmark broadenings result in similar intensity distributions, no line shift or asymmetry is predicated for the latter and none is observed experimentally. As with Lorentz broadening, modern theory may be used to calculate $\Delta\nu_R$ directly without recourse to experimental collision cross-sections, but variations of detail occur in the formulae depending on the theory used and the range of application.

At the level of atom concentration encountered in AAS and AFS, resonance broadening is usually of little importance despite the high values of σ_R for most elements. For example, it is uncommon to introduce into flame cells solutions exceeding a concentration of 1 M, yet even this would result[10] in resonance broadening half-widths of only ca. 10^{-5} nm. Similarly, the atom concentration within most commonly used light sources would be too low to produce significant resonance broadening. However, $\Delta\nu_R$ may not be negligible in some of the intense spectral sources used for AFS, as these are often run at high power and produce sufficient atom concentrations to give severe self-absorption even over a very small path length.

(*v*) *Stark and Zeeman broadenings*

These well-known types of spectral line broadening are usually of little importance in analytical AAS and AFS. Stark broadening results from the splitting of the electronic levels of an atom due to the presence of strong, non-uniform electric fields or large densities of moving electrons or ions. It is negligible in flames and is also considered small in hollow-cathode lamps and electrodeless discharge tubes under the conditions employed in AAS and AFS. It becomes important, however, in arcs and sparks and may be a major factor in determining their relative unsuitability for use as sharp-line sources.

Zeeman broadening is a similar splitting effect observed in the presence of a strong magnetic field. Obviously the effect will be obtained only by deliberate application of such a field, so that it is of no importance here except to mention that it might be applied to produce a deliberate shift in a source or absorption line to obtain better overlap of the two.

(vi) Quenching broadening

It was mentioned earlier that radiationless transitions from the excited state of a resonance line are usually of no importance. While this may be true for a discharge within an inert gas, such as in most low-pressure spectral sources, quenching collisions can occur in flames as the result of the presence of foreign gas molecules with vibrational levels very close to the excited state of the resonance line. This may produce line broadening owing to the reduction in the lifetime of an excited atom by the same mechanism as natural broadening. The frequency distribution will have a collision rate dependence of the same type as Lorentz broadening and the half width, $\Delta\nu_Q$, for a resonance line may be determined from equation (2-19) if σ_L^2 is replaced by the quenching cross-section σ_Q^2. Values of σ_Q^2 observed experimentally are discussed in Section 4.5.4 and it will be seen that in certain flames σ_Q^2 may be sufficiently large for $\Delta\nu_Q$ to make a significant contribution to the overall broadening of many lines used in AAS and AFS.

(vii) Self-absorption broadening

This type of broadening is only of importance for lines that have their lower level at or near the ground state so that self-absorption within the source is easily induced. It is of particular relevance to analytical AAS and AFS, however, as by their very nature these techniques are mainly concerned with those spectral lines giving strong absorption. Indeed, it would appear[11,12] that self-absorption is the most serious of all the factors causing line broadening in sources used for AAS and AFS. This arises not because self-absorption itself multiplies the line width by a large factor, as it usually only increases the width by at most a factor of two or three, but because this multiplication factor is applied to the sum of all the other broadening factors. Thus self-absorption may often cause more serious line broadening of resonance lines than any other single broadening effect.

Self-absorption results in broadening of the spectral line simply because the maximum absorption occurs at the centre of the line; a greater *proportional* amount of radiation is therefore lost from the centre of an emission line than from its wings. Thus the shape of the line changes as well as its intensity, as illustrated in Fig. 2-4. The curves show the effect of increasing optical density on (a) Gaussian and (b) Lorentzian profiles for a source in which the absorbing atom concentration is uniform throughout, but the concentration of emitting atoms decreases linearly to zero between the centre and edge of the source. The effect of a given degree of self-absorption varies considerably with the model chosen for the spectral source,[13] but Fig. 2-4 is representative in that the relative increase in line width is usually somewhat greater for a collision-broadened profile than for a Doppler-broadened profile. In both cases, very high levels of self-absorption can result in self-reversal of the profile at its centre* so that a minimum is observed. Quite apart from any line width considerations, this condition can have an adverse effect on AAS and

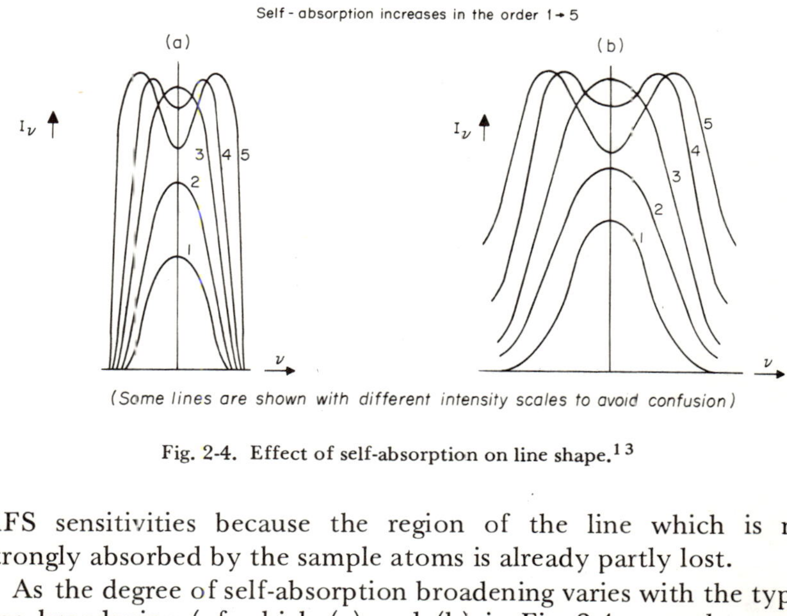

Fig. 2-4. Effect of self-absorption on line shape.[13]

AFS sensitivities because the region of the line which is most strongly absorbed by the sample atoms is already partly lost.

As the degree of self-absorption broadening varies with the type of line broadening (of which (a) and (b) in Fig. 2-4 are only extreme cases), it is not practical to give even an approximate expression for an "absorption half-width". It is only possible to calculate a value at a particular level of absorption, and for a specific model of the atomic distribution in the source, by considering the effect of the absorption on the intensity at each point in the line profile. This must be done by representing the variation with frequency of emission intensity and absorption by complex mathematical functions

* Except in the special cases where the concentration of emitting atoms is uniform throughout the source or actually increases towards the edge.

of the type mentioned in Sections 2.5.3 and 3.2.1. Some typical results are plotted in Fig. 2-5, which shows the variation of the half-width ratio (self-absorbed/non-self-absorbed) with the degree of self-absorption, p. This is defined such that the doubling of p means a doubling of the self-absorption, and so on. The three cases shown are all for a uniform distribution of absorbing atoms through the source and for the following distribution of emitting atoms: (1) concentration is uniform throughout the source; (2) concentration is

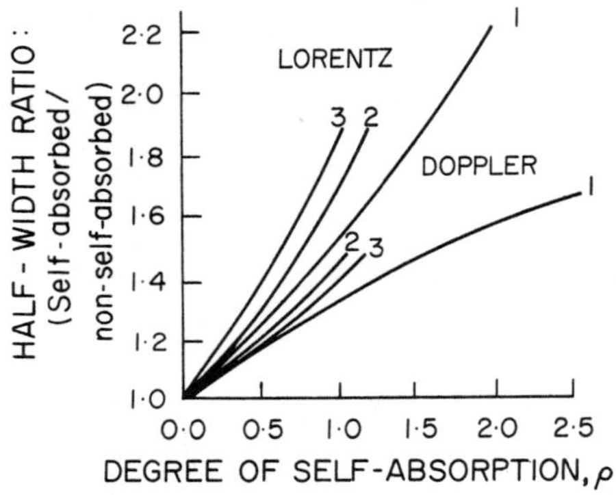

Fig. 2-5. The increase of half-width with self-absorption.[13]

zero at all points except the centre of the source; (3) concentration decreases uniformly between the centre and edge of the source.

(*viii*) *Hyperfine structure effects*

In the discussion so far it has been assumed that, although each separate spectral line isolated by a normal monochromator is not truly monochromatic, it can at least be represented by a single, uniform profile. In practice, when many spectral lines are examined under high resolution they are found to be made up of several closely spaced and probably partially overlapping profiles. These components, known as the hyperfine structure (hfs) of the line, are the result of small variations in the atomic energy levels due to the interaction of a nuclear spin with the spins of the electrons or to the presence of several isotopes. Thus, in many cases the "single" line will be quite complex, being made up of components due to several abundant isotopes of the element with the possibility of each of these components being further subdivided owing to the nuclear spin of the isotope. Isotopes with both atomic weight and number even (i.e. even numbers of both protons and electrons) have no nuclear spin and hence no extra subdivision of their line.

Each hfs component of a line must be considered as an individual spectral line with its own profile broadened by factors discussed above. Furthermore, the equations given in the next chapter for the calculation of absorption at a line must also be applied individually to each component. This would present no problems if the hfs components of both the emission and absorption lines were completely separated, i.e. if there was no overlap of their spectral profiles.* Unfortunately, the hfs components are very closely spaced so that the widths of the components are comparable to their separation in most common spectral sources. Unless the overlap is complete, so that the line may be regarded with fairly good accuracy as a single profile, any attempt to calculate the total line profile, after carrying out the individual line broadening calculations described above, is extremely complicated. Usually only an approximate graphical solution can be attempted.

The effect of hyperfine structure on absorption measurement is frequently neglected when, for example, line broadening estimates are made for the sharp line spectral sources used for peak absorbance measurements. This is partly because the effects on sensitivity are relatively small with most elements, and partly because the effect cannot be influenced (from the point of view of improving sensitivity) by practical measures. Thus, estimates of other broadening factors are still of some use even if the true line width due to the hfs apparently makes the Doppler, Lorentz, etc. widths appear somewhat irrelevant. However, it should be remembered that for certain aspects of absorption measurements the hfs can have important consequences and must not be overlooked (see Section 3.5.2).

2.5.3 The Observed Line Profile and Its Measurement

As already described, the profile of a line emitted by atoms in a spectral light source or absorbed in an absorption cell results from a combination of the factors discussed in Section 2.5.2. Unfortunately, this combination cannot be carried out by some simple addition process because the profiles are represented by complex mathematical functions. The methods used are too specialized to discuss here, but a simplified summary should prove useful in understanding

* In this particular context, the question of whether the spectrometer used is capable of resolving the components is irrelevant.

the effect of the different broadening factors on the complete line. The same techniques are also required in the derivation of expressions that relate the measured absorption to atomic concentration (see Section 3.2.1).

Suppose the variation in intensity with frequency for two particular types of broadening can be represented by functions of the type $I(x) = f(x)$ and $I(x) = \phi(x)$ where "x" is defined in terms of the central frequency of the profile; i.e. for any frequency ν, $x = (\nu - \nu_0)$. It may be shown that if both of these factors act at a point on the profile where $x = \nu'$, the resultant intensity at that point is given by

$$I(\nu') = \int_{-\infty}^{+\infty} f(x)\phi(\nu' - x)\, dx = F(\nu') \tag{2-20}$$

This process may be repeated for any number of broadening processes. The difficulty is that exact integration of the complex functions (i.e. mathematical expressions) represented by f, ϕ, F, etc. is impossible and approximate numerical methods must be used. These are greatly simplified if the functions can be combined in some way before attempting to carry out the integration. To do this the functions representing individual broadening factors are placed into two classes.

(*a*) *Gaussian functions* are of the form

$$f(x) = c\, e^{-x^2/\beta_2^2} \tag{2-21}$$

where c and β_2 are constants. Doppler broadening results in a function of this type.

(*b*) *Dispersion* (*Lorentzian*) *functions* are of the form

$$f(x) = c/(1 + x^2/\beta_1^2) \tag{2-22}$$

where c and β_1 are constants. Natural and Lorentzian broadening and the other types of broadening via collisions result in functions of this type.

If f and ϕ are both Gaussian and have the constants β_2' and β_2'', it may be shown that F is also Gaussian and has a constant given by

$$\beta_2^2 = (\beta_1')^2 + (\beta_2'')^2 \tag{2-23}$$

Similarly, if f and ϕ are both dispersion functions, we obtain a constant for the dispersion function, F, by

$$\beta_1 = \beta_1' + \beta_1'' \tag{2-24}$$

However, if f is Gaussian and ϕ dispersion (or vice versa), then F is known as a *Voigt function* and has both β_1 and β_2 parameters. It will obviously be a function of the form

$$F(x) = (c\ e^{-x^2}/\beta_2^2)/(1 + x^2/\beta_1^2) \tag{2-25}$$

Furthermore, if two Voigt functions are combined in the same way the result is another Voigt function whose β_1 and β_2 parameters are obtained from the two formulae given above.

Thus any symmetrical* line profile can be represented by a Voigt function. This is most important because the various half-widths are constants (for a particular profile) equivalent to the β_1 and β_2 parameters. Hence the two groups of half-widths may be added independently to give two parameters ($\Delta\nu_{\text{Gauss.}}$ and $\Delta\nu_{\text{D isp.}}$) which define the Voigt profile representing the actual line. Tables of the function have been calculated by evaluating the integral so that the total half-width of the Voigt profile may be obtained (see, for example, references 14 and 15).

If line widths are to be determined experimentally, it is necessary first to use a spectrometer of sufficiently high resolving power, and secondly to analyse the recorded profile by a reversal of the above procedure. This must now allow for a further line broadening factor—that due to the instrument used to observe the line, as no spectrometer can have an infinitely high resolving power. Most of the factors causing instrumental broadening may be represented by Gaussian or dispersion functions so that the *instrumental profile* will itself be a Voigt function. Although spectrometers of sufficiently high resolution are readily available,[16,17] such practical measurements are made difficult by the problems of profile analysis—since factors such as self-absorption, line asymmetry and shift and hyperfine structure add further complications. For this reason, accurate experimental values of line widths directly of interest in AAS and AFS are relatively few. It also explains why absorption techniques requiring actual measurements of absorption line profiles have seen little practical application (see Section 3.5.1). With the current interest in absolute analysis by AAS, the accurate estimation of actual line widths and the ability to record absorption line profiles remain important problems of concern to the analytical chemist.

* Hence the effects of collisional shift, hyperfine structure, etc. must be excluded from this treatment.

REFERENCES

1. See, for example, H. G. Kuhn, "Atomic Spectra", Longmans, Green, London, 1962.
2. See, for example, R. Mavrodineanu and H. Boiteux, "Flame Spectroscopy", Wiley, New York, 1965, Ch. 17.
3. A. G. Gaydon and H. G. Wolfhard, "Flames, Their Structure, Radiation and Temperature", 3rd ed., Chapman and Hall, London, 1970, Ch. 9.
4. A. G. Gaydon, "The Spectroscopy of Flames", Chapman and Hall, London, 1957, pp. 66-72.
5. H. E. White and A. Y. Eliason, *Phys. Rev.*, 1927, **29**, 790. (The tables are reproduced in several textbooks on atomic spectra; e.g. in ref. 1 above.)
6. W. L. Wiese, "Present status of our knowledge of atomic transition probabilities", *Proc. Xth Colloq. Spectros. Inter., Maryland, 1962,* pp. 37-56.
7. C. Th. J. Alkemade, *Applied Optics,* 1968, **7**, 1261.
8. R. G. Breene, "The Shift and Shape of Spectral Lines", Pergamon Press, Oxford, 1961.
9. N. P. Penkin and L. N. Shabanova, *Opt. Spectrosc.,* 1968, **25**, 446.
10. I. Rubeska and V. Svoboda, *Anal. Chim. Acta,* 1965, **32**, 253.
11. A. I. Bodretsova, B. V. L'vov and V. I. Mosichev, *J. Appl. Spectrosc.,* 1966, **4**, 149.
12. G. F. Kirkbright and M. Sargent, *Spectrochim. Acta,* 1970, **25B**, 577.
13. T. Tako, *J. Phys. Soc. Japan,* 1961, **16**, 2016.
14. J. T. Davies and J. M. Vaughan, *Astrophys. J.,* 1963, **137**, 1302.
15. D. G. Hummer, *Mem. Roy. Astron. Soc.,* 1965, **70**, 1.
16. S. Tolansky, "High Resolution Spectroscopy", Methuen, London, 1947.
17. P. Jacquinot, *Reports Progr. Phys.,* 1960, **23**, 268.

CHAPTER 3

Theory of Atomic Absorption Measurements

3.1 INTRODUCTION

The opportunity has been taken in several places in the preceding pages to point out the important consequences of finite spectral line-widths on the theory and practice of AAS. In order to discuss these further we find it necessary to define several parameters that may be used to represent atomic absorption and to discuss at some length the conditions under which they correspond to experimentally determined quantities. It will be seen that, in the most widely used form of AAS that has evolved, the theoretical complications, and their consequent experimental difficulties, have been reduced to the minimum for the application of atomic absorption measurements to chemical analysis. Unfortunately, restriction of the discussion to correspond to this minimum can result in considerable misunderstanding and misinterpretation of anomalies that apparently occur when the limits of the approximations are approached. Moreover, it becomes more difficult to appreciate the reasons for the choice of the absorption techniques used at present by analytical chemists and to visualize the problems involved with the development of these techniques into a sophisticated means of absolute analysis.

In view of this, we shall discuss at some length three quantities: the *absorption coefficient*; the *total absorption factor*; and the *absorbance*. The conditions under which these quantities may be used will be indicated and the boundaries of these conditions will be related to the "abnormalities" that are found in practical measurements; for example, the curvature of calibration graphs, which ideally should be straight lines. It will be seen that at the present time the widespread use of AAS is based mainly on the straightforward interpretation of *absorbance* measurements made with a sharp-line spectral source and that the other two quantities are of much less importance to the analytical chemist. Finally, the possibilities of developing AAS into an absolute analytical technique rather than into one that relies on calibration standards will be discussed.

3.2 THE ABSORPTION COEFFICIENT

3.2.1 Definition of the Absorption Coefficient, k_ν

The absorption coefficient at a discrete frequency, ν, is defined by the equation

$$I_\nu = I_\nu^0 \ e^{-k_\nu L} \tag{3-1}$$

where I_ν^0 and I_ν are the initial and final intensities of radiation of frequency ν passed through an absorption cell of length L. A plot of k_ν against frequency represents the profile of the absorption line, which is of the same form as that shown in Fig. 2-3 for an emission line. Thus, values of k_ν may only be measured directly by using a monochromating device able to resolve the profile of the absorption line; i.e. one that has a spectral bandpass considerably less than the line width.

k_ν may be expressed as a function of frequency for a general line profile by using equation (2-20). Substituting the expressions given to represent Gaussian and Lorentzian functions, and replacing the particular point ν' with a general frequency ν, we obtain

$$k_\nu = C \cdot \int_{-\infty}^{\infty} \frac{e^{-x^2/\beta_2^2}}{1 + (\nu - \nu_0 - x)^2/\beta_1^2} \cdot dx \tag{3-2}$$

This may be rearranged as

$$k_\nu = C \cdot \frac{\beta_1^2}{\beta_2^2} \int_{-\infty}^{\infty} \frac{e^{-(x/\beta_2)^2}}{(\beta_1/\beta_2)^2 + (\nu - \nu_0 - x)^2/\beta_2} \cdot dx \tag{3-3}$$

Equation (3-3) is considerably simplified by writing

$$y = x/\beta_2, \quad \omega = (\nu - \nu_0)/\beta_2 \quad \text{and} \quad a = \beta_1/\beta_2$$

Hence, we obtain

$$k_\nu = Ca\beta_1 \int_{-\infty}^{\infty} \frac{e^{-y^2}}{a^2 + (\omega - y)^2} \cdot dy \tag{3-4}$$

Such an expression for k_ν was originally developed from the basic theories of line broadening by Voigt,[1] and a comparison with equation (3-4) allows the three constants describing the Gaussian and Lorentzian functions to be related to the half-widths of the absorption profile:

$$\beta_1 = \frac{\Delta\nu_c}{2}; \qquad \beta_2 = \frac{\Delta\nu_D}{\sqrt{2 \ln 2}} \tag{3-5}$$

and

$$C = \frac{2k'_0}{\beta_1 \pi} = \frac{2k'_0}{\pi \Delta\nu_c} \tag{3-6}$$

where $\Delta\nu_c$ is the sum of the half-widths of the Lorentzian-type functions (i.e. $\Delta\nu_N$, $\Delta\nu_L$, $\Delta\nu_Q$ and $\Delta\nu_R$ of Section 2.5.2), $\Delta\nu_D$ is the Doppler half-width (i.e. the Gaussian function half-width) and k'_0 is an ideal* quantity representing the peak absorption coefficient which would be obtained with an absorption line *broadened only by the Doppler effect.*

* k'_0 is obviously an ideal quantity as it is impossible to obtain an absorption line without at least the Lorentzian contribution from natural broadening.

After substitution of these relations, equation (3-4) becomes

$$k_\nu = \frac{k'_0 a}{\pi} \int_{-\infty}^{\infty} \frac{e^{-y^2}}{a^2 + (\omega - y)^2} \cdot dy \tag{3-7}$$

which is the form of the *Voigt expression* most widely quoted following its derivation and use by Mitchell and Zemansky.[2] Equation (3-7) relates k_ν to the number of absorbing atoms, which is included in k'_0 (see Section 3.2.2), and is necessary if equation (3-1) is to be used for theoretical calculations of the variation of light absorption with concentration (*curves of growth*). For this purpose equation (3-7) must be evaluated either by using computer calculated tables of the integral,[3,4] or by replacing it with an approximate analytic expression. Several of these have been suggested, and the most recent[5,6] can give quite accurate results over a wide range of line-widths and absorption values. For example, equation (3-8)* gives k_ν/k'_0 with a maximum error of only +2·5% and −5% over most conditions likely to be required.

$$\begin{aligned} k_\nu = k'_0 & \left[1 - \frac{\Delta\nu_c\sqrt{\ln 2}}{\Delta}\right] \exp\left[-2{\cdot}772\left(\frac{\nu-\nu_0}{\Delta/\sqrt{\ln 2}}\right)^2\right] \\ & + \left[\frac{\Delta\nu_c\sqrt{\ln 2}}{\Delta}\right] \frac{1}{1+4\left(\dfrac{\nu-\nu_0}{\Delta/\sqrt{\ln 2}}\right)^2} + 0{\cdot}016\left[1 - \frac{\Delta\nu_c\sqrt{\ln 2}}{\Delta}\right] \\ & \times \left[\frac{\Delta\nu_c\sqrt{\ln 2}}{\Delta}\right] \left\{\exp\left[-0{\cdot}4\left(\frac{\nu-\nu_0}{\Delta/\sqrt{\ln 2}}\right) 2{\cdot}25\right] \right. \\ & \left. - \frac{10}{10 + \left[\dfrac{\nu-\nu_0}{\Delta/\sqrt{\ln 2}}\right]^{2\cdot25}}\right\} \end{aligned} \tag{3-8}$$

Δ is the observed half-width of the absorption line (resulting from all broadening effects).

3.2.2 *The Integrated Absorption Coefficient, K*

Although its dependence on frequency complicates the use of k_ν, integration of k_0 with respect to ν over the whole line results in the most useful quantity, K, which is independent of frequency. It may

* Equation (3-8) is obtained by using the notation of this book in equation (5) of reference 6.

be shown (Mitchell and Zemansky,[2] p. 95) that K is equivalent to the quantity I_{ab}/I_0 as defined in equation (2-10). Hence we obtain

$$K = \int k_\nu \, d\nu = \frac{h\nu_0}{c}(B_{ji}N_j - B_{ij}N_i) \tag{3-9}$$

where ν_0 is the frequency at the centre of the line and the small variation of ν throughout the line has been neglected.

In many absorption cells, including flames, the value of N_i is quite small compared with that of N_j (see Section 2.4.2) so that, to a good approximation, we may simplify equation (3-9) to

$$K = \frac{h\nu_0}{c} B_{ji}N_j \tag{3-10}$$

Substitutions for B_{ji} may be made from equations (2-2) to (2-6) giving

$$K = \frac{\lambda_0^2}{8\pi}\frac{g_i}{g_j} A_{ij}N_j \tag{3-11}$$

or

$$K = \frac{\pi e^2}{mc} f_{ji}N_j \tag{3-12}$$

These equations are important as they show that K may be represented by a simple, linear relationship with N_j and that this relationship is quite independent of whatever physical processes are responsible for the formation of the absorption line. Unfortunately, it is difficult to make direct use of this fact to measure atomic populations by absorption because K can only be measured by taking the area under the actual absorption profile. It is important to remember that the value of the integrated absorption coefficient so obtained *does not* correspond to the absorption measured with a continuum source and a conventional monochromator (see Section 3.3).

Mitchell and Zemansky[2] also give an alternative integration of k_ν which shows that

$$K = \frac{2\sqrt{\ln 2}}{\sqrt{\pi}\,\Delta\nu_D} k_0' \tag{3-13}$$

where k'_0 is the peak absorption coefficient for a purely Doppler broadened line, as explained in Section 3.2.1. Thus, after substituting for K, we obtain

$$k'_0 = \frac{2\sqrt{\ln 2}}{\sqrt{\pi}\Delta\nu_0} \frac{\lambda_0^2}{8\pi} \frac{g_i}{g_j} A_{ij} N_j \qquad (3\text{-}14)$$

This equation is necessary for the calculation of curves of growth of absorption with atom concentration.

3.3 THE TOTAL ABSORPTION FACTOR, A_T

3.3.1 The Definition of A_T

In view of the difficulties already mentioned in measuring k_ν and K, it is necessary to have some other indication of how much radiation is actually absorbed during an experiment. The total absorption factor, A_T, may be calculated from a knowledge only of the amounts of light entering and leaving the absorption cell but, unlike k_ν and K, it depends on both the emission line from the source and the absorption line in the cell. Nearly all analytical AAS measurements are directly concerned with A_T as the variables normally used in AAS (percent absorption and absorbance) are both closely related to A_T.

As the total absorption factor is defined in terms of the total amount of light passing through the cell at a particular spectral line, the intensities used are integrated with respect to the complete absorption line. Hence A_T is given by

$$A_T = \frac{\text{Absorbed radiation intensity}}{\text{Incident radiation intensity}} = \frac{\Delta I}{I_0} \qquad (3\text{-}15)$$

where $\Delta I = I_0 - I$, where I is the amount of radiation leaving the cell. The *percent absorption* commonly used in AAS is defined as $[(\Delta I/I_0) \times 100]$, so that for most purposes it may be considered as an alternative notation for values of A_T.

Although A_T is easily measured experimentally, it is as difficult to relate it theoretically to the number of absorbing atoms as when k_ν is employed. In fact, it is related by expressing A_T in terms of k_ν and evaluating the integral approximately for special cases of interest in AAS. Now, by definition

$$I_0 = \int I_\nu^0 \, d\nu \quad \text{and} \quad I = \int I_\nu \, d\nu$$

so that we may rewrite equation (3-15) as

$$A_T = \frac{\int I_\nu^0 \, d\nu - \int I_\nu^0 \, e^{-k_\nu L} \, d\nu}{\int I_\nu^0 \, d\nu} \tag{3-16}$$

by using equation (3-1). In order to proceed further it is necessary to assume that I_ν^0 will always be integrated over the same limits as k_ν is; i.e. to impose the condition that the integration will be made over the source line-width if it is narrower than the absorption line, or the absorption line-width if it is narrower than the source line. Equation (3-16) may then be rewritten as

$$A_T = \frac{\int I_\nu^0 \, d\nu \int (1 - e^{-k_\nu L}) \, d\nu}{\int I_\nu^0 \, d\nu}$$

$$= \frac{I_0 \int (1 - e^{-k_\nu L}) \, d\nu}{I_0}$$

i.e.

$$A_T = \int (1 - e^{-k_\nu L}) \, d\nu \tag{3-17}$$

The limits of this integration are considered as a number of special cases, depending on the width of the narrower of the two lines.

3.3.2 *Evaluation of A_T for a Wide Source Line*

This situation applies if the background source emits a continuum or a very broad, self-absorbed line such as is obtained from an arc or flame. As the absorption line is much narrower than that from the source, we may assume that the total absorbed radiation will be given by integration of equation (3-17) across the entire absorption line, i.e. from "zero to infinity". This is quite easy for small or large values of the optical density, $k_\nu L$ (i.e. of the absorption).

(a) Small $k_\nu L$

In this case equation (3-17) may be simplified* to

$$A_T = \int_0^\infty k_\nu L \, d\nu = L \int_0^\infty k_\nu \, d\nu = LK \tag{3-18}$$

* Since $(1 - e^{-x}) \to x$ as $x \to 0$.

Hence, from equation (3-12),

$$A_T = \frac{\pi e^2}{mc} f_{ji} N_j L \tag{3-19}$$

i.e. A_T will increase linearly with N_j (and therefore with the concentration of absorbing atoms*) at low optical densities.

(*b*) *Large* $k_\nu L$

When the optical density becomes sufficiently large, the central region of the line shows complete absorption. Further increases in optical density then result only in changes at the edge of the line profile, as shown in Fig. 3-1. Thus, the form of the line profile (i.e.

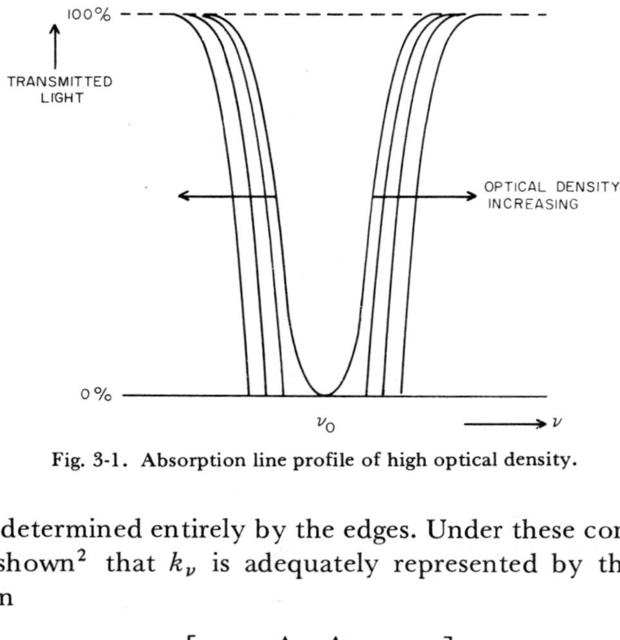

Fig. 3-1. Absorption line profile of high optical density.

of k_ν) is determined entirely by the edges. Under these conditions, it may be shown[2] that k_ν is adequately represented by the analytic expression

$$k_\nu = \left[\frac{\Delta\nu_0 \Delta\nu_c}{4\sqrt{\pi}\sqrt{\ln 2}(\nu - \nu_0)^2}\right] k'_0 \tag{3-20}$$

Hence the complete integration of equation (3-17) is possible without undue difficulty, the resultant expression for A_T being

$$A_T = \left[\frac{k'_0 \Delta\nu_D^2 a\sqrt{\pi}L}{\ln 2}\right]^{1/2} \tag{3-21}$$

* But see Section 3.5.3.

After substitution for k'_0 using equation (3-13), we obtain

$$A_T = \left[\frac{2\pi e^2 \Delta\nu_D a f_{ji} N_j L}{mc\sqrt{\ln 2}}\right]^{1/2} \tag{3-22}$$

i.e. at high optical densities, A_T is proportional to the square root of the concentration of absorbing atoms.

3.3.3 Evaluation of A_T for a Sharp Line Source

This is the case when a low-current spectral lamp is used as the light source, as in most analytical AAS measurements. As the source line width, $\Delta\nu_s$, is small when compared with the width of the absorption line, we may assume that absorption will occur over the full width of the source line. Thus, the integration limits of equation (3-17) will be zero and $\Delta\nu_s$. In practice, the usual Gaussian function is converted into a triangular one and the actual integration limit is $\Delta\nu_s\sqrt{\pi}/2\sqrt{\ln 2}$.

A further consequence of using a sharp line source is that we are only concerned with the very centre of the absorption line. Thus it is valid to assume that k_ν is a constant and equal to the peak absorption coefficient of the line. This is not equal to k'_0 since the line is not purely Doppler-broadened. However, to a good approximation we may write $k_{(\nu=0)} = bk'_0$, where b is a constant of proportionality. Hence, we obtain

$$A_T = (1 - e^{bk'_0 L}) \int_0^{\frac{\Delta\nu_S\sqrt{\pi}}{2\sqrt{\ln 2}}} d\nu \tag{3-23}$$

(a) Small $k'_0 L$

Using $(1 - e^{-x}) = x$ for small values of x we obtain

$$A_T = bk'_0 L \frac{\Delta\nu_s\sqrt{\pi}}{2\sqrt{\ln 2}} \tag{3-24}$$

after performing the integration in equation (3-23). Substitution for k'_0 as before gives

$$A_T = \frac{b\Delta\nu_s}{\Delta\nu_D}\frac{\pi e^2}{mc} f_{ji} N_j L \tag{3-25}$$

i.e. A_T varies linearly with N_j.

(*b*) *Large* k'_0L

As we are now concerned only with the line centre, high optical densities will result in very strong absorption. Thus, to a good approximation, $e^{-bk'_0L} = 0$. Hence, on integration, equation (3-23) becomes

$$A_T = \frac{\Delta\nu_s\sqrt{\pi}}{2\sqrt{\ln 2}} \tag{3-26}$$

i.e. at high optical densities the absorption reaches a constant value independent of N_j.

3.4 THE ABSORBANCE, A, FOR A SHARP LINE SOURCE

An expression for the absorbance, A, may be evaluated without the need to assume limiting values of the optical density, as is required to estimate A_T. On the other hand, the derivation is only simple if a narrow source line width is assumed, i.e. if we assume that k_0 has a constant value. Thus, we restrict ourselves to the use of A with such a source; the fact that the derivation does not require an assumption based on the extreme values of $k_\nu L$ makes this expression extremely useful (see Section 3.5).

The absorbance is defined by

$$A = \log_{10}\left(\frac{I_0}{I}\right) \tag{3-27}$$

where I_0 and I are the incident and transmitted intensities at the absorption cell.

A is related to A_T by

$$A = \log_{10}\left(\frac{1}{1 - A_T}\right) \tag{3-28}$$

Thus, there is no simple relationship between the two except at low values of A_T ($< 0{\cdot}10$) when a graph of A against A_T is, to a good approximation, a straight line.

Substitution into equation (3-27) of

$$I_0 = \int_0^{\Delta\nu_s} I_\nu^0 \,.\, d\nu$$

and

$$I = \int_0^{\Delta\nu_s} I_\nu \, d\nu = \int_0^{\Delta\nu_s} I_\nu^0 \, e^{-k_\nu L} \, d\nu$$

gives

$$A = \log_{10} \left[\frac{\int_0^{\Delta\nu_s} I_\nu^0 \, d\nu}{\int_0^{\Delta\nu_s} I_\nu^0 \, e^{-k_\nu L} \, d\nu} \right] \tag{3-29}$$

If we now assume that over the limits of integration (i.e. over the narrow region at the centre of the absorption line defined by the source line-width) k_ν is a constant and equal to bk_0' as before, we may take the exponential term outside of the integration. Hence,

$$A = \log_{10} \left[\frac{1}{e^{-bk_0' L}} \right]$$

$$= \log_{10} \left(e^{bk_0' L} \right)$$

i.e.

$$A = 0{\cdot}434 \, bk_0' L \tag{3-30}$$

Substitution for k_0' as before gives

$$A = 0{\cdot}434 \, b \, \frac{2\sqrt{\ln 2}}{\Delta\nu_D \sqrt{\pi}} \frac{\pi e^2}{mc} f_{ji} N_j L \tag{3-31}$$

Thus a plot of absorbance against N_j should be rectilinear over a wide concentration range (but see Section 3.5).

3.5 PRACTICAL ATOMIC ABSORPTION METHODS

3.5.1 Historical Approach

Atomic absorption spectroscopy has been used as a tool of fundamental research for many years, particularly in astrophysical work and for the determination of spectral constants, such as transition probabilities and oscillator strengths. Such applications require techniques that readily allow a theoretical analysis of the results, and, as mentioned several times in earlier chapters, these methods are too complex and too restricted in application to be of much use for routine chemical analysis.

From a theoretical standpoint, the most satisfactory means of applying absorption measurements is the use of equation (3-10) in conjunction with a determination of the integrated absorption coefficient, K. For this purpose the light source is a lamp whose emission intensity is of known spectral distribution, usually a continuum. The transmission of the absorption cell is then measured at a number of points on the absorption line and equation (3-1) is used to calculate k_ν at each of these frequencies. After k_ν has been plotted as a function of frequency, K corresponds to the area under the curve. Although this method has been used quite often by physicists, the experimental difficulties are great when conventional pressures and temperatures are used in the absorption cell. Under these circumstances the overall width of the absorption line will usually be only a very small fraction of a nanometre. Thus a spectrometer of extremely high resolving power is required to measure k_ν at different points on the line profile. Even when such an instrument is available (e.g., a Fabry-Perot interferometer) its use presents many practical problems. In certain cases, the problem may be considerably reduced by measuring k_ν directly with a sharp line spectral source whose emission line may be shifted slightly to allow measurements at different points on the absorption line profile. The Zeeman scanning technique, in which a strong magnetic field is applied to the spectral lamp to shift the frequency of the emission line, has been used to demonstrate this.[7] Another, perhaps more obvious solution, which has been used a number of times for theoretical studies, is to operate an absorption cell at extremely high foreign (i.e. non-absorbing) gas pressures. This produces a very wide absorption line that may be effectively resolved by a conventional spectrophotometer. Such experiments, extended to changes in temperature as well as in pressure, have proved[8] that K is quite independent of conditions in the absorption cell.

The problems of resolving the absorption line profile have also been avoided for practical applications of AAS by measuring the total absorption factor, A_T. However, as we have seen in Section 3.3, this is only simply related to concentration at very low absorption values, when it is proportional to N_j, and at very high absorption values, when it is proportional to $\sqrt{N_j}$. In the intermediate region which, in most experimental arrangements, covers a wide range of atomic populations of interest, the relationship between A_T and N_j cannot be expressed in simple analytic terms. Instead, it is necessary to compute theoretical curves of growth (of A_T against N_jL) using the Voigt expression with assumed values of the "a" parameter (see Section 3.2.2) and to compare them with experimental results.[9] Obviously, such techniques are not necessary for straightforward chemical analysis, but the lack of a constant relationship between A_T and N_j in this important region greatly reduces the usefulness of AAS with a continuum source (since it is A_T that is being measured). At very high absorption levels the $\sqrt{N_j}$ region may be passed and A_T increases more slowly than expected, eventually reaching a limiting value. This results from the fact that under these circumstances most of the effect of increasing absorption is apparent at the edges of the absorption line (as explained earlier and shown in Fig. 3-1). The narrow monochromator bandpass usually used for AAS measurements with a continuum may then mean that these changes are not detected.

Several other sophisticated techniques have also been suggested[10] in which, instead of using a continuum source to determine A_T, a line spectral source is used to measure the total absorption (the so-called "total line absorption factor A_L"). However, the applicability of these methods is again rather limited and they have not been widely used.

3.5.2 The Walsh Method of Peak Absorbance Measurements

As we have seen, most of the traditional atomic absorption methods are unsatisfactory for routine use as analytical techniques. The acceptability of AAS for such use came only after the suggestion by Walsh[11] that measurement of the absorbance, A, at the peak of the absorption line profile would give a linear relationship with concentration over a wide range of absorbance values (see equation (3-31)). In theory, such measurements should have the same freedom from the effects of changes in the absorption line profile as those of the integrated absorption coefficient, K, without the serious experimental difficulties that arise when K is measured (as it is only necessary to measure the total absorption of a sharp line spectral source).

In practice, the technique works sufficiently well with both flame and other atom cells to allow its widespread routine use. Furthermore, because measurements are made in the spectral region of strongest absorption, there is a high effective "contrast" of the absorption cell with respect to the light source. Thus, this method can also give greater sensitivity than, say, measurement of A_T with a continuum source. However, under actual experimental conditions the relationship between A and N_j is not as simple as it would appear from the discussion given above, and the range over which the absorbance varies linearly with N_j is often much smaller than might be expected. There are several possible reasons for this, not all of which will necessarily apply in every case.

1. The half-width of the emission line from the source is not always sufficiently small relative to that of the absorption line to justify the assumption that a constant value of k_ν (equal to bk'_0) may be used in the evaluation of A. This fact is not always appreciated, as it would seem that the very low pressures and the low temperatures in such sources as the hollow cathode lamp must always give a line width much narrower than that in the usual flame absorption cells. However, unless these lamps are run at low currents self-absorption of the resonance lines is easily produced for many elements and, as we have seen in Section 2.5.2(vii), this can lead to quite serious broadening of the lines. When low-temperature atom sources (such as the graphite filament) are used, the source line width may not be negligible for absorbance measurements even in the absence of self-absorption.

2. In flame absorption cells, and others with relatively high foreign gas pressures, the Lorentz shift of the absorption line peak will be considerable (see Section 2.5.2(iii)). As the source emission line is unaffected the absorbance measurement will no longer be a true peak value. This can result in a reduction of the expected absorbance by as much as 25%, depending on the absorption line profile and the extent of the shift. Willis[12] has recently attempted practical correction of absolute absorbance measurements for the effects of absorption line shift.

3. The important effects of the hyperfine structure (hfs) of many spectral lines on their effective width has already been discussed in Section 2.5.2(viii). However, our derivation of an expression for the absorbance* has assumed the existence of a single, isolated resonance line. Thus if the equation is to be rigorously applied we must use it for each individual component of the emission and absorption line rather than for the line as a whole. This always results in a lower absorbance than would be obtained with a single line, the error

* And, indeed, for all of the other absorption quantities used to which this source of error also applies.

depending on the number of hfs components and the relative values of the component separation, the line widths and (if any) the absorption line shift. The linearity of the calibration graphs is also reduced. Both effects have been discussed in some detail by L'vov[10] who shows that the errors are minimized when the line-widths are large enough to make the hfs separation negligible. Of course, if the emission line is too wide we merely introduce a different error into the absorbance measurements. Thus, a common example will be the situation in which the absorption line is sufficiently broadened to be considered as a single component but the hfs of the emission line is important. Some values calculated for this case by L'vov are shown

TABLE 3-1
Relative absorbance values calculated for two-component and single-component emission lines at different levels of absorption[10]

	Absorbance for a two-component emission line/Absorbance for a single-component emission line					
Ratio of hfs component separation to line-width	2:1	1:1	1:2	1:3	1:4	1:5
Absorbance for a single-component line						
0·1	0·36	0·87	1·0	1·0	1·0	1·0
0·5	0·35	0·75	0·93	0·97	0·99	1·0
1·0	0·34	0·65	0·86	0·92	0·96	0·98
1·5	0·32	0·57	0·81	0·89	0·94	0·96
2·0	0·30	0·54	0·76	0·86	0·92	0·94

in Table 3-1; the further simplifying assumptions were made that the emission line consists of just two components with a pure Lorentz profile for the absorption line. It can be seen that in serious cases the absorbance may be reduced to one-third of the expected value. Practical corrections for the effect of hfs on absorbance values have recently been described by Willis.[12]

4. Throughout our discussion we have related absorbance values* to the ground state population, N_j, with the implied assumption that N_j is directly proportional to the sample concentration. In practice there may be a more complex relationship owing to compound formation and dissociation in the absorption cell. This problem is particularly important with flame cells and is discussed for them in Chapter 6.

* And the other quantities.

3.5.3 The Possibility of Absolute Analysis by Atomic Absorption Spectroscopy

Numerous equations have been derived which, at first sight, seem to provide useful relations between atomic concentration and absorption measurements. The preceding discussion (particularly in Section 3.5.2), however, has also illustrated that in some cases the use of the equations is restricted by the extreme practical difficulties in making the required measurements, whereas in others the derivation of the expressions has involved approximations that limit their applications to certain idealized situations. It should also be mentioned that the equations include constants, such as transition probabilities, that have been determined with only poor precision for many elements and spectral lines. Finally, the expressions themselves still provide only a value of N_j in the absorption cell; in many instances relating this to the sample concentration is itself a major problem.

In view of these facts, the present situation with regard to absolute analysis may be summarized as follows. By carrying out carefully and precisely a complex set of measurements it is possible to determine with quite good accuracy the absolute concentrations of a few elements that meet fairly well the approximations required by the theoretical expressions. Furthermore, if only poor accuracy is sufficient, the measurements may be extended to a wider range of applications. Indeed, several usable techniques have been reported recently. Most interest has centred on the use of these techniques to determine the free atom fractions of various elements in the flames commonly used as absorption cells in AAS. The techniques usually involve direct application of equation (3-19) by measuring the total absorption (A_T) with a continuum source[12-15] or direct application of equation (3-31) by measuring the absorbance, A with a sharp line source.[12,15-18] It is assumed that by working at low absorption levels with carefully chosen experimental conditions, the effect of the approximations made in order to derive the theoretical relations between signal and concentration will not be significantly large in comparison with the experimental errors. However, even after allowance has been made for errors of theory and/or technique it is found that there is a consistent discrepancy between the results obtained with the two techniques. The most thorough comparison of the two techniques is that of Willis[12] who suggests that more realistic results will be obtained when a continuum source is used unless corrections are made for the effects of hyperfine structure and line shift on the peak absorbance method. He also points out the serious uncertainties that occur in many of the "*f*-values" needed in either technique.

Thus, we may conclude that the effort recently devoted to the subject has considerably clarified the nature of the problem and has indicated the most promising areas for attempts to improve the techniques. While this should make it possible to obtain reasonable results by using great care and a realistically chosen problem, it seems almost certain that these techniques will see little practical analytical application for some time to come. Compared with the simplicity, accuracy and range of application of conventional AAS using calibration by standards, they all remain impracticable for routine use.

REFERENCES

1. W. Voigt, *Münch. Ber.*, 1912, p. 603.
2. A. C. G. Mitchell and M. W. Zemansky, "Resonance Radiation and Excited Atoms", Cambridge Univ. Press, Cambridge, 1934.
3. D. W. Posener, *Australian J. Phys.*, 1959, 12, 184.
4. G. D. Finn and D. Mugglestone, *Mon. Not. R. Astr. Soc.*, 1965, **129**, 221.
5. B. H. Armstrong, *J. Quant. Spec. Radiat. Transfer,* 1967, 7, 61.
6. E. E. Whiting, *J. Quant. Spec. Radiat. Transfer,* 1968, **8**, 1379.
7. T. Hollander, B. J. Jansen, J. J. Plaat and C. T. J. Alkemade, *J. Quant. Spec. Radiat. Transfer,* 1970, **10**, 1301.
8. J. C. Strijland and A. J. Nanassy, *Physica,* 1958, 24, 935.
9. R. B. King, B. R. Parnes, M. N. Davis and K. H. Olsen, *J. Opt. Soc. Am.*, 1955, **45**, 350.
10. B. V. L'vov, "Atomic Absorption Spectrochemical Analysis", Hilger, London, 1970, Ch. 1.
11. A. Walsh, *Spectrochim. Acta,* 1955, 7, 108.
12. J. B. Willis, *Spectrochim. Acta,* 1971, **26B**, 177.
13. L. De Galan and J. D. Winefordner, *J. Quant. Spec. Radiat. Transfer,* 1967, 7, 251.
14. P. J. Th. Zeegers, W. P. Townsend and J. D. Winefordner, *Spectrochim. Acta,* 1969, **24B**, 243.
15. J. B. Willis, *Spectrochim. Acta,* 1970, **25B**, 487.
16. S. R. Koirtyohann and E. E. Pickett, *Proc. XIIIth Colloq. Spectroscop. Intern., Ottawa, 1967,* p. 270.
17. C. S. Rann, *Spectrochim. Acta,* 1967, **23B**, 827.
18. L. De Galan and G. F. Samaey, *Anal. Chimica Acta,* 1970, **50**, 39.

CHAPTER 4

Theory of Atomic Fluorescence Measurements

4.1 INTRODUCTION

As absorption is the initial process that must occur before fluorescence can be observed, the expressions relating atomic fluorescence intensity to atom concentration are, naturally, similar to those already derived in Chapter 3 for atomic absorption. They are, however, more complex. This is partly due to the extra process involved; we must now account for the proportion of the absorbed energy which is re-emitted as fluorescence and allow for the possibility that the fluorescence radiation will suffer self-absorption before it can be measured. As the absolute intensity of the source radiation is of importance in AFS, any loss of this radiation due to self-absorption before it reaches the region of the cell from which fluorescence is measured will also be important. The other complicating factor is the number of different types of fluorescence that are possible. The yield of fluorescence obtained with a given absorption of radiation will vary with the type of fluorescence transition, and the effect of re-absorption of fluorescence will depend on the population of the lower level of this transition.

In view of these difficulties no attempt will be made to derive complete expressions for the fluorescence intensity in a single step. Instead, we shall first relate an ideal fluorescence intensity to the total absorption factor, A_T, derived in Chapter 3. Such a quantity has limited practical value since it is dependent on the assumptions that all absorption of the source radiation occurs within the region of the fluorescence cell actually used for measurements, and that none of the fluorescence radiation is lost by re-absorption before it can be detected. However, it will serve to illustrate the basis of more rigorous expressions without introducing many confusing correction terms. This will be followed by a discussion of the different categories of atomic fluorescence accompanied, where appropriate, by the expressions required to evaluate the fluorescence yield factor in a particular case. Finally we shall see how practical fluorescence measurements differ from the ideal case, and compare the analytical curves derived from the more rigorous expressions that may be developed with those which would be expected from the initial simple treatment.

4.2 AN IDEAL FLUORESCENCE INTENSITY

An expression for the fluorescence intensity may be obtained with little difficulty in terms of the total absorption factor, A_T, derived at some length in the previous chapter, by making the following assumptions:

1. That the particular fluorescence transition considered is excited by absorption of energy at only a single frequency.*
2. That the entire fluorescence cell is within the solid angle viewed by the detector; i.e. there will then be no region of the cell where the incident radiation is absorbed without exciting fluorescence which is measured by the detector. If this were not the case the effective incident intensity with respect to the fluorescence measurements would be less than the value where the exciting beam enters the cell (and which is used in the calculation).

* If more than one absorption line contributes to a particular fluorescence line, it is necessary merely to apply the equation separately to each absorption line and sum the results.

3. That no fluorescence emission is lost by reabsorption in the cell.

In this case, the integrated fluorescence intensity in a direction perpendicular to the exciting light beam is given by

$$I_F = I_0 w \frac{\Omega}{4\pi} A_T \phi \qquad (4\text{-}1)$$

where I_0 is the radiant flux (expressed as energy per unit time per unit area of the fluorescence cell face on which it is incident) which excites the fluorescence under consideration; w is the width of the exciting beam of radiation; Ω is the solid angle over which the excited fluorescence is detected and measured (4π is simply the total number of radians over which fluorescence is emitted from the cell); A_T is the total absorption factor for the spectral line at which the fluorescence is excited (and therefore depends on the absorption path length through the cell as well as atom concentration); ϕ is the *fluorescence yield* (or power, or efficiency). It is the fraction of the absorbed photons which is re-emitted as fluorescence radiation.

The value of ϕ for a particular fluorescence transition will depend on the type of fluorescence (i.e. on the number of alternative paths by which the atom may return to the ground state) and on the *quenching of fluorescence* which occurs in the cell. This process is the deactivation of excited atoms by collisions with foreign atoms or molecules in the fluorescence cell, rather than as a result of a transition involving the emission of radiation. It produces an overall reduction in the fluorescence intensity measured rather than a change in the shape of calibration graphs. However, depending on the species present in the cell, quenching may result in values of ϕ far below unity, and this process is a very important factor in the selection of a suitable cell for fluorescence measurements. Quenching processes are discussed in detail in Section 4.5. The effects on ϕ of different types of fluorescence are given in the following section.

By using the equations for A_T obtained in Chapter 3, expressions for the "ideal" fluorescence intensity may be derived using equation (4-1) for the same four special cases. Thus, for a continuum source we see from equations (3-19) and (3-22) that I_F should be proportional to N_j at low optical densities and to $N_j^{1/2}$ at high optical densities. Similarly, equations (3-25) and (3-26) show that, for a sharp line source, I_F should again be proportional to N_j at low optical densities but should be independent of N_j at high concentrations. The degree of relevance of these findings to practical situations where the fluorescence intensity can no longer be considered as "ideal" will be discussed in Section 4.4.

4.3 TYPES OF ATOMIC FLUORESCENCE

The main types of atomic fluorescence process are summarized in Fig. 4-1 which shows diagrammatically the transitions involved. The

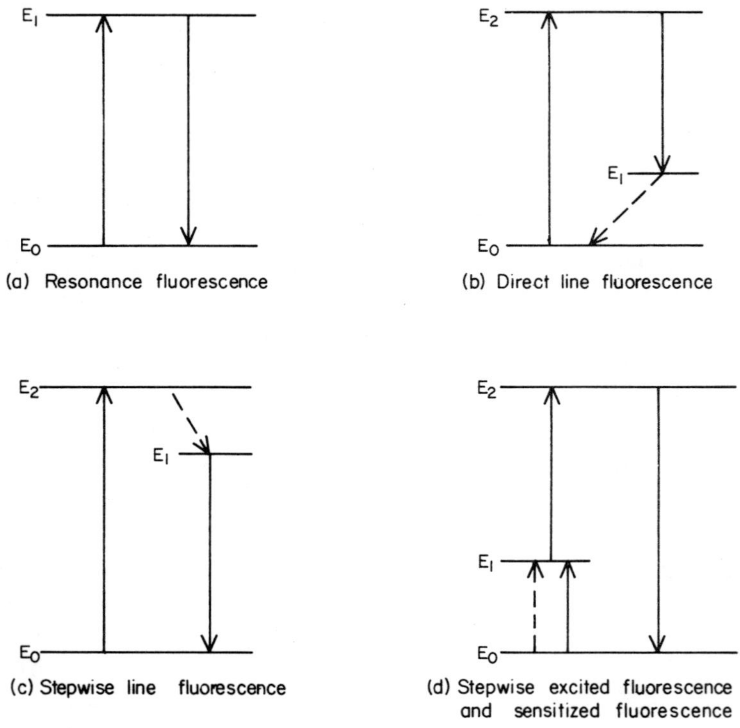

Fig. 4-1. The main types of atomic fluorescence and their electronic transitions.

energy levels are indicated by numerical subscripts used in ascending order for increasing energy. This system is used in preference to the notation of Chapter 2 to avoid confusion when the Einstein transition probabilities are applied, as in the situation discussed here the values of B_{ji} and B_{ij} applying to a particular fluorescence need not share two common energy levels. However, the distinction between B_{ji} and B_{ij} will still be clear even with this notation. Dashed lines are used on the diagrams to represent radiationless transitions (e.g. deactivation by collisions). Each of the types shown on the diagram is discussed below in more detail, although not all of them have actually been applied to analytical AFS.

4.3.1 Resonance Fluorescence

Resonance fluorescence occurs when the atom re-emits a spectral line of the same wavelength as that used for excitation. Many of the AFS measurements made by analytical chemists involve this type of fluorescence. Pure resonance fluorescence is observed only when no other types of fluorescence can occur simultaneously from the same excited state, i.e. when the selection rules allow no other transition from the excited state considered to a lower level. An example of pure resonance fluorescence without the presence of an intermediate

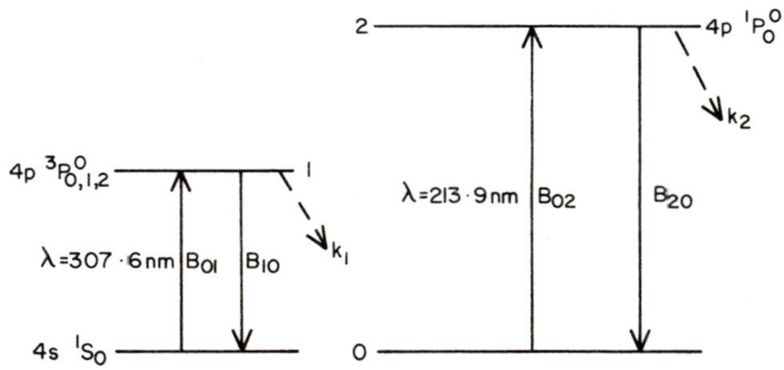

Fig. 4-2. Pure resonance fluorescence of zinc.

level is shown in Fig. 4-2 for the 213·86 nm and 307·59 nm lines of zinc. Obviously, pure resonance fluorescence will not be observed unless there is a singlet ground state.

The yield, ϕ, of pure resonance fluorescence is easily evaluated as it is merely the ratio of the numbers of stimulated emission transitions without and with quenching of atoms in the excited state. Hence, using the Einstein transition probabilities (see Section 2.3.2) and defining as k the probability per second that an excited atom is deactivated by a quenching collision, we obtain for the zinc example shown in Fig. 4-2:

$$\phi_{307\cdot 59} = \left(\frac{B_{10}}{B_{10} + k_1}\right) \quad \text{and} \quad \phi_{213\cdot 86} = \left(\frac{B_{20}}{B_{20} + k_2}\right) \tag{4-2}$$

The ease of determining ϕ when the resonance fluorescence is accompanied by fluorescence of another type may vary with the form that the latter takes. A simple example is shown for thallium in

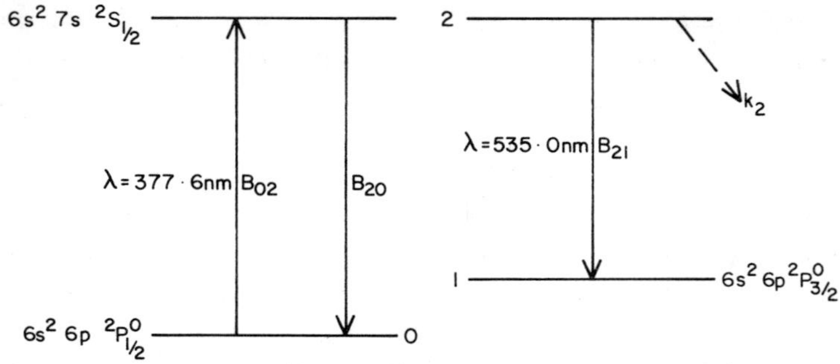

Fig. 4-3. Fluorescence from the lower thallium excited state.

Fig. 4-3. In this case excitation at the 377·6 nm resonance line produces resonance fluorescence accompanied by a further emission at 535·0 nm. In this case the yield of resonance fluorescence is given by

$$\phi_{377\cdot 6} = \frac{B_{20}}{(B_{20} + B_{21} + k_2)} \tag{4-3}$$

4.3.2 Direct Line Fluorescence

Direct line fluorescence is observed when transitions between the excited state of the resonance line and a lower intermediate level are not forbidden by the selection rules. Thus, direct line fluorescence will always occur at a longer wavelength than that of the resonance line which excites it (and which is observed simultaneously as resonance fluorescence). This situation will arise if the ground state is a multiplet, as we have already seen for thallium in Fig. 4-3. In this simple example the expression for ϕ is again easily evaluated, being*

$$\phi_{535\cdot 0} = \frac{B_{21}}{(B_{20} + B_{21} + k_2)} \tag{4-4}$$

There are other situations which can result in direct line fluorescence. For example, the ^4P excited state of the antimony resonance line lies above the ^2D state as shown in Fig. 4-4. As a result, excitation at 217·6 nm produces direct line fluorescence at 267·1 nm

* It should be remembered that ϕ can be defined in terms of energy rather than numbers of photons; it is then necessary to incorporate a ratio of line wavelengths whenever the absorption and fluorescence wavelengths differ as in this example (as the energies of the absorbed and emitted photons will then be different).

and 277·0 nm. Evaluation of an expression for the yield at each line is again quite straightforward. Although the direct line fluorescence is normally less intense than the resonance fluorescence at the corresponding absorption wavelength, it may well be more intense than resonance fluorescence observed at another, weaker, absorption line. Thus, in certain cases, depending on experimental conditions, the use of direct line fluorescence gives complete freedom from the effects of scattered source radiation without resulting in an unacceptable decrease in the fluorescence intensity received.

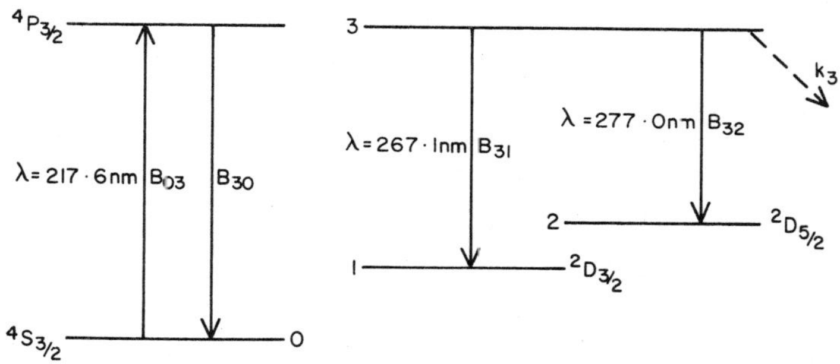

Fig. 4-4. Resonance and direct line fluorescence of antimony.

4.3.3 *Stepwise Line Fluorescence*

When an atom is initially excited to a higher excited state by absorption of radiation at a ground state line, it may undergo deactivation (often by a collision or other radiationless process) to a lower excited state rather than return directly to the ground state. If a transition is permitted from this level to one still lower (which will often be the ground state), stepwise line fluorescence is observed as shown in Fig. 4-1(c). For example, if sodium atoms are excited only by the 589·0 nm *D* line (this would be achieved by filtering the incident source radiation with a monochromator or a narrow-band interference filter), stepwise line fluorescence is observed at 589·6 nm provided suitable species are present to give energy transfer by collisions. This is illustrated in Fig. 4-5. The quenching rates are shown with the same notation as previously, but the vertical transitions with rates k_{12} and k_{21} represent the rates of collisional transfer of energy between the two excited states. In this particular case the splitting of the $^2P^0_0$ level is so small that k_{12} is relatively large and there is said to be good *mixing* between the $^2P^0_{1/2}$ and

$^2P^0_{3/2}$ states. When the sodium atoms are in the presence of a species which gives efficient transfer of energy by collisions, the mixing becomes fast compared to all other radiative or radiationless transitions. The populations of these two excited states are then in partial equilibrium and their ratio is practically the same as for thermal equilibrium and obeys the Boltzmann law. This allows a simple derivation of a direct expression for the ratio of fluorescence intensities concerned,[1] but its application is limited to special cases of stepwise line fluorescence such as this where the level separation is extremely small.

It will be appreciated that although the species present with the sodium atoms must give efficient collisional mixing between the

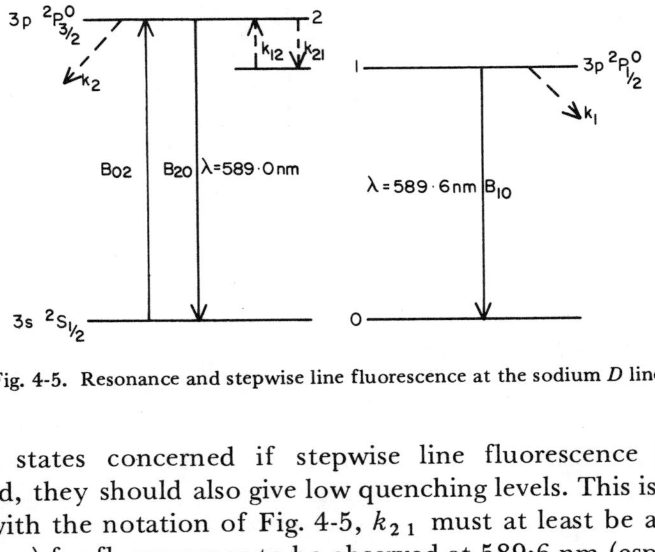

Fig. 4-5. Resonance and stepwise line fluorescence at the sodium *D* lines.

excited states concerned if stepwise line fluorescence is to be observed, they should also give low quenching levels. This is essential since, with the notation of Fig. 4-5, k_{21} must at least be as large as $(k_2 + B_{20})$ for fluorescence to be observed at 589·6 nm (especially as quenching with rate k_1 will also occur at this lower excited state). As a result this particular example is not observed when an air-hydrogen flame is used as the fluorescence cell because nitrogen is an efficient quencher (see Section 4.5.3) but can be seen in an oxygen-argon-hydrogen flame because argon gives low quenching while still retaining efficient mixing capability for the two $^2P^0$ states of sodium.

When the intermediate excited level is at much lower energy than the initial excited state it may be assumed that any mixing between the two levels is negligible; i.e. energy can be transferred to the lower level by collisional deactivation but is unlikely to return via the reverse process. However, because of the larger energy gap the deactivation may not occur in a single stage and the exact path of stepwise line fluorescence may be difficult to visualize. An example of this is shown in Fig. 4-6, again for sodium. Excitation in which only the ground state doublet at 330 nm is used produces not only resonance fluorescence at the same wavelengths but also stepwise line fluorescence at the *D* lines. It is believed that the required deactivation from the $4p\ ^2P^0$ to the $3p\ ^2P^0$ levels probably occurs via the $3d\ ^2D$ and $4s\ ^2S$ intermediate levels (not shown in the diagram) and may also give other fluorescence lines at longer wavelengths. Obviously, under these conditions, any attempt to derive an expression for the fluorescence yield at a particular

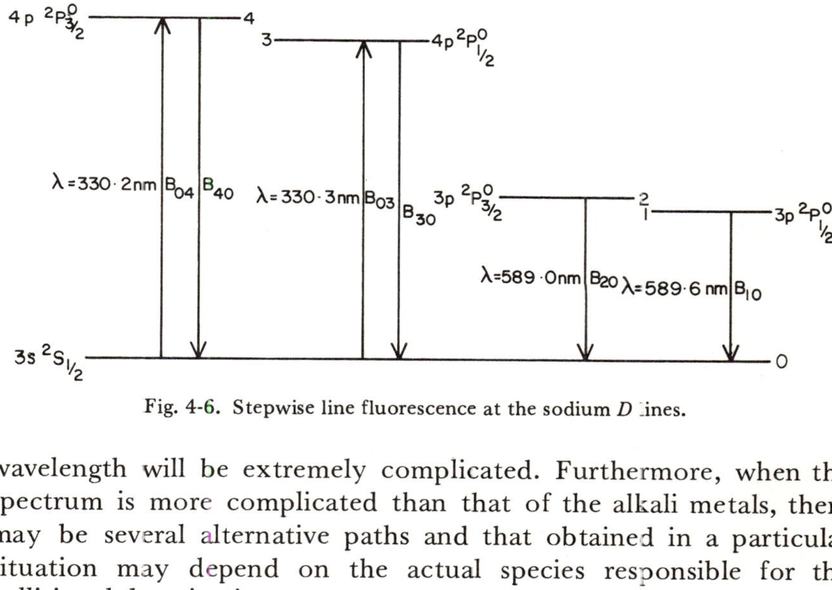

Fig. 4-6. Stepwise line fluorescence at the sodium *D* lines.

wavelength will be extremely complicated. Furthermore, when the spectrum is more complicated than that of the alkali metals, there may be several alternative paths and that obtained in a particular situation may depend on the actual species responsible for the collisional deactivation.

The situation can also be complicated by an energy level configuration intermediate between the types occurring in the previous two examples; i.e. where the lower excited level is neither sufficiently close to the higher one to assume an equilibrium distribution between the two nor sufficiently far below to assume that re-excitation by collision is negligible. Such an example occurs with lead, excitation at 283·3 nm to the $7s\ ^3P_1^0$ level producing stepwise line fluorescence at 368·3 nm from the $7s\ ^3P_0^0$ level. The technique needed to evaluate the rate of collisional re-excitation from $^3P_0$ to $^3P_1$ has been discussed by Jenkins[2] and is obviously complicated by the simultaneous emission of resonance and direct line fluorescence from the $7s\ ^3P_1^0$ level.

4.3.4 *Stepwise Excitation Fluorescence*

Stepwise excitation fluorescence is really the converse of stepwise line fluorescence, i.e. it occurs as the result of a stepwise absorption of energy by the atom in order to reach a particular excited state. A simple example is seen when the first level reached from the ground state is populated thermally owing to the relatively high temperature of a flame fluorescence cell. Absorption of a line involving a transition from this level can then produce excitation of fluorescence. This is seen with the thallium energy levels shown in Fig. 4-3. At sufficiently high temperatures the $6s^2\ 6p\ ^2P_{3/2}^0$ level is populated enough to give quite strong absorption at 535·0 nm so that stepwise excited fluorescence is observed at both 535·0 nm and 377·6 nm. Obviously, this type of stepwise excited fluorescence will be observed only when the lower level of the absorption line is relatively close to the ground state unless an extremely hot flame is used. In any case, the intensity is unlikely to be sufficiently intense for analytical use in most instances.

Another well-known mechanism for stepwise excited fluorescence involves the use of a very intense excitation source. An excited level reached by absorption of a ground state line may then itself act as the lower level for further absorption of another line to reach a much higher excited state. For example, absorption of the mercury 253·7 nm resonance line may be followed by absorption of the 435·8 nm line to reach the $7s\ ^3S_1$ level from which stepwise excited fluorescence is observed. Again, this is likely to be of low intensity even with an extremely bright excitation source.

Stepwise excitation fluorescence may also be observed as the exact converse of stepwise line fluorescence, i.e. absorption of the excitation radiation raises atoms to (say) the lower level of a close doublet and mixing by collisional processes involving other species present results in a population of the upper level. Thus, a fluorescence line from this level is observed, the intensity depending on the separation of the levels and the efficiency of the collisional mixing (or coupling) provided by the species concerned. A typical example will occur with the sodium levels shown in Fig. 4-5, excitation at 589·6 nm only resulting in fluorescence both at 589·6 nm (resonance) and 589·0 nm (stepwise excited) even though the latter transition involves photons of higher energy than provided by the absorption transition. Fluorescence lines with this characteristic have recently been observed for antimony, bismuth and arsenic in flame cells and referred to as thermally assisted fluorescence[3] since the flame species involved in the mixing process have considerable thermal energy. The example of bismuth is illustrated in Fig. 4-7, excitation at 206·2 nm to the $^4P_{5/2}$ level resulting in fluorescence at 293·8 nm and 262·8 nm from the $^2P_{3/2}$ level which lies 0·12 eV higher. It can be seen that, as mentioned earlier, any attempt to write an expression for the yield of stepwise fluorescence would involve some difficulty (especially as only some of the levels are shown in most of the simplified Grotrian diagrams given here).

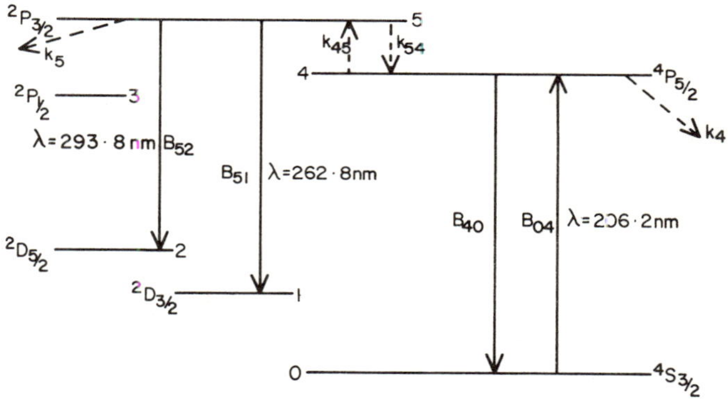

Fig. 4-7. Stepwise excited fluorescence from bismuth $^2P_{3/2}$ level.

4.3.5 Sensitized Fluorescence

Sensitized fluorescence* is obtained when an atom which has been excited by normal absorption of resonance radiation is able to transfer this energy to another atom during a collision, thus raising the second atom to an excited level from which a radiative transition is possible. In order to obtain sufficient collisional excitation of this kind a high concentration of the donor atoms is needed; sensitized fluorescence is thus unlikely to be observed from flame cells (especially as the energy is more likely to be lost by normal quenching collisions with the flame species). However, it has been observed in closed cells containing metal vapours at relatively high pressure. For example, if such a mixture of mercury and thallium atoms is strongly irradiated with mercury resonance radiation at 253·7 nm it is possible to observe sensitized fluorescence of thallium at the 377·6 nm and 535·0 nm lines.

* Sometimes referred to as photosensitized fluorescence.

4.4 CALCULATIONS OF ACTUAL FLUORESCENCE INTENSITY

It is now necessary to examine the relevance of the simple expression for fluorescence intensity derived in Section 4.2 and to discuss some recent attempts to obtain wider agreement with practical measurements. Before doing this it is useful to consider some of the assumptions made earlier in terms of the dimensions of the "standard" fluorescence cell shown in Fig. 4-8. This could be achieved in practice by making measurements with parallel light beams in order to obtain the rectangular geometry for the atom cell. As calculations usually assume conditions of uniform atomic

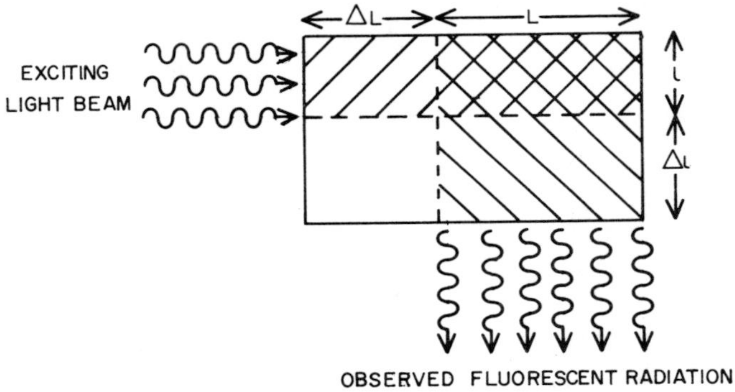

Fig. 4-8. Dimensions of a standardized fluorescence cell.

concentration, temperature, etc. the use of a flame cell would require that the standard cell of Fig. 4-8 be situated at the centre of a flame-shielded flame (see Chapter 6). The assumption that no incident radiation is lost by absorption before it can excite useful (i.e. detected) fluorescence would be met by a cell with $\Delta L = 0$. If the condition that no fluorescence be lost by re-absorption is to apply generally, it is also necessary that $\Delta l = 0$ and l be as small as possible so that the path length over which re-absorption can occur is minimized.

In practice it is obviously impracticable to make l sufficiently small for re-absorption to be negligible under all conditions, as the fluorescence intensity produced also depends on l (i.e. on the width of the exciting light beam). Thus equation (4-1) may be applied only when re-absorption of fluorescence is negligible; i.e. (a) at low atom concentrations, and (b) at all concentrations for direct line fluorescence (and sometimes the other types of non-resonance fluorescence) when the lower level of the fluorescence transition is far enough above the ground state for its equilibrium population to be extremely small. The analytical curves predicted by equation (4-1) will then apply and are shown in Fig. 4-9. The four asymptotes shown correspond to the four limiting cases discussed in Section 4.2; the two full curves are only obtained by numerical solution of the Voigt integral for all concentrations, as explained in Section 3.2.1. It can be seen that the limiting cases give a useful indication of the shape of the full curves. As most fluorescence measurements use resonance fluorescence, however, equation (4-1) is really only of practical interest at low concentrations. Nevertheless, AFS is a trace analysis technique and the linear dependence of fluorescence on (low) concentrations under all conditions is of considerable importance.

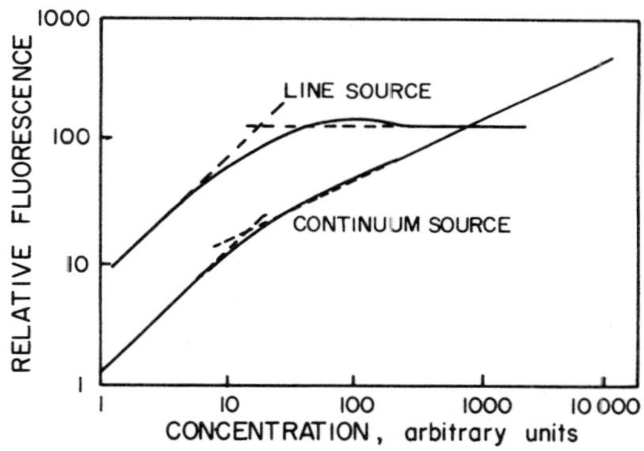

Fig. 4-9. Ideal analytical curves assuming no re-absorption of fluorescence (applicable to direct line fluorescence and certain other examples of non-resonance fluorescence).

Extension of equation (4-1) to wider concentration ranges under practically useful conditions* has been described by Hooymayers[4] for $\Delta L = 0$ and his general relation has recently been adapted to $\Delta L \neq 0$ by Zeegers and Winefordner.[5] The latter have also tested the equation experimentally and found good agreement between actual and calculated analytical curves. We shall derive the basic equation without attempting to evaluate the complex integral involving exponential terms of the absorption coefficient, k_ν.

Consider a thin layer of the standard fluorescence cell in the direction of the exciting radiation, with width dx and distance x from the edge BDF of the excitation region, as shown in Fig. 4-10.

* However, the derivation is carried out only for a single absorption line as with the AAS equations of Chapter 3, so that the important effect of hfs is not taken into account.

Let I_ν be the incident radiation flux of the exciting light beam expressed as energy per unit time at frequency ν per unit area and per unit frequency interval. Then from equation (3-1) the remaining radiation flux at the plane CD will be $I_\nu \exp(-k_\nu \Delta L)$. Hence, using equation (3-1) again for the region between CD and EF, the absorbed radiation at frequency ν will be proportional to $I_\nu \exp(-k_\nu \Delta L)$ $[1 - \exp(-k_\nu L)]$. If the third dimension of the cell (i.e. perpendicular to the plane of the paper in Fig. 4-10) is of length l', the

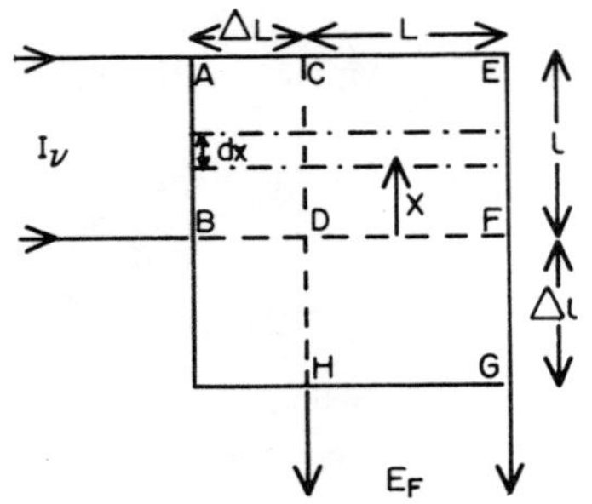

Fig. 4-10. Symbols used in the derivation of an expression for the actual fluorescence radiation E_F (see text for explanation).

integrated energy absorbed per second in the thin layer considered can be written as

$$E_A = l' \,\mathrm{d}x \int_0^\infty I_\nu \exp(-k_\nu \Delta L)\ [1\text{-}\exp(-k_\nu L)]\,\mathrm{d}\nu \qquad (4\text{-}5)$$

which we will abbreviate to

$$E_A = l' \,\mathrm{d}x\ \theta$$

Now from the definition of the fluorescence yield factor, ϕ, the fluorescence radiation emitted by the thin layer considered in a direction perpendicular to the exciting light beam and through a small solid angle Ω is simply

$$\mathrm{d}F'_{\mathrm{em}} = \frac{\Omega}{4\pi} E_A \,\phi = \frac{\Omega}{4\pi} l' \,\mathrm{d}x\ \theta\phi \qquad (4\text{-}6)$$

This will be reduced by reabsorption while travelling to the edge BDF of the excitation region to a value

$$\mathrm{d}F_{\mathrm{em}} = \frac{\Omega}{4\pi} l' \,\mathrm{d}x\ \theta\phi \exp(-k_{\nu'} x) \qquad (4\text{-}7)$$

again from equation (3-1) and where ν' is the frequency of the fluorescence line. Hence the total fluorescence radiation at frequency ν' leaving the excitation region along the line DF is obtained by integration of equation (4-7) for values of x between 0 and l, i.e.,

$$
\begin{aligned}
F_{em} &= \int_0^l \frac{\Omega}{4\pi} l'\theta\phi \exp(-k_{\nu'}x)\,dx \\
&= \frac{\Omega}{4\pi} l'\theta\phi \frac{1}{k_{\nu'}} [\exp(-k_{\nu'}x)]_0^l \qquad (4\text{-}8) \\
&= \frac{\Omega}{4\pi} l'\phi\theta \frac{1}{k_{\nu'}} [1 - \exp(-k_{\nu'}l)]
\end{aligned}
$$

Now by further application of equation (3-1), the radiation reaching the edge of the cell at HG after further re-absorption will be equal to $F_{em} \exp(-k_{\nu'}\Delta l)$. Hence, E_F, the integrated fluorescence energy leaving the cell per second along the edge HG, is given by

$$E_F = \frac{\Omega}{4\pi} l'\theta\phi \int_0^\infty \frac{1}{k'_\nu} [1 - \exp(-k_{\nu'}l)] \cdot [\exp(-k_{\nu'}\Delta l)]\,d\nu' \quad (4\text{-}9)$$

After substitution of the full expression for θ and multiplication and rearrangement* of the exponential terms within each integral, we obtain

$$
\begin{aligned}
E_F = \frac{\Omega}{4\pi} l'\phi \int_0^\infty I_\nu [\{1 - \exp[-k_\nu(L + \Delta L)]\} - \{1 - \exp(-k_\nu \Delta L)\}]\,d\nu \\
\times \int_0^\infty \frac{1}{k_{\nu'}} d_{\nu'} \times \int_0^\infty [\{1 - \exp[-k_{\nu'}(l + \Delta l)]\} - \{1 - \exp[-k_{\nu'}\Delta l]\}]\,d\nu'
\end{aligned}
\qquad (4\text{-}10)
$$

But $\int_0^\infty \frac{1}{k_{\nu'}} d_{\nu'}$ may be written as $1/\int_0^\infty k_{\nu'}\,d\nu$ which is simply (see Section 3.2.2) I/K where K is the integrated absorption coefficient for the fluorescence line. Hence using equation (3-11) with λ' as the wavelength of the fluorescence line, A its transition probability, g_0

* Using $e^a e^b = e^{a+b}$.

and g_1 the statistical weights of the lower and upper states and N_0 the population of the lower state, we obtain finally

$$E_F = \frac{\Omega}{4\pi} l'\phi \frac{8\pi}{(\lambda')^2 A N_0} \frac{g_0}{g_1}$$

$$\times \int_0^\infty I_\nu \{1 - \exp[-k_\nu(L + \Delta L)]\} - \{1 - \exp[-k_\nu \Delta L]\} \, d\nu$$

$$\times \int_0^\infty [\{1 - \exp[-k_{\nu'}(l + \Delta l)]\} - \{1 - \exp[-k_{\nu'}\Delta l]\}] \, d\nu' \qquad (4\text{-}11)$$

This equation has been evaluated by Hooymayers[4] and Zeegers and Winefordner[5] for resonance fluorescence in which $\nu = \nu'$ and N_0 is equal to the ground state population of the sample element. Both found it still necessary to assume extreme cases in which the source line is either much wider than the absorption line (continuum source) or much narrower (sharp line source) in order to avoid the difficulty of defining the excitation source line profile and assigning a particular value to it for each experiment. In both reports the authors have calculated for these two special cases the full analytical curves by numerical solution of the Voigt integrals occurring in equation (4-11). They have also derived asymptotic expressions for the limiting cases of low and high concentrations using approximations for the integrals similar to those discussed in Section 3.3 for the evaluation of A_T.

Since the Voigt expressions for k_ν depend on the profile of the absorption line, as expressed* by the so-called "damping constant", a (see Section 3.2.1), the analytical curve obtained by evaluation of equation (4-11) will obviously vary with the value of a assigned to the absorption line as well as with the dimensions of the fluorescence cell. This is illustrated in Fig. 4-11 which shows curves calculated by Hooymayers[4] for different values of a with the same fluorescence cell in which $\Delta L = 0$, $L = 4l$ and $\Delta l = 3l/2$. It can be seen that the changes with different values of a are considerable. Furthermore, at high concentrations the curves differ considerably from those calculated using the "ideal" equation (see Fig. 4-9) even though part of the possible deviation from ideality has still not been included (i.e. ΔL is

* From equation (3-5) it is seen that $a = \frac{\beta_1}{\beta_2} = \frac{\Delta\nu_C}{\Delta\nu_D} \cdot (\ln 2)^{1/2}$ where $\Delta\nu_C$ and $\Delta\nu_D$ are defined in Section 3.2.1.

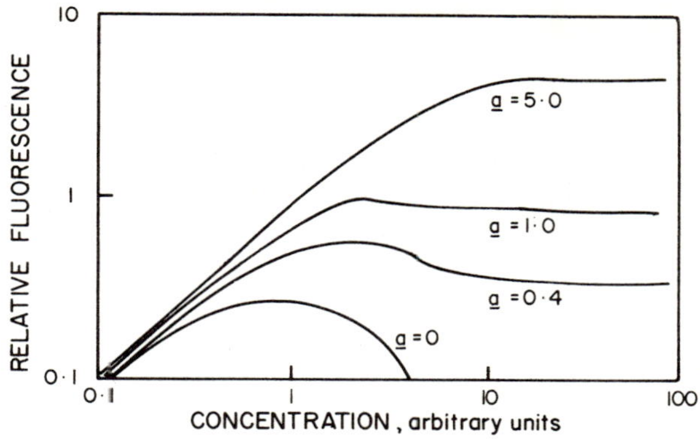

(a) Continuum source

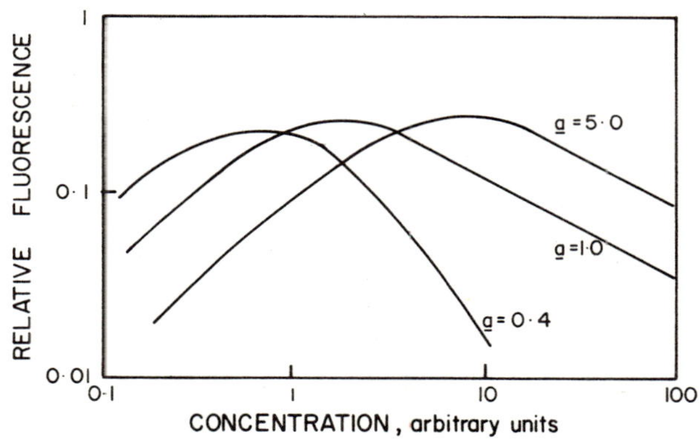

(b) Sharp line source

Fig. 4-11. Variation of calculated analytical curves with the *a*-parameter [ref. (4)] (single spectral line, fluorescence cell dimensions: $\Delta L = 0$, $L = 4l$, $\Delta l = 3l/2$).

still zero); i.e. the continuum source curves reach a plateau instead of continuing to rise in proportion to $(N_j)^{1/2}$ whereas those for a sharp line source fall steadily instead of reaching a plateau.

Zeegers and Winefordner[5] found that for the particular example of the magnesium resonance line at 285·2 nm, their experimental conditions corresponded to an *a* value of 0·4. Their calculated curves for a sharp line source and an absorption line with *a* equal to 0·4 are shown in Fig. 4-12 for several different configurations of the fluorescence cell. It is seen that the departures from ideality increase considerably when ΔL is increased from zero. Similar effects are observed with a continuum source.

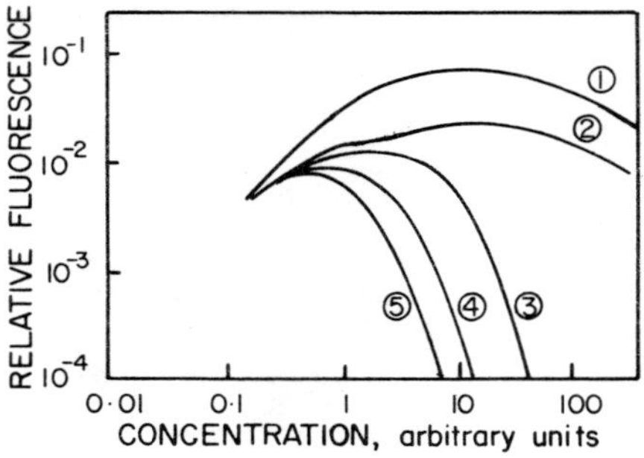

(a) L = l = 5 in all cases. ① ΔL = 0, Δl = 0; ② ΔL = 0, Δl = 9·5; ③ ΔL = 1, Δl = 9·5; ④ ΔL = 3, Δl = 9·5; ⑤ ΔL = 5, Δl = 9·5

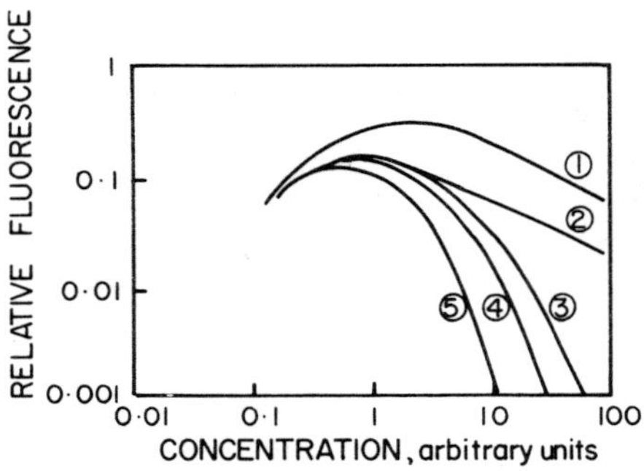

(b) L = 10, l = 5 in all cases. ① ΔL = 0, Δl = 0; ② ΔL = 0, Δl = 9·5; ③ ΔL = 0·5, Δl = 9·5; ④ ΔL = 1, Δl = 9·5; ⑤ ΔL = 3, Δl = 9·5

Fig. 4.12. Variation of calculated analytical curves with fluorescence cell dimensions for a sharp line source and a single absorption line with *a*-parameter equal to 0·4 [ref. (5)].

4.5 QUENCHING OF FLUORESCENCE

The quenching of fluorescence was mentioned in Section 4.2 with an indication of its important effect on the fluorescence yield observed in many practical situations. We shall now discuss in more detail the mechanism by which the process occurs and give some examples of individual quenching processes and of practical measurements which are particularly relevant to AFS. Some basic concepts necessary for a proper understanding of the principles involved will also be mentioned briefly. However, for a more thorough discussion of excitation and de-excitation processes and related topics the reader is referred to a recent, comprehensive review by Alkemade and Zeegers[6] and to the works of Laidler and Shuler,[7] Laidler,[8] and

Shuler *et al.*[9] Many of the basic principles involved are discussed concisely in the textbooks of Laidler,[10] and Mavrodineanu and Boiteux.[11]

4.5.1 Basic Concepts

As we have seen, the fluorescence process involves the absorption of photons from a primary exciting light beam to produce an atom (or molecule) in an excited state. Up to this point in this text the main concern has been the de-excitation of this state by the re-emission of radiation (fluorescence). As there is a finite probability of such a transition occurring (the Einstein coefficient B_{ij}; see Section 2.3.2), however, there will also be a finite lifetime for the excited state before its excess energy is removed by this process. During this period it is possible for an excited species to be de-excited by other *non-radiative* processes as a result of collisional interaction with other species present in its environment. The resultant reduction in observed fluorescent power is generally referred to as *quenching* of the fluorescence. The effectiveness of quenching under particular circumstances will thus depend on the rate at which collisions occur, the type of non-radiative process involved and the efficiency of this process in removing energy from the excited species.

The fraction of absorbed photons actually re-emitted as fluorescence radiation is known as the *fluorescence yield factor, ϕ*. This is defined by the Stern-Volmer formula as

$$\phi = \frac{B_{ij}}{\text{Total probability per second of de-excitation}} \qquad (4\text{-}12)$$

The total probability of de-excitation is simply the summation of B_{ij} with the *rates* of all non-radiative processes contributing to the quenching. For example, two straightforward but important examples of de-excitation processes are (a) inelastic collisions involving transfer of energy

$$A^* + B \rightleftharpoons A + B$$

and (b) reactions in which the quenching species dissociates after receiving the excess energy from the excited atom or molecule

$$A^* + CD \rightleftharpoons A + C + D$$

In these processes the superscript * is used to indicate a particle in an excited state. The symbol $\rightleftharpoons$ is used in each example as, by the

Principle of Microreversibility, the processes can proceed in either direction, that from right to left in (a) and (b) being excitation processes for A*. The *rate, r*, of each reaction is defined as the number of particles A* quenched per second per unit volume by the process. It is generally expressed in the form

$$r_1 = k_1 [A^*] [B]$$
$$r_2 = k_2 [A^*] [CD] \tag{4-13}$$

for processes (a) and (b) respectively. k_1 and k_2 are the *rate constants* or rate coefficients for the two processes. The probabilities of a particle A* being quenched in each of these examples are thus $r_1/[A^*]$ and $r_2/[A^*]$ respectively. Hence, in a fluorescence experiment in which quenching occurs only by the processes (a) and (b) the yield factor is given by

$$\phi = \frac{B_{ij}}{k_1 [B] + k_2 [CD] + B_{ij}} \tag{4-14}$$

It can be seen from equation (4-14) that the maximum value of ϕ is unity when no quenching occurs. Note also that the effects of different quenching species are directly additive via the products of their concentrations and the rate constants. This property allows experimental measurement of the rate constants of quenching for complex situations, such as those encountered in flame cells. If only one quenching species were present its rate constant would follow directly from the measured value of ϕ, if it is assumed that its concentration and the value of B_{ij} are known. When n different quenching species act simultaneously in the flame, it is necessary to measure ϕ for each of a set of n (or more) isothermal flames having different but known compositions. All of the unknown rate constants may then be obtained from the straightforward solution of the set of simultaneous equations that results. This technique has been used to obtain the flame quenching values quoted in Section 4.5.3, although, as explained below, it is generally more useful to give the results in a form other than that of the direct rate constants.

4.5.2 Reaction Rate Theory

The fundamental relationship used in reaction rate studies is the empirical rule generally known as the *Arrhenius* law. This is usually written in the form

$$k_n = A e^{-E/kT} \tag{4-15}$$

and thus relates the rate constant, k_n, of a reaction to the absolute temperature, T. A and E are constants referred to respectively as the *frequency factor* and the *activation energy* of the reaction. Since we are concerned with quenching processes that occur as the result of collisions between gaseous molecules, it is useful to interpret equation (4-15) in terms of collision theory, i.e. in terms of the number of collisions occurring between the atoms or molecules concerned. This is most simply achieved by using the *classical collision theory*, but at the expense of treating the colliding particles as hard spheres. Thus, the quantized structure of the atoms and molecules is ignored and the results calculated must be expected to differ from experimental determinations and those predicted by the *absolute reaction rate theory* which takes the internal structure into account. This more sophisticated theory is really necessary to interpret fully the quenching of excited atoms in flames, as the most important processes are known to involve conversion of the internal energy of the molecules (see Section 4.5.3). The subject is too complex to discuss here but a fairly straightforward treatment has been given by Laidler.[10] We shall, however, derive the reaction rate formulae obtained with the simple classical theory as they are still able to provide a useful format within which most measurements of quenching of atomic fluorescence are reported and interpreted.

The number of binary collisions occurring per second per unit volume in a gas mixture containing two species, A and B, considered to be hard spheres is

$$Z_{A,B} = [A]\,[B]\,\pi(r_A + r_B)^2\,\bar{v}_{rel} \qquad (4\text{-}16)$$

where r_A and r_B are the radii of the species A and B, and $\bar{v}_{rel}$ is their average relative velocity. $Z_{A,B}$ is sometimes referred to as the *collision number*, although most authors prefer to reserve this term for the number of collisions made by *one* of the species; i.e. for $Z_{A,B}/[A]$ or $Z_{A,B}/[B]$. The reciprocal of this quantity is known as the *collisional lifetime*, τ, of the particle. Thus for species A we have

$$\tau_A = \frac{[A]}{Z_{A,B}} = \frac{1}{[B]\,(r_A + r_B)^2\,\bar{v}_{rel}} \qquad (4\text{-}17)$$

Values of τ are directly related to the spectral line widths obtained as a result of collisional or quenching broadening (see Section 2.5.2). A typical value of τ for a metal atom in a flame at 1 atm and 2500 K is about 10^{-10} second. The number of ternary collisions (i.e. between a metal atom and two flame molecules) under the same conditions is three or four orders of magnitude smaller than the number of binary collisions so that they will have a negligible effect on the value of τ.

The term $\pi(r_A + r_B)^2$ in equation (4-16) is known as the gas kinetic cross section for the process considered since it is a function of the quantity $(r_A + r_B)/2$ which, in the kinetic theory, would be the minimum distance between the two hard spheres representing A and B. In practice it is impossible to assign a definite diameter to an atom or molecule and it is customary to define a *collisional cross section* in terms of equation (4-16) *for a particular process.* Thus, the collisional cross section represents the value of $(r_A + r_B)^2$ which gives the collision frequency ($Z_{A,B}$) measured by observing the rate of the process under consideration. The collisional cross section of an atom or molecule will, therefore, vary with the process it is undergoing. It is usually written in the form $\sigma^2 = (r_A + r_B)^2$, but an alternative notation of the type, $S = \pi(r_A + r_B)^2$, is also used. As the symbol S is frequently replaced by σ, care is needed, when cross section data reported in the literature are used, to ascertain whether the values quoted include the factor π.

The collision frequency may be related to the Arrhenius equation in terms of the classical theory by writing the rate of a reaction between the particles A and B as

$$r_n = Z_{A,B}\, e^{-E/kT} = [A]\,[B]\,\pi\sigma^2 \bar{v}_{rel}\, e^{-E/kT} \qquad (4\text{-}18)$$

i.e. the rate is the number of particles colliding per second and having a joint energy E in excess of the mean energy. By comparison with equation (4-13), we see that the rate constant for the reaction is given by

$$k_n = \pi\sigma^2 \bar{v}_{rel}\, e^{-E/kT} \qquad (4\text{-}19)$$

The average relative velocity of the two particles is defined as

$$\bar{v}_{rel} = \left(\frac{8kT}{\pi\mu}\right)^{1/2} \qquad (4\text{-}20)$$

where μ is the reduced mass of the two particles; i.e.

$$\frac{1}{\mu} = \left(\frac{1}{m_A} + \frac{1}{m_B}\right)$$

Thus, we obtain finally

$$r_n = [\mathrm{A}]\,[\mathrm{B}]\sqrt{\frac{8\pi kT}{\mu}}\,\sigma^2\,\mathrm{e}^{-E/kT} \tag{4-21}$$

and

$$k_n = \sqrt{\frac{8\pi kT}{\mu}}\,\sigma^2\,\mathrm{e}^{-E/kT} \tag{4-22}$$

In the application of this theory to the interpretation of quenching measurements, equation (4-22) is frequently used to rewrite the Stern-Volmer formula in terms of the collisional cross section rather than the rate constant k_n (cf. equation (4-14)). Thus, measurements of ϕ for a particular atom may conveniently be tabulated as the quenching cross sections* observed with different quenching species. However, with the type of quenching measurements of interest in AFS work there is little point in considering an activation energy for the reaction and the inclusion of the exponential term in equation (4-22) is unnecessarily cumbersome. As a result of this it is convenient to define *an effective quenching cross section*, σ_Q^2 or S_Q, such that

$$k_n = \sqrt{\frac{8\pi kT}{\mu}}\,\sigma_Q^2 = \sqrt{\frac{8kT}{\pi\mu}}\,S_Q \tag{4-23}$$

It will be noted that effective cross sections defined in this way are functions of temperature (unless the activation energy, E, is zero) and that they will automatically include any factors required to account for possible deviations of the actual value of k_n from the simple classical theory. Thus σ_Q^2 or S_Q represent a useful means of reporting experimental quenching values (see Section 4.5.3).

4.5.3 Types of Quenching Process and Experimental Results

The de-excitation processes by which quenching of excited metal atoms may occur are summarized below together with an indication of their relative importance in AFS. Experimental measurements of quenching cross sections are given where possible but relatively few data have been reported in the literature; in fact, the values given in Table 4-1 represent most of the results available at the present time.

* It is customary when discussing quenching processes to speak of quenching cross sections rather than collisional cross sections and this convention will generally be followed here.

TABLE 4-1
Experimentally determined quenching cross sections

	Quenching cross section,[a] σ_Q^2, Å^2									
	Quenching species									
Atom (excited level)	H_2	H_2O	O_2	N_2	CO	CO_2	Ar	Other[b]	Temperature, K	Reference
Li ($2^2P_{1/2,3/2}$)	5·2	1·9	—	6·8	12·6	9·2	≤0·3	—	1400	(*c*)
Na ($3^2P_{1/2,3/2}$)	2·9	0·5	12·3	7·0	11·9	17·0	≤0·1	—	1400	(*d*)
	2·6	0·7	10·8	6·7	13·0	15·9	—	—	2000	(*e*)
	5·1	—	—	12·9	—	—	—	3·8(HD) 3·2(D_2)	390	(*f*)
K ($4^2P_{1/2,3/2}$)	1·0	0·9	15·5	5·6	12·4	21·4	≤0·2	—	1400	(*g*)
	1·1	0·8	15·6	6·1	14·0	21·0	—	—	1900	(*h*)
K ($4^2P_{1/2}$)	—	—	—	11·2	—	—	—	—	350	(*i*)
K ($4^2P_{3/2}$)	—	—	—	12·4	—	—	—	—	350	(*i*)
K ($5^2P_{1/2,3/2}$)	19·1	4·8	—	14·3	—	—	—	—	2000	(*h*)
Rb ($5^2P_{1/2,3/2}$)	0·6	1·3	25·0	6·1	11·8	24·0	≤0·3	—	1400	(*g*)
	1·1	1·3	26·4	8·0	—	—	—	—	2000	(*j*)
Rb ($5^2P_{1/2}$)	—	—	—	11·8	—	—	—	—	300	(*k*)
Rb ($5^2P_{1/2}$)	1·9	—	—	18·5	—	—	—	1·9(HD), 1·0(D_2), <0·3(CH_4), 0·6(CD_4), 44·3(C_2H_4), 0·6(C_2H_6)	340	(*l*)
Rb ($5^2P_{3/2}$)	—	—	—	11·5	—	—	—	—	300	(*k*)

Rb ($5^2P_{3/2}$)	1·0	—	—	13·7	—	—	—	1·6(HD), 1·6(D_2), 1·0(CH_4), 1·0(CD_4), 30·2(C_2H_4), 1·9(C_2H_6)	340	(*l*)
Cs ($6^2P_{1/2,3/2}$)	1·8	5·6	—	25·1	—	—	—	—	1400	(*g*)
Cs ($6^2P_{1/2}$)	—	—	—	24·5	—	—	—	< 0·6(CH_4)	350	(*m*)
Cs ($6^2P_{3/2}$)	—	—	—	21·9	—	—	—	—	350	(*m*)
Tl ($7^2S_{1/2}$)	0·03	1·8	13·2	6·4	13·6	32·5	⩽ 0·1	—	1400	(*n*)
Pb ($6p7s\ ^3P_1$)	0·4	8·0	15·0	5·7	13·0	29·0	≈ 0·0	≈ 0·0(He)	1400	(*o*)
Hg ($6\ ^3P_1$)	8·0	—	20·0	<0·3	6·5	5·0	—	—	300	(*p*)
	7·8	—	19·3	0·23	6·9	3·2	—	7·2(D_2), 0·28($^{15}N_2$), < 0·001(Xe)	296	(*q*)

[a] Values are effective experimental cross sections. Also the factor π is omitted even where it was included in the original source of the data (see text).

[b] The quenching species is shown in parentheses after the value of σ_Q^2.

[c] D. R. Jenkins, *Proc. Roy. Soc.*, 1968, **A306**, 413.

[d] Ibid., 1966, **A293**, 493.

[e] H. P. Hooymayers and P. L. Lijnse, *J. Quant. Spec. Radiat. Transfer*, 1969, **9**, 995

[f] B. P. Kibble, G. Copley and L. Krause, *Phys. Rev.*, 1967, **159**, 11.

[g] D. R. Jenkins, *Proc. Roy. Soc.*, 1968, **A303**, 453.

[h] H. P. Hooymayers and C. Th. J. Alkemade, *J. Quant. Spec. Radiat. Transfer*, 1966, **6**, 847. Certain of the values used differ from those given in this reference having been revised subsequent to publication of the paper.

[i] D. A. McGillis and L. Krause, *Can. J. Phys.*, 1968, **46**, 25.

[j] H. P. Hooymayers and P. L. Nienhuis, *J. Quant. Spec. Radiat. Transfer*, 1968, **8**, 955.

[k] J. A. Bellisio, P. Davidovits and P. J. Kindlmann, *J. Chem. Phys.*, 1968, **48**, 2376.

[l] E. S. Hrycyshyn and L. Krause, *Can. J. Phys.*, 1970, **48**, 2761.

[m] D. A. McGillis and L. Krause, *Can. J. Phys.*, 1968, **46**, 1051.

[n] D. R. Jenkins, *Proc. Roy. Soc.*, 1968, **A303**, 467.

[o] Ibid, 1969, **A313**, 551.

[p] J. P. Carrat, D. Casalta, J. L. Cojan and J. Hamel, *J. Physique*, 1966, **27**, 608.

[q] J. S. Deech, J. Pitre and L. Krause, *Can. J. Phys.*, 1971, **49**, 1976.

As will be seen, most of the data apply to process A(b)(ii) (see below), which is by far the most important, and are for elements giving the simplest spectra (i.e. mainly the alkali metals). Data are also included on cross sections for the mixing of excited levels (see below) as this process is closely related to quenching and the results may prove useful in certain fluorescence studies. For a more comprehensive discussion of the de-excitation processes and the literature data, the reader is referred to the review of Alkemade and Zeegers.[6]

A. Physical processes

(*a*) *Involving conversion of translational energy*, i.e. the excess electronic energy of the excited metal atom is converted into translational energy of the colliding species, no internal degrees of freedom of the latter being involved in the process.

(*i*) *Quenching by collisions with free atoms:*

$$A^* + B \rightleftharpoons A + B$$

This process is mainly of interest from the point of view of quenching by rare gas atoms. Values shown in Table 4-1 for σ_Q^2 are all less than *ca.* 0·3 Å^2 and, within the accuracy of the determinations, may be taken as almost zero. Indeed, measurements by Copley *et al.*[12] for Na($3^2P_{1/2,3/2}$) using metal vapour in a sealed bulb at low temperature ($\leqslant 200°C$), gave values of $<10^{-2}$ Å^2 for He, Ne, Ar, Kr and Xe. This is to be expected because transfer of kinetic energy must occur between the valence electron of the excited atom and the rare gas atom; the transfer is extremely inefficient owing to the great difference in mass of the colliding species.

(*ii*) *Quenching by collisions with free electrons:*

$$A^* + e^- \rightleftharpoons A + e^-$$

In view of the much smaller mass and higher velocity of the quenching electrons, this process should be much more efficient than (*i*) above. Direct measurements are not available but studies of the reverse excitation process[13] indicate that σ_Q^2 may be of the same order as that obtained with the major quenching process A(b)(ii) (see below). This is unlikely to be of any consequence in present applications of AFS in which flame atom cells are used, because the free electron concentration outside the primary reaction zone is relatively low in most of the flames employed. However, the process may become significant if attempts are made to use as atom cells devices such as plasmas which contain far higher electron populations.

(*iii*) *Mixing of excited states by collisions with free atoms:*

$$A^* + B \rightleftharpoons A^\dagger + B$$

This process involves the conversion, as the result of a collision, of one excited state of the atom to another (rather than to the ground state as in quenching). It has been widely observed for the doublets of the alkali metals where the energies of the two excited states are relatively close. Cross section data are shown in Table 4-2 for both processes involving collisions with other metal atoms such as

$$Na(^2P_{3/2}) + Na(^2S_{1/2}) \rightleftharpoons Na(^2P_{1/2}) + Na(^2S_{1/2})$$

and those involving collisions with rare gas atoms such as

$$Na(^2P_{3/2}) + Ar \rightleftharpoons Na(^2P_{1/2}) + Ar$$

It will be noted that the cross sections are large for Na and K, but are of the same order as quenching cross sections by process A(a)(i) for Rb and Cs where the energy difference between the doublet levels is rather larger.

(*b*) *Involving conversion of internal energy*, i.e. the excess electronic energy of the excited metal atom is converted on collision to electronic energy within another atom or molecule or to vibrational/rotational energy of the colliding molecule.

(*i*) *Quenching by collisions with other atoms:*

$$A^* + B \rightleftharpoons A + B^*$$

This can only occur with a large cross section when the two atoms have excited states of very similar energy (the slight difference during transfer being compensated for by their translational energy). It is simply the process of sensitized fluorescence described in Section 4.3.5 and, as explained there, is unlikely to be of any practical consequence in AFS work owing to the relatively low concentrations of metal atoms present in the atom cells employed.

(*ii*) *Quenching by collisions with molecules:*

$$A^* + BC \rightleftharpoons A + BC$$

As mentioned earlier, this is by far the most important process for consideration of quenching in AFS applications, and most of the data in Table 4-1 relate to it. The ability of many of the simple molecules found in flame gases to quench efficiently a wide variety

TABLE 4-2
Experimentally determined mixing cross sections

Atom (excited level)	Mixing cross sections,[a] $Å^2$										Temperature, K	Reference
	Mixing species											
	He	Ar	Na	K	H_2	N_2	HD	D_2	CH_4	Others[b]		
Na ($3\,^2P_{3/2} - 3\,^2P_{1/2}$)	—	—	—	—	13·4	24·2	14·0	16·6	24·5	25·8(CD_4), 30·6(C_2H_2), 30·0(C_2H_4), 30·2(C_2H_6)	390	(*c*)
Na ($3\,^2P_{3/2} - 3\,^2P_{1/2}$)	14·3	17·8	—	—	—	—	—	—	—	—	400	(*d*)
Na ($3\,^2P_{3/2} - 3\,^2P_{1/2}$)	13·1	20·6	—	—	—	—	—	—	—	—	400	(*e*)
Na ($3\,^2P_{3/2} - 3\,^2P_{1/2}$)	—	—	90	—	—	—	—	—	—	—	550	(*d*)
Na ($3\,^2P_{1/2} - 3\,^2P_{3/2}$)	—	—	—	—	25·5	45·8	26·8	31·2	47·1	48·1(CD_4), 58·0(C_2H_2), 56·6(C_2H_4), 58·0(C_2H_6)	390	(*c*)
K ($4\,^2P_{3/2} - 4\,^2P_{1/2}$)	—	—	—	—	16·9	21·0	15·6	15·9	—	—	350	(*f*)
K ($4\,^2P_{3/2} - 4\,^2P_{1/2}$)	13·0	7·0	—	79·5	—	—	—	—	—	—	360	(*g*)

Rb ($5\ ^2P_{3/2} - 5\ ^2P_{1/2}$)	—	—	—	—	—	2·2	—	—	—	—	300	(*h*)
Rb ($5\ ^2P_{3/2} - 5\ ^2P_{1/2}$)	0·032	$2{\cdot}5 \times 10^{-4}$	—	—	—	—	—	—	—	—	300	(*i*)
Rb ($5\ ^2P_{3/2} - 5\ ^2P_{1/2}$)	—	—	—	—	4·8	7·3	8·0	9·5	13·4	12·1(CD_4), 10·2(C_2H_4), 24·5(C_2H_6)	340	(*j*)
Rb ($5\ ^2P_{1/2} - 5\ ^2P_{3/2}$)	—	—	—	—	3·5	5·1	5·7	7·0	9·5	8·9(CD_4), 7·3(C_2H_4), 18·1(C_2H_6)	340	(*j*)
Cs ($6\ ^2P_{3/2} - 6\ ^2P_{1/2}$)	—	—	—	—	14·0	8·0	10·2	8·9	6·7	—	310	(*k*)
Cs ($6\ ^2P_{3/2} - 6\ ^2P_{1/2}$)	8×10^{-5}	—	—	—	—	—	—	—	—	—	300	(*i*)
Cs ($6\ ^2P_{3/2} - 6\ ^2P_{1/2}$)	—	$1{\cdot}6 \times 10^{-5}$	—	—	—	—	—	—	—	—	310	(*g*)

[a] Values are effective experimental cross sections. The factor π is omitted even where it was included in the original source of the data (see text).

[b] The quenching species is shown in parentheses after the value of the mixing cross section.

[c] M. Stupavsky and L. Krause, *Can J. Phys.*, 1968, **46**, 2127.

[d] J. Pitre and L. Krause, ibid., 1967, **45**, 2671.

[e] L. A. Jordan and P. A. Franken, *Phys. Rev.*, 1966, 142, 20.

[f] D. A. McGillis and L. Krause, *Can. J. Phys.*, 1968, **46**, 25.

[g] L. Krause, *Appl. Optics*, 1966, **5**, 1375.

[h] J. A. Bellisio, P. Davidovits and P. J. Kindlmann, *J. Chem. Phys.*, 1968, **48**, 2376.

[i] A. Gallaher, *Phys. Rev.*, 1968, 172, 88.

[j] E. S. Hrycyshyn and L. Krause, *Can. J. Phys.*, 1970, **48**, 2761.

[k] D. A. McGillis and L. Krause, *Can. J. Phys.*, 1968, **46**, 1051.

of excited atoms is generally attributed to the ease with which the electronic energy of the atoms is converted to vibrational energy within the molecule. Although the large number of vibrational levels available in each molecule for "matching" to the atomic levels of many different metals must contribute to this effect it should be remembered that the mechanisms involved are complex; in general the proximity of a given vibrational level of a given molecule to the excited state level of the atom concerned is not necessarily a good indication of the size of the quenching cross section.

(*iii*) *Mixing of excited states by collisions with molecules:*

$$A^* + BC \rightleftharpoons A^\dagger + BC$$

Comparision of this process with A(a)(iii) is analogous to that of quenching by atoms or molecules; the existence of only a small energy difference between the mixed levels is no longer necessary for a large cross section and reference to the small number of data included in Table 4-2 will show that values of Rb and Cs are now of the same order as those for Na and K. It should be noted that, in the presence of molecules, quenching and mixing of the excited atomic states will occur simultaneously and at similar rates. This effect has not been taken into account for most of the data in Table 4-1 where results are given for overall values of σ_Q^2 for the unresolved alkali metal doublets. Although this has little practical effect on the application of the results to AFS work, it may be of considerable importance in fundamental fluorescence studies.

B. Chemical processes

Very little information is available concerning chemi-quenching, although processes of the type

$$A^* + BC \rightleftharpoons A + B + C$$

must make some contribution to the overall quenching observed for certain species present in flame gases. Zeegers[14] has estimated the quenching due to dissociation of H_2O from the rate constant of the reverse excitation process

$$H + OH + K\ (4^2S) \rightleftharpoons H_2O + K\ (5^2P)$$

and finds for an air-hydrogen flame at about 2000 K that the ratio of chemical to total quenching is *ca.* 0·1. Thus it would seem that the process is of little practical interest in analytical AFS.

REFERENCES

1. C. Th. J. Alkemade, in R. M. Dagnall and G. F. Kirkbright (eds.), "Atomic Absorption Spectroscopy", Butterworths, London, 1970, pp. 73-98.
2. D. R. Jenkins, *Proc. Roy. Soc. Lond.*, 1969, **A313**, 551.
3. R. Smith, in J. D. Winefordner (ed.), "Spectrochemical Methods of Analysis", Wiley-Interscience, New York, 1971, Ch. 4, pp. 235-282.
4. H. P. Hooymayers, *Spectrochimica Acta*, 1968, **23B**, 567.
5. P. J. Th. Zeegers and J. D. Winefordner, *Spectrochimica Acta*, 1971, **26B**, 161.
6. C. Th. J. Alkemade and P. J. Th. Zeegers, in J. D. Winefordner (ed.), "Spectrochemical Methods of Analysis", Wiley-Interscience, New York, 1971, Ch. 1, pp. 3-125.
7. K. J. Laidler and K. E. Shuler, *Chem. Rev.*, 1951, **48**, 153.
8. K. J. Laidler, "The Chemical Kinetics of Excited States", Clarendon Press, Oxford, 1955.
9. K. E. Shuler, T. Carrington and J. C. Light, *Appl. Opt.*, 1965, Suppl 2, 81.
10. K. J. Laidler, 'Reaction Kinetics, Vol. 1: Homogeneous Gas Reactions", Pergamon Press, Oxford, 1963.
11. R. Mavrodineanu and H. Boiteux, "Flame Spectroscopy", Wiley, New York, 1965.
12. G. Copley, B. P. Kibble and L. Krause, *Phys. Rev.*, 1967, **163**, 34.
13. I. P. Zapesochnyi, *High Temp.*, 1967, **5**, 6.
14. P. J. Th. Zeegers, "Recombination of Radicals and Related Effects in Flames", PhD. Thesis, Utrecht, 1966.

CHAPTER 5

Spectral Light Sources

5.1 INTRODUCTION

The spectral light source forms an essential component for both atomic absorption and fluorescence spectrometry, and the optimization of its performance is of major importance if the capabilities of the two techniques are to be fully exploited. In consequence, a major part of the effort devoted to improvement of AAS and AFS instrumentation in recent years has been concerned with the light source. The results of this effort may be seen, for example, in the hollow-cathode lamps supplied as light sources with most commercial atomic absorption spectrophotometers; present-day lamps show many-fold improvements over those available for the early commercial spectrophotometers with respect to both their critical performance characteristics and their operating and storage lives. Development of light sources continues to be of major interest in this field and further improvements of the lamps described here are probable, particularly in the case of the high-intensity sources required for AFS.

5.1.1 Classification of Sources

The division of spectra into line, band and continuum types has been explained in Section 2.1. Spectral lamps used in AAS and AFS are usually referred to as line or continuum sources, the latter being applied to both true continua and those lamps emitting many closely spaced lines or bands over a wide wavelength range. The line and continuum categories may be further subdivided according to whether the lamp emits radiation *continuously* with respect to time or in short, specific bursts of energy at regular intervals, i.e. in *pulses.* Sources that are merely modulated for use with tuned detection systems (see Chapter 10) are generally classed as continuous. Thus, at the present time most sources used in AAS and AFS fall into this group. However, true pulsed operation is possible* with several types of lamps and its ability to increase the absolute intensity available from a given lamp may soon result in more widespread application of the technique.

The light sources may be further classified according to their method of construction, independently of the output characteristics. Most common in AAS and AFS are permanently† *sealed-off* lamps. These offer considerable advantages of simplicity, ease of use and good reproducibility, especially if the source requires a particularly low or high operating pressure. Many types of sealed-off lamp may be produced in a similar *demountable* form. Thus, by using, for example, removable electrodes or a tap in place of a permanent seal for the envelope, alterations may be made at any time after manufacture. This greatly facilitates investigations of the optimization of various lamp parameters or reduces the cost of providing line sources for a number of elements required only occasionally in the

* and sometimes essential, e.g. with many types of lasers.

† Obviously it may be possible to dismantle and re-condition some types of sealed-off lamp if suitable facilities are available.

laboratory. Finally, several types of emission source may be classified as *consumable* as they require continuous replacement of the element of which the spectrum is required. For example, flames, arcs, plasmas and flow-through microwave discharges are described below as sources for AAS and AFS. Although such types are usually very versatile, it will be seen that they have many disadvantages which have limited their application.

Although many types of light source can be used for AAS and AFS, most are only briefly summarized here. The greater part of this chapter is devoted to a detailed description of the hollow-cathode lamp (HCL), the electrodeless-discharge lamp (EDL), the vapour-discharge lamp (VDL) and several continuum sources. These account for most of the published work in AAS and AFS and almost all analytical applications. In each case the principle and construction of the device is discussed together with its operating requirements and the characteristics of its spectral output.

5.1.2 Requirements of Light Sources Used in AAS and AFS

Some essential requirements of spectral light sources were mentioned briefly in Chapter 1 and emphasized to a certain extent in the theoretical discussion of the following chapters. We shall list the most important factors here before proceeding to a discussion and comparison of individual light sources.

For use with both AAS and AFS, it is important that the light source emit a steady uniform level of radiation to obtain an analytical signal of low noise level allowing good precision of measurement and good detection limits. In many cases it is also important that the source should show little long-term drift in emission intensity as most fluorescence spectrometers and many commercial absorption spectrometers have only single-beam operation and frequent adjustment of instrumental settings is inconvenient. In view of the need for convenience and ease of use with commercial atomic absorption spectrometers designed for large numbers of routine analyses, the spectral lamp used should also require the minimum of maintenance and adjustment to obtain optimum performance. This requirement usually necessitates the use of a permanently sealed-off source.

The characteristics of the spectra produced by sharp line sources are also important. They should be "clean", with the minimum interference to the analytical line due to the presence of spectral lines of the filler gas or impurities in the lamp. This is achieved by the careful choice of the filler gas and the use of a getter in the lamp (see Section 5.2.3). The intensity of the analytical line must also be high for optimum results. In AFS it is essential that the maximum intensity is obtained, since the signal intensity (and hence the analytical signal) will be directly proportional. Even though such high intensity is not needed for AAS, a reasonably high level is required to permit low signal-noise levels and good precision to be obtained. With AAS, however, the profile of the source line is even more important than its intensity, as explained in Chapter 3. Thus, the source must be designed to avoid long path-lengths through the excited vapour; this helps to avoid self-absorption of the resonance lines which is a major source of their broadening. Similarly, the avoidance of high temperatures, pressures and current densities within the source is also an important aid to obtaining the sharpest possible source line for AAS measurements (see Section 2.5.2). With both AAS and AFS sources it is essential that the emitted resonance radiation should not pass through a cloud of cool vapour (of the analyte element) between leaving the excitation region and reaching the exit window of the lamp; this would produce self-reversal of the resonance lines, removing the central region required for both peak absorbance measurements and the efficient excitation of fluorescence.

In view of the need for sealed-off sources mentioned above and the large numbers of relatively expensive commercial lamps required in many laboratories, it is essential that such lamps should have long operating and storage lives. In many cases this has been achieved by careful attention to minor constructional details. For example, metal electrodes and silica optical windows should be fused directly to glass lamp bodies rather than attached by cemented joints. Optical windows should always be positioned so that transmission will not be reduced by formation of an opaque coating on the inner surface, particularly where volatile compounds are used or there is a possibility of the lamp being used at currents above the recommended value. Similarly, constructional materials likely to be attacked by the compound within the lamp should be avoided. For instance, the alkali metals are known to attack silica and certain types of glass. For hollow-cathode lamps of low-melting-point elements, accidental spillage from the cathode should be avoided by careful cathode construction or use of a higher melting compound or amalgam. Finally the overall construction of the lamps should be as strong as possible avoiding such obvious hazards as thin glass necks and joints.

5.2 HOLLOW-CATHODE LAMPS

5.2.1 Introduction

The hollow-cathode lamp is the most widely used of the many devices available for the production of fine line spectra and since its introduction in 1916 by Paschen[1] it has seen extensive application as a source of intense, sharp spectral lines. Several factors have been responsible for this popularity, viz: (a) lines may be produced over the whole optical region from the infrared to the vacuum ultraviolet; (b) intense spectra may be excited using only low currents; (c) extremely sharp lines are quite readily obtained; (d) the lamps and their associated power supplies are relatively simple and inexpensive to construct; (e) they give a stable, noise-free output over long periods of time. In spite of its widespread use for fifty years, the hollow-cathode lamp only became generally available commercially after the introduction of atomic absorption spectroscopy as a routine analytical method. More lamps have probably been manufactured since then than during the whole of the previous period; all commercial AAS instruments available at the present time are designed for use with these sources.

During the period from 1916 to 1955, only flow-through hollow-cathode lamps were used. This resulted from the requirement of most workers for very pure spectra which could not be obtained with sealed-off devices owing to subsequent out-gassing of the various components. There were also problems of leaking at window and electrode seals and of filler gas adsorption (see below). However, in 1955 Crosswhite, Dieke and Legagneur[2] produced a successful sealed-off iron lamp using a getter material (see below) and Walsh[3] suggested the HCL as a suitable source for AAS on the basis of the availability of a sealed device. As a result of later improvements and the less stringent requirements of spectral purity for sources used in routine AAS, most commercial lamps provided as AAS sources have been of the permanently sealed type. In view of this we shall consider the construction of these devices in detail and discuss flow-through types only briefly from the point of view of the special type of demountable HCL (see Section 5.2.6(a)).

Many different designs of HCL have been described in the literature, but the most common can generally be placed into one of two classes. The Schüler[4] lamp design uses water or liquid-air cooling of the cathode to give extremely sharp lines for fundamental applications such as ultra-high resolution studies of hyperfine structure. When line width is less critical, or for studies of high-energy lines, a hot hollow-cathode has generally proved more useful; this allows a simpler lamp design and much higher intensity. Provided that the lamp current is kept relatively low, the resonance lines are sharp enough for peak absorption measurements and all sealed-off lamps used for AAS are of this type. Thus cooled-cathode lamps will not be described here,* although they may be of some use for isotope analysis by AAS.[6,7]

5.2.2 Mechanism of the Hollow-Cathode Lamp

In general terms the hollow-cathode lamp consists of a sealed tube containing an anode, a cylindrical hollow cathode made of, or somtimes lined with, the metal whose spectrum is required, and an inert gas at a pressure of about 1 to 5 torr (depending on the gas used). When a potential of 500 to 1000 volts is applied between the electrodes a discharge of the carrier gas strikes and appears as a glowing positive column. However, at the pressure used (and for a range of 2-3 torr about it) the discharge concentrates into the hollow cathode, giving higher current densities than those normally attained in the discharge and hence intense illumination of the cathode space. The appearance of the cathode at this stage is shown in Fig. 5-1. The spectrum of the carrier gas obtained in the negative glow region shows no Stark line broadening (see Section 2.5.2) in spite of its high intensity because the potential drop between the electrodes with a steady discharge is usually only 250-300 volts. However, as the current is increased the glow occupies a larger area of the inner surface of the cathode; once the whole surface is covered a further increase in current produces an increase in the potential difference and the freedom from Stark broadening may no longer hold. With most cathode designs the current at which this occurs will usually be far in excess of other limitations to the current imposed by requirements for lamps used as AAS sources.

The low cathode-potential fall is due to the small wastage of positive ions from the hollow cathode. Thus, in conjunction with the stream of carrier gas ions directed into the cylinder by the applied field this produces a high concentration of ions within the discharge. Following acceleration by the field these ions bombard the inner surface of the cathode giving a strong sputtering action as they

* The reader is referred to the book by Tolansky[5] for further details of this type of lamp.

displace atoms from the surface. The sputtered metal atoms accumulate inside the cathode and are excited to give an emission spectrum by second-order collisions with the high concentration of excited inert gas atoms and ions. The vapour cloud of the metal atoms must be sufficiently concentrated if its spectrum is to predominate over that of the carrier gas. However, this concentration is still relatively very low so that problems of self-absorption or even resonance broadening (see Section 2.5.2) should not interfere with the excitation of an intense metal spectrum.

The sputtering mechanism of the hollow-cathode lamp explains its ability to produce intense metal spectra even when cooled to temperatures at which the vapour pressure of the metal is negligibly small. Under these conditions it is particularly important that

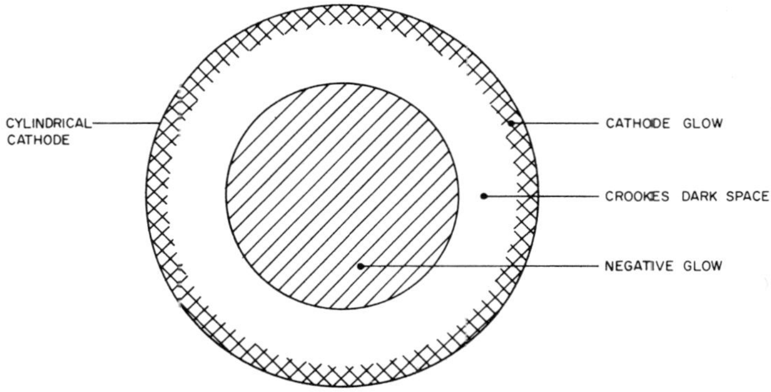

Fig. 5-1. Hollow-cathode carrier-gas discharge obtained on striking.

efficient sputtering should be obtained; it may be necessary to employ argon as the inert carrier gas rather than neon or helium because it is a more effective sputtering agent owing to its greater atomic weight.* When a hot hollow cathode is used, as in AAS, the thermal vapour pressure of many metals may be considerable so that the metal pressure approaches that of the carrier gas. The metal vapour discharge may then be partly or completely self-sustaining and a very large increase in intensity of its spectrum is observed. Many of the lamps used in AAS are run close to this condition and careful adjustment of the current (i.e. vapour pressure) is important to avoid serious self-absorption and other line-broadening effects. Lamps run largely in metal vapour tend to give mostly lines of fairly low excitation energy (i.e. "arc lines") so that mainly the first spectrum is observed with few ion lines. If a "spark spectrum"

* Argon 39·95, Neon 20·18, Helium 4·00.

containing many ion lines is required it is necessary to ensure excitation by collisions with an inert carrier gas having a high excitation potential, such as helium (19·7 volts).

5.2.3 Construction of Hollow-Cathode Lamps

It is proposed to discuss in some detail the main components of commercially available lamps rather than to describe a practical method of lamp construction. However, in a laboratory equipped

TABLE 5-1
Major manufacturers of hollow-cathode lamps

Atomic Spectral Lamps Pty. Ltd., 23–31 Islington Street, Melbourne, Australia.
Cathodeon Ltd., Church Street, Cambridge, England.
FIVRE, Via Panciatichi 70, Firenze-Castello, Italy.
Hitachi Ltd., 1–4 Marunouchi, Chiyoba-Ku, Tokyo, Japan.
Dr. Kern und Sprenger GmbH., Post Box 751, Florenz Sartorius Strasse 5, Gottingen, Germany.
Perkin Elmer Corp. Instrument Division, Norwalk, Connecticut 06852, U.S.A.
Quarzlampen GmbH., Hanau/Main, Hoehensonnenstrasse, Germany.
Rank Precision Industries Ltd., 98 St. Pancras Way, London, N.W.1., England.
Westinghouse Electric Corporation, Electronic Tube Division, P.O. Box 284, Elmira, New York 14902, U.S.A.

with a suitable vacuum system and glass- and metal-working facilities, there should be no real difficulty in the construction of sealed lamps similar to those available from lamp manufacturers but at considerably lower cost. Practical instructions for laboratory manufacture have been published by Jones and Walsh[8] and by Bodretsova *et al.*[9] Unfortunately, details of the most recent improvements in lamp design have not always been published owing to commercial considerations. Although many useful hints can be obtained from the following pages and from the catalogues published by the larger lamp manufacturers,* it may prove necessary to carry out many measure-

* The major Western manufacturers of hollow-cathode lamps are listed in Table 5-1.

ments with each element in order to optimize parameters such as cathode dimensions and filler-gas pressure.

(a) Lamp envelope and window

The early commercial sealed hollow-cathode lamps were constructed to a design similar to that shown in Fig. 5-2 and were generally made as large as possible. The large volume was intended to increase the useful life of the lamp by minimizing the effects of leakage of air into the tube, adsorption of the filler gas onto particles sputtered from the cathode and liberation of gaseous impurities by the electrodes during use. The optical window was placed on a side

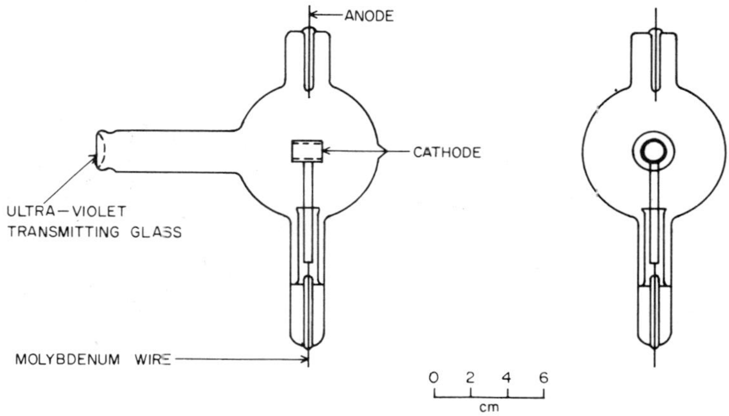

Fig. 5-2. Early design of hollow-cathode lamp for commercial atomic absorption spectrometers.

arm some distance from the cathode further to reduce the possibility of deposition of sputtered metal atoms onto it (with consequent loss of transmission). The envelope was made of thin electronic grade (molybdenum) glass with thermal expansion properties similar to those of the leads to the electrodes so that a vacuum-tight seal could readily be produced.

Recent improvements in lamp technology (see below) have led to the production of lamps of much smaller volume and of a more convenient design similar to that shown in Fig. 5-3. The lamp illustrated has an optical window of 25·4 mm diameter (similar to the older design), yet an overall length of only 165 mm and a maximum diameter of about 38 mm. This small size is particularly useful when several lamps are kept on "warm-up" to allow rapid interchange of elements; it also greatly reduces the storage space required for a large number of lamps. The metal-to-glass seals at the base of the lamp are protected by a plastic cover cemented to the glass, electrical connections being made via a special "lamp socket" built into this base. Use of these plastic bases, which in some designs incorporate a special fitting allowing direct attachment to the spectrometer, has greatly reduced the possibility of accidental breakage compared with the older style of lamp.

The elongated shape for the lamp body is used because it is still necessary to place the optical window a considerable distance from the cathode to avoid loss of transmission due to metal deposition.

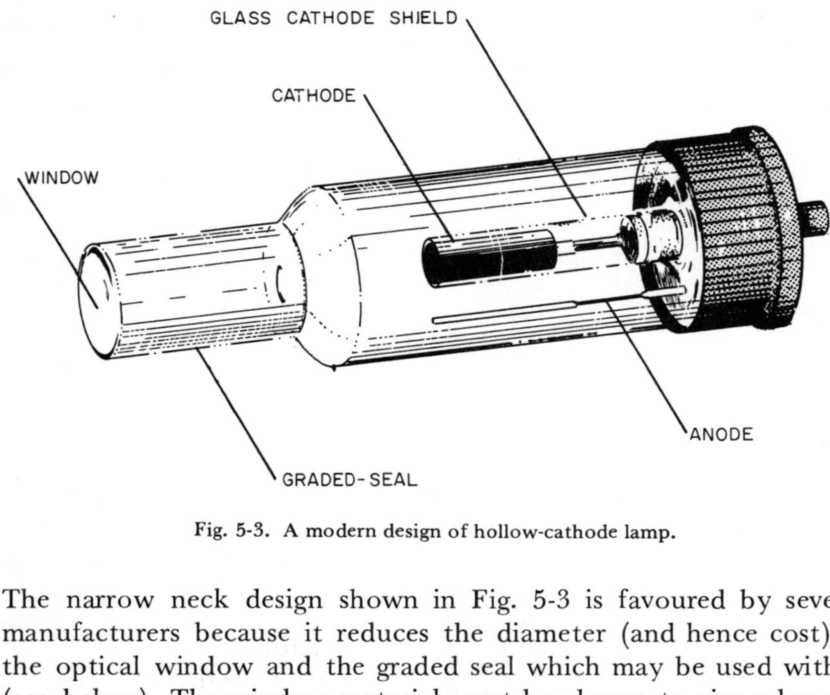

Fig. 5-3. A modern design of hollow-cathode lamp.

The narrow neck design shown in Fig. 5-3 is favoured by several manufacturers because it reduces the diameter (and hence cost) of the optical window and the graded seal which may be used with it (see below). The window material must be chosen to give adequate transmission at the wavelength of the resonance line(s) required from a particular lamp. Transmission characteristics for several window materials used in commercial lamps are shown in Fig. 5-4. Corning 7740 Pyrex glass gives good transmission above 300 nm and is used for lamps of a number of elements, including the alkali metals and many rare earth metals. It has the added advantage of being easy to seal to the electronic grade glass normally used for the lamp envelopes. However, Pyrex is unsuitable for lamps of the majority of elements, which have their best absorption lines below 300 nm and either silica or an ultraviolet transmitting glass must be used as the window material. As can be seen from Fig. 5-4, the latter is inferior to silica at all wavelengths below about 320 nm and gives very poor transmittance below 250 nm. In spite of this it has been used by several manufacturers because it can be sealed directly to the glass envelope of the lamp, whereas silica cannot (owing to the strain imposed by the large difference between the expansion coefficients of the two materials).* Early silica window lamps used a cement seal (e.g. epoxy resin) but this was rather unsatisfactory because of early

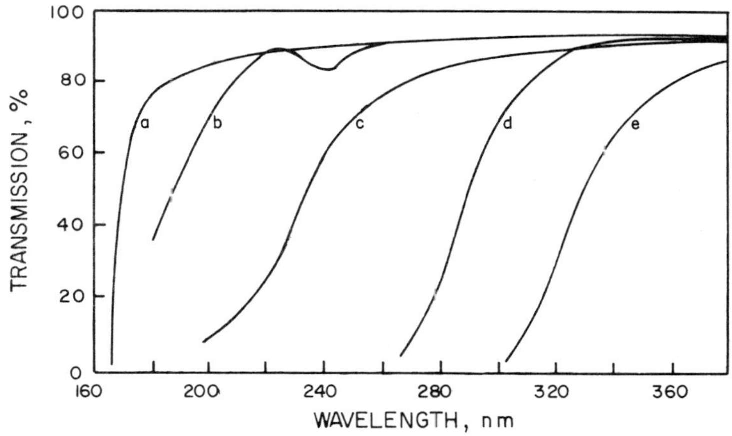

Fig. 5.4. Transmission characteristics of window materials. (a) Suprasil (optical quality silica); (b) commercial silica; (c) Corning 9741 (uv transmitting glass); (d) Corning 7740 (pyrex glass); (e) Corning 7720.

leakage at the seal and also because it was not possible to obtain adequate degassing of the lamp blanks by heating them to a sufficiently high temperature. Most manufacturers now prefer to use silica windows properly sealed to the envelope via a graded seal. These are simply open-ended cylinders produced by the glass manufacturers and having silica at one end and a suitable glass at the other. The joint between the two is achieved by a special process interposing material of thermal expansion carefully graded between the two extremes. The seals are extremely reliable without being prohibitively expensive, and lamps of this type are now normally used whenever Pyrex windows are unsatisfactory.

* Lamp envelopes of silica would be far too expensive to manufacture for normal AAS use and the silica-to-metal seal required at the electrodes also presents difficulty.

(b) Electrode configuration

Recent improvements in the design of the cathode and the other electrodes and shields associated with it have been of similar importance to those of the lamp envelope. As a result, high-quality lamps have been extended from a relatively small range of "favourable" elements to most of the elements for which AAS is a useful analytical technique. Many different cathode shapes have been suggested, some of which are shown in Fig. 5-5, but most manufacturers seem to have standardized on the narrow-bore cylindrical type (h). Those with rectangular apertures ((a), (b), (c)) give effectively a greater output of light since most spectrometers have as the dispersive element a monochromator with a rectangular entrance slit. However, there appear to be stability problems owing to concentration of the cathode glow at sharp corners. The cup-shaped designs ((c), (d)) have been quite widely used for low-melting elements as they facilitate retention of the metal within the cathode. However, most manufacturers now prefer the use of higher melting alloys or compounds in the ordinary narrow-bore cathode (h). The spherical cavity (f) is the most efficient from the electrical point of view but has not been widely used, presumably owing to difficulties of construction. Whatever the shape of the cathode, the dimensions of the hollow are quite critical if optimum performance is to be obtained (although this seems to have been overlooked in the design of some early commercial lamps). Mitchell[10] has shown that at any filler-gas pressure the ratio of resonance line intensity to unwanted ion line intensities decreases if the diameter of the bore is increased or its length decreased (see Fig. 5-6). The filler-gas pressure required also depends on the bore diameter, an approximate relationship being $d = c/p$ where p is the gas pressure (torr), d the diameter (cm) and c a constant varying between 0·5 and 10, depending on the cathode material, the type of filler gas and the operating temperature.

The cathode material also has an important effect on the operation of the lamp. Obviously it must be a reasonably good electrical conductor and in view of the fairly high operating temperature* of the types used in AAS should not melt or deform during operation. A reasonably low vapour pressure is also important to avoid rapid evaporation onto the interior surfaces of the lamp. Elements operated as liquids in the cup-type cathodes described

* 400-500°C may be reached near the maximum recommended operating current.

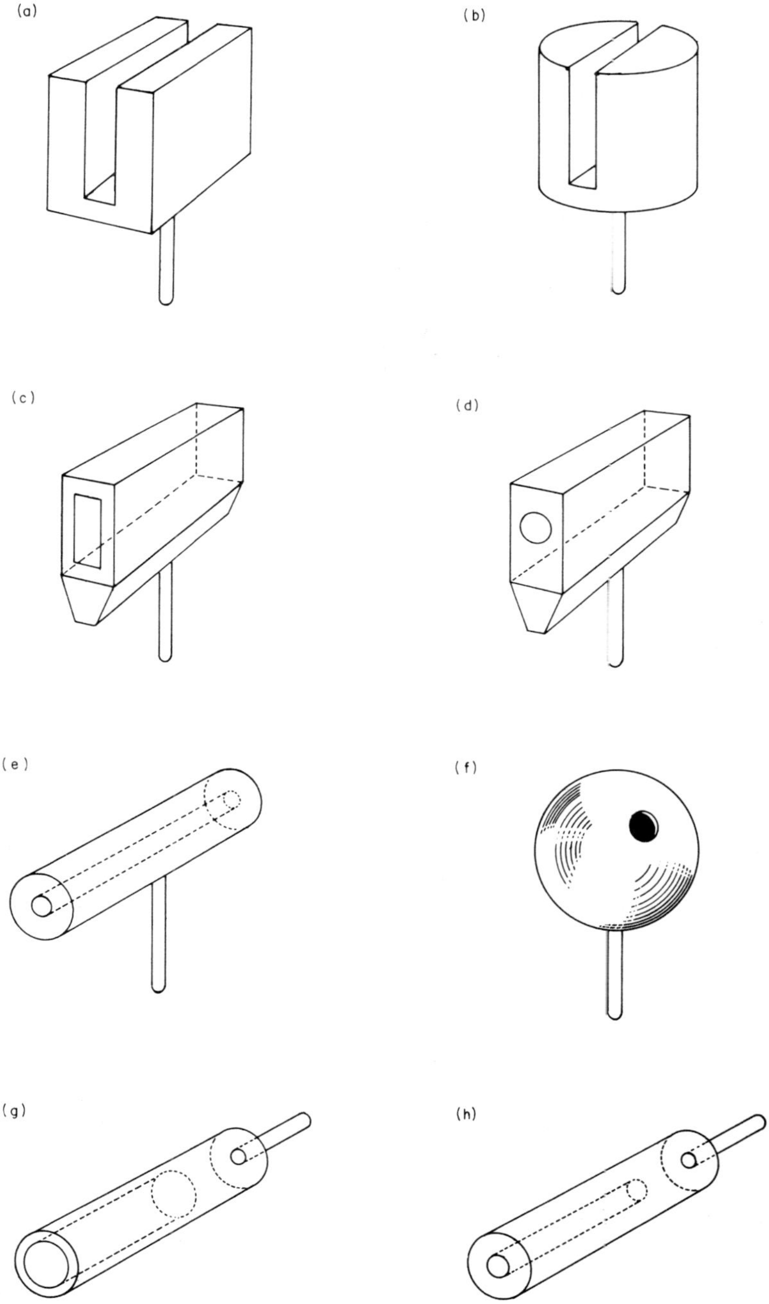

Fig. 5-5. Possible hollow-cathode shapes.

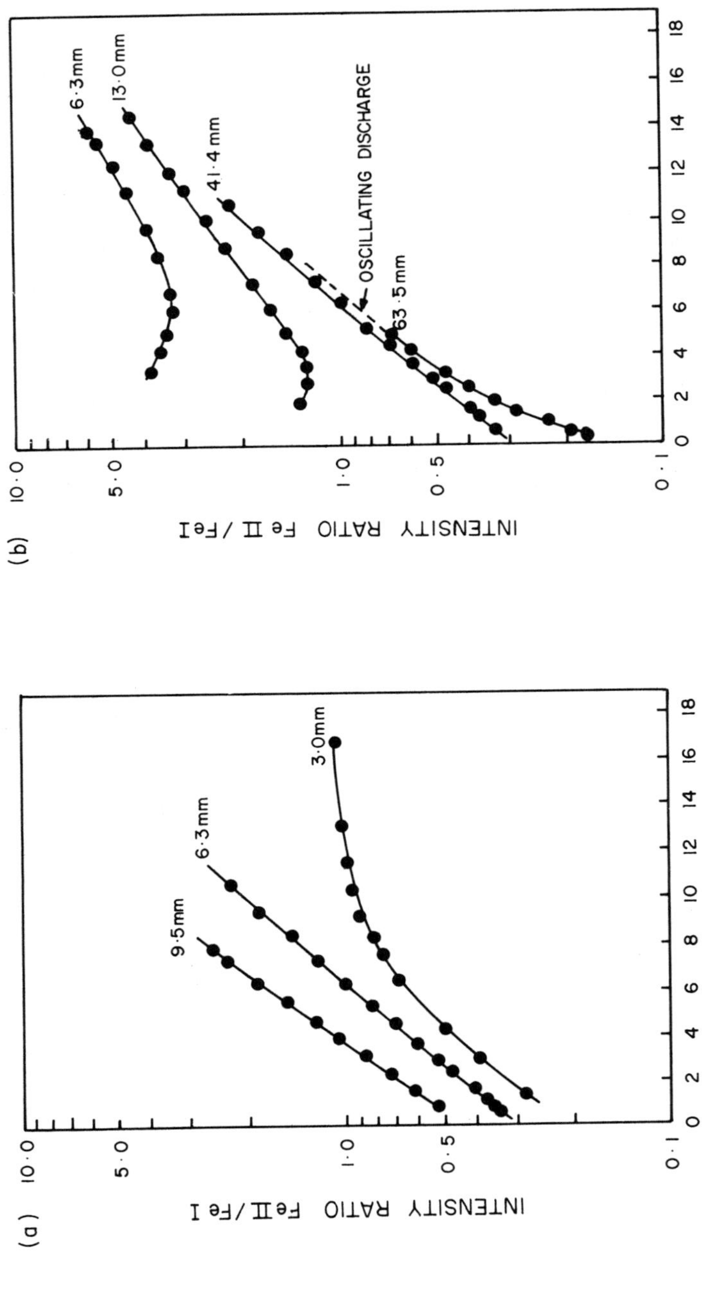

Fig. 5-6. Ratio of intensities Fe II 274·93/Fe I 273·36 at different neon pressures for a hollow-cathode at 160 mA.[10] (a) For different bore diameters, bore length 41·4 mm; (b) for different bore lengths, bore diameter 6·3 mm.

above must, of course, have relatively low vapour pressures even when molten (e.g. Ga, Pb, Sn). With many of the common metals it is possible to use the simple procedure of machining the cathode from a solid rod of the pure element. When this is not practicable owing to the cost or the physical nature of the element, a cathode shell of copper, brass, aluminium or iron may be fitted with an appropriate lining. At one time such a lining was frequently very thin and produced from foil, electrolytic plating, or deposition of a suitable salt coating. However, the sputtering process rapidly depletes such thin linings, exposing the base metal and leading to fluctuations and a gradual reduction in output intensity. Thus more satisfactory results are obtained with a fairly thick lining of a suitable alloy or salt. In fact, recent improvements to a number of "difficult" lamps have involved careful selection of the matrix used to contain the analyte element so that the rate at which it is sputtered may be closely controlled. The difficulties of producing varied matrices as alloys have been reduced considerably by using the techniques of powder metallurgy; this involves thorough mixing of the dry powders followed by sintering to produce a solid cathode lining of the correct dimensions. Strong "blocks" may be produced from even quite high melting powders by adding a lower melting bonding powder such as aluminium or copper. These techniques have also been important in the production of multi-element lamps (see Section 5.2.6(e)).

Considerable improvements have also been made to the anode design and position, although it would now seem that a quite simple arrangement works well. Some possible configurations which have been used in commercial lamps are shown in Fig. 5-7. Although types (a) and (b) were used for early commercial lamps, they are now less popular owing to the inconvenience of the side seal and connection for the anode. Types (c) and (d) are both currently available in commercial lamps, (c) having a circular anode concentric with the end of the cathode and mounted on an insulated stem. With type (d) the whole anode may be exposed in some lamps, whereas other manufacturers prefer to insulate the anode as far as the end of the cathode. These three possibilities seem to be a matter of individual preference, most manufacturers using the same design whatever the element; in practice there is probably little advantage to any one of them (except that the straight wire anode may be cheaper to produce). The selection of the anode material is also fairly straightforward, a metal such as tungsten usually being preferred. At one time titanium or zirconium was used by some manufacturers

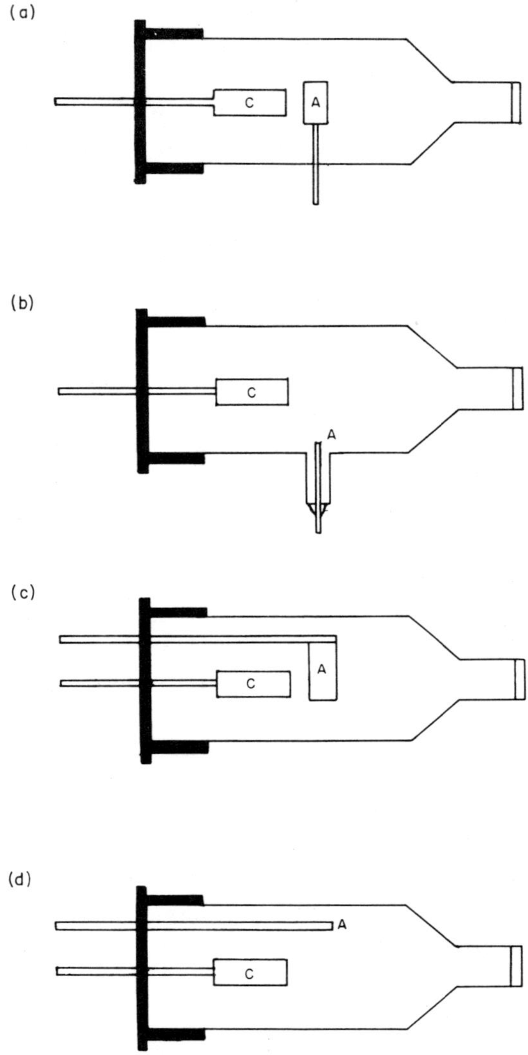

Fig. 5-7. Positions of the cathode and anode in hollow-cathode lamps.

because they actively absorb gaseous impurities in the lamp and thus act in the same way as the uranium getters incorporated in the first sealed-off lamps with much simpler lamp construction. However, it is now recognized that for the requirements of AAS light sources a getter is unnecessary if lamps are carefully constructed with adequate precautions in the selection and degassing of the materials used.

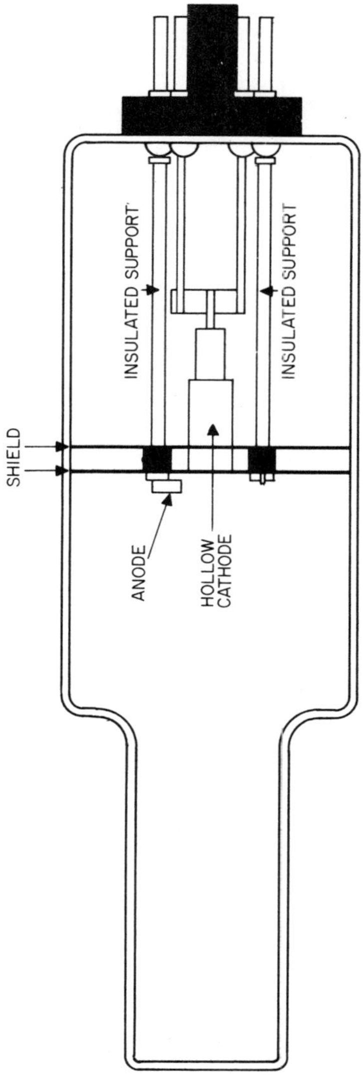

Fig. 5-8. Commercial hollow-cathode lamp with simple disc electrode shields.

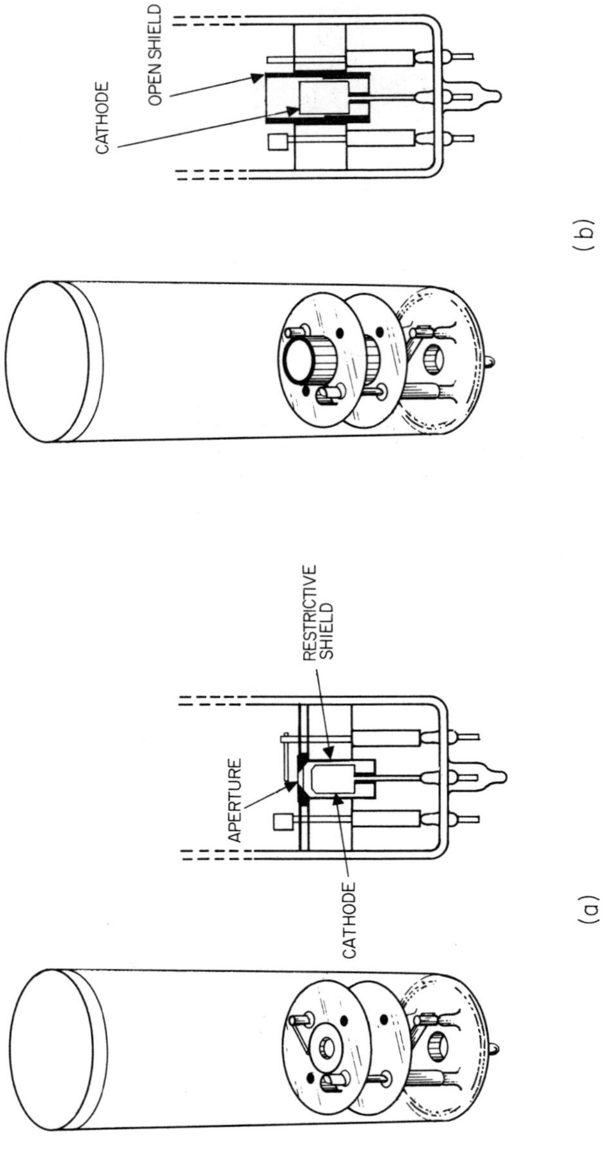

Fig. 5-9. Elaborate hollow-cathode lamp shields. (a) Restrictive type; (b) open type.

Ideally, the construction of the electrodes should be such that the energy of the discharge is concentrated within the cathode, maintaining the cloud of metal vapour within a restricted area and at a uniform level of excitation. In practice this is not always possible and a cloud of weakly excited atoms may form at the mouth of the cathode or just outside it, leading to serious self-absorption of the resonance lines and a possible shortening of lamp life. To reduce this problem many commercial lamps incorporate some form of cathode shield. A simple glass shield used by Atomic Spectral Lamps Pty. Ltd. and mainly intended to prevent discharges to the outside of the cathode is shown in Fig. 5-3. A disc shield used in conjunction with a circular anode by Pye Unicam Ltd. is shown in Fig. 5-8. Ceramic shields have been used in some lamps for difficult elements. Those designed by the Perkin-Elmer Corporation are illustrated in Fig. 5-9. The restrictive type is recommended for molten or volatile metals such as lead or bismuth, whereas the open shield is used for the less volatile metals such as titanium or germanium. Manning and Vollmer[11] report that both types of shield help to restrict the metal vapour cloud to a region in front of the cathode, concentrate the discharge in the cathode cavity and reduce filler-gas cleanup.

(*c*) *Filler gas*

As mentioned earlier, the mechanism of the hollow-cathode discharge requires a low pressure of an inert gas within the lamp. The properties of the rare gases relevant to their use for this purpose are summarized in Table 5-2. The efficiency of the filler-gas ions in sputtering metal atoms from the cathode depends not only on their mass but also on their velocity. Thus, with the strength of the potential field normally present in the hollow cathode, argon is generally more effective than the heavier krypton or xenon. In view of its very low mass helium is extremely poor in its sputtering ability and is not usually used in hollow-cathode lamps of the type required for AAS. Neon, however, has a sufficiently high mass to act as an efficient sputtering agent and from a practical point of view is comparable with argon in this respect. It should also be remembered that with the uncooled lamps used in AAS, the operating temperature of the cathode may be sufficiently high to give a significant vapour pressure of a volatile element. Thus, the results obtained with a particular inert gas often vary considerably between involatile and volatile metals. Furthermore, a meaningful comparison between the different gases can only be made if the pressure is carefully optimized in each case; a change of only a few torr will often produce a very large change in the intensity of the emitted metal resonance lines.

Detailed observations of the effects of different gases and pressures have been made by Crosswhite *et al.*[2] (Fe), Mitchell[10] (Fe and U) and Bodretsova *et al.*[9] (Al, Bi, Ca, Cd, Co, Cr, Cu, Fe, Mg, Mo, Ni, Pb, Sb, Sn, Ti and Zn). Although generalized conclusions can be misleading, some comments on the results of Bodretsova *et al.* are of interest. They found that similar variations of metal line intensity with filler-gas pressure occur with He, Ne, Ar and Xe. When the gas pressure is gradually reduced with lamps of involatile metals, the resonance line intensity increases to a peak and then decreases again for most gas/metal combinations. With volatile metals the intensity generally continues to increase until the pressure is too low to maintain the discharge. In both cases the optimum pressure for neon is around 3 torr whereas that for argon is about 1 torr. The action of the other gases with different metals is less consistent, although helium generally shows similar behaviour to that of neon, and xenon to that of argon. The highest intensities were usually observed with neon and argon, although xenon gave a considerably greater intensity in a few cases (e.g. Al, Ni and Sb). Neon resulted in a considerably more intense output than that of argon for several of the involatile metals and in most other cases the two gases produced comparable metal line intensities at the optimum pressures; only in one or two examples (e.g. Ca and Pb) did argon result in a significantly higher intensity than that of neon.

TABLE 5-2

Properties of hollow-cathode filler gases

Gas	Atomic weight	Ionization potential (eV)	Excitation potential (eV)
Helium	4·00	24·580	19·818
Neon	20·18	21·559	19·619
Argon	39·95	15·755	11·548
Krypton	83·80	13·996	9·915
Zenon	131·30	12·127	8·315

In practice, only argon and neon are used as the filler gas in commercial lamps. Although neon will often produce a higher emission intensity of the metal lines, it is not an automatic choice because the spectral line emission of the inert gas itself must be considered. As explained in Section 12.3.1(c), any of these background lines lying close to a metal resonance line may lie within the spectral bandpass observed by the monochromator during an absorption measurement. As they are not absorbed by the sample atoms a loss of sensitivity and linearity of calibration graphs results. The hollow-cathode spectrum of neon is shown in Fig. 5-10 and that of argon in Fig. 5-11. The spectra consist mostly of lines of Ne(I), Ne(II), (particularly below 450 nm), Ar(I) and Ar(II) (particularly below 400 nm), although some impurity lines may be identified with both elements. When reference is made to these diagrams it should be remembered that many weaker lines of the spectra are not included. However, these will be of negligible intensity compared with those of the metal resonance lines in properly constructed lamps. In fact many of the less intense lines appearing in Figs. 5-10 and 5-11 will also be of little consequence under most circumstances. It should also be noted that in order to include these lines and retain a reasonably small diagram the higher peaks have been compressed by the recording technique while the most intense lines have been limited to an arbitrary maximum height.

When preparing a particular lamp reliable results may only be obtained if tests are carried out using both neon and argon at their optimum pressures (see above). It is then possible to examine closely the spectral region of interest, i.e. that within a few nanometres of the most sensitive resonance line, in order to decide which rare gas gives the most favourable signal-to-background ratio for the resonance line. In a few cases the choice may actually depend on the spectral bandwidth of the spectrometer in which the lamp will eventually be used. However, some general conclusions may still be drawn from Figs. 5-10 and 5-11. Argon emits rather fewer lines than neon does in the regions between 325 and 350 nm and between 580 and 700 nm, whereas neon is somewhat better between 370 and 520 nm. There is little to chose between the two in the regions below 325 nm and between 350 and 370 nm and 520 and 580 nm.

In practice, most commercial lamp manufacturers now use neon wherever possible, reserving argon for the special cases where it gives a considerably less intense background spectrum around the resonance line. The number of these special cases varies with different manufacturers and there is little doubt that personal preference still plays a part in the choice of gas. Argon was used almost exclusively in the early commercial lamps, probably because only the background spectrum was considered when the choice was made (rather

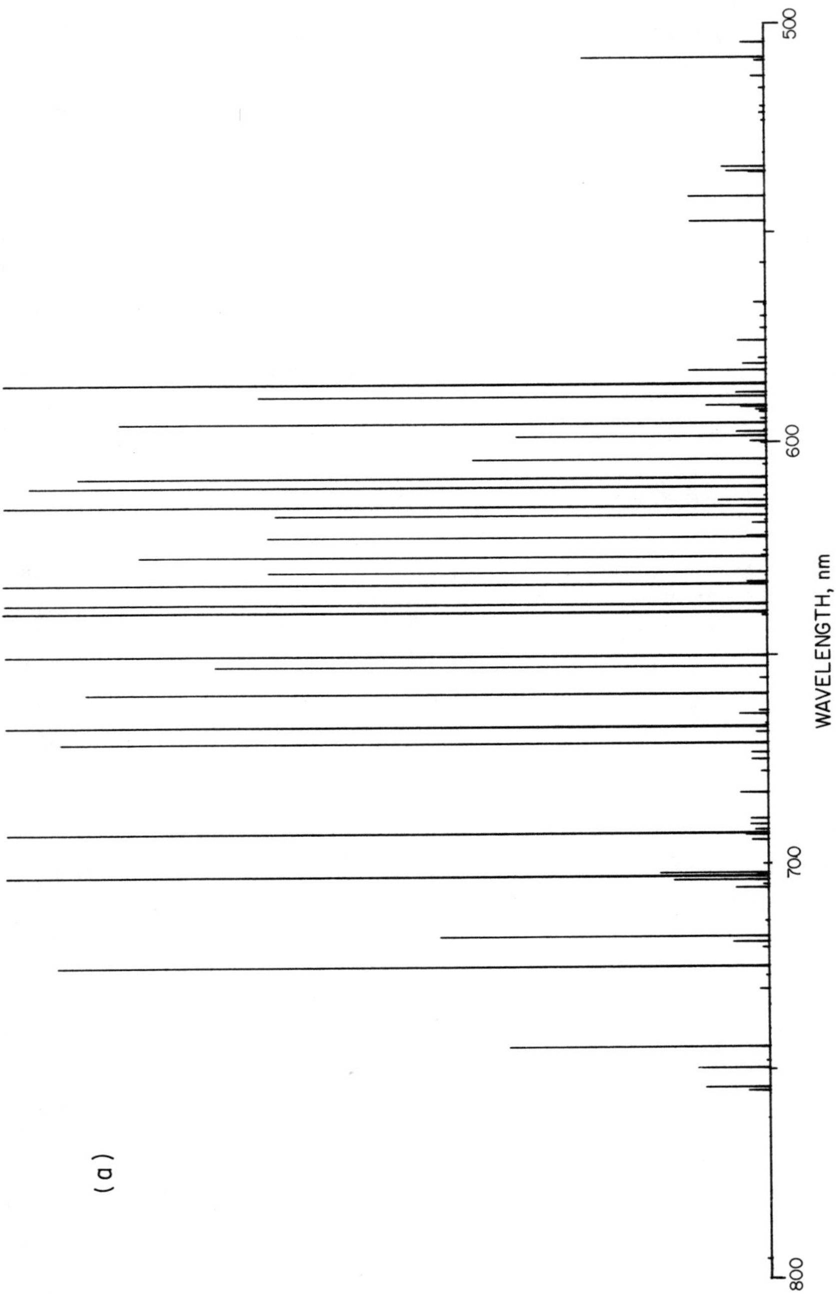
500
600
700
800
WAVELENGTH, nm
(a)

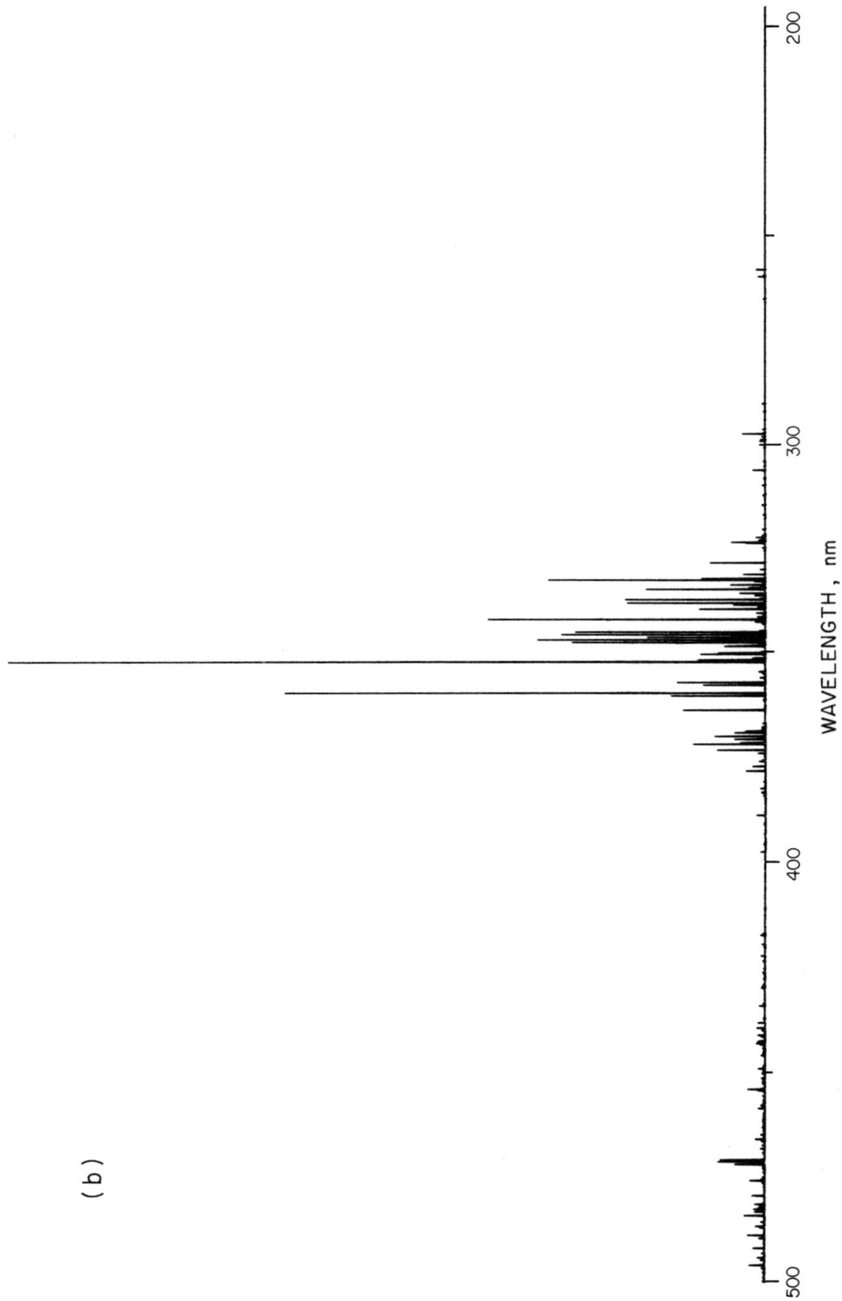

Fig. 5-10. Spectrum of neon in a hollow-cathode discharge. (a) 800-500 nm; (b) 500-200 nm. (The intensity scale from 500 to 200 nm is scale expanded 10x compared with that for 800-500 nm.)

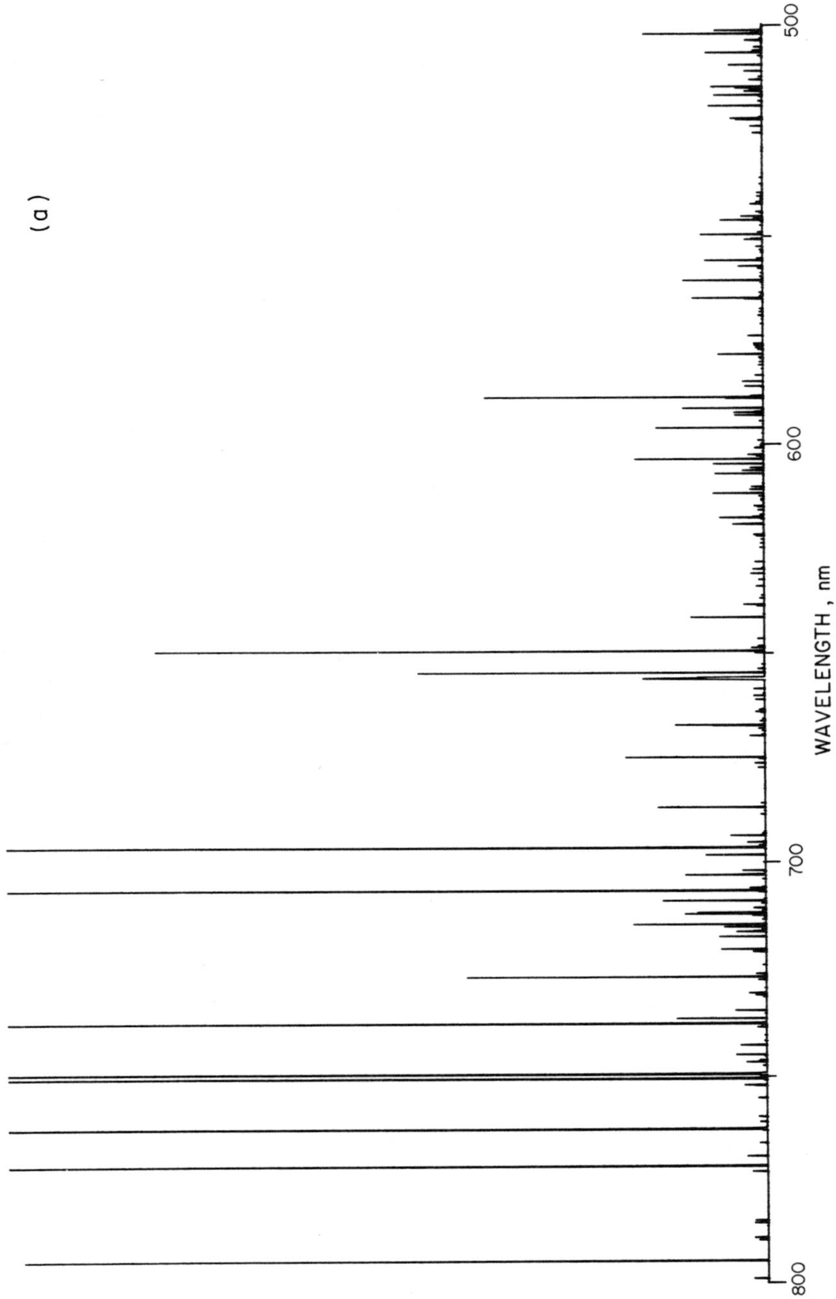
(a)
500
600
700
800
WAVELENGTH, nm

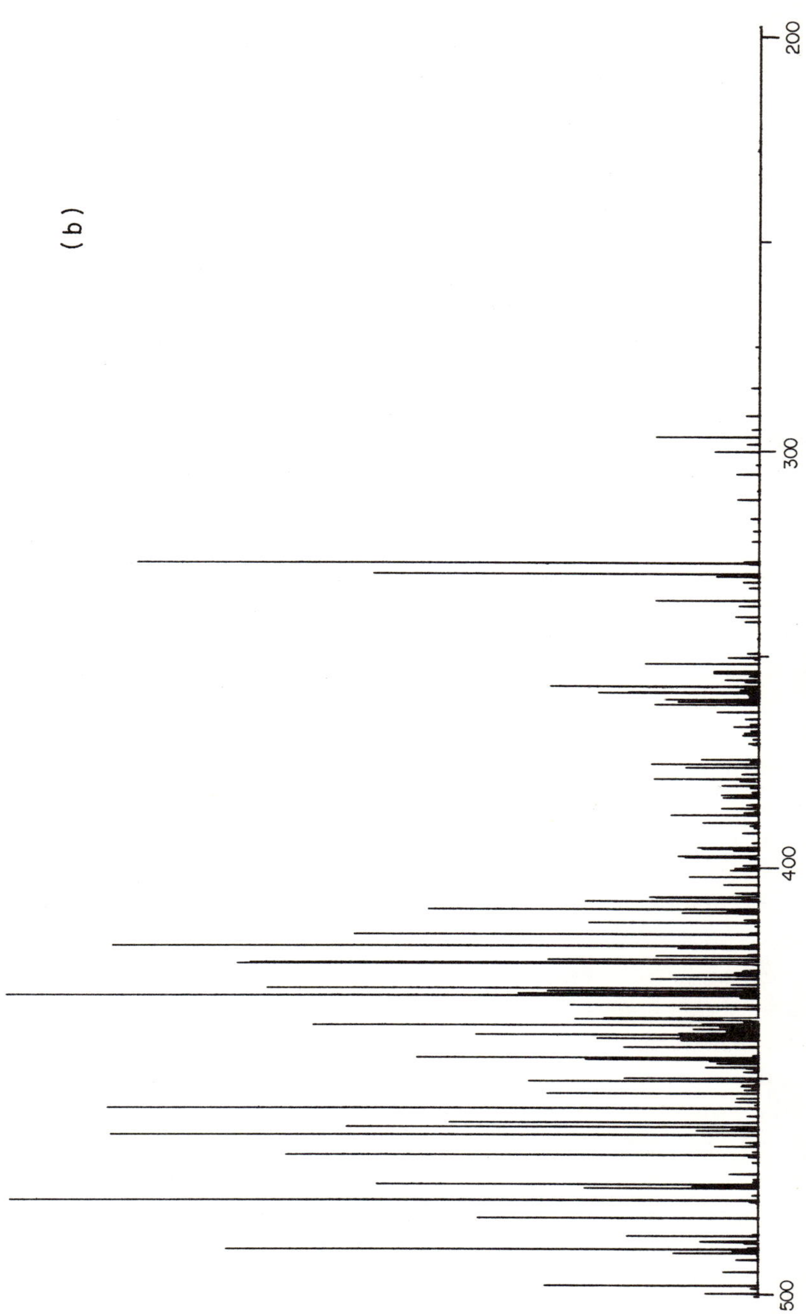

Fig. 5-11. Spectrum of argon in a hollow-cathode discharge. (a) 800-500 nm; (b) 500-200 nm.

than including also the actual intensity obtained from the metal resonance lines) and argon is seen to be as good as or even better than neon in the very important AAS ultraviolet region (i.e. below 370 nm). High-purity argon is also considerably cheaper than neon and its choice may have seemed natural at that time.

5.2.4 Operation of Hollow-Cathode Lamps

(*a*) *Power requirements*

A stable light source is required in both AAS and AFS to allow convenient routine operation without frequent adjustments of the spectrometer. For AAS in particular, an extremely stable source is needed to allow measurement of very small changes in its intensity at the detector (i.e. of very small absorption signals). The HCL is inherently a stable source compared with some other discharge sources owing to the design of its cathode and the fact that it operates in the glow-discharge region where the voltage across the lamp electrodes depends much more on the construction than on the lamp current. However, in most cases it is able to give sufficient stability for use in AAS only when operated from a specially designed and relatively sophisticated power supply unit. Where precise measurements are needed such a unit may be required to maintain a current stability better than 0·1% over a relatively wide range of voltage and current.

The operating voltage with most HCL designs lies in the range 150 to 300 volts and is relatively independent of the current; usually a range of less than 50 volts will cover all currents at which the lamp can be satisfactorily operated. The maximum current permissible with the conventional, sealed-off, uncooled lamps is usually about 40 or 50 mA; higher currents will overheat the cathode causing unstable operation and, in severe cases, permanent damage. In practice, most lamp manufacturers recommend a much lower maximum current to restrict the shortening of the lamp life as a result of the higher rates of metal volatilization and filler-gas "clean-up". As the starting voltage of the lamp is generally rather higher than the operating voltage, and with some designs may exceed it by as much as 150 volts, it is therefore necessary to protect the lamp* against current surges by placing a load resistor in series with it. In many cases it is convenient to place the load resistor in series with another variable resistor which may be used to adjust the operating current of the

* and the power supply unit if it is capable of supplying only small currents.

lamp, rather than to adjust it by altering the supply voltage. If the supply voltage is V_1 and the operating voltage and current selected for the lamp are V_0 and i_0 respectively, the value R_1 of the resistor required may be calculated from $R_1 = (V_1 - V_0)/i_0$. This should be clear from Fig. 5-12. For example, a 500-volt supply would be quite adequate to start and operate most commercial lamps. The load resistor required to limit to 50 mA the current of a lamp whose operating voltage is 250 volts at this current would then be 5 kΩ. Similarly, if it were necessary to operate this lamp at 10 mA (at

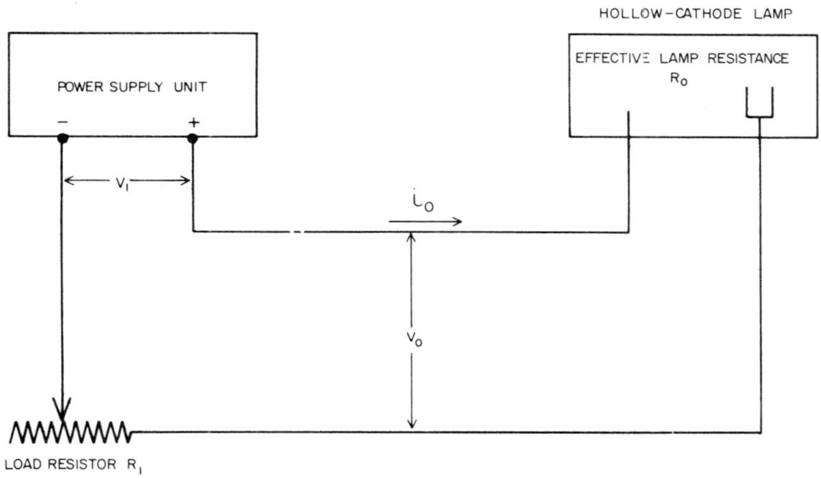

FOR LAMP: $V_0 = R_0 i_0$

FOR CIRCUIT: $V_1 = (R_0 + R_1)\, i_0$

HENCE: $V_1 = R_1 i_0 + V_0$

or $R_1 = \dfrac{V_1 - V_0}{i_0}$

Fig. 5-12. Connection of a hollow-cathode lamp to the power supply via a load resistor.

which its operating voltage may be, say, 220 volts) a total load resistance of 28 kΩ would be needed. Thus, with a 500-volt supply a load resistance variable from about 5-50 kΩ would be sufficient for most purposes. Then even if the effective lamp resistance should drop to zero owing to a fault the maximum current through the circuit would be limited to 100 mA.

In AAS applications the lamp will be used with a power supply that minimizes the effects of voltage fluctuations in the ac mains.

Hence the major source of instability in the lamp output will be fluctuations in its voltage/current characteristics which occur as a result of changes within the lamp itself. For example, the nature of the hollow-cathode surface and the plasma within it will be changed by the sputtering and vaporization of the metal during operation of the lamp. If a constant voltage is being applied to the electrodes the current at which the lamp is operating will change causing a fluctuation in the intensity of the light output. Thus, ideally, the lamp should be operated with a *current*-stabilized power unit which is able accurately to maintain the current supplied to the lamp at a preset value even when quite large changes occur in the voltage/

TABLE 5-3
Effects of supply voltage on the operating stability of a hollow-cathode lamp used with a non-current-stabilized power supply unit

Supply voltage V_1	Load resistor R_1 ($k\Omega$)	Effective lamp resistance R_0 ($k\Omega$)	Potential drop across lamp V_0 (volts)	Lamp operating current i_0 (mA)
500	30	20	200	10
500	30	22	211·5	9·62
1000	80	20	200	10
1000	80	22	215·5	9·81
1500	130	20	200	10
1500	130	22	217·0	9·87

current characteristics. Most of the more expensive commercial atomic absorption spectrometers now use this type of lamp supply unit.

Where only a more simple voltage-stabilized supply is available, it is desirable to have the supply voltage (and hence the load resistor) as high as possible even though 500 volts would be adequate to run most lamps. This will then minimize the change in the current that occurs when, say, the voltage needed to sustain the discharge varies. An example to illustrate this point is given in Table 5-3. It is assumed that the load resistor (R_1) has been chosen for each of three supply voltages (V_1) to give a lamp current (i_0) of 10 mA with a discharge which "normally" has a potential drop (V_0) of 200 volts at this point. The lamp current is then calculated for each of these supply voltages and load resistors after the voltage current characteristic has changed so that the lamp would now operate at 10 mA with an applied voltage V_0 of 220 volts; i.e. the "effective resistance", R_0, of the lamp is assumed to have changed from 20 kΩ to 22 kΩ.* It can be seen that whereas a 3·8% change in the lamp operating current occurs with the 500-volt supply it is only 1·3% with the 1500-volt supply. If the lamp had been operated direct from a 200-volt supply, the current would have dropped to 200/22 = 9·1 mA, i.e. a 9% change.

It has been assumed in the preceding discussion that the hollow-cathode lamp is operating on direct current whereas, as explained elsewhere, almost all AAS and AFS spectrometers use a modulated light source with a synchronous (ac) amplifier. This is achieved in some spectrometers by placing a rotating sector (or chopper) in front of the light source. The effective intensity of the lamp will be reduced (for example, by 50% if the sector contains equal numbers of the same size open and closed segments) but this is not a serious disadvantage in AAS as the ac amplifier also gives an improvement in the stability and noise level of the signal. However, a much more convenient and effective solution to the problem is obtained by modulation of the current supplied to the lamp† and this system is used in many commercial spectrometers. Some care is then needed in the selection of the modulation frequency, waveform and peak voltage to obtain satisfactory results. The effects occurring are illustrated in Fig. 5-13. Diagram (a) shows the result of using sine-wave modulation (such as would be obtained with half-wave rectification) when the peak voltage of the supply is just high enough to start the discharge of the lamp. The lamp voltage will follow the supply voltage until the peak when the discharge initiates and the voltage of the lamp falls to a value depending on its characteristics. This value will remain fairly steady until the supply voltage has fallen again and the discharge extinguishes when the lamp voltage will again follow that of the supply until firing occurs again on the next cycle. As in dc operation the current flowing while the discharge is present will depend on the supply and lamp voltages and any impedance in series. Hence it will drop somewhat during the discharge period as the supply voltage is falling.

Referring again to diagram (a), it can be seen that under these conditions the average current supplied to the lamp is very much less than the peak current. In fact with this particular example the ratio

* A large change in lamp characteristics is needed to give a clear illustration; in practice, a hollow-cathode lamp in good condition would show much smaller fluctuations.

† Modulation should not be confused with pulsed operation which can be used to obtain a large increase in the output intensity of a hollow-cathode lamp. Pulsed operation is discussed in Section 5.2.4(c).

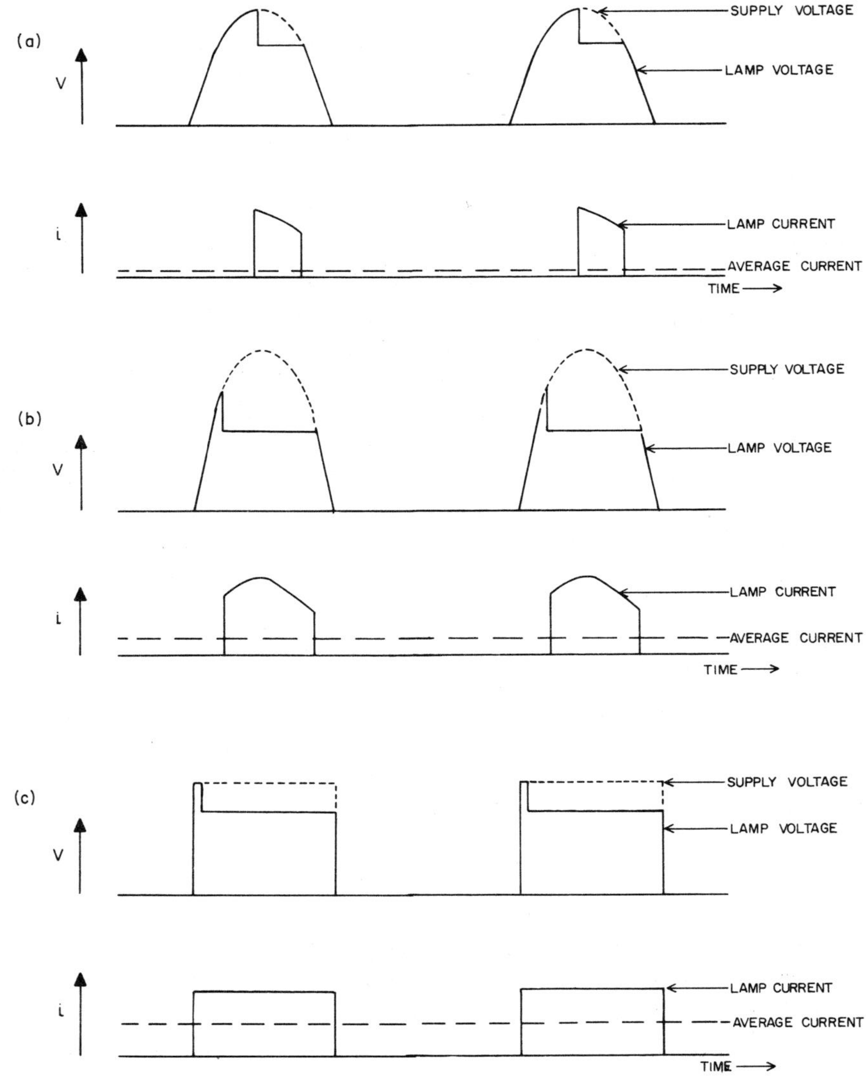

Fig. 5-13. Schematic illustration of the effects of modulation of hollow-cathode lamp power supply.

is about 10 : 1 so that if an average current of 10 mA were needed to obtain sufficient output, peak currents of around 100 mA would be necessary. Unlike dc excitation at this level, the high peak current would not lead to overheating of the cathode or serious self-absorption but it would require a higher level of ionization in the lamp and hence increase filler-gas clean-up. Thus, operation of the lamp in this way may lead to a considerable shortening of the lamp life. The peak-to-average current ratio may be reduced for a particular waveform simply by increasing the peak voltage as shown in diagram (b) of Fig. 5-13. Obviously, the discharge will now start earlier in the cycle and will be maintained for a larger and larger part of the cycle as the peak voltage is increased further. Thus, by using a sufficiently high peak voltage it would eventually be possible to maintain the discharge for almost a half-cycle, giving a ratio of just over 2 : 1. In practice, it is more satisfactory to modulate the power supply in the form of a square wave as shown in diagram (c). A peak-to-average current ratio of 2 : 1 will then always be obtained with any voltage high enough to fire the discharge (although it is advantageous to use a high supply voltage and series impedance as with dc operation).

Quite apart from the need for modulation in AAS and AFS spectrometers, the operation of the hollow-cathode lamp in this way with a current higher than the apparent average current actually improves the output intensity. This is because the relationship between intensity, I, and current, i, has been shown to be not linear but of the form $I = C \cdot i^n$ *in the absence of self-absorption.*[2,9] C and n are constants for a particular cathode and filler gas; n usually has a value between 2 and 3 for metal lines excited by neon or argon.* The degree of self-absorption in most lamps will depend mainly on the average current so that operation of a particular lamp with a square-wave-modulated supply at an average current of 10 mA will actually give an increase in output over the same lamp operated at 10 mA dc. This should be compared with the effect of a rotating sector which must always reduce the effective intensity of the lamp. Thus, the use of this device for modulation should be avoided in a spectrometer designed for the hollow-cathode source.

(b) *Warm-up and lamp life*

Although most hollow-cathode lamps show a steady change in their spectral output during the useful lamp life, this is not usually of any significance during the normal periods of continuous use which in most laboratories will not exceed a few hours. However, nearly all lamps do exhibit a "warm-up" period during which the intensity

* A much smaller degree of enhancement would be inferred if the effect of a high peak current were calculated from plots of intensity against dc or average current supplied by manufacturers, since in many cases these are seriously distorted by self-absorption at higher currents.

drifts quite rapidly until the lamp has reached an equilibrium condition with respect to the cathode temperature and the population of analyte atoms. This effect will be observed each time the lamp is operated and may last only a few minutes or up to an hour. In view of this, many spectrometers incorporate an extra, unstabilized lamp power supply which allows one or more lamps to be run and stabilized at approximately the correct current until required for analysis of a sample. Such a device is not normally supplied with double beam spectrometers as the detection system in these compensates for any drift in intensity. This may not always be desirable, since lamps in which the metal atom population is highly temperature-dependent usually show the longest warm-up periods. These lamps may be operated under self-absorption conditions so that the actual resonance line width (and hence the peak absorption sensitivity) may change during the warm-up of the lamp; double-beam operation will not compensate for this.

The shelf-life of a lamp will depend mainly on the reliability of construction, particularly at any joints and seals; with modern techniques it may be considered as indefinite for most lamps in fairly routine use. The operating life is limited mainly by the adsorption of filler-gas onto sputtered metal atoms. The effects of loss of material from the cathode and degassing of lamp materials as they become hot during operation are generally much less important in the latest commercial lamps which are very carefully designed and constructed to minimize these factors. Although neon appears to show more rapid adsorption than the other gases this is not a critical factor in its use as a filler gas. Much more important is the choice of electrode configuration to minimize the effect of gas pressure, since this will obviously show a gradual reduction during the life of the lamp. With most elements it is then possible to obtain a useful operating life of several thousand hours at currents up to the maximum recommended by the manufacturer. It is usually possible to operate the lamp beyond this period but with reduced stability and intensity of the metal lines (although some lamps actually show an increase in output near the end of life).

(*c*) *Pulsed operation*

A major factor limiting the output intensity of the hollow-cathode sources used in AAS and AFS is, as mentioned previously, their dependence on the same discharge to atomize the sample element and to excite the atoms. Thus, any attempt to increase the level of excitation by intensifying the discharge must also produce a higher concentration of metal atoms in the cathode region with the consequent problems of self-absorption and broadening of resonance lines and shortened lamp life. One possible solution to this problem, suggested by Bodretsova *et al.*,[12] is the operation of the hollow cathode with a high-frequency power supply (e.g. 60 MHz instead of the usual dc or a few Hz ac). Although such a discharge is extremely efficient in terms of excitation it produces a lower potential drop and hence a lower energy of the ions bombarding the cathode surface than in a dc discharge. As a result, the same level of spectrum excitation may be obtained with much less sputtering of the electrode metal compared with that produced in the dc-operated lamp. Thus Bodretsova *et al.* found that, for a given degree of self-absorption and broadening of the resonance line, use of a high-frequency power supply produced a 100-fold improvement in intensity for easily sputtered metals such as magnesium and a 25-fold improvement for those metals, such as aluminium, that were difficult to sputter.

An even more impressive improvement in the intensity obtained from a standard hollow-cathode lamp was reported by Dawson and Ellis[13] using true pulsed operation of the lamp at much lower frequencies. This is a logical extension of the effect observed when the lamp power supply is modulated (see above); it simply involves increasing the pulse interval (i.e. the ratio of the period between pulses to the pulse duration) from unity to give a peak-to-average current ratio somewhat greater than the value of 2 : 1 obtained with normal square-wave modulation. It is then possible to use relatively high (peak) excitation currents while avoiding excessive thermal vaporization of the metal since the effective (average) current is only a few milliamperes. However, there is a useful limit to the value of the peak current that can be used with a particular metal (even when the effective current is kept constant by making the pulses shorter and shorter) since sputtering of the cathode material depends only on the amplitude of the pulses and eventually the stage will be reached where this process alone is able to produce a sufficiently high concentration of metal atoms for self-absorption and broadening of resonance lines to occur. Nevertheless, Dawson and Ellis[13] were able to obtain several hundred-fold improvements in observed intensity from a number of lamps without an apparent increase in self-absorption or line broadening. In fact, a later and more extensive study by Katskov *et al.*[14] has shown that the actual improvement is probably somewhat smaller than this, although still considerable.

They compared the results obtained for several lamps using normal dc operation and a power supply giving square-wave pulses at a frequency of 300 Hz with a pulse interval (see above) of 5. A Fabry-Perot interferometer was used to observe under high spectral resolution the broadening and self-absorption of the metal resonance lines. It was found that, provided the peak current of the pulses is not made too high as explained above, the "quality" of the lines is similar when the mean current of the pulsed power supply is the same as the direct current. Thus, it was possible to make a direct comparison between the intensities obtainable with the two modes of operation, as shown in Table 5-4. It can be seen that pulsed operation gives an improvement varying from about 20-fold to

TABLE 5-4
Improvements in hollow-cathode lamp intensity using a pulsed power supply[15]

Lines (nm)	i_{mean} (mA)	i_{peak} (mA)	$\left[\frac{I\ \text{pulsed}}{I\ \text{dc}}\right]_{\text{meas.}}$	n	$\left[\frac{I\ \text{pulsed}}{I\ \text{dc}}\right]_{\text{calc}}$
Ag 328·1	8	40	23	1·9	21
Al 309·3	20	100	55	2·5	56
Co 240·7	10	50	36	2·3	41
Cr 357·9	20	100	102	2·9	107
Mo 313·3	30	150	21	1·9	21
Ni 341·5	15	75	20	1·8	18
Pd 247·6	10	50	132	3·0	125
Rh 343·5	10	50	44	2·3	41
Si 251·6	30	150	26	2·0	25

130-fold depending on the metal. This variation is due to the different value of n for different metals in the equation $I = C \, . \, i^n$ (see Section 5.2.4(a)); the ratio of pulsed to dc intensity should be given simply by $(i\ \text{pulsed}/i\ \text{dc})^n$ and under conditions where self-absorption is insignificant it should be possible to calculate the ratio using a constant value of n for a particular lamp. This is illustrated in the last column of Table 5-4 using values of n measured by Bodretsova *et al.*[9] for conventional dc excited lamps; there is excellent agreement between the measured and calculated ratios.

More recently Mitchell and Johansson[16] have used pulsed hollow-cathode lamps as atomic fluorescence sources. As very sharp source lines are much less important than in AAS, they were able to use high peak currents without undue consideration of the relatively high atom concentrations which would consequently be sputtered. Their measurements were made with rather shorter pulses than those described above; for example, a pulse current of 190 mA corresponded to a mean current of only 8·5 mA, and they suggest that extremely high pulse currents could be obtained in this way for good fluorescence sensitivity without serious problems of overheating or self-reversal of the light source. Such short pulses also allow fluorescence measurements of very brief duration without any loss of fluorescence intensity compared with measurements made in the normal way. This technique has been used in an atomic fluorescence spectrometer for "simultaneous" multielement analysis, the different lamps being pulsed sequentially so that apparently continuous measurement of each element is obtained.

5.2.5 Spectral Characteristics of Hollow-Cathode Lamps

(*a*) *Line broadening and self-reversal*

The breadth and degree of self-absorption of the metal resonance lines is, as mentioned previously, one of the most important characteristics of the hollow-cathode lamp from the point of view of its use as a light source in AAS. Unfortunately, as a result of the experimental difficulties involved, relatively few measurements have been made of the line profiles that are of direct interest to AAS applications. There seems to be general agreement[9,12,17-24] that in the absence of self-absorption both metal and filler-gas lines are broadened mainly as a result of the Doppler effect. In both cases collisional (Lorentzian) broadening is the next major factor producing a line-width which generally amounts to a few percent of the Doppler width, although a similar contribution by resonance broadening has been noted[21] for some filler-gas lines and this effect may also occur with some metal lines if the atom population within the cathode increases greatly at higher operating currents. Very much larger Lorentzian line-widths have apparently been observed for certain lines of Ar(II) and Ne(II),[21,24] but these appear to result from an anomalous effect which is of no interest to AAS applications.

As the Doppler width can be readily calculated from the source temperature (see Section 2.5.2(ii)), there would seem to be little difficulty in estimating the width of a non-self-absorbed metal line simply by measuring the operating temperature of the cathode with, for example, a suitable thermocouple. However, there is considerable disagreement in the reported literature as to whether the observed Doppler temperature* is actually equal to the operating temperature

* i.e. the temperature calculated from the measured Doppler half-width using equation (2-18).

of the cathode.* This situation is complicated by the fact that measurements of the line-width are difficult and subject to considerable errors in many cases; as $\Delta\nu_D$ is proportional to $T^{1/2}$ the uncertainty in the estimated Doppler temperature may be quite large. Thermocouple measurements of the cathode wall temperature in typical modern lamps have shown a range of *ca.* 300 to 800 K at currents of about 20 to 90 mA for an iron cathode[17] and 350 to 430 K at currents of 5 to 15 mA with a calcium oxide/tungsten/titanium cathode.[23] However, the corresponding Doppler temperatures measured at the calcium 422·67 nm resonance line were reported by Yasuda[17] to be far higher (*ca.* 1300 to 2700 K for currents of 30 to 90 mA), whereas Bruce and Hannaford[23] found them virtually identical to the thermocouple values. Similarly, argon and neon lines were found by Gofmeister and Kagan[19,20] to give Doppler temperatures of 320-700 K in the current range 20-100 mA, and this is again close to the expected cathode temperature.

A much more extensive study of this problem has been made by Kreye and Roesler.[21,24] The conclusions drawn from their own line-width measurements and from their collation of ancillary data may be summarized as follows: (i) The Doppler temperature observed for the neutral filler-gas lines (i.e. ArI and NeI) is generally a good indication of the actual gas temperature. (ii) The doublet lines of ArII and NeII give similar Doppler temperatures but quartet lines show anomalously low values for reasons which are not of immediate concern. (iii) The gas temperature observed with argon is consistently about twice that observed for neon with similar lamps operated at the same discharge current. (iv) The Doppler temperature observed for a metal line is somewhat larger than that of the filler-gas lines unless there is a large interval, τ_e, between excitation and emission of the line, i.e. unless the transition probability is small. (v) It is suggested that this latter effect is due to there being insufficient time for metal atoms with high energy to come to equilibrium with the gas temperature before emission because of the relatively long interval between metal/gas atomic collisions (*ca.* 20 ns).

Some Doppler temperatures for hollow-cathode lines[21,24] are listed in Table 5-5 together with values of τ_e for the metal lines. Allowing for the general uncertainty in such measurements it can be seen that there is fair agreement between the ArI and NeI

* This should not be confused with the *excitation* temperature of the spectral lines which depends on the energy provided by the excitation mechanism and in a hollow-cathode lamp is generally greatly in excess of the gas or vapour temperature.

temperatures and the thermocouple measurements reported by Yasuda[17] (argon-filled lamp) and Bruce and Hannaford[23] (neon-filled lamp). The temperatures observed for Au 479·2 and Fe 372·0 are also comparable, both of these lines having large values of τ_e, whereas those of the other metal lines with smaller values of τ_e are obviously too high to be representative of the gas temperature within the cathode. It should be noted that the value of τ_e for the calcium 422·67 nm line is about 5·4 ns.[23] Thus, if the hypothesis of Kreye and Roesler is valid, it would be expected to show a high Doppler

TABLE 5-5
Doppler temperature for hollow-cathode lamps[21,24]

Element[a]	Line(s)[b]	Current (mA)	Doppler temp. (K)	τ_e (ns)
Ar I	451·1 and 470·2 nm	10–25	700–1100	—
Ar I	451·1 nm	15	1200	—
Ar II	Doublets	10–25	700–1100	—
Ar II	Doublets	15	1070	—
Ar II	Quartets	10–25	400–800	—
Ar II	Quartets	15	800	—
Ne I	347·3 nm	15	640	—
Ne II	Doublets	15	590	—
Ne II	Quartets	15	350	—
Au I	312·3 nm	10–25	1500–1900	25
Au I	479·2 nm	10–25	1000–1300	70
Au I	242·8 nm	—	4700	6·5
Au I	267·6 nm	—	4700	17
Fe I	248·3 nm	—	3900	0·17
Fe I	372·0 nm	—	690	75

[a] All metal lines are for lamps with argon as filler gas.
[b] Where more than a single line is indicated temperatures have been averaged from several comparable lines.

temperature as observed by Yasuda[17] rather than a value characteristic of the cathode temperature.[23] In any event, until further data are available it would seem unwise to assume narrow Doppler line widths characteristic of the cathode temperature for many metal resonance lines, as those of interest in AAS and AFS usually exhibit large transition probabilities (i.e. small values of τ_e).

In practice, the question of whether or not the Doppler widths of the metal resonance lines are actually characteristic of the cathode temperature is not of paramount importance in AAS applications. In either case, the possible increase in Doppler width with rising lamp temperature is insufficient for most elements to explain the serious reduction in peak absorbance values frequently observed with increasing lamp current. This is almost certainly due[9,17,18,22,23] to a rather larger increase in self-absorption broadening which occurs as a result of the higher concentration of metal atoms in the hollow-cathode bore when the operating current is increased. Both the sputtering process and thermal vaporization contribute to the enhanced production of metal atoms, their relative importance depending on the lamp design and the volatility of the analyte

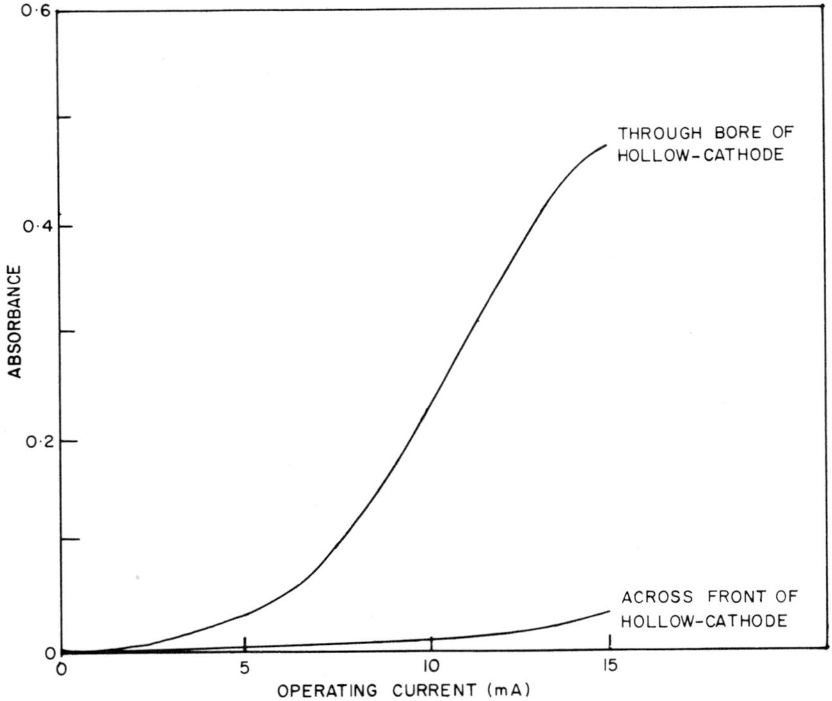

Fig. 5-14. Absorbance measured at the 422·67 nm resonance line within a calcium hollow-cathode lamp.[23]

element. A major contribution to the improvements in modern hollow-cathode lamps has been the reduction in this effect when larger lamp currents are used to increase the intensity of the resonance line. Particularly important is the limitation of the formation of a cloud of "cool" metal atoms in front of the cathode as these will give the most serious self-absorption of the ground-state metal lines of interest in AAS and AFS. Figure 5-14 shows measurements by Bruce and Hannaford[23] of the absorbance observed through the bore and across the front of the cathode of a commercial calcium lamp. Although it was necessary to modify the cathode to allow the beam from a second source lamp to pass through, this is not expected to affect the significance of the results.

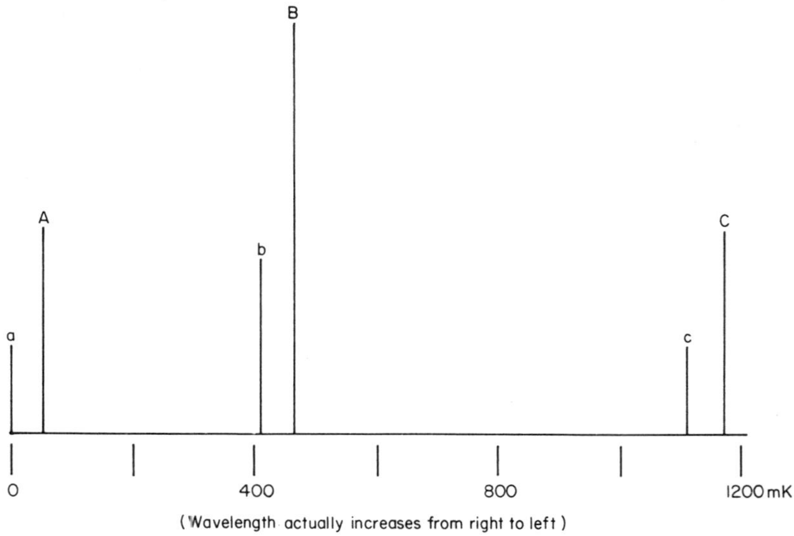

(a) Actual hyperfine structure; the heights of the lines represent the theoretical relative intensities of the components.

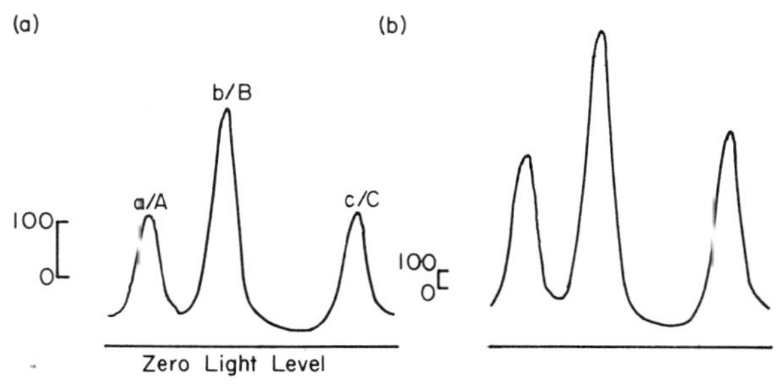

(b) Fabry-Perot spectrometer scan of the profile emitted by a hollow-cathode lamp operated at (a) 15 mA and (b) 25 mA.

Fig. 5-15. Hyperfine structure of the thallium 377·6 nm line.

The manufacturer's recommended operating current for this lamp (based on measurements of AAS sensitivity) is 4 to 5 mA, and the concentration of metal atoms within the cathode rises rapidly if this value is exceeded. The much smaller absorbance by atoms outside the cathode is an indication of efficient lamp design.

More direct evidence of the effects of self-absorption has also been obtained by actual observation of the line profiles. Analysis of the profile recorded for the calcium 422·67 nm line to deconvolute the contributions of Doppler and self-absorption broadening has confirmed[17,23] that the latter is much more important as an explanation of the increase in half-width with current, even at quite modest currents. However, as mentioned above, such measurements are difficult and involve considerable uncertainty. A much simpler

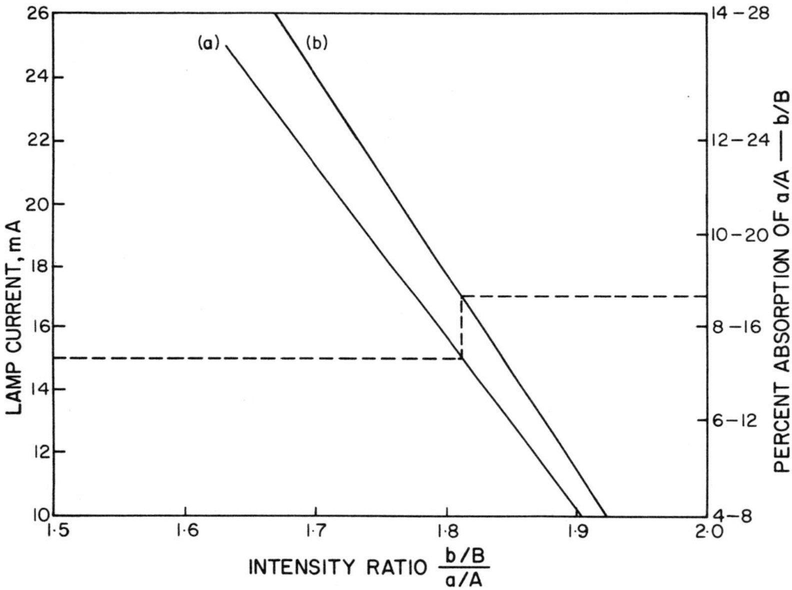

Fig. 5-16. Variation of the hfs intensity ratio of the thallium 377·6 nm line. (a) Observed experimentally for a hollow-cathode lamp at different operating currents; (b) calculated for different degrees of self absorption.

qualitative assessment of the effects of self-absorption may be made by observing the relative intensities of the hyperfine components of the line[9,12,18,22] (see Section 2.5.2(vii)). Provided a high resolution spectrometer is available this technique is easily applied to most metal lines of interest* without any need to modify the source. Special cases may give additional information for particular elements, but in general the technique relies on the fact that any differences in intensity of hfs components will depend only on their relative transition probabilities (since the energy levels are very similar). Thus, if one peak is twice as intense as another the degree of

* Except, of course, for a few elements such as calcium which show no resolvable hfs.

absorption at these lines will be in approximately the same ratio and a rapid change in the intensity ratio occurs as self-absorption increases. The effect is clearly illustrated by the thallium 377·6 nm line,[22] the hyperfine structure of which is shown in Fig. 5-15(a). In practice, the line-width obtained from an ordinary hollow-cathode lamp results in the six components appearing as three peaks, as shown in a Fabry-Perot interferometer scan of the line (Fig. 5-15(b)). The variation with lamp current of the intensity ratio of the *a/A-b/B* peaks (which is theoretically 2·0 in the absence of self-absorption) is shown in Fig. 5-16. Also shown is an approximate calculated plot of the variation in the intensity ratio with different levels of self-absorption. This gives an estimate of the degree of self-absorption at a particular operating current. For example, as indicated in Fig. 5-16, the self-absorption of the *b/B* peak is about 17% at a current of 15 mA (the recommended lamp current is actually 20 mA).

(*b*) *Line interference*

The practical effects of non-absorbed lines appearing within the spectral bandpass of an atomic absorption spectrometer set to the wavelength of the analyte absorption line are discussed in Chapter 12; we are concerned here with the HCL characteristics producing this type of interference. It is most commonly observed when lines of the filler gas lie close to the analyte line, although the effect has also been noted for non-absorbed (i.e. non-ground-state) lines of the metal atom itself and for ionic lines of the metal. The first type is probably the easiest to avoid, or at least to reduce in severity, simply by careful choice of the gas and the lamp pressure (see Section 5.2.3(c)). Figure 5-17 shows the spectrum observed in the vicinity of the chromium 357·9 nm resonance line, (a) in an argon-filled lamp and (b) in a neon-filled lamp of modern design. The effect is such that with the argon-filled lamp the less strongly absorbed 359·4 nm chromium line actually appeared to be more sensitive. Less often it may be possible to reduce the effects of non-absorbed metal lines by a similar change in lamp construction which may enhance the resonance line intensity or reduce ionic line intensities. The latter effect is illustrated for the nickel 232·0 nm resonance line in Fig. 5-18; the drastic reduction in intensity of the nearby 231·6 nm ion line between diagrams (a) and (b) was obtained simply by reducing the pressure of the neon filler gas. It should be noted that in some cases a straightforward improvement in resonance line intensity is useful in reducing the interference as it allows a narrower spectral

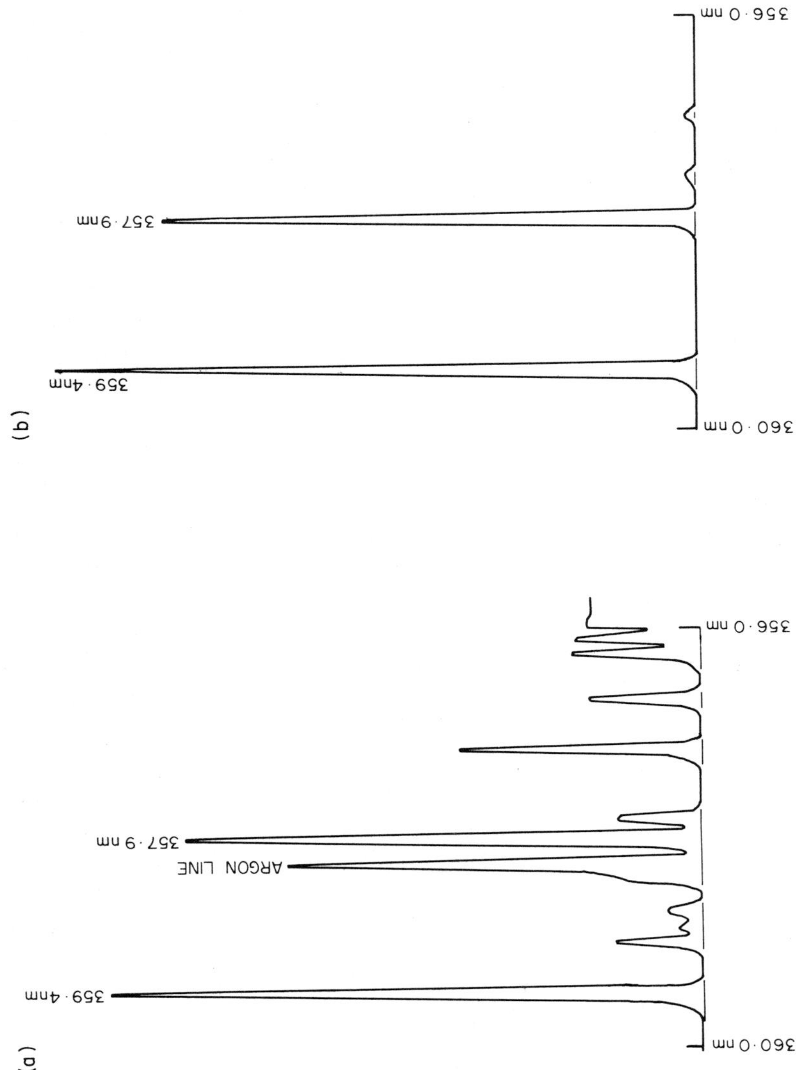

Fig. 5-17. Filler gas line interference at the chromium 357·9 nm line. (a) Spectrum from an older argon-filled lamp; (b) spectrum from a modern neon-filled lamp.[25]

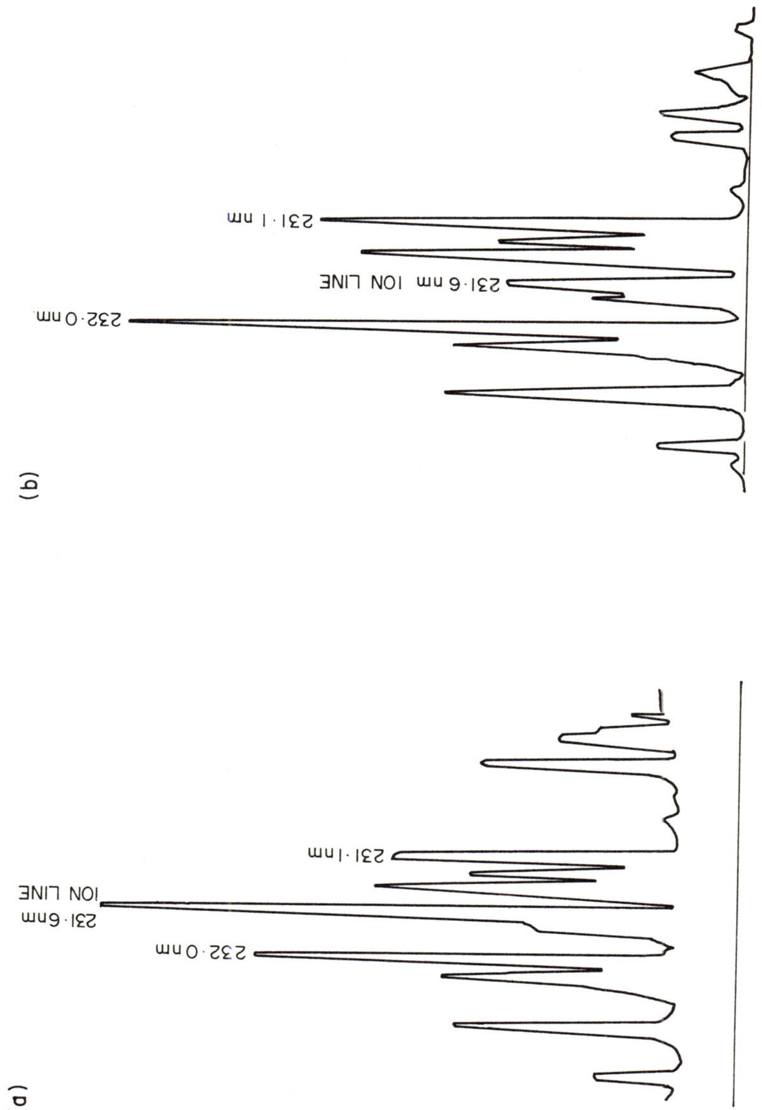

Fig. 5-18. Metal ion line interference at the nickel 232·0 nm line. (a) Spectrum from a lamp containing too high a pressure of neon; (b) spectrum from a lamp containing the correct neon pressure.[25]

bandpass to be employed while maintaining a high signal-to-noise ratio.

(c) *Signal-to-noise ratio*

The signal-to-noise ratio is an important factor in the selection of a source for both AAS and AFS as it limits the smallest analytical signal that can be observed and thus has a direct effect on the detection limit. A hollow-cathode lamp in good condition is inherently stable and, as discussed in Section 5.2.4(a), the effect on

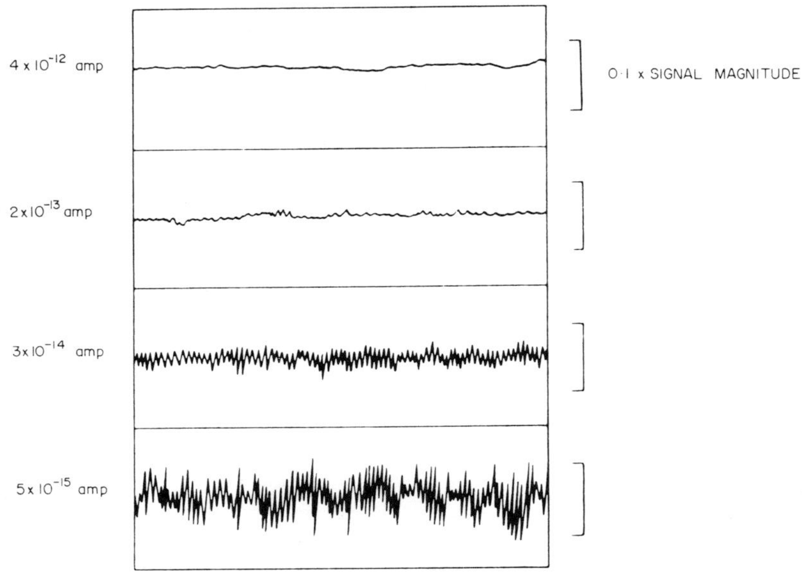

Fig. 5-19. Signal recordings of the same lamp output at different levels of incident intensity on the photomultiplier.[15]

the output intensity of any fluctuations that may occur within the lamp can be minimized by careful construction of the power supply unit. Early studies of the application of the lamps to AAS[26,27] suggested that quite high noise levels were present and that they could not be removed or reduced by using a more stable power supply. However, it has been shown[9] that these workers almost certainly measured fluctuations not in the lamp output but in the detection circuit, probably the shot noise of the photomultiplier. This effect is illustrated in Fig. 5-19 which shows a series of signals recorded from a neon-filled iron hollow-cathode lamp operated at 50 mA (i.e. the output of the lamp is the same in each case). The intensity actually received at the detector was reduced in stages and it can be seen that there is an apparent increase in the "noise level of the source". As the fluctuation is about 0·1% of the signal at the highest intensity (diagram (a)), the lamp noise cannot exceed this level. Thus, a fairly high resonance line intensity is important in AAS even though the intensity has no direct effect on the sensitivity. This is particularly so for elements which have their strongest absorption lines in regions of low photomultiplier sensitivity.

5.2.6 Special Types of Hollow-Cathode Lamp

(a) Demountable lamps

Demountable hollow-cathode lamps are not usually supplied with commercial atomic absorption spectrometers and have seen relatively little application compared to the sealed types, although they have

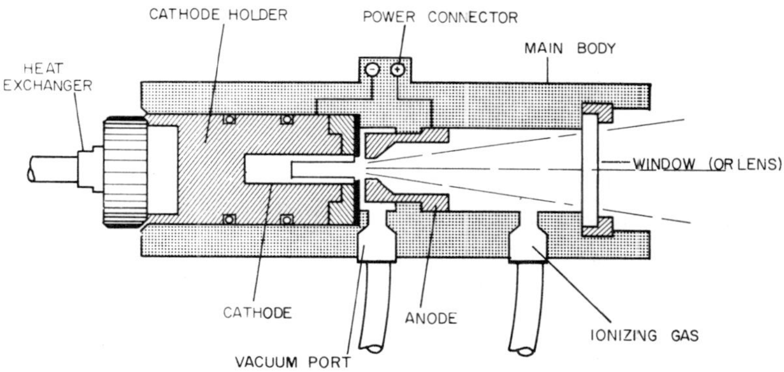

Fig. 5-20. Glomax demountable hollow-cathode lamp.

been described by a number of workers.[26-32] Recently, a commercial device has been introduced* which is of suitable size to replace the standard sealed-off lamps in most spectrometers without extensive modifications. The lamp (Fig. 5-20) is intended to be as versatile as possible with easily interchanged cathodes and windows and a variable anode-to-cathode distance. It is also possible to water-cool the cathode via a heat-exchanger attached to the cathode support and this, in conjunction with the continuous flow of fresh filler gas during operation, allows the use of a somewhat higher current than with a conventional lamp. As all operating parameters can also be optimized for a particular application it should be

* Glomax Lamp, Barnes Engineering Co., Stamford, Connecticut.

possible to use this lamp as a high-intensity source for AFS as well as for specialized applications in AAS. However, for routine analyses it offers little economic advantage over conventional lamps, unless occasional determinations of many elements are likely to be required (the costs would be similar for about ten elements assuming that suitable vacuum and gas control systems are already available in the laboratory; if this were not the case the figure would exceed twenty elements).

(b) High-intensity lamps

Several lamp manufacturers refer to their standard hollow-cathode sources as "high-intensity lamps", but in the present context we are concerned with several elaborate modifications which have been suggested to allow operation of a sealed lamp at greatly increased resonance line intensity without the loss of the sharp profile. In practice, for most elements this means increasing the level of excitation without producing a higher concentration of metal atoms within the cathode; i.e. without introducing the problem of serious self-absorption.

The first and the most widely known solution to this problem was suggested by Sullivan and Walsh.[33] They used the conventional cathodic discharge to produce metal atoms at the optimum pressure but excited the cloud of atoms in front of the cathode with a second low-voltage discharge between two auxiliary electrodes. This not only gives a direct increase in the resonance line intensity prior to the onset of self-absorption but, in view of the low excitation energy available in the low-voltage auxiliary discharge, may actually enhance the resonance line intensity relative to other metal lines, filler-gas lines and ionic lines. Hence, the increased intensity of the resonance line is usually accompanied by a further improvement in the line-to-background intensity ratio. When compared with the conventional type of HCL the lamp also gives improved AAS sensitivity and calibration linearity for a number of elements. This results from the avoidance of self-absorption by a cloud of cool atoms in front of the cathode, a phenomenon often observed for the more volatile elements with early lamps of conventional design. The Sullivan and Walsh lamp, however, shows less pronounced improvement over the most modern types of conventional HCL and its use has recently been less popular. This restriction of its application has been hastened by the need to operate the lamp with a second power supply for the auxiliary discharge, especially as the small voltage drop (20-30 V) requires the use of currents up to 500 mA and hence a relatively expensive power pack.

Two attempts have been made to improve further the original Sullivan and Walsh design, one aim being to produce even sharper resonance lines for sensitive AAS measurements[34] and the other to provide an extremely intense source for atomic fluorescence.[35] Van Gelder's narrow line source[34] uses an open-ended, cylindrical hollow cathode, the discharge being struck through this cylinder between an ordinary cathode and a perforated disc anode. Metal atoms are sputtered from the inside of the hollow cathode as in a conventional lamp and the resultant resonance radiation leaves the lamp via the perforations of the anode. The discharge is constricted by a glass capillary to ensure (a) that sputtered metal atoms are restricted to an area within the cylinder where excitation can occur, and (b) that radiation is viewed only from the axis of the discharge where the excitation of sputtered atoms is most intense (i.e. where the concentration of ground-state metal atoms giving self-absorption is at its lowest).

Lowe's atomic fluorescence source[35] also uses a common anode and an open-ended cylinder for the hollow cathode. However, in this case the anode is placed behind the hollow cathode and the auxiliary (booster) cathode to one side of it so that there is no restriction of the resonance radiation leaving the lamp. A conventional, modulated discharge is struck between the anode and the hollow cathode while a high current (400 mA) dc discharge is obtained between the anode and the booster cathode. A set of restricting baffles ensures that this latter discharge can only occur through the hollow cathode (whereas in the original Sullivan and Walsh design the auxiliary discharge passes only across the front of the hollow cathode). This arrangement provides maximum excitation of metal atoms sputtered by the primary discharge, and for some elements the lamp was found to give a more than ten-fold increase in fluorescence intensity with apparently only a small increase in line-width compared with the original design.

Human and Butler[36] have described a high-intensity hollow-cathode lamp in which the auxiliary discharge is operated at microwave frequency (2450 MHz);* this has the advantage of needing only a simple wire antenna mounted within an ordinary

* Note that this device is quite different from both the radiofrequency HCL's described in Section 5.2.4(c) and the electrodeless microwave lamps described in Section 5.3.

hollow cathode. In use the ordinary modulated hollow-cathode discharge is used to sputter metal atoms while 10-15 watts of microwave power are applied to the antenna to boost the excitation of these atoms without increasing the normal concentration of metal atoms. It was found that at comparable absorbance sensitivities (i.e. comparable line-width/self-absorption) the auxiliary discharge was able to increase the resonance line intensity by up to 15 times.

(*c*) *Multi-element lamps*

Multi-element hollow-cathode lamps which emit resonance lines of two or more elements are useful as they can reduce the lamp cost per element and, provided suitable combinations are available, allow easier and more convenient operation in many routine analytical problems. Early multi-element lamp designs used several separate cathodes within one source,[8,37] but there were problems of alignment and of switching between the cathodes and this type has seen little application. However, several designs using a single cathode have been more successful and some of these are available commercially.

The most convenient technique with a single cathode is to construct it from a suitable alloy, but performance often deteriorates rapidly owing to preferential sputtering of one of the elements (e.g. of zinc when brass is used for a copper-zinc lamp[38]). A better approach is to press individual rings of various metals (or compounds) into a single cathode.[39,40] However, only certain combinations can be used together and the operating current must be carefully chosen to give a stable output for all elements rather than being optimized for each one. Successful multi-element lamps have been developed by the Perkin-Elmer Corporation[41-44] using sintered cathodes of mixed metal powders or intermetallic compounds. A number of lamps, containing up to seven elements, have been constructed; great care must be taken, however, in the choice of elements and materials used in a single lamp to ensure a similar useful life for each element and to avoid spectral interferences. It is also important that the group of elements should often require analysis together in a particular laboratory as these lamps often have the disadvantages of reduced intensity and limited operating conditions compared to conventional lamps. For example, the seven-element Al–Ca–Cu–Fe–Mg–Si–Zn lamp may be useful for routine blood serum analysis.

5.3. ELECTRODELESS-DISCHARGE LAMPS

5.3.1 Introduction

Although less widely used by spectroscopists than the hollow-cathode lamp, high-frequency electrodeless discharges have been recognized for many years as excellent sources of sharp-line spectra. They have several useful properties for high-resolution work and other fundamental studies, namely ease of construction, absence of electrode contamination, maintenance of the discharge at very low vapour pressures, and the need for only small amounts of the element or compound to be studied. However, their application to atomic absorption spectroscopy has been almost insignificant in comparison with the widespread use of the hollow-cathode lamp. This is undoubtedly due partly to the selection of the hollow-cathode lamp by Walsh and his co-workers (for reasons that are still largely valid) as the most suitable spectral source for their original analytical AAS work. With most elements a source line of the intensity and sharpness required for AAS is obtained with good stability and reproducibility more easily by using a hollow-cathode lamp rather than an electrodeless-discharge lamp in its present state of development. However, this is not generally the case with the type of source line required for atomic fluorescence spectroscopy, and it seems possible that the extensive development of the microwave-excited electrodeless-discharge lamp for this purpose may also lead to lamps which are useful alternatives to the HCL for routine AAS applications.

Electrodeless discharges may be excited at both *radio frequencies* (100 kHz to 100 MHz) and *microwave frequencies* (>100 MHz). The latter are a much more recent introduction as suitable power supplies have been available only during the last twenty or twenty-five years. In spite of this, most of the work by analytical chemists with these sources has used microwave excitation. This is partly due to the ready availability of suitable power supply units (originally inexpensive medical diathermy generators but more recently commercial units designed for spectroscopic applications) and partly because the properties of the microwave lamps (see below) were more appropriate for the very high output intensities required by the two groups of analytical chemists* originally concerned with the application of electrodeless-discharge lamps as spectral sources for atomic fluores-

* Winefordner and co-workers in the United States and West and co-workers in the United Kingdom; references to the work of both these groups will be found in the following pages.

cence spectroscopy. Radio-frequency (RF) discharges, however, have been used for fundamental fluorescence studies by many physicists because of the high stability obtainable; RF discharges may be also more suitable for obtaining resonance lines with very low self-absorption (see below).

Radio-frequency lamps have the advantage of a "skin effect" if the discharge conditions are carefully chosen. This concentrates the discharge near the tube walls so that the emitted radiation travels a negligible distance through vapour within the lamp and self-absorption and self-reversal of the resonance lines are minimized. On a more practical level, the construction of suitable RF power units is extremely well-documented and a current-stabilized supply equivalent to those used for the hollow-cathode lamp may be easily produced from widely available components (see Section 5.3.4). In contrast to this, a true "skin effect" is not normally obtained with microwave discharges and the construction of microwave generators and components involves more specialized electronic engineering. The problem of obtaining a fully stabilized unit giving stability comparable to that of hollow-cathode lamp supplies has been an important restriction on the spectroscopic application of EDL sources (see Section 5.3.5). For many applications, however, microwave excitation has two major advantages over RF excitation: both the useful operating life-time of lamps[45] and the efficiency of conversion of input power into emitted radiation[46,47] increase significantly at higher frequencies. The restricted life of RF excited lamps appears to result from absorption of the metal onto the tube walls, but even RF lamps show adequate working lives if frequencies towards the higher end of the RF range are used.[45] The rate of increase of lamp efficiency with frequency is difficult to assess, as although many different frequencies have been used for RF excitation, most work on microwave excitation has been carried out at 2450 MHz owing to the availability of commercial generators operating at this frequency. However, Jacobsen and Harrison[46] observed a sixteen-fold increase in intensity for 40 watts power when the excitation frequency was increased from 150 MHz to 3000 MHz, and Forrester *et al.*[47] reported a lamp output of 0·025 watt cm^{-2} steradian^{-1} for 95 watts input at 2450 MHz compared with 0·002 watt cm^{-2} steradian^{-1} for 1000 watts input at 70 MHz.

Although the ancillary equipment required for excitation is somewhat different for RF and microwave excitation, the basic requirements for lamp construction are rather similar. The sources are simply glass or silica envelopes containing appropriate quantities of an inert gas and the element (or one of its volatile compounds) to be excited. In view of this, constructional techniques are discussed first here, followed by separate sections dealing with the operation and characteristics of RF and microwave lamps respectively. As microwave-excited sources appear at present to be of more interest to analytical chemists these are dealt with in more detail. The reader interested in RF-excited sources as alternative absorption sources to hollow-cathode lamps or for specialized fluorescence work will, however, find adequate notes and references to serve as an introduction to the subject.

5.3.2 *Construction of Electrodeless-Discharge Lamps*

Electrodeless-discharge lamp sources suitable for both RF and microwave excitation are available commercially.* In the experience of the authors, however, manufacturers have not (at least until very recently) attempted to produce lamps to meet the specific requirements of analytical chemists, and superior sources are often produced in the laboratory, especially when very high intensity is required for sensitive atomic fluorescence measurements. As the fabrication techniques are also relatively simple, owing to the absence of electrodes, these are described below in sufficient detail for the reader wishing to undertake lamp construction in his own laboratory. This requires only simple glass-working facilities for the preparation of lamp bulbs and a small vacuum line for the filling procedure.

(*a*) *Preparation of lamp bulbs*

The most widely used materials for lamp bulbs are Pyrex glass and (for elements emitting their resonance lines below about 300 nm) transparent silica tubing. The latter is available in suitable spectroscopic grades (such as "Vitreosil" and "Suprasil") from, for example, Jencons Ltd. in the U.K., and Hanovia Lamp Division, Engelhard Industries in the U.S.A. Although both Pyrex and silica are inert to attack by RF or microwave discharges under most conditions, this is not the case with lamps of alkali and alkaline earth elements which frequently show a rapid and irreversible attack on the lamp walls.[48,49,50] The attack appears to be least serious for the heavier alkali metals (e.g. rubidium[48]) on Pyrex, but in other cases it is only

* For example, RF lamps have been supplied by Varian and microwave lamps by Ophthos Instrument Co., and Scintillonics-Interface Inc. in U.S.A. and EDT Supplies Ltd., EMI Electronics Ltd., Thermal Syndicate Ltd. and Southern Instruments Ltd. in U.K.

possible to use the lamps for a very short experiment. In certain cases it is possible to reduce the rate of attack by forming a protective layer of borate glass on the inside of the lamp as the metals will not react by displacement of boron owing to the much greater stability of the B—O bond compared to the Si—O bond. This may be achieved with Pyrex (borosilicate glass) by washing the inner surface of the lamp with hydrofluoric acid solution[49,50] to remove a layer of silicon and leave a borate glass surface. Alternatively, both Pyrex[48] and silica[50] may be internally glazed with a protective layer of borate glass by wetting the lamp bulbs with a saturated solution of boric acid followed by heating to a high temperature. Neither method is very effective for obtaining a long operating life, as in the former case the layer is very thin and in the latter the protective surface eventually cracks. A much more effective solution to the problem[50] involves the use of commercially available* two-ply glass having an outer layer of soda glass and an inner ply of silica-free glass intended to resist attack by sodium vapour. However, this glass is more difficult to work and seal than are conventional types and although it gives long-lived lamps with magnesium its transmission is very low at the 285 nm resonance line.

A variety of shapes have been used for the lamp bulbs but some of the more simple designs seem adequate for most purposes. Early work with RF discharges favoured cylinders one or two centimetres in diameter and perhaps 10 cm long, although bulbs with capillary sections (to produce a "thread" of discharge of great intensity) and very long cylinders (up to 2 m) were also common.[5] More recently, spherical bulbs have been popular, especially for lamps of the alkali metals,[15,48,51,52] and they allow the maximum possible output intensity while retaining the advantage of a wall discharge. Bulbs ranging in diameter from about 1 to 3 cm have been used. Frequently, cylindrical or spherical RF lamps are encased in heater units (see below) but when these are not required (i.e. for volatile elements) it is often useful to seal the lamp into a Pyrex or silica cover to protect it from draughts as the vapour pressure of a volatile metal can fluctuate significantly, even with small temperature changes.

Most work with microwave discharges has been with simple lamps made by sealing at both ends a relatively narrow, thick-walled Pyrex or fused silica tube, as shown in Fig. 5-21(a). The bulbs are usually between 3 and 10 cm long with an internal diameter of

* Glass Tubing and Components Ltd., Chesterfield, England; type X93.

between 5 and 15 mm. Forrester *et al.*[47] examined several more elaborate designs with additional bulbs and constrictions but found no advantages for most elements. However, they did suggest that the greatest intensity from a high-power microwave discharge could be obtained by constricting it to a diameter of about 1 mm. As even a silica capillary of this diameter would melt and seal itself under these conditions, it was necessary to use a tube of 1 mm x 4 mm cross-section. This design has not been widely used, partly because ordinary tubing is more widely available and partly because of the severe loss of line sharpness when the discharge is operated in this mode. Protective cases have not been widely used with microwave lamps as some protection from draughts is usually afforded by the cavities used for excitation (see Section 5.3.4(b)). However, the use

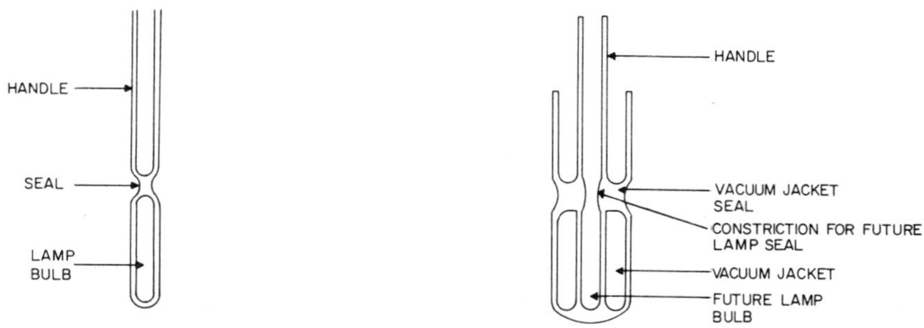

Fig. 5-21. Microwave electrodeless discharge lamps. (a) Finished lamp of simple, conventional design; (b) lamp blank with evacuated outer jacket already in place.

of a vacuum jacket around the bulb has recently been found advantageous for some elements[53,54] and allows a higher and more uniform operating temperature.

Some care is needed in the final preparation of lamp bulbs for both RF and microwave discharges to remove any trace contamination. Even when the spectra of such impurities are unimportant for a particular application, they may cause a significant reduction in the lamp intensity and even a serious shortening of the useful operating life. Thus it is important to clean the bulb blank thoroughly, for example by boiling it in a concentrated nitric/sulphuric acid mixture (1 : 1 v/v) and then rinsing in distilled water and drying. Subsequently, care must be taken not to touch the outside of the bulb with the hands, particularly if silica is used, as traces of oil or grease from the skin may be deposited and impair its operation. For this reason a handle permanently attached to the bulb is useful. For example, the microwave lamp shown in Fig. 5-21 simply has a short piece of the original tubing left in place above the seal. When preparing blank bulbs, the point at which the final seal is to be made should be constricted as much as possible to minimize the heating required when the lamp is completed. This is particularly important with silica because overheating breaks down some of the silica and the oxygen produced may contaminate or even quench the discharge. In view of this, vacuum-jacketed microwave lamps should have the jacket sealed around the blank bulb[54] (as shown in Fig. 5-21(b)) rather than onto the finished lamp. The final stage of blank preparation is degassing to avoid subsequent outgassing during lamp operation. This is achieved by heating the bulb under vacuum. For obvious reasons the process must be carried out just prior to filling and is therefore discussed below in that context.

(b) The vacuum line

A simple vacuum line design is quite suitable for preparing most types of sealed-off electrodeless-discharge lamps of interest for AAS and AFS applications and the ability to achieve high vacuum conditions is not essential. It should be remembered, however, that although sealed lamps are operated at relatively high pressure (typically in the range 0·1 to 10 torr for microwave-excited lamps), a very much better vacuum is desirable to achieve effective degassing of the lamp walls. As filling of the lamps involves flushing with gas and repetitive pumping down at pressures in the region of 1 torr, a two-stage rotary pump affords a good compromise between efficient pumping in this region and a sufficiently high vacuum for degassing (typically 0·001 torr for a modern pump in good condition). Some workers have suggested[55] that a vacuum better than 10^{-6} torr is required if the lamps are to be degassed sufficiently for use as high-stability sources. It would then be necessary to use an ancillary oil or mercury diffusion pump for the degassing stage of the lamp preparation. Under these circumstances great care is needed to avoid any traces of contamination during construction and use of the vacuum line if long pump-down times are to be avoided.

A basic system suggested for preparation of sealed-off lamps is shown diagrammatically in Fig. 5-22. The cold trap, T_1, is essential, not only to avoid damage to the pump if volatile materials (e.g. iodine) are used in lamp preparation, but also to avoid contamination of the finished lamps by pump oil (the amount of oil vapour diffusing back from most commercial pumps is more than sufficient to cause detectable contamination). A simple "cold-finger" trap cooled in liquid air or nitrogen is sufficient and should be made readily demountable for cleaning if large quantities of volatile materials are used in the system. The second cold trap, T_2, is also desirable, especially if the vacuum gauges used contain mercury which would otherwise almost certainly contaminate the lamps to some extent. This trap is also a useful aid to lamp purity when ordinary commercial grade compressed gases are used for lamp preparation, as these often contain traces of oil and hydrocarbon vapours.

The filler gas is added to the system via a needle valve, NV, which should be of the precision, low-throughput type designed for vacuum use. Such a valve provides the most convenient means of adding gas to a pumped-down lamp in order to reach exactly a particular lamp pressure. It also allows an exact pressure to be set while continuous

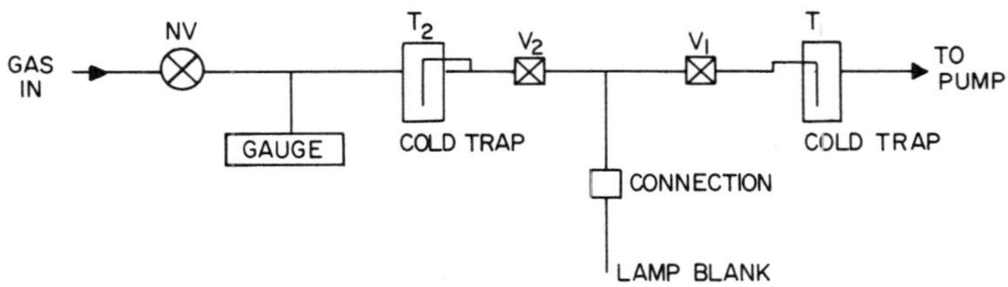

Fig. 5-22. Schematic representation of a vacuum system for the preparation of sealed electrodeless discharge lamps.

pumping proceeds; this may be useful at some stages of lamp preparation or to test particular lamp characteristics. The lamp and its connection to the vacuum line may be isolated by two valves, V_1 and V_2. These are situated to allow removal and replacement of lamps without the need to admit air to a large part of the system, as the line is intended for fairly rapid preparation of sealed lamps. However, this does mean that the lamp is not isolated from the valves by any form of trap. Thus, unless greaseless valves are available, great care must be taken during their assembly. The high-vacuum silicone greases or a more traditional type, such as Apiezon N, are suitable, but the smallest possible quantity should be used to avoid any excess entering the line; this would act as a sink for impurities and may itself act as a lamp contaminant. With Apiezon N it is useful, especially if high vacuum is to be used during degassing, to degas the grease before its application. This is easily achieved by placing some of the grease into a glass tube which is heated in hot water while under vacuum; after cooling under vacuum the grease should be used as soon as possible.

The means of attaching the lamp blank to the vacuum system also requires some consideration. Sealing of the finished lamp is generally easier if the lamp is mounted vertically rather than horizontally. Obviously, if solid materials are to be added to a blank of the type shown in Fig. 5-21(a) it must be suspended vertically rather than mounted above the line. Attachment of the lamp blank to the line by rubber pressure tubing should be avoided as an almost inevitable source of contamination. Some workers favour direct sealing between the two, but this may present problems when only limited glass-working skill is available; furthermore, the lamp cannot be moved during the sealing process. Use of a cone- or ball-type ground joint is more convenient but still requires the sealing of part of the joint to the optical quality silica used for most lamps. Perhaps the best compromise is the use of an O-ring sealing, high-vacuum metal union. Hand-tightened components suitable for attachment to glass tubing are widely available and relatively inexpensive. They may be repeatedly resealed as no swaging of joints is necessary and the best types are quite adequate even for high vacuum (10^{-6} torr). If the O-ring is lightly greased the lamp is easily rotated to facilitate the final sealing.

The vacuum gauge indicated schematically in Fig. 5-22 may in practice take one of several forms, depending on the particular lamp preparation technique and the justifiable expense. The simplest suitable gauges are the rotatable miniature McCloud types. These have the disadvantage of precluding continuous reading of the pressure but they do give correct direct measurement with all gases and vapours and are relatively inexpensive. Mercury contamination does not occur when the gauge is isolated by a cold trap, as mentioned above. In most systems two of these gauges would be necessary, one reading *ca.* 0·001 to 1 torr and the other *ca.* 0·1 to 10 torr. Most of this range could also be covered by using the largest type of *precision* barometric gauges (not the simple types used in rough vacuum systems). These are relatively expensive and rather cumbersome, but have the advantage of continuous reading while still showing identical response to all gases and vapours. Remote reading electrical gauges are often preferred despite the need of different calibrations for each gas used in the system and the risk of serious errors when vapours are present. Care should be taken to select a commercial unit that gives adequate precision at the pressures needed for filling the lamps (for example, many Pirani gauges are unreliable above *ca.* 1 torr). A suitable electrical gauge (such as the Penning type) is generally the best choice as a monitor when high-vacuum degassing is to be used. Careful positioning of the gauge head is then required if a true indication is to be obtained of whether the lamp has reached, say, 10^{-6} torr, as errors are easily made at this degree of vacuum.

(c) *Materials used in lamp preparation*

Helium, neon and argon have all been used as the filler gas, although argon seems to be the most popular. It is readily available; it produces a large, stable discharge over at least* the pressure range 0·1 to 10 torr and thus allows the pressure to be optimized for the vaporization characteristics of the material used in a particular lamp. Furthermore, serious spectral interference due to its background spectrum has not been widely reported in AAS and AFS applications. As with early work on the hollow-cathode lamp (see Section 5.2.3(c)), however, direct comparisons between the three gases are rarely found in the literature. Thus, it may be worthwhile to investigate each of the gases for a particular application whenever optimum performance is essential. Ultra-high-purity (research grade) gases are unnecessary except for special applications (for example, where it is necessary to exclude the other inert gases or oxygen or nitrogen in order to obtain an extremely pure spectrum); generally "high-purity" cylinder gases are quite adequate for lamps prepared for AAS and AFS, especially if added to the lamp via a suitable trap to remove any volatile impurities.

The form in which an element is added to the lamp is of vital importance for successful lamp preparation. As there is no sputtering mechanism of the type found in the hollow cathode, the material used must produce the correct vapour pressure of the element at the operating temperature of the lamp. In many cases the range of vapour pressures for optimum performance is small and a high proportion of erratic or poor results in lamp preparation may be traced to this cause (see Section 5.3.5). Most metals will not have the correct vapour pressure at the "normal" temperature of an RF or microwave discharge and two courses of action are open: either the lamp itself may be cooled or heated, or a more suitable form of the metal may be used at the preparation stage. The former solution has

* Sealed microwave-excited lamps may prove difficult to initiate at pressures below about 1 torr.

been widely used for RF discharges but less so with microwave excitation (see Section 5.3.4 and 5.3.5), although the latter does offer the possibility of quite wide changes in temperature by careful choice of gas pressure, microwave cavity and operating power. Careful choice of the compound used has been widely advocated for the preparation of microwave lamps.[56-58]

It is not possible to give an exact criterion based on vapour pressure for the selection of the best compound of each metal; there is no real substitute for practical experience of preparing lamps for an actual analytical application. However, as a very general guide it can be said that materials having a vapour pressure of about 1 torr at 200-400°C will usually prove suitable. Most microwave-excited lamps described in the literature have used either the pure metal or one of its halides, or sometimes a mixture of the two. Generally the iodide is the most popular (owing to ease of handling) with the chloride as an alternative when its vapour pressure is more suitable; few lamps using the metal bromide have been reported. Although very high purity chemicals (for example, the so-called "spectroscopic grade") are unnecessary for the applications described here, it is important to avoid sealing into the lamp bulb impurities such as water or free iodine. For this reason, many workers prefer to prepare the halides when required, actually *in situ* in the lamp bulb or in an adjacent part of the vacuum line. This is particularly easy with the iodides which can often be made simply by heating a small quantity of pure iodine with a slight excess of the metal. Chlorides are best prepared by heating the metal in a stream of chlorine gas.

(*d*) *Lamp assembly*

This final stage of lamp construction involves four important steps: the cleaned and prepared lamp bulb is attached to the vacuum line and degassed; a small quantity of the metal or a suitable compound is added to the bulb; the correct pressure of filler gas is introduced; the bulb is then permanently sealed off by melting the glass or silica. Each of these steps must be carried out in the correct manner if a lamp giving optimum performance and extended operating life-time is to be obtained.

The complexity of the degassing procedure varies greatly with lamp preparation methods described by different workers. Although considerations of the operating temperature of the EDL, the required working life and spectral purity are important factors, the method of adding the metal to the lamp after degassing must be taken into account before deciding how elaborate the process should be. If the lamp is to be re-opened to the atmosphere after degassing, as recommended by West and co-workers,[58] it is obviously pointless to carry out a prolonged process at high temperatures and under high vacuum. Similarly, if degassing is to be carried out with the sample actually *in situ*, as suggested by Corliss and co-workers,[56] prolonged heating of the whole lamp is not feasible. Under these conditions it is probably adequate to flush the lamp several times with the filler gas at a pressure of a few torr while sparking the bulb with a Tesla coil. The gas discharge will then scrub water vapour and gases from the surface of the walls. After pumping the bulb down using a conventional rotary pump (i.e. to a pressure in the range 0·1 to 0·001 torr), the bulb is heated with a blow torch for a few minutes to drive off impurities situated deeper in the walls. If possible the softening point of the glass or silica should be approached. The procedure of Corliss *et al.* (where the sample is already in the lamp) is probably preferable to subsequent demounting and this process must then be carried out in two stages, heating half the tube at a time while the other half is wrapped in wet asbestos tape so that the sample remains condensed. The impurities driven from the walls* should then be flushed out with fresh filler gas.

To obtain the highest spectral purity, discharge stability and long operating life-time, a more elaborate degassing method than that described above should be used. This is also most important when EDL sources are to be operated at high temperatures to obtain discharges from relatively involatile substances such as compounds of the rare earths.[57] High-vacuum conditions (better than 10^{-6} torr) attainable with an ancillary diffusion pump are then necessary. If degassing is effected solely by heating, the lamp should be held near its softening point for 1-2 hours and it is then much more convenient to use a small demountable furnace for heating instead of a torch. Alternatively, Gleason and Pertel[55] suggest that the scrubbing process described above should be carried out with great care at 5×10^{-7} torr; they also find that excitation of a microwave discharge in a few torr of argon helps the process and gives a good indication of the extent of degassing (pure argon gives a violet plasma which becomes milky white when excessive water vapour is present on the walls of the lamp).

When extensive degassing under high vacuum is to be carried out it is also necessary to resort to a more elaborate means of introducing

* Together with oxygen formed by actual decomposition of silica at high temperatures.

the solid sample than simply placing it in a lamp bulb of the type shown in Fig. 5-21. Tomkins and Fred[57] suggested sealing the solid under vacuum into a capillary tube placed in the vacuum line. This tube may subsequently be moved into the degassed lamp and broken open by the judicious use of iron slugs and magnets. An easier solution to the problem is the use of a lamp blank with one or more side-arms, such as that shown in Fig. 5-23. This particular configuration was used by Gleason and Pertel[55] for a microwave-excited mercury lamp, but the principle is readily applied to many other applications with only minor modifications. The tube R_3 is the

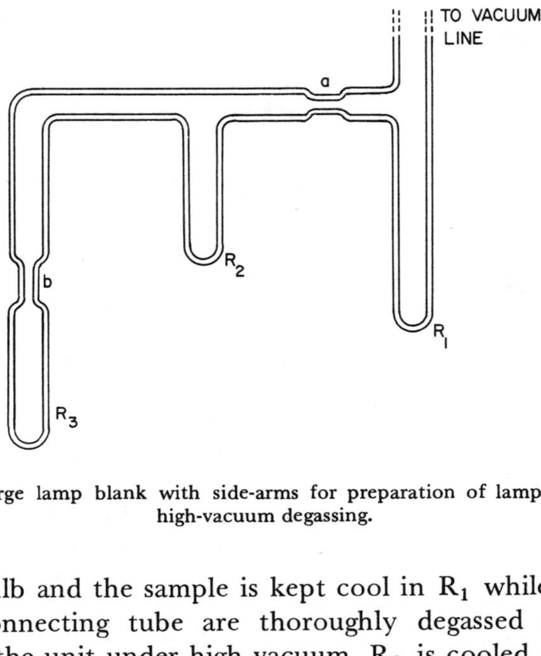

Fig. 5-23. Discharge lamp blank with side-arms for preparation of lamps with elaborate high-vacuum degassing.

final lamp bulb and the sample is kept cool in R_1 while R_3 and R_2 and their connecting tube are thoroughly degassed as described above. With the unit under high vacuum, R_2 is cooled and some of the sample is driven into it by warming R_1. The absence of impurities from R_2 and R_3 is then checked, with further argon scrubbing if necessary, before adding gas at the correct pressure to the unit and sealing at the point "a". Finally, the correct amount of sample is driven into R_3 by heating R_2 before sealing the finished lamp bulb at point "b".

Whichever method is used to introduce the sample into the lamp bulb, several general precautions should be observed. In the first place, it is not sufficient to leave a lump of solid or a few grains of powder in the bulb; the material must be heated and then recondensed on the lamp wall to degas it and drive off any water vapour which may be present. This is usually best achieved using a microburner as very little heat is required for the small amounts of solid involved. When the solid has been added directly to the final lamp bulb it is best to repeat the process several times, driving the material up and down the walls. If an iodide is prepared *in situ* it is essential to drive off any excess free iodine. Although only a few milligrammes of solid are normally added initially, even this is more than is needed (successful small microwave lamps have been prepared with as little as 0·1 microgramme of sample). Thus only a very narrow band of material needs to be left on the wall, the excess being driven out of the bulb by the heating process. There has been some discussion in the literature over the importance of the amount of material used, although until recently no attempt was made at a systematic study. For example, Worden *et al.*[59] found that curium lamps showed self-reversal when 1 mg was used but not with 100 μg; Silvester and McCarthy[60] reported that cadmium lamps gave greater intensities when containing 5 mg of solid than when 35 or 50 mg were used. On the other hand, Winefordner and co-workers[61] observed no systematic correlation between lamp performance and amounts of material in the range 0·23-10 mg.

In fact, the importance of the quantity of material depends to a large extent on the mode of lamp operation.* If the vapour pressure in an RF or microwave-excited lamp is carefully controlled by thermostating the operating temperature of the lamp with external heating (or sometimes cooling), the amount of unvaporized material left will have little effect. A number of workers appear intuitively to have obtained this condition in microwave-excited lamps where the discharge is the sole source of heat, by critical adjustment of the discharge condition, the position of the lamp in the microwave cavity, the excitation power, and the flow of cooling air. However, under these circumstances it is easier to obtain *reproducible* results if the lamp is operated so as to vaporize completely all material present in the bulb, and the amount of material added during construction then becomes of major importance. This is confirmed by a recent study reported by Cooke, Dagnall and West.[62] In order to obtain high precision and reproducibility when adding minute quantities of material to the lamps, they used the technique of preparing lamps with 5 or 10 μl aliquots of aqueous solutions. The water was

* This topic is discussed further in Sections 5.3.4 and 5.3.5.

removed under vacuum prior to subliming the material onto the lamp wall and then proceeding as described above. Provided suitable solutions can be prepared, this technique should prove most useful, especially when using a lamp blank with side-arm of the type discussed above. The material could be deposited on to the limb R_1 using a separate rough vacuum line before proceeding with the

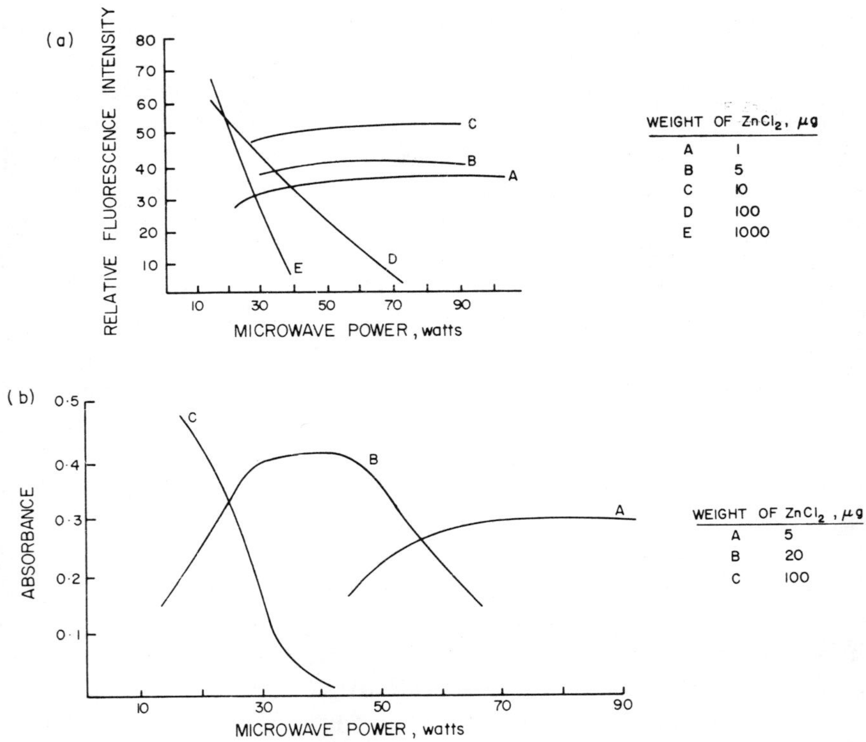

Fig. 5-24. Effect of microwave power on performance of zinc electrodeless discharge lamps containing different weights of $ZnCl_2$.[62]

degassing and material transfer stages in the same way as with a more usual solid sample. The importance of adding a small, carefully controlled quantity (by whatever technique) to microwave lamps operated with complete vaporization is illustrated in Fig. 5-24. This shows the atomic fluorescence and atomic absorption signals obtained using microwave-excited zinc lamps containing small amounts of $ZnCl_2$ over a range of operating power. It can be seen that self-absorption (indicated by loss of absorption sensitivity) followed by self-reversal (indicated by loss of fluorescence sensitivity) is observed at progressively lower microwave power as the amount of material is increased.

Once the correct amount of sample has been placed in the degassed lamp bulb the filler gas may be added at the required pressure. Again, careful optimization for a particular application is necessary and the only general guide that can be given is that between 2 and 5 torr of argon is often a useful pressure for first consideration. If many tests are to be made to prepare a high-performance lamp it is useful to test the discharge on the vacuum line (possibly with a flow of gas at the chosen pressure), or alternatively it may be more convenient to use a demountable lamp fitted with a tap[60] for preliminary experiments. When the finished lamp is finally sealed off, care is still required to avoid unsuccessful results; it is most important to avoid excessive heating of the glass or silica. If the lamp walls have not been thoroughly degassed, sufficient impurities may be driven into the bulb to contaminate or even quench the discharge (the lamp seal is unlikely to reach such a high temperature during normal operation). An equally important effect occurs with silica lamps, irrespective of the efficiency of the degassing, and has frequently been overlooked. Owing to the high melting point of the silica, some decomposition inevitably occurs during the sealing operation and oxygen may be driven into the lamp. If a large area of silica is heated for a long period the amount of oxygen may be sufficient to quench the discharge completely, especially when unusually low filler-gas pressures are used. In fact, many "inexplicable" failures or instabilities of lamps may originate from this source.

5.3.3 Operation and Characteristics of Radio-Frequency-Excited Electrodeless-Discharge Lamps

Electrodeless-discharge lamps may be excited at radio frequencies in two modes of operation. Many spectroscopic applications have used a type of RF glow discharge for which two ring electrodes are mounted on the outside of a cylindrical lamp at a suitable distance apart.[5] It is possible to maintain these discharges over a very wide pressure range (about 10^{-3} to 10^{2} torr), so that the metal vapour pressure may often be optimized without the need for an additional pressure of inert gas to support the discharge. This may be an advantage for some applications, but such sources have not been widely applied as sources for absorption or fluorescence measurements, although L'Vov[15] has used a mercury AAS source of this type operated at 145 MHz.

More recent interest for absorption and fluorescence applications has been shown in an annular type of discharge excited by positioning the lamp within the field of the coil of an RF generator. The discharge is then maintained only over a very narrow pressure range and the lamp is filled with an inert gas at the correct pressure (e.g. 1 or 2 torr of argon). Some care is needed in the choice of characteristics for the RF oscillator circuit to obtain the best results. For example, it is important to obtain a clean waveform without spurious oscillations over the whole range of currents to be used. This facilitates maintenance of a true skin discharge and also ensures that the available power is used mainly to heat the lamp envelope and to excite low energy lines (rather than unwanted non-resonance lines from higher energy levels). Similarly, although the discharge is maintained by the RF magnetic field, sufficient RF electric field must be present to initiate it. There is also a minimum frequency for any given lamp below which a discharge will not start (depending on the mean free path of an electron in the vapour for each cycle of the applied field). As the lamps have a strong negative coefficient of resistance, the oscillator should approach as closely as possible the characteristics of a constant power source; in most cases a further worthwhile improvement in lamp stability results from using a current-stabilized supply for the oscillator valves.

Lamps of this type have been described for a fairly wide range of elements, both with and without auxiliary heaters. Bell, Bloom and Lynch[48] prepared spherical lamps of 10 mm diameter for the alkali metals and operated them in the assembly shown in Fig. 5-25.* The lamp is mounted under a glass cover immediately above the exciting oscillator with a common case. Heat from the thermionic valves is thus used as an auxiliary source for the lamp and the entire package acts as an efficient heat sink to stabilize the lamp operating temperature. The temperatures shown in the diagram are the equilibrium values reached at the recommended operating current for a rubidium lamp. Although this system provides a simple, compact lamp unit, it has the disadvantage of operating properly only at the current for which the package was designed (the corresponding shell temperatures for Cs, K and Na are 40, 80 and 120°C, respectively). Thus adjustment of the lamp output by changing the excitation current is not recommended. To overcome this problem while still

* This design formed the basis of the Varian commercial RF lamp.

maintaining good temperature stability (for precise fluorescence measurements with the alkali metals), Atkinson, Chapman and Krause[49] designed a cylindrical lamp in which the base is used as a metal reservoir and is mounted in a heater block. This allows accurate independent control of the operating current and the metal vapour pressure; the discharge alone maintains the base reservoir at about 70°C and this may be gradually increased to about 200°C using the heater (which is thermostated to give stable operation). On the other hand, other workers have reported sufficient stability of spherical RF lamps for their use as sources in AAS simply by covering the bulb and coil to prevent convection currents and exclude draughts, the lamp temperature being adjusted via the excitation power. Thus L'vov and co-workers[15] have prepared convenient AAS sources for Rb, Cs, K, Na, Zn, Cd, Hg, Se and Te, while Ivanov and co-workers,[63,15] have extended the range to include As, Bi, Ca, Cu, Fe, Ga, In, Mg, Mn, Mo, P, Pb, S, Sn and Tl (owing to the low volatility of Fe, Mn, Mo and Cu it was necessary to prepare the lamps using $FeCl_2$, $MnCl_2$, MoO_3 and CuI_2). Apparently, some of these lamps are now available commercially in the USSR.[15]

The designers of the RF lamps described above have all reported that in their particular applications these sources offer a considerable improvement over more conventional alternative devices. For example, Atkinson *et al.*[49] compared their alkali metal RF lamps with commercial vapour-discharge lamps (see Section 5.4.1) as sources for sensitive atomic fluorescence measurements. From the observations at the 766·5 nm resonance line of potassium (Fig. 5-26) it is evident that the integrated intensity measured with a normal spectrometer is about twice as great for the RF lamp and shows fairly similar increases with both lamps. However, the simultaneous change in the halfwidth of the line (shown in the second diagram) is quite modest for the RF lamp, whereas there is a very sharp increase in line-width for the vapour-discharge lamp. This has a considerable effect on the observed fluorescence intensity, as illustrated in Fig. 5-27. When the potassium vapour pressure in the fluorescence cell is below 10^{-2} torr the vapour-discharge lamp produces no significant fluorescence, whereas the RF lamp is most efficient at low vapour pressures. In fact, at very low vapour pressures the absorption line is narrower than the incident exciting line from the RF lamp. Thus, as the vapour pressure in the cell is increased the broadening of the absorption line gradually matches its width to that of the exciting

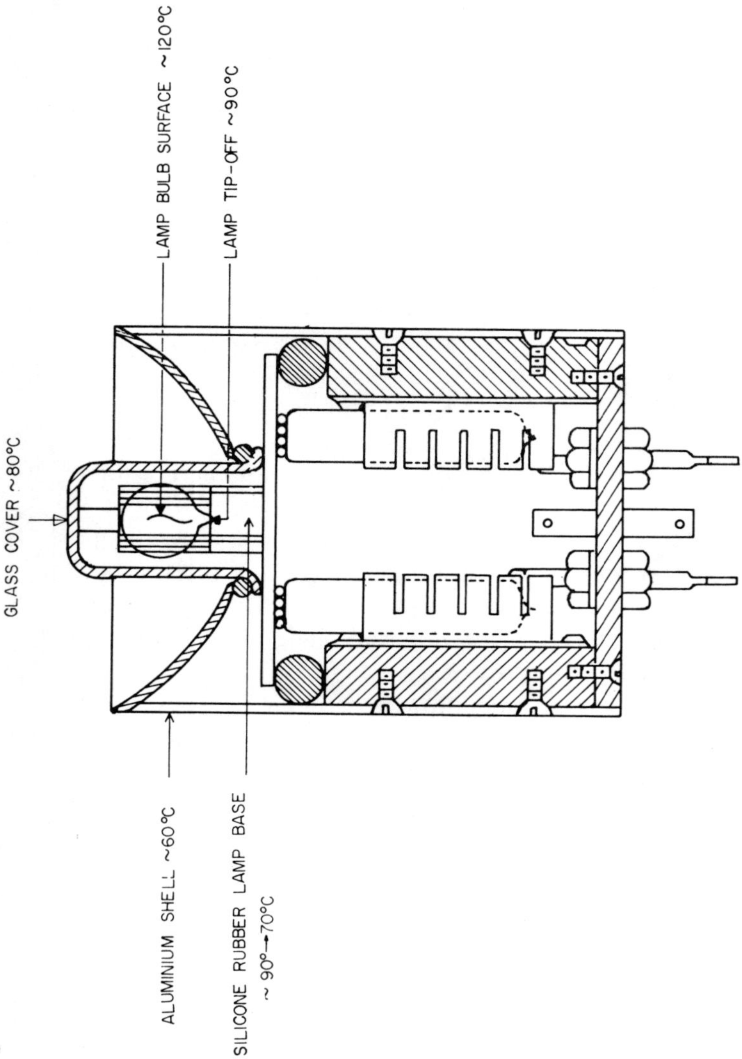

(a) Assembly used for a Rb lamp

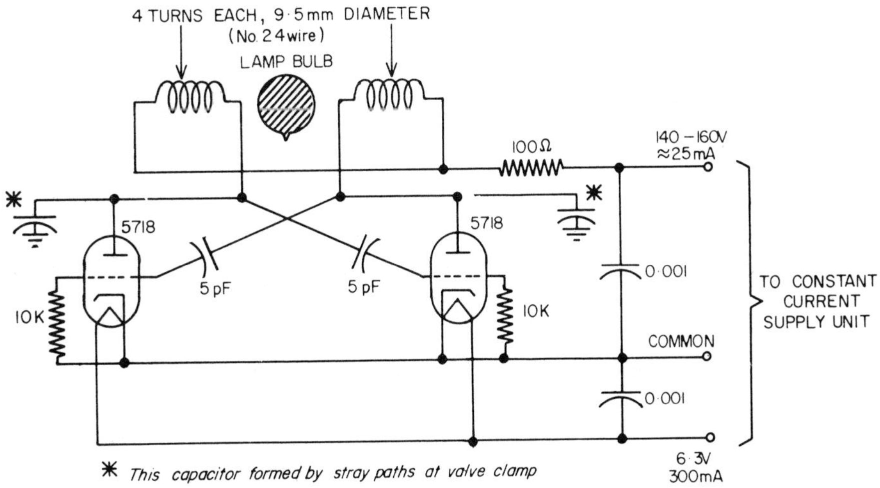

(b) Oscillator circuit

Fig. 5-25. Compact radio frequency exciter assembly used with alkali metal electrodeless discharge lamps.[48]

line and a small maximum in the fluorescence intensity is obtained. The second peak for the RF lamp (as well as that for the vapour-discharge lamp) results from increasing self-absorption in the fluorescence cell at higher potassium concentrations (see Section 4.4).

In a similar fashion L'Vov[15] has compared the spectral characteristics of the spherical RF lamps with those of hollow-cathode lamps from the point of view of their use as sources for AAS. He reports that, for the cadmium 228·8 nm resonance line, equal line-widths for the two sources are obtained when the intensity

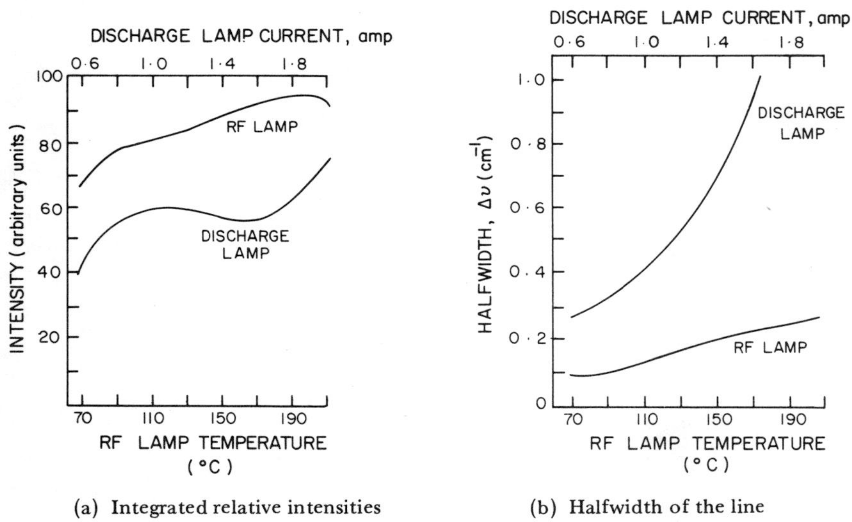

Fig. 5-26. The comparative effect of operating conditions on the potassium 776·5 nm resonance line for a vapour discharge lamp and a radio frequency electrodeless discharge lamp.[49]

from the RF lamp is about 1000 times that obtained with the hollow-cathode lamp. By reducing the current to the RF lamp it was possible to improve the line sharpness considerably while maintaining a greater intensity than is available from the hollow-cathode lamp. Similarly, Ivanov *et al.*[63] reported similar AAS sensitivities for a number of metals using both hollow-cathode and spherical RF lamp sources, but the source intensity was considerably better in the latter case (e.g. a 200-300-fold improvement at the Cu 324·7 nm line). It was found that the background spectrum in these sources using xenon as filler gas was of no significance (e.g. 0·1 to 0·2% of the line intensity at the Sb 231·1 nm and Pb 283·3 nm lines). L'Vov[15] suggests that even less background emission is obtained with argon as filler gas and that lamp stability and noise are no worse than for the hollow-cathode source when the working conditions are correctly chosen.

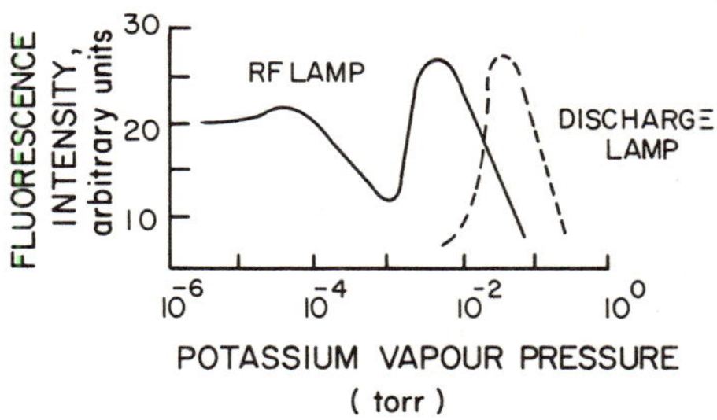

Fig. 5-27. Variation of potassium fluorescence intensity with vapour pressure, excitation at the 776·5 nm line with a radio frequency electrodeless discharge lamp or a vapour discharge lamp. (The curves are normalized to show equal peak intensities.)

5.3.4 Operation and Characteristics of Microwave-Excited Electrodeless-Discharge Lamps

(*a*) *Power supplies*

The construction of a microwave generator is, as mentioned above, more complex than the assembly of a dc or RF power source and requires a considerable knowledge of specialized microwave techniques. This situation has had two main consequences for the development of microwave excitation of electrodeless-discharge lamps. In the first place, although many reports are to be found in the literature describing the designs of the lamps themselves and even of suitable microwave cavities, there are very few descriptions of complete microwave generators or specialized modifications of interest to the spectroscopist. Secondly, almost all reported applications of microwave-excited lamps have used commercially available generators which are either actual medical diathermy units or "spectroscopic power supplies" produced by minor modifications to such designs. Thus, with but few exceptions, "microwave" excitation may be interpreted as excitation at the standard 2450 MHz frequency generated by the magnetron valve used in these units. However, some early generators were modified from surplus radar equipment: Jacobsen and Harrison[46] used excitation at 3000 MHz, and recently Kikuchi[64] has operated a lamp at 35 GHz (i.e.

35,000 MHz) using a klystron valve source. The following discussion deals exclusively with lamps and equipment operating at 2450 MHz.

Although medical diathermy units are readily available and relatively cheap, they may sometimes be unsuitable for use with discharge lamps intended as sources in AAS or AFS. The most serious problem is the frequent lack of any form of stabilization for the power supply to the magnetron valve; as a small percentage change in the anode voltage may alter the microwave output by 50 watts or more, such a generator can show somewhat erratic behaviour. Furthermore, the rectification of the high-voltage supply to the magnetron is sometimes rather unsatisfactory, leading to considerable mains frequency modulation of the microwave output. This may cause problems in certain spectrometers when the radiation from the lamp is chopped for use with a modulated detection system whose frequency is mains-derived. Generators sold specifically for spectroscopic use generally have a more sophisticated rectification circuit and this problem is less likely to arise. In many cases some form of stabilization for the magnetron power supply is also introduced, often by applying the voltage to the magnetron via series resistors. As explained for hollow-cathode lamp supplies (see Section 5.2.4(a)), this will minimize the change in anode current (and hence microwave power) resulting from a given fluctuation in the supply voltage. Even so it may still be advantageous to operate the generator from a voltage-stabilized power supply unit (although one with a sufficiently high power rating may not be readily available in many laboratories) and such a device should be particularly useful with generators intended for medical use.

Unfortunately, the change of magnetron power with voltage is so large that even these precautions may prove inadequate for the very stable light sources required for AAS and AFS work; a supply voltage stabilized to 0·1 per cent could still give output fluctuations of several watts for a generator with 100 to 200 watts full power rating. There also remains the effect on the anode current of changes in the load applied to the magnetron (caused by variations in discharge and lamp characteristics; see below). As the magnetron's power *vs* current characteristic is much steeper and almost a linear variation,* it would seem that more satisfactory results should be obtained by using a current-stabilized circuit for the magnetron supply. Practical considerations make the direct regulation of the magnetron current difficult, however, and until recently the authors were not aware of

* A useful rough guide for the magnetron output is that an anode current of 1 mA represents a microwave power of about 1 watt.

any commercial units or published circuits that achieve it at reasonable cost. Brandenberger[65] has pointed out a method which avoids this difficulty and has described a relatively simple modification to a commercial microwave generator. As the basic circuits of many of these units are quite similar, it should be possible to apply Brandenberger's modification without serious difficulty to currently available designs. The additional expense should be worthwhile for many applications as Brandenberger observed the output of a helium-discharge lamp operated at 50-60 watts to have a stability (long term and short term for ±5% changes in mains voltage) of about 0·1% as well as a total rms ripple of about 0·1%. It should be remembered, however, that even a current-stabilized device cannot compensate for the changes in lamp condition which are often responsible for the fluctuations in output observed with tubes containing metal vapour as well as the inert gas (see below).

As explained for the hollow-cathode source (see Section 5.2.4), when modulation of the lamp output is required use of a rotating chopper is almost always less satisfactory than direct electronic modulation of the lamp power supply. This is particularly true when electrodeless-discharge lamps are to be used as high-intensity fluorescence sources (as the chopper entails a considerable loss of the available radiation) and several commercial microwave generators are now available with facilities for modulation of the microwave output at a suitable frequency. This is achieved by modulation of the anode current of the magnetron oscillator. A relatively simple modification[66] to produce this effect involves superimposing an ac component on the dc potential of the anode using a transformer. However, this has the disadvantages[54] of working efficiently only with low-frequency sinusoidal waveforms (owing to the use of the transformer) and of frequently giving a distorted waveform on the output as a result of the relatively high dc current carried by the secondary winding of the transformer. More satisfactory results may be obtained when an adequate stabilized power supply is used for the magnetron simply by adding the desired waveform to the reference element of the circuit using a low-power signal generator. In this way it is possible to obtain reliable modulation with any waveform and at high frequencies (>100 kHz if necessary). It should be remembered, however, that with any system of modulation the electrodeless discharge lamp itself and possibly the actual emission line observed may have a considerable effect on the ease with which 100% modulation of the emitted radiation is attainable at a particular

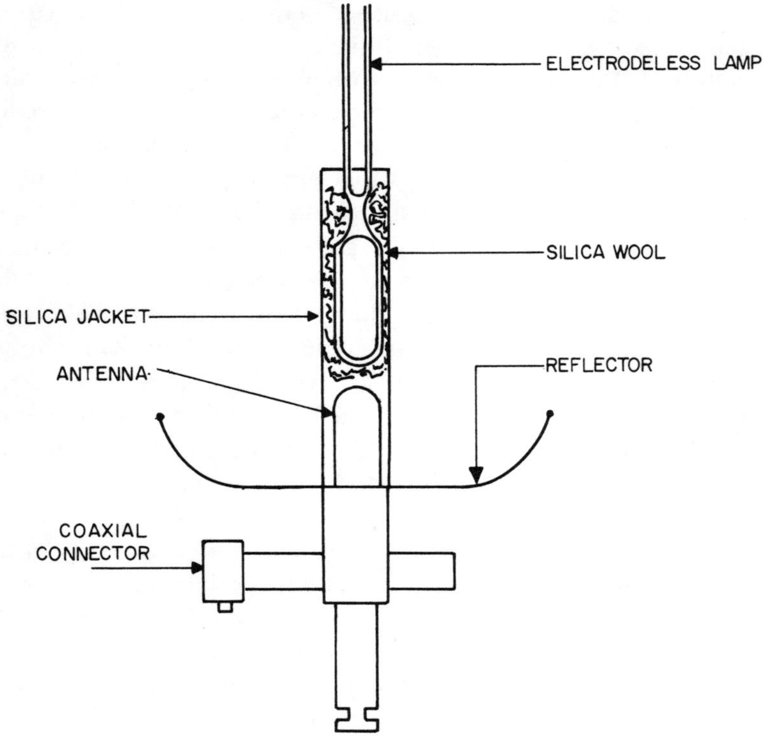

(a) Open A-type antenna with thermal insulation of the lamp body.[61]

Fig. 5-28. Operation of microwave excited electrodeless lamps with an antenna device.

frequency. For example, Phillips[67] was able to obtain 100% modulation above 150 kHz for a helium lamp, but only up to about 110 kHz for iodine-argon and mercury-helium lamps. At low frequencies (less than 100 Hz) it may be necessary to set the depth of modulation* considerably below 100% to maintain the discharge, owing to the finite lifetime of the ions in the discharge.

(*b*) *Transfer of power to the discharge*

Reliable operation of any microwave generator can only be obtained when the power is efficiently transferred to the gas in the discharge. This is achieved either by focusing the microwave radiation onto the lamp, using an antenna device with a reflector or by placing the lamp in a cavity whose resonant frequency is tuned to 2450 MHz to increase the field strength in the gas. In recent years relatively few workers have used antennae (see for example refs. 53,

* Modulated microwave generators usually have provision for continuous variation of the depth of modulation from 0 to 100%.

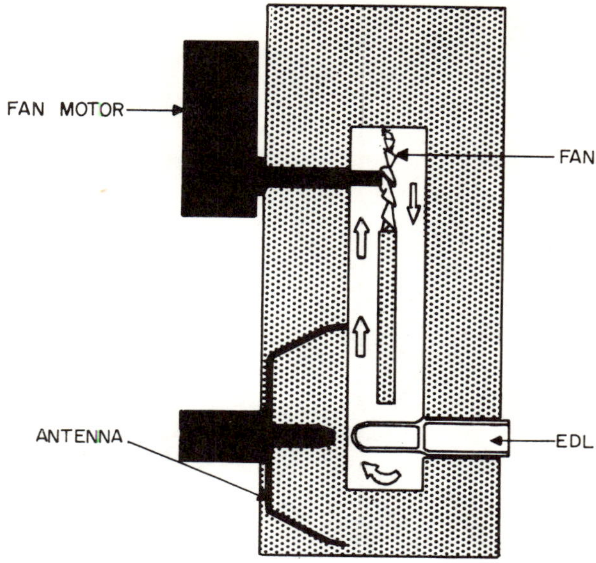

(b) A-type antenna incorporated in a temperature-stabilized lamp housing.

55, 60 and 61), an example of which is illustrated in Fig. 5-28(a). Antennae appear to be less efficient than the best of the resonant cavities and as a result of the exposed position of the lamp it is often difficult to maintain the lamp at a sufficiently high temperature to vaporize metals or their compounds (hence the insulating jacket shown in Fig. 5-28(a)). This problem appears to have been overcome in a commercial unit* which incorporates an A-type antenna in a temperature-controlled lamp housing (see Fig. 5-28(b)). It is claimed that use of a small forced air oven attached to the unit allows the lamp temperature to be maintained within ±0·1°C of a predetermined optimum value, ensuring optimum high intensities with good stability and lamp life. Furthermore, the diffuse microwave output obtained from the antenna is said to avoid the need for critical tuning and positioning of the EDL inherent in many designs of resonant cavity (see below); this simplifies operation and improves the stability of the discharge.

Most reports of microwave-excited discharge lamps concern the use of resonant cavities, usually one of a range of "standard" designs. The most popular of these are available commercially (often from

* Scintillonics-Interface Inc., Fort Collins, Colorado 80521, U.S.A. Based on the design of Winefordner and co-workers.

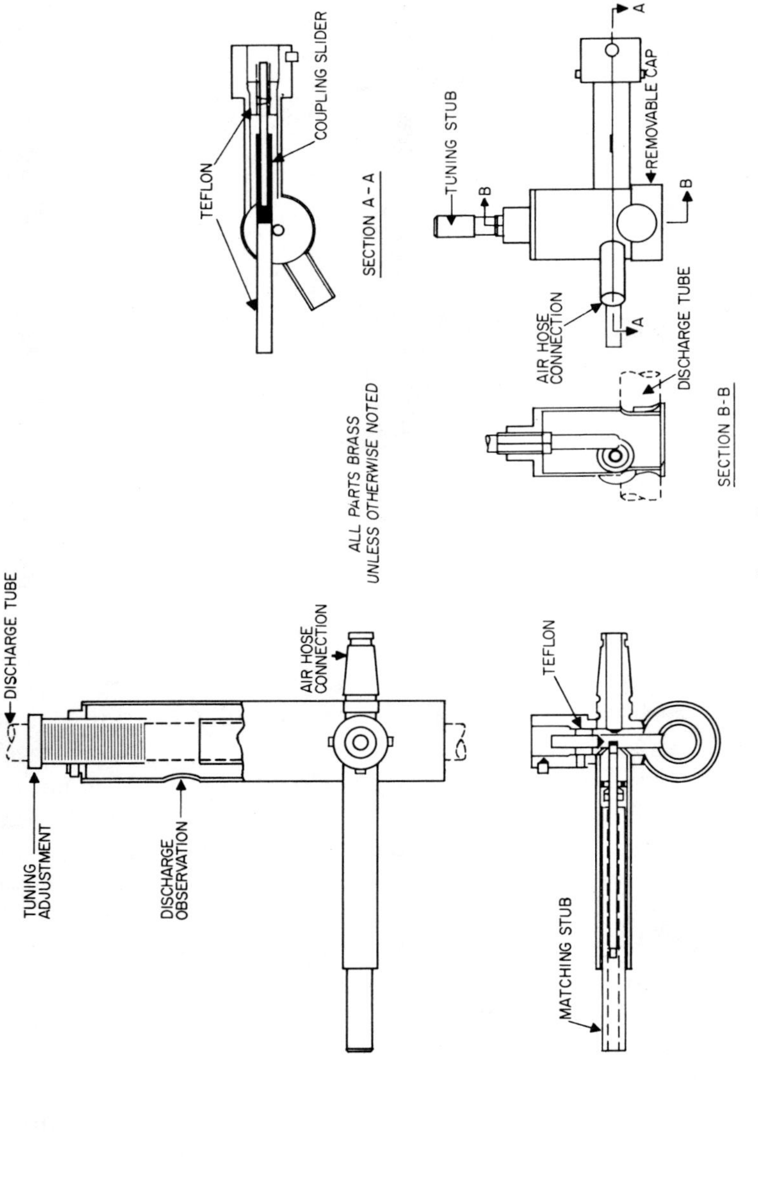

(a) Broida-type $\frac{3}{4}$-wave cavity.[68]

(b) Evenson-type $\frac{1}{4}$-wave cavity.[68]

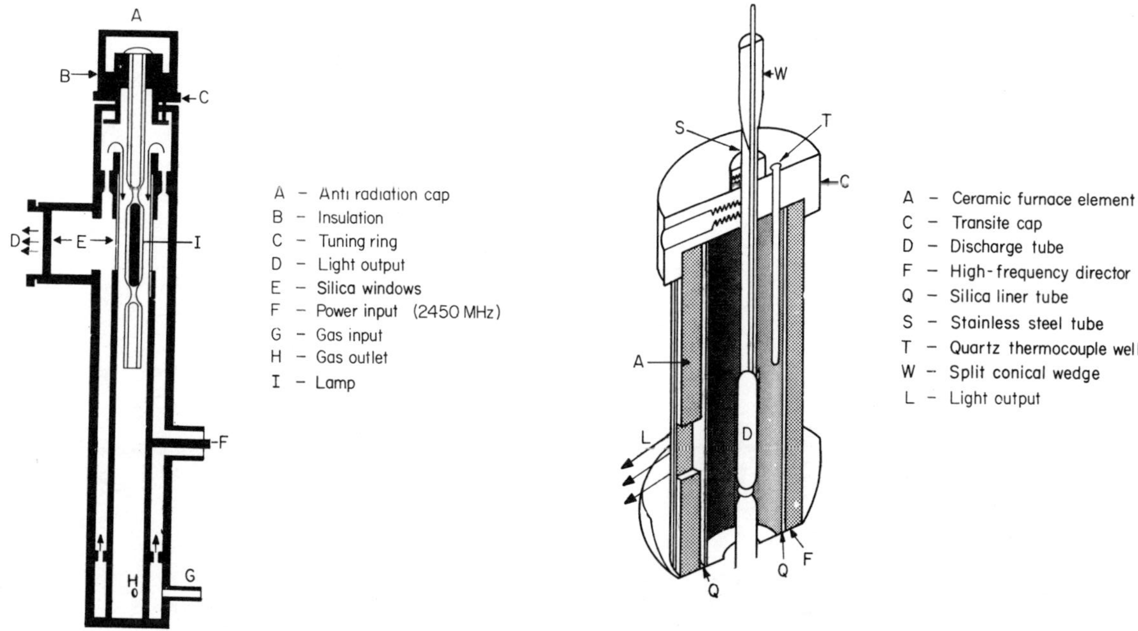

(c) Commercial $\frac{3}{4}$-wave cavity specifically designed for use with AAS and AFS sources (*EMI Electronics Ltd.*)

(d) Furnace assembly incorporating a cavity for high temperature operation of discharge lamps[57]

Fig. 5-29. Some types of cavity used in microwave excitation of electrodeless discharge lamps.

the manufacturer or distributor of a microwave generator) but they are all fairly straightforward, and construction to an existing design should present no serious difficulties. The main features of a number of the "standard" designs have been reviewed by Fehsenfeld *et al.*[68] These workers point out that the presence of a discharge within the cavity changes both the resonant frequency and the load presented by the cavity to the microwave generator. Furthermore, the properties of the discharge vary with the nature of the species present and the pressure. Hence, the most efficient designs provide both a tuning adjustment (to allow the resonant frequency of the cavity to be adjusted to the microwave frequency) and a coupling adjustment (for matching the cavity load to the generator) with sufficient degrees of variation for a wide range of discharge conditions. In some designs (including some available commercially) the coupling adjustment may give an inadequate range, or be omitted altogether. However, Schrenk *et al.*[69] have described the construction of a simple impedance-matching transformer which can be inserted between an existing cavity and input cable.

Two cavities that have been particularly widely used are illustrated in Fig. 5-29(a) and 5-29(b). They are respectively, the Broida-type $\frac{3}{4}$-wave cavity and the Evenson $\frac{1}{4}$-wave cavity. Of the two, the latter gives more efficient transfer of microwave power to the discharge over a wide range of pressures but has the disadvantage of requiring critical tuning to achieve this performance and sometimes giving arcing between the tuning stub and the coupler at this setting. McCarroll[70] has found that both these problems can be eliminated or made less serious by the addition of extension tubes $\lambda/2$ (or more) long to both sides of the cavity along the length of the discharge tube. Although both cavities have provision for forced air cooling of the lamp, the open form of the Evenson design makes it considerably more susceptible to fluctuating temperatures in the lamp than is the Broida cavity. This can prove a serious problem in obtaining good lamp stability for certain elements, since significant fluctuations in the population of metal atoms in the discharge may result. For this reason many workers have preferred to use the Broida cavity for excitation of lamps used as sources in AAS and AFS. Even so the thermal stability may be inadequate for some applications and the modified $\frac{3}{4}$-wave cavity shown in Fig. 5-29(c) should afford a further improvement. The lamp tube has a silica shield immediately adjacent to it inside the cavity, radiation leaves the cavity via a second silica window mounted on the outer wall. Thus, hot or cold air can be passed around the lamp in a carefully controlled manner using the small inlet and outlet tubes. Adjustment of the lamp power and the air flow rate (together with the actual input temperature of the air in difficult cases) allows a stable optimum temperature to be maintained for a wide range of lamps used in AAS and AFS. The cavity provides for tuning of the resonant frequency by adjustment of the coaxial gap, as in the Broida-type, but no coupling stub is fitted and in certain cases it may be advantageous to use an external matching device as mentioned above. Unlike most other designs used to date, the lamp is mounted in the cavity and held in place by a locating ring fitted to it rather than lamp and cavity being independently clamped. This should reduce the risk of relative movement between the two with the consequent possibility of significant changes occurring in the tuning of the cavity; with some critically tuned cavities vibrations in the apparatus can actually cause an unstable output via this mechanism.

As mentioned earlier, microwave-excited lamps of nonvolatile metals are usually constructed using a suitable halide to allow a sufficient atom concentration to be obtained at the "normal" operating temperature of the discharge. When this is not possible additional heating of the lamp is required. The unit shown in Fig. 5-29(d) consists of a small furnace placed on top of a microwave director and was designed by Tomkins and Fred[57] for operation of rare-earth iodide lamps at 750-800°C. The main problems with this type of high-temperature operation are those related to additional lamp outgassing and provision of sufficient cooling of insulation material at the connector between the antenna or cavity and the microwave input cable.

(c) *Characteristics of the discharge*

Although the main requirements* for a light source in AAS or AFS have been largely satisfied by the characteristics of microwave discharges of the permanent gases, these elements are of limited interest for analytical applications of the technique. The situation with discharges emitting the metal resonance lines is, unfortunately, somewhat less satisfactory. However, the problems mentioned here are now widely recognized and it seems likely that work currently in hand will soon result in much more reliable operation of metal

* As mentioned several times elsewhere in this book, these include easy starting with little warm-up time, high intensity (especially for AFS), narrow resonance line profiles (especially for AAS) and a stable and reproducible output.

microwave lamps. Thus, new lamps of many elements require a "running-in" period of several hours, especially those prepared using the metal and free iodine. To obtain a reproducible output on subsequent operation of the lamp, careful positioning and tuning of lamp and cavity are necessary together with a period of about 5 to 30 minutes for warm-up and stabilization during which further adjustment may be needed. Most properly constructed lamps allow initiation of the discharge (with a Tesla vacuum tester) while cold, but for a few elements it may be necessary to warm the lamp first. This may also prove helpful should a well-used lamp prove difficult to start.

Unless a lamp has deteriorated or been improperly prepared, the discharge should emit a "clean" metal spectrum. Examples of improper preparation are the presence of an iodine continuum due to a residue of free iodine in the finished lamp, and an intense mercury spectrum resulting from the use of mercury vacuum gauges without adequate trapping. Provided the correct pressure of filler gas has been used, it should be possible to obtain the metal spectrum almost completely free of background inert gas lines. In practice, the results obtained with many metals even for a well-made lamp depend very much on the way in which it is operated. The authors have found it useful to distinguish two distinct modes of operation; these also form a realistic framework into which it is possible to fit the other desirable characteristics of a lamp. In the so-called *high-intensity mode*, the inert-gas lines are usually extremely weak in comparison with those of the metal spectrum and can be completely ignored during most measurements with the metal resonance lines. This mode of operation, which is usually obtained at relatively high microwave power, also gives an extremely intense metal spectrum and with many lamps the output stability and reproducibility are limited only by those of the microwave equipment itself. Unfortunately, for very many elements the high-intensity mode of operation is also characterized by severe self-absorption broadening, or even serious self-reversal, of the metal resonance lines. Hence a lamp operated in this way may prove completely unsatisfactory as a source for AAS and, contrary to expectations from measurements of its intensity with a conventional spectrometer, give quite poor fluorescence sensitivity.

On the other hand, in the *low-intensity mode* (usually obtained at very low operating powers) there will frequently be virtually no self-absorption of metal resonance lines and the line profile is extremely sharp as the discharge conditions minimize other forms of line broadening (e.g. Doppler, pressure and Stark broadening). However, it is now found that the background inert-gas spectrum is often quite intense in comparison with the metal spectrum and may cause spectral interference with resonance lines of some elements. Furthermore, many lamps prove difficult to operate reproducibly under these conditions, frequently showing unacceptably poor stability, and the authors have found that in some cases the intensity of a metal resonance line may be no better than that obtained with similar line width and superior stability from a hollow-cathode source. The change in line-width (i.e. self absorption) when going from the low- to the high-intensity mode is illustrated in Fig. 5-30 for the thallium 377·6 nm resonance line. This diagram shows the hyperfine structure of the line, as observed with a high-resolution Fabry-Perot interferometer, under five different operating levels. As a result of the poor stability of this particular lamp in the low-intensity mode, the spectrometer was scanned rapidly while observing the spectra on an oscilloscope (each photograph shows the spectrum repeated several times; the peaks may be identified by comparison with Fig. 5-15). The very severe self-reversal with an actual decrease in the useful intensity at higher power is clearly illustrated.

The characteristics of the metal microwave-excited lamps would seem to be quite adequate for AAS and AFS sources if it were possible to operate them in the high-intensity mode at a relatively high level of excitation (i.e. input power) while maintaining minimal self-absorption of the resonance lines. At present, with the commonly used methods of preparation and operation the use of high power results in a high lamp temperature and thus in many cases a high vapour pressure of free metal atoms within the lamp. There are several ways in which it may be possible to use a high level of microwave power while maintaining a low (and, in fact, preferably the optimum) concentration of metal atoms. For example, improved methods of cooling the lamps may allow the use of high power while maintaining a lamp temperature which is sufficiently low *and* reproducible without hot or cold spots to reduce stability. On the other hand, it may prove possible to add precisely the correct amount of material to the lamp so that operation at high temperature with complete vaporization of a relatively volatile substance gives optimum performance without self-absorption. Given an even lamp temperature this would certainly give reproducible operation of the lamps. Finally, with some elements of low volatility

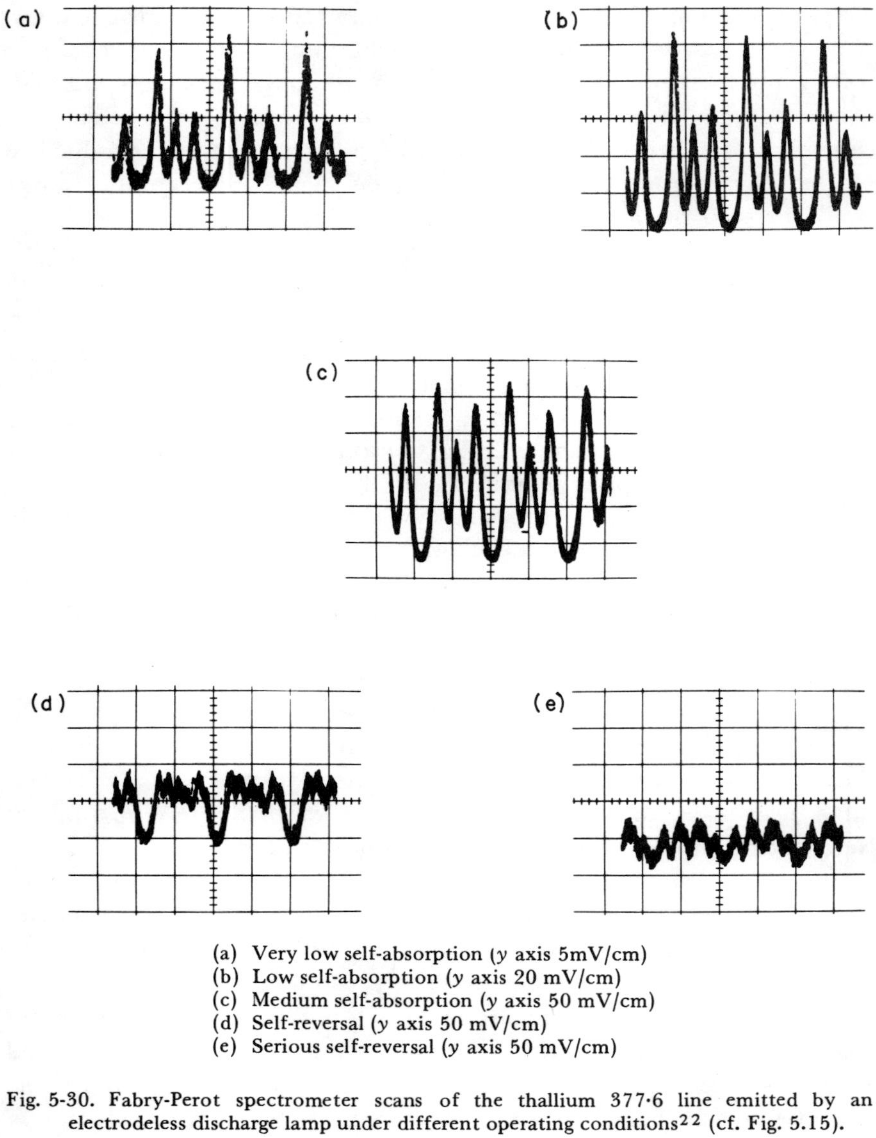

(a) Very low self-absorption (y axis 5mV/cm)
(b) Low self-absorption (y axis 20 mV/cm)
(c) Medium self-absorption (y axis 50 mV/cm)
(d) Self-reversal (y axis 50 mV/cm)
(e) Serious self-reversal (y axis 50 mV/cm)

Fig. 5-30. Fabry-Perot spectrometer scans of the thallium 377·6 line emitted by an electrodeless discharge lamp under different operating conditions[22] (cf. Fig. 5.15).

(or using their compounds of sufficiently low volatility) it may be possible to obtain good results by heating the lamp well above the temperature attained with the discharge alone. The exact vapour pressure required could then be selected and accurately maintained by control of the furnace temperature. The rather greater Doppler width of the resonance lines thus obtained would not be a serious problem for many applications, especially with a lamp to be used as an AFS source.

5.4 OTHER LIGHT SOURCES

As mentioned at the beginning of this chapter, we shall discuss the remaining types of spectral source of interest in AAS and AFS only briefly. Some, such as the vapour-discharge lamp, were of interest in early work but are unlikely to see much further application for analytical purposes. Others, such as the laser, may well become important alternatives to the hollow-cathode lamp and electrodeless-discharge lamp as the result of future technical improvements. Several other possible sources which are unlikely ever to be of more than isolated interest for practical analytical applications will also be listed.

5.4.1 Vapour-Discharge Lamps

Vapour-discharge lamps form simple, robust sources of line spectra of the volatile elements (Cd, Cs, He, Hg, K, Na, Ne, Rb, Tl, and Zn) and for many years were the most popular commercially available* types for purposes such as spectrograph calibration. The commercial lamp shown in Fig. 5-31 is simply an inner discharge tube surrounded by an envelope to protect it from damage and draughts. The discharge tube is usually silica, but a protective layer inside a special glass is required for alkali metals to prevent attack on the surface. The envelope is usually of non-ultraviolet transmitting glass and it is then necessary to cut an opening to view the resonance lines of Zn, Cd, Hg. The discharge tube contains a small amount of the required metal, an inert gas at the correct pressure, and two tungsten electrodes for the discharge. The design shown in the diagram also incorporates starting electrodes operating via a thermistor which switches them off when the lamp reaches its normal operating temperature. These lamps may be operated with either ac or dc although Wotan lamps are intended for operation at 220-240 V ac via a choke to regulate the current passed. However, use of a stabilized power unit is almost essential for atomic fluorescence (or absorption) applications and a method of modulating the output has been described.[71]

The recommended operating current of the commercial lamps, to give stable, efficient operation is *ca.* 1·0 to 1·5 amp (depending on

* Osram G.m.b.H. ("Wotan" lamps), Munich, Germany. Phillips Gloeilampenfabriek N.V., Eindhoven, Netherlands.

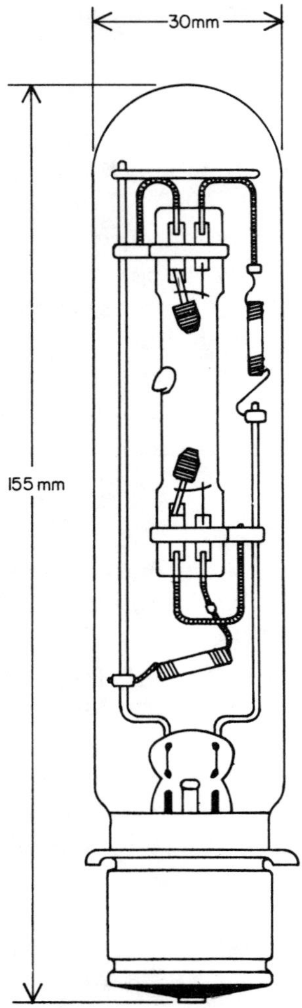

Fig. 5-31. Construction of a commercial vapour discharge lamp.

the element). However, under these conditions the resonance lines are found to be strongly self-absorbed or even self-reversed.[22,72] This is illustrated in Fig. 5-32 for the hfs components of the thallium 377·6 nm resonance line.* It can be seen that reduction of the current reduces the self-absorption and several workers have been able to use these lamps as sources for atomic fluorescence spectroscopy.[73-83]

* These spectra should be compared with those shown in Figs. 5-15 and 5-30 for the hollow-cathode and electrodeless-discharge lamps respectively.

The lines are still somewhat broad and the thallium lamp was found[22] to give similar line widths to those from a hollow-cathode lamp only if operated at 0·3 to 0·5 amp, when the intensity also becomes comparable. As the stability is also rather poor at these low currents (compared with the HCL source) these sources have seen little use for absorption work.[84-86]

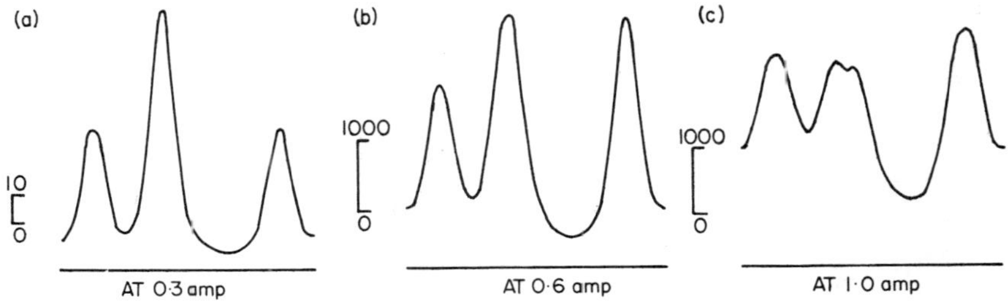

Fig. 5-32. Fabry-Perot spectrometer scans of the thallium 377·6 nm line emitted by a vapour discharge lamp at different operating currents[22] (cf. Fig. 5.15).

5.4.2 *Continuum Sources*

The continuum source is sometimes claimed to have an important advantage over all line sources for AAS and AFS work in that a single continuum lamp is sufficient for most elements. Unfortunately, as explained in earlier chapters, serious problems arise in both absorption and fluorescence applications. Thus, the sensitivity for AAS will only approach that obtained with a sharp line source if a high resolution monochromator is used, and the calibration graphs also show two distinct forms over the normal concentration range. In fluorescence work, the intensity of these sources often appears deceptively high. This is partly because the best absorption lines in some cases lie in a region of the uv where some currently available sources are well below their peak output.* Furthermore, the intensity *per unit wavelength* of most continuum sources is quite low compared with that of a line source at a strong emission line. As the bandpass of a monochromator used in AFS may easily exceed the absorption line-width by a factor of 100, conventional measurements of the source intensity with this equipment will indicate a far higher value than is actually "seen" by the absorption line of the analyte

* Prugger[87] has reported the variation of spectral radiance with wavelength for a number of continuum sources.

element. This effect is particularly serious when fluorescence measurements are made under conditions producing light scatter in the fluorescence cell. As scatter may be detected over the whole bandpass of the monochromator, the scatter signal will be greatly enhanced relative to the fluorescence intensity when changing from a line to a continuum source.

As a result of these problems, continuum sources have seen much less application to AAS and AFS than might otherwise be expected. Several reports have, however, appeared in the literature over the years and some of the specialized applications seem quite promising. The conventional tungsten filament lamp is really of insufficient intensity for either technique, although the more recently introduced tungsten iodide lamps with silica envelopes have been examined as absorption sources.[88] Similarly, several workers[89-91] have attempted to use the well-known hydrogen and deuterium discharge lamps as absorption sources. These are low-pressure lamps giving a far more intense continuum than does a tungsten filament between *ca.* 170 and 400 nm, with a peak output at approximately 230 nm (the deuterium is about 30% more intense than hydrogen at this wavelength). Even so their intensity is too low to allow precise atomic absorption measurements with the narrow monochromator bandpass which is frequently required.

More satisfactory results using a continuum source have been reported for the high-pressure xenon arc by a number of workers, both for AAS[92-96] and AFS.[97-104] These lamps as used in spectroscopy are usually small thick-walled silica bulbs containing two tungsten electrodes a few millimetres apart and xenon gas at a pressure of several atmospheres. Thus an extremely small bright discharge is obtained which may virtually be treated as a point source. Most of the applications mentioned above use this type of lamp with power ratings in the range 50 to 100 watts. Many other types are available for industrial purposes at power ratings of up to 20 kW, but their shape and size (and operating requirements) generally make them unsuitable for laboratory spectroscopic purposes. In any case, the power units used with such powerful lamps would generally be inadequate for the more stringent requirements of spectroscopic applications. In fact, there are many reports of work with much smaller lamps having suffered from the use of power supplies not intended for such applications. Satisfactory results in AAS or AFS can only be expected if the xenon lamp is operated with a specially designed unit giving high current stability (0·1% or better even at maximum output) and low current ripple on the dc output (this is important when synchronous detection systems are used). Operation in this way also helps to avoid sudden shifts of the arc inside the lamp, which can sometimes cause erratic changes in the lamp intensity. Suitable commercial units are available with outputs of up to 500 or 1000 watts. Some of these give only one fixed output for a lamp of a particular rating, but it is possible to obtain supplies giving a wide range of current adjustment and these are much more useful. The use of a pulsed power supply for the xenon arc has also been suggested.[105,106] As with other light sources already described, this technique allows a relatively small spectroscopic lamp to be run at currents well above its continuous rating, thus increasing the output intensity. For example, Klein[106] was able to operate a xenon lamp rated at 2·2 kW with 10 kW pulses.

5.4.3 Lasers

The unique properties that distinguish lasers from other optical sources may be summarized briefly as:

(i) The extremely coherent and unidirectional beam of radiation that is obtained.
(ii) The large amount of energy available over a small spectral region.
(iii) The extremely small area into which the light beam may be concentrated.

These properties offer two possibilities that are of considerable interest with respect to spectral sources for AAS and AFS. For absorption work, the coherent nature of the light beam should allow the use of much longer path lengths (either directly or by use of multiple reflection techniques through a small cell) without the usual problems of light scatter in the cell and insufficient intensity at the detector. This could result in large improvements in atomic absorption sensitivities without any other changes in current techniques. For fluorescence work, the extremely high energy that may be concentrated over the width of a resonance line and within a small area of a sample cell should produce tremendous improvements in sensitivity over anything that is possible with currently available conventional light sources.

Lasers have now been used quite widely for a number of years and their application to absorption and fluorescence studies of molecules in the gas phase has been reported in the literature. However, only very recently have any attempts been made to apply the advantages mentioned above to atomic absorption and fluorescence measurements. The explanation of this delay is quite simple. Most of the commercially produced lasers were of two types, either solid state (in which the active material is a solid such as ruby, glass, Nd-YAG, etc.) or gas (in which the active material is a permanent gas such as He/Ne, Ar, Kr, Xe, CO_2, N_2 etc.). In both cases laser action is obtained at one or perhaps several spectral lines which are characteristic of the active material. Thus, the coincidence of a laser line with a metal resonance line suitable for AAS or AFS would be quite fortuitous and rather unlikely. As a result, no attempts appear to have been made to produce a laser source for use at metal resonance lines by choosing the appropriate active material in a conventional laser. Several metal vapour lasers have been developed and the He/Cd laser is available commercially. The metal ion lines emitted by these lasers are of no direct interest in AAS or AFS and their application would still depend on the fortuitous coincidence of a laser line with a metal resonance line.

This situation has been dramatically altered by the advent of *tunable* lasers of several different types (see below).* These devices permit the laser output (in the form of a "line" whose half-width may be pre-selected and made extremely narrow if necessary) to be tuned through a given wavelength range and hence peaked at any discrete wavelength within that range. Thus, a tunable laser may be used as a sharp line source at *any* metal absorption line lying within its tuning range. Quite apart from any advantages of using a laser source as such, these devices seem to be the perfect sources for AAS and AFS; i.e. a single sharp line source which (like a continuum) can be used with many different elements and which allows the half-width of the source line to be varied over a wide range. As a result, it seems likely that once the initial drawbacks of these devices have been overcome (see below), the tunable laser will become an extremely important source in AAS and AFS. In view of this rather more space has been given to the topic than may appear warranted by the number of applications so far reported.

Perhaps the most serious deterrent to the use of a laser source for AAS and AFS at the present time is the extremely high cost of a suitable tunable laser for most elements; quite reasonably priced

* The term tunable as used here should not be confused with some conventional commercial lasers in which it is possible to *select* laser action at one of several spectral lines available from the active material by making an adjustment to the cavity.

commercial units are widely available but their wavelength range lies too far into the visible region for use at the absorption lines of many elements. Some of these cheaper units also give a pulsed rather than a continuous output. Their operating life may be limited if left running for an extended period. Together with the short duration of the pulses this would require some modifications to normal AAS and AFS technique.

Four distinct types of tunable laser have been reported at the time of writing, but interest in the subject is considerable and other techniques may be introduced. The four types are:

Organic dye lasers. Many commercial models are available, some being relatively simple and cheap. The principle of operation is quite straightforward and components are easily purchased for assembly of a complete system in the laboratory. Most reports of tunable laser applications actually refer to organic dye lasers. The present state of the art is summarized below.

Parametric oscillator lasers. Only one or two commercial models are available and the principle and construction is somewhat more complex than for the dye lasers. However, there is a possibility that a more versatile tunable uv output could be more readily obtained by this technique and some details are mentioned below.

Diode lasers. At present tunable diode lasers are in the development stage and give only an infra-red output. Hence they are of no further interest here.

Spin-flip Raman lasers. The comment on diode lasers also applies to these devices.

Dewey[107] has written an excellent review of all these devices with thorough coverage of the literature up to 1971, particularly for dye lasers. Part of his discussion is also relevant to AAS and AFS. Schlossberg and Kelley[108] have described in simple terms the mechanisms of the tunable lasers mentioned above and have summarized the performance achievable to the time of writing.

(*a*) *Organic dye lasers*

The development of dye lasers and early work have been reviewed by Dewey[107] and Kagan *et al.*,[109] while Bradley[110] has described some of the most recent developments and Webb[111] has discussed the mechanism of dye lasers and summarized the "state of the art".

Dye laser action is based on the optical pumping of an active material consisting of a dilute solution of an organic (dye) compound whose absorption and emission spectral bands do not overlap.

Absorption of light from the pumping source excites these molecules to excited singlet energy levels and the re-emitted fluorescence radiation produces a laser action in the cavity under suitable conditions. In practice, very fast radiationless transitions occur within the excited singlet states so that all lasing transitions occur from the bottom vibrational level of the first singlet state to the vibrational levels of the ground state, giving a broad continuous band of emission which may be between *ca.* 20 and 200 nm wide. The exact position of the band of laser radiations will depend on the structure of the dye, the concentration of the solution used, and features of the cavity (such as its path length). The wavelength ranges of some of the most important groups of laser dyes are shown in Fig. 5-33. It can be seen that the range from about 340 nm to 900 nm is fairly well covered, but the uv region of most interest in AAS and

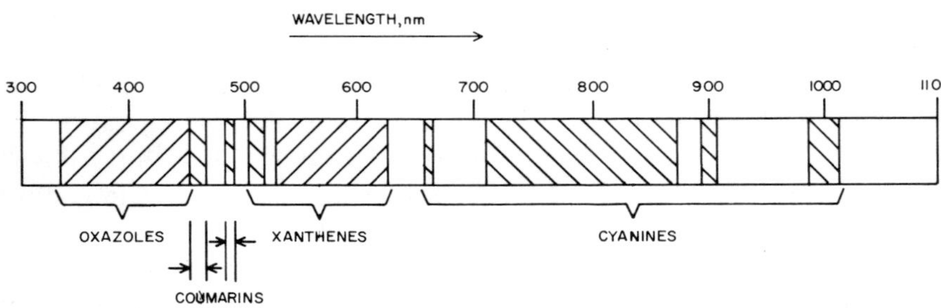

Fig. 5-33. Spectral ranges covered by the main groups of laser dyes.[110]

AFS is not available. Thus, the discovery of organic dyes giving laser emission further into the uv would be an important development for the direct use (see below) of tunable dye lasers for the atomic absorption and fluorescence measurements of many metals.

Early dye lasers were obtained using a pulsed conventional laser as the pumping source, the characteristics of the pumping pulse being quite critical for a successful dye laser. Subsequent investigation of the possibilities of other organic dyes and the development of special flashlamps with very short pulse rise times has allowed pumping of dye lasers without using an expensive auxiliary laser. Many relatively inexpensive commercial tunable lasers operating on this principle are now available. It should be noted that as a result of the more critical requirements for flashlamp pumped dyes some of the spectral regions shown in Fig. 5-33 may not be completely covered by this type of laser. Worden and Gough[112] have surveyed the wavelengths available from fifty-nine flashlamp pumped dyes. To avoid thermal gradients in the dye cell due to the heat transferred during pumping, it is often necessary continuously to circulate dye through the cell at a fairly high flow rate. This is particularly so with flashlamp pumping and/or when high repetition rates are required for the pulsed output, but with even a simple circulation system the thermal problem is no longer the factor limiting the maximum pulse rate and it does not prevent continuous (CW) operation of dye lasers. Until recently it was thought that CW dye lasers were not possible for a different reason. Strong intersystem crossing occurs between the energy levels of these organic molecules and leads to conversion of the excited singlet-state molecules (formed by pumping) in to excited triplet-state molecules. These have a much longer lifetime, so that an appreciable population builds up in the dye solution. As the absorption band of these molecules overlaps the singlet fluorescence band, strong absorption of the dye output can occur and eventually lead to quenching of the laser action. This also explains the need for a short pulse rise time and duration in the pumping source. However, the triplet state can easily be quenched by adding a suitable species to the dye (e.g. oxygen or cyclooctatetraene) and this technique allowed first long pulse operation of certain dyes and recently CW operation with solutions of Rhodamine 6G dye in water.[113,114] Obviously, an intense continuous source is required to pump such lasers and at present this is only possible using a very expensive argon ion laser. Thus, application to routine problems is unlikely until the overall cost can be reduced.

As mentioned earlier, the spectral width of the laser band obtained from the dye varies with different compounds but is always large compared with the width of an atomic spectral line. In some early work[109] there was a tendency to select dyes giving a fairly narrow band and to tune the laser by producing a shift in the peak of this band. This could be achieved by changing factors such as the dye concentration, the solution temperature, or the path length through the dye cell. Subsequently it was found that insertion of a spectral dispersive element into the laser cavity caused the entire output to fall into a single mode characterized by the wavelength-setting element and with a line-width equal to the spectral bandpass of the element. This is easily achieved by, for example, placing a prism between the mirrors of the cavity or by replacing one mirror with a diffraction grating. It is then possible to obtain a relatively narrow spectral line which can be tuned across the entire band over which a particular dye will lase, and it is desirable to find dyes giving as wide a band as possible. Lines narrower than 0·1 nm are produced without difficulty using a diffraction grating, but even this may be inadequate for sensitive absorption measurements. However, much sharper lines may be produced *without serious loss of laser intensity* by inserting a Fabry-Perot etalon into the cavity to reduce further the spectral bandpass. Coarse tuning is then obtained via the main grating (or, quite often, an echelle grating or another Fabry-Perot etalon of greater bandpass), while the final exact matching of the line to the desired wavelength is achieved by rotating the etalon slightly.

Spectral line-widths less than 0·001 nm can be produced in this way using just an echelle grating and one Fabry-Perot etalon.[115] By including a telescope inside the cavity to ensure that a wide, collimated beam of light is incident on each dispersive element, Hänsch[116] recently obtained a linewidth of 0·0004 nm with relative freedom from drift due to small movements of the optical components. Use of an additional Fabry-Perot interferometer external to the laser cavity further reduced the line-width to about 8×10^{-6} nm.[117] Although an extremely narrow line such as this would require considerable expertise in assembly and operation of the complete "laser unit", it can be seen that it is already possible to obtain tunable laser lines narrower than those available from most hollow-cathode lamp sources used in AAS without difficulty and using only commercially available lasers and optical components. In fact, wavelength drift of the line (from its setting on a metal resonance line) may prove more of a problem than its actual width, but even this should not be serious with the specially designed optics now available. The possibility of wavelength drift is also completely avoided in a quite different technique of locking the dye laser output to atomic resonance lines suggested by Bölger and Waysenfeld.[118] They were able to obtain "mirrors" giving high reflection coefficients only at metal resonance lines simply by generating the correct vapour pressure of an alkali metal inside a small glass cell with a wedged window. This cell could then be used as one mirror of a dye laser with an output in the correct spectral range, causing most of the laser radiation to be locked onto the frequencies of, say, the sodium or rubidium resonance lines. Thus the dye laser output is simultaneously narrowed to a line-width suitable for absorption methods and tuned to the correct wavelength without the need for any elaborate auxiliary optical components.

Although the progress in dye laser techniques described above is adequate to allow these devices to be successfully applied as light sources in both AAS and AFS (see below), one further problem remains before they can be considered as universal sources suitable for the determination of most metals. This is the availability of a completely tunable output below the present uv limit of about 360 nm and down to *ca.* 200 nm. Extension of the range of direct dye laser output below 360 nm requires the development of both suitable dyes fluorescing in this region and sufficiently intense pumping sources of even shorter wavelength. Development of pumping sources for this region should not prove difficult, as suitable flashlamps are already available for pulsed operation and many advances have recently been made in ultraviolet laser techniques[119] needed for CW pumping. Unfortunately, predictions of the likelihood of discovery of organic dyes that can be used for routine uv laser operation are not practicable at the time of writing and we can only await future developments.

However, an alternative method of obtaining a tunable uv output from a dye laser is currently receiving considerable attention and seems likely to prove most useful in the near future. This involves doubling the frequency (i.e. halving the wavelength) of laser radiation by generating second-harmonic radiation in a suitable non-linear optical material such as ADP or KDP crystals. The technique was demonstrated as long ago as 1965 and is now widely used to obtain additional spectral lines from fixed wavelength lasers, including many commercial systems. As each particular laser wavelength requires a crystal especially cut to give phase matching for normal incidence of the laser beam, application of the technique to tunable laser outputs is less satisfactory. However, it has been possible to tune dye laser outputs through relatively small uv ranges (e.g. 290 to 300 nm[120] and 280 to 290 nm[121]) using a single crystal provided the crystal is simultaneously tuned to the correct phase-matching angle for each frequency. Thus, in this way many uv metal resonance lines of interest in AAS and AFS could be covered using a single dye laser and a relatively small selection of specially cut frequency-doubling crystals. Recently Gabel and Hercher[122] reported a CW, frequency-doubled, dye laser radiation tunable from 260 to 235 nm. Unlike the previous techniques they placed the frequency doubling crystal (SHG) actually inside the dye laser cavity so that only a uv laser output was observed beyond the end mirrors. It is not clear whether the entire uv range was obtained using a single crystal or whether separate adjustment of the crystal was needed during tuning. Yet another technique currently under active investigation not only avoids any problem of resetting the device during tuning of the dye laser but also triples the fundamental frequency instead of just doubling it. This technique replaces the specially cut crystal with a relatively simple cell containing the vapour of a volatile metal (e.g. the alkali metals) and, when necessary, an inert gas to phase match the refractive indices at the input and output wavelengths.[123,124,125] Once the practical difficulties are overcome (e.g. a tendency to deposit metal films on the optical windows) it should be possible to obtain from a dye laser radiation tunable in the very important range 180 to 300 nm. Paradoxically, coverage of wavelengths *above* 300 nm may prove more difficult as this will require operation of the dye laser in the infrared region above 900 nm where currently available dyes are less satisfactory.

Some applications of dye lasers which are of direct interest to AAS and AFS have already been reported in the literature. Kuhl *et al.*[126] determined sodium vapour concentrations down to 2 ppb in an enclosed cell by absorption using a pulsed dye laser. They averaged each measurement from a number of laser shots and recorded the incident and transmitted intensity for every shot to reduce the effects of variations in the laser output. Their source line (narrowed using a Fabry-Perot interferometer) had a width of about 5×10^{-4} nm and could be set to the desired wavelength with an accuracy of about $2{\cdot}5 \times 10^{-4}$ nm. Bowman, Gibson and Sandford[127,128] have measured sodium concentrations in the upper atmosphere by using a dye laser to excite fluorescence radiation at heights as great as 100 km. Of more direct interest, Denton and Malmstadt[129] determined barium in a conventional flame cell, using a pulsed dye laser to excite fluorescence. To avoid problems of laser instability and irreproducibility it was necessary to use a relatively wide source "line" (compared with the barium absorption line in the flame) and to take data only when the laser output fell within a preselected range. It was also necessary to use an ultrasonic nebulizer producing very small droplets to avoid serious scattering problems in the flame used. A detection limit of 1 ppm Ba was obtained. Fraser and Winefordner[130,131] have carried out an extensive study of atomic fluorescence in flame cells using a pulsed dye laser pumped with a nitrogen laser as a tunable sharp line source. They measured detection limits for the elements Al, Ca, Cr, Co, Ga, In, Fe, Mn, Mo, Ni, Sr, Ti and Tl in either air-hydrogen or nitrous oxide-acetylene flames using an optical and electronic system which was rather more elaborate than that normally used for AFS work. In general terms, it was found that, with respect to detection limits and linearity of analytical curves, the system was equivalent to a conventional system using a range of sharp line sources for all elements with strong absorption lines between 360 and 650 nm, but was considerably more expensive. A major problem was the high level of scattered radiation (from the laser) received as a background to resonance fluorescence, and for a number of elements better detection limits were actually obtained by measuring at less intense non-resonance fluorescence lines. This work is treated in more detail in Chapter 14.

(*b*) *Parametric oscillator lasers*

The basic component of this type of tunable laser is a non-linear optical crystal similar to those mentioned earlier for frequency-doubling applications. Under the correct conditions the crystal is able partially to convert a high-intensity pumping source into signal and idler waves whose frequencies sum to that of the incident wave. By placing the crystal between two wavelength selective mirrors to form an optical cavity, it is possible to obtain a laser output at the signal frequency. Of particular importance is the dependence of the signal and idler frequencies on the refractive indices of the crystal. Thus, the frequency of the output may be varied by altering the temperature, rotation, or electric field of the crystal so as to vary its refractive indices, and a tunable laser is obtained. Generally a sufficient intensity is only obtained using a laser pump source, and pulsed operation is used in most cases. Continuous (CW) operation is possible, but operation is more difficult, the output is less stable, and careful design is essential.

Until recently, it was not possible to obtain a wide, continuous tuning range with one pump laser and one crystal. A commercial system,* however, is now available which provides coverage from 0·5 μm to 3·7 μm (540 to 3700 nm) using a Q-switched ND : YAG laser as pump source for a $LiNbO_3$ crystal. This range coverage is, however, only obtained by using different lines from the pump laser as the actual pump wavelength, depending on the region required. It is also necessary to use a number of different sets of special mirrors to cover the complete range quoted. Tuning is achieved by varying the temperature of the crystal using a special oven. This is claimed to give a frequency stability and resettability equal to the band-width (1 to 2·5 cm^{-1} unless an additional line-narrowing Fabry-Perot etalon is used). There appear to be no reported

* Chromatix, Mountain View, California, U.S.A.

applications of this type of system to AAS or AFS, partly because of the cost of these devices and partly because the available wavelengths are too long for the absorption lines of most elements. The frequencies that are available, however, would seem almost ideal for tripling to provide a tunable source in the uv region of most interest for AAS and AFS. Thus, should the metal vapour devices mentioned earlier become available for routine application, tunable parametric oscillator lasers may prove as useful as their organic dye counterparts.

5.4.4 Miscellaneous Sources

As mentioned previously, there are many types of spectral light source other than those already described here which could be applied to atomic absorption and fluorescence measurements, but very few have been considered for routine analytical work. For the sake of completeness we shall mention one or two for which AAS or AFS applications have been reported in the literature, mostly as solutions to rather non-routine problems.

Most interest has been shown in consumable sources, which are more versatile for research purposes than the other types described above. Arcs, sparks and plasmas are widely used in emission spectrography and spectrometry, but the resonance lines emitted are generally too broad for high sensitivity to be obtained in AAS and it is often difficult to obtain sufficient intensity for AFS without introducing problems of self-reversal. Since sources such as plasmas tend to be rather noisy, it is not possible to reduce these disadvantages by using high scale expansion of signals. In spite of these difficulties, Strasheim and his co-workers have made comparative tests of several of these sources, describing the use for AAS of a time-resolved spark,[132,133] and gas-stabilized dc arcs.[134,135] They reported that good precision and the ease of multi-element analysis often compensated for the much lower sensitivity (about one-third lower) compared with hollow-cathode lamp sources. A high-frequency plasma torch has also been used[136] as a source for AAS.

Flames may also be used as primary light sources for atomic absorption and fluorescence, although again sensitivity will be rather poor. It is necessary to use a hot flame at atmospheric pressure in order to obtain sufficient emission intensity from most elements, so that Doppler and Lorentz (collision) broadening will be severe. The need for quite high atom concentrations in the flame (to obtain sufficient intensity) often leads to self-absorption of resonance lines, introducing further difficulties. This is particularly true when a flame is used as the primary light source for atomic fluorescence; although the line-width is of much less importance than for absorption measurements it is necessary to obtain extremely intense emission for good AFS sensitivity. This requires a high atom concentration and results in severe self-absorption and self-reversal. Matters may be improved somewhat by using a double ("flame-shielded") flame with emitting atoms present only in the centre flame; this avoids the region of cool absorbing atoms normally present at the edge of the flame. Thus, flames offer similar advantages and disadvantages to arcs and plasmas plus the facts that their composition and operating conditions are generally uniform and well established and that no expensive power supply units are required. Flames were used as the spectral source by Alkemade and Milatz[137] for their first work on atomic absorption and since then they have been used several times for special purposes.[138-141]

REFERENCES

1. F. Paschen, *Ann. der Physik,* 1916, **50**, 901
2. H. M. Crosswhite, G. H. Dieke and C. S. Legagneur, *J. Opt. Soc. Am.,* 1955, **45**, 270.
3. A. Walsh, *Spectrochim. Acta,* 1955, **7**, 108.
4. H. Schüler, *Z. Physik,* 1926, **35** , 323.
5. S. Tolansky, "High Resolution Spectroscopy", Methuen, London, 1947; Chapter 4.
6. J. A. Goleb and Y. Yokoyama, *Anal. Chim. Acta,* 1964, **30**, 213.
7. J. A. Goleb, *Anal. Chem.,* 1963, **35**, 1978.
8. W. G. Jones and A. Walsh, *Spectrochim. Acta,* 1960, **16**, 249.
9. A. I. Bodretsova, B. V. L'Vov, E. N. Pavlovskaya and V. K. Prokof'ev, *J. Appl. Spectroscopy,* 1965, **2**, 62.
10. K. B. Mitchell, *J. Opt. Soc. Am.,* 1961, **51**, 846.
11. D. C. Manning and J. Vollmer, *Atomic Absorption Newsletter,* 1967, **6**, 38.
12. A. I. Bodretsova, B. V. L'Vov and V. I. Mosichev, *J. Appl. Spectroscopy,* 1966, **4**, 149.
13. J. B. Dawson and D. J. Ellis, *Spectrochim. Acta,* 1967, **23A**, 565.
14. D. A. Katskov, G. C. Lebedev and B. V. L'Vov, *Zh. Prikl. Spektrosk.,* 1969, **10**, 215.
15. B. V. L'Vov, "Atomic Absorption Spectrochemical Analysis", Hilger, London, 1970.
16. D. G. Mitchell and A. Johansson, *Spectrochim. Acta,* 1970, **25B**, 175.
17. K. Yasuda, *Anal. Chem.,* 1966, **38**, 592.
18. D. K. Davies, *J. Applied Phys.,* 1967, **38**, 4713.
19. V. P. Gofmeister and Yu. M. Kagan, *Opt. Spectros.,* 1968, **25**, 185.
20. *Idem,* ibid., 1969, **26**, 371.
21. W. C. Kreye and F. L. Roesler, *J. Opt. Soc. Am.,* 1970, **60**, 1100.

22. G. F. Kirkbright and M. Sargent, *Spectrochim. Acta,* 1970, **25B**, 577.
23. C. F. Bruce and P. Hannaford, ibid., 1971, **26B**, 207.
24. W. C. Kreye, *J. Opt. Soc. Am.,* 1971, **61**, 833.
25. R. J. Reynolds, *Laboratory Methods,* 1970, **81**, 197.
26. S. R. Koirtyohann and C. Feldman, *Developments in Applied Spectros.,* 1964, **3**, 180.
27. J. I. Dinnin and A. W. Helz, *Anal. Chem.,* 1967, **39**, 1489.
28. G. I. Goodfellow, *Applied Spectros.,* 1967, **21**, 39.
29. G. Rossi and N. Omenetto, *Applied Spectros.,* 1967, **21**, 329.
30. R. E. Popham and W. G. Schrenk, *Applied Spectros.,* 1968, **22**, 192.
31. G. Rossi and N. Omenetto, *Talanta,* 1969, **16**, 263.
32. G. D. Propster, paper presented at Society for Applied Spectroscopy Meeting, Chicago, May 11, 1971.
33. J. V. Sullivan and A. Walsh, *Spectrochim. Acta,* 1965, **21**, 721.
34. Z. van Gelder, *Applied Spectros.,* 1968, **22**, 581.
35. R. M. Lowe, *Spectrochim. Acta,* 1971, **26B**, 201
36. H. G. C. Human and L. R. P. Butler, *Spectrochim. Acta,* 1970, **25B**, 647.
37. A. Walsh, L.S.U. Intern. Symp. Modern Methods of Analytical Chemistry, Baton Rouge, January 1962.
38. W. T. Elwell and J. A. F. Gidley, *Anal. Chim. Acta,* 1961, **24**, 71.
39. H. Massmann, *Z. Instrumenteck.,* 1963, **71**, 225.
40. L. R. P. Butler and A. Strasheim, *Spectrochim. Acta,* 1965, **21**, 1207.
41. C. Sebens, J. Vollmer and W. Slavin, *Atomic Absorption Newsletter,* 1964, **3**, 165.
42. D. C. Manning, D. Trent and J. Vollmer, ibid., 1965, **4**, 234.
43. J. Vollmer, C. Sebens and W. Slavin, ibid., 1965, **4**, 306.
44. F. J. Fernandez, D. C. Manning and J. Vollmer, ibid., 1969, **8**, 117.
45. W. F. Meggers and F. O. Westfall, *J. Research Nat. Bur. Standards,* 1950, **44**, 447.
46. E. Jacobsen and G. R. Harrison, *J. Opt. Soc. Am.,* 1949, **39**, 1054.
47. A. T. Forrester, R. A. Gudmundsen and P. O. Johnson, ibid., 1956, **46**, 339.
48. W. E. Bell, A. L. Bloom and J. Lynch, *Rev. Sci. Instr.,* 1961, **32**, 668.
49. R. J. Atkinson, G. D. Chapman and L. Krause, *J. Opt. Soc. Am.,* 1965, **55**, 1269.
50. K. M. Aldous, R. M. Dagnall and T. S. West, *Anal. Chim. Acta,* 1969, **44**, 457.
51. R. G. Brewer, *Rev. Sci. Instr.,* 1961, **32**, 1356.
52. R. A. Franz, *Rev. Sci. Instr.,* 1963, **34**, 589.
53. K. E. Zacha, M. P. Bratzel, J. D. Winefordner and J. M. Mansfield, *Anal. Chem.,* 1968, **40**, 1733.
54. K. M. Aldous, D. Alger, R. M. Dagnall and T. S. West, *Laboratory Practice,* 1970, **19**, 587.
55. W. S. Gleason and R. Pertel, *Rev. Sci. Instr.,* 1971, **42**, 1638.
56. C. H. Corliss, W. R. Bozman and F. O. Westfall, *J. Opt. Soc. Am.,* 1953, **43**, 398.
57. F. S. Tomkins and M. Fred, *J. Opt. Soc. Am.,* 1957, **47**, 1087.
58. R. M. Dagnall and T. S. West, *Applied Optics,* 1968, **7**, 1287.
59. E. F. Worden, R. G. Gutmacher and J. G. Conway, *Applied Optics,* 1963, **2**, 707.
60. M. D. Silvester and W. J. McCarthy, *Spectrochim. Acta.* 1970, **25B**, 229.
61. J. M. Mansfield, M. P. Bratzel, H. O. Norgordon, D. N. Knapp, K. E. Zacha and J. D. Winefordner, *Spectrochim. Acta,* 1968, **23B**, 389.

62. D. O. Cooke, R. M. Dagnall and T. S. West, *Anal. Chim. Acta,* 1971, **54**, 381.
63. N. P. Ivanov, L. V. Minervina, S. V. Baranov, L. G. Pofralidi and I. I. Olikov, *J. Anal. Chem. (USSR),* 1966, **21**, 998.
64. T. T. Kikuchi, *Applied Optics,* 1972, **11**, 687.
65. J. R. Brandenberger, *Rev. Sci. Instr.,* 1971, **42**, 1535.
66. R. F. Browner, R. M. Dagnall and T. S. West, *Anal. Chim. Acta,* 1969, **45**, 163.
67. L. F. Phillips, *Rev. Sci. Instr.,* 1971, **42**, 1078.
68. F. C. Fehsenfeld, K. M. Evenson and H. P. Broida, *Rev. Sci. Instr.,* 1965, **36**, 294.
69. W. G. Schrenk, S. E. Valente and K. E. Smith, *Applied Spectros.,* 1972, **26**, 108.
70. B. McCarroll, *Rev. Sci. Instr.,* 1970, **41**, 279.
71. F. M. Hamm, T. L. Martin and P. B. Zeeman, *Anal. Chem.,* 1970, **43**, 490.
72. B. J. Russell, J. P. Shelton and A. Walsh, *Spectrochim. Acta,* 1957, **8**, 317.
73. J. D. Winefordner and R. A. Staab, *Anal. Chem.,* 1964, **36**, 165.
74. *Idem,* ibid., 1964, **36**, 1367.
75. J. M. Mansfield, J. D. Winefordner and C. Veillon, ibid., 1965, **37**, 1049.
76. G. I. Goodfellow, *Anal. Chim. Acta,* 1966, **35**, 132.
77. R. M. Dagnall, T. S. West and P. Young, *Talanta,* 1966, **13**, 803.
78. R. M. Dagnall, K. C. Thompson and T. S. West, *Anal. Chim. Acta,* 1966, **36**, 269.
79. R. S. Hobbs, G. F. Kirkbright, M. Sargent and T. S. West, *Talanta,* 1968, **15**, 997.
80. N. Omenetto and G. Rossi, *Spectrochim. Acta,* 1969, **24B**, 95.
81. T. J. Vickers and R. M. Vaught, *Anal. Chem.,* 1969, **41**, 1477.
82. P. D. Warr, *Talanta,* 1970, **17**, 543.
83. D. R. Jenkins, *Spectrochim. Acta,* 1970, **25B**, 47.
84. W. Slavin, D. J. Trent and S. Sprague, *Atomic Absorption Newsletter,* 1965, **4**, 180.
85. D. C. Manning, D. J. Trent, J. Sprague and W. Slavin, ibid., 1965, **4**, 255.
86. D. J. Trent, D. C. Manning and W. Slavin, ibid., 1965, **4**, 335.
87. H. Prugger, *Spectrochim. Acta,* 1969, **24B**, 197.
88. V. A. Fassel and V. G. Mossotti, *Anal. Chem.,* 1963, **35**, 252.
89. N. P. Ivanov and N. A. Kozyreva, *Zh. Analit. Khim.,* 1964, **19**, 1266.
90. V. L. Ginsberg and G. P. Satarina, *Zavodsk. Lab.,* 1965, **31**, 249.
91. N. P. Ivanov and N. A. Kozyreva, *Zavodsk. Lab.,* 1965, **31**, 566.
92. V. G. Mossotti and V. A. Fassel, *Spectrochim. Acta,* 1964, **20**, 1117.
93. V. A. Fassel, V. G. Mossotti, W. E. L. Grossman and R. N. Kniseley, ibid., 1966, **22**, 347.
94. M. Margoshes and M. M. Darr, *NBS Tech. Note,* 1965, **272**, 18.
95. L. de Galan, W. W. McGee and J. D. Winefordner, *Anal. Chim. Acta,* 1967, **37**, 436.
96. C. W. Frank, W. G. Schrenk and C. E. Moloan, *Anal. Chem.,* 1967, **39**, 534.
97. R. M. Dagnall, K. C. Thompson and T. S. West, *Anal. Chim. Acta,* 1966, **36**, 269.
98. D. W. Ellis and D. R. Demers, *Anal. Chem.,* 1966, **38**, 1943.
99. C. Veillon, J. M. Mansfield, M. L. Parsons and J. D. Winefordner, *Anal. Chem.,* 1966, **38**, 204.
100. D. L. Manning and P. Heneage, *Atomic Absorption Newsletter,* 1968, **7**, 80.

101. M. S. Cresser and T. S. West, *Anal. Chim. Acta*, 1970, **51**, 530.
102. *Idem, Spectrochim. Acta*, 1970, **25B**, 61.
103. D. H. Cotton and D. R. Jenkins, ibid., 1970, **25B**, 283.
104. M. P. Bratzel, R. M. Dagnall and J. D. Winefordner, *Anal. Chim. Acta*, 1970, **52**, 157.
105. A. V. Sheklein and V. A. Popov, *Zh. Nauchn. i. Prikl. Fotogr. i. Kinematogr.*, 1964, **9**, 192.
106. L. Klein, *Applied Optics*, 1968, **7**, 677.
107. C. F. Dewey, "Excitation of gases using wavelength-tunable lasers", in "Modern Optical Methods in Gas Dynamic Research", p. 221, ed. D. S. Dosanjh, Plenum Press, 1971.
108. H. R. Schlossberg and P. L. Kelley, *Physics Today*, July 1972, **25**, 36.
109. M. R. Kagan, G. I. Farmer and B. G. Huth, *Laser Focus*, September 1968, p. 26.
110. D. J. Bradley, "Recent developments in dye lasers and their applications"; paper presented at "Electro-Optics '71", Brighton, England, March 1971.
111. J. P. Webb, *Anal. Chem.*, May 1972, **44**, 30A.
112. J. T. Worden and L. Gough, *Appl. Phys. Letters*, 1971, **19**, 345.
113. D. G. Peterson, S. A. Tuccio and B. B. Snavely, *Appl. Phys. Letters*, 1970, **17**, 245.
114. A. Dienes, E. P. Ippen and C. G. Shank, *IEEE J. Quantum Electronics*, 1972, **8**, 388.
115. D. J. Bradley, A. J. F. Durrant, G. M. Gale, M. Moore and P. D. Smith, *IEEE J. Quantum Electronics*, 1968, **4**, 707.
116. T. W. Hänsch, *Applied Optics*, 1972, **11**, 895.
117. T. W. Hänsch, A. Shahin and A. L. Schawlow, *Phys. Rev. Letters*, 1971, **27**, 707.
118. B. Bölger and C. H. Weysenfeld, "Locking of organic dye laser frequency to atomic resonance lines"; paper presented at VII Inter. Quantum Electronics Conf., Montreal, May 1972.
119. W. F. Hagen, *Industrial Research*, 1972, **14**, 48.
120. B. G. Huth, G. I. Farmer, L. M. Taylor and M. R. Kagan, *Spectroscopy Letters*, 1968, **1**, 425.
121. D. J. Bradley, J. V. Nicholas and J. R. D. Shaw, *Appl. Phys. Letters*, 1971, **19**, 172.
122. C. Gabel and M. Hercher, "A continuous tunable source of coherent uv radiation"; paper presented at VII Inter. Quantum Electronics Conf., Montreal, May 1972.
123. J. F. Young, G. C. Bjorklund, A. H. Kung, R. B. Miles and S. E. Harris, *Phys. Rev. Letters*, 1971, **27**, 1551.
124. S. E. Harris and R. B. Miles, *Appl. Phys. Letters*, 1971, **19**, 385.
125. G. C. Bjorklund, D. S. Carson, S. E. Harris, A. H. Kung, R. B. Miles and J. F. Young, "Generation of ultraviolet radiation in phase-matched metal vapours"; paper presented at VII Inter. Quantum Electronics Conf., Montreal, May 1972.
126. J. Kuhl, G. Marowsley and R. Torge, *Anal. Chem.*, 1972, **44**, 375.
127. M. R. Bowman, A. J. Gibson and M. C. W. Sandford, *Nature*, 1969, **221**, 456.
128. M. C. W. Sandford and A. J. Gibson, *J. Atm. Terr. Phys.*, 1970, **32**, 1423.
129. M. B. Denton and H. V. Malmstadt, *Appl. Phys. Letters*, 1971, **18**, 485.
130. L. M. Fraser and J. D. Winefordner, *Anal. Chem.*, 1971, **43**, 1693.
131. *Idem.*, ibid., 1972, **44**, 1445.
132. A. Strasheim, *Nature*, 1964, **196**, 1194.

133. A. Strasheim and H. G. C. Human, *Spectrochim. Acta,* 1968, **23B**, 265.
134. H. G. C. Human, L. R. P. Butler and A. Strasheim, *Analyst,* 1969, **94**, 81.
135. *Idem,* paper presented at 15th Colloq. Spect. Intern, Madrid, 1969.
136. S. Greenfield, P. B. Smith, A. E. Breeze and N. M. D. Chilton, *Anal. Chim. Acta,* 1968, **41**, 385.
137. C. T. J. Alkemade and J. M. W. Milatz, *J. Opt. Soc. Amer.,* 1955, **45**, 583.
138. D. C. Manning and W. Slavin, *Atomic Absorption Newsletter,* No. 8, November 1962.
139. R. K. Skogerboe and R. A. Woodriff, *Anal. Chem.,* 1963, **35**, 1977.
140. C. S. Rann, *Spectrochim. Acta,* 1968, **23B**, 827.
141. G. F. Kirkbright and S. J. Wilson, *Analyst,* 1970, **95**, 833.

CHAPTER 6

Flames

6.1 INTRODUCTION

This chapter is concerned with the use of a flame as the atom cell in analytical atomic absorption and fluorescence spectrometers. The flame is, at present, by far the most widely used method of obtaining a population of sample (analyte) atoms in commercial atomic absorption instruments and will be discussed here in some detail. Non-flame cells, which are gradually increasing in importance, are described in Chapter 7, while Chapter 8 deals with nebulization devices, which are used almost exclusively to introduce sample solutions into the flame as a fine aerosol. In the present chapter our only concern with samples will be the process by which a population of free analyte atoms is obtained from the aerosol introduced into the flame. This topic is discussed in Section 6.5. Prior to this we shall deal briefly with the general properties and characteristics of flames, followed by a more detailed survey of the particular flames that have been most widely used in AAS and AFS.

An exact description of the nature of a flame is difficult; it may, however, be considered as the gaseous product of highly exothermic reactions, usually having a relatively high temperature and emitting radiation. We shall be concerned only with flames in which all the reactants are in gaseous form and in which one of the reactants is always oxygen. This statement requires some further explanation. Thus, the restriction to gaseous reactants excludes flames such as those obtained from finely divided solids or oil sprays but not those obtained from the use of liquid or solid materials which are vaporized prior to their combustion at a burner. Similarly, the requirement of oxygen as a reactant precludes, for example, flames of fluorine with hydrogen or hydrocarbons, but not the use of oxidants such as nitrous oxide. The reacting gases may also be diluted by the presence of an inert gas such as nitrogen or the rare gases. Although these diluents take no direct part in the chemical reactions they greatly influence the characteristics of the flame obtained as a result of the reactions. Finally, all of the flames discussed are of the *stationary* type obtained in a stable form at a burner, as opposed to those *propagating* through a tube or combustion chamber, or due to an explosion. As a result, the correct choice of burner is almost as important as the selection of the reactant gases and is considered during the description of each particular flame.

A wide range of chemical substances may be used to obtain a flame, even within the restrictions made above, but only relatively few have been utilized for flame spectroscopy and hence for AAS and AFS applications. Of these, the most important are:

Fuels—hydrogen, coal gas, propane, acetylene and butane.
Oxidants—air, oxygen, oxygen + inert gas and nitrous oxide.

The properties of the flames obtained by reaction of combinations of these gases depend not only on the basic gases but also on the way in which the combustion occurs. Thus it is necessary before proceeding further to consider a notation by which the nature of a particular flame can conveniently be described.

6.1.1 Classification of Flames

All flames of the type of interest here involve a *flow* of gases, both in a cool state prior to combustion at the burner and in the hot, burning gases moving through the flame itself. Thus, a basic classification of flame types may be made in terms of the state of flow; this may be described as either *laminar* or *turbulent.*

In a turbulent flow of gas, it is possible to identify specific regions (called turbulence balls) in which all flow is in a particular direction, but the regions themselves have random velocities and may collide with each other and sometimes mix. Thus, flow occurs in directions other than the average flow direction of the gas as a whole, whereas with laminar flow all volume elements move parallel to each other. In laminar flow, transport phenomena within the gas may be described in terms of molecular motion (e.g. diffusion), but under turbulent conditions the transport is due to the motion of the turbulence balls.

For analytical applications the flames used are relatively small and the region of interest is usually fairly close to the burner. Thus, a laminar flame can be obtained only if the burner design produces a laminar flow in the unburnt gases leaving it. The converse is not necessarily true and a turbulent flame may be obtained even when the burner has produced laminar gas flow. This is particularly so if the flow is close to the borderline of turbulence when any relatively large fluctuations in the gas may cause the flame to oscillate between the laminar and turbulent forms. Although turbulent flames have been used in analytical spectroscopy (see Section 6.4.2) they provide much less well-defined conditions for atomization than a laminar flame and are much more noisy (acoustically as well as with respect to the signals being measured). Hence a majority of the flames described here are of the laminar type. The conditions necessary to obtain laminar flow at the burner are discussed later.

A further division of flame types may be made according to whether the gases are *premixed* before leaving the burner or whether some, or all, of the gases are allowed to mix only on reaching the point of ignition. For convenience, this latter type will sometimes be referred to as *unpremixed* flames, but in practice only two distinct forms are likely to be encountered in AAS and AFS applications. The most obvious is the *diffusion flame* in which only the fuel gas leaves the burner, subsequently mixing with the surrounding air which acts as the oxidant to support a flame. A more versatile modification involves the supply of both fuel and oxidant via a burner designed so that mixing can only occur by diffusion at the flame itself. A further development of this concept gives the second type of unpremixed flame in which the burner outlets are designed to give forced mixing of the gases and result in a turbulent flame. It will be seen subsequently that in many cases the logical choice of one particular classification for an actual flame may appear difficult. For example, under almost all conditions a premixed flame burning in air is actually surrounded by a diffusion flame in which partly combusted flame gases burn further with the air as oxidant (see Section 6.2.2), but the complete flame is invariably referred to as premixed.

In view of the overlap encountered in practice between the fundamental groups described above, it is useful to add here a further classification based simply on the flames likely to be encountered in the laboratory and to mention the most important characteristics of the flames in each group. The notation used was chosen because it provides an indication of the nature of the flames and because the flames thus described will be readily recognized by those concerned with the analytical applications of flame spectroscopy.

(a) *Premixed laminar flames* are obtained by mixing the fuel and oxidant gases at the base of the burner or in a special mixing chamber and ensuring that the correct mixture leaves the top of the burner with laminar flow. They produce a high temperature with relatively low background emission and noise; a wide variety of flame shapes is possible for different applications, and the temperature and chemical composition of the different flames are well documented and easily reproduced. There is the possibility of an explosive "flash-back" of the flame into the burner but this may be minimized and most gas mixtures burn safely on a properly designed burner (see Section 6.3.1). The majority of flames currently used in analytical AAS and AFS are of this type.

(b) *Diffusion flames in air* have either only a fuel gas or a fuel gas mixed with inert diluent (usually nitrogen or argon) supplied to the burner. This is combusted at the burner outlet as it comes into contact with the surrounding air. As the only mixing with oxidant occurs by diffusion, the flames are fuel-rich and hydrocarbon gases produce very luminous and sooty flames. Hence, hydrogen is the most commonly encountered fuel for spectroscopic applications. The flames are relatively cool and give exceptionally low background emission; they were therefore widely used in early atomic fluorescence studies. They have the disadvantage of always experiencing some mixing with the nitrogen of the atmosphere which may act as an inert diluent (important from the point of view of fluorescence quenching). Although nitrogen-hydrogen and argon-hydrogen flames are employed in AAS work, premixed hydrogen-oxygen + inert gas flames are now more commonly employed in AFS work.

(c) *Diffusion flames with additional oxidant* have both fuel and oxidant supplied via a burner whose outlets are shaped so that nearly all mixing occurs by diffusion above the burner head. The flames are similar to type (b) except that it is possible to vary the type and quantity of oxidant, which may also be used to nebulize solutions.

(d) *Unpremixed turbulent flames* are a logical development of type (c); the fuel and oxidant are still supplied separately to the burner head but the outlet jets are shaped to give a forced mixing of the gases. This produces a much hotter flame than can be obtained by diffusion alone but like ordinary diffusion flames still ensures complete freedom from the possibility of flash-back. For this reason such flames were once popular for flame spectroscopy. They exhibit many disadvantages, however, such as high noise levels, lack of distinct flame zones and uniform areas of known characteristics, high background emission, and difficulties of obtaining a flame shape suitable for absorption measurements.

(e) *Premixed turbulent flames* have all the disadvantages of type (d) together with the possibility of flash-back; they have seen little use.

6.1.2 Requirements of Flames used in Atomic Absorption and Fluorescence Measurements

A number of general requirements should be considered when a flame atom cell is to be used in AAS or AFS. Some of these are a matter of common sense, while others are more concerned with the burner and its ancillary equipment than with the actual choice of flame. Thus, safety is an important factor, especially with commercial instruments that may be used by an operator whose experience of flames is limited to use of a laboratory Bunsen burner. As a result the gases to be used should be restricted to those that remain reasonably safe during handling and storage. Even if the possibility of flame flash-back cannot be completely eliminated (owing to the use of premixed flames), the burner design should minimize this risk and also its possible consequences. For some applications the choice of flame may be restricted by the cost and availability of compressed gases in certain areas or by industrial or legal restrictions. For example, the use of compressed hydrogen may be prohibited in some buildings.

Whichever flame is chosen as the atom cell for AAS or AFS, it must meet a number of general operating requirements. Thus it is important that the characteristics of the flame should remain stable during use and be reproducible from day to day. Similarly, in most applications good precision requires stable, noise-free signals accompanied by reasonably low levels of background radiation and noise from the flame itself. It is also important that in each application the flame gives sufficient breakdown of the sample, both to produce an adequate population of free analyte atoms for good sensitivity and also to prevent interference effects resulting from the nature of the sample matrix. The ability of the flame to accept a variety of sample solvents may also be important in certain applications. For example, cool diffusion flames may be unsuitable for nebulization of certain organic solvents owing to the formation of excessive carbon in the flame.

Certain specific requirements of AAS or AFS must also be taken into account. For example, a major factor in obtaining sensitive fluorescence measurements is the choice of a flame containing low populations of those species that most readily quench excited metal atoms before their excess energy can be re-emitted as a fluorescence signal. Although such a choice may seem relatively simple, the situation becomes more complicated if, for example, the same flame must also have a high temperature and the ability to break down stable oxides. Similarly, no matter what the other requirements, a flame used as a fluorescence cell must permit minimal scattering of the incident source radiation as in many applications this will be indistinguishable from the fluorescence signal.

Absorption measurements are less exacting from this point of view, but may impose other specialized requirements upon the flame employed. For example, many widely used optical systems are such that good absorption sensitivity can only be obtained if the conventional circular flame is replaced by a long, flat flame. The flame chosen must then be capable of stable, safe combustion on a burner of this form. Absorption by the flame species can also be an important consideration. Some common flames absorb strongly in certain spectral regions used for absorption measurements, and this can limit precision, either because the spectral source is not sufficiently intense to make this loss of light negligible or because the flame absorption actually contributes noise to the absorption signal.

6.2 PROPERTIES OF PREMIXED LAMINAR FLAMES

This section is intended to give only a brief discussion of flame properties, sufficient for a proper understanding of the selection of flames and burners for analytical applications. For more detailed treatment the reader is referred to a number of textbooks, such as those by Gaydon and Wolfhard,[1] Lewis and Von Elbe,[2] Fristrom and Westenberg,[3] Gaydon,[4] Fenimore[5] and Mavrodineanu and Boiteux.[6] References are given to detailed sources of information on specific points. Although much of this section is not applicable to unpremixed flames (for example, limits of inflammability and burning velocity have little meaning in analytically useful diffusion flames) some of the information given on sources of flame radiation and methods of temperature measurement is also relevant to the other flame types and will not be repeated in Section 6.4.

6.2.1 Combustion Characteristics of Premixed Flames

The propagation of a stable premixed flame using a particular combination of gases and burner depends basically on three factors: the limits of inflammability, the ignition temperature and the burning velocity. Typical values of these properties for some common flame gas mixtures are given in Table 6-1 which, for convenience, also includes the stoicheiometric reactions and maximum flame temperatures. The data should only be used as an approximate guide as large variations occur in the values reported in the literature for some of the properties, often because of varying experimental conditions or different measurement methods.

If combustion is initiated at a particular point in a mixture of fuel and oxidant, it will spread to the bulk of the mixture and form a flame only if the percentage of fuel in the mixture falls between certain limits. These are known as the *limits of inflammability*. They vary greatly for any given combination of gases with experimental factors such as the unburnt gas temperature, the pressure and the combustion equipment used. Flames used in analytical spectroscopy are mostly well within the limits, but they may become more significant in applications where particularly fuel-lean or fuel-rich flames are needed in order to obtain optimum sensitivity or freedom from interferences (see Section 12.2). For each gas mixture within the limits of inflammability there is a minimum temperature at which spontaneous propagation of combustion through the mixture

TABLE 6-1

Combustion characteristics of some premixed laminar flames at atmospheric pressure

Fuel	Oxidant	Stoicheiometric reaction	Energy released (kcal)	Limits of inflammability at room temp. (vol %)			Minimum ignition temp.		Maximum burning vel.		Maximum flame temp.			
											Calculated		Experimental	
				Lower	Upper	Notes	(K)	Notes	cm/s	Notes	K	Notes	K	Notes
Hydrogen	Air	$H_2 + \frac{1}{2}O_2 + 2N_2 \rightarrow H_2O + 2N_2$	57·8	4	75	*a*	803	*b*	310	*c*	2373	*a*	2318	*d*
	O_2	$H_2 + \frac{1}{2}O_2 \rightarrow H_2O$	57·8	4	94	*a*	723	*b*	1400	*c*	3083	*a*	2933	*d*
	$(O_2 + Ar)^m$	$H_2 + \frac{1}{2}O_2 + 2Ar \rightarrow H_2O + 2Ar$	57·8						335	*e*			2250	*h*
	$(O_2 + He)^m$	$H_2 \frac{1}{2}O_2 + 2He \rightarrow H_2O + 2He$	57·8						518	*e*				
	N_2O	$H_2 + N_2O \rightarrow H_2O + N_2$	77·5						390	*d*	2920	*i*	2880	*l*
	NO	$H_2 + NO \rightarrow H_2O + \frac{1}{2}N_2$	79·4				1593	*a*	55	*a*	3113	*a*	3093	*d*
	NO_2	$H_2 + \frac{1}{2}NO \rightarrow H_2O + \frac{1}{2}N_2$	61·8						200	*a*	2933	*d*	1823	*d*
Acetylene	Air	$C_2H_2 + 5/2O_2 + 10N_2 \rightarrow 2CO_2 + H_2O + 10N_2$	300·1	2·5	80	*j*	623	*b*	158	*c*	2523	*j*	2500	*j*
	O_2	$C_2H_2 + 5/2O_2 \rightarrow 2CO_2 + H_2O$	300·1	2·0	95	*j*	608	*b*	1140	*c*	3341	*j*	3160	*j*
	$(O_2 + Ar)^m$	$C_2H_2 + 5/2O_2 + 10Ar \rightarrow 2CO_2 + H_2O + 10Ar$	300·1						228	*f*				
	$(O_2 + He)^m$	$C_2H_2 + 5/2O_2 + 10He \rightarrow 2CO_2 + H_2O + 10He$	300·1						336	*f*				
	N_2O	$C_2H_2 + 5N_2O \rightarrow 2CO_2 + H_2O + 5N_2$	401·5	2·2	67	*j*			160	*d*	3152	*a*	2990	*k*
	NO	$C_2H_2 + 5NO \rightarrow 2CO_2 + H_2O + 5/2N_2$	408·1						87	*d*	3363	*d*	3368	*d*
	NO_2	$C_2H_2 + 5/2NO_2 \rightarrow 2CO_2 + H_2O + 5/4N_2$	320·2						135	*d*				
Methane	Air	$CH_4 + 2O_2 + 8N_2 \rightarrow CO_2 + 2H_2 + 8N_2$	191·8	5·3	15	*a*	918	*b*	45	*c*	2222	*a*	2148	*b*
	O_2	$CH_4 + 2O_2 \rightarrow CO_2 + 2H_2O$	191·8	5·1	61	*a*	918	*b*	450	*c*	3010	*a*	2950	*b*
	$(O_2 + Ar)^m$	$CH_4 + 2O_2 + 8Ar \rightarrow CO_2 + 2H_2O + 8Ar$	191·8						96	*g*	2531	*g*		
	$(O_2 + He)^m$	$CH_4 + 2O_2 + 8He \rightarrow CO_2 + 2H_2O + 8He$	191·8						130	*g*	2253	*g*		
Town gas	Air	Town gas + $0{\cdot}98O_2 + 3{\cdot}9N_2 \rightarrow CO_2 + H_2 + 3{\cdot}9N_2$	108·8	9·8	24·8	*b*	560	*b*	55	*l*	2113	*l*	1980	*l*
	O_2	Town gas + $0{\cdot}98O_2 \rightarrow CO_2 + H_2O$	108·8	10·0	73·6	*b*	450	*b*			3073	*l*	3013	*l*
Propane	Air	$C_3H_8 + 5O_2 + 20N_2 \rightarrow 3CO_2 + 4H_2 + 20N_2$	530·6	2·1	9·4	*b*	510	*b*	82	*l*	2198	*l*	2198	*l*
	O_2	$C_3H_8 + 5O_2 \rightarrow 3CO_2 + 4H_2O$	530·6				490	*b*			3123	*l*	3123	*l*
Butane	Air	$C_4H_{10} + 13/2O_2 + 26N_2 \rightarrow 4CO_2 + 5H_2O + 26N_2$	687·9	1·9	8·4	*b*	490	*b*	82·6	*l*	2203	*l*	2168	*l*
	O_2	$C_4H_{10} + 13/2O_2 \rightarrow 4CO_2 + 5H_2O$	687·9				460	*b*			3183	*l*	3173	*l*
Ethane	Air	$C_2H_6 + 7/2O_2 + 14N_2 \rightarrow 2CO_2 + 3H_2O + 14N_2$	378·8	3·0	12·5	*b*	530	*b*	85·6	*b*				
	O_2	$C_2H_6 + 7/2O_2 \rightarrow 2CO_2 + 3H_2O$	378·8	4·1	50·5	*b*	500	*b*						
Cyanogen	Air	$(CN)_2 + O_2 + 4N_2 \rightarrow 2CO + 5N_2$	126·7	7·6	38·0	*b*			90				2603	*b*
	O_2	$(CN)_2 + O_2 \rightarrow 2CO + N_2$	126·7						270	*d*	4813	*d*	4640	*d*

[a] A. G. Gaydon and H. G. Wolfhard, "Flames, Their Structures, Radiation and Temperature", Chapman and Hall, London, 3rd ed., 1970.
[b] R. Mavrodineanu, *Spectrochim. Acta*, 1961, 17, 1016.
[c] G. E. Andrews and D. Bradley, *Combustion and Flame*, 1972, **18**, 133.
[d] J. B. Willis, *Applied Optics*, 1968, 7, 1295.
[e] G. H. Morgan and W. R. Kane, *Proc. 4th Intern. Symp. on Combustion, 1952*, p. 313, Williams, Wilkins and Co., Baltimore, 1953.
[f] C. E. Mellish and J. W. Linnett, ibid., p. 407.
[g] W. H. Clingman, R. S. Brokaw and R. N. Pease, ibid., p. 310.
[h] R. F. Browner and J. D. Winefordner, *Anal. Chem.*, 1972, **44**, 247.
[i] J. B. Willis, V. A. Fassel and J. A. Fiorino, *Spectrochim Acta*, 1969, **24B**, 157.
[j] A. Williams and D. B. Smith, *Chem. Revs.*, 1970, **70**, 267.
[k] J. B. Willis, J. O. Rasmuson, R. N. Kniseley and V. A. Fassel, *Spectrochim. Acta*, 1968, **23B**, 725.
[l] R. Mavrodineanu and H. Boiteux, "Flame Spectroscopy", Wiley, New York, 1965.
[m] Mixture equivalent to air.

can be initiated. This is known as the *ignition temperature.* For obvious reasons, the combinations of gases used in flame spectroscopy have ignition temperatures well above room temperature at all mixture strengths within their limits of inflammability. Thus, the flame can be initiated only by supplying heat energy in the form of another flame, a spark, etc. Although the ignition temperature may, in some cases, vary with mixture strength by as much as several hundred degrees, this is unlikely to present any problems and generally any mixture that will support a flame can be ignited simply by applying a match or lighted taper.

Once a combustible gas mixture has been ignited, a flame results from propagation of the reaction by thermal conductivity between adjacent layers of the mixture. The rate at which this occurs is known as the *burning velocity* and, as with the other factors, it varies greatly with experimental conditions. The values shown in Table 6-1 (mostly maximum values obtained at or near the stoicheiometric proportions) illustrate the wide range of burning velocities obtained with different gas mixtures. For each combination of gases there is also a large variation of burning velocity with the mixture strength. This is illustrated in Fig. 6-1 for methane flames with oxygen/inert gas mixtures. The burning velocity is, from the analytical point of view, the most important of the three factors because it has a direct effect on the range of premixed gas mixtures that can be burnt safely and stably at a particular burner. If a stable, stationary flame is to be maintained on the burner, the vertical rise velocity of the gases through the burner head must be carefully matched* to the burning velocity (i.e. the speed with which the flame tends to propagate back into the burner). Should the gases travel at too high a velocity, the burning velocity will be unable to maintain a flame at the burner head and the flame will "lift-off" and eventually extinguish (although it is often possible to maintain a rather unstable flame several inches above a burner). Thus, careful consideration of burning velocities is necessary during burner design; a unit working satisfactorily with, say, town-gas/air flames would not be safe with an acetylene-oxygen mixture. Similarly, very lean or rich mixtures of particular gases may burn safely under conditions where the stoicheiometric mixture could be dangerously close to flash-back owing to its much higher burning velocity. In view of the variation in reported burning velocities it is, therefore, important not to use any literature values (including those in Table 6-1) for critical applications without first checking their reliability and the conditions under

* This does not imply a simple equality; see Section 6.3.1.

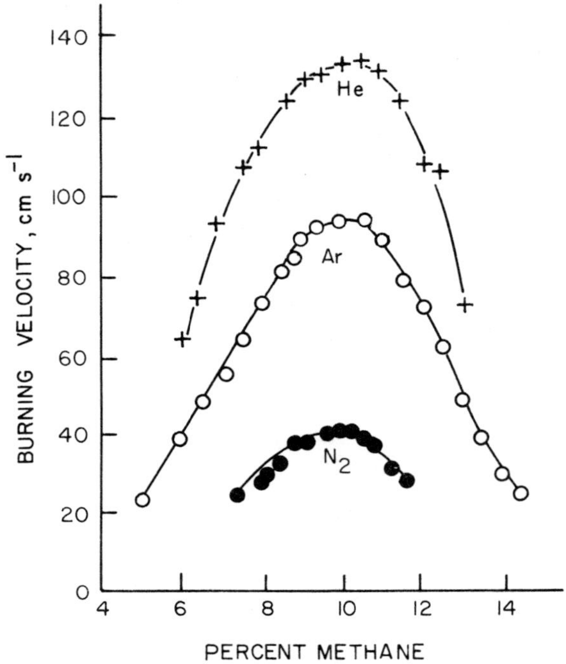

Fig. 6.1. Variation of burning velocity with mixture strength for flames of methane with "airs" in which the inert gas is nitrogen, argon, or helium.[7]

which they were obtained. Andrews and Bradley[8] have discussed in detail this situation and recommended the measurement methods most likely to give reliable values as well as critically reviewing some of the available data. Table 6-1 includes selected values of the burning velocities, but no attempt has been made at complete coverage of the literature and some of the figures are of unknown reliability.

6.2.2 Structure of Premixed Flames

The gases leaving the top of the burner tube enter a *preheating zone* where they are raised to their ignition temperature by the heat of the existing flame. Surrounding this zone is the brightly luminous *primary reaction zone* whose outer surface forms the flame front. As the preheating zone is non-luminous, distinction from the reaction zone is usually difficult by straightforward observation of the flame and the two regions are often referred to as simply the "primary zone". For a circular burner the flame front is approximately conical in shape, this shape being such that the burning velocity always equals the component of the gas velocity perpendicular to its surface; i.e. referring to Fig. 6-2(a), $S = v \cos \alpha$. A major contribution to the final shape of this surface is made by the velocity distribution of the gas mixture as it leaves the burner. As mentioned earlier, laminar flow of this gas is essential to obtain a laminar flame and the dimensions of the burner tube and the flow rates of the gases are selected such that Poiseuille flow is obtained. The velocity is then not uniform but assumes an approximately parabolic distribution, having a maximum along the central axis and falling virtually to zero along the walls of the tube.

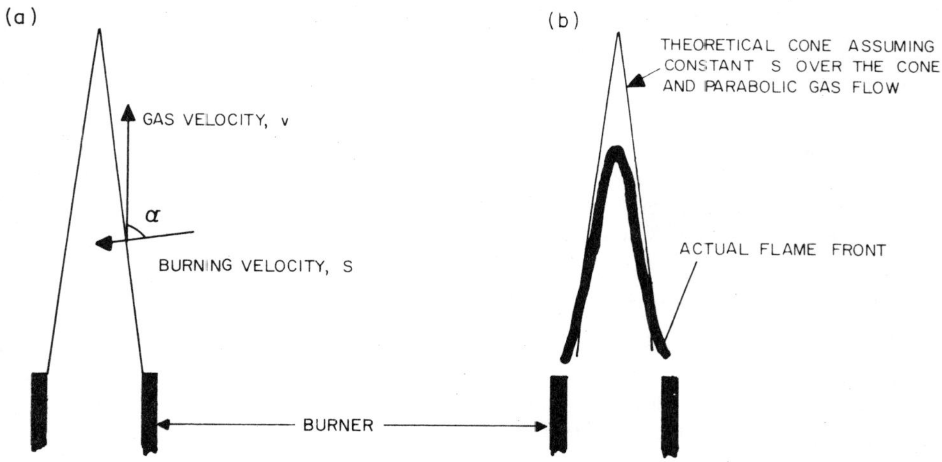

Fig. 6-2. Shape of the flame front for a circular burner. (a) Relation between gas velocity and burning velocity: (b) comparison of the actual and theoretical shapes of a propane-oxygen flame.[1]

In practice the conical shape expected from parabolic flow is distorted, as shown in Fig. 6-2(b), by several factors. Thermal expansion of the gases occurs as they pass through the flame front and this leads to a lower flow rate at the tip, with consequent rounding. Furthermore, the increased radius of curvature of the flame front causes an increase in burning velocity near the tip (owing to increased heat transfer and diffusion of active species) and this produces further rounding of the tip. The heating, expansion and acceleration of the gases at the flame front also causes a pressure difference across it, producing a back-pressure which distorts the parabolic velocity distribution as the gases leave the burner. At the same time the burning velocity is reduced to near zero at the rim of the burner owing to the loss of heat and the chain-carrying radicals at the wall. As a result the conical shape is considerably distorted outwards so that the flame front actually overhangs the burner and a small dead space is observed between the top of the burner and the base of the flame. The back-pressure of the flame is sufficient to prevent diffusion of air through this space and into the cone.

In flames burning at atmospheric pressure the primary reaction zone is extremely thin (usually only a fraction of a millimetre). Thus the residence time of the gases in this region is far too short for the reaction between fuel and oxidant to reach completion or for thermodynamic equilibrium to be established. Thus conditions in the primary reaction zone are not well established and as high noise levels and intense background radiation (see Section 6.2.3) are also present this region is seldom employed in flame spectroscopy. This situation quickly changes as the partially combusted gases and the flame radicals pass from the reaction zone into the region known as the *interconal zone*. This is much broader than the reaction zone and may approach a cylindrical shape owing to expansion and increased rise velocity of the gases as they pass through the flame front. The nature of the interconal zone for particular fuel/oxidant combinations varies considerably with the mixture strength (i.e. fuel: oxidant ratio), as this will affect the composition of gases passing from the primary reaction zone. In a typical hydrocarbon flame the gases will be made up of molecules such as CO, H_2, CO_2, H_2O, O_2 or N_2 and radicals such as H, OH, C_2, CH and CN. As no fresh reactants are added at this stage, equilibrium is usually established fairly quickly via recombination of radicals and dissociation of the combustion products. In most flames presently used in analytical absorption or fluorescence measurements this region of thermal equilibrium begins almost immediately outside the reaction zone and extends for some distance through the flame. There is some evidence for the persistence of greater than equilibrium concentrations of radicals such as H or OH in the interconal zone of some of these flames but this is of no consequence for most analytical applications.* Thus, data are available on the temperature and composition of the interconal zone which permit prediction of the manner in which dissociation, ionization and excitation equilibria will control the behaviour of samples introduced into the flame. As a result, this region of premixed flames is frequently the most useful as the volume to be viewed for AAS or AFS measurements.

* See Section 6.5.3, however.

The hot, partly combusted gases of the interconal zone come into contact with the oxygen of the surrounding air and further reactions can then occur. These reactions give rise to the *secondary reaction zone* which is responsible for the well-known blue outer mantle of hydrocarbon flames with air. The participation of the atmospheric air is a somewhat complex process; both diffusion and entrainment of the air, due to the upward motion of the hot gases, are involved. In many respects the secondary reaction zone is useful because part of the extra energy liberated serves to reduce the rate at which the temperature of the interconal zone would otherwize fall with increasing distance from the reaction zone (see Section 6.2.5). The secondary zone also tends to stabilize the flame on the burner and allows a wider range of gas mixtures to be used with a particular burner than might otherwise be possible. Unfortunately, with many of the flames used in analytical spectroscopy the reactions of the interconal zone are accompanied by strong radiation (see Section 6.2.3) which frequently exhibits an appreciable noise component. Thus, optical measurements of species in the interconal zone may be impeded in certain regions of the spectrum. The secondary zone may be removed in part simply by excluding the surrounding air using the technique of *flame separation.* However, this inevitably results in a reduction of the temperature of the interconal zone and may only be advantageous in certain circumstances (see Section 6.3.2).

6.2.3 Radiation from Flames

Discussion of the radiation from flames involves two approaches. First, we may list the species from which it originates and compare the intensities and types of radiation from different regions of the flame. This is mainly a straightforward matter of quoting recorded information. Secondly, the excitation mechanisms responsible for radiation from different species and different parts of the flame must be interpreted. Here the situation is much more confused and there is a dearth of experimental data and rigorous theoretical interpretation.

The most intense radiation from the flame originates in the primary reaction zone, as might be expected, and is mainly attributable to molecular band spectra. However, atomic lines (due to C and to metal atoms if these are added to the flame) and a continuum background (due to incandescent carbon particles or non-quantized processes) are also observed. The species producing the molecular spectra obviously vary with the gases used, but some of the most common are C_2, CH, CHO, OH, CN, NO and NH.

Radiation from the interconal zone is much less intense owing to the existence of low populations in this region of the flame of many of the species giving strong band spectra. Probably the most intense emission from this region is due to OH. However, there are one or two notable exceptions, especially when fuel-rich flames are used; higher concentrations of carbon-containing species may then persist into the interconal zone. For example, the interconal zone of the fuel-rich nitrous oxide-acetylene flame exhibits a bright red emission owing to the presence of a high concentration of CN radicals. The secondary reaction zone also gives much less intense radiation than the primary zone. The main emission observed in this zone arises from OH radicals whose radiation is usually much more intense than in the interconal zone. The characteristic blue colour of the outer cone of hydrocarbon flames is caused by a continuum arising from the reaction.

$$CO + O \rightarrow CO_2 + h\nu$$

A similar reaction with NO produces the whitish yellow-green colour at the tip of some hydrogen flames. The orange colour of the outer cone of these flames is mainly due to vibration-rotation bands of H_2O (and also of OH).

Radiation from flames may, in the first instance, be separated into two categories—thermal and non-thermal. The latter is often referred to simply as chemiluminescence without further explanation, but this can be misleading; the reader will appreciate that all energy and radiation from flames is essentially chemical in origin.

With thermal excitation, a species is raised to its excited state by virtue of the high temperature of its surroundings before emitting radiation. In other words, energy is transferred between flame species and converted to electronic excitation by collision processes. This energy is that responsible for the heat of the flame and released during the reactions of the fuel and oxidant. It is contained in the flame species by virtue of their motions (both internal and external) and should not be confused with that in those flame species which are actually formed in an excited state. Gaydon and Wolfhard[1] have suggested that energy transfers of this type, which result in electronic excitation, are unlikely to be due to the translational motion of the flame species but more probably occur via the rotations and vibrations of the molecular species.

The total radiation from a given region of the flame can only be called thermal if complete thermal equilibrium is present. (This is discussed in detail in Section 6.2.4.) However, in flame spectroscopy it is more usual to refer to the emission from each individual species. It is then possible to indicate whether a minority of species are giving non-thermal radiation when the rest of the hot gases are, for practical purposes, in thermal equilibrium. By convention, a particular emission is referred to as "non-thermal" if more than a small amount is produced as such, although obviously some part of it (however small) will still be thermal in origin. The best indication of non-thermal origin for a particular emission is to use that radiation to determine a spectroscopic temperature and compare it with the true equilibrium temperature. (This is discussed further in the following sections.)

Excitation of species sufficiently to emit radiation may be produced by many factors other than the high temperature of their surroundings. For example, the intense emission from discharges results from the presence of large numbers of fast-moving electrons. However, in flames by far the most common source of non-thermal radiation is *chemiluminescence.* This arises when the emitting species is formed directly in an excited state as the result of a chemical reaction. For example, chemiluminescence in the flame gases themselves may result from reactions such as

$$CH + O_2 \rightarrow CO + OH^*$$

or

$$O + ClO_2 \rightarrow ClO + O_2^*$$

(* denotes an electronic excited state).

Similar processes may be obtained with metals added to the flame:

$$C + MO \rightarrow CO + M^*$$

or

$$H + H + M \rightarrow H_2 + M^*$$

(This second example is really an induced chemiluminescence, not a true one, since M is not chemically modified by the reaction.) It is also possible that instead of an excited flame species losing its excess energy by radiation as chemiluminescence, it could transfer it to another flame species or a metal atom by collision

$$\text{e.g.} \quad OH^* + M \rightarrow OH + M^*$$

The emission from the metal atom in a case like this is often termed chemiluminescence, but "energy transfer" would be more accurate.

In many cases although non-thermal emission from a flame can be identified as such, it may not be clear whether it is due to true chemiluminescence, induced chemiluminescence or an energy transfer. It is then much more satisfactory to term the radiation "non-thermal" or "overexcitation" rather than "chemiluminescent".

Discussion of the regions of the flame giving thermal and non-thermal radiation will be considered below in the section concerned with equilibria in flames.

6.2.4 Equilibrium in Flames

Three types of equilibrium may be considered in flames: chemical, thermal and radiational. The most significant failure to reach equilibrium will be through incomplete combustion. However, the main chemical reactions involved in flames are normally completed in a very short time and chemical disequilibrium is generally restricted to the reaction zone. As explained above, an exception may occur in some flames when greater than equilibrium concentrations of certain radicals persist well into the interconal zone, but this is of importance in only relatively few examples relevant to analytical applications (see Section 6.5.3). Thus, in many premixed flames it is possible to calculate the flame gas composition, assuming complete equilibrium, and obtain a very good approximation to the actual situation outside the reaction zone. The minor chemical changes which give rise to phenomena such as chemiluminescence will be considered in conjunction with thermal equilibrium as they have a much greater effect on the radiation from the flame gases than on their composition.

When thermal equilibrium exists in a region of the flame, the radiation emitted by all species will be a function of the flame temperature. Conversely, if radiation from the flame is used to measure its temperature the same value will always be obtained. Even if we assume chemical equilibrium, thermal equilibrium does not correspond to a state of complete equilibrium, which would also require radiative equilibrium. The latter is only achieved if all radiation emitted by the flame is replaced by absorption of light. This is only true for a black body and clearly the situation in a flame is usually very far from this. However, Gaydon and Wolfhard[1] have shown that in flames at atmospheric pressure, the flame species experience far more deactivations by collision than by emission of light. Thus, radiation of light has only a very small effect on the thermal equilibrium. (This is probably not true in very low pressure flames where radiation may cause a significant decrease in flame temperature.)

As might be expected, most overexcitation in flames originates from the primary reaction zone, since the greatest release of energy occurs there and thermal equilibrium cannot be expected. Radicals such as OH, CH, C_2 and HCO are well known for their non-thermal emission from this region of the flame. Similarly, overexcitation of metal atoms added to hydrocarbon flames occurs mainly in the reaction zone. At the other extreme, the outer cone, a few centimetres above the reaction zone in premixed flames, is a region where for most purposes thermal equilibrium has been achieved (minor variations do occur, because as discussed below some equilibria, such as ionization, take longer to achieve than others). The observed radiation is then for all practical purposes entirely thermal. Between these two well-defined regions is a transition zone where overexcited particles undergo sufficient collisions to reach equilibrium. Thus, it is often possible to observe some non-thermal radiation from the interconal zone. The extent and height to which it is observed will vary with the flame and the species considered. For example, C_2 and CH chemiluminescence will persist to a greater height in a fuel-rich flame. OH chemiluminescence generally persists much longer than that from metal atoms.

It should be clear from the foregoing that the thermal departures from equilibrium are of the greater concern in flame spectroscopy and that these are of more interest in measurements of emission spectra than in absorption or fluorescence work. Some comments are included here, however, in view of their relevance to problems such as the reliability of temperature measurements.

Thermal equilibrium is best studied by considering the equilibrium of individual flame species; their state of equilibrium is directly dependent on the number of collisions they have undergone since formation in the reaction zone. Equilibrium of this type is indicated by whether the particular energy of a given species is equilibrated according to the Maxwell-Boltzmann Law. In its simplest form this merely states that if a species has possible energy levels E_0, E_1, E_2, etc., then the number of atoms/molecules with these energies will be proportional to

$$e^{-E_0/kT},\ e^{-E_1/kT},\ \text{etc.}$$

Each type of energy requires a different number of collisions to reach this state of equilibrium, since different degrees of freedom are involved. The approximate numbers are:

Translational	10
Rotational	10^3
Vibrational	10^5
Excitation and dissociation	10^7
Ionization	10^9

Thus, if all species start their life in the flame out of equilibrium, translational energy will be the first class to reach equilibrium *within its own class*, i.e. to obey a Maxwell-Boltzmann distribution. This will be followed by rotational energy, and so on, until after sufficient collisions equilibrium is achieved between all classes and the flame gases are in thermal equilibrium. The result of the finite number of collisions which have occurred by any point in the flame is that we may be able to obtain, for example, an OH rotational temperature in the reaction zone because at that point the rotational energies (i.e. the rotational spectral lines) obey the Maxwell-Boltzmann Law. If an attempt were made to obtain the OH vibrational temperature at the same point it might be found that the vibrational energies did not obey the Law. In this case, it would be difficult to solve the equations, graphs, etc. used to calculate the temperature from the vibrational spectral lines. However, the OH rotational temperature that was obtained would bear no relation to the "actual" temperature of the flame–it would be only a function of the reactions forming the OH, the point of measurement, etc.

In the region of the flame where most species have suffered enough collisions to equilibrate all types of energy, it is usually acceptable to speak of "thermal equilibrium" even if this is not strictly true. For example, it may be possible to measure the same temperature using a thermocouple or the spectral lines of a metal atom added to the flame but to obtain a much higher OH rotational temperature. This would mean that some OH chemiluminescence was still being observed owing to a persistent reaction, but the amount of energy involved would not be enough to make any difference to the practical level of equilibrium of metal atoms added to the flame. Similarly, the ionization of, say, potassium added to the flame might not have reached equilibrium because of the high number of collisions required. Thus, a temperature measured using the ionization distribution in the Maxwell-Boltzmann Law might be erroneous.

In most cases, the energy involved would be too small to prevent the flame gases as a whole having an "actual" equilibrium temperature.

6.2.5 Flame Temperature

The usual concepts of temperature and conventional methods of measuring high temperatures are not, in general, applicable to flames. Devices such as thermocouples, resistance thermometers and optical pyrometers can only be used with special precautions and for certain restricted applications. The most widely used, and the most informative, means of measuring flame temperature are spectroscopic, i.e. they utilize radiation emitted or absorbed by discrete particles present in the flame gases or added to them.

It should be clear from the discussion of Sections 6.2.3 and 6.2.4 that unless the whole flame is in thermal equilibrium we cannot expect the temperature to be the same in all regions. Further, unless each region of the flame studied is in complete thermal equilibrium, different spectroscopic methods will indicate different temperatures. Thus, it is possible to talk of translational, rotational, electronic, etc. temperatures. As long as their limitations are realized, such temperatures are still very useful. Indeed, in many cases they provide more information about the flame processes than can be obtained under equilibrium conditions.

Most spectroscopic temperatures are measured by applying the Maxwell-Boltzmann Law to the populations of the energy levels of the species concerned. Thus they are often referred to as "population" temperatures. This applies even to the translational temperature, which depends on the distribution of kinetic energy and ultimately on the Maxwellian distribution of the velocities of the atoms or molecules.

Although a wide variety of spectroscopic techniques have been reported as methods of temperature measurement, relatively few have been used routinely for obtaining data on the "actual" flame temperature, such as the values given in Table 6-1. Perhaps the most well-known is the method of sodium line reversal. Briefly, this involves adding sodium to the flame which is then viewed through a spectroscope against a bright background continuum source. If the latter is operated to be hotter or colder than the flame, the sodium *D* lines will appear respectively either as dark lines in absorption or as bright lines standing out against the background. A reversal point is obtained when both source and flame are at the same temperature. The source may be a black body or one of known emissivity (such as a standard tungsten filament lamp) and is readily calibrated for a range of temperatures using a pyrometer. Thus the flame temperature is found simply by noting the lamp current at which a reversal point is obtained. This technique is discussed in considerable detail by Gaydon and Wolfhard (ref. 1, Chapter 10). It is extremely reliable for low and medium temperature flames (up to about 2600 K) when a tungsten strip-filament lamp can be used as the source, but some difficulty has been experienced in finding equally accurate alternative sources for flames of higher temperature. The problem of finding a suitable source can be avoided by calculating the temperature simply from the relative emission intensities of spectral lines of metal atoms added to the flame.[9,10] This method also involves difficulties, however, notably the problem of obtaining sufficiently accurate values of the transition probabilities of the metal lines and of measuring emission intensities free from errors caused by self-absorption. A technique using relative atomic absorption intensities has been suggested[11] to overcome the latter problem. A recently suggested atomic fluorescence technique[12] avoids both self-absorption and transition probability errors but requires a background light source as for line-reversal measurements. Flame temperatures have also been widely measured using the band spectra of molecular species present in the flame gases themselves; such values are generally referred to as rotational or vibrational temperatures, depending on the type of spectral band used for the measurement. However, measurements made of emission spectra (see, for example, ref. 4, Chapter 8) generally suffer from self-absorption errors, while measurements in absorption[13] present considerable experimental difficulty. In any case, the temperatures obtained are frequently characteristic of the state of equilibrium of the species used rather than giving a true indication of the "actual" flame temperature (see previous sections).

As mentioned previously, it is difficult to visualize an actual temperature for the primary reaction zone of a premixed flame owing to the absence of equilibrium. In most cases, the true maximum temperature of the flame, and the value of most interest in analytical applications, is attained in the interconal zone, a short distance beyond the flame front. The temperature then falls steadily with increasing height in the flame. The rate of this change depends on the type of flame, the mixture strength used and to some extent even the design of the burner. A uniform temperature will persist for the greatest volume in a flame which is shielded by a second, concentric flame. Conversely, a rapid fall of temperature may be observed in flames where the use of the flame separation technique (see Section 6.3.2) removes the secondary reaction zone from around the interconal zone. Similarly, a more rapid fall-off of temperature would be expected in the thin, flat flames commonly used in atomic absorption measurements (see Section 6.3.1) than at the centre of a more conventional circular flame. The effect of separation on the temperature of a nitrous oxide-acetylene flame burning on an atomic absorption burner is illustrated in Fig. 6-3 for stoicheiometric and also fuel-rich and fuel-lean flames. The much more uniform temperature distribution within a properly flame-shielded flame is illustrated in Fig. 6-4 for an air-acetylene flame of the mixture strength giving maximum flame temperature.

The marked effect of the fuel:oxidant ratio on flame temperature is clearly seen from Fig. 6-3. It is generally assumed that the highest flame temperature in the interconal zone is observed with a stoicheiometric mixture. In practice, the maximum is obtained either with richer mixtures if the fuel is highly endothermic (as it releases heat on decomposition) or with leaner mixtures if the fuel is highly exothermic (and absorbs heat on decomposition). Thus, hydrogen flames show their maximum temperature very close to the stoicheiometric point, whereas with acetylene flames the maximum is observed on the rich side. This latter effect is illustrated for an air-acetylene flame in Fig. 6-5. In fact, the relatively small changes in flame temperature with mixture ratio observed in the lower region of the interconal zone have little practical effect on absorption or fluorescence measurements. The accompanying changes in flame composition may, however, be of considerably greater importance, such as with the determination of refractory elements in a nitrous oxide-acetylene flame (see Section 6.3.3).

As the interconal zones of the premixed laminar flames discussed here approximate fairly closely to a state of chemical and thermal equilibrium, it is possible to calculate from basic thermo-chemical data the composition of the flame gases obtained from a particular fuel/oxidant mixture and hence, by making a heat balance, the flame temperature itself. Both aspects of this problem have been discussed in considerable detail by Gaydon and Wolfhard (ref. 1, Chapter 12) who also tabulate some relevant data and results. It should be pointed out that their technique of calculation of composition using a set of flame reactions and equilibrium constants is ideal for manual calculations but may be more conveniently replaced for computer

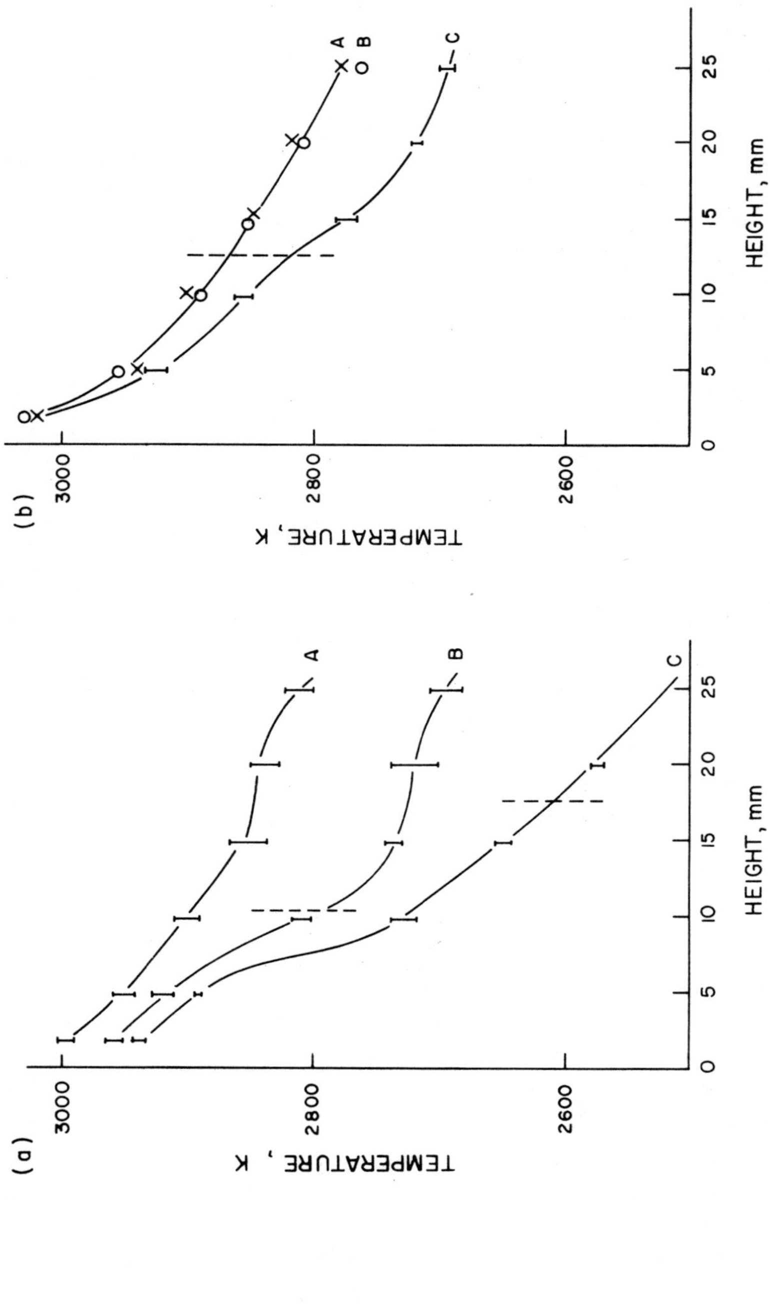
(b)
TEMPERATURE, K
3000
2800
2600
A
B
C
0
5
10
15
20
25
HEIGHT, mm
(a)
TEMPERATURE , K
3000
2800
2600
A
B
C
0
5
10
15
20
25
HEIGHT, mm

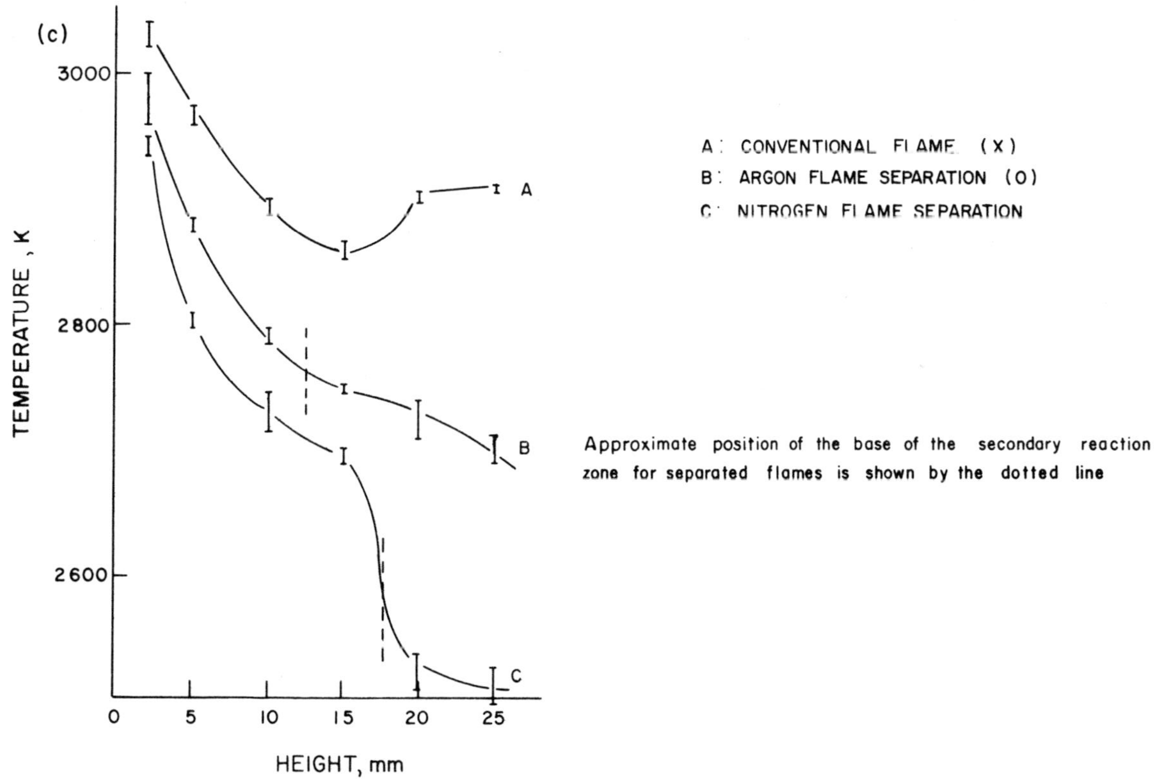

Fig. 6-3. The effect of flame separation on the variation of temperature with height above the burner for the nitrous oxide-acetylene flame.[14] (a) Stoicheiometric flame : $N_2O/C_2H_2 = 2$; (b) fuel-lean flame : $N_2O/C_2H_2 = 3$; (c) fuel-rich flame : $N_2O/C_2H_2 = 1{\cdot}8$.

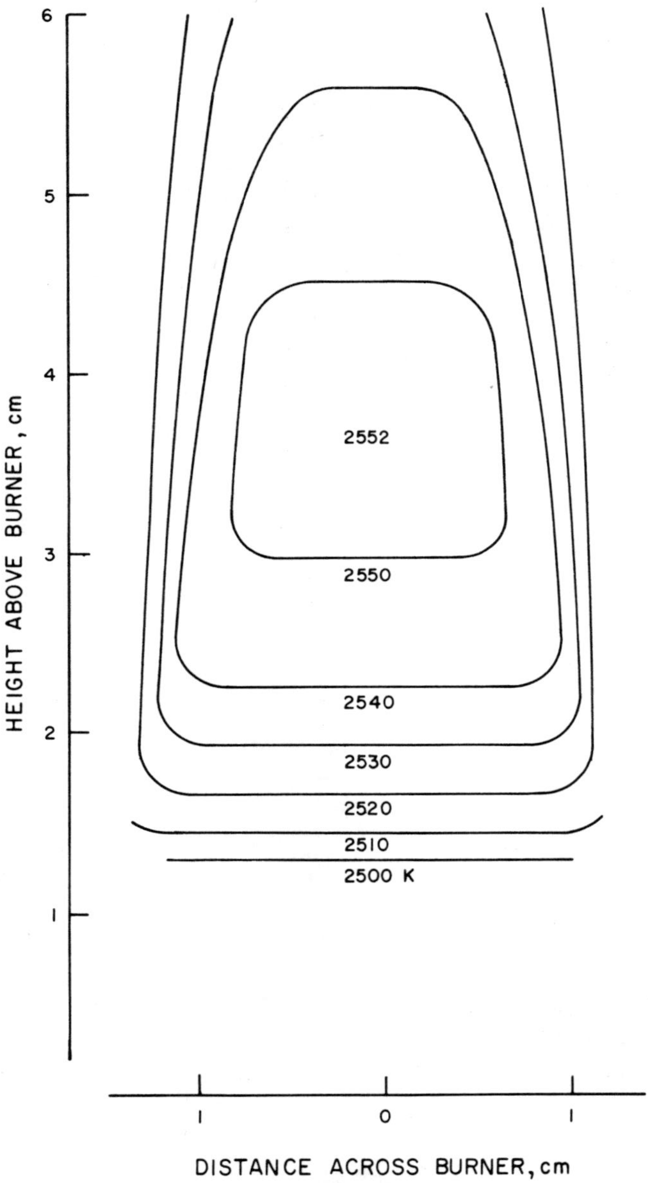

Fig. 6-4. Temperature distribution within a flame-shielded air-acetylene flame.[15] (Air : acetylene = 10·1 : 1.)

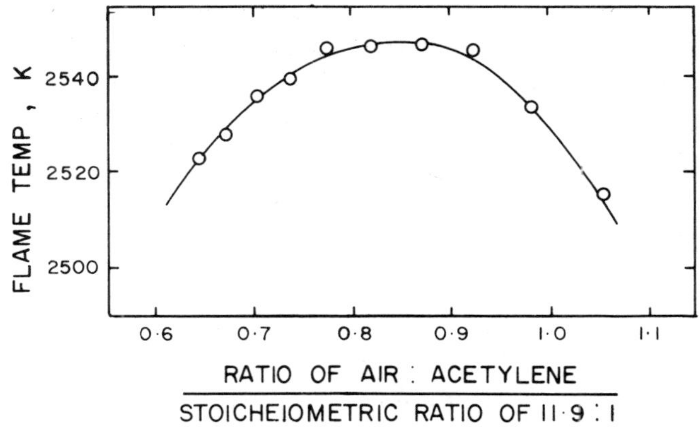

Fig. 6-5. Effect of fuel ratio on maximum temperature of an air-acetylene flame.[15]

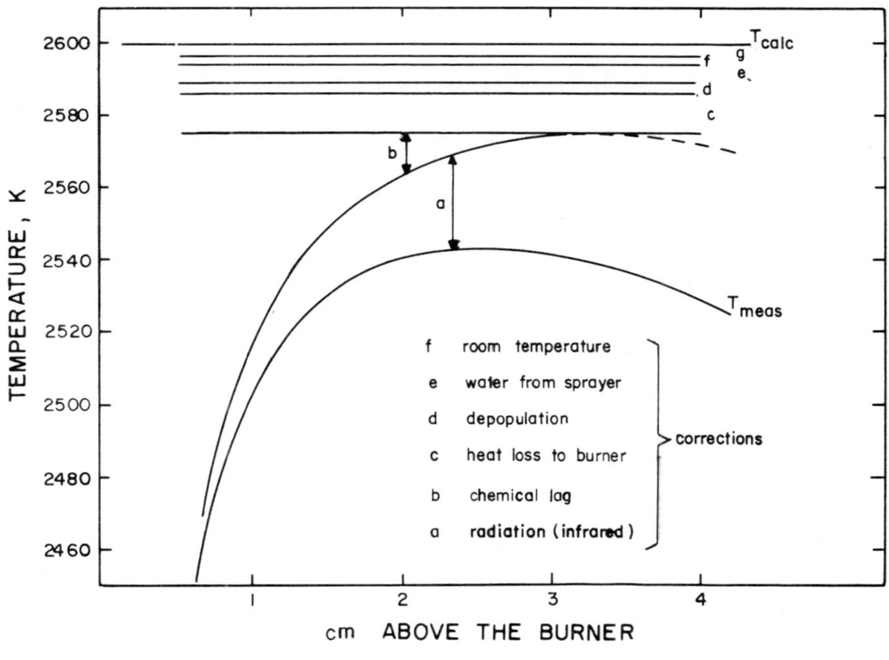

Fig. 6-6. Factors contributing to the difference between calculated and experimental temperatures of an air-acetylene flame.[18] Corrections: (a) radiation loss from the flame between top and height of measurement; (b) incomplete equilibration of the gases: (c) net heat loss to the burner (minus heat restored to the flame); (d) a difference between gas temperature and measured excitation temperature (caused by depopulation of the upper level of the measuring line 589 nm, solution 5000 ppm Na); (e) cooling effect of water sprayed into the flame; (f) effect of room temperature on flame temperature. The remaining discrepancy (g) may be attributed to the differences between the temperature scales TTS–IPTS.

calculations by the method of free energy minimization[16,17] which requires no assumptions of flame reactions or knowledge of equilibrium constants. Reference to Table 6-1 will show that maximum experimental temperatures of practical flames generally fall well below the values calculated using these techniques. Nevertheless, calculated temperatures can provide a useful guide, especially when reliable experimental data are not available. A large part of the difference may be accounted for by loss of heat to the surrounding atmosphere, and flame-shielding of the important region of the flame can result in a considerable improvement of the correlation between the two values. Snelleman[18] has studied in detail the remaining sources of temperature loss and has been able to estimate correlation factors for an air-acetylene flame. The factors contributing to the difference between calculated temperature and the reversal temperature of a flame-shielded air-acetylene flame are summarized in Fig. 6-6.

6.3 APPLICATION OF PREMIXED LAMINAR FLAMES TO ATOMIC ABSORPTION AND FLUORESCENCE SPECTROSCOPY

The premixed flames that have been most widely employed in analytical flame spectrometry are those of air with coal (town) gas, propane or acetylene. Although adequate for much AAS work, these flames provide for efficient atomization for only about 35 elements; in these flames the sensitive determination of a large group of elements which form thermally stable oxides is not possible. Furthermore, these flames are relatively unsuitable for sensitive atomic fluorescence measurements as they contain high concentrations of species that exhibit large quenching cross-sections (see Section 4.5). As a result of these limitations considerable research in this area of flame spectroscopy has been carried out and has resulted in the availability of a far wider range of flames for routine AAS and AFS measurements. When supported at suitable burners these flames allow the sensitive determination of the majority of the metallic, and some non-metallic, elements by AAS and AFS. In this section the most important aspects of modern burner design are discussed and the characteristics of those premixed flames that are used in routine analytical applications are described.

6.3.1 Burner Design

Any burner to be used for analytical spectroscopy with premixed, laminar flames must meet a number of requirements:

(i) There must be complete mixing of the fuel and oxidant gases.

(ii) The flow of these gases must be laminar as they leave the top of the burner.
(iii) There must be minimum loss of analytical sample within the burner, absence of sample contamination owing to burner corrosion by acidic solutions, and ease of cleaning.
(iv) Burner port(s) must be designed to give a stiff, stable flame with a suitable gas flowrate, and to prevent flash-back of the flame into the burner.
(v) Any flash-back must be contained within the burner.
(vi) The burner should run for long periods without overheating or clogging.
(vii) The flame obtained should be of a convenient size and shape, quiet, and usable over a wide range of mixture strengths.

In practice the design of the burner head and its port(s) is the most critical, and the construction of the main body of the burner is usually straightforward. If possible the fuel and oxidant gases should be mixed only within the burner itself and the volume of mixed gases should be kept small to reduce the consequences of any explosion resulting from misuse of the burner. Alternatively, some form of flame trap to quench combustion may be incorporated within the burner.

Usually, sufficient care is taken to ensure complete mixing of the fuel and oxidant gases. For example, the fuel may be passed through a jet to induce turbulence as it meets the other gases. In this case it may be necessary to ensure that the burner body is sufficiently long and narrow to permit re-establishment of laminar flow if the burner head design itself is not able to achieve this. For a particular gas mixture, laminar flow in the burner tube depends on its velocity and the width of the tube. The state of flow is conveniently expressed by its Reynolds number, R_e, where

$$R_e = v \,.\, l/\alpha \qquad (6\text{-}1)$$

and v is the average gas velocity and α is its kinematic viscosity (i.e. viscosity/density). For a circular tube, l is equal to the diameter of the tube but with a rectangular burner of dimensions $a \times b$, $l = 2ab/(a+b)$. In many cases the rectangular burner is such that $a \gg b$ and then $l \approx 2b$. For laminar flow the value of R_e must be below about 2300 and the dimensions of the burner and the required (volume) flow rate through it must be chosen accordingly. Even then, pure laminar flow will only be established if the burner tube is sufficiently long, an approximate value for the minimum length, L_m, being

$$L_m = 0{\cdot}05\, R_e \,.\, l \qquad (6\text{-}2)$$

The burner body is usually machined from stainless steel as this is strong and durable and avoids most problems of contamination caused by corrosion with acidic sample solutions. However, glass[19] and plastic[20] have also been used, and recently some commercially available burners for AAS have been produced from titanium. This avoids some of the residual contamination problems experienced with stainless steel, but is also far easier to machine and weld. As a result burner production on a large scale is less costly and even burners that require a head with a long, narrow slit (see below) can be produced in one piece without the need for screws, gaskets, etc. If possible, the inside of the burner should be polished and free from corners, edges and depressions likely to trap any large droplets left in the nebulizer spray and result in contamination between consecutive samples. This problem is particularly serious when the burner body or adjacent metal pipes remain sufficiently cold during operation (owing to the flow of cold gas through them) to produce significant condensation of the sample mist. These low temperatures may occur either when the burner head is cooled too efficiently by a water jacket (see below) or when the material used (e.g. glass) gives very poor conduction of heat from a relatively hot burner head. In the latter case it may be necessary to warm the burner body using some form of heating jacket. Care must be taken not to raise the gas temperature in the burner to too high a value, as in some cases the burning velocity increases significantly with preheating temperature and the designed safety margin of the burner with a particular gas mixture may be seriously reduced.

The factors involved in designing an efficient burner head have been clearly analysed by Fiorino *et al.*[21] Briefly, the friction of the walls of the burner port(s) reduces the upward velocity of the gas stream so that in order to obtain a sufficiently high velocity to balance the burning velocity the area of the port(s) must produce an average stream velocity several times greater than the burning velocity. In practice, as the flow velocity tends to zero at the burner wall, the flame would always strike back in this region if it were not for the quenching effect of the rim and wall of each port. This effect results from the ability of the walls to remove heat and chain-carrying species from the gas; the distance for which this ability extends from the wall is known as the *quenching distance*.

The quenching distance, like the burning velocity, is a characteristic feature of each flame-gas mixture. Thus, the diameter of the burner port(s) must be such as to simultaneously ensure adequate quenching for a particular gas mixture and give the correct gas velocity for the volume of that mixture normally passed through the burner. In practice this gas velocity is frequently well above the minimum needed to obtain a stable balance of the burning velocity, both to act as a safety margin and to provide a fairly "stiff" flame which is less prone to fluctuations such as those caused by draughts. This safety margin is particularly necessary with some of the hotter flames which have very small quenching distances as they are often supported at burners where quenching might prove inadequate should some flame instability initiate a flash-back. Under these conditions, a worthwhile improvement in quenching ability can be obtained by cooling the burner head with a water jacket.[22] It has also been claimed that water cooling eliminates slow drift of analytical signals under certain conditions[23] and reduces the rate of build-up of carbon deposits on burners used with some fuel rich hydrocarbon flames.[24] Excessive cooling, however, may result in deposition in the burner ports of solid material from the nebulized samples as well as causing condensation of sample spray in the body of the burner. To avoid this problem while still obtaining an increased safety margin, some manufacturers have produced burner heads with air-cooling fins, as illustrated in Fig. 6-7.

The use of a single, relatively large circular burner port, as in the well-known Bunsen burner, is really only suitable for slow-burning gas mixtures such as town gas with air. Many early analytical flame spectroscopy experiments were, however, carried out using this type of burner and Mavrodineanu and Boiteux[6] have described burners of this type suitable for premixed air-acetylene (see Fig. 6-8) and oxygen-acetylene flames. As relatively little protection against flash-back is provided by quenching, these burners usually incorporate some form of safety device, the most common being a combination of jet (for the fuel gas) and venturi (for the total gas mixture) near the base of the burner as shown in Fig. 6-8. The burner port usually takes the form of either a set of small holes or a narrow slot. The former is most commonly obtained by drilling a set of holes in a solid plate (see Fig. 6-9). In this way any required flame size and shape is readily obtained, and varying the number and diameter of the holes allows easy optimization of gas velocity (and quenching distance) for any required gas flow rate. If the holes can be made

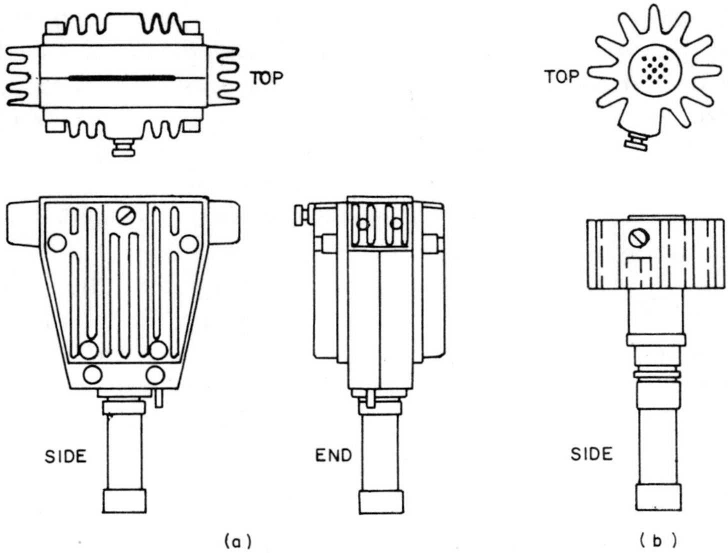

Fig. 6.7. Commercial burners for (a) absorption and (b) fluorescence measurements with air-cooling fins (Evans Electroselenium Ltd.).

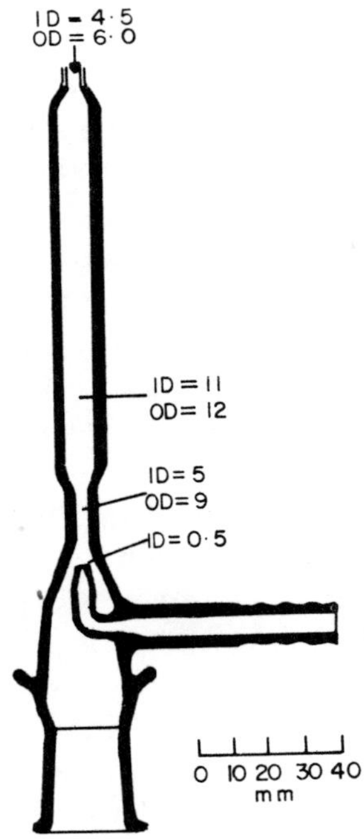

Fig. 6-8. Premixed air-acetylene burner using a single, large diameter port and a jet and venturi safety device.[6]

sufficiently deep an improved laminar gas flow will be obtained (see above), although it is difficult to drill a large number of narrow deep holes in a hard metal such as stainless steel.

To avoid this problem, Padley and Sugden[26] have devised a burner* in which the ports take the form of short lengths of stainless

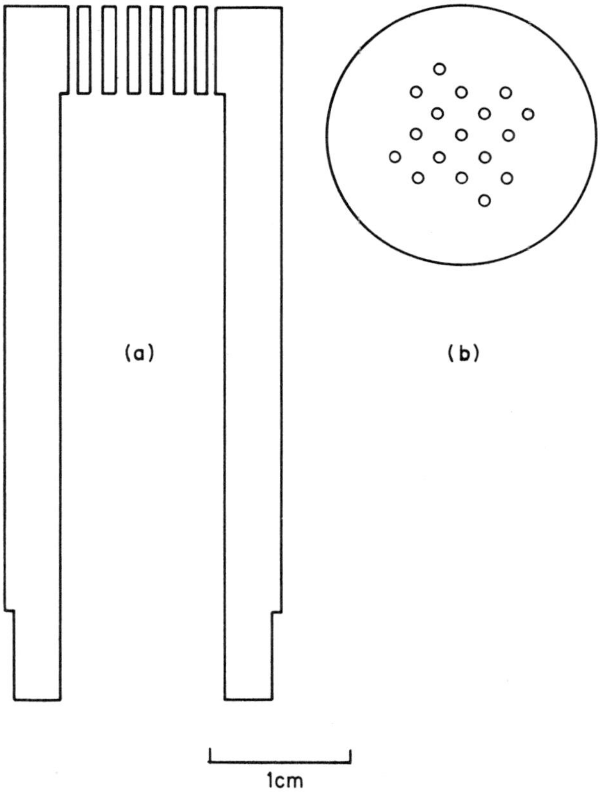

Fig. 6-9. Premixed air-acetylene burner using an array of seventeen holes (0·75 mm diameter) drilled directly into the stainless steel head.[25] (a) Side view; (b) top view.

steel capillary tubing suitably mounted in the burner head. Production of a large number of extremely fine holes several centimetres long then presents no problems and the burner head gives excellent laminar gas flow and quenching properties. This type of burner has not been widely used in analytical flame spectroscopy, partly because the usual techniques of assembly involve soldering many capillary tubes to a support, but mainly because water-cooling the burner head to prevent damage by overheating is then necessary. It

* A modified version of this burner is shown in Fig. 6-17.

was considered that, under the conditions encountered during analysis, fine capillaries cooled in this way would be particularly prone to clogging. Recently, however, Aldous *et al.*[27] described a capillary burner constructed by bonding the tubes into a block of high-temperature-resistant epoxy resin.* If the stainless steel tubes, which have low heat conductivity, are left to project several millimetres above the top surface of the resin (and of the metal burner case fitted around the block) no overheating of the epoxy resin occurs even without water cooling, provided a reasonably high flow rate of gas is used through the burner. Thus, the whole burner head equilibrates to a moderate temperature and it has been claimed that no clogging occurs even when a solution containing 10% glucose is nebulized for ten minutes.

Burner heads with a single port in the form of a long narrow slot giving a "flat" flame have been widely used in burners fitted to commercial atomic absorption spectrometers. Burners of this type have the advantage of being easy to machine into a plate of any thickness so that laminar flow is readily attained. In most cases the burner plate is produced in several parts and spacers or gaskets are used to obtain the desired slot width (see Fig. 6-10), but modern machining techniques and the use of titanium burners also allow a narrow slot to be cut in a single piece of metal without difficulty. Slot burners are easier to clean than those consisting of a matrix of holes and produce a more homogeneous flame. With some gas mixtures, however, it may prove more difficult in designing a slot burner to obtain the correct combination of quenching effect and gas velocity for the required gas flow rate. In view of this, rectangular burners have also been designed with several parallel slots[28,29] which are narrower than would normally be used. Apart from improving the quenching efficiency this also allows a considerably broader flatter flame than is possible with a single slot and this may be advantageous with the optical systems employed in some spectrometers. A disadvantage of using very narrow slots is that they are more prone to clogging, and considerable difficulty has been experienced with carbon deposition on the slits of burners used with hot fuel-rich nitrous oxide-acetylene flames owing to partial degradation of the fuel. This problem is surmounted on several commercially available burners without resorting to water cooling (which introduces other problems, as discussed above) by the use of raised edges on the burner surface, as shown in Fig. 6-11. Apparently any carbon deposit

* A burner constructed in this way is shown in Fig. 6-14.

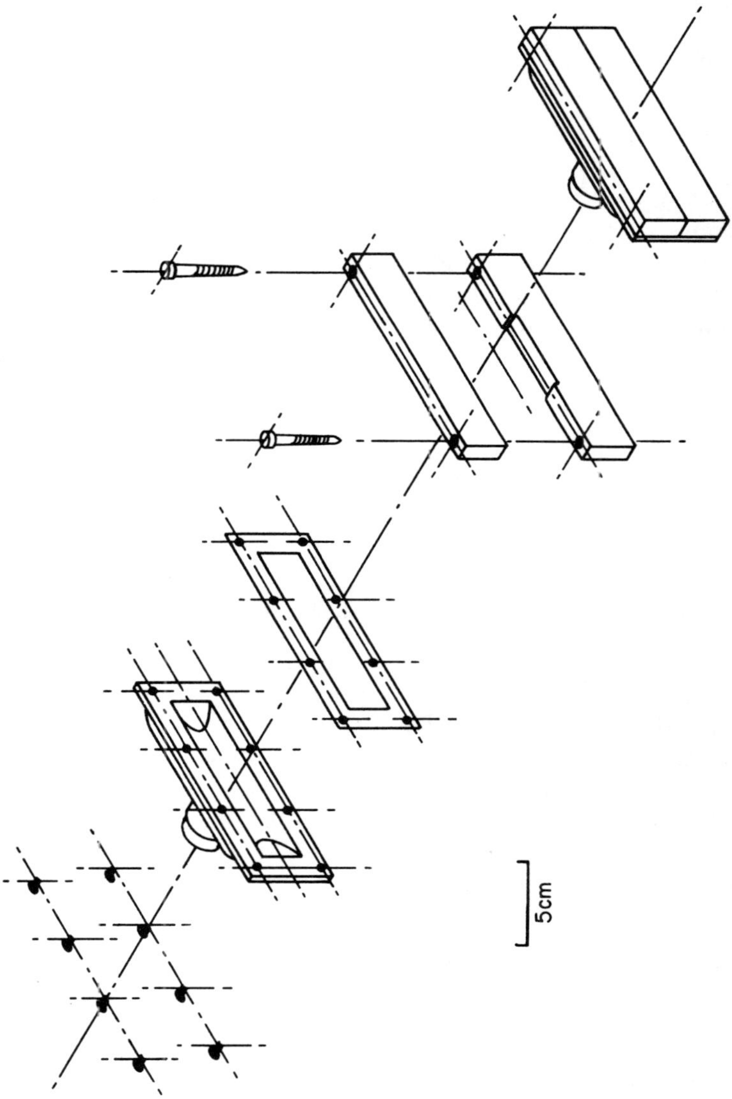

Fig. 6-10. Construction of a commercial, long slot, atomic absorption burner for premixed nitrous oxide-acetylene flames (Techtron Pty. Ltd.).

formed is then quickly burnt away at the high temperature reached on the raised edges and by the additional flow of entrained air over them as a result of the shape of the burner surface. This flow of air also improves the stability of the flame if the grooves in the burner surface are correctly positioned.

Circular burners have also been constructed with a single, narrow slot as the burner port. This is readily achieved by using a metal tube with a solid insert supported at its centre. The width of the slot may be changed simply by interchanging the centre pieces of the burner. In an early design[30] the centre piece was held in position by three small screws or pins passed through the outer tube near the top of the burner. This has the disadvantage of disturbing the laminar gas flow and the flame shape and it is also difficult to locate the central plug accurately. A more recent design, illustrated in Fig. 6-12,

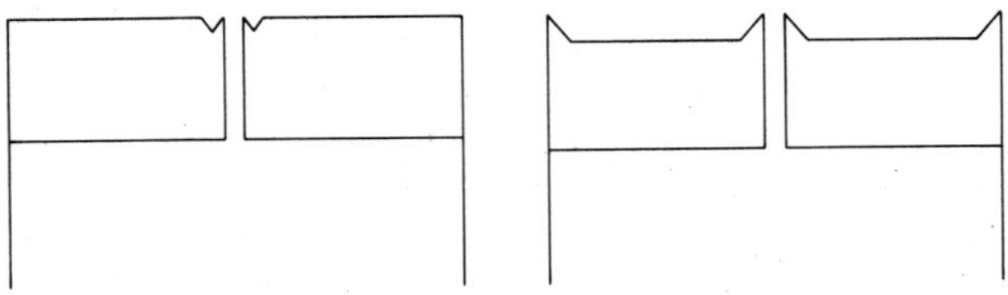

Fig. 6-11. Surface configurations used with narrow slot long-path (AAS) burners to minimize carbon deposition from fuel-rich nitrous oxide-acetylene flames.

requires only a single-threaded hole mounted near the base of the burner, the central insert being supported on a long stem which screws directly into this hole. The brass insert shown in the diagram was intended to conduct heat away from the central insert, but is unnecessary under most operating conditions and would be a disadvantage when nebulizing corrosive solutions.

Although a very narrow burner port is needed for adequate quenching with fast-burning flames, it is possible to support even oxygen-acetylene flames on a single port of reasonable width without using excessively high gas flow rates. Alternatively, more slow-burning gas mixtures may be supported on a wider port than would normally be safe. This is possible by making use of the nozzle effect. For our purposes a nozzle can be considered as a tube (or duct of any shape) with converging sides which smoothly accelerate the flow of a gas through it. Provided the ratio of the inlet and outlet areas is greater than 4 : 1 (ref. 3, p. 104) such a nozzle gives a much more uniform velocity profile with a far higher boundary velocity than would be obtained in an ordinary tube with similar average gas flow and exit port diameter. Furthermore, laminar flow is obtained even though the gas velocity leaving the port may be extremely high; Reynolds numbers as great as 40,000 may be obtained (ref. 1, p.15). Thus a safe, stable, laminar flame may be supported under flow conditions where conventional burner ports would give a non-laminar flame and be prone to flash-back (as the low boundary velocity would require much more effective quenching). By making use of

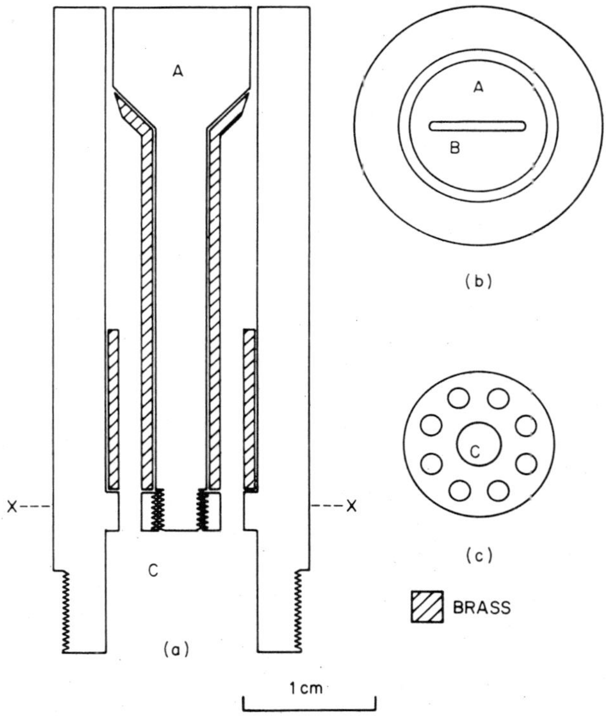

Fig. 6.12. A circular nitrous oxide-acetylene burner head using a single narrow slot.[25] (a) Side view; (b) top view; (c) section through x–x. A, Centre piece; B, screwdriver slot; C. threaded hole for centre piece.

this effect, Fiorino *et al.*[21] were able to support an oxygen-acetylene flame on a single slot 7·62 cm long and 0·25 mm wide provided the burner plates were water-cooled. Similarly, many of the single port, circular burners mentioned earlier also make use of a nozzle.[6]

In the foregoing discussion we have described burners which give either a circular flame or a long flat flame. The effect of flame geometry on atomic fluorescence measurements has already been discussed at some length in Chapter 4 and the reader will appreciate that the necessary compromise between optimum excitation conditions and optimum collection of fluorescence radiation justifies the general use of a circular burner. It may, however, be possible to obtain further improvement by using a burner whose dimensions have been chosen to match the characteristics of a particular optical system and a capillary tube burner of the type described above could easily be constructed with any desired shape. The almost universal use of the longest possible flat flame for absorption measurements is often justified on much less satisfactory grounds. A common explanation is simply that, according to Beer's law, "the longer the flame the greater the absorption". This is true in solution spectrophotometry, where doubling the path length means simply doubling the amount of sample in the light beam, but is quite incorrect in AAS if more or less constant sample uptake and gas flow rates are assumed. The optimum flame geometry is merely that which places as many sample atoms as possible in the path of the light beam per unit time as it travels from the source to the detector. In practice, most modern detection systems are based on a narrow entrance slot (giving a narrow solid angle and optical path) and since the light beam from a hollow-cathode lamp is also usually rather narrow the optimum flame shape may, indeed, be rectangular. No light must reach the detector, however, without having passed through the flame *or the region of the flame in which a population of free analyte atoms is obtained* as this would reduce the absorption sensitivity. The use of a long slot burner may, therefore, result in a flame that is too thin to give optimum sensitivity, particularly for those elements only efficiently atomized in the interconal zone of the flame (e.g. refractory elements in a fuel-rich nitrous oxide-acetylene flame). Thus, even for absorption measurements both the width and the length of the flame must be considered and it is again advantageous to construct a burner that exactly matches the optical system in use. It should also be remembered that, even when all sample atoms are placed in the path of the light beam, there is no advantage in obtaining a larger flame by using higher gas flow rates unless the rate of sample uptake is correspondingly increased. Without this there will merely be a greater "dilution" of analyte atoms in the flame gases.

6.3.2 Separated and Sheathed Flames

As explained in Section 6.2, atomic absorption or fluorescence measurements with premixed, laminar flames are mostly confined to the well-characterized interconal zone of the flame. With most burner designs in routine use the laminar flow characteristics are not sufficiently good to prevent at least some entrainment into the interconal zone of the atmospheric air which supports the secondary reaction (diffusion) zone. As a result, quenching of fluorescence radiation may occur in the interconal zone of a flame with which great care has been taken to avoid introduction of a highly quenching species by careful choice of fuel and oxidant gases (e.g a hydrogen-oxygen-argon flame). Similarly, in a flame chosen for its ability to break down refractory oxides (e.g. a fuel-rich nitrous oxide-acetylene flame) the additional oxygen may re-oxidize some of the free metal atoms or reduce the concentration of those species responsible for the reducing action of this region of the flame. Furthermore, any radiation involved in the absorption or fluorescence measurements must travel through the secondary zone. Thus, in fluorescence work, species present in this relatively cool region may increase the scattered radiation received at the detector, whereas in absorption molecular species may absorb sufficiently at the wavelength of measurement to cause an interference effect. Even when a modulated source and detection system are used for either technique, the relatively intense radiation emitted at some wavelengths by the outer zone and received at the detector may be registered as noise and cause deterioration in the value of the detection limit attainable. Hence, in many cases it would be useful if the part of the interconal zone used in AAS or AFS could be isolated from the effects of the surrounding atmosphere or of the secondary reaction zone. This may be achieved by the use of flame sheathing and/or flame separation.

A separated flame is one in which the secondary reaction zone is physically moved some distance from the primary zone to produce a greatly extended interconal zone between the two. This separation may be achieved by surrounding the lower part of the flame with either a solid silica wall or a wall of an inert gas. In either case the flame must remain stable and adequately attached to the burner, and it must be possible to maintain a stable secondary zone at the desired height above the primary zone.

A sheathed flame is one in which the flame is protected in some way from the surrounding atmosphere without physically separating its zones. Examples include a sheath of filtered air (to protect the flame from atmospheric contaminants), a sheath of oxygen (to increase the temperature at the edge of the flame), and a sheath of another flame (to protect species normally attacked by atmospheric oxygen or quenched by atmospheric nitrogen or to maintain a more uniform temperature within the interconal zone of the main flame).

In practical terms, air or oxygen sheathing has seen little routine application to atomic absorption or fluorescence measurements, although the former would be useful wherever analyses need to be carried out in an atmosphere contaminated by a significantly high concentration of the analyte element. Oxygen sheathing is really of interest only in flame emission spectroscopy where an increase in flame temperature is particularly useful, although Gilbert has suggested (ref. 31, p. 177) that on certain burners oxygen sheathing actually increases the ability of an oxygen/acetylene flame to atomize refractory metals by allowing stable combustion of a more fuel-rich flame than is usually possible. In analytical AAS and AFS more use has been made of separated flames or flame-sheathed flames in which the nebulized sample solution is supplied only to the central flame. Either technique can be used to obtain many of the improvements in flame performance mentioned above and the final choice may be somewhat arbitrary. The main differences between the two techniques are that separation reduces the background radiation from flame species (whereas sheathing may actually increase it) and that sheathing increases the flame temperature (whereas separation actually reduces it). Thus, a flame sheath may be more satisfactory for improving atomization of refractory elements but a separated flame would often prove preferable for the measurement of extremely weak fluorescence signals.

The traditional method of flame separation is with a Smithells-type separator as used for many years in fundamental studies of flame structure. This is simply a glass or silica tube, sometimes fitted with optical windows, which is attached to the top of the burner to prevent access of atmospheric oxygen and to provide a stable auxiliary port at which the separated secondary reaction zone may burn. Although this type of separator has been applied to both absorption and fluorescence determinations by Hingle *et al.*[32,33] it has a number of disadvantages. These include the limited life of the separator with high-temperature flames, the deposition of soot on the sides or on optical windows with fuel-rich flames or when organic solvents are nebulized, the limited range of fuel:oxidant ratio for stable separation, and the problem of scattered radiation in fluorescence measurements.

In general, more satisfactory results are obtained when separation is produced by flowing a laminar stream of an inert gas such as nitrogen or argon around the outside of the flame. An adequate laminar gas flow may be obtained without excessive lengths of tubing if the inert gas is passed through a short length of a matrix containing a large number of narrow bore holes. This is produced either by tightly packing alternate layers of corrugated and plain stainless steel strips[24,25,34] or by using a large number of stainless steel capillary tubes.[27] Burners for fluorescence (circular) and absorption (rectangular) measurements constructed using both these techniques are shown in Figs. 6-13 and 6-14 respectively. The circular device in Fig. 6-13(a) is intended to fit around the air-acetylene and nitrous oxide-acetylene burner heads shown in Figs. 6-9 and 6-12. The rectangular burner shown in Fig. 6-13(b) is fitted with a narrow slit for combustion of nitrous oxide-acetylene flames but the same burner may be used for air-acetylene with a burner plate containing a row of small boles.[35] Both burners shown in Fig. 6-14 may be used for air-acetylene, air-hydrogen and nitrous oxide-acetylene flames simply by varying the number and diameter of the capillary tubes.

An obvious application of separated flames to atomic absorption spectroscopy was described by Kirkbright *et al.*[35] who determined a number of refractory oxide forming elements in nitrogen or oxygen separated nitrous oxide-acetylene flames. They reported between 15 and 50% improvements in sensitivity on separation as a result of the more favourable atomization conditions thus obtained. A more unexpected effect of flame separation is the considerable increase in transparency obtained below 200nm for air and nitrous flames with acetylene.[35-39] Absorption of radiation by the atmosphere in the region immediately below 200 nm is mainly due to oxygen molecules and the improved flame transparency on separation is thought to be due to removal of the oxidizing outer flame mantle and the lack of oxygen species in the interconal zone.[36,37] The extent of this improvement is illustrated in Fig. 6-15 for the air-acetylene flame at 193·7 nm (arsenic absorption line) and in Fig. 6-16 for the nitrous oxide-acetylene flame at 180·7 nm (sulphur absorption line). As a result, it has been possible to obtain better detection limits for arsenic and selenium in both air-acetylene[35] and nitrous oxide-acetylene[36] flames. Further benefits, obtained by use of the separated nitrous oxide flame with a modified spectrometer which excludes air from most of the optical path, have been the direct determination of iodine at 183·0 nm,[37] sulphur at 180·7 nm[38] and phosphorus at 177·5 nm.[39] Furthermore the high transparency and extended volume of absorbing atoms in this separated flame has

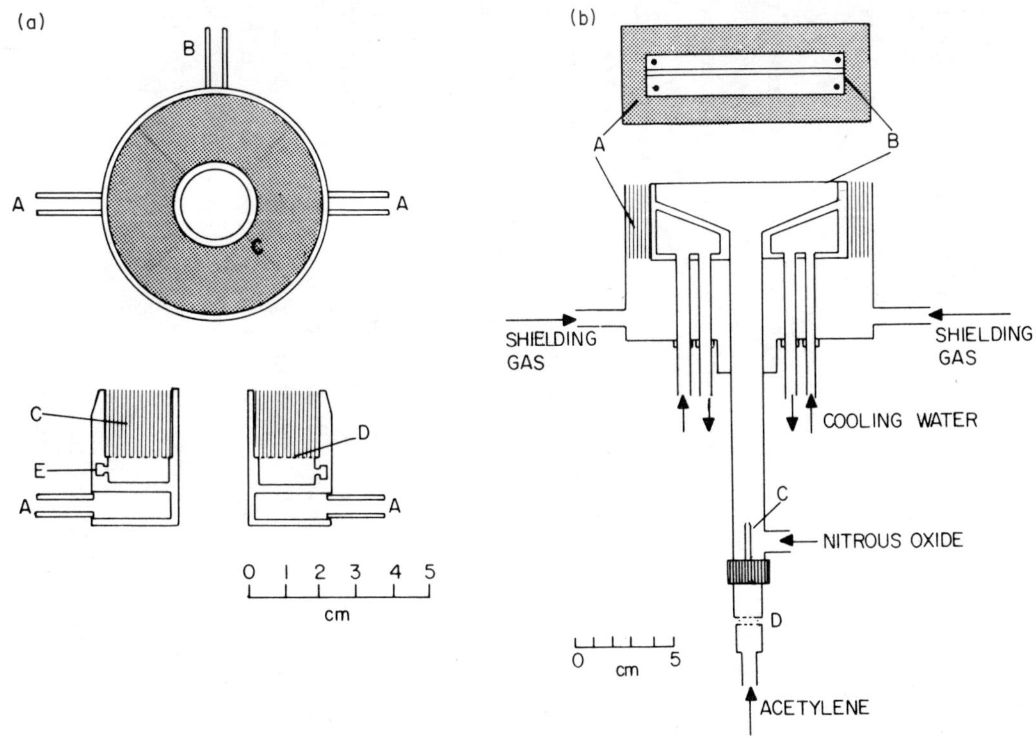

Fig. 6-13. Burners for flame separation using a laminar flow of an inert gas. (a) Circular burner;[25] A, water pipes; B, nitrogen-argon inlet; C, laminar flow matrix; D, support for matrix; E, nitrogen-argon distribution channel. (b) Rectangular burner;[24] A, laminar flow matrix; B, removable burner plate; C, acetylene jet; D, anti-blowback valve.

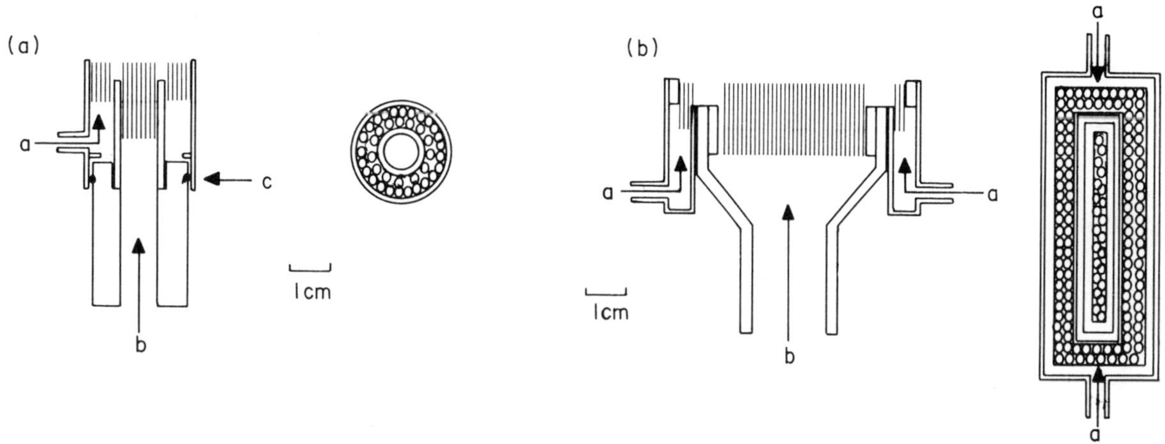

Fig. 6.14. Capillary burners for flame separation using a laminar flow of an inert gas.[27] (a) Circular burner; (b) rectangular burner. a, Nitrogen-argon inlet; b, premixed flame gases; c, O-ring seal for removable outer gas shielding device.

allowed the determination of arsenic and selenium with a triple pass optical system.[40]

Rather more applications of separated flames have been reported for atomic fluorescence measurements. Most reported improvements on flame separation appear to result from a reduction in flame

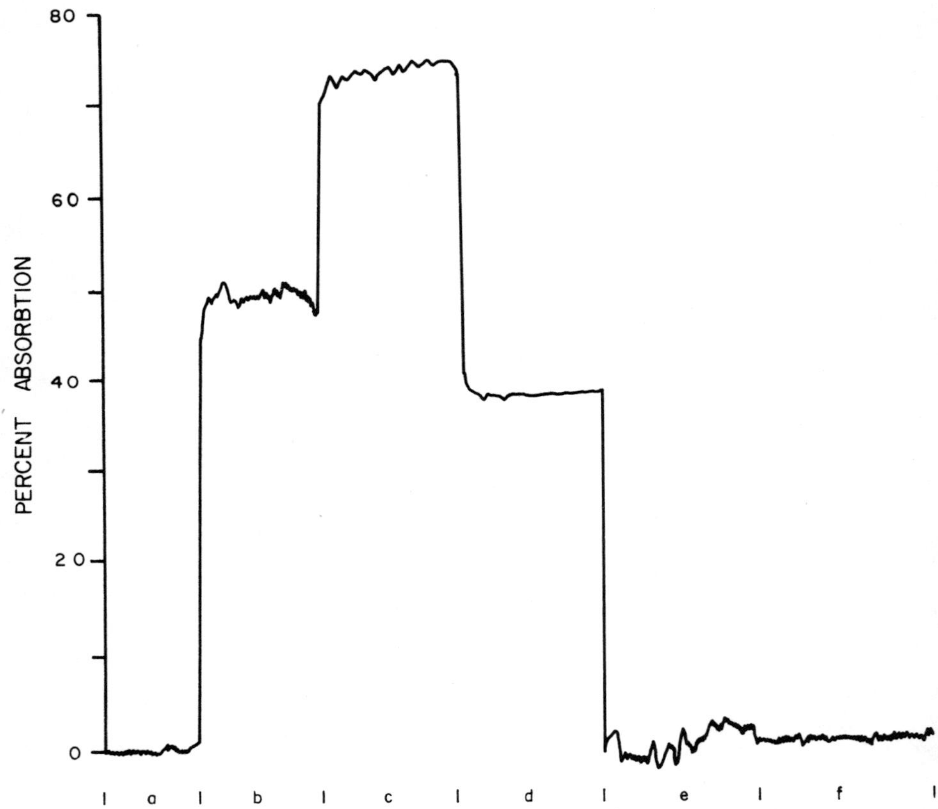

Fig. 6-15. Background absorbance in an air-acetylene flame at the arsenic 193·7 nm line.[35] (a) No flame; (b) conventional flame, optimum fuel flow for arsenic determinations; (c) conventional flame, same fuel flow as (d); separated flame, optimum fuel flow for arsenic determination; (e) conventional flame, absorbance set to zero (by increasing gain); (f) separated flame, absorbance set to zero.

background radiation (and hence signal noise) rather than from less serious quenching by the flame gases. Thus, the most promising applications were with unmodulated detection systems for the air-acetylene flames[34,41] and with simple non-dispersive spectrometers for both the air-acetylene[42] and nitrous oxide-acetylene[43] flames. Some improvement in detection limits has been reported, however, for the separated air-acetylene flame even with conventional, modulated spectrometers[41,44-48] and under these same conditions separation of the nitrous oxide-acetylene flame appears to give a very useful improvement in fluorescence sensitivity[49-51] as a

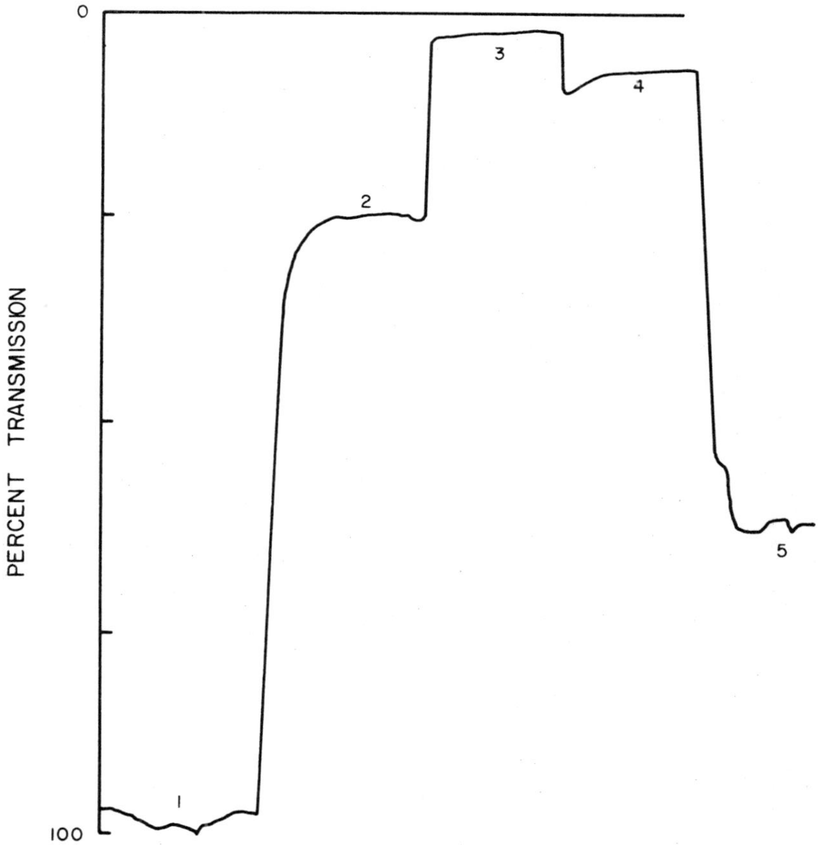

Fig. 6-16. Background absorbance in a nitrous oxide-acetylene flame at the sulphur 180·7 nm line.[38] 1, No flame, nitrogen separation gas only (transmission set to 100%); 2, no flame, separation gas off (air only in light path across burner); 3, air-acetylene flame; 4, unseparated nitrous oxide-acetylene flame, fuel-rich; 5, Nitrogen separated nitrous oxide-acetylene flame, optimum acetylene flow rate for maximum sensitivity for sulphur.

result of the very high background radiation and noise from the conventional flame.

There appear to have been no reported studies by analytical chemists of the effect of flame separation (with argon) on the yield of fluorescence radiation in flames where exclusion of entrained air may be expected to have a significant effect on the concentration of highly quenching species (i.e. hydrogen-argon-oxygen flames; see

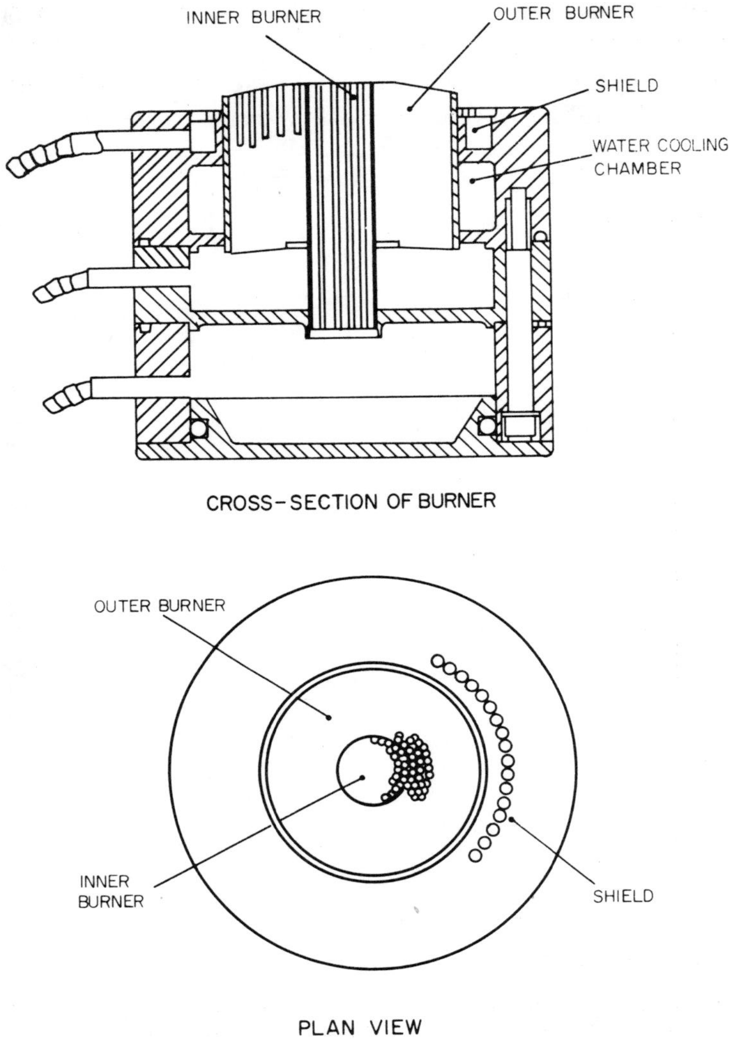

Fig. 6.17. Capillary burner for application of flame sheathed hydrogen-oxygen flames to atomic fluorescence measurements.[51] The nebulized sample is supplied only to the inner burner and the outer flame is surrounded by a further sheath of inert gas as an additional precaution against entrainment of air.

Chapter 4 and Section 6.3). However, considerable use of an outer flame sheath to exclude entrained air has been made in fundamental fluorescence studies under these circumstances. Jenkins[52] has used the modified Padley and Sugden capillary burner shown in Fig. 6-17 to study the effect of flame composition on fluorescence intensity in hydrogen-oxygen flames diluted with different inert gases. He con-

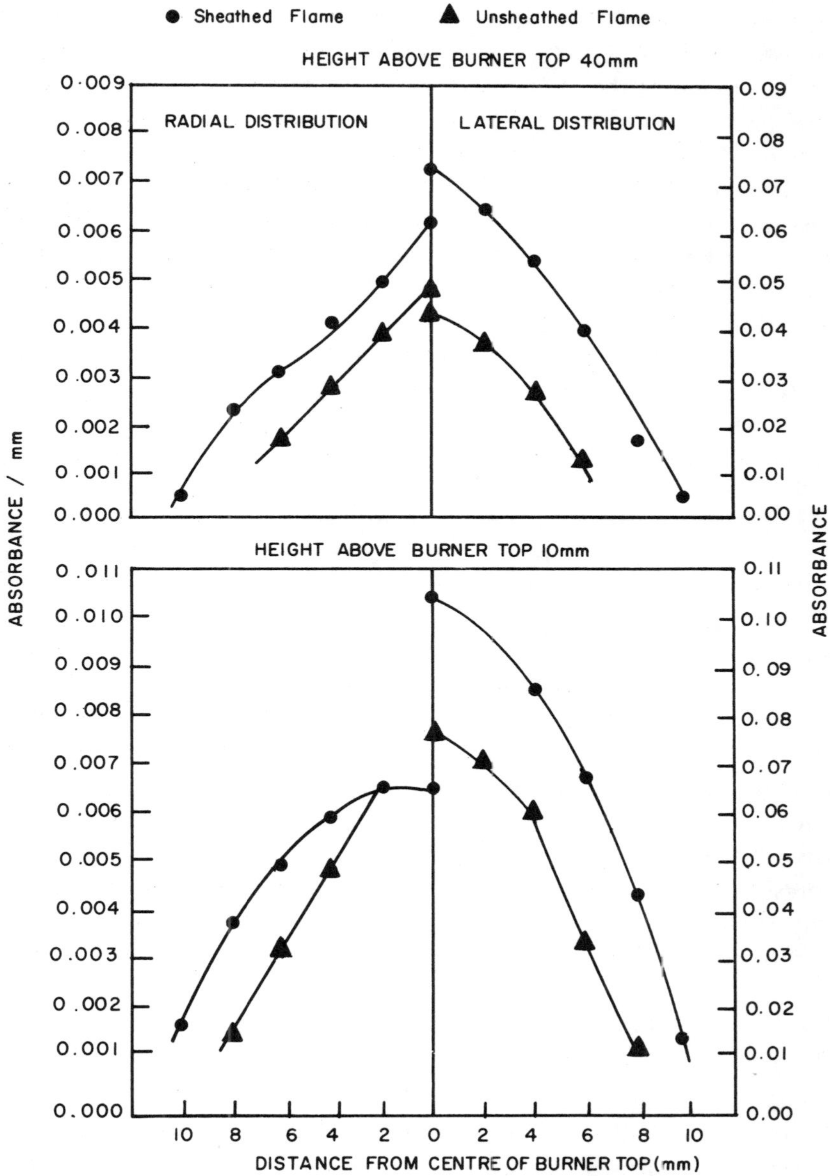

Fig. 6.18. Lateral and radial distribution of calcium atoms in a premixed air-acetylene flame with and without a sheath flame.[55]

cluded that if the maximum fluorescence intensity were to be obtained in argon-diluted flames, the protection of the outer flame against air entrainment was essential over most of the flame height. However, very similar results could be obtained without this protection close to the burner head as a result of the extremely laminar flow provided by this type of capillary burner. The situation may well be different for many burners used in routine AFS measurements and use of a flame sheath (or separation with a flow of argon) might result in higher fluorescence intensity close to the burner.

Few other applications of flame-sheathed flames to either AAS or AFS appear to have been reported at the present time although several workers have used these flames for special purposes, such as the measurement of free atom fractions and studies of absolute absorption techniques (see Sections 3.5.3 and 11.4.3). Of particular interest in this respect is a study by Willis[53] in which absorption signals in a long-path flame were compared with and without a flame sheath. He observed that the distributions of sodium, barium and magnesium atoms in an air-acetylene flame were virtually unchanged on sheathing although the total number of barium atoms at any particular height increased significantly, presumably as a result of the more fuel-rich flame obtained by prevention of air entrainment. Similarly, the distribution of magnesium atoms in a nitrous oxide-acetylene flame on the same burner was virtually unchanged on sheathing with an air-acetylene flame. An atomic fluorescence burner based on a design used for theoretical studies by Alkemade and co-workers[54] and with provision for flame sheathing has recently been described by Slevin *et al.*[55] The effect of the flame sheath in improving the atomization efficiency of calcium in a premixed air-acetylene flame is illustrated by the absorbance measurements shown in Fig. 6-18 and the sheath was also found to enhance the fluorescence sensitivity of tin determinations. The advantage of flame sheathing a premixed oxygen-hydrogen-argon flame for AFS determinations of arsenic was also investigated with this burner.[56]

6.3.3 Premixed Flames Used for Routine Analysis

Prior to the introduction of atomic absorption as an analytical technique, commercial flame-emission spectrometers were used principally for alkali metal analyses. Generally these instruments were supplied with burners suitable for premixed flames of air with either town gas (giving a flame temperature of *ca.* 1800°C) or propane (*ca.* 1900°C), although provision was also made in many of the more expensive spectrometers for combustion of the hotter air-acetylene flame (*ca.* 2200°C) which had been introduced in the 1930s by Lundegardh as a "high-temperature" source suitable for the excitation of a wide range of elements. With this background it was perhaps natural that an air-town gas flame should have been used with the introduction of AAS as an analytical technique[57] and that it was rapidly superseded by air-propane and air-acetylene flames as the application of the method became more widespread. Unfortunately, the emphasis on AAS as a "universal" method of trace metal analysis in contrast to the quite limited routine applications of its emission predecessor, soon revealed the inadequacy of these flames for this purpose. The efficiency of free atom formation, upon which AAS sensitivity depends, is a function of flame composition and temperature and is quite low for many elements even in the air-acetylene flame. Furthermore, chemical interferences are prevalent, owing to the relative stability of many compounds in these flames. As a result, it was found that sensitive determinations by AAS were possible for only about 35 elements, and even for these many routine analyses were subject to interference problems. The determination of the rare-earth elements and many others forming stable, refractory oxides was not possible except at relatively high concentrations.

In view of this limitation of the air-acetylene flame considerable experimental effort has been devoted during the past decade to the development of premixed flames able to (a) provide efficient atomization for most elements, (b) reduce problems of chemical interference, and (c) be suitable for routine use. These efforts have met with such success that the air-town gas and air-propane flames are now rarely used; the air-acetylene flame has been replaced by the nitrous oxide-acetylene flame for many elements. A similar success has been achieved in the development of specialized flames for AFS. Development here has been along different lines, and via other flames, in view of the requirement for a medium that gives minimal background radiation, noise and quenching but still minimizes chemical interferences. It seems, however, that this latter important requirement of freedom from interference effects, and the development of intense spectral line sources for AFS, may lead to the adoption of the ubiquitous nitrous oxide-acetylene flame for routine analytical AFS applications as well as for AAS.

(a) Air-acetylene flame

The premixed air-acetylene flame has seen such wide application to routine analytical AAS that even a summary of the available literature reports is impossible here. It provides a stable, reproducible absorption cell which is reasonably safe to use, gives low signal and acoustic noise levels, and produces sufficient atomization of many

TABLE 6-2

Free atom fractions in premixed air-acetylene flame

Reference	58	59	59	60	61	53	53	53
Notes	*a, e*	*b, f*	*b, g*	*a, c, h*	*a, i*	*a, c, j*	*a, c, j*	*b, d, j*
Element								
Ag	0·66				0·70		*k*	
Al	< 0·00001	< 0·0001	< 0·0001		< 0·00005			
Au		0·26	0·21		0·40	0·63	0·76^k	
Ba	0·0011	0·0019	0·0034	0·0009	0·0018	0·0026	0·0031	0·0013
Be					0·00005			
Bi					0·17			
Ca	0·14	0·069	0·052	0·066	0·07	0·070	0·083	0·070
Cd	0·50	0·77	0·80		0·38			
Co	0·41	0·13	0·023		0·28			
Cr	0·064	0·19	0·53	0·065	0·071			
Cu	0·98	0·38	0·40	1	0·88	1	1·19^k	0·51
Fe	0·66			0·38	0·84			
Ga	0·16				0·16			
In	0·67	0·13	0·10		0·67			
K	0·25				0·32	0·28	0·34	0·38
Li	0·20				0·12	0·26	0·31	0·14
Mg	0·59	1·09	1·05	0·64	1·06	0·84	1	1
Mn	0·45	0·33	0·33	0·59	0·62	0·70	0·83	0·37
Na	1·00	0·68	0·63		1·04	0·53	0·63^k	0·57
Pb	0·44				0·77			
Rb						0·16	0·19	0·16
Si					< 0·001			
Sn	< 0·0001	0·003	0·061		0·043	0·078	0·092	0·060
Sr	0·13	0·087	0·068	0·075	0·063			
Ti					< 0·001			
Tl	0·36				0·52			
V		< 0·0004	< 0·004		0·15			
Zn	0·45	0·62	0·66			1·10	1·31	0·66

[a] Values measured by integrated (continuum) absorption technique.
[b] Values measured by peak (sharp line) absorption technique and less reliable for many elements.
[c] Values relative to an assumed value of 1 for copper.
[d] Values relative to an assumed value of 1 for magnesium.
[e] 10·2 cm slot burner, air: acetylene = 5·6:1.
[f] 5 cm slot burner, air: acetylene = 7·0:1.
[g] 5 cm slot burner, air: acetylene = 5·36:1.
[h] Rectangular burner, with flame-shielded flame, air: acetylene = 9·0:1.
[i] 5 cm slot burner, air: acetylene = 9·85:1.
[j] 7·6 cm slot burner, air: acetylene = 7·85:1 except for Ba, Sn, where a "richer" flame was used.
[k] More recent measurements by Willis[62] show that these elements are completely atomized in a flame shielded flame, air: acetylene = 7·15:1.

analytical samples for sensitive, interference-free determinations. Unfortunately, as mentioned above, its atomization efficiency has proved inadequate for the sensitive determination of many other common elements and its use does not allow interference-free analyses of some practical samples of the 35 analyte elements that can be determined. The free atom fractions (i.e. the efficiency with which atoms of an analyte element are formed on nebulization of aqueous solutions of simple salts) obtainable in this flame for a number of elements have been determined by a number of workers and are shown in Table 6-2. The large variation in values obtained by different workers for particular elements reflects errors in the various measuring techniques, the use of quite different experimental conditions and the different theoretical expressions used for calculation by different workers. It is, nevertheless, possible to see clearly the distinction between elements that are easy and difficult to atomize. Although only a few elements are completely atomized (i.e. have a free atom fraction of unity), sufficient sensitivity would be obtained for most applications if the fraction were greater than *ca.* 0·1 or 0·2.

In addition to providing an indication of the ease with which measurements may be made for a particular element, the values of the free atom fractions also serve as a guide to the extent to which optimization of flame conditions is important. For example, Fig. 6-19(a) shows that for a readily atomized element such as copper the stoicheiometry of the flame and the region used for absorption measurements have relatively little effect on the signal obtained. In contrast, these parameters become of greater importance for an element of lower free atom fraction such as calcium (Fig. 6-19(b)) and quite critical for an element such as tin which is difficult to atomize (Fig. 6-19(c)). It should be remembered when selecting the flame conditions for difficult elements that as a result of the mechanism of the atomization process (see Section 6.5) the optimum position in the flame may vary as the air:acetylene ratio is altered. This is illustrated for molybdenum in Fig. 6-20, for which the largest signal is obtained in a very rich flame.

In comparison with AAS, relatively few atomic fluorescence measurements have been reported in the air-acetylene flame. This has resulted partly from the desire to obtain the highest possible fluorescence detection limits, for which lower temperature hydrogen flames are more attractive as a result of their very low background emission and noise (see Section 6.3.3). The air-acetylene flame is also unsuitable in this respect because of the high concentration of

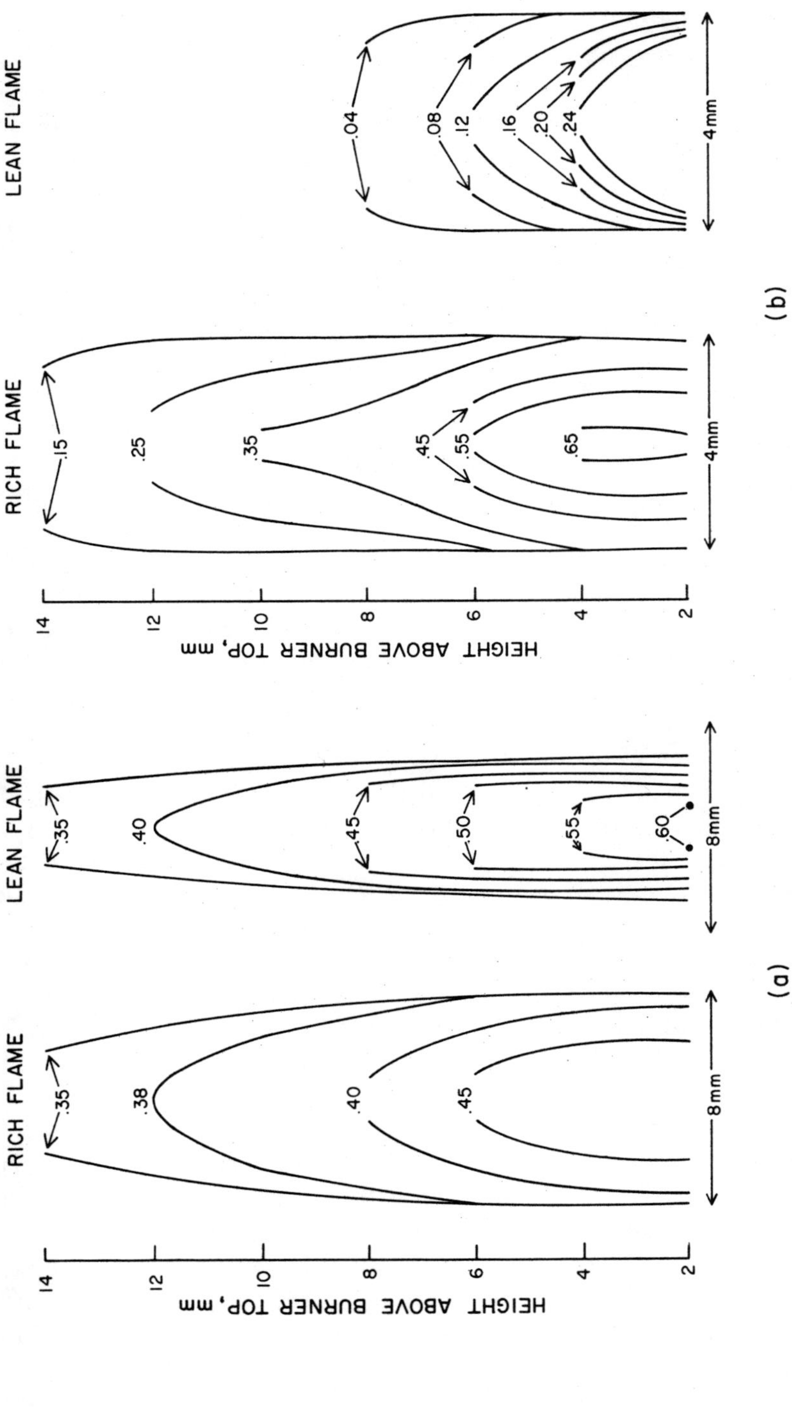
RICH FLAME
LEAN FLAME
RICH FLAME
LEAN FLAME
HEIGHT ABOVE BURNER TOP, mm
HEIGHT ABOVE BURNER TOP, mm
14
12
10
8
6
4
2
.35
.38
.40
.45
.35
.40
.45
.50
.55
.60
.15
.25
.35
.45
.55
.65
.04
.08
.12
.16
.20
.24
8mm
8mm
4mm
4mm
(a)
(b)

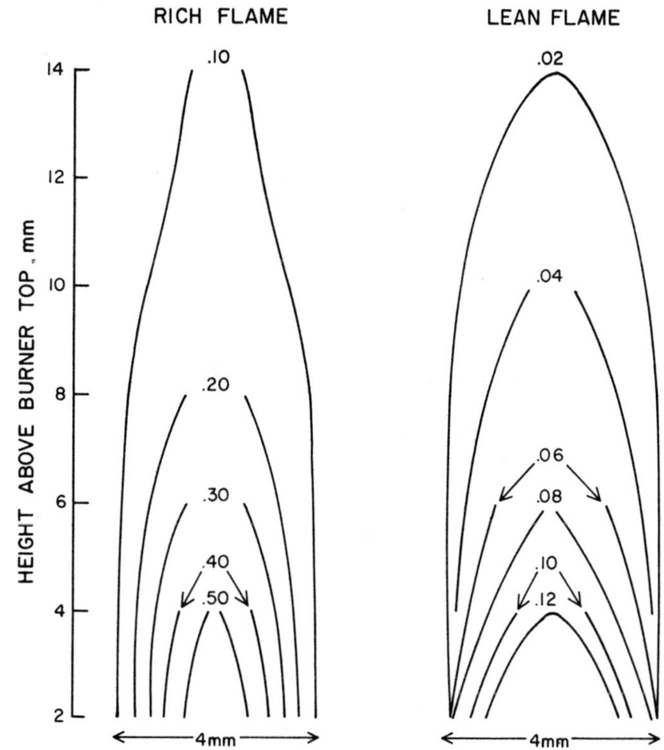

(c)

Fig. 6-19. Absorption profiles in a 10 cm air-acetylene flame under fuel-rich (air : acetylene = 5·65 : 1) and fuel-lean (air : acetylene = 9·45 : 1) conditions.[63] (a) Copper, 324·7 nm, 1 ppm; (b) calcium, 422·7 nm, 20 ppm; (c) tin, 286·3 nm, 500 ppm (rich-flame) and 5000 ppm (lean flame).

fluorescence quenching species (particularly N_2 and CO and possibly CO_2 in fuel-lean flames) compared with flames such as those obtained with hydrogen-oxygen-argon mixtures (see Section 4.5). In recent years, however, it has been realized (as with AAS) that in practical situations efficient atomization and freedom from interferences in AFS can often be obtained for many more elements and samples by using the hotter air-acetylene flame. Thus, several AFS

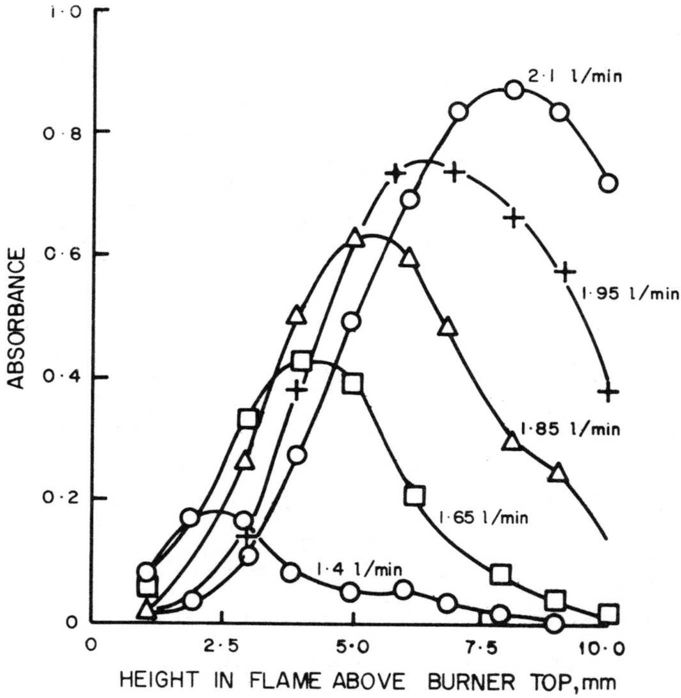

Fig. 6-20. Effect of flame stoicheiometry and height of measurement on absorbance of 100 ppm molybdenum in an air-acetylene flame.[64] (The air flow was constant at 5·5 l/min while varying the acetylene flow as shown.)

studies have now been reported using this flame. Many of these applications have used the technique of flame separation in order to reduce the flame background emission and these have already been described in Section 6.3.2. Provided a conventional spectrometer of good resolution and a modulated source and ac detection system is used, however, even flame separation is not essential and the conventional air-acetylene flame is able to give sufficient fluorescence sensitivity for the determination of many elements (see Table 6-3).

TABLE 6-3

Atomic fluorescence detection limits in an air-acetylene flame[44]

Element	λ(nm)	HCL current, (mA)	Detection limit, ppm	
			Separated	Unseparated
Ag	328·1	45	0·02	0·08
As	193·7	30	nd[a]	nd
Au	242·8	40	7·0	14
Bi	306·8	40	55	220
Ca	239·9	45	nd	nd
	422·7	45	0·05	0·1
Cd	228·8	25	0·1	0·15
Co	240·7	40	1·7	2·5
Cr	357·9	40	5·7	14
	425·4	40	0·5	1·0
Cu	324·7	40	0·05	0·14
Fe	248·3	40	0·15	0·45
Hg	253·7	24	60	120
Mn	279·5	40	0·05	0·13
Ni	232·0	40	0·5	1·0
Pb	283·3	40	3·0	6·0
	405·7	40	1·0	2·3
Pd	247·6	40	nd	nd
	340·4	40	1·0	3·5
Pt	265·9	40	60	90
Sb	231·1	40	nd	nd
Se	196·1	18	180	180
	204·0	18	45	45
Sn	286·3	40	75	300
	303·4	40	25	100
Te	214·3	40	20	30
Tl	276·8	40	7·0	14
	377·6	40	1·2	3·0
Zn	213·8	40	0·10	0·13

[a] nd = not detected.

(*b*) *Nitrous oxide-acetylene flame*

The nitrous oxide-acetylene flame was little more than a curiosity, of interest only to combustion chemists, prior to its introduction by Willis[65] in 1965 as a possible atom cell for AAS. It is now widely used, however, for routine applications and its use is essential for the determination of some elements by AAS. The advent of this flame has almost doubled the number of elements that may be determined and has resulted in the elimination of many of the chemical interferences that were previously encountered for a number of common sample matrices when an air-acetylene flame was employed.

The nitrous oxide-acetylene combination was originally chosen for its ability to give a flame almost as hot as oxygen-acetylene but with a very much lower burning velocity. This latter feature allows the

TABLE 6-4

Free atom fractions in the premixed nitrous oxide-acetylene flame

Reference	59	59	53	53	61
Notes	*b, d*	*b, e*	*a, c, f*	*b, c, f*	*a, g*
Element					
Ag					0·57
Al	0·29	0·29	0·28	0·15	0·13
Au	0·17	0·16	0·48		0·27
B					0·0035
Ba	0·11	0·074	0·20	0·08	0·17
Be					0·095
Bi					0·35
Ca	0·50	0·34	0·94	0·76	0·52
Cd	0·67	0·60			0·56
Co	0·13	0·11			0·25
Cr	1·00	1·02			0·63
Cu	0·54	0·49	0·68	0·38	0·66
Fe					0·83
Ga					0·73
In	0·45	0·37			0·93
K					0·12
Li			0·30	0·32	0·34
Mg	1·00	1·07	1	1	0·88
Mn	0·38	0·39	0·52	0·38	0·77
Na	0·31	0·32	0·23	0·44	0·97
Pb					0·84
Si					0·06
Sn	0·24	0·35	0·52	0·34	0·82
Sr	0·98	0·57			0·26
Ti			0·17	0·17	0·11
Tl					0·55
V			0·51	0·17	0·32
Zn	0·54	0·49	0·61	0·56	

[a] Values measured by integrated (continuum) absorption technique.

[b] Values measured by peak (sharp line) absorption technique and less reliable for many elements

[c] Values relative to an assumed value of 1 for magnesium.

[d] 5 cm slot burner, nitrous oxide : acetylene = 2·01:1.

[e] 5 cm slot burner, nitrous oxide : acetylene = 1·82:1.

[f] 7·6 cm slot burner, nitrous oxide:acetylene optimized within the range (1·73-1·87):1 for each element.

[g] 5 cm slot burner, nitrous oxide:acetylene = 2·14:1.

flame to be supported at slot burners as long as 100 mm while retaining safe operation by relatively inexperienced operators. It was further noted during early work[9,66,67] with this flame that the characteristic red, interconal zone obtained under slightly fuel-rich conditions (see Section 6.2.3) is extremely effective for the atomization of elements that form stable, refractory oxides without the need to resort to the unpleasant and inconvenient "sooty" flame previously used to obtain reducing conditions with other gas mixtures. The reducing nature of this zone appears to result[17,67] from the presence in the interconal zone of a substantial concentration of gaseous carbon-containing species (e.g. C, CN, CH and HCN) and very low concentrations of oxidizing species. With careful optimization of flame and burner conditions the interconal zone extends over a relatively large volume and allows extremely sensitive absorption measurements even for elements forming stable oxides, while low noise levels (i.e. good detection limits) are retained.

The effectiveness of this flame compared with the air-acetylene flame for elements such as Al, Be, Ti and V can be seen by comparing the free atom fractions shown in Table 6-4 for the nitrous oxide-acetylene flame with those of Table 6-2. A typical absorption profile for a "refractory" element (titanium) is illustrated in Fig. 6-21. It can be seen that the flame volume in which a strong absorption signal is obtainable is considerably enlarged in the rich flame. The confinement of efficient atomization to a central region of the flame occurs only for such difficult elements; in other cases the adjustment of flame-operating conditions is much less critical. This is apparent from Fig. 6-22 which shows envelopes of absorption for a number of metals, i.e. the contours along which the absorption signal has fallen to 1% of the maximum value obtained at the centre of the flame. As mentioned above, even when the correct flame region is chosen, successful determination of the refractory elements requires exact adjustment of the fuel:oxidant ratio. This is strikingly illustrated in Fig. 6-23(a) which shows the effect of this parameter on free atom fractions calculated theoretically for equilibrium conditions. Figure 6-23(b) shows the simultaneous change in the concentrations of major flame species; the critical point is coincident with a sudden decrease in concentration of the reducing species (HCN, C_2H and unburnt acetylene) followed by a rise in the level of oxidizing species (OH and O). It should be noted that these plots are for a constant temperature of 2900 K so that the situation is not strictly comparable with a real flame where the temperature will change with the fuel:oxidant ratio.

Few applications of the nitrous oxide-acetylene flame to AFS measurements have been reported as, like the air-acetylene flame, it contains high concentrations of efficient quenching species (particularly N_2 and CO) and this flame also exhibits more intense background emission than the air-acetylene flame. Nevertheless, the

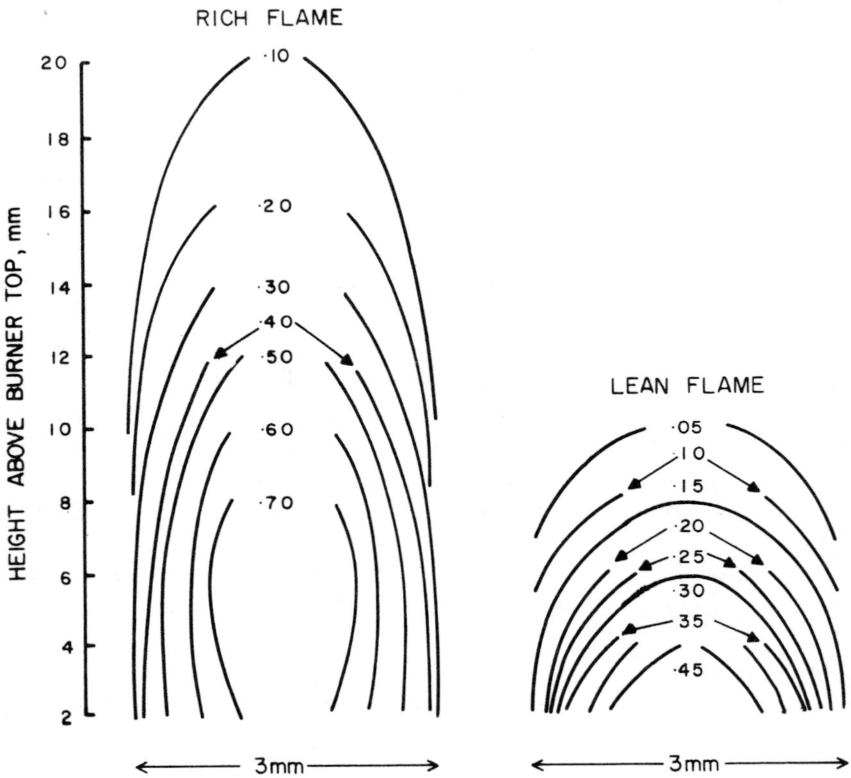

Fig. 6-21. Absorption profiles in a 5 cm nitrous oxide-acetylene flame under fuel-rich (nitrous oxide : acetylene = 1·1 : 1) and fuel-lean (nitrous oxide : acetylene = 1·15 : 1) conditions for titanium (365·3 nm, 2500 ppm).[63] (The red interconal zone extends to a height of 40-50 mm in the rich flame and 4-6 mm in the lean flame.)

difficulty of atomizing refractory elements in other flames more suitable for fluorescence studies has prompted several studies of the feasibility of using this flame as a fluorescence cell.[43,49-51] In each case flame separation as described in Section 6.3.2 was utilized to minimize flame noise and background from the secondary reaction zone. Useful results can be obtained for those refractory oxide-

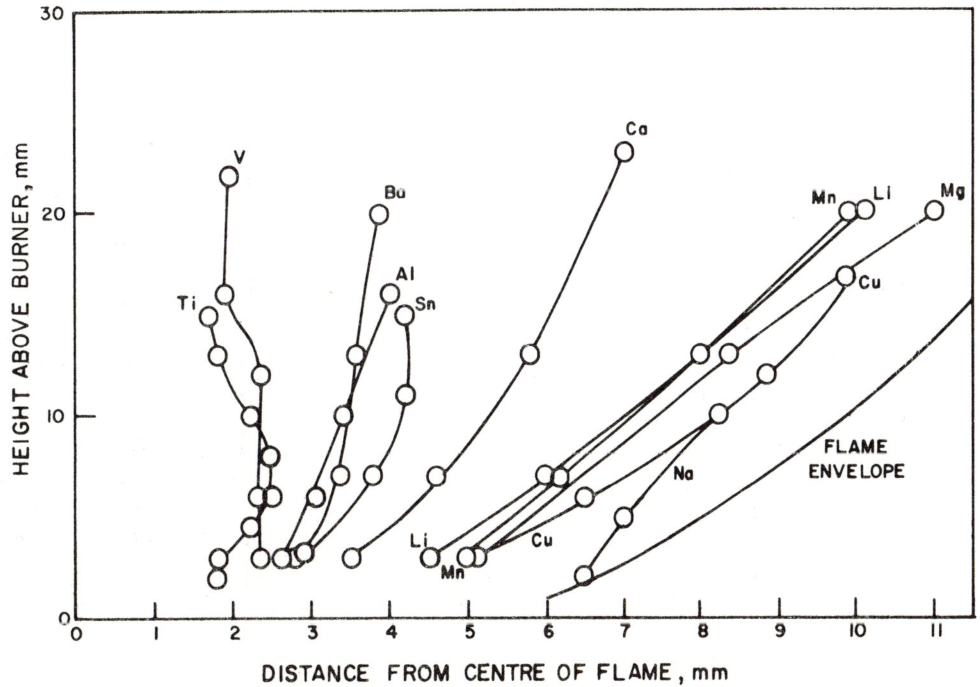

Fig. 6-22. Envelopes of metal absorption in a nitrous oxide-acetylene flame (nitrous oxide : acetylene = 1·87 : 1).[53] (The points correspond to absorbance values equal to 1% of the maximum value obtained at the centre of the flame.)

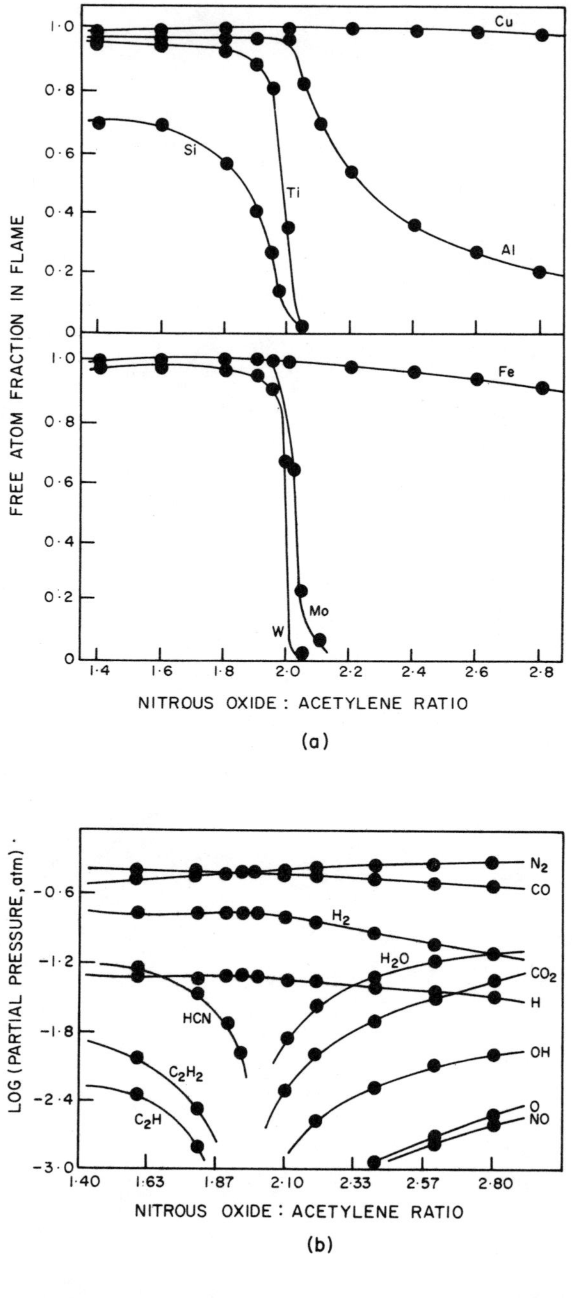

Fig. 6-23. Calculated effect of fuel : oxidant ratio for a nitrous oxide-acetylene flame under isothermal conditions.[68] (a) On metal-free atom fractions; (b) on partial pressures of major flame species.

forming elements that have strong fluorescence lines below *ca.* 300 mm. There is also the possibility that the flame may be useful to minimize chemical interference effects while retaining reasonable fluorescence sensitivity for other elements having their fluorescence wavelengths in this more favourable spectral region (e.g. magnesium[43]). The nitrous oxide-acetylene flame has also been used as an atom cell in the atomic fluorescence measurements made by Winefordner and co-workers[69-71] using a tunable laser as the light source (see also Sections 5.4.3 and 14.4). As a result of the intense radiation available from the laser, the low-duty cycle of the pulsed source and the sophisticated electronics of the detection system used, they were able to overcome some of the difficulties mentioned above and utilize the efficient atomization offered by this flame to determine a number of refractory oxide-forming and rare-earth elements with high sensitivity.

(*c*) *Oxygen-acetylene flame*

Premixed oxygen-acetylene flames burning on small single-jet burners have been used in analytical flame emission spectroscopy (see, for example, ref. 6). The small flame supported at this type of burner is unsuitable as an absorption cell and does not appear to have been investigated for fluorescence measurements. Analytical AAS and AFS measurements have been reported in unpremixed turbulent oxygen-acetylene flames and are described separately in Section 6.5. As a result of the difficulty of burning large premixed oxygen-acetylene flames safely on the type of burner commonly employed for AAS and AFS work (see above) there appear to have been few reports of their application to the former technique and none to the latter. A rectangular burner producing a premixed flame 76 mm in length has, however, been investigated as an absorption cell for refractory elements by Fiorino *et al.*[21] Fassel and co-workers have conducted a comprehensive study of the general properties of oxygen-acetylene flames (see ref. 72 for details). It would seem that the oxygen-acetylene flame gives very similar results to the slightly cooler nitrous oxide-acetylene flame. Advantages in the use of this flame are the lower cost of oxygen compared with that of nitrous oxide and the reduced intensity of the CN emission bands, which in the nitrous oxide flame may cause high noise levels in several narrow spectral regions even when absorption or fluorescence measurements are made with modulated source and detection systems.

(d) Hydrogen flames

A widely recognized feature of all premixed hydrogen flames is the low background emission intensity which they exhibit. The background emission is less intense than that of the acetylene flames discussed above, even for the hot hydrogen flames obtained by direct combustion with oxygen or nitrous oxide, and very much less intense than this for the cooler flames obtained using air or oxygen plus an inert diluent. This would be a significant advantage for atomic fluorescence measurements and useful for absorption work. Furthermore, the great advantage of the low degree of fluorescence quenching observed in premixed oxygen-argon-hydrogen flames has already been described in some detail in Chapter 4 and is discussed further elsewhere in this chapter. Unfortunately, a disadvantage of all hydrogen flames for practical AAS and AFS applications is their limited ability to atomize some analyte elements and vaporize some types of sample matrix. As a result, analytical sensitivity by either absorption or fluorescence is unsatisfactory for many elements and even those readily atomized may be subject to many chemical interferences. Thus, although these flames have been widely used in fundamental work* and to obtain exceptionally good fluorescence detection limits for favourable elements in pure solutions, their range of application to practical analyses with real samples is somewhat limited.

The stoicheiometric oxygen-hydrogen flame has an extremely high burning velocity—actually greater than that of the oxygen-acetylene flame (see Table 6-1). As mentioned above, it is also less effective for atomization than the oxygen-acetylene or nitrous oxide-acetylene flames in spite of having a similar temperature to the latter, owing to the absence of carbon-containing reducing species and the presence of high concentrations of oxidizing species even under very fuel-rich conditions.[68] Thus, no analytical applications of these premixed flames have been reported.† The burning velocity of the oxygen-hydrogen mixture is considerably lower under very fuel-rich conditions or in the presence of inert diluents such as argon or nitrogen. When using these slow-burning mixtures on simple burners, however, great care must be taken to avoid accidental combustion of a dangerous (i.e. lean or stoicheiometric) gas mixture. The nitrous oxide-hydrogen flame can be burnt safely over a wide range of mixture strengths owing to its low burning velocity compared to that

* A number of references to this work will be found elsewhere in this chapter.

† Applications of the corresponding unpremixed flames are described in Section 6.4.2.

of the oxygen-hydrogen flame (see Table 6-1). The poor atomization efficiency in this flame for many elements has been confirmed both by theoretical calculations of flame composition[68,73] and by experiment.[61,74] Free atom fractions for a range of elements are shown in

TABLE 6-5

Free atom fractions in premixed nitrous oxide-hydrogen and air-hydrogen flames[61]

Element	N_2O-H_2 flame[a]	Air-H_2 flame[b]
Ag	0·72	0·85
Al	< 0·0001	< 0·0001
Au	0·43	0·54
B	< 0·001	< 0·001
Ba	0·0046	0·005
Be	0·0004	< 0·0001
Bi	0·26	0·63
Ca	0·036	0·15
Cd	0·62	0·37
Co	0·28	0·21
Cr	0·042	0·31
Cu	0·92	0·96
Fe	0·91	0·82
Ga	0·16	0·45
In	0·61	1·07
K	0·20	0·40
Li	0·094	0·14
Mg	0·97	0·87
Mn	0·54	0·75
Na	0·90	1·06
Pb	0·93	1·03
Si	< 0·001	< 0·002
Sn	0·059	0·38
Sr	0·039	0·17
Ti	< 0·003	< 0·002
Tl	0·61	0·68
V	0·0081	0·018

[a] 5 cm long slot burner, hydrogen: nitrous oxide = 1·72:1.
[b] 5 cm long slot burner, hydrogen: air = 1·12:1.

Table 6-5. Furthermore, the high concentration of nitrogen in the burnt gases gives rise to quenching of atomic fluorescence (i.e. low fluorescence quantum yields) similar to that given by the nitrous oxide-acetylene flame. Hence, the only advantage of the use of this flame lies in its relatively low background emission; the flame has not been applied to practical analysis.

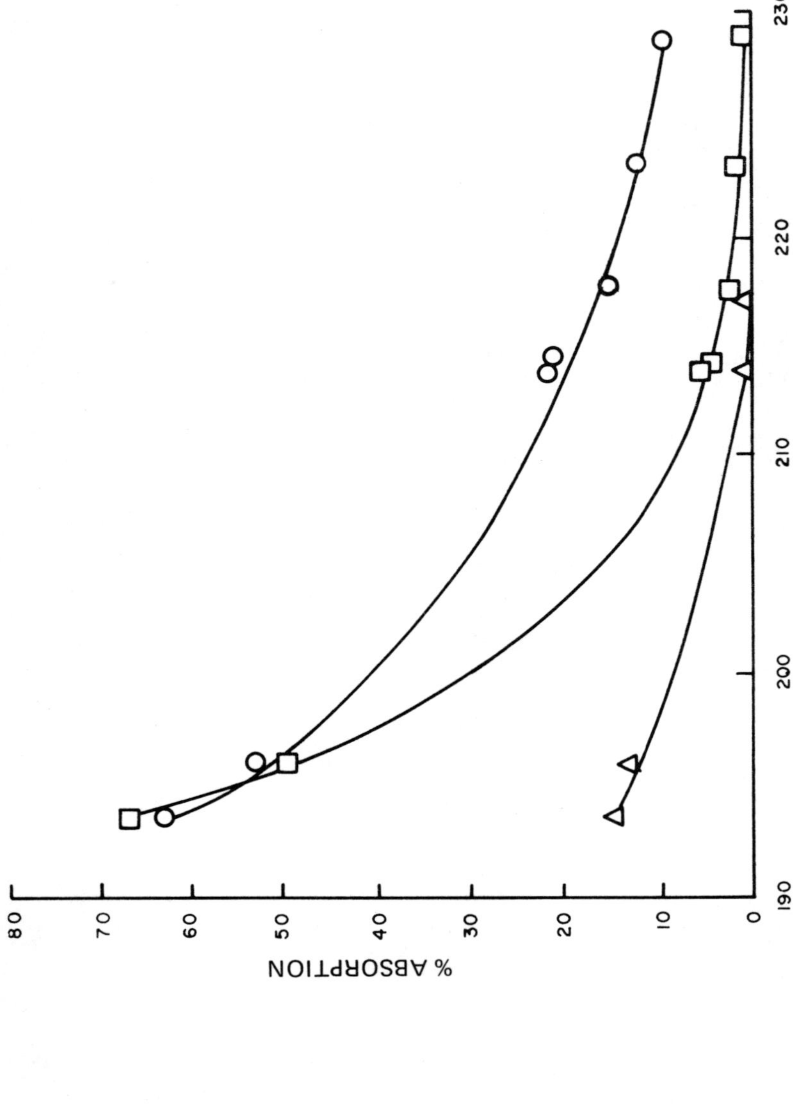

Fig. 6-24. Background flame absorption below 230 nm for air-acetylene (○), air-hydrogen (□) and argon-hydrogen (△) flames.[75]

A number of analytical studies of the use of cooler hydrogen flames with air or oxygen-argon mixtures have been reported; many of these employ diffusion flames, however, and are described in Section 6.4. The premixed air-hydrogen flame can be burnt on a standard long path air-acetylene burner and is recommended by some instrument manufacturers for a few absorption determinations where special conditions apply. Provided sufficient atomization is achieved

TABLE 6-6

Free atom fractions in a premixed oxygen-argon-hydrogen flame[80]

Flame gas mixture[a]	Height in flame(mm)[b]	Cu	Mg	Mn	Sr[c]
1	20·8	0·60	0·052	0·17	0·094
2	20·8	0·80	0·28	0·26	0·012
3	20·8	0·71	0·024	0·016	0·001
4	20·8	0·78	0·004	0·005	0·0002
1	35·3	0·73	0·095	0·18	0·080
2	35·3	0·80	0·29	0·23	0·005
3	35·3	0·72	0·021	0·013	0·0002
4	35·3	0·68	0·004	0·003	0·00007
1	57·0	0·97	0·10	0·19	0·079
2	57·0	0·94	0·22	0·19	0·001
3	57·0	0·80	0·016	0·01	0·0001
4	57·0	0·78	0·002	0·004	0·0004

[a] $Ar:O_2:H_2$ flow rates (l/min) are

1 13·8:0·70:6·84
2 13·8:2·24:6·84
3 13·8:3·44:6·84
4 13·8:3·95:6·84.

[b] A cylindrical flame with a concentric outer flame sheath was used.
[c] Ionization was not suppressed for these values.

(see Table 6-5) and chemical interferences do not occur, the very low flame background absorption and emission is always an advantage. This effect is enhanced for elements determined at resonance lines below 200 nm (As, Se) because background absorption by flame species is lower in this flame than in acetylene flames. Argon-diluted flames exhibit even lower background absorption in this wavelength region, as illustrated in Fig. 6-24. An exceptional useful application of the air-hydrogen flame in AAS is to the determination of tin. Tin is more efficiently atomized in this flame than in the air-acetylene flame,[66] possibly owing to the formation of an intermediate tin hydride which is readily dissociated. Both premixed air-hydrogen[76,77,78] and argon-oxygen-hydrogen[79] flames have been applied to AFS measurements. Some free atom fraction data obtained in the latter flame are shown in Table 6-6. The highest fluorescence yield will be obtained with the argon flame but full benefit of the reduced quenching is obtained only if measurements are made near the base of a flame with good laminar flow characteristics, or if an outer sheath is used to prevent entrainment of atmospheric nitrogen.[52] Quenching will also be minimized by using the stoicheiometric flame diluted with as much argon as possible,[81] but this requirement is often incompatible with efficient atomization (see Table 6-6).

6.4 OTHER FLAME ATOM CELLS USED IN ANALYTICAL SPECTROSCOPY

Although the majority of AAS and AFS applications involve the use of laminar premixed flames, a few other flame types have been investigated as possible alternatives. Some of those available are discussed briefly here together with reports of more recent improvements or new applications. For convenience, the long-path cells which may be obtained by directing a flame along an absorption tube are also described here.

6.4.1 Laminar Diffusion Flames

Laminar diffusion flames have been studied to a lesser extent than their premixed counterparts, partly because of the difficulty of defining fundamental characteristics that can easily be measured. As a result few data are available on the basic properties of these flames. The exact nature of such flames varies with the type of burner used, as this will effectively change the gas mixture that actually burns for a given supply of combustants. Some current views on the basic properties of these flames have been summarized by Gaydon and Wolfhard[1] but the subject is too diffuse to discuss here. Suffice it to say that a large area of the flame may be regarded as a reaction zone; unlike the reaction zone of premixed flames, however, this region is not an area that exhibits intense spectral radiation or noise and any departures from equilibrium are minor by comparison. This situation arises from the gradual mixing and combustion of the fuel and oxidant, in contrast to the gases of premixed flames having a high chemical potential which is rapidly released in the reaction zone. In a diffusion flame a preheating zone exists on each side of the reaction zone in which the temperatures of the fuel and oxidant are increased. There may also be a region of hot flame gases in which combustion has been completed. In practice, however, the slow rate of mixing may mean that any such region will be a considerable distance from the burner, except in special cases such as the flat diffusion flames obtained from adjacent thin streams of fuel and oxidant.

As a result of the relatively low intensity of combustion in diffusion flames and the extended area over which it occurs, heat loss from these flames is considerable. Thus the temperature in many cases will be low, especially if aqueous solutions are nebulized into the flame in spectroscopic applications. This can change the degree to which combustion moves to completion compared with premixed flames. For example, with diffusion flames of almost all hydrocarbons solid carbon particles are formed and continue to burn as they move upwards from the burner. In many cases combustion may not be completed at all and unburnt carbon will leave the flame as soot. From an analytical point of view the presence of carbon particles in the flame gases is of far greater importance than simply evidence of incomplete or slow combustion; it results in a characteristic luminous yellow flame, which is unsuitable for spectroscopic measurements whether by emission, absorption or fluorescence techniques. Thus we are concerned here only with the laminar diffusion flames of hydrogen. A luminous flame is then no longer obtained and combustion will also proceed at a rather greater rate owing to the more rapid diffusion of the hydrogen fuel. In fact one of the major advantages of the use of hydrogen laminar diffusion flames for AAS or AFS applications is their extremely low background emission in both the visible and ultraviolet regions.

Most analytical applications of laminar hydrogen diffusion flames have involved the use of atmospheric air as the oxidant gas; i.e. a laminar stream of hydrogen is supplied through a burner and mixes with the surrounding air by diffusion to give a combustible mixture. There would be little additional difficulty in surrounding the hydrogen stream by a similar laminar flow of oxygen or perhaps nitrous oxide to obtain a hotter flame. Such a device would be more versatile and still remain free from the flash-back danger inherent in premixed flames, but there appear to have been no applications of this type of flame to AAS or AFS. A probable explanation lies in the use of the air diffusion flames mostly because of their very low temperature and the superiority of premixed flames if a hotter flame is required. Indeed a feature common to many AAS or AFS studies with air/hydrogen diffusion flames has been the simplicity of the burner required. In many cases hydrogen is passed through the type of simple cylindrical or rectangular Meker-type burner supplied with many flame emission or absorption spectrophotometers. Under these circumstances the implicit assumption made above that all mixing of hydrogen and air occurs by diffusion may not be completely true. The gas stream leaving the burner may already have some degree of turbulent motion owing to too high a Reynolds number in the burner tube or else turbulence may occur within a short distance above the outlets. Under these conditions some mixing may occur by entrainment of air. In many cases a flame is also obtained whose base is visibly a short distance above the top of the burner and air-entrainment in this region will then be appreciable. From the point of view of application to AAS and AFS this situation is, however, of little consequence. The flames retain the low temperature and background emission characteristics of "laminar diffusion flames" and the analytical chemist will probably notice little difference from, say, the flame obtained with a carefully designed capillary burner producing truly laminar flow conditions.

The disadvantages of air/hydrogen diffusion flames should be apparent from our earlier discussion of their premixed counterparts. Thus, low atomization efficiency and high susceptibility to chemical interferences may arise as a result of the lower temperature, and the effect of sample nebulization in this respect will be even more serious unless care is taken to remove all large droplets from the sample aerosol. The need to nebulize using the hydrogen flow may also be regarded as a disadvantage. This problem has frequently been avoided by using a flow of nitrogen or argon for nebulization, the inert gas and sample mist being added to the hydrogen prior to passage through the burner. An even cooler flame is then obtained and this has proved an advantage in several reported applications,[82,83] e.g. in the determination of arsenic and selenium at their resonance lines below 200 nm where the flame background is low compared to those of other flames. For fluorescence studies argon is preferred to nitrogen[84] in order to reduce quenching by the flame gases. In spite of their simplicity and safe operation these entrained air flames have seen little application to the direct analysis by nebulization of solutions of real samples.

6.4.2 Turbulent, Unpremixed Total-consumption Flames

As explained in Section 6.1.1, the low temperature and inefficient combustion of laminar diffusion flames can be overcome, without losing their freedom from flash-back dangers, by forced mixing of the oxidant and fuel gases using a burner with specially shaped outlet jets. Unfortunately this forced mixing of the high velocity gas streams also produces a flame which is extremely turbulent, exhibits high acoustic noise, and to a certain extent shows the high background radiation associated with the reaction zone of premixed flames. In spite of these and other disadvantages described below a considerable number of applications of these flames to AAS and AFS have been reported. This may be attributed to the almost exclusive use of one particular type of burner which provides a uniquely compact, simple and inexpensive complete "flame atom cell system". No other types of turbulent unpremixed burner or flames will be discussed here.

The burner system in question is usually known as the total-consumption nebulizer-burner and was introduced by Gilbert.[85] Several variations of the original design are available commercially, but that shown in Fig. 6-25 is representative. The oxidant gas is passed through a nozzle which surrounds a fine capillary tube. This capillary is extended by a slightly wider tube which protrudes from the base of the burner and can be dipped into a beaker of sample solution. The motion of the oxidant past the capillary outlet causes a pressure drop and solution is drawn along the capillary into the gas stream and nebulized.* The oxidant jet is surrounded by a second concentric nozzle through which the fuel gas is passed. Forced mixing of oxidant, fuel and sample aerosol is obtained and a high-temperature turbulent flame may be burnt with complete safety even for stoicheiometric oxygen/acetylene and oxygen/hydrogen mixtures. No problem is experienced with overheating of the burner even though large flames are usual with the small, uncooled burner shown in Fig. 6-25. This results from the small "dead" space always present between the tip of the burner and the base of the flame. Combustion under these conditions is probably stabilized by the presence at the jet of a small diffusion flame obtained from part of the fuel gas and entrained atmospheric air. There are no problems of matching gas and burning velocities so that the same unit may be used without modification for air/hydrogen or even argon-hydrogen/entrained air flames.

* The mechanism of the nebulization process is explained in detail in Section 8.2.

The nebulizer-burner has the advantage of simplicity and of supplying all of the nebulized sample solution to the flame (as explained in Section 8.2, use of a remote nebulizer with premixed flames inevitably results in the loss of a considerable fraction of the

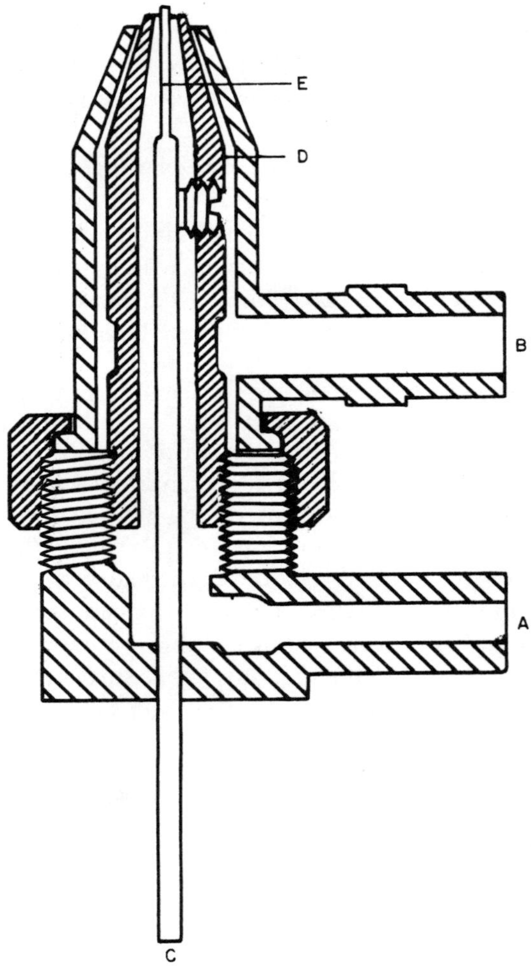

Fig. 6-25. Total consumption nebulizer burner.[85] A, Oxidant inlet; B, fuel inlet; C, capillary tube for sample solution; D, fuel nozzle; E, oxidant nozzle.

sample aerosol). On the other hand it also adds directly to the disadvantages of the turbulent flame already mentioned. Thus the large volume of liquid drawn into the flame may give appreciable cooling, reducing the temperature to a level well below that observed with equivalent premixed flames.[86] This problem can be overcome, however, by restricting the solution uptake rate.[87] A more serious disadvantage is the relatively large fraction of the sample that enters the flame as large droplets.[88] As explained in Section 6.5.1 these are incompletely evaporated by the flame and pronounced chemical interferences may occur for many practical sample matrices. Several improvements to the original design have been reported to reduce some of these disadvantages. These include use of an oxygen sheath[89] to increase flame temperature and stability when nebulizing solutions, devices to improve the reproducibility of liquid sample volumes added to the flame,[86,90] and addition of a small graphite premixing channel[91] which gives a quieter more stable flame. The burners have also been used with separate premixing chambers[92,93] but the unit then becomes a total-consumption nebulizer-burner giving flames similar to those described in Section 6.2 and 6.3 rather than a turbulent flame device.

Although the original Gilbert burner was intended for application to flame emission spectroscopy, several workers have applied this type of burner to both AAS and AFS. Apart from problems already noted, their application to the former technique is complicated by the incompatibility (for high absorption sensitivity) of the short flame path length with the optical systems of commercial atomic absorption spectrometers designed for rectangular flames. The difficulty has been overcome by using several identical burners in line or by passing the light beam through the same flame several times (with mirror and/or prism systems) to increase the optical path length. Both systems have a number of operating problems (e.g. misalignment, increased signal noise, difficulty of obtaining identical flame conditions at three separate burners, increased complexity and expense, etc.) and have seen little application in recent years. An alternative approach is the use of a long-path absorption tube along which the flame is directed. This is described separately in Section 6.4.3.

A number of applications of the total-consumption nebulizer-burner to AFS have also been reported for relatively cool, turbulent hydrogen flames.[78,94-98] The turbulent oxygen/acetylene flame has also been investigated and found to be relatively unsuitable.[99,100] In fact Pearce *et al.*[101] found that for a number of elements in these turbulent flames there was little variation in fluorescence yield between hydrogen-argon/entrained air, hydrogen/oxygen and acetylene/oxygen flames. Thus, the advantage of the cooler hydrogen flames must lie in their lower flame noise and possibly more efficient atomization of certain elements. It is sometimes suggested that these cooler flames at total-consumption burners are more likely to exhibit scattering interferences from solid particles in the path of the light beam. Recent measurements[78,102] show, however, that this is not normally a problem except, perhaps, where large amounts of releasing agent are added to samples to suppress chemical interferences.

6.4.3 Long-Path Flame-Absorption Cells

Long-path absorption systems based on non-flame methods of atomization (i.e. using furnaces etc.) are described in Section 7.2. A number of devices have been reported, however, in which a conventional flame is directed along a steel, ceramic or silica tube.[103-111] This relatively simple technique can, under suitable conditions, provide an extremely long path length in which absorbing analyte atoms exist in the hot flame gases. In principle it should be possible to use any flame in this way, but most of the reported applications appear to be based on the use of total consumption nebulizer-burner units. Hydrogen flames with air or oxygen are also the most commonly used; acetylene flames may show a prohibitively high level of background noise and absorption over such an extended optical path. In practice, even hydrogen flames produce relatively high signal noise so that detection limits are often not as good as would be expected from sensitivity measurements reported in the literature. The effective signal : noise ratios are also reduced by the need to use a relatively narrow optical beam for passage through the tube, as this results in less light being available at the photodetector.

Although dramatic increase in the attainable sensitivity (for 1% absorption) is possible with this technique, a number of other disadvantages which are sometimes overlooked may occur. Perhaps the most important of these is the relatively low temperature and atomization efficiency of the flame gases within the tube, particularly with hydrogen flames. Thus, chemical interferences frequently occur with real sample solutions and even with pure solutions only a few elements have a sufficiently long lifetime under these conditions to show the full benefit of, say, a 50 cm path length. This is illustrated in Fig. 6-26 where it can be seen that the absorbance of Fe and Co (measured across the tube) falls off very rapidly with distance from the entrance compared with that of the easily atomized Pb, Ag and Cu. The problem can be minimized to a certain extent by placing the tube in a high temperature furnace[111] or by heating it electrically. In fact, this precaution also reduces the tendency of such devices to show memory effects and signal drift. This may occur when the tube walls gradually become coated with deposits produced from sample materials introduced into the flame. Drift may occur when the absorption tube is constructed from polished silica. Thus, part of the "zero" signal reaching the detector is due to light that has been reflected several times within the tube,

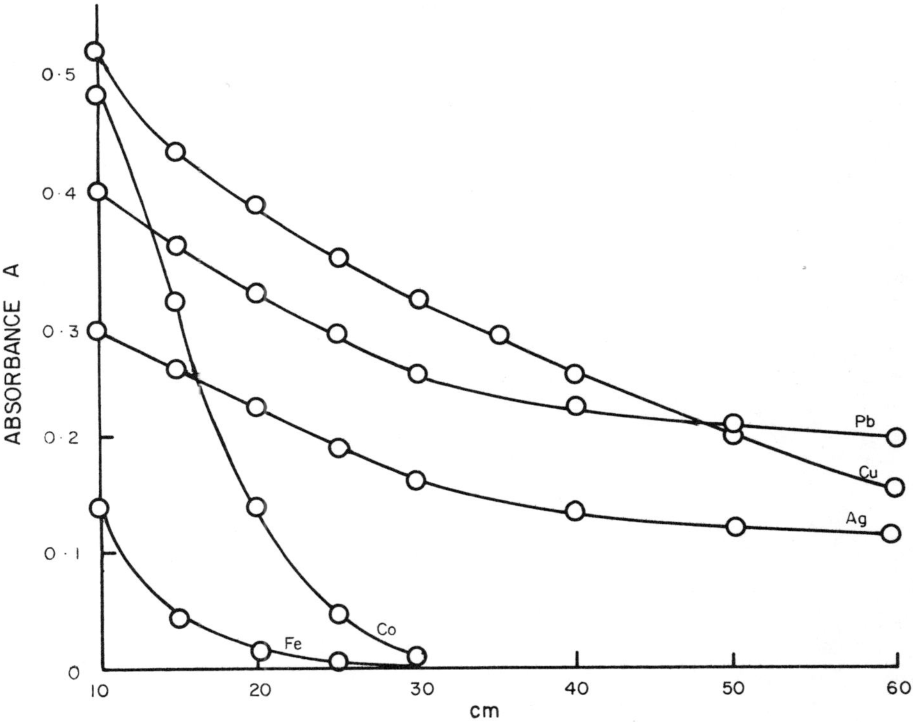

Fig. 6-26. Variation of absorbance with position in the absorption tube for a long-path device using an air-hydrogen flame.[110] (Absorbance values at each distance indicated were measured perpendicular to the tube axis.)

and any change in the reflection coefficient owing to deposition of solid material will be recorded as a drifting baseline.

Some of the problems mentioned above were overcome by Hingle *et al.*[111] who replaced the usual turbulent hydrogen flames with a premixed air-acetylene flame supplied from an indirect nebulizer system. The long-path burner (see Fig. 6-27) used by these workers employed the technique of flame separation described in Section 6.3.2 to avoid the high flame-absorption and noise which would otherwise result from such a long path-length through an air-

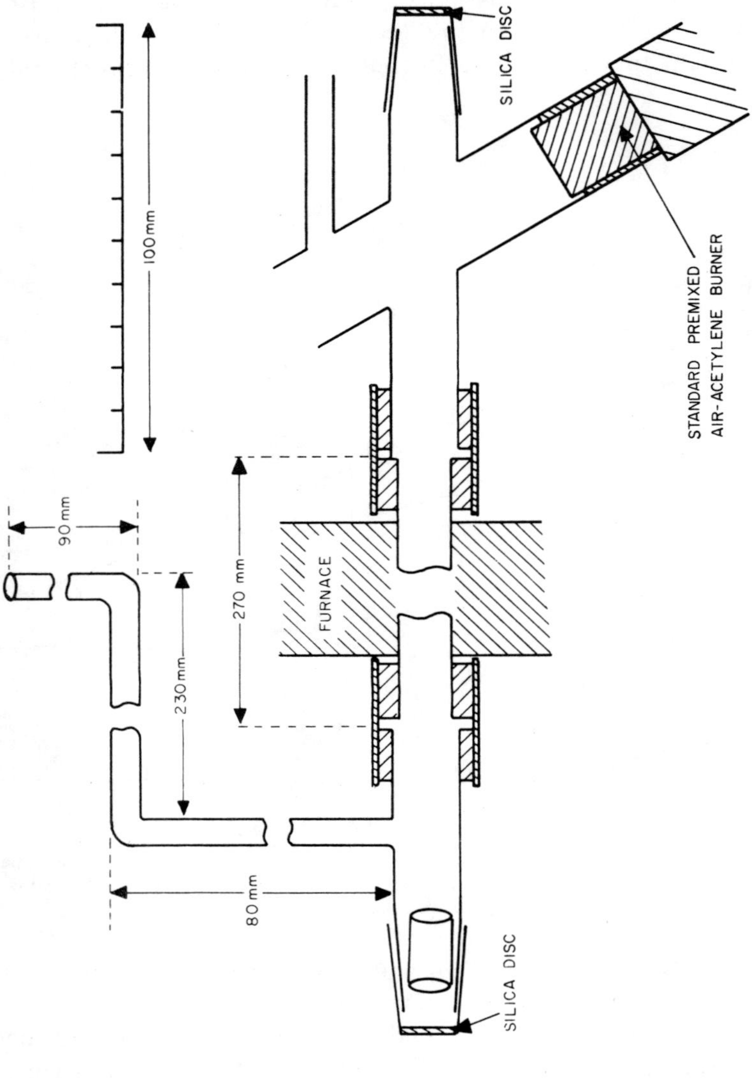

Fig. 6-27. Long-path absorption cell using a premixed, separated air-acetylene flame.[111]

acetylene flame. They reported good sensitivities for elements that are difficult to atomize as well as for easy atomized ones, and improved precision, stability and freedom from interference compared to some of the devices described above.

6.5 ATOMIZATION OF NEBULIZED SAMPLES IN THE FLAME

The formation in the flame of a population of free metal atoms from the analytical sample added to it is a complex process involving several distinct stages. The nature of this process must be fully understood if full use is to be made of the potential of the flame atom cell during absorption and fluorescence measurements. Although it will be possible to obtain a signal from almost any sample simply by adding it to the "recommended" flame, better performance with respect to both sensitivity and freedom from interferences can almost invariably be obtained by ensuring that optimum conditions are obtained at each stage of the atomization process. Practical suggestions to this end will be found in Chapter 11; for the present we are concerned mainly with the atomization mechanism which forms the basis of such recommendations. As the majority of samples are added to the flame as a nebulized mist (aerosol) we shall consider the atomization process to begin from the moment the solution leaves the nebulizer jet.

6.5.1 Desolvation of the Sample Solution

Vaporization of the sample cannot begin until all solvent has been removed from the spray droplets that enter the flame. With the indirect nebulizers mostly used with premixed flames (see Section 8.2.1) evaporation of solvent will begin in the spray chamber and continue in the preheating zone of the flame. This forms an important contribution to the desolvation process which must be completed as soon as possible in the flame if efficient atomization is to be obtained in the interconal region used for absorption and fluorescence measurements. Alkemade[112] has shown that the total time required for complete desolvation of a droplet is proportional to its initial surface area and that in a typical flame water droplets of diameter $\leqslant 1$ μm will be completely dry before they have travelled 0·03 mm through the flame, whereas droplets >30 μm in diameter must travel at least 30 mm. Thus, it is important that the nebulizer should produce a spray of extremely fine droplets; large droplets may be allowed to fall out of the sample mist in the spray chamber as they will probably make little contribution to the analytical signal and may have an adverse effect by reducing the flame temperature. This latter effect is particularly well known in total-consumption nebulizer burners (see Section 6.4.2) in which the design is such that a large proportion of an aqueous sample solution reaches the flame in the form of droplets greater than 25 μm in diameter.[88] Here much of the sample will not be atomized until a height of several centimetres is reached in the flame.

The use of a spray chamber in conjunction with a nebulizer producing a large fraction of droplets larger than, say, 10 μm is not very satisfactory as it merely wastes part of the sample and results in poorer sensitivity. Some improvements may be achieved in favourable cases by replacing the water by an organic solvent which evaporates more readily. In certain cases the correct choice of an organic solvent can also result in production of a better spray (i.e. smaller droplets) by the nebulizer owing to the different viscosity and surface tension of the liquid sample. As a result, mixtures of water and organic solvent may sometimes give better results than using pure solvent, even though the latter is more volatile. A more satisfactory way of ensuring complete desolvation is by use of a nebulizer that ensures a uniform spray of small droplets even with aqueous samples. This can be achieved with ultrasonic nebulizers (see Section 8.3.1), but these are more complex and expensive than the simple pneumatic type. An alternative solution is the use of a heated spray chamber (see Section 8.2.1(d)). By use of infrared heating and a condenser it is possible to remove most of the solvent in the spray chamber so that the "spray" entering the flame consists principally of dry particles. Thus, vaporization of the sample aerosol may commence as it enters the flame.

6.5.2 Vaporization of the Sample Matrix

The rate of vaporization of the dry aerosol particles will obviously have an important effect on the formation of free analyte atoms low in the flame and consequently on the absorption or fluorescence sensitivity obtained. Unfortunately, in many cases a quantitative discussion of the topic is not practical. This arises because often the compound finally obtained as a vapour in the flame is not identical to the metal salt present in the sample solution that was nebulized. For a given metal salt the composition at the vapour stage may depend on factors such as the solvent used, concomitant elements present in the sample, reactions occurring in the solid or molten state (either amongst the original constituents or with the flame gases) and partial decomposition dependent on the flame temperature. As the melting or boiling points of the compounds obtainable from a particular salt may vary widely, the rate of volatilization can also be drastically altered by apparently minor changes in technique. This situation accounts for many interferences and apparent anomalies in sensitivity reported in the literature. In fact, many of the practical examples discussed in Chapter 12 can readily be placed into one of several simple classes of compound formation.

(a) The most straightforward case is a pure solution of a compound which is desolvated and vaporized unchanged so that the species obtained in the flame is indeed just a vapour of the original sample material. Many anhydrous halides (e.g. NaCl) fall into this category and analysis obviously presents no problems (except in a few cases for compounds of exceptionally low volatility).

(b) If the compound of the analyte element contains oxygen, breakdown may often occur to form the metal oxide which frequently has a high melting point or is difficult to dissociate. For example[113]

$$Al(NO_3)_3 \rightarrow Al_2O_3 \text{ (m.pt. 2320 K)}$$

This effect can also occur with molecules readily associated with water molecules, e.g.[114]

$$MgCl_2 . H_2O \rightarrow MgO \text{ (m.pt. 3070 K)}$$

Oxide formation by these processes has been a major obstacle to efficient atomization of many elements and was mainly responsible for the introduction of high-temperature reducing flames to routine AAS applications (see Section 6.3.3).

(c) When an acid is present in the sample solution, it may react with the original compound on desolvation and form a completely different species. This explains the reduced sensitivity of calcium determinations when phosphoric acid is added to samples.[115] The melting point of the phosphate formed (1940 K) is somewhat higher than that of the compound present without phosphoric acid. The reverse effect may also be obtained and is frequently used to improve the sensitivity obtained with refractory elements. Thus, in the determination of aluminium by AAS, improved sensitivity is obtained when aluminium sample solutions containing hydrofluoric acid are nebulized.[116] The melting point of the fluoride formed (1310 K) is considerably lower than that of the aluminium oxide which would otherwise be formed (see above). Similarly[117] addition of HF to zirconium samples results in the production of $ZrOF_2$, which is rather more volatile then the ZrO otherwise formed in the flame.

(d) A similar effect may be obtained with some metals by forming an organic complex with a suitable ligand. For example[118] addition of oxine to magnesium sample solutions increases the rate of vaporization of magnesium because the complex is carried into the flame where it decomposes to release magnesium, preventing the usual formation of stable magnesium oxide. The use of this technique to improve AAS sensitivities for a range of elements in the air-acetylene flame is described in Chapter 12.

(e) Unexpected compounds of the analyte element may also need to be volatilized in the flame as a result of interelement effects between species present in the sample matrix. Thus[115] the presence of aluminium in calcium samples may depress the calcium signal owing to compound formation. Calcium aluminate (m.pt. 1820 K) is much less volatile than the calcium compound that would otherwise be formed. The converse effect may also be observed. Thus, it has been suggested[114] that the enhancement of magnesium signals by the addition of calcium to the sample matrix can be explained by formation of an interelement compound showing greater volatility.

Although not involving compound formation, a related effect occurs when a large excess of another metal compound is present in the sample matrix with the analyte compound; the latter will be dispersed as a trace component in the particles formed after desolvation. Hence, if the concomitant salt is less volatile, its slower rate of vaporization will govern the release of the analyte element into the flame. Conversely, if the bulk of the sample matrix is more easily volatilized, it will actually improve vaporization of the analyte compound which will be left as much finer particles than would otherwise be obtained after the nebulization and desolvation processes.

In practice, quantitative treatment of the rate of vaporization of a given compound in the flame is only straightforward when the flame temperature is higher than the boiling point. The situation is then analogous to the boiling of a water droplet (as in the desolvation stage) and the rate of vaporization is directly proportional to the radius of the particle, with the total time for evaporation dependent on its initial surface area.[112] Obviously the rate will also depend on the difference between the flame temperature and the boiling point of the compound. When, as is often the case, the desolvated particles do not reach their boiling point in the flame used, the rate of vaporization also depends on other factors. Although the original size of the particle is still the major factor in deciding the total time of evaporation, other important considerations include the saturation pressure of the vapour at the surface of the particle (in the previous case this pressure would, of course, be close to 1 atm for all compounds), the density of the particle, the molecular weight of the compound and the rate of diffusion of the vapour into the ambient flame gases. It is possible to derive approximate expressions for the rate and time of evaporation by making a number of simplifying assumptions in the case of molten particles[113] (i.e. where the flame temperature exceeds the melting point of the compound), but the situation is more complex for vaporization of solid particles (which may often be inhomogeneous or of irregular shape). In any case, calculations are likely to involve considerable errors as average values must be assumed for a number of the quantities included in the expressions. Hence, experimental evidence is likely to be of more utility in deciding whether vaporization is complete under a particular set of flame conditions.

It should be remembered that although the melting and boiling points of the material affect the *rate* of vaporization, it is not essential that they lie below the flame temperature for complete evaporation to be obtained. The final partial pressure of the analyte element in the flame is usually very low and seldom exceeds *ca.* 10^{-3} atmospheres.[119] This low pressure is usually much less than the saturation pressure of the compound in the flame and vaporization must ultimately reach completion. Unsatisfactory results may, however, still be observed if the velocity of the flame gases is too high to allow complete vaporization before the analyte compound has passed out of the flame region used for measurements. Thus, the height of observation can considerably affect the results obtained, particularly with refractory compounds or cool flames. This type of effect is usually most pronounced in turbulent flames supported at total-consumption nebulizer-burners (see Section 6.4.2) as these flames show a higher rise velocity and larger dry aerosol particle size than laminar, premixed flames used in conjunction with indirect nebulizers.

6.5.3 Formation of Free Analyte Atoms

Many of the molecules introduced into the flame gases during the vaporization stage will quickly break down, as complex molecules are not stable at the high temperature of the flame. This does not, however, mean that most of the analyte element is available in the flame gas as free atoms immediately after vaporization. Some or all of the metal introduced may react either with a flame species or a constituent of the nebulized solution to form a molecule in the gaseous state. These are mostly diatomic species, in particular the metal monoxides, although under suitable conditions fluorides or chlorides may be present. The monohydroxides of elements such as the alkaline earths are also stable and an appreciable proportion of these elements may be present in this form, particularly in hydrogen flames where even dihydroxides such as $Sr(OH)_2$ may occur. As a result of this possibility of compound formation, the fraction of the analyte element present in the flame as free atoms will vary greatly with flame conditions and is often much less than unity. Thus, unlike the vaporization stage, we are no longer concerned with the rate at which the process proceeds to completion but with the degree to which compounds of the analyte metal are dissociated.

If we consider the concentration of the metal M to depend on the dissociation of a compound MX, the *degree of dissociation*, β, is defined* as

$$\beta = \frac{[M]}{[M] + [MX]} \tag{6-3}$$

Provided that three species are in a state of equilibrium, i.e.

$$MX \rightleftarrows M + X \tag{6-4}$$

their concentrations will be related through the law of mass action, as for any other chemical reaction, to a dissociation constant, K_d:

$$K_d = \frac{[M]\,[X]}{[MX]} \tag{6-5}$$

thus β may be expressed simply in terms of K_d and [X]. Furthermore, for reactions in the gas phase K_d may be calculated in terms of the temperature and fundamental constants of the species involved using formulae derived from statistical mechanics.† Hence we obtain an expression of the form

$$\beta^{-1} = 1 + \frac{[X]}{f(T)\exp(-D_0/kT)} \tag{6-6}$$

* [ ] denotes the concentration of the enclosed species.

† The reader is referred to a standard textbook on the subject, such as that by Fowler and Guggenheim.[120]

where D_0 is the dissociation energy of the molecule MX, k is the Boltzmann constant, T is the flame temperature and $f(T)$ is a temperature-dependent parameter incorporating the statistical weights and partition functions of the species involved. It can be seen that for a given value of [X] the value of β will depend mainly on the exponential term, i.e. on the dissociation energy of MX and on the flame temperature. Thus the stability of a metal compound in the flame is characterized principally by D_0, some values of which are shown in Table 6-7 for species of interest in AAS and AFS. In general terms, molecule formation is unlikely in flames of analytical interest where the D_0 value is less than about 3·5 eV; dissociation becomes difficult, however, for D_0 values greater than *ca.* 5–6 eV.

In most cases, the formation of MX has a negligible effect on the value of [X] so that β is independent of the actual metal concentration introduced into the flame. This arises because X is usually a flame gas constituent which will be present at a partial pressure much greater than that of the metal atom under most circumstances of analytical interest. Even when X is a species introduced with the sample solution the same situation will apply, as almost invariably significant compound formation will persist in the flame only if X is a bulk constituent of the sample solution. As a result, interference effects which may alter the nebulization rate, or give rise to incomplete vaporization or partial ionization, can only change the total value of [M] and [MX]; they do not alter their ratio, i.e. the value of β.

As discussed in Section 6.2.4, the flame gases of interest are usually (but not always) in a sufficient state of chemical equilibrium outside the reaction zone for the mass action law to be applied directly to the dissociation process. This situation generally applies to laminar premixed hydrocarbon flames and hot ($>$*ca.* 2500 K) hydrogen flames under stoicheiometric or fuel-rich conditions but not to fuel-lean flames or to the cooler hydrogen flames. Such departures from equilibrium are due to the presence in the interconal zone of excess (i.e. greater than equilibrium) concentrations of long-lived radicals formed in the reaction zone. The turbulent unpremixed flames described in Section 6.4.2 are particularly likely to show above equilibrium radical concentrations in the region used for analytical measurements as this is, in effect, part of the reaction zone.[31,124] Equilibrium laws obviously cannot be applied to the degree of dissociation in the presence of excess free radicals which affect (either favourably or adversely) the dissociation reaction(s).

TABLE 6-7

Dissociation energies of metal compounds important in AAS and AFS[a]

	D_0			D_0	
Molecule	kcal mole^{-1}	eV molecule^{-1}	Molecule	kcal mole^{-1}	eV molecule^{-1}
AgCl	74	3·2	GeCl	81	3·5
AgF	83·8	3·6	GeF	115	5·0
AgO	46	2	GeO	158·8	6·78
AlCl	117	5·1	InCl	104	4·5
AlF	158	6·85	InF	123	5·35
AlH	67	2·9	InO	76	3·3
AlO	106	4·6	KCl	100·5	4·36
AsO	113	4·9	KF	117	5·07
AuCl	81	3·5	KOH*	85	
BCl	118	5·1	LiCl	113	4·9
BF	181	7·85	LiF	137	5·95
BO	184	8·0	LiO	81	3·5
BaCl	115	5·0	LiOH*	105	
BaF	115	5·0	MgCl	81	3·5
BaO	133	5·75	MgF	105·5	4·5
BaOH*	114		MgO	94	4·1
$Ba(OH)_2$*	213		MnCl	85·3	3·7
BeCl	93	4·0	MnF	120	5·2
BeF	136	5·9	MnO	96	4·1
BeO	106	4·6	MoO	116	5·0
BiCl	72·3	3·13	NaCl	98	4·25
BiF	61	2·65	NaF	114	4·9
BiO	71	3·1	NaOH*	80	
CaCl	105		NiCl	88	3·8
CaF	125	5·4	NiF	88	3·8
CaO	100	4·3	NiO	97	4·2
CaOH*	104		PF	104	4·5
Ca(OH)	204		PO	148	6·4
CdCl	48·8	2·1	PbCl	71	3·1
CdO	<88	$<3·8$	PbF	69	3
CeO	185	8·03	PbO	89·3	3·87
CrCl	86·5	3·75	RbCl	102	4·4
CrF	92	4	RbF	120	5·2
CrO	101	4·3	RbOH*	87	
CsCl	105	4·55	SbCl	85	3·7
CsF	123	5·33	SbF	104	4·5
CsOH*	90		SbO	92	4
CuCl	83	3·6	SeO	100	4·3
CuF	81	3·5	SiCl	104	4·5
CuO	95	4·1	SiF	115	5·0
FeCl	83	3·6	SiO	187	8·1
FeO	99	4·3	SnCl	74	3·2
GaCl	114	4·94	SnF	90	3·9
GaF	140	6·1	SnO	124·5	5·40
GaO	69	3·0	SrF	126	5·45

TABLE 6-7 (*continued*)

Molecule	D_0 kcal mole^{-1}	D_0 eV molecule^{-1}	Molecule	D_0 kcal mole^{-1}	D_0 eV molecule^{-1}
SrO	97	4·2	WO	157	6·8
SrOH*	103		YCl	81	3·5
$Sr(OH)_2$*	202		YO	168·5	7·3
TaO	195	8·4	YbO	122	5·3
TeO	90	3·9	ZnCl	49	2·1
ThO	196	8·5	ZnO	65	2·8
TiO	166	7·2	ZrO	181	7·8
TlCl	87·5	3·8			
TlF	104	4·5			
VO	147·5	6·4			

[a] All data taken from recommended values tabulated by Gaydon[121] except for those marked * which were taken from Cotton and Jenkins.[122,123]

However, in many cases the reactions involving the metal atom proceed sufficiently rapidly in both directions for a state of "pseudo-equilibrium" to be assumed and the mass action law may then be applied to the individual reaction using the actual radical concentrations.[52]

Whereas under true equilibrium conditions the degree of compound dissociation may be calculated by thermodynamic techniques (see, for example, refs. 1, 16, 17, 68, 125), in the presence of excess radicals calculations can only be performed using experimentally determined values of the radical concentrations. Such measurements have mostly been carried out for laminar, premixed hydrogen flames in which the excess radicals are H, O and OH (see, for example, refs. 5, 52, 126-129). It should be remembered that the population of free metal atoms required for AAS or AFS measurements will only be significantly affected by excess radical concentrations if those radicals directly control the equilibrium of the metal compound. This is easily checked in a laminar, premixed flame showing uniform conditions (i.e. at the centre of a flame-shielded flame and reasonably close to the burner). Normally the absorption or fluorescence signal observed from the metal atoms will be independent of height in the flame. When the dissociation is affected by excess radicals, however, the signal will change with height owing to the gradual decay of the radicals as the flame gases approach equilibrium. For example, Li atoms in a flame form a stable hydroxide by the reaction

$$Li + H_2O \rightleftarrows LiOH + H$$

so that the population of Li atoms is dependent on the hydrogen radical concentration, the value of β increasing with M. Thus, in

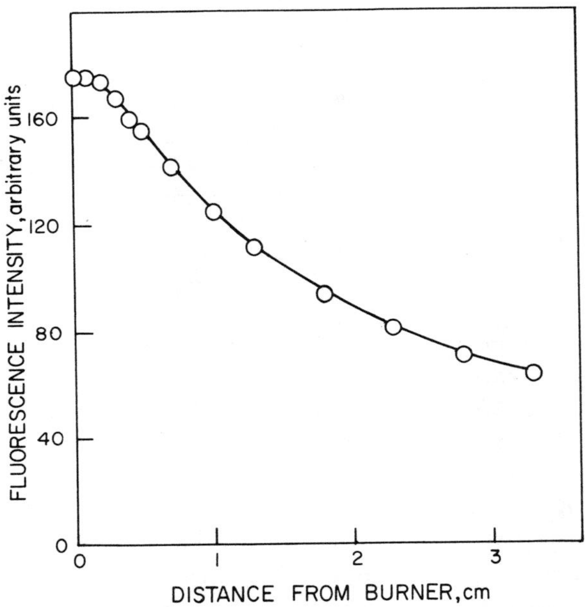

Fig. 6-28. Intensity of lithium 670·7 nm fluorescence in a hydrogen-oxygen-argon premixed flame at 1600 K.[52]

cool hydrogen flames used for AFS measurements the lithium fluorescence signal decreases with height above the burner as shown in Fig. 6-28 owing to a similar decrease in the number of hydrogen radicals. On the other hand, it has been shown, for example, that the formation of stable magnesium oxide is governed by the reactions

$$Mg + OH \rightleftarrows MgO + H$$

$$H + H_2O \rightleftarrows H_2 + OH$$

These are equivalent to the reaction

$$Mg + H_2O \rightleftarrows MgO + H_2$$

and as both $[H_2O]$ and $[H_2]$ are virtually constant the value of β for formation of magnesium atoms is almost independent of height even in hydrogen flames showing high excess concentrations of H and OH.[52]

6.5.4 Ionization of Analyte Atoms

AAS and AFS measurements require a population of free, *neutral* metal atoms. A loss of analytical sensitivity will result if the atoms formed by the molecular dissociation process are "depleted" by ionization:

$$M \rightleftharpoons M^+ + e^- \tag{6-7}$$

Under equilibrium conditions the concentrations of these species are related by *Saha's Law* which is exactly analogous to the mass action law of molecular dissociation (see Section 6.5.3). Hence we obtain for the *degree of ionization,* δ, an expression of the same form as equation (6-6), i.e.

$$\delta^{-1} = 1 + \frac{[e^-]}{f(T)\,.\exp\,(-E_i/kT)} \tag{6-8}$$

where E_i is the ionization energy of the atom. As before, variation of δ at given $[e^-]$ will depend mainly on the exponential term; $f(T)$ is again a parameter which depends on fundamental constants (for the atom and ion of the element) and which shows a small temperature variation. The value of E_i varies widely between different elements (from about 3·9 to 10·5 eV) and consequently the degree of ionization varies widely in a particular flame. The extent of this variation can be seen clearly from Fig. 6-29 which shows the degree of ionization calculated using the Saha Law at 2500 K (i.e. the air-acetylene flame temperature) and 3200 K (somewhat hotter than the nitrous oxide-acetylene flame). The different plots at each temperature indicate the effect of the factor $f(T)$ between different types of ionization transitions (i.e. different groups of elements).

In many cases the ionization process shows an important difference from the molecular dissociation discussed in Section 6.5.3. For dissociation the values of β were generally independent of the metal concentration because, for example, the concentration of oxygen atoms in most flames is not significantly affected by the addition of metal atoms at the concentrations encountered in analytical work. In the flame regions of interest, however, the electrons present may often originate principally from the analyte element itself; some important exceptions to this generality are discussed below. Hence when significant ionization occurs the value of δ will vary with the concentration of the analyte element in the sample solution, decreasing as the concentration increases. As a result, calibration plots in AAS or AFS for the neutral atom will deviate from linearity and bend *away* from the concentration axis. This concentration effect has a further consequence in that the degree of ionization is also influenced by molecular dissociation in

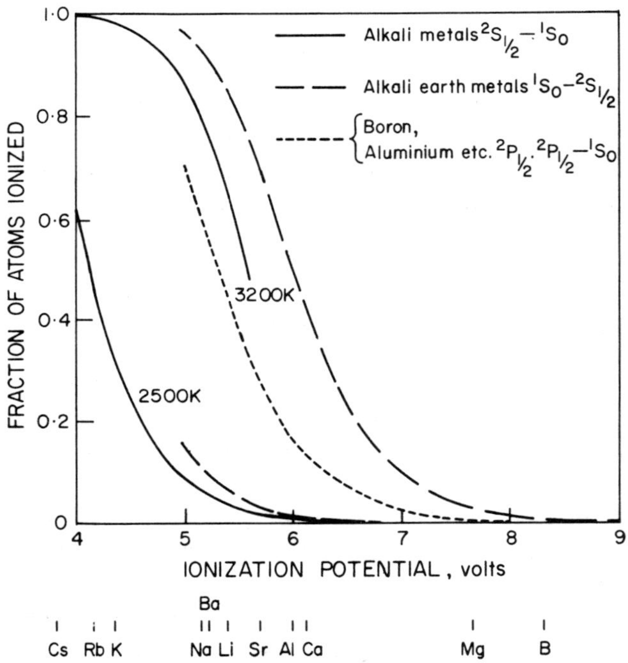

Fig. 6-29. Calculated degree of ionization for various metals at 2500 K and 3200 K assuming a metal concentration of 1 part in 10^6 of the flame gases.[64]

the flame. Thus, when the element exists partly as molecules, owing to incomplete dissociation, the value of δ for free atoms will be higher (as the concentration of free atoms is lower) but the total element concentration added to the flame will actually be affected less by ionization. Furthermore, under equilibrium conditions the value of β should remain constant with height in the flame (see Section 6.5.3) but the actual number of free metal atoms may change owing to other factors, such as incomplete vaporization (see Section 6.5.2). As a result the value of δ may vary with height in the flame even within a region which shows virtually isothermal conditions. The possibility of ionization of molecular species containing the analyte element is often also overlooked; this would compete with the dissociation process of neutral molecules and affect the equilibria controlling both free atom formation and the subsequent ionization of atoms. For example, in some hydrogen flames the concentrations of ions such as $CaOH^+$ and $BaOH^+$ may actually exceed those of Ca^+ and Ba^+.

Although the concentration dependence of ionization is caused by the lack of electrons in the flame produced other than by direct ionization of analyte atoms, this same effect readily allows suppression of virtually all ionization of the analyte. Thus, the addition of a large excess of an element which is as easily (or, preferably, even more easily) ionized as the analyte atoms will raise the concentration of electrons in the flame and shift the equilibrium given by equation (6-7) to the left for the original metal. Referring to equation (6-8), it is apparent that any ionization of the original element will be negligible for practical purposes when $[e^-] \geqslant 100\, f(T) \exp(-E_i/kT)$. This is easily achieved for most analyte element/flame combinations used in AAS or AFS by the addition of excess potassium (ionization potential 4·34 eV) to the sample solution. Under these circumstances improved sensitivity and linear calibrations will result in the air-acetylene flame for elements of ionization potential up to *ca.* 5·5 eV and in the nitrous oxide-acetylene flame for those with ionization potentials as high as *ca.* 6·5 or 7·0 eV. It should be remembered that if ionization suppression is not used, the same effect may cause an interference in practical determinations; i.e. if the sample contains a more easily ionized concomitant element not present in the standard solutions used for calibration. Ionization may also be suppressed when a sufficiently high electron concentration is produced by formation of natural flame ions but, as explained below, this is not the case in most situations (except perhaps at very low metal concentrations).

In the presence of natural flame ions it would be necessary to allow for the electrons thus formed, when using the equation given above to evaluate the degree of ionization of an added metal. In view of the high ionization potentials of all such flame species, negligible equilibrium concentrations would be expected in normal flames. In practice, natural ions are not found in hydrogen flames but appear to be formed at levels greatly exceeding the equilibrium concentration in the reaction zone of hydrocarbon flames (see, for example, refs. 1,

2, 5, 127, 130, 131). Although a wide variety of ions has been observed the most abundant, and the only one carried into the interconal zone in significant quantities, is H_3O^+. This is probably formed and removed by the mechanism[127]

$$CH + O \rightarrow CHO^+ + e^-$$

$$CHO^+ + H_2O \rightarrow H_3O^+ + CO$$

$$H_3O^+ + e^- \rightarrow H_2O + H$$

Calculations of the degree of ionization based on the Saha Law for a metal added to the flame would be higher than the actual value for any flame in which this departure from equilibrium is significant. Furthermore, the concentration of H_3O^+ ions will gradually decrease with increasing height in the flame, so that a corresponding increase in the value of δ should be observed and result in a variation with height of absorption or fluorescence signals. It should be noted here that even in the absence of natural flame ions, ionization equilibrium may not be achieved for a considerable height in the flame. This arises partly from the relatively large number of collisions required to establish ionization equilibrium (see Section 6.2.4) and partly from the need to form free metal atoms before the ionization can begin. The height required to establish equilibrium will increase with increasing ionization potential. Sodium, for example, requires a time period equivalent to a height of about 10 cm in a typical air-hydrogen flame.[132] Equilibrium appears to be established unexpectedly quickly for the alkaline-earth metals; this may be due to the effect of a chemi-ionization process involving MOH^+ ions.[112]

REFERENCES

1. A. G. Gaydon and H. G. Wolfhard, "Flames, Their Structure, Radiation and Temperature", Chapman and Hall, London, 3rd ed., 1970.
2. B. Lewis and G. Von Elbe, "Combustion, Flames and Explosions of Gases", Academic Press, New York, 2nd ed., 1961.
3. R. M. Fristrom and A. A. Westenberg, "Flame Structure", McGraw-Hill, New York, 1965.
4. A. G. Gaydon, "Spectroscopy of Flames", Chapman and Hall, London, 1957.
5. C. P. Fenimore, "Chemistry in Premixed Flames", Pergamon Press, Oxford, 1964.
6. R. Mavrodineanu and H. Boiteux, "Flame Spectroscopy", Wiley, New York, 1965.
7. W. H. Clingman, R. S. Brokeau and R. N. Pease, *Fourth Symposium (International) on Combustion*, Williams and Wilkins, Baltimore, 1953, p. 310.
8. G. E. Andrews and D. Bradley, *Combustion and Flame*, 1972, **18**, 133.

9. G. F. Kirkbright, M. K. Peters, M. Sargent and T. S. West, *Talanta,* 1968, **15,** 663.
10. G. F. Kirkbright, M. Sargent and S. Vetter, *Spectrochim. Acta,* 1970, **25B,** 465.
11. R. F. Browner and J. D. Winefordner, *Anal. Chem.,* 1972, **44,** 247.
12. N. Omenetto, R. F. Browner, J. D. Winefordner, G. Rossi and P. Benetti, *Anal. Chem.,* 1972, **44,** 1683.
13. H. J. Kostkowski and H. P. Broida, *J. Opt. Soc. Am.,* 1956, **46,** 246.
14. G. F. Kirkbright, and S. Vetter, *Spectrochim. Acta,* 1971, **26B,** 505.
15. W. Snelleman and J. A. Smit, *Metrologia,* 1968, **4,** 123.
16. W. B. White, S. M. Johnson and G. B. Dantzig, *J. Chem. Phys.,* 1958, **28,** 751.
17. J. E. Chester, R. M. Dagnall and M. R. G. Taylor, *Anal. Chim. Acta,* 1970, **51,** 95.
18. W. Snelleman, *Metrologia,* 1968, **4,** 117.
19. V. C. O. Schuller and G. S. James, *J. South African Inst. Min. Metall.,* 1962, **62,** 786.
20. L. R. P. Butler, ibid., 1962, **62,** 796.
21. J. A. Fiorino, R. N. Kniseley and V. A. Fassel, *Spectrochim. Acta,* 1968, **23B,** 413.
22. R. Friedman and W. C. Johnson, *J. Appl. Phys.,* 1950, **21,** 791.
23. R. Lockyer and G. E. Hammes, *Analyst,* 1959, **84,** 385.
24. G. F. Kirkbright, M. Sargent and T. S. West, *Talanta,* 1969, **16,** 1467.
25. D. N. Hingle, G. F. Kirkbright, M. Sargent and T. S. West, *Lab. Practice,* 1969, **18,** 1069.
26. P. J. Padley and T. M. Sugden, *Proc. Royal Soc. Ser. A,* 1958, **248,** 248.
27. K. M. Aldous, R. F. Browner, R. M. Dagnall and T. S. West, *Anal. Chem.,* 1970, **42,** 939.
28. E. A. Boling, *Spectrochim. Acta,* 1966, **22,** 425.
29. J. V. Sullivan and A. Walsh, ibid., 1966, **22,** 1843.
30. G. F. Kirkbright, A. Semb and T. S. West, *Talanta,* 1968, **15,** 441.
31. P. T. Gilbert, *Proc. Xth Colloq. Spectros. Inter.,* pp. 171-215, Spartan Books, Washington, 1963.
32. D. N. Hingle, G. F. Kirkbright and T. S. West, *Talanta,* 1968, **15,** 199.
33. *Idem, Analyst,* 1968, **93,** 522.
34. R. S. Hobbs, G. F. Kirkbright, M. Sargent and T. S. West, *Talanta,* 1968, **15,** 997.
35. G. F. Kirkbright, M. Sargent and T. S. West, *Atomic Absorption Newsletter,* 1969, **8,** 34.
36. G. F. Kirkbright and L. Ranson, *Anal. Chem.,* 1971, **43,** 1238.
37. G. F. Kirkbright, T. S. West and P. J. Wilson, *Atomic Absorption Newsletter,* 1972, **11,** 53.
38. G. F. Kirkbright and M. Marshall, *Anal. Chem.,* 1972, **44,** 1288.
39. *Idem,* ibid., 1973, **45,** 1610.
40. G. F. Kirkbright, L. Ranson and T. S. West, *Spectrosc. Letters,* 1972, **5,** 25.
41. R. S. Hobbs, G. F. Kirkbright and T. S. West, *Talanta,* 1971, **18,** 859.
42. P. L. Larkins, *Spectrochim. Acta,* 1971, **26B,** 477.
43. P. L. Larkins and J. B. Willis, ibid., 1971, **26B,** 491.
44. R. F. Browner and D. C. Manning, *Anal. Chem.,* 1972, **44,** 843.
45. V. Sychra and J. Matousek, *Talanta,* 1970, **17,** 363.
46. T. L. Martin, F. M. Hamm and P. B. Zeeman, *Anal. Chim. Acta,* 1971, **53,** 437.

47. G. F. Kirkbright, L. Ebdon and T. S. West, ibid., 1969, **47**, 563.
48. *Idem, Talanta,* 1970, **17**, 965.
49. R. M. Dagnall, G. F. Kirkbright, T. S. West and R. Wood, *Anal. Chim. Acta,* 1969, **47**, 407.
50. *Idem, Analyst,* 1970, **95**, 425.
51. *Idem, Anal. Chem.*, 1970, **42**, 1029.
52. D. R. Jenkins, *Spectrochim. Acta,* 1970, **25B**, 47.
53. J. B. Willis, ibid., 1970, **25B**, 487.
54. C. Van Trigt, T. J. Hollander and C. T. J. Alkemade, *J. Quant. Spectry. Radiat. Trans.*, 1965, **5**, 813.
55. P. J. Slevin, V. I. Muscat and T. J. Vickers, *Appl. Spectry.*, 1972, **26**, 296.
56. T. J. Vickers, P. J. Slevin, V. I, Muscat and L. T. Farias, *Anal. Chem.*, 1972, **44**, 930.
57. B. J. Russell, J. P. Shelton and A. Walsh, *Spectrochim. Acta,* 1957, **8**, 317.
58. L. de Galan and J. D. Winefordner, *J. Quant. Spectros. Radiat. Transfer,* 1967, **7**, 251.
59. S. R. Koirtyohann and E. E. Pickett, paper presented at XIII Colloquium Spectroscopicum Internationale, Ottawa, June 1967.
60. P. J. M. Zeegers, W. P. Townsend and J. D. Winefordner, *Spectrochim. Acta,* 1969, **24B**, 243.
61. L. de Galan and G. F. Samaey, ibid., 1970, **25B**, 245.
62. J. B. Willis, ibid., 1971, **26B**, 177.
63. C. L. Chakrabarti, M. Katyal and D. E. Willis, ibid., 1970, **25B**, 629.
64. J. B. Willis, in "Analytical Flame Spectroscopy", ed. R. Mavrodineanu, Macmillan, London, 1970.
65. *Idem, Nature,* 1965, **207**, 715.
66. M. D. Amos and J. B. Willis, *Spectrochim. Acta,* 1966, **22**, 1325.
67. G. F. Kirkbright, M. K. Peters and T. S. West, *Talanta,* 1967, **14**, 789.
68. L. L. Reigle, W. J. McCarthy and A. C. Ling, *J. Chemical and Engineering Data,* 1973, **18**, 79.
69. L. M. Fraser and J. D. Winefordner, *Anal. Chem.*, 1971, **43**, 1693.
70. *Idem,* ibid., 1972, **44**, 1444.
71. N. Omenetto, N. N. Hatch, L. M. Fraser and J. D. Winefordner, ibid., 1973, **45**, 195.
72. T. G. Cowley, V. A. Fassel and R. N. Kniseley, *Spectrochim. Acta,* 1969, **24B**, 157.
73. J. E. Chester, R. M. Dagnall and M. R. G. Taylor, *Analyst,* 1970, **95**, 702.
74. J. B. Willis, V. A. Fassel and J. A. Fiorino, *Spectrochim. Acta,* 1969, **24B**, 157.
75. J. E. Schallis and H. L. Kahn, *Atomic Absorption Newsletter,* 1968, **7**, 75.
76. J. D. Winefordner and R. A. Staab, *Anal. Chem.*, 1964, **36**, 1367.
77. D. M. Armentrout, ibid., 1966, **38**, 1235.
78. M. P. Bratzel, R. M. Dagnall and J. D. Winefordner, ibid., 1969, **41**, 713.
79. R. F. Browner, R. M. Dagnall and T. S. West, *Anal. Chim. Acta,* 1969, **46**, 207.
80. D. S. Smyly, W. P. Townsend, P. J. Th. Zeegers and J. D. Winefordner, *Spectrochim. Acta,* 1971, **26B**, 531.
81. D. R. Jenkins, ibid., 1967, **23B**, 167.
82. R. M. Dagnall, K. C. Thompson and T. S. West, *Atomic Absorption Newsletter,* 1967, **6**, 117.
83. H. L. Kahn and J. E. Schallis, ibid., 1968, **7**, 5.
84. M. B. Denton and H. V. Malmstadt, *Anal. Chem.*, 1972, **44**, 1813.
85. P. T. Gilbert, "The design of flame emission and absorption sources", Eastern Analytical Symposium, November 1969, New York.

86. *Idem,* A. S. T. M. Special Technical Publications, 1959, No. 269, p. 73.
87. L. de Galan and J. D. Winefordner, *J. Quant. Spectros. Radiat. Transfer,* 1967, **7,** 703.
88. J. A. Dean and W. J. Carnes, *Anal. Chem.,* 1962, **34,** 192.
89. P. T. Gilbert, Beckman Bulletin No. 3063 (1961), Fullerton, California.
90. J. W. Robinson and R. J. Harris, *Anal. Chim. Acta,* 1962, **26,** 439.
91. R. N. Kniseley, A. P. D'Silva and V. A. Fassel, *Anal. Chem.,* 1963, **35,** 910.
92. V. G. Mossotti and M. Duggan, *Applied Optics,* 1968, **7,** 1325.
93. M. P. Bratzel, R. M. Dagnall and J. D. Winefordner, *Anal. Chem.,* 1969, **41,** 1527.
94. D. W. Ellis and D. R. Demers, ibid., 1966, **38,** 1943.
95. M. P. Parsons and J. D. Winefordner, ibid., 1966, **38,** 1595.
96. M. P. Bratzel and J. D. Winefordner, *Anal. Letters,* 1967, **1,** 43.
97. J. M. Mansfield, M. P. Bratzel, H. O. Norgorden, D. O. Knapp, K. E. Zacha and J. D. Winefordner, *Spectrochim. Acta,* 1968, **23B,** 389.
98. K. E. Zacha, M. P. Bratzel, J. D. Winefordner and J. M. Mansfield, *Anal. Chem.,* 1968, **40,** 1733.
99. J. D. Winefordner and R. A. Staab, ibid., 1964, **36,** 165.
100. J. W. Robinson and C. J. Hsu, *Anal. Chim. Acta,* 1968, **43,** 109.
101. S. J. Pearce, L. de Galan and J. D. Winefordner, *Spectrochim. Acta,* 1968, **23B,** 793.
102. N. Omenetto, L. P. Hart and J. D. Winefordner, *Applied Spectroscopy,* 1972, **26,** 612.
103. K. Fuwa and B. L. Vallee, *Anal. Chem.,* 1963, **35,** 942.
104. K. Fuwa, P. Pulido, R. McKay and B. Vallee, ibid., 1964, **36,** 2407.
105. S. R. Koirtyohann and C. Feldman, *Developments in Applied Spectroscopy,* 1964, **3.**
106. E. J. Agazzi, *Anal. Chem.,* 1965, **37,** 365.
107. C. L. Chakrabarti, J. W. Robinson and P. W. West, *Anal. Chim. Acta,* 1966, **34,** 269.
108. T. V. Ramakrishna, J. W. Robinson and P. W. West, ibid., 1967, **37,** 20.
109. I. Rubeska, ibid., 1968, **40,** 187.
110. I. Rubeska and B. Moldan, "Atomic Absorption Spectrophotometry", Iliffe, London, 1969.
111. D. N. Hingle, G. F. Kirkbright and T. S. West, *Talanta,* 1968, **15,** 199.
112. C. Th. J. Alkemade, in "Analytical Flame Spectroscopy", ed. R. Mavrodineanu, Macmillan, London, 1970.
113. B. V. L'Vov, "Atomic Absorption Spectrochemical Analysis", Hilger, London, 1970.
114. D. J. Halls and A. Townshend, *Anal. Chim. Acta,* 1966, **36,** 278.
115. R. Herrmann and C. T. J. Alkemade, "Chemical Analysis by Flame Photometry", Interscience, New York, 1963.
116. K. Konopicky and W. Schmidt, *Z. Anal. Chem.,* 1960, **174,** 262.
117. J. B. Willis, *Applied Optics,* 1968, **7,** 1295.
118. F. J. Wallace, *Analyst,* 1963, **88,** 259.
119. C. T. J. Alkemade, in "Flame Emission and Atomic Absorption Spectrometry", Vol. 1, ed. J. A. Dean and T. C. Rains, Dekker, New York 1969.
120. R. H. Fowler and E. A. Guggenheim, "Statistical Thermodynamics", Cambridge, 1949.
121. A. G. Gaydon, "Dissociation Energies and Spectra of Diatomic Molecules", Chapman and Hall, 3rd ed., London, 1968.
122. D. H. Cotton and D. R. Jenkins, *Trans. Farad. Soc.,* 1968, **64,** 2988.
123. *Idem,* ibid., 1969, **65,** 1537.

124. V. A. Fassel, R. B. Myers and R. N. Kniseley, *Spectrochim. Acta,* 1963, **19,** 1187.
125. P. W. J. Boumans, "Theory of Spectrochemical Excitation", Plenum, New York, 1966.
126. T. M. Sugden, *Trans. Farad. Soc.,* 1956, **52,** 1465.
127. *Idem, Annual Revs. Phys. Chem.,* 1962, **13,** 369.
128. P. J. T. Zeegers, "Recombination of Radicals and Related Effects in Flames", Thesis, Utrecht, 1966.
129. D. R. Jenkins and T. M. Sugden, "Radicals and molecules in flame gases", in "Flame Emission and Atomic Absorption Spectroscopy", Vol. 1 (Chap. 5), eds. J. A. Dean and T. C. Rains, Dekker, New York, 1969.
130. T. Hollander, "Self-Absorption, Ionization and Dissociation of Metal Vapour in Flames", Thesis, Utrecht, 1964.
131. E. M. Bulewicz, *Combustion and Flame,* 1967, **11,** 297.
132. D. E. Jensen and P. J. Padley, *Trans. Farad. Soc.,* 1966, **62,** 2140.

CHAPTER 7

Non-flame Absorption and Fluorescence Cells

This chapter describes briefly some of the non-flame atom cells that have been used in AAS and AFS. The structure and properties of the flame atom cells most commonly employed in AAS and AFS are detailed elsewhere in this volume (Chapter 6). Devices of extended path-length, where a flame is employed for the introduction of the sample, are also discussed in Chapter 6. Techniques in which samples are introduced into flames by methods other than direct nebulization of solutions are discussed in Chapter 8.

7.1 GENERAL CONSIDERATIONS

Although the flames discussed in Chapter 6 have been very widely used as atom cells in AAS and AFS, they do have disadvantages for some analytical work. The volumes of the sample solutions required may frequently be larger than those available, and flame cells are not usually convenient for the direct atomization of solid samples. In some locations it may be inconvenient to use high-pressure cylinders of support and fuel gas and, in closed automated systems where no operator is in attendance, it may not be possible to use a flame as the atom cell. Furthermore, the atom concentration attainable in flame cells is limited by the dilution effect of the high flowrate of unburnt gas used to support the flame and to transport relatively small volumes of sample solution to the flame. The atom concentration is also limited by the expansion of the flame gas on combustion. Precise control over the chemical environment in flame cells cannot be achieved; the degree of control of chemical composition of the burnt gases that is obtained by variation of the fuel : oxidant concentration ratio is accompanied by simultaneous changes in the flame temperature and its spectral absorption and emission characteristics.

Thus, for the analysis of small liquid or solid samples, and for the determination of trace amounts of many elements in larger samples, it would be advantageous to achieve a higher concentration of atoms in a smaller cell volume than is possible with flames employing solution nebulization techniques. This may be accomplished with non-flame atom cells. Winefordner, for example, in a consideration of a typical graphite cell technique, has demonstrated that a substantially higher peak concentration of atoms may be expected in such a cell than in a typical flame.[1] This gain factor results directly from avoidance of the limiting sample dilution effects that occur in flame cells. In order to assist the formation and maintenance of a high free atom fraction for the analyte element for AAS or AFS it is also an advantage to be able to use a chemically inert gaseous atmosphere; also, in AFS, selection of a suitable inert supporting gas may give greater fluorescence efficiency as a result of reduced quenching of radiationally excited atoms. With suitably designed non-flame cells a considerable advantage over flames in the recorded noise that originates from the atom cell should result.

It is apparent, therefore, that non-flame devices may find widespread application in AAS and AFS, provided that they are comparable to flame cells in simplicity and reliability.

7.2 FURNACES

The use of high-temperature furnaces in AAS/AFS offers the

possibility of obtaining a high concentration of absorbing atoms within a well-defined volume, with very low background emission and noise. Several furnace devices have been reported in the literature and the most well-known designs will be discussed here.

7.2.1 Low-Temperature Furnaces

These devices usually operate at temperatures less than *ca.* 1500°C, and there may be significant memory effects between samples together with more serious chemical interferences than are encountered in flames. These disadvantages may be outweighed, however, by the potentially high AAS sensitivity for some elements as a result of the long path lengths that are possible.

Mislan[2] has described the construction and performance of an atomic absorption spectrophotometer which employs a 36-cm silica tube of internal diameter 2·5 cm heated to a maximum temperature of 1250°C by a wire-wound resistance furnace. Sample solutions are transferred to the absorption tube via a conventional indirect nebulizer-spray chamber assembly. The device was employed for the determination of cadmium and excellent detection limits were obtained. Hudson[3] has made AAS measurements on sodium vapour obtained in a stainless-steel absorption cell heated by a resistance wire. Atomic absorption studies using furnaces have also been made by Vidale,[4] Choong and Loong-Seng[5] and Tomkins and Ercoli.[6]

Black, Glenn, Bratzel and Winefordner[7] have reported the design and operation of a simple platinum tube furnace for AFS. In this device the sample solution is nebulized continuously with argon through a compact electrically heated furnace (maximum temperature 1600°C) to effect atomization. The analyte vapour emerging from the furnace was employed for AFS with electrodeless discharge-tube sources. The analyte atomization efficiency and the severity of interferences observed in the determination of cadmium were found to depend on the manner in which the furnace design controlled the heat transfer efficiency to the nebulized sample. Two designs of furnace were described by these workers. This type of furnace was shown to be effective for the determination of relatively volatile elements by AFS; the advantages of this non-flame technique include its simplicity for sample introduction, good precision attainable with continuous sample uptake and relative freedom from chemical interferences. The volume of sample required, however, may be greater than for other non-flame methods.

7.2.2 The L'Vov Graphite Cuvette

A number of devices that employ graphite furnaces of the type used by King[8] have been constructed for analytical use in both AAS and AFS. The earliest and best-known of these is the furnace device first described by L'Vov in 1959.[9] During the past ten years L'Vov and his co-workers have undertaken extensive AAS studies with various models of the original graphite curvette.[10-15] This work has been reviewed by L'Vov in his textbook[16] and elsewhere.[17]

A diagram of the basic graphite tube assembly employed recently

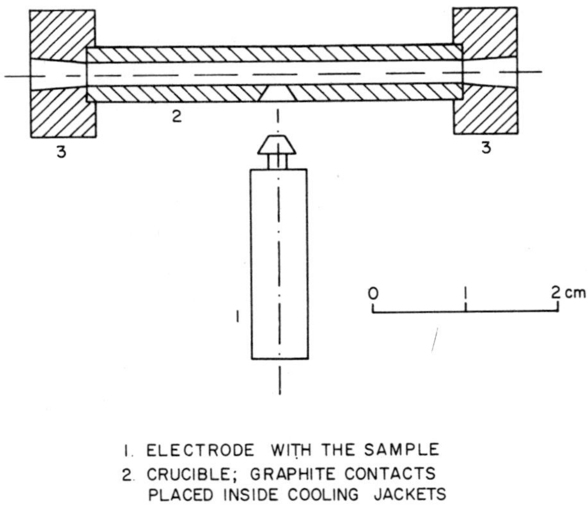

Fig. 7-1. The atomization system of the L'Vov furnace.[17]

is shown in Fig. 7-1. Graphite cylinders 30-50 mm in length and of internal diameter 2·5-5 mm (depending on the analytical requirements) are used. The furnace may be heated electrically by an alternating current from a 4 kW 220/10V transformer. The temperature to which the tube is heated is regulated by changing the voltage to the primary circuit of the transformer. In early forms of this furnace the inside wall of the tube was lined with tantalum or tungsten foil to eliminate vapour diffusion through the porous walls. More recently, preference has been given to tubes lined both inside and outside with a layer of pyrographite. This has low gas permeability, high heat conductivity and resistance to oxidation, so that it precludes vapour diffusion through the walls and ensures uniform heating of the tube and longer lifetime.

The sample is introduced into the tube on an auxiliary carbon rod electrode 6 mm in diameter. The tip of this rod is shaped to fit an orifice in the tube wall. In operation, the sample to be analysed may be placed on this electrode in solution or powder form. Five of these auxiliary electrodes are positioned on a turntable device below the graphite tube. When the samples are in position the complete assembly is closed and purged with nitrogen or argon, and the tube is heated to the desired temperature. This requires only 20-30 seconds. The appropriate auxiliary electrode is then moved into place on the turntable below the tube and raised to fit the orifice in the tube wall. The electrode is then heated electrically by an independent current which is obtained from a separate transformer (1 kW, 220/15 V). The circuit to the electrode is completed via the main tube so that heating can only occur when the two units are in contact. The sample is rapidly vaporized giving a pulsed signal, the auxiliary electrode is lowered away from the furnace, and the turntable is rotated to position the next electrode and its sample beneath the furnace. In the original form of the L'Vov furnace the pulse vaporization of the sample was achieved by using a dc arc between an electrode placed in the crucible and the auxiliary electrode mounted under the crucible.

The pulse vaporization method with the graphite furnace was employed by L'Vov using the auxiliary equipment shown in Fig. 7.2. The spectrophotometer has two channels and permits the simultaneous recording of absorption for two elements, one of which may be an internal standard. A continuum source (hydrogen lamp) is used to correct for interference caused by non-selective molecular absorption and scattering of source radiation. The high-intensity atomic line sources used may be either high-frequency electrodeless discharge lamps or hollow-cathode lamps operated from a pulsed power supply. These two types of source provide intensities two or three orders of magnitude greater than that of the conventional hollow-cathode lamp. The radiation emitted by the line sources of the two elements is combined by means of a semi-reflecting mirror before being modulated and merged with the continuum source radiation, using a rotating mirror chopper.

Table 7-1 gives data on the absolute sensitivity obtained in the L'Vov furnace[17] and shows the amounts of different elements actually measured (M_{meas}), the corresponding experimental conditions, and extrapolations of these values to 1% absorption for a 2·5 mm tube (M_{extrap}). The attainable precision claimed for the

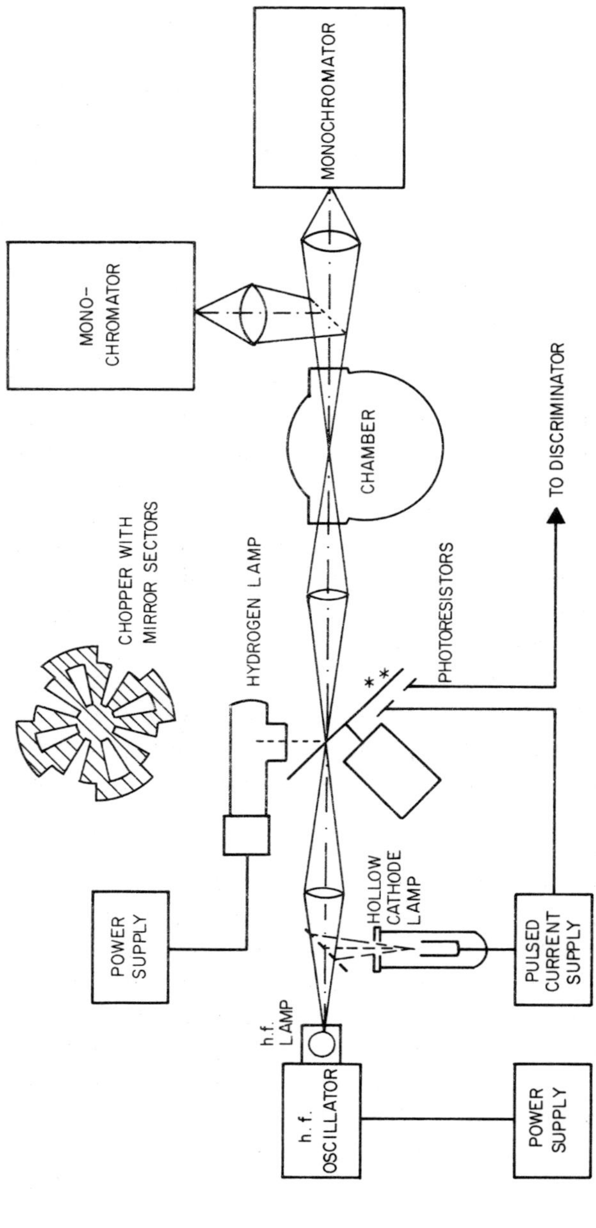

Fig. 7-2. Spectrophotometer used by L'Vov.[17]

TABLE 7-1

Absolute sensitivity attained for various elements with the L'Vov furnace[17]

Line, nm	Diameter, mm	Pressure of argon, atm	Temperature, °C	D_{meas}	M_{meas}, g	M_{extr}, g
Ag 328·1	2·5	2	1800	0·020	$5{\cdot}0 \times 10^{-13}$	1×10^{-13}
Al 309·3	4·5	1	2100	0·03	$2{\cdot}5 \times 10^{-11}$	1×10^{-12}
Au 242·8	2·5	2	1700	0·28	$7{\cdot}0 \times 10^{-11}$	1×10^{-12}
B 249·8	2·5	2	2400	0·105	$5{\cdot}0 \times 10^{-9}$	2×10^{-10}
Ba 553·5	3·0	1	2200	0·05	$1{\cdot}0 \times 10^{-10}$	6×10^{-12}
Be 234·9	4·5	6	2400	0·17	$2{\cdot}6 \times 10^{-12}$	3×10^{-14}
Bi 306·8	2·5	2	1800	0·031	$2{\cdot}5 \times 10^{-11}$	4×10^{-12}
Ca 422·7	2·5	2	2300	0·244	$2{\cdot}5 \times 10^{-11}$	4×10^{-13}
Cd 228·8	1·2	1	1500	0·014	$6{\cdot}0 \times 10^{-14}$	8×10^{-14}
Co 240·7	2·5	2	2200	0·020	$7{\cdot}5 \times 10^{-12}$	2×10^{-12}
Cr 357·9	2·5	2	2200	0·125	$5{\cdot}0 \times 10^{-11}$	2×10^{-12}
Cs 852·1	2·5	2	1900	0·072	$6{\cdot}5 \times 10^{-12}$	4×10^{-13}
Cu 324·8	2·5	2	2100	0·044	$6{\cdot}3 \times 10^{-12}$	6×10^{-13}
Fe 248·3	2·5	2	2100	0·11	$2{\cdot}5 \times 10^{-11}$	1×10^{-12}
Ga 287·4	2·5	2	2100	0·12	$2{\cdot}5 \times 10^{-11}$	1×10^{-12}
Hg 253·7	2·5	2	700	0·026	$4{\cdot}5 \times 10^{-10}$	8×10^{-11}
In 303·9	2·5	1	1900	0·08	$8{\cdot}0 \times 10^{-12}$	4×10^{-13}
K 404·4	2·5	2	1800	0·074	$6{\cdot}3 \times 10^{-10}$	4×10^{-11}
Li 670·8	3·0	1	1900	0·05	$5{\cdot}0 \times 10^{-11}$	3×10^{-12}
Mg 285·2	4·5	2	1800	0·10	$3{\cdot}0 \times 10^{-12}$	4×10^{-14}
Mn 279·5	2·5	2	2000	0·048	$2{\cdot}5 \times 10^{-12}$	2×10^{-13}
Mo 313·3	2·5	2	2500	0·08	$5{\cdot}0 \times 10^{-11}$	3×10^{-12}
Ni 232·0	2·5	2	2200	0·012	$2{\cdot}5 \times 10^{-11}$	9×10^{-12}
Pb 283·3	2·5	2	1900	0·063	$3{\cdot}0 \times 10^{-11}$	2×10^{-12}
Pd 247·6	2·5	2	2100	0·05	$5{\cdot}0 \times 10^{-11}$	4×10^{-12}
Pt 265·9	2·5	2	2300	0·086	$2{\cdot}5 \times 10^{-10}$	1×10^{-11}
Rb 780·0	2·5	2	1900	0·030	$7{\cdot}5 \times 10^{-12}$	1×10^{-12}
Rh 343·5	2·5	2	2300	0·034	$6{\cdot}3 \times 10^{-11}$	8×10^{-12}
Sb 231·1	2·5	2	2000	0·040	$5{\cdot}0 \times 10^{-11}$	5×10^{-12}
Se 196·1	2·5	2	1600	0·092	$2{\cdot}0 \times 10^{-10}$	9×10^{-12}
Si 251·6	2·5	2	2250	0·23	$2{\cdot}7 \times 10^{-12}$	5×10^{-14}
Sn 286·3	2·5	2	2000	0·022	$1{\cdot}0 \times 10^{-11}$	2×10^{-12}
Sr 460·7	3·0	1	2200	0·05	$2{\cdot}0 \times 10^{-11}$	1×10^{-12}
Te 214·3	2·5	2	2000	0·030	$7{\cdot}6 \times 10^{-12}$	1×10^{-12}
Ti 365·3	2·5	2	2500	0·054	$5{\cdot}0 \times 10^{-10}$	1×10^{-11}
Tl 276·8	2·5	2	1800	0·01	$2{\cdot}5 \times 10^{-12}$	1×10^{-12}
Zn 213·8	4·5	4	1500	0·04	$1{\cdot}0 \times 10^{-12}$	3×10^{-14}

graphite tube technique operated with small (2-5 μl) liquid samples corresponds to a relative rms error in a single measurement of 5-8%. This arises from the error in sample application, which may be 1-2·5%, and from the error caused by the recording circuit when pulses with a very short rise time (a few tenths of a second) are registered by a comparatively slow recording potentiometer (1 sec for full scale deflection).

In addition to extensive use of the graphite tube furnace for the determination of traces of metallic and non-metallic elements having resonance lines above 200 nm, L'Vov and Khartsyzov[16] have examined the possibility of its application to the determination of sulphur, phosphorus and iodine in the ultra-violet region between 170 and 190 nm. Using high-frequency discharge lamp sources, the conventional argon-purged graphite furnace, lithium fluoride windows and lenses, and a vacuum grating monochromator, they obtained sensitivities (for 1% absorption) of 3×10^{-12} g P at 177·5 nm, 3×10^{-11} g I at 183·0 nm and 1×10^{-10} g S at 180·7 nm. L'vov and co-workers have also made use of the favourable properties of the graphite furnace to study the Lorentz widths of resonance lines,[17,18] to determine absolute values of oscillator strengths[13,14,17,18] and to estimate atomic diffusion coefficients.[19]

7.2.3 *The Massman Furnace*

Massman[20] has described a graphite cuvette system for use in both AAS and AFS. The furnace arrangements employed are shown in Figs. 7-3 and 7-4. The graphite tube used for AAS is 55 mm long, 6·5 mm internal diameter and 1·5 mm wall thickness. An orifice 2 mm in diameter is cut into the centre of the tube length to permit insertion of the sample with a micropipette. For AFS a cup-shaped graphite cuvette 40 mm long, 6·5 mm internal diameter and 1·5 mm wall thickness is used. The source radiation enters at the open top of the cuvette, and the fluorescence radiation is viewed via a slit cut into the wall. The sample is inserted into the cuvette via its open top.

The tube or cuvette may be heated within a few seconds to temperatures as high as 2600°C, a power supply being used that is capable of supplying up to 400 A. The absorption tube is purged with argon, but the optical path through the tube is open to the atmosphere. Emission from the fluorescence cuvette, which is similarly purged with argon, is viewed via a silica window in the side of the chamber. Modulated radiation from hollow-cathode lamp sources was employed for both techniques but high current levels were necessary with fluorescence measurements to obtain sufficiently intense emission. In order to avoid rapid deterioration of the hollow-cathode lamps of volatile elements the high current was used for only 10 to 15-second intervals.

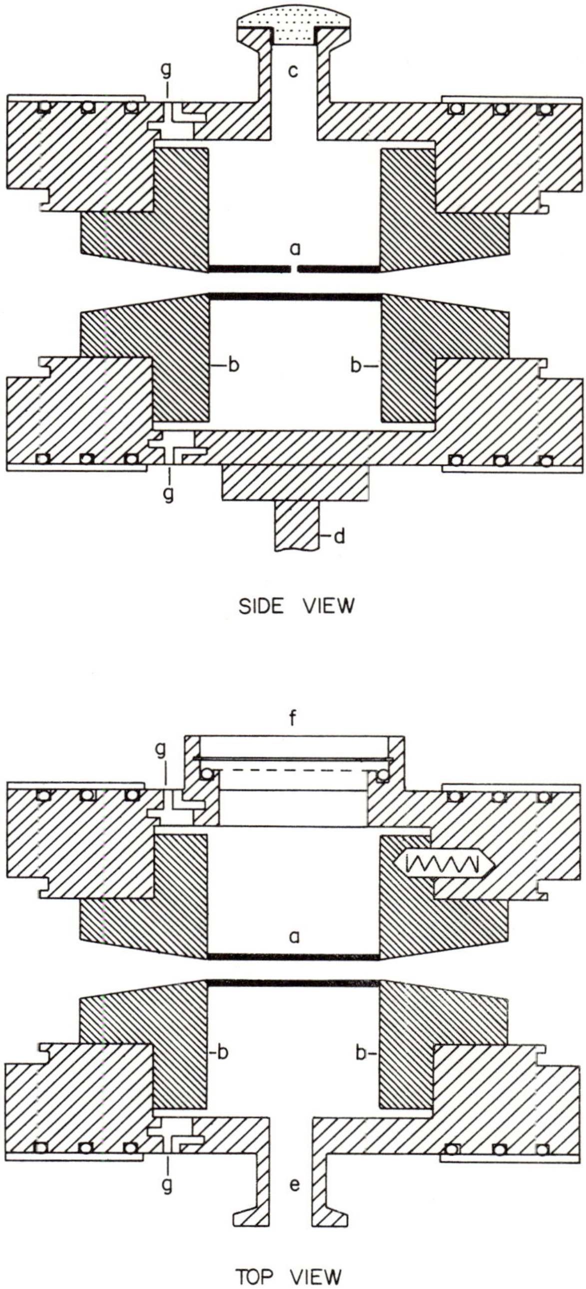

Fig. 7-3. Argon chamber with graphite cuvette for atomic absorption analysis.[20] a, graphite tube; b, steel holders; c, sample inlet port; d, mounting holder; e, argon inlet; f, observation window; g, insulation.

The sample solution volumes employed are between 5 and 200 μl for AAS work and 5 to 50 μl for AFS. Solid samples of up to 1 mg weight were also reported to be acceptable with this apparatus; for larger samples unacceptable background absorption effects are encountered. However, in some cases, correction for non-selective background absorption in AAS may be effected by using a two-channel spectrometer and monitoring the absorption of a nearby elemental non-resonance or filler-gas line which is emitted from the source but is not absorbed by the analyte element.[21]

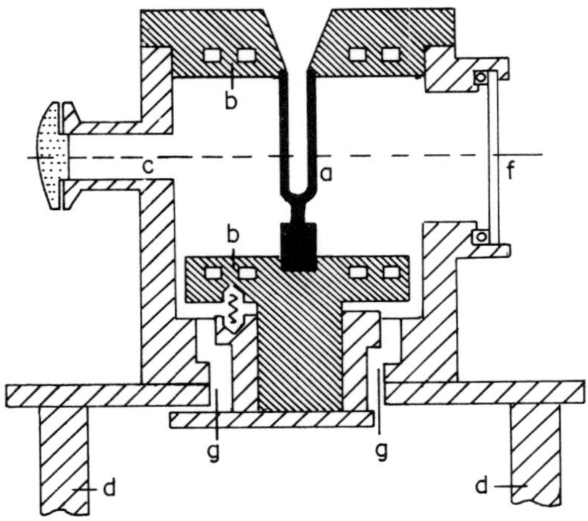

Fig. 7-4. Argon chamber with graphite cuvette for atomic fluorescence analysis.[20] a, graphite cuvette; b, steel holders; c, argon flow; d, mounting holder; e, quartz window for viewing fluorescence radiation; f, insulation.

Table 7-2 shows detection limits obtained by Massman for 16 elements by atomic absorption and nine elements by AFS. Only for zinc and cadmium were clearly superior detection limits obtained by AFS. For other elements the available source intensities were insufficient to permit low detection limits by AFS. The analytical precisions achieved in AAS and AFS were similar when the dissolved solid content of sample solutions was 2% or less and not more than 100 and 30 μl sample volumes were taken for AAS and AFS determinations respectively. Standard deviations between 4 and 12%, depending on the matrix element and its concentration, were obtained. Figure 7-5 shows the scatter of fluorescence intensities observed for the determination of silver; 5 μl samples containing 0·01 ppm silver produce in this case a relative standard deviation of 8%. Higher precision is obtainable by use of an internal standard element whose absorption is measured simultaneously with a two-channel spectrometer.

A graphite tube furnace similar to that described by Massman, and of comparable performance, has recently been introduced as an accessory to a commercially available atomic absorption spectrophotometer.[22] The determination of copper and strontium in milk, which is difficult by AAS without sample pretreatment when flame cells are used, was chosen for preliminary studies to illustrate the application of the device. The recorded signal during the analysis sequence in this determination is shown in Fig. 7-6. A 25 μl sample

TABLE 7-2

Detection limits obtained by AAS and AFS with Massman-type graphite cuvette[20]

Element	Line, nm	Detection limit, AAS, g	Detection limit AFS, g
Zn	213·86	8×10^{-13}	4×10^{-14}
Cd	228·80	2×10^{-12}	$2{\cdot}5 \times 10^{-13}$
Ag	328·07	8×10^{-13}	$1{\cdot}5 \times 10^{-12}$
As	189·04	6×10^{-10}	—
Se	196·09	2×10^{-9}	—
Sb	231·15	1×10^{-10}	2×10^{-10}
Fe	248·33	2×10^{-11}	3×10^{-9}
Hg	253·65	2×10^{-10}	—
Tl	276·79	4×10^{-11}	2×10^{-9}
Mn	279·48	8×10^{-12}	—
Pb	283·31	1×10^{-11}	$3{\cdot}5 \times 10^{-11}$
Mg	285·21	5×10^{-13}	$3{\cdot}5 \times 10^{-12}$
In	303·94	2×10^{-10}	—
Bi	306·77	2×10^{-10}	—
Cu	324·75	1×10^{-11}	$4{\cdot}5 \times 10^{-10}$
Na	588·99	7×10^{-12}	—

of milk is introduced into the furnace and the tube is operated at low current to evaporate the water, giving an absorption signal (A) as steam is driven off. The power is then slowly increased to char the organic matter, the smoke produced also causing non-specific absorption (B). The power is continuously increased until all the organic material has been burnt off. Full power is then applied to bring the tube to its chosen final temperature, and the atomic absorption signal for the analyte element is recorded. The signal in Fig. 7-6 was produced by 25 μl of 0·05 g/ml copper, which corresponds approximately to the normal concentration in milk. The use of a continuum source for correction for non-specific background

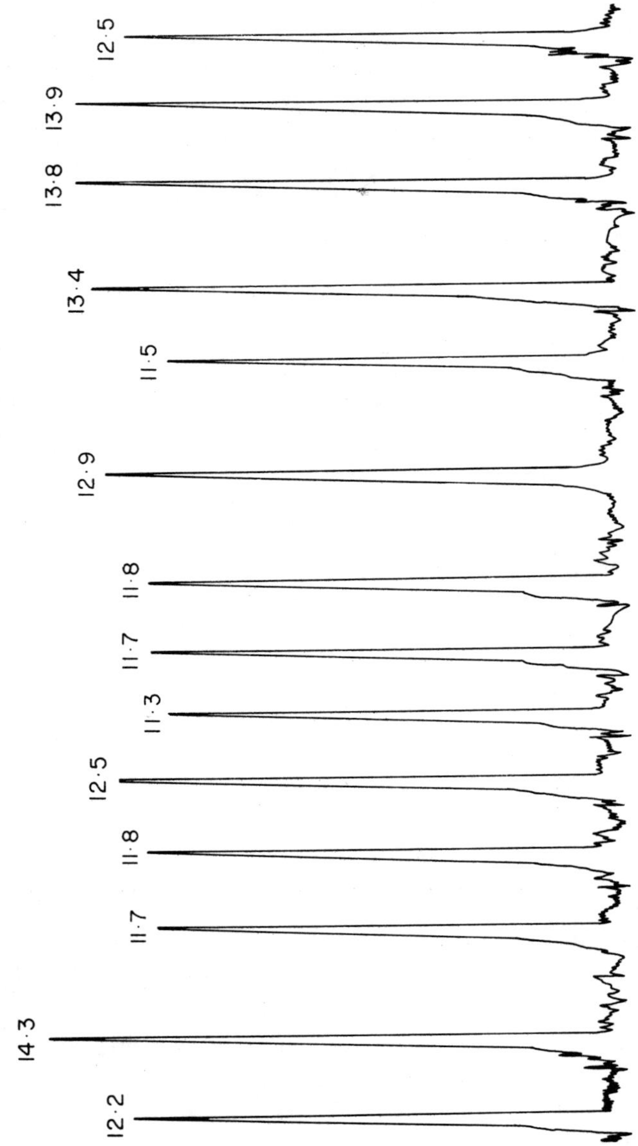

Fig. 7-5. Illustration of the reproducibility attainable in AFS with the Massman graphite curvette.[20] Each peak was obtained from a 5 ml sample of 0·01 ppm silver solution.

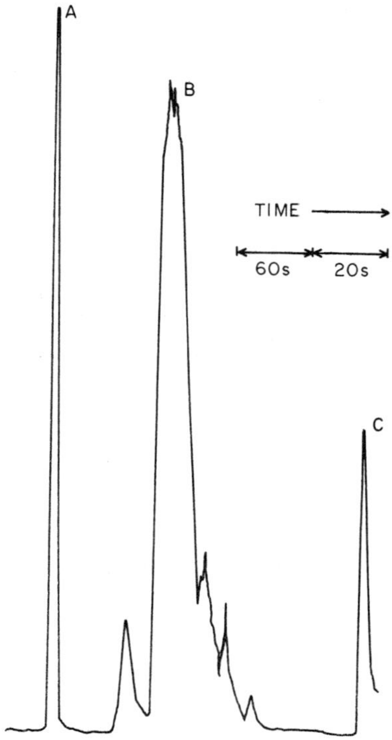

25 μl samples were used and both drying (A) and charring (B) sequences are shown as well as the atomic absorption signal (C)

Fig. 7-6. Analysis of copper in milk with a graphite furnace.[22]

absorption[23,24] is valuable with this technique, as it also indicates whether any analyte element is being lost in the low power sample preparation operations. Some recent applications of this graphite tube furnace technique may be found in Chapter 13.

Pickford and Rossi[25] have developed a system which permits the automatic repetitive operation of a Massman furnace with solution samples. Automation of the sample delivery step permits higher reproducibility to be obtained than when manual introduction of samples is effected with a hypodermic syringe.

7.2.4 Other Graphite Tube Devices

Woodriff and co-workers[26-28] have developed a graphite tube furnace device for AAS which employs a tube 150 mm in length and 7 mm in diameter heated by a current from an arc welder. The sample is introduced into the tube on a carbon cup inserted through

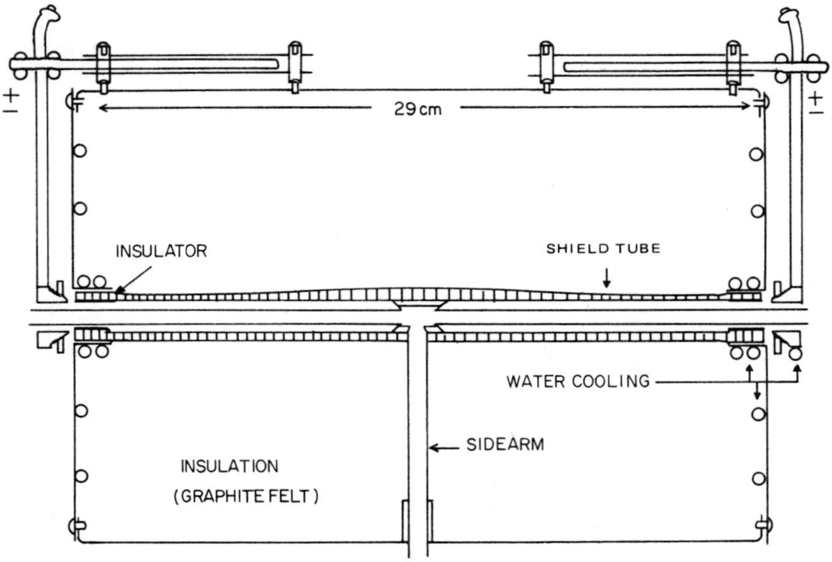

Fig. 7-7. Design of the Woodriff furnace.

TABLE 7-3

Absolute sensitivity and detection limits obtained for various elements with the Woodriff-type furnance[26]

Element	Line, nm	Furnace temperature, °C	Sensitivity g, per 1 per cent absorption	Detection limit, g[a]
Ag	328·1	1265	8×10^{-12}	$1{\cdot}2 \times 10^{-11}$
Cd	228·8	1200	9×10^{-12}	$1{\cdot}3 \times 10^{-11}$
Zn	213·9	1200	9×10^{-12}	$3{\cdot}9 \times 10^{-11}$
Pb	283·3	1200	1×10^{-11}	$5{\cdot}7 \times 10^{-11}$
Li	670·8	1200	1×10^{-11}	3×10^{-11}
Cu	324·7	1550	9×10^{-11}	$3{\cdot}2 \times 10^{-10}$
Mn	279·5	1200	8×10^{-11}	$2{\cdot}7 \times 10^{-10}$
Mg	285·2	1200	8×10^{-11}	$2{\cdot}5 \times 10^{-11}$
Fe	248·3	2200	1×10^{-10}	$2{\cdot}8 \times 10^{-10}$
Al	309·2	1550	2×10^{-10}	$5{\cdot}4 \times 10^{-10}$
Ni	232·0	2200	2×10^{-10}	$7{\cdot}3 \times 10^{-11}$
Dy	421·2	2200	1×10^{-10}	$2{\cdot}2 \times 10^{-10}$
Ho	410·4	2200	9×10^{-11}	$3{\cdot}3 \times 10^{-10}$
Er	400·8	2200	1×10^{-10}	$3{\cdot}7 \times 10^{-11}$
Ca	422·7	1200	1×10^{-10}	$5{\cdot}3 \times 10^{-10}$

[a] Amount producing absorbance value equal to twice experimental standard deviation of the blank.

the side-arm of the tube, or it may be nebulized and carried in with an inert gas. Figure 7-7 shows the furnace design, and Table 7-3 illustrates the sensitivities obtained with this apparatus for 15 elements. No significant matrix effects were observed from the presence of relatively large amounts of Al, Cr, Cu, Fe, Ni, Mn, Zn, and Mg on the atomic absorption signal obtained for 10^{-9} g of silver.

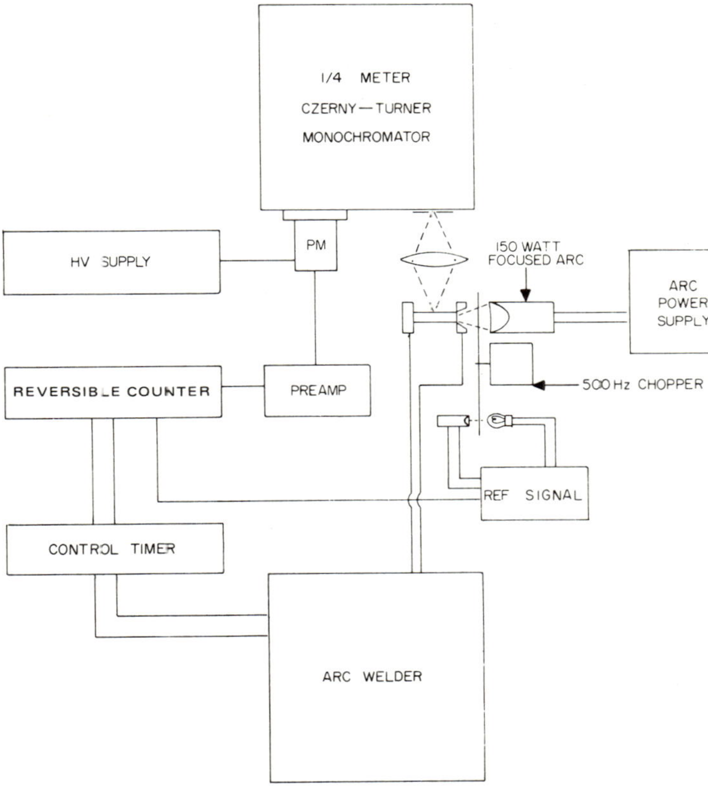

Fig. 7-8. Instrumental assembly for use with a graphite cell system in atomic fluorescence spectroscopy.[1]

Winefordner[1] has described the construction of a graphite cell system for AFS studies. This is shown schematically in Fig. 7-8. The graphite tube is heated by an electric arc welder of 300A output, and sample solutions are introduced into the cell by means of a hypodermic syringe inserted through a rubber septum in the front of the cell housing. A low flow-rate of argon is used to provide a continuously flushed cell. Radiation from a 150-W xenon arc lamp, modulated at 500 Hz by a mechanical chopper, is used as the excitation source. A photon-counting system is employed to register the fluorescence emission after its detection via a 25-cm grating monochromator and photomultiplier. Analytical results obtained with this system have not yet been published.

7.3 FILAMENTS AND OPEN CELLS

7.3.1 Wire Filament Devices

Atomization devices in which a wire loop or a sample boat is introduced into the hot flame gases above the primary reaction zone of premixed flames are described in Chapter 8. Several devices of the same type in which the sample loop is heated electrically have been described. In these open devices the atomic vapour released from the filament passes through an unconfined volume in the absorption or fluorescence light path and transient analytical signals are obtained. It may be difficult to achieve freedom from matrix effects when a large temperature gradient exists between the hot filament and the cooler volume above it in which AAS or AFS measurements are made. Owing to the unconfined nature of the analytical cell volume in these devices, however, the tendency towards memory effects with some analyte atoms may be less than with furnace devices.

Ulfvarson,[29] and Brandenburger and Bader,[30,31] were able to detect nanogram amounts of mercury by collecting it as an amalgam on a wire and then heating the wire to vaporize the mercury into the optical path of an atomic absorption spectrophotometer. Brandenburger has also employed this technique with a closed system in which the mercury vapour is confined within an absorption tube after vaporization from the filament.[32,33] The filament technique has also been applied to the AAS determination of Cd, Zn, Pb, Te, Cu, Ag, Au and Pt after electrolytic deposition on a wire filament.[31,33]

Bratzel, Dagnall and Winefordner[34] have described the application in AFS of a similar filament technique to that employed by Brandenburger and Bader. In this device sample solutions are vaporized from a platinum loop which is positioned in a flowing stream of inert shielding gas. The vaporized analyte element is swept away from the wire loop and irradiated by light from an electrodeless discharge lamp. The atomic fluorescence obtained is measured with a dc electrometer system. Typical absolute detection limits reported are Cd, 10^{-15} g; Hg, 10^{-8} g; and Ga, 10^{-7} g.

7.3.2 The Graphite Filament Cell

West and Williams[35] have described a device in which a graphite filament, 2 mm in diameter and *ca.* 20 mm in length, is heated to 2000 to 2500°C within 5 seconds by passage of a current of *ca.* 100 A at 5 V. The filament is mounted on water-cooled stainless steel electrodes, the assembly being housed in an argon-filled chamber fitted with quartz windows. Small liquid samples (5 μl) are placed on a depression in the filament. The original device was used for the detection of silver and magnesium by AAS and AFS. Amounts of 10^{-10} g of these elements produced 1% absorption in AAS, while in AFS 10^{-16} g Mg and 3×10^{-11} g Ag were detectable.

An improved design of the device, in which inert gas shielding of the filament eliminates the need for a closed chamber assembly, is shown in Fig. 7-9. The limits of detection and maximum determinable amounts observed for a range of elements by AFS with this type of assembly are given in Table 7-4. Typical sensitivity values (for 1% absorption) and detection limits for AAS work with a similar device are shown in Table 7-5.

As with other non-flame cells in which pulse atomization is effected, it is important that the response time of the electronic circuitry employed is fast enough to permit the variation of the AAS or AFS signal with time to be recorded without distortion. Under these conditions either peak-height or peak-area methods of measurement may be employed. Additionally, with the filament device, and in contrast to the L'Vov or Massman furnaces, no further energy is available in the free space above the filament after the vaporization pulse to prevent condensation of the atomic vapour. The decay of the atomic population is therefore promoted by condensation of the analyte atoms with their own species or with the atoms or molecules formed from concomitant elements present in the sample. The effective lifetime of the free atoms may therefore be considerably shorter than when a furnace system is used. Hence, it is necessary to view the atomic vapour close to the filament if interelement effects are to be minimized.

Figure 7-10 illustrates the manner in which the absorbance obtained for Cu, Ni, Pb and Al decreases as the height of observation above the filament is increased.[41] Interference effects from volatile matrix elements have been observed with the carbon filament device

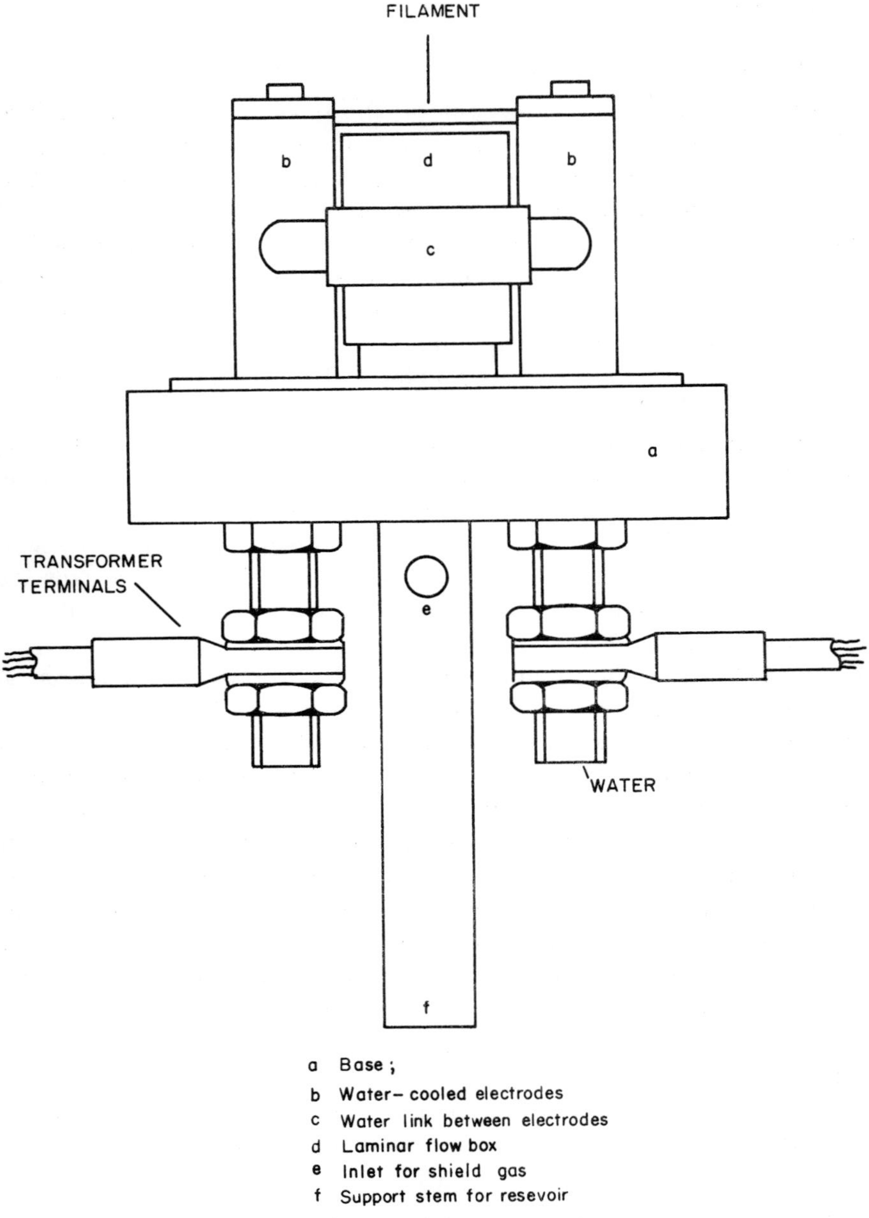

Fig. 7-9. Filament atom reservoir for use in open atmosphere.[36]

to be generally more serious than in flame systems; interference effects from refractory-oxide-forming elements, however, appear to be significantly less troublesome with this device than in flames. This most probably results from the fact that interelement effects

TABLE 7-4

Detection limit and upper determination limit data in AFS carbon filament atom reservoir of West *et al.*

Element	Limit of detection,[a] g	Maximum determinable amount, g	Concentration in 1 μl samples, ppm	Reference
Mg	10^{-12}	1×10^{-9}	0·001	37
Ag	10^{-12}	2×10^{-9}	0·001	37
Pb	10^{-11}	$1{\cdot}5 \times 10^{-7}$	0·01	37
Zn	2×10^{-14}	4×10^{-10}	0·00002	37
Bi	10^{-11}	1×10^{-8}	0·01	37
Tl	5×10^{-11}	2×10^{-9}	0·05	37
Ga	5×10^{-11}	1×10^{-8}	0·05	37
Cd	$1{\cdot}5 \times 10^{-13}$	—	0·00015	41
Au	4×10^{-12}	—	0·0008[b]	38
Sb	10^{-9}	3×10^{-8}	1	39
Cu	10^{-12}	4×10^{-9}	0·001	39
Mn	5×10^{-12}	2×10^{-9}	0·005	40
Co	2×10^{-11}	6×10^{-9}	0·02	39
Ni	5×10^{-12}	5×10^{-9}	0·005	39

[a] Signal : Noise = 2:1.
[b] 5 μl sample.

TABLE 7-5

Atomic absorption sensitivity data obtained with carbon filament atom reservoir of West *et al.*

Element	Sensitivity g/1% abs	Detection limit, ppm (g)	Reference
Cu	$3{\cdot}3 \times 10^{-11}$	5×10^{-11}	41
Pb	7×10^{-12}	5×10^{-11}	41
Mn	5×10^{-11}	5×10^{-11}	42
Ni	$2{\cdot}4 \times 10^{-10}$	3×10^{-10}	41
Al	6×10^{-10}	1×10^{-9}	41

occur in the vapour phase above the filament on condensation of those (more volatile) elements that readily vaporize from it, whereas it is difficult to vaporize more refractory elements from the filament.

Amos[43] has utilized a modified form of the graphite rod device of West *et al.* in which a hole is drilled across the diameter of the rod to

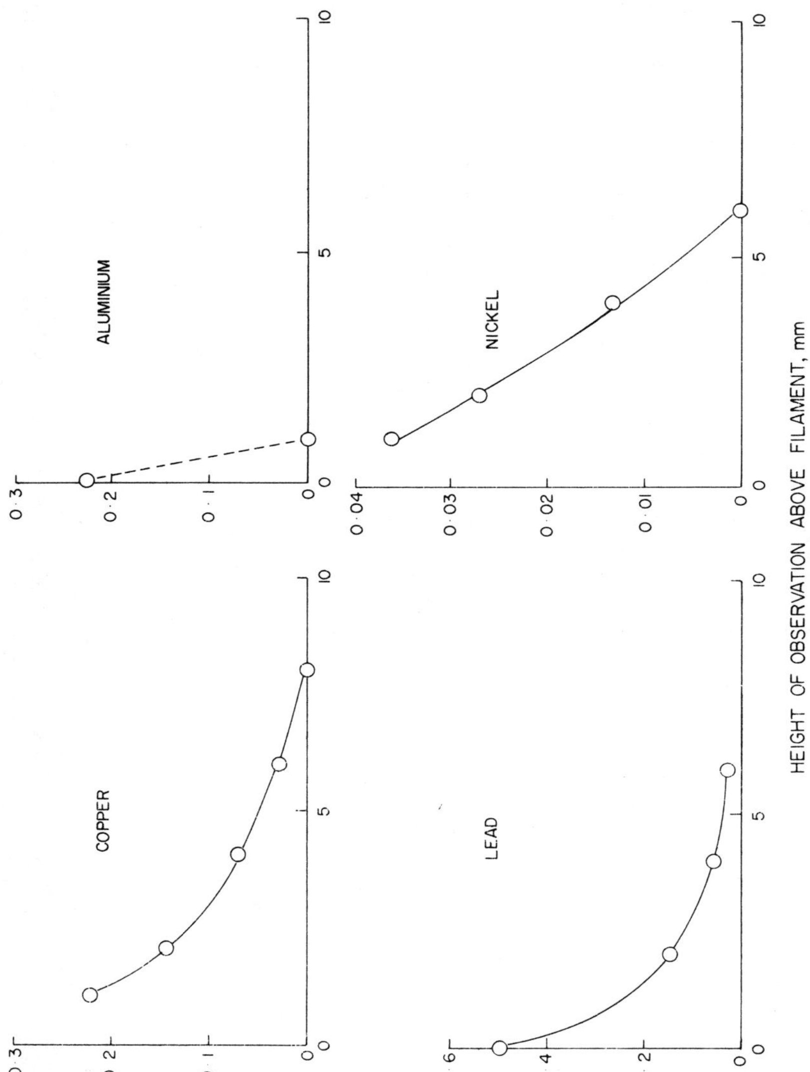

Fig. 7-10. Variation of absorbance with height of observation above carbon filament.[41]

form a cavity into which the sample liquid (0·5 to 1 μl) can be introduced. Amos *et al.*[44] have compared this device with the carbon rod atomizer for AAS and AFS, and have developed methods for the direct analysis of lead in blood and urine. The replacement of the argon or nitrogen shield gas for the carbon rod device by hydrogen has been described; in operation the glowing rod then ignites the hydrogen to form a diffusion flame surrounding the rod. This modification assists atomization of less volatile elements and tends to minimize inter-element effects. Considerably less serious spectral and chemical interferences have been observed in the determination of lead in biological materials, for example, when the rod is surrounded by a diffusion flame in this way. Winefordner *et al.*,[45] in the development of a method for the determination of copper in serum by AAS using graphite rod atomization, have described the use of an outer argon-hydrogen flame sheath to surround the argon shield used to provide an inert atmosphere for operation of the filament. A number of other applications of the graphite filament technique are tabulated in Chapter 13.

7.3.3 *Sample Boat Cells*

Donega and Burgess[46] have described a device similar to that used by West and co-workers, but in which sample boats, 50 mm long and 6 mm wide, are cut from graphite sheet, or tantalum or tungsten foil. The sample boat is heated electrothermally with a current of 30-50 A at 12 V which is sufficient to heat the boat to ~2200°C in less than 0·1 second. The sample boat is supported on two copper rods mounted via insulators in a brass baseplate fitted with a central quartz window and an outer O-ring seal. The whole assembly is fitted into one end of a quartz tube 50 mm in diameter, with a second quartz window sealed to the other end. The chamber may be purged with inert gas and used at operating pressures between 1 and 760 torr. The complete assembly is shown schematically in Fig. 7-11.

The device may be used with liquid samples up to 50-100 μl in volume. The mode of operation for AAS is similar to that employed by West and co-workers for their carbon filament atom cell. Table 7-6 shows some sensitivities obtained with this device. Little data concerning the effect of matrix elements on the sensitivity obtained has been reported. The method of operation at low pressure should result in low background absorption interference, in the same manner as reported by Massman for the hot hollow-cathode technique (see Section 7.4.3), but this may be achieved only at the

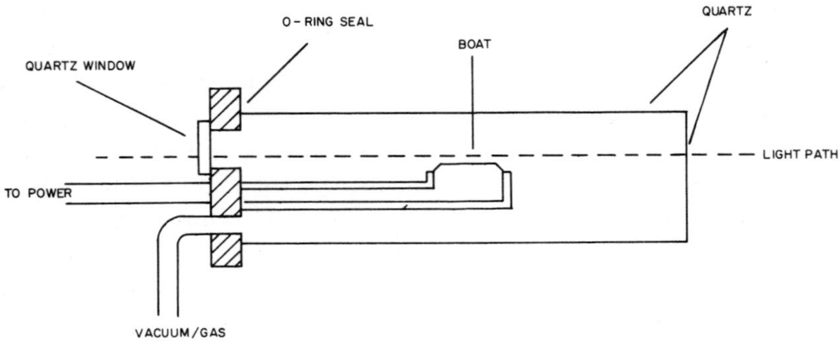

Fig. 7-11. Absorption chamber and filament assembly used by Donega and Burgess.[46]

TABLE 7-6

Absolute sensitivities reported for filament technique of Donega and Burgess[46]

Element	Wavelength, nm	Sensitivity, g	Operating conditions[a]
Al	309·3	4×10^{-10}	1
Cr	359·4	6×10^{-12}	2
Cu	324·8	3×10^{-12}	2
Mn	279·5	3×10^{-12}	2
Mo	313·3	3×10^{-9}	3
Ni	341·5	3×10^{-9}	2
Pt	265·9	3×10^{-7}	2
Si	251·6	3×10^{-9}	2
Na	589·0	3×10^{-11}	4
V	318·4	6×10^{-10}	1

[a] (1) Tantalum boat, 300 torr hydrogen. (2) Tantalum boat, 300 torr argon. (3) Tungsten boat, 1 torr argon. (4) Graphite boat, 100 torr argon.

expense of short residence time in the effective absorption volume and the requirement of a fast-response detector system. Several commercial tantalum ribbon atomizers of the type described by Donega and Burgess are available; several applications of these devices to trace analysis with microlitre liquid samples are listed in Chapter 13.

7.4 CATHODE SPUTTERING CELLS

The sputtering action which is responsible for the production of atomic vapour within a hollow-cathode lamp may be exploited for the atomization of samples for AAS. Several methods of placing the sample in the cathode have been devised.

7.4.1 *Cathodes Fabricated from the Sample*

Walsh and workers[47,48] constructed a sputtering chamber in which metal samples, machined to the shape of an open-ended, hollow cathode, were clamped. The chamber was then evacuated, filled with argon at the required pressure, and a discharge initiated as in a conventional hollow-cathode source. The sputtering chamber was fitted with silica windows and could be placed in the light path of an atomic absorption spectrophotometer. The determination of Ag and P in copper and Si in aluminium and steel with this device was reported.[49] A disadvantage of the technique is the need to prepare the cathode from the sample itself; this tends to restrict the technique to metal samples.

7.4.2 *Use of Liquid Samples*

In order to avoid the restriction of the Walsh cell, Goleb and Brody[50] evaporated sample solutions on to the inner wall of an aluminium hollow-cathode. This was then used as part of a water-cooled demountable hollow-cathode lamp with a continuous-flow low-pressure system. Only small sample volumes were required, and 1 μg quantities of Na, Ca, Mg, Si and Be were detectable; considerable interelement effects were observed. Goleb[51] has also applied this device as an atom source for the isotopic analysis of uranium.

Ivanov *et al.*[52] employed a graphite hollow cathode in a sputtering cell; in this way they achieved some of the advantages of the graphite cuvette technique in addition to those gained from the sputtering cell technique. Sample solutions were evaporated directly onto the walls or onto a fine molybdenum wire which was then placed along the central axis of the cathode.

Gandrud and Skogerboe[53] have studied the use of a simple hollow-cathode discharge cell for the detection of nanogram amounts of Ag, As, Ca, Cd, Hg, Pb, Se, and Zn. In operation, sample solutions are evaporated on graphite or aluminium discs which are then placed on a brass rod cathode. The cell is then evacuated, filled with argon at 3 torr and the discharge initiated. The transient absorption signal is then recorded over the period of sample vaporization; the tendency towards arcing, which is caused by pressure rise due to the outgassing of the cell and sample, is avoided by continuous pumping to maintain the low pressure of argon. Kirkbright and Wilson have recently described the use of a modified form of this device, shown in

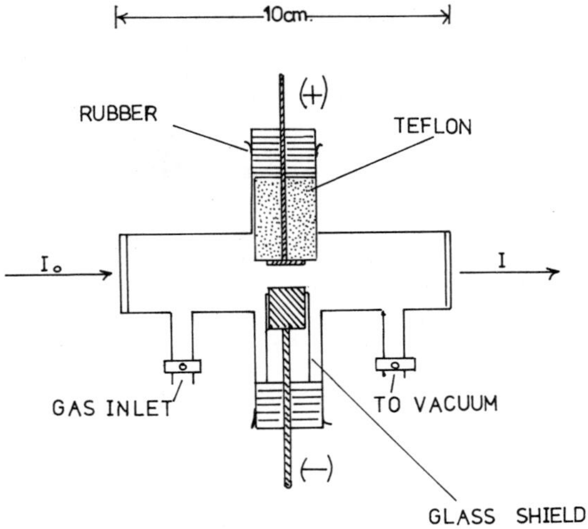

Fig. 7-12. Modified form of discharge cell of Skogerboe and Gandrud[53] used for the determination of sub-microgram amounts of iodine by direct AAS at 183·0 nm.[54]

Fig. 7-12, for the direct determination of iodine by AAS at 183·0 nm.

7.4.3 Solid Samples in a Hot Hollow Cathode

Massman[55] has described the use of a hot hollow-cathode assembly for AAS determinations with solid samples. In this device a graphite tube (30 mm in length and 7 mm internal diameter) is supported on a small cylindrical graphite electrode which passes through the wall of the cathode tube. This carrier electrode also holds the solid sample in a depression 2 mm in diameter and 3 mm deep bored out of its end; the dimensions of the depression may be varied depending on the sample size. The cathode assembly is mounted on a 2 mm diameter molybdenum rod whose insulated base passes through the water-cooled base plate of the metal housing of the assembly. This housing acts as the anode and is at ground potential. The compartment is fitted with quartz windows and may be purged with argon. The discharge is then operated at an argon pressure in the region 1 to 10 torr at a power of up to 1 kW.

When hollow-cathode cells are used in AAS, continuum radiation emitted by the cathode, and the spectrum emitted by the sample itself, might be expected to seriously interfere with the absorption measurements. With cool cathodes this interference is difficult to avoid, since the atomic vapour persists only as long as the discharge current is applied. Massman pointed out, however, that this difficulty may be overcome with the hot cathode device, as under suitable operating conditions atomic vapour may persist after the discharge current has been allowed to fall to zero. When the graphite hollow cathode discharge is operated with a 50 Hz half-wave current, therefore, the discharge, which creates an atomic vapour, is present for only one-half of the operating period. During the alternate discharge-free half-period the atomic vapour may still persist, so that its absorption of the primary source radiation may be measured without interference.

In the instrumental assembly described by Massman a rotating sector, driven in phase with the discharge current cycle, is placed between the cell and the monochromator slit. This sector thus transmits the primary source radiation only during a prescribed period (just less than one-half cycle) while the discharge current is not applied. Hence, a detector operated in phase with this sector receives no radiation originating from the atom cell. The primary hollow-cathode lamp radiation is modulated at 450 Hz by a rotating sector placed between the source and the graphite cathode cell. When the peak absorption signal is measured, linear calibration is possible only over a narrow working range; e.g. 1-10 ppm silver in a 30 mg sample of lead. When integration of the signal is performed, it is possible to achieve improved linear range; with integration, for example, 1–300 ppm silver in lead then yields a linear working curve. In the analysis of Pb, Zn or Al samples (30 mg) coefficients of variations of 8 to 15% were obtained. In the determination of silver in lead with integration of the absorption signal a standard deviation of 4% was achieved.

The hot graphite hollow cathode cannot be heated above 2000°C, so that only samples whose evaporation proceeds sufficiently rapidly below this temperature can be analysed. The method is more suitable for the determination in milligram samples of low concentrations of relatively volatile elements such as lead, zinc or aluminium than for the determination of small absolute amounts of analyte element. This results from the short residence time of the atomic vapour in the absorption volume at the low pressure employed. The vaporization rate of samples having metals such as Fe, Ni, and Cu as the major constituents is too slow to allow their effective analysis by this technique. When these elements are present as impurities in more volatile matrices, however, the evaporation rate may be sufficiently high to permit their determination. Table 7-7 shows the detection limits obtainable by AAS with this device for the determination of various elements in 30 mg samples of relatively volatile samples. These are compared with the detection limits obtained for the same samples by emission spectrography with a hot hollow cathode and a medium quartz spectrograph.[56] Even with large sample weights (100 mg) of Pb, Zn, Sb or Al, less than 1% interference is encountered owing to background absorption at wavelengths greater than 220 nm.

TABLE 7-7

Detection limits obtained with hot hollow cathode for the determination of various elements in solid metal samples[54,55]

		Atomic absorption method		Atomic emission method	
Element	Sample, 30 mg	Line, nm	Det. limit, ppm	Line, nm	Det. limit, ppm
Ag		328·07	0·3	328·07	0·1
Sb		231·15	50	252·85	1·0
Zn		213·86	0·05	213·86	0·5
Cu	Lead	324·75	2·8	324·75	0·3
Cd	Lead	228·80	0·01	228·80	0·2
Mg	Zinc	285·21	1·2	—	—
Mn	Zinc	279·48	2·2	—	—
Cr	Aluminium	357·87	10	—	—

From this point of view the hot graphite hollow cathode operated at low pressure may be superior for large samples to graphite furnaces operated at atmospheric pressure.

7.5 OTHER TYPES OF ATOM CELL

7.5.1 Arcs, Sparks and Plasmas

The use of the dc arc for sample atomization in AAS has been described by several workers.[57,58,59] Robinson has reported that 20% absorption was obtained for aluminium when a spark was used to atomize the sample;[60,61] the solution was nebulized in the conventional fashion and the aerosol was passed between two electrodes between which the spark was struck.

There have also been several reports of the use of RF or microwave plasma sources to atomize for samples AAS. Wendt and Fassel[62,63] employed an induction-coupled plasma together with a triple-pass optical system to obtain large absorbance signals for AAS.

Friend and Diefenderfer[64] investigated the possibility of using a plasma jet for the determination of refractory elements. The RF plasma has been applied to AAS by at least three groups of workers.[65,66,67] These plasma systems, some of which have previously been used extensively for analytical emission spectroscopy, show great promise for AAS. The high energy available should permit considerable freedom from matrix interference effects, and the possibility of transferring samples to the plasma with low inert gas flow-rates should result in high atomic concentration for the analyte element and great sensitivity. Greater control over the chemical environment can be achieved with plasmas than with flames and, consequently, the lifetime of free ground state atoms may be greater in these atom cells.

7.5.2 Lasers

The use of lasers offers the possibility of atomization for the direct analytical examination of solid surfaces.[68,69,70] The heat produced at a solid when a focused laser beam strikes it may vaporize the solid material over a small area of the surface. This technique has been employed in spark-emission spectroscopy; a spark is passed across electrodes placed just above the surface so that it passes through the plume of atoms produced by laser vaporization. In atomic absorption spectroscopy this direct sampling and atomization method should provide for low emission background and high sensitivity for small solid samples. In the laser microprobe-atomic absorption instrumental assembly devised by Mossotti, Laqua and Hagenah[68] a pulsed laser system is employed to produce rapid heating and vaporization from the sample surface. The atomic absorption in the transient atomic vapour cloud which is produced is then measured with a fast response system.

7.5.3 Solid Propellant Technique

Venghiattis[71] has proposed a technique for the direct conversion of solid samples to atomic vapour in which the powdered sample is mixed with a solid-propellant powder; the mixture is then compressed and ignited. The atomic vapour of the analyte element in the gases of the "flame" that is produced then passes into the absorption light path of a conventional atomic absorption spectrophotometer. Calibration may be performed by the similar technique of mixing standard samples with the propellant mixture.

The technique has been applied to the determination of trace elements such as Au, Ag, Cu, Pb, and Zn in ores. Although it may be difficult to achieve better than 5% precision without very careful sample preparation, this technique may find considerable application for analyses in the field or on-site where the transportation or use of gas cylinders, solutions and chemicals is inconvenient.

REFERENCES

1. J. D. Winefordner, "Atomic Absorption Spectroscopy", eds. G. F. Kirkbright and R. M. Dagnall, Butterworths, London, 1971, p. 37.
2. J. P. Mislan, Paper presented at 7th Conf. Anal. Chem. in Nuclear Technol., Gatlinberg, Tenn., 1963.
3. R. D. Hudson, *Phys. Rev.*, 1964, **135**, 1212.
4. G. L. Vidale, Space Science Lab. Aerospace Operation, General Electric T.I.S. Report R 605D330, 1960.
5. S. P. Choong and W. Loong-Seng, *Nature,* 1964, **204**, 276.
6. F. S. Tomkins and B. Ercoli, *Applied Optics,* 1967, **6**, 1299.
7. M. S. Black, T. H. Glenn, M. P. Bratzel and J. D. Winefordner, *Anal. Chem.*, 1971, **43**, 1769.
8. A. S. King, *Astrophys. Journal,* 1908, **27**, 353.
9. B. V. L'Vov, *J. Eng. Phys.*, 1959, **2**, (2), 44.
10. B. V. L'Vov, *J. Eng. Phys.*, 1959, **2**, (11), 56.
11. B. V. L'Vov, *Spectrochim. Acta,* 1961, **17**, 761.
12. B. V. L'Vov, *Zavodsk. Lab.*, 1962, **28**, 931.
13. B. V. L'Vov, *Opt. i. Specktrosc.*, 1965, **19**, 507.
14. B. V. L'Vov, and G. V. Pljusk, *Proc. 15th Symp. Spectroscopy, Minsk, 1963,* Vol. 2, p. 159. Acad. Sci. USSR, 1964.
15. B. V. L'Vov and G. G. Lebedev, *Zh. Prikl. Spectroscop.*, 1967, **7**, 264.
16. B. V. L'Vov, "Atomic Absorption Spectrochemical Analysis", Hilger, London, 1970.
17. B. V. L'Vov, *Spectrochim. Acta,* 1969, **24B**, 53.
18. B. V. L'Vov and G. G. Lebedev, *Symp. Theoret. Spectry., Yerevan, 1966.*
19. G. I. Nikolayev and V. B. Aleskovskii, *Zhur. Techn. Fiz.*, 1964, **34**, 753.
20. H. Massman, *Spectrochim. Acta,* 1968, **23B**, 215.
21. H. Massman, *Z. Anal. Chem.*, 1967, **225**, 203.
22. D. C. Manning and F. Fernandez, *Atomic Absorption Newsletter,* 1970, **9**, 65.
23. S. R. Koirtyohann, *Anal. Chem.*, 1965, **37**, 601.
24. H. L. Kahn, *Atomic Absorption Newsletter,* 1968, **7**, 40.
25. C. J. Pickford and G. Rossi, *Analyst,* 1972, **97**, 647.
26. R. Woodriff, R. W. Stone and A. M. Held, *Appl. Spectroscopy,* 1968, **22**, 408.
27. R. Woodriff and G. Ramelow, *Spectrochim. Acta,* 1968, **23B**, 665.
28. R. Woodriff and R. W. Stone, *Applied Optics,* 1968, **7**, 1337.
29. U. Ulfvarson, *Acta Chem. Scand.*, 1967, **21**, 641.
30. H. Brandenburger and H. Bader, *Helv. Chim. Acta,* 1967, **50**, 1409.
31. H. Brandenburger, *Chimia,* 1968, **22**, 449.
32. H. Brandenburger and H. Bader, *Atomic Absorption Newsletter,* 1967, **6**, 101.
33. H. Brandenburger and H. Bader, *Atomic Absorption Newsletter,* 1968, **7**, 53.
34. M. P. Bratzel, R. M. Dagnall and J. D. Winefordner, *Atomic Absorption Newsletter*, 1969, **48**, 197.

35. T. S. West and X. K. Williams, *Anal. Chim. Acta,* 1969, **45**, 27.
36. J. F. Alder and T. S. West, *Anal. Chim. Acta,* 1970, **51**, 365.
37. R. G. Anderson, I. S. Maines and T. S. West, *Anal. Chim. Acta,* 1970, **51**, 355.
38. J. Aggett and T. S. West, *Anal. Chim. Acta,* 1971, **55**, 349.
39. D. Alger, R. G. Anderson, I. S. Maines and T. S. West, *Anal. Chim. Acta,* 1971, **57**, 271.
40. L. Ebdon, G. F. Kirkbright and T. S. West, *Anal. Chim. Acta,* 1972, **19**, 1301.
41. R. G. Anderson, H. N. Johnson and T. S. West, *Anal. Chim. Acta,* 1971, **57**, 281.
42. L. Ebdon, G. F. Kirkbright and T. S. West, *Anal. Chim. Acta,* 1971, **58**, 39.
43. M. D. Amos, *American Lab.*, 1970, 33.
44. M. D. Amos, P. A. Bennett, K. G. Brodie, P. W. Y. Lung and J. P. Matousek, *Anal. Chem.*, 1971, **43**, 211.
45. M. Glenn, J. Savory, L. Hart, T. Glenn and J. D. Winefordner, *Anal. Chim. Acta,* 1971, **57**, 263.
46. H. M. Donega and T. E. Burgess, *Anal. Chem.*, 1970, **42**, 1521.
47. W. G. Jones and A. Walsh, *Spectrochim. Acta,* 1960, **16**, 249.
48. B. M. Gatehouse and A. Walsh, *Spectrochim. Acta,* 1960, **16**, 602.
49. A. Walsh, *Proc. Xth Colloq. Spectrosc. Int., 1962,* 127.
50. J. A. Goleb and J. K. Brody, *Anal. Chim. Acta,* 1963, **28**, 457.
51. J. A. Goleb, *Anal. Chem.*, 1963, **35**, 1978.
52. N. P. Ivanov, M. N. Gusinsky and A. D. Jesikov, *Zh. Analit. Khim.*, 1965, **20**, 1133.
53. B. W. Gandrud and R. K. Skogerboe, *Appl. Spectroscopy*, 1971, **25**, 243.
54. G. F. Kirkbright and P. J. Wilson, *Anal. Chim. Acta,* 1974, **68**, 462.
55. H. Massman, *Spectrochim. Acta,* 1970, **25B**, 393.
56. H. Massman, *IXth Colloq. Spectr. Int., 1961,* Vol. II, 170, 1962.
57. M. Marinkovic, B. Bojovic and D. Pesic, *XIIIth Colloq. Spect. Int.*, Hilger and Watts, London, 1967, p. 1181.
58. U. I. Belyaev, L. M. Ivantsov, A. V. Karyakin, P. H. Phi and V. V. Shemet, *Zh. Analit. Khim.*, 1968, **23**, 508.
59. T. Kantor and L. Erdey, *Spectrochim. Acta, 1969,* **24B**, 283.
60. J. W. Robinson, *Anal. Chim. Acta,* 1962, **27**, 465.
61. J. W. Robinson, *Ind. Chemist,* 1962, **38**, 226, 362.
62. R. R. Wendt and V. A. Fassel, *Anal. Chem.*, 1965, **37**, 920.
63. R. R. Wendt and V. A. Fassel, *Anal. Chem.*, 1966, **38**, 337.
64. K. E. Friend and A. J. Diefenderfer, *Anal. Chem.*, 1966, **38**, 1763.
65. C. Veillon and M. Margoshes, *Spectrochim. Acta,* 1968, **23B**, 503.
66. S. Greenfield, P. H. Smith, A. E. Breeze and N. M. D. Chilton, *Anal. Chim. Acta,* 1968, **41**, 385.
67. H. Brandenburger, *Chimia,* 1968, **22**, 449.
68. V. G. Mossotti, K. Laqua and W. D. Hagenah, *Spectrochim. Acta,* 1967, **23B**, 197.
69. S. D. Rasberry, B. F. Schribner and M. Margoshes, *Appl. Optics,* 1967, **6**, 81, 87.
70. A. V. Karyakin and V. A. Kaigorodov, *Zh. Analit. Khim.*, 1968, **23**, 930.
71. A. A. Venghiattis, *Atomic Absorption Newsletter,* 1967, **6**, 19.

CHAPTER 8

Introduction of Liquid Samples into Flame Atom Cells

8.1 INTRODUCTION

The quantitative analytical application of AAS and AFS using a flame as the sample cell rests on the uniform and efficient transfer of the samples to the flame. Flame sample cells provide best reproducibility when supplied with liquid samples. In the early literature of analytical flame emission spectroscopy several techniques were described for the direct introduction of solid or liquid samples into premixed flames; in these techniques the sample was placed on a platinum wire,[1] loop or gauze disc[2] or on a paper "spill"[3] which was introduced into the flame above the primary reaction zone for vaporization to proceed directly. Although for some elements high sensitivity was obtainable with these techniques, the attainable precision was frequently poor and serious solute vaporization interferences and memory effects were encountered. The introduction of liquid samples to flames as a fine aerosol mist was first described by Gouy.[4] This technique, known as *nebulization*, has become the most widely employed in modern instrumental systems for quantitative analytical work in AAS and AFS. The word *nebulizer* has now virtually replaced the older inaccurate term *atomizer* to describe the device employed to produce the sample aerosol; the word atomizer is now used to describe the system, in this case a flame, in which free atoms are generated from the analytical sample. The most commonly encountered form of nebulizer is the *pneumatic nebulizer.* This may be used in an *indirect* or *direct* mode of operation in conjunction with the burner assembly employed. This chapter describes the construction and operational characteristics of each of these types of nebulizer and also some less common types of non-pneumatic nebulizer devices.

8.2 PNEUMATIC NEBULIZERS

In pneumatic nebulizer devices the liquid sample is introduced into a high velocity jet of the oxidant (or inert) gas employed in the combustible mixture used to form the flame. For *indirect* nebulization the aerosol produced is then mixed with the fuel-gas (and frequently additional oxidant or inert gas) and passed via a mixing/expansion chamber to the burner base. In direct nebulizers the aerosol/oxidant mixture is brought into contact with the fuel-gas as soon as the aerosol is formed. The mixture is then ignited to form a turbulent flame at a combined nebulizer-burner unit with no mixing or expansion chamber.

8.2.1 Indirect Nebulizers and Expansion Chambers

The most commonly used type of pneumatic nebulizer employed with a mixing chamber to form an indirect nebulizer assembly is shown in Fig. 8-1. In the simple *right-angle nebulizer* (see Fig. 8-2(a)) a high velocity gas stream is passed through a duct and creates a suction pressure drop along a second duct at 90° to it. This is connected to a capillary tube placed in the sample solution and liquid is drawn into it.

Although this type of nebulizer is simple to construct, it is relatively inefficient and difficult to align precisely. The most frequently used type is the *concentric nebulizer*, shown in Fig. 8-2(b), in which the jet for the sample liquid is surrounded by the concentric outlet for the nebulizing gas. Glass, quartz, plastic, stainless steel, platinum and silver nebulizer units have all been recommended for analytical work, but the most common nebulizers

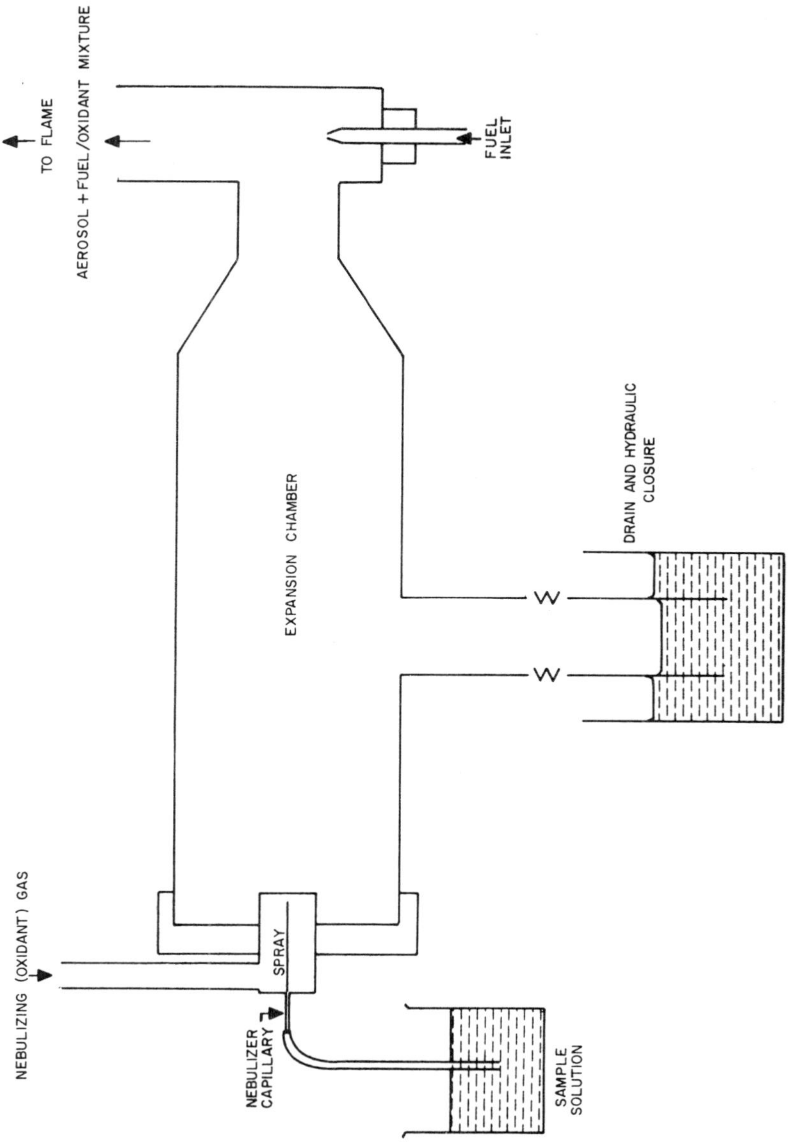

Fig. 8.1. Typical indirect nebulizer and expansion chamber assembly.

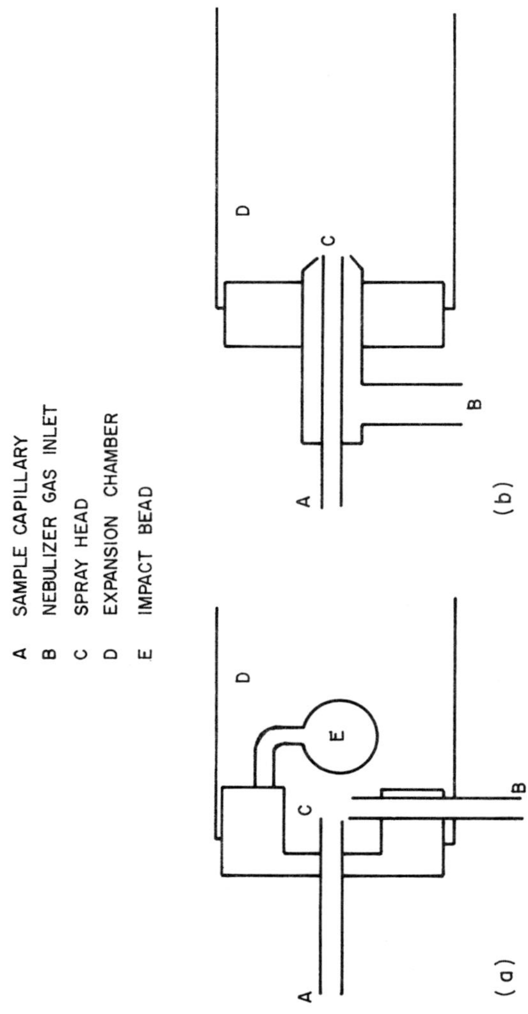

Fig. 8-2. Indirect nebulizers, (a) right-angle type, (b) concentric type.

used on modern flame spectrophotometers utilize platinum-iridium alloy (90 : 10) capillaries mounted in stainless steel gas supply inlets.

(a) Primary nebulization

The high velocity gas stream passing the capillary carrying the liquid to be nebulized creates a pressure drop at its outlet. The liquid is thus drawn out of the capillary and "shattered" into many droplets by the gas stream. The energy required for the production of droplets in this way is proportional to the surface tension of the sample liquid and the flow rate and inversely proportional to the diameter of the droplets produced. The required energy is very small, however, compared with the total kinetic energy available in the gas stream in concentric nebulizers. The solution flow rate obtained may be predicted from the Poiseuille equation to be

$$Q_d = \frac{\pi P r^4}{8 L \eta} \tag{8-1}$$

Where P is the pressure drop along the capillary, r and L are the radius and length of the capillary respectively and η is the viscosity of the liquid sample.

For a given nebulizer assembly the pressure drop created by the high velocity gas stream depends on the dimensions of the annulus and on the operating pressure of the gas stream. The solution flow rate therefore increases with the operating pressure of the nebulizing gas. Above a certain value for the operating pressure, however, it is usually found that no further substantial increase in the aerosol concentration produced in the gas is obtained. This effect is caused by the formation of a greater fraction of large droplets which do not remain suspended as an aerosol when high gas flow rates are employed. Additionally the use of high nebulizing gas flow rates requires a greater fuel-gas flow rate to maintain the desired flame stoicheiometry; unnecessarily large flames may then be produced.

The pressure drop along the capillary, and therefore the solution flow rate, depends quite critically on the position of the sample capillary with respect to the outlet of the duct carrying the gas stream. This is preset for optimum overall performance in some commercial instrumental systems. A variety of adjustable nebulizer assemblies are available, however, in which it is possible to control the solution flow rate independent of the operating pressure by variation of the position of the sample capillary in the gas stream. The use of this type of nebulizer permits the optimization of the solution flow rate (and drop-size distribution) for best signal : noise ratio at high or low concentrations or greatest freedom from solute vaporization interferences in AAS or AFS.

Although the mechanism by which the column of sample liquid emerging from the capillary is broken down into droplets has been investigated by several workers,[5-8] the processes which occur are complex and no universally accepted theory has been proposed. One of the simplest mechanisms postulated is that as the liquid column emerges from the capillary it is set into oscillation by the gas stream and filaments of liquid are drawn out from the surface. This process

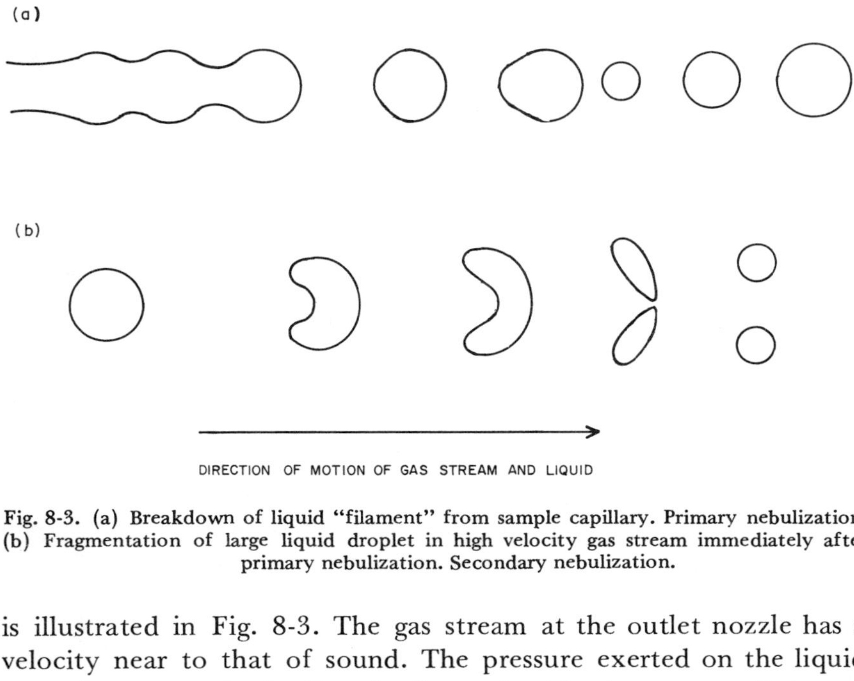

Fig. 8-3. (a) Breakdown of liquid "filament" from sample capillary. Primary nebulization. (b) Fragmentation of large liquid droplet in high velocity gas stream immediately after primary nebulization. Secondary nebulization.

is illustrated in Fig. 8-3. The gas stream at the outlet nozzle has a velocity near to that of sound. The pressure exerted on the liquid may thus come both directly from the kinetic energy of the gas stream and via shock waves generated at the edges of the nozzle. The gas stream breaks the liquid column into droplets which are then subjected to rapid acceleration. The pressures acting radially and tangentially on these droplets to cause acceleration in the gas stream oppose the internal pressure due to the surface energy which acts to hold the droplet together. The dynamic pressure approximates to $D(V_g - V_d)^2$, where D is the density of the nebulizer gas and V_g and V_d are the velocities of the gas and droplet respectively. Thus the pressure is greatest when the velocity of the gas is large compared to that of the droplet, i.e. before the droplet has experienced appreciable acceleration. At this point, immediately after the liquid column has been fragmented, this dynamic pressure may considerably exceed the internal droplet pressure, $4S/d$, where S is the surface energy and d is the droplet diameter, particularly for any relative large droplets formed initially. The droplets may then fragment further until they become stable in the gas stream at a sufficiently small diameter (i.e. high internal pressure) and high velocity (i.e. small relative velocity, $V_g - V_d$). This droplet fragmentation process is illustrated in Fig. 8-3(b).

Within a short time (distance) after leaving the nebulizer gas outlet the droplets achieve a velocity similar to that of the gas stream and may form a stable aerosol. The aerosol contains droplets of a range of different diameters below the approximate maximum permitted by the above considerations of external dynamic pressure *vs* internal pressure due to surface energy. The *drop-size distribution* is controlled by the operating pressure and alignment at the nebulizer, and this in turn has an important effect on the efficiency of the nebulization process and the vaporization in the flame.

In an attempt to establish the manner in which the mean droplet diameter, d_0, produced at a pneumatic nebulizer is controlled by the experimental conditions and physical properties of the solution nebulized, Nukiyama and Tanasawa[5] developed the expression

$$d_0 = \frac{585}{V_g - V_d}\left(\frac{S}{\rho}\right)^{0\cdot 5} + 597\left(\frac{\eta}{S\rho^{0\cdot 5}}\right)^{0\cdot 45}\left(\frac{10^3 . Q_d}{Q_g}\right)^{1\cdot 5} \quad (8\text{-}2)$$

where ρ is the density of the liquid (g cm^{-3}), S is the surface tension (dyne cm^{-1}), V_g and V_d are the velocities of the gas and liquid (ms^{-1}) respectively, η is the liquid viscosity (g cm^{-1} s^{-1}) and Q_d and Q_g are the liquid and gas flow rates (cm^3 min^{-1}) respectively. This empirical formula applies only under specific conditions to a particular concentric nebulizer used at subsonic velocities. It does, however, allow some general observations to be made concerning the relative importance of the operating conditions on the magnitude of d_0.

Thus, as the solution flow rate is usually much smaller than the gas flow rate, the second term of equation (8-2) has frequently little influence on d_0. Under these conditions the viscosity of the sample liquid has little effect, and d_0 is inversely proportional to the gas flow velocity and is influenced by the surface tension and density of the liquid sample. The mean droplet size increases, however, at high solution flow rates, Q_d, when the second term of equation (8-2) becomes significant. As the solution flow rate is inversely proportional to the viscosity (equation 8-1), the contribution of the second term in equation (8-2) to d_0 also varies inversely with viscosity. An example of this effect may be seen at high solution flow rates in the use of protective agents of high viscosity (e.g. glycerol) to provide smaller droplet size for minimization of solute vaporization influences (see Chapter 12.2.2).

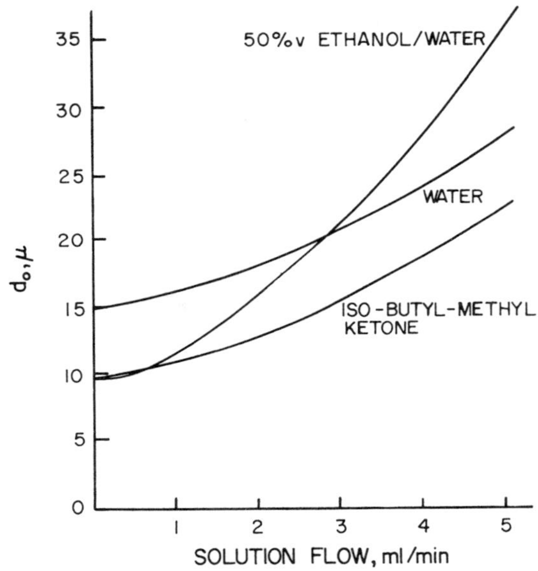

Fig. 8-4. Theoretical curves from equation (8-2) for variation of mean droplet diameter, d_0, with solution flow rate at pneumatic nebulizer (from ref. 9).

By inserting into equation (8-2) approximate values typical of a pneumatic nebulizer (V_g = 333 m/s; $V_d \sim$ 1 m/s, Q_g = 8·5 litres/min) and the known physical constants of a variety of solutions used in analytical flame spectroscopy, Stupar and Dawson[9] have recorded the data shown in Fig. 8-4 for variation in d_0 with solution flow rate. Although no allowance is made for the effect of other processes, such as solvent evaporation, on d_0, the curves obtained illustrate the effect obtainable with organic solvents whereby a significant reduction in droplet diameter may be realized. It is also clear from these curves that this effect may be quite critically dependent on the solution flow rate itself. Thus at low flow rates the use of an ethanol-water mixture leads to a decrease in d_0 compared to that for water, whereas at high flow rates the reverse is true. The values of d_0 shown in Fig. 8-4 also appear to be of the same order of magnitude as those obtained in practice with many pneumatic nebulizers. Figure 8-5 shows the drop-size distribution reported by Stupar and Dawson[9] for a pneumatic nebulizer. From this distribution it is clear that the largest number of droplets have a diameter of about 5 μ, whereas most of the sample volume is contained in droplets of about 20 μ diameter, the mean (volume/surface) diameter d_0, is found to

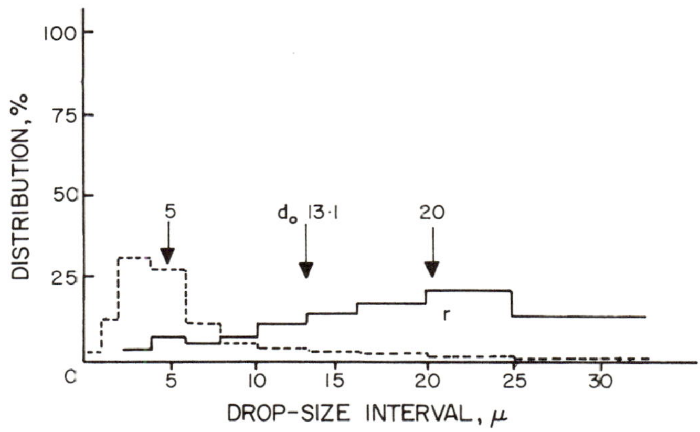

Fig. 8-5. Drop-size distribution from a pneumatic nebulizer (from ref. 9). —— Volume distribution; --- numerical distribution.

be 13 μ. A value of 9 μ for d_0 has been calculated by Willis[10] for a pneumatic nebulizer using a liquid flow rate of 1 ml/min.

The experimental determination of d_0 for pneumatic nebulizers has been shown to depend on the nebulizer construction and the extent of solvent evaporation in the spray chamber. Koirtyohann and Pickett[11] have reported a value of 8·6 μ for d_0 using the nebulizer from a commercial atomic absorption spectrophotometer. Under rather different conditions, using a pneumatic nebulizer fitted with an impact bead (see below) and a different method of particle collection for the measurement of d_0, Willis[10] has obtained a value of 2 μ; with both fixed and variable liquid flow rate nebulizers between 65 and 88% of the particles formed had a diameter of less than 2 μ depending on the flow rate used.

(b) Secondary nebulization

Even though a small value of d_0 may be achieved at an indirect nebulizer, a very high fraction of the total volume nebulized (rather than the number of droplets) is usually contained in very large droplets which are deposited on the spray-chamber walls. In order to avoid wastage of these large droplets, which do not reach the flame, several devices have been described to permit their further fragmentation. Thus if the aerosol initially generated is allowed to impinge on a barrier surface placed a few millimetres from the outlet jet, further droplet break-up may occur when large drops strike this surface.[10,12] The use of a spherical "impact bead" (see Fig. 8-2) in this way has been adopted on some commercial equipment. Although the

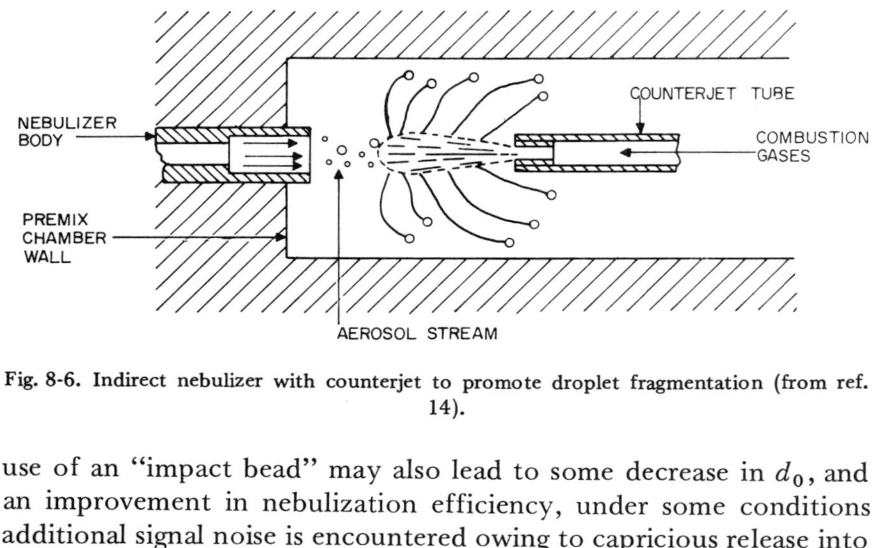

Fig. 8-6. Indirect nebulizer with counterjet to promote droplet fragmentation (from ref. 14).

use of an "impact bead" may also lead to some decrease in d_0, and an improvement in nebulization efficiency, under some conditions additional signal noise is encountered owing to capricious release into the gas stream from the impact bead of large liquid droplets which may be formed on it.

An alternative device to provide for droplet break-up has been proposed by Feldman[13] and adopted commercially.[14] This is shown schematically in Fig. 8-6. An additional gas jet is positioned within the expansion chamber on the same axis as the primary nebulizer jet and is supplied with premixed fuel and oxidant. This is directed in a laminar stream towards the aerosol produced at the primary jet and acts as a barrier surface to rapidly decelerate and fragment the larger droplets in the stream (in the same manner as for the primary nebulization the tendency to droplet break-up depends on the external dynamic pressure). A two to three fold gain in overall nebulizer efficiency (and therefore sensitivity in AAS) has been observed with this type of device for a range of elements in AAS with both air-acetylene and nitrous oxide-acetylene flames.

(*c*) *Expansion chamber*

Glass or plastic expansion chambers are most commonly employed for transfer of the aerosol to the burner base. These are generally fitted with a drain tube which has a U-tube or other means of maintaining a hydraulic closure, and in some instruments an additional side tube is fitted to the chamber to permit the supply of an auxiliary flow of oxidant gas. The fuel-gas may be introduced into the oxidant/analyte mist in the expansion chamber to ensure thorough mixing before the gas passes into the burner. For reasons of safety, however, particularly when a glass expansion chamber is employed, the fuel-gas may sometimes be introduced into the oxidant/analyte mist after it has passed to the base of the burner head. The dimensions of the chamber affect the fraction of the nebulized sample which reaches the flame, the analytical signal reproducibility and the time required after commencement of nebulization before maximum analytical signal is obtained. The use of a large volume chamber decreases the fraction of nebulized sample reaching the flame and lengthens the time required to establish a steady signal. It may, however, produce a damping effect on fluctuations in aerosol density which occur owing to poor nebulizer design or pressure variations in the nebulizer gas supply; this damping effect leads to improvement in flame stability and signal noise levels. The use of a small volume expansion chamber, on the other hand, permits rapid attainment of the maximum signal and speeds up the analysis as the "wash" time is also decreased, while a slight improvement in analytical sensitivity may also be obtained owing to its somewhat higher efficiency.

The drop-size distribution and mean droplet diameter, d_0, produced initially at the nebulizer jet (with or without the use of an impact bead or counterjet device) is significantly affected by the processes which occur in the expansion chamber before the aerosol reaches the flame. The two principal processes involved are the collection of the larger droplets on the walls of the expansion chamber and the loss of solvent from droplets by evaporation. When low sample flow rates are employed extensive evaporation of the solvent can occur before the droplets reach the flame. In the smallest droplets desolvation may be virtually complete, so that the flame receives a relatively "dry" aerosol. The extent to which solvent evaporation can proceed, however, is governed by the saturated vapour pressure of the solvent at the temperature of the expansion chamber and by the volume of the chamber. With high solution flow rates the gas mixture rapidly becomes saturated with solvent and the degree of desolvation which can occur is thereby reduced. Additionally, when the liquid flow rate is increased the drop size may increase beyond the point at which solvent evaporation can lower the value of d_0 sufficiently to produce a stable aerosol; there will then be an increasing loss of solution by deposition on the walls of the chamber.

From calculations of the rate of loss of surface from a stationary spherical droplet in still, dry air, using the Langmuir equation, it is possible to show that even droplets as large as 50 μ might evaporate completely during their residence time in the expansion chamber.[15] As pointed out by Herrman and Alkemade, however, several factors can modify this conclusion obtained using the ideal model. Thus the rate of evaporation may be significantly reduced when the droplets cool during evaporation, whereas the evaporation may proceed faster than calculated owing to the motion of the droplets with respect to the gas. Additionally, the evaporation rate would be decreased owing to the fact that the air does not remain dry after evaporation begins; the decreasing droplet diameter and increasing solute concentration may also affect the saturated vapour pressure of the solvent at the chamber temperature. It is probable that the maximum total loss of evaporation from the aerosol produced at the type of pneumatic indirect nebulizer used in AAS and AFS with a nebulizer gas flow rate of 7 – 10 l/min is *ca.* 0·05 – 0·1 g per minute.

A further process which may occur in the expansion chamber is the recombination of droplets to form larger droplets which deposit on the walls of the chamber or fall out of the stream to waste. The probability of droplet recombination in this way depends on N^2, where N is the number of droplets per unit volume. The effect may not be negligible at high liquid flow rates; it has been estimated that the value of d_0 is increased significantly by recombination only when the gas flow rate, Q_g, exceeds the liquid flow rate, Q_d, by less than a factor of 5000.[5,16]

(*d*) *Heated expansion chambers*

The extent to which solvent evaporation proceeds in the expansion chamber can be appreciably increased by preheating the gas used for nebulization[17,18] or by use of a heated chamber.[19,20,21] Although the former technique may be simpler to arrange and is capable of producing a significant improvement in solvent evaporation, difficulties may arise when the nebulizer assembly is heated by the gas. The adjustment of the sample capillary in the gas duct is temperature-dependent and difficult to maintain when the assembly is supplied with heated gas; the sample may also become warm as the sample capillary is heated and this can give rise to difficulties owing to bubble formation or sample degradation.

The use of a heated expansion chamber has proved more popular in AAS than preheating the nebulizer gas. A typical arrangement of this type is shown in Fig. 8-7. In the device illustrated, infrared radiation is employed to provide heat to the aerosol,[20] although other nebulizer/expansion chamber assemblies which utilize direct heating by an electric coil[21] or by high frequency energy have been described. A large gain in efficiency of transfer of the aerosol to the flame can be attained using these devices. In order to avoid the transfer of unwanted vaporized solvent to the flame a water-cooled condenser is used after the heating stage. A gain in sensitivity of approximately ten-fold has been obtained in AAS for a wide range of elements.[20,21] In a commercially available heated nebulizer/expansion chamber assembly the infrared radiation is supplied by a 400 W heater positioned around a Pyrex expansion chamber in a manner which minimizes conductive heating of the chamber wall. The heater is backed by a cylindrical reflector to provide intense heating within the chamber to evaporate the solvent before droplets can strike the walls. This arrangement minimizes the drift caused by the cooling effect of droplets striking the walls[20] which is experienced with nebulizers in which the energy for evaporation is provided by conductive heating of the chamber walls.

Table 8-1 shows some AAS sensitivities and detection limits obtained by Venghiattis[21] using a heated chamber burner and some comparable values attainable with a conventional indirect nebulizer. In addition to the improved (1% absorption) sensitivity obtained owing to greater solute transfer efficiency, a similar gain in detection limits is observed. This effect of the heated chamber has been attributed to the production of smaller solute particles and the delivery of a more uniform aerosol to the flame by break-up of the larger particles. As the volume of the heated expansion chamber was greater than that of the chamber used in the conventional assembly, however, it is also possible that the improved signal : noise ratios

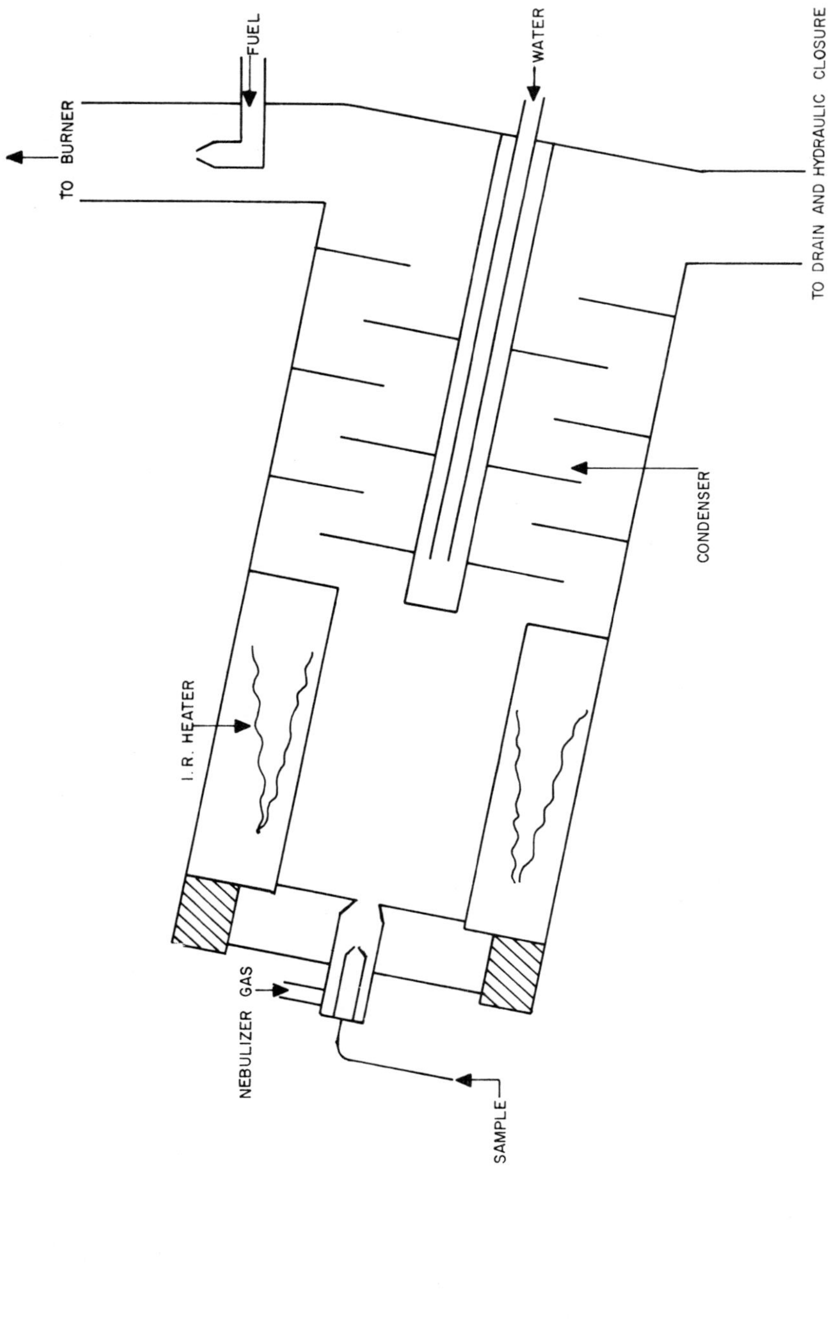

Fig. 8-7. Indirect nebulizer with heated expansion chamber and water-cooled condenser.

observed were partly the result of damping of the fluctuations in aerosol density reaching the flame; a greater proportion of the partly or completely desolvated larger droplets, which would otherwise arrive at the flame irregularly to cause "noise", would be expected to "fall-out" from the gas stream in the larger expansion chamber.

The use of indirect nebulizers with heated expansion chambers is primarily of benefit when additional sensitivity is sought. One of the most serious disadvantages, however, is that for samples of high dissolved solid content significant "memory" effects are frequently experienced because of deposition of sample on the chamber walls. Owing to the tendency for these deposits to become "baked" onto the walls excessively long wash-times may be required between sample measurements to eliminate this effect.

TABLE 8-1

Sensitivities and detection limits in AAS using a heated expansion chamber and a conventional cold expansion chamber (from ref. 21)

Element	Sensitivity, ppm/1% abs		Detection limits, ppm		Flame
	Heated	Cold	Heated	Cold	
Ag	0·008	0·08	0·0005	0·005	Air-C_2H_2
Al	0·3	1·1	0·015	0·1	N_2O-C_2H_2
Ca	0·01	0·08	0·001	0·002	Air-C_2H_2
Cu	0·008	0·1	0·0005	0·005	Air-C_2H_2
Fe	0·02	0·15	0·002	0·005	Air-C_2H_2
Mg	0·001	0·008	—	—	Air-C_2H_2
Mn	0·01	0·08	0·005	0·003	Air-C_2H_2
Pb (217·0 nm)	0·02	0·25	—	—	Air-C_2H_2
Pb (283·3 nm)	—	—	0·004	0·01	Air-C_2H_2
Si	0·4	1·2	0·02	0·1	N_2O-C_2H_2
Ti	—	—	0·04	0·2	N_2O-C_2H_2
Zn	0·0015	0·04	0·0005	0·002	Air-C_2H_2

(e) *Nebulizer/expansion chamber efficiency*

The *efficiency* of indirect nebulizer and expansion chamber assemblies, defined as the fraction of the nebulized analyte which passes into the flame, is usually low. With aqueous solutions an efficiency in the range 1 to 10% is commonly obtained. When organic solvents are employed, however, the efficiency may be as high as 20%. This *organic solvent effect* results from an increase in the yield of sample reaching the flame owing to the influence of the surface tension, viscosity and vapour pressure of the organic solvent in decreasing the mean drop size and increasing the rate of evaporation of solvent from the droplets. An estimate of the efficiency of a nebulizer/expansion chamber assembly may be obtained by measurement of the rate of collection of unused nebulized sample in the drain tube from the nebulizer and of the sample uptake rate at the nebulizer capillary. The difference between these values represents roughly the rate of delivery of aerosol mist to the burner and this may be expressed as a percentage of the sample uptake rate to obtain the percentage efficiency. This procedure does not take into account evaporation of solvent from the droplets or loss of solute by deposition of solid particles in the burner connections or the burner itself. The value obtained, which is commonly termed the *nebulizer efficiency*, may, therefore, not accurately express the rate of introduction of the analyte element into the flame.

8.2.2 Direct Nebulization into Unpremixed Flames

Turbulent unpremixed flames used in analytical flame spectroscopy are usually supported at integral nebulizer-burner units so that the sample aerosol is produced pneumatically and introduced directly into the fuel-gas and oxidant at the point where they mix and are ignited. The reader is referred to Chapter 6 for discussion of the structure and properties of the flames formed at this type of burner; we are concerned here only with the nebulization process and the manner in which it is affected by the experimental operating conditions.

(*a*) *Aspiration*

As with indirect nebulizers, the liquid flow rate (sample uptake rate) into total consumption nebulizer-burner units is controlled by the pressure drop along the capillary (see equation (8-1)). It thus depends on the operating gas pressure at the concentric outlet jet. As the nebulizing gas is ignited immediately after leaving the outlet jet, it is even more difficult with these devices than with indirect nebulizers and premixed flames to control the sample uptake rate and flame gas flow rates independently. The other variables in the Poiseuille equation (equation (8-1)) also apply to these units, i.e. the sample uptake rate will depend on the capillary diameter and length and the viscosity of the solution.

The sample uptake rate may have a significant effect on the

temperature attained by unpremixed flames at nebulizer-burner units. This results from the fact that a much greater fraction of the solution reaches the flame. Hence the heat required for desolvation may constitute an appreciable fraction of the total heat available from the combustion, particularly with aqueous solutions. Thus, for example, a temperature decrease of 600°C has been reported for an oxygen/hydrogen flame at this type of burner with a sample uptake rate of 6 ml/min.[22,23] In addition to the temperature effect high sample flow rates result in a large fraction of the nebulized sample volume being formed as large droplets. For both of these reasons it may usually be preferable to operate the integral nebulizer-burner units at relatively low sample solution flow rates. Several methods are available to accomplish this while maintaining an adequate gas flow to produce a flame of the required dimensions. For example, the flow rate may be reduced by the use of a long, narrow, capillary to the sample solution; in practice this may not be satisfactory with solutions of high dissolved solid content which tend to block the capillary. Alternatively the sample flow rate may be reduced by controlling its delivery to the capillary via a motor-driven syringe or peristaltic pump.

(*b*) *Drop-size distribution*

The drop size distribution at direct nebulizer-burner units is controlled by the same factors as those discussed in Section 8.2.1 for primary nebulization with indirect pneumatic nebulizers. In addition, there may also be a tendency for the turbulence created near to the capillary outlet when the fuel and oxidant gas are mixed to give rise to some degree of secondary nebulization, i.e. droplet break-up as the drops are subjected to rapid changes in direction and velocity. This may, in fact, occur to some extent for the largest droplets which are not rapidly desolvated at the flame front.

As with indirect nebulizers a large fraction of the sample volume is contained in the larger droplets. While these might fall out of the stream before reaching the flame with indirect nebulizer systems, in integral nebulizer-burner units they are injected into the base of the turbulent flame. A proportion of these larger droplets are deflected out, or fall from, the flame before they are desolvated and are therefore wasted. Even if the droplets remain in the flame, desolvation and vaporization may not be completed until reaching a considerable height. This is partly a result of the large droplet size and lack of evaporative loss of solvent before entering the flame, but a further factor is the high rise velocity in the hot turbulent flames such as oxygen/hydrogen and oxygen/acetylene supported at integral nebulizer-burner units. The severity of solute vaporization interferences encountered is, however, still proportional to the mean droplet diameter as with indirect nebulizers. The use of low sample flow rates may thus lead to a reduction in these interferences by producing a decrease in d_0 and simultaneously give rise to less severe solvent cooling of the flame. Nebulizer-burner units are frequently supplied with samples in organic solvents of low surface tension (such as ethanol or acetone) in order to achieve similar effects without necessitating a greatly reduced input of sample solution into the flame.

8.3 NON-PNEUMATIC NEBULIZATION

Simple indirect pneumatic nebulizers of the type described in Section 8.2 have found widespread application in analytical flame spectroscopy as a result of several attractive features; these include simplicity, ease of operation, reproducibility, small sample volume requirements and little tendency to memory effects. Several fundamental disadvantages do, however, also exist with pneumatic nebulization. Thus the aerosol produced contains droplets of widely varying diameter (1 to 50 μ) and the sample transfer efficiency is usually low. The sensitivity obtained in AAS or AFS in the flame is limited by the low aerosol density attainable; this may amount to a maximum of only *ca.* 0·03 g of solution per litre of gas owing to the large gas flow rate required for nebulization. The nebulizer performance varies with the viscosity and surface tension of the solution, temperature of the nebulizing gas and solution and the vapour pressure of the solvent. Sample solutions of high salt content may cause problems by blocking the nebulizer jet outlet.

In view of these problems, there has been much interest in the development of alternative methods of nebulization which overcome at least some of the disadvantages of the pneumatic devices. The following sections describe briefly some of the non-pneumatic techniques and devices which have been used.

8.3.1 Ultrasonic Nebulization

Aerosol formation by ultrasonic techniques has been shown to be useful in analytical atomic spectroscopy as an alternative to pneumatic nebulization.[9,24-29] Figure 8-8 (from ref. 9) illustrates two nebulizer assemblies which have been used to provide aerosols for atomic absorption spectroscopy. Ceramic crystal transducers are employed to break up the liquid; the power is transferred from the vibrating crystal to the solution via a velocity transformer (Fig. 8-8(i)) or by use of a coupling liquid with a focusing device (Fig.

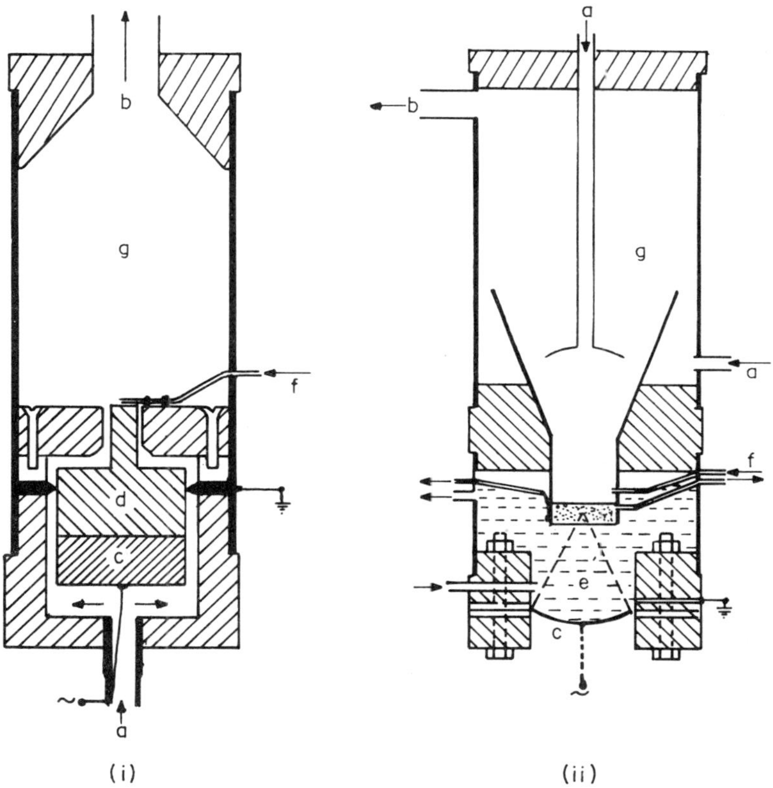

Fig. 8-8. Ultrasonic indirect nebulizers (from ref. 9): (i) velocity transformer, (ii) focusing bowl. a, air; b, aerosol; c, ceramic crystal transducer; d, velocity transformer; e, coupling medium (water); f, sample solution, g, spray chamber.

8-8(ii)). Various vibration frequencies in the range 20 kHz to 5·4 MHz have been employed by different workers.

It has been postulated that standing waves are formed on the surface of the sample liquid and lead to the rupture of sections of the surface bounded by nodal lines to form droplets.[30,31,32] The mean droplet diameter in the aerosol is predicted from these considerations to be inversely proportional to the square of the frequency of the ultrasonic energy employed and also to depend on the surface tension and viscosity of the liquid sample. Cavitation, viscosity, vapour pressure and the ultrasonic power employed are probably more important in determining the rate of ultrasonic nebulization than is the mean droplet diameter.[9] The approximate drop-size distributions produced from ultrasonic devices operating at different frequencies are shown in Fig. 8-9. It is evident that ultrasonic nebulizers are capable of producing more homogeneous aerosols than those produced by conventional pneumatic nebulizers (cf. Fig. 8-5).

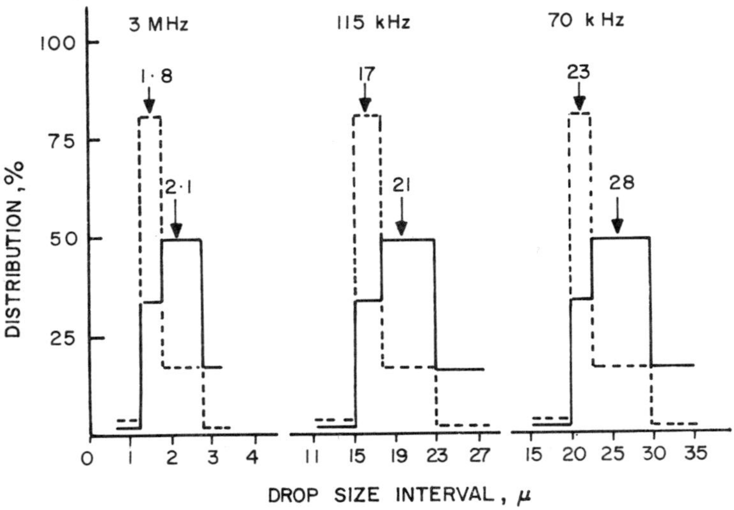

Fig. 8-9. Drop-size distribution from ultrasonic nebulizers (from ref. 9). ------ Numerical distribution; —— volume distribution.

Also, as d_0 depends on the operating frequency, it is necessary to employ frequencies greater than *ca.* 500 kHz if the value of d_0 is to be smaller than, or comparable to, that obtained with pneumatic devices. Stupar and Dawson,[9] have reported that unless high frequencies and high operating power are used to generate fine aerosols (~5 μ diameter) at sample flow rates of 1-5 ml/min, the sensitivity and solute vaporization interference obtained when these devices are used in AAS may be worse than that attainable with a pneumatic nebulizer. These workers demonstrated that the interference of aluminium on magnesium was related to the estimated mean diameter of the dried particles in the flame. Figure 8-10 summarizes their results; the more serious interferences encountered with the ultrasonic nebulizer may be attributed to the relatively low operating frequencies available.

The potential advantage of ultrasonic nebulization for AAS and AFS lies in the possibility of more precise control of droplet size (to

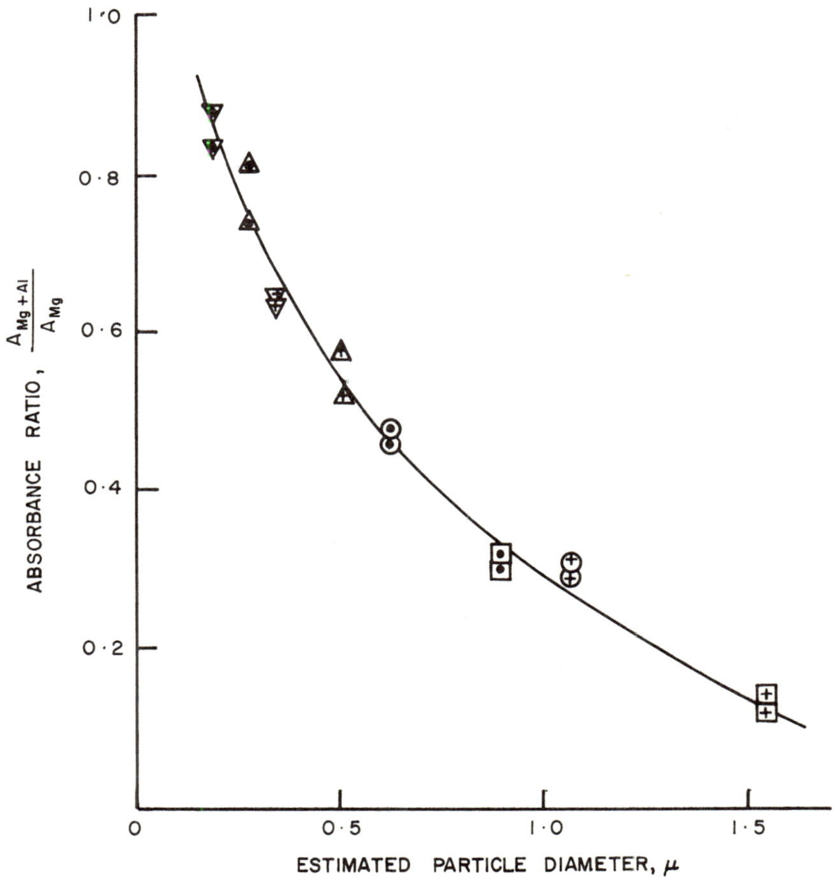

Fig. 8-10. Experimental curve of variation of the absorbance ratio for magnesium (mixed solution/pure solution) with estimated mean diameter of dry particle (from ref. 9). •, 20 ppm Al; +, 100 ppm Al; △, pneumatic (~0·10 ml/min); ▽, pneumatic (~3·5 ml/min); ○, ultrasonic (115 kHz); □, ultrasonic (70 kHz).

increase sample transfer efficiency and minimize solute vaporization interferences) and aerosol density (by independent control of flow rate of gas used to sweep aerosol into the flame). A practical disadvantage is their tendency towards "memory" effects which leads to difficulties with time taken for washing out of sample solutions after measurement. In spite of this and the greater cost and complexity of ultrasonic devices compared to pneumatic nebulizers, the fundamental advantages which they offer will undoubtedly lead to the development of apparatus which is more suitable for routine analytical practice.

8.3.2 Other Techniques

A considerable amount of work has been undertaken with non-pneumatic nebulizers other than those utilizing ultrasonic energy. Centrifugal nebulizers, in which the sample solution is brought into contact with a rapidly spinning disc, have been investigated.[33,34] Walton and Prewett[35] reported the formation of a uniform spray of droplets from 3 to 15 μ in diameter with this type of nebulizer and several other workers have also used nebulizers of this type. Several forms of electric and electrostatic nebulizers have been described. In these devices the sample is nebulized by the direct action of a spark at the solution surface,[36,37] or by the production of a very high electrostatic field at the liquid surface.[38,39] In the latter case, the field is applied between a metal capillary carrying the sample and a counter electrode, to give rise to forces which effect the nebulization. Although this particular technique may appear attractive, the rate of aerosol formation depends quite critically on the dipole moment of the liquid molecules so that in practical analytical samples this effect may give rise to a type of "physical" interference.

Droplet formation produced by "effervescence" resulting from the passage of small gas bubbles through a solution, or from liberation of gas bubbles in the solution, was one of the earliest techniques employed for the introduction of liquid samples into flames. In the latter case the gas bubbles may be generated chemically[40] or electrolytically.[41] The bubbles entrain the liquid as they emerge through the surface and may then be passed to the flame in a current of air. The reproducibility of the rate of aerosol generation is poor with these techniques, however, and they are not applied in modern analytical flame spectroscopy.

8.4 DISCRETE SAMPLING DEVICES

In Section 8.1 it was mentioned that early in the development of analytical flame emission spectroscopy various techniques were described for the direct introduction of solid or liquid samples into flames without nebulization.[1-3] In order to examine small liquid or solid samples directly by flame AAS or AFS without dilution several techniques for the direct introduction of such samples into flames have recently been described. In each of these techniques a known weight or volume of the sample is placed into a container, or onto a probe, and this is then inserted directly into the flame gases above the primary reaction zone. The sample is then vaporized rapidly and produces a short-lived atomic population for AAS or AFS. A transient analytical signal is obtained; after sample insertion the signal increases as the rate of vaporization increases and then decreases as the sample available for vaporization becomes exhausted. The peak atomic concentration achieved in the flame by this technique may be greater than that produced by conventional nebulization of an aqueous solution into the flame; the effective sensitivity (for 1% absorption), therefore, may be increased. Also, as the devices employed usually accept very small liquid or solid samples (often less than 100 μl or 100 μg) the absolute detection limit which is attainable is usually considerably lower than that produced with solution samples and conventional pneumatic nebulization. The transient analytical signal may be recorded from either the peak height or the peak area.

8.4.1 The Boat-in-Flame Technique

Kahn, Peterson and Schallis[42] have described a technique in which a small solid or liquid sample is placed into a tantalum boat (50 mm in length) which is then inserted into an air-acetylene flame supported at a Boling triple-slot burner. The device used to place the boat in the flame may also be used to position it close to the flame initially to desolvate liquid samples or to dry ash solid samples before vaporization in the flame itself. The technique functions effectively for elements such as Ag, As, Cd, Hg, In, Pb, Se and Tl which are relatively volatile and easily vaporized at the temperature achieved in the boat (somewhat less than the flame temperature). The detection limits reported for these elements are listed in Table 8-2. Figure 8-11 shows typical recordings obtained in the AAS determination of silver by this technique. It can be seen that the absorption signal obtained represents the absolute weight of analyte element in the sample rather than its concentration. The entire sample atomization process is usually complete within only a few seconds after the insertion of the boat into the flame; the amplification and recorder system employed, therefore, must have a fast response time ($\leqslant$1 sec).

TABLE 8-2
Detection limits obtained for AAS with boat-in-flame technique[43]

Element	Detection limit, ppm
As	0·02
Bi	0·003
Ag	0·0001
Cd	0·0001
Hg	0·02
Pb	0·001
Se	0·01
Te	0·01
Tl	0·001

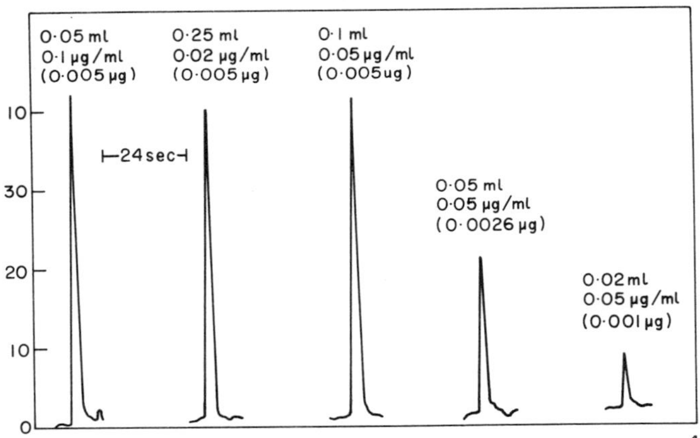

Fig. 8-11. Absorption signals for silver with boat-in-flame technique.[43]

Interference effects may be more pronounced with this device than in a flame with ordinary nebulization; the method of standard additions, or close matching or sample and standard composition, is therefore recommended. Kahn and Sebestyen[44] have used this device for the determination of lead in blood and urine.

8.4.2 Tube-in-Flame Techniques

Robinson[45] in early AAS studies demonstrated the increase in absorption sensitivity possible by passing the flame gases containing the analyte atoms into a T-piece shaped adaptor; the residence time of the analyte atoms in the light beam is then increased as they are made to travel along the tube rather than pass rapidly out of the beam as in a conventional flame. This technique may also be used in conjunction with discrete sampling devices of the sampling boat type described above. A longer residence time of the transient population of atoms produced on sample vaporization is then obtained in the optical path.

White[46] has described the use of a technique in which small liquid samples may be placed onto a platinum wire loop which is then placed in the flame gases. As the sample vaporizes it passes, together with some flame gases, through a hole cut into a nickel tube half way

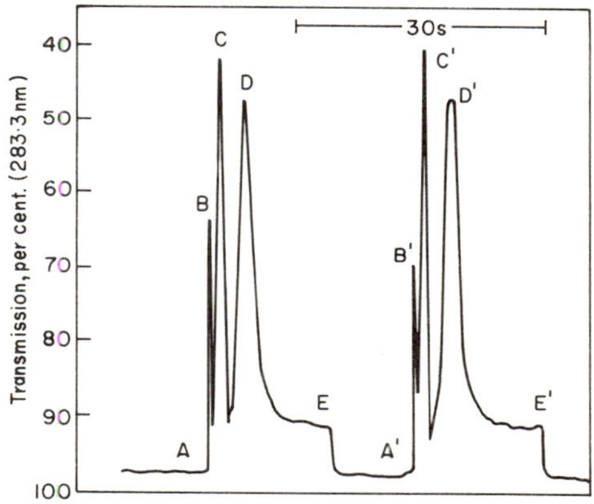

Fig. 8-12. Absorption signals at 283·3 nm obtained with duplicate 10-μl samples of blood containing 100 μg of lead per 100 ml (from ref. 47). (This was the physiological concentration and not the result of an added lead standard.) Oxidation procedure and experimental conditions as stated. A and A′, crucible introduced into the flame below the absorption tube; B and B′, C and C′ absorption signals from the combustion-products of the samples; D and D′, lead atomic absorption signals; and E and E′, crucible removed from the flame.

along its length. The tube is aligned in the flame so that the light beam from the source passes along its central axis. An attainable precision of *ca.* 5% was claimed for this device for elements which are relatively volatile; detection limits of 1×10^{-10} g for Cd and 1×10^{-9} g for Pb were obtained. An advantage of this technique compared to the boat-in-flame technique[42] is that the nickel tube yields an absorption cell of well-defined volume so that variations in sensitivity with boat position in the light beam are minimized.

Delves[47] has described a device which combines the best features of both the boat-in-flame technique and the absorption tube technique of White. In this technique, now referred to as the "Delves Cup Technique", the tantalum boat of Kahn *et al.*[42] is replaced by a nickel micro crucible (10 mm o.d., 5 mm deep, 0·006 in. metal foil) to minimize intra-boat variations in sensitivity and a nickel absorption tube is used above the boat to minimize variation of absorption with observation height in the air-acetylene flame and provide increased residence time. Delves applied this technique to the direct determination of lead in 10 μl samples of whole blood. The blood samples are transferred to the crucible and dried at 140°. A small volume of 100 volume hydrogen peroxide is then added and the crucible is reheated at 140° for 1-2 minutes to effect partial oxidation of the blood before introducing the crucible into the flame. Figure 8.12 shows traces of the absorption signals obtained for lead at 283·3 nm by this technique. By this method 1×10^{-10} g of Pb gave 1% absorption and the standard deviation was 4% at the 3 ng level. Thirty-nine blood samples were analysed in the preliminary assessment of the technique; the results were found to give good correlation with those obtained by an automated spectrophotometric procedure for the determination of lead using dithizone. Several other workers have also reported the application of this technique to the determination of Pb in blood.[48-51] The advantages offered by the technique should undoubtedly result in its widespread usage.

REFERENCES

1. R. Bunsen, *Ann. Chem. Pharm.*, 1859, **111**, 257.
2. J. M. Eder and E. Valenta, *Klasse Kais. Akad. Wiss. Wien.*, 1893, 467.
3. H. Ramage, J. H. Sheldon and W. Sheldon, *Proc. Roy. Soc.*, 1933, **B113**, 308.
4. G. L. Gouy, *Ann. Chim. Phys.*, 1879, **18**, 5.
5. S. Nukiyama and Y. Tanasawa, *Trans. Soc. Mech. Eng., Japan*, 1939, **5**, 62, 68.
6. R. A. Castelman, *J. Res. Nat. Bur. Standards*, 1931, **6**, 369.
7. R. W. Lane, *Ind. Eng. Chem.*, 1951, **43**, 1312.
8. D. M. Bitron, *Ind. Eng. Chem.*, 1955, **47**, 23.
9. J. Stupar and J. B. Dawson, *Applied Optics*, 1968, **7**, 1351.
10. J. B. Willis, *Spectrochim. Acta*, 1967, **23A**, 811.
11. S. R. Koirtyohann and E. E. Pickett, *Anal. Chem.*, 1966, **38**, 1087.
12. R. Mavrodineanu and H. Boiteux, "Flame Spectroscopy", Wiley, New York, 1965.
13. F. J. Feldman, *American Laboratory*, August 1969, p. 47.
14. Instrumentation Laboratories Inc., Lexington, Mass., U.S.A.
15. R. Herrman and C. T. J. Alkemade, in "Chemical Analysis by Flame Photometry", Interscience, New York, 1963, p. 24.
16. B. Lewis, R. N. Pease and H. S. Taylor, "Combustion Processes", Princeton Univ. Press, 1956.
17. A Strasheim, L. R. P. Butler and E. C. Maskew, *J. S. African Inst. Min. Net.*, 1962, **62**, 796.

18. R. A. G. Rawson, *Analyst,* 1966, **91,** 630.
19. J. E. Allan, 4th Australian Spec. Conf., Canberra, 1963.
20. A. Hell, W. F. Ulrich, N. Shifrin and J. Ramirez-Munoz, *Appl. Optics,* 1968, **7,** 1317.
21. A. A. Venghiattis, *Appl. Optics,* 1968, **7,** 1313.
22. J. A. Dean, Kresge-Hooker, *Record of Chem. Progress,* 1961, **22,** 179.
23. W. H. Foster Jnr. and D. N. Hume, *Anal. Chem.,* 1959, **31,** 2028.
24. H. Dunkin, G. Pforr, W. Mikkeleit and K. Geller, *Z. Chem.,* 1963, **3,** 196.
25. C. D. West and D. N. Hume, *Anal. Chem.,* 1964, **36,** 412.
26. R. H. Wendt and V A. Fassel, *Anal. Chem.,* 1965, **37,** 920; 1966, **38,** 338.
27. W. J. Kirsten and G. O. Bertilson, *Anal. Chem.,* 1966, **38,** 648.
28. H. C. Hoare, R. A. Mostyn and B. T. Newland, *Anal. Chim. Acta,* 1968, **40,** 181.
29. J. Spitz and G. Uny, *Appl. Optics,* 1968, **7,** 1345.
30. R. J. Lang, *J. Acoust. Soc. Amer.,* 1962, **34,** 6.
31. K. Bisa, K. Dirnagl and R. Esche, *Siemens Z.,* 1954, **8,** 341.
32. E. G. Lierke and G. Griesshammer, *Ultrasonics,* 1967, **5,** 224.
33. E. Beckman and P. Waentig, *Z. phys. Chem.,* 1910, **68,** 385.
34. K. R. May, *J. Appl. Phys.,* 1949, **20,** 932.
35. W. H. Walton and W. C. Prewett, *Proc. Phys. Soc.,* 1949, **B62,** 341.
36. G. A. Hemsalech, *Phil. Mag.,* 1917, **33,** 1.
37. W. Herschel, *Z. Physik,* 1928, **47,** 147.
38. H. Staubel, *Naturwiss.,* 1953, **40,** 337.
39. H. Staubel, *Mikrochim. Acta,* 1955, 329.
40. T. Török, *Z. Anal. Chem.,* 1939, **116,** 29; 1940, **119,** 120.
41. E. Beckman, *Z. Electrochem.,* 1899, **5,** 327.
42. H. L. Kahn, G. E. Peterson and J. E. Schallis, *Atomic Absorption Newsletter,* 1968, **7,** 35.
43. "Analytical Methods for Atomic Absorption Spectrophotometry", Perkin-Elmer Corp., Norwalk, Conn., U.S.A., 1971.
44. H. L. Kahn and J. S. Sebestyen, *Atomic Absorption Newsletter,* 1970, **9,** 33.
45. J. W. Robinson, *Anal. Chim. Acta,* 1962, **27,** 465.
46. R. A. White, Brit. Non-Ferrous Metals Res. Assoc., Research Report A 1707, Sept. 1968.
47. H. T. Delves, *Analyst,* 1970, **95,** 431.
48. F. J. Fernandez and H. L. Kahn, *Atomic Absorption Newsletter,* 1971, **10,** 1.
49. A. A. Cernik and M. H. P. Sayers, *Brit. J. Ind. Med.,* 1971, **28,** 392.
50. D. G. Mitchell, K. M. Aldous and F. J. Ryan, *Proc. Pittsburgh Conf. on Anal. Chem. and Appl. Spectrosc., Cleveland, Ohio, March 5-9, 1973,* paper no. 148.
51. *Idem,* ibid., paper no. 86.

CHAPTER 9

Wavelength Selection

9.1 INTRODUCTION

The majority of atomic absorption or fluorescence measurements require some form of wavelength selection in the radiation that provides the signal to be processed by the spectrometer. This arises because the radiation used for the measurements is frequently accompanied by unwanted light of other wavelengths. This unwanted radiation may be simply stray daylight, a complex spectrum emitted by the light source or atom cell, or may arise from the weaker absorption lines of the analyte element which would reduce the sensitivity of AAS measurements at the main wavelength for that element (see Chapter 11). Under certain conditions, and particularly with fluorescence measurements, it is possible to use electronic methods (i.e. phase-sensitive amplifiers or solar-blind photomultipliers, as discussed in Chapter 10) to avoid detection of the unwanted light and allow useful measurements to be made without wavelength selection. Sensitive absorption measurements made in this way are, however, only possible for those elements whose atoms exhibit simple line spectra. Even in fluorescence measurements it may be difficult to obtain high signal : noise ratios for many elements by these techniques. Thus, in most reported applications of AAS and AFS, an optical monochromator device is used to preselect the radiation that is allowed to reach the photomultiplier. In commercial instrumentation the most widely used device is the traditional optical prism or grating monochromator which may be used to select any single spectral line (or, at least, a closely spaced group of lines) within a chosen wavelength range. The much simpler optical filters still commonly found in flame photometers may also sometimes be used, particularly for a number of fluorescence applications. Finally, several less widely known devices such as "resonance detectors" and "selective resonance modulation" have been applied to AAS and AFS. Each of the above techniques is described briefly in this chapter and the emphasis is placed on topics that are directly relevant to analytical spectrometry rather than on details of principle or construction. The reader is referred to specialized textbooks[1,2,3] for a more comprehensive discussion of the subject.

9.1.1 Classification of Instruments

The routine application of atomic absorption and fluorescence techniques requires only a relatively elementary knowledge of the optical instruments utilized. The particular requirements of these techniques have, however, given rise to a terminology which differs in some respects from that used in many general courses or textbooks which in other respects form an excellent introduction to the subject. Thus, it is useful to consider AAS and AFS instrumentation in perspective with respect to spectroscopic apparatus in general.

The classical method of wavelength separation involves *dispersion* of the incident radiation so that an observer may view, individually, light of any particular wavelength within the range covered by the instrument. In its simplest form such an instrument is known as a *spectroscope* and allows visual observation of the spectrum produced by the dispersing device. A permanent record may be obtained by using a *spectrograph* in which the observer's eyepiece is replaced by a camera whereby the spectrum can be photographed to produce the well-known pattern of lines on a photographic plate or film. To avoid the need to use the photographic process if just a record of light intensity at a few of these lines is required, the camera may be replaced by a *quantometer*. This replaces the photographic emulsion with a mask which allows a fine slit to be aligned with any line of the spectrum. A photocell placed behind this slit is then used to observe variations in the intensity of that line. The final development of this principle produces the *monochromator* in which a single slit and photocell is maintained in a fixed position while the spectrum is *scanned* across it by moving the dispersive element to alter the angle at which light of a particular wavelength emerges.

The spectroscope is obviously of little use for measurement of atomic absorption or fluorescence signals, while the use of a spectrograph is both inconvenient and time-consuming; it would also be somewhat impractical for routine absorption measurements which would require individual recording of line intensity with and without a sample in the atom cell. A spectrograph used with a quantometer is quite suitable for regular measurements at just one wavelength but is inconvenient if frequent wavelength changes are required. It is possible to use a quantometer with a fixed detector at each absorption line required,[4] but this becomes expensive when many wavelengths are provided and may be impractical when several closely spaced lines are to be used. Thus the monochromator is at present the most satisfactory means of obtaining a versatile, direct-reading optical system for AAS or AFS. All direct-reading photoelectric instruments (i.e. monochromator or spectrograph + quantometer) are conveniently described as *spectrometers* or *spectrophotometers*.

Each of these classical spectroscopic instruments conforms to a more or less standard arrangement of components:

(1) An entrance aperture whose image is observed separately in the spectrum for each dispersed wavelength present in the incident radiation. The most widely used aperture is a narrow rectangle, i.e. a slit. Hence the general association of a "spectrum" with the pattern of fine lines which is observed or photographed.
(2) A collimating device to produce a parallel beam of radiation from that passing through the aperture.
(3) The dispersing element which separates this radiation by giving each different wavelength a different emergent angle.
(4) A system of lenses or mirrors to focus the different wavelengths as a sharp image of the entrance aperture.

(5) A device which allows these images to be viewed, photographed or photo-electrically recorded.

The dispersing element may take a number of forms, some of which are more common than others, and the particular type used is often employed as the basis of a classification of spectroscopic instruments given above. The original dispersing element used by Newton was a transparent *prism* which, because its refractive index varies with wavelength, transmits each particular wavelength at a different angle. The prism is still employed but has to a large extent been replaced by the *diffraction grating* in which light is deflected by a pattern of fine slits at different angles according to its wavelength. This results from the wavelength-dependence of the positions and intensities of the interference patterns obtained. The interference of light may also be used in numerous other devices to obtain a dispersive element; for historical reasons these are generally referred to as *interferometers.* One or two of these devices have been applied to AAS or AFS but no further mention will be made of them here. They are best considered from our point of view merely as rather specialized types of spectrometer.

There remains a further group of wavelength separating devices which do not fall into the above classification as they do not employ a dispersive element to obtain a spectrum. Thus, we describe below several applications to AAS or AFS of devices in which the required wavelength is isolated by *absorption* of all other incident radiation before the light beam reaches the photodetector. The most widely used of these devices is the simple optical filter which depends for its action on the (molecular) absorption bands of liquid, solid or sometimes gaseous materials.* Other, more modern, instruments actually use atomic absorption by the very elements whose radiation is to be isolated as a means of detecting individual wavelengths (see Section 9.3).

9.1.2 *Characteristics of Instruments*

Several important characteristics are common to all classical (i.e. dispersive) spectroscopic instruments and are widely used for comparisons of performance. These refer to the *effectiveness* of the instrument for dispersing different wavelengths and to the *efficiency*

* It is emphasized that this does not include the so-called "interference filters" (see Section 9.3.1) which are simple dispersive devices; they are, in fact, miniature interferometers.

with which it carries out this process, i.e. the instrument would be 100% efficient only if the intensity of each particular wavelength reaching the detector was identical to its original value as a component of the undispersed incident beam

The ability of a spectrometer to spread out different wavelengths is termed its *dispersion*. This may be measured either in terms of the *angular dispersion* of the separated beams of light as they emerge from the dispersive device or in terms of the *linear dispersion* of the separated images of the entrance aperture when they are brought to a focus at the exit aperture. Thus, if a device produces a change $\Delta\theta$ in angle for a wavelength difference $\Delta\lambda$, the angular dispersion is $\Delta\theta/\Delta\lambda$. Similarly, if two images for wavelengths separated by $\Delta\lambda$ are focused a distance Δx apart, the linear dispersion will be $\Delta x/\Delta\lambda$. A useful approximation when the changes considered are not too large is that

$$\Delta x = f\Delta\theta \tag{9-1}$$

where f is the focal length of the optical system and is assumed to be reasonably independent of wavelength.

Equation (9-1) suggests that it should be possible to obtain separate spectral lines at the detector for any two wavelengths, no matter how small the difference between those wavelengths, simply by constructing an instrument of sufficiently large focal length. This is true, however, only if it is possible to produce an infinitely narrow entrance slit, failing which the images of the slit will become broader as the focal length is increased to obtain greater separation (of their centres). Quite apart from practical considerations such as the amount of light able to enter the instrument through an extremely narrow slit, all images are in fact diffraction patterns and have a finite minimum size which depends on the wavelength of the light as well as the construction of the instrument. Thus there is a definite limit to the ability of a particular device to separate closely spaced adjacent wavelengths; this is usually termed the *resolving power*. It is defined as $\lambda/\Delta\lambda$ where $\Delta\lambda$ is the smallest wavelength difference (at wavelength λ) which allows separation (i.e. resolution) into two distinct spectral lines.

This definition of resolving power requires qualification. Each wavelength entering the instrument is, as already explained in Chapter 2, actually a narrow region of the type shown in Fig. 2-3 in which the intensity is a maximum at the centre and gradually decreases towards zero at the edges. If the instrument was able to resolve wavelength differences far smaller than the width of this region, the spectral line obtained on, say, a photographic plate would show the same distribution of intensity measured against distance along the spectrum. When this is not the case, the line obtained will have a similar appearance but the distribution will be largely a function of the instrument rather than the original wavelength distribution emitted by the source of light. This second situation will apply with the majority of light source/spectrometer combinations of interest in AAS and AFS. In any event, the variation of intensity across the width of the line means that the resolving power cannot be considered as simply a limit on one side of which two adjacent lines

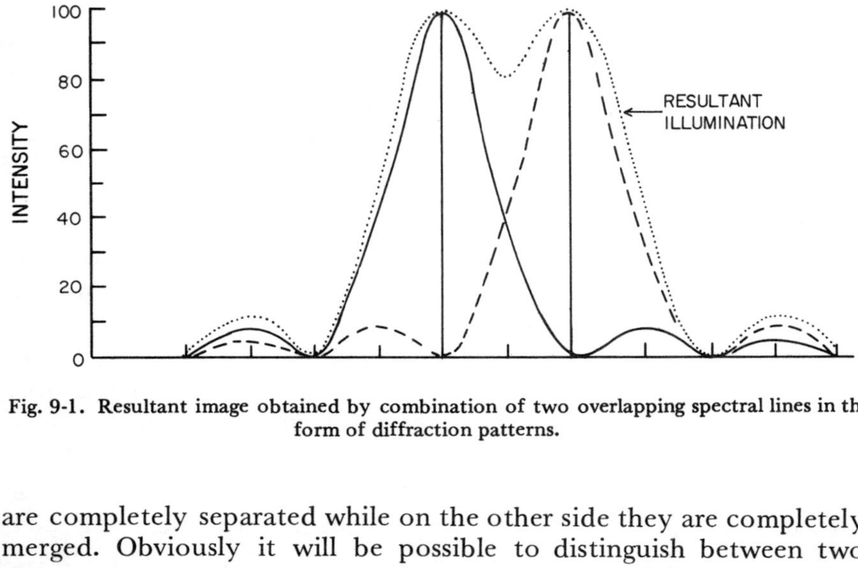

Fig. 9-1. Resultant image obtained by combination of two overlapping spectral lines in the form of diffraction patterns.

are completely separated while on the other side they are completely merged. Obviously it will be possible to distinguish between two adjacent peaks even when there is considerable overlap of the lines and it is necessary to decide on a practical criterion of just how close the lines can be before they can no longer be considered as resolved. Usually this is based on the fact that each observed image is actually a diffraction pattern, the "line" being the main central peak of this pattern with a succession of progressively weaker peaks on either side. The *Rayleigh criterion* states that two lines of equal intensity are just resolved if the maximum of the central peak of one line falls on the first minimum of the other. The intensity midway between the two peaks is then equal to 81% of the peak intensity. This is illustrated in Fig. 9-1. The criterion does not apply when the peaks are of unequal intensity, but it is then usual to consider the lines as resolved if at any point between the peaks the intensity falls to 81% of that of the less intense maximum.

Although the resolving power bears a closer relation to practical conditions than the dispersion of an instrument, Rayleigh's criterion is really only useful when applied to very high-quality spectrometers. The actual resolution is affected by many instrumental factors such as the actual design of the instrument (assuming a given dispersion), the width of the entrance slit, the quality of the optical components, etc. Thus, in many cases the instrument may be incapable of achieving its theoretical resolving power. In both analytical atomic absorption and fluorescence spectroscopy it is much more common to refer to the capability of a monochromator in terms of its experimentally determined *bandpass*. This is simply the width of the transmitted peak as observed at the detector (i.e. on a photographic plate or on a recording obtained photoelectrically while scanning the monochromator slowly across a narrow spectral line). It is generally defined in the same way as the half-width of a spectral line (see Section 2.5), i.e. the bandpass is the width of the observed peak in wavelength units between the two points where the transmitted intensity is equal to half that at the maximum. The bandpass will obviously decrease with the width of the entrance slit, so that for monochromators it is usual to quote a value either at a given slit-width or at the smallest practicable width. The bandpass is a particularly useful concept for AAS and AFS work because not only does it give a clear picture of the capability of an instrument but it may also be applied to non-dispersive devices such as optical absorption filters.

The efficiency with which a monochromator is able to carry out the wavelength selection process is usually termed its *transmission*. This is defined as the percentage of the incident radiation that is passed by the device at the wavelength selected. As with the bandpass it is a convenient, practical term which may be applied both to dispersive and non-dispersive systems. With a monochromator transmission depends both on the design and construction of the instrument and upon the manner in which it is used. Thus, as with any optical system, the maximum radiant power that can be passed will depend on the relative values of the focal lengths and apertures of the mirrors or lenses. Similarly the amount of radiation lost during each reflection or transmission step within the device will depend on the quality of the components. Of particular importance is the width of the entrance slit; an increase not only increases the bandpass but also the transmission of the instrument. The slit and its illumination are, in fact, major factors in deciding the transmission; the relations between these factors are relatively complex. This arises because the brightness observed at the detector depends not only on the intensity of the radiation but also on the solid angle over which energy is collected. As a result, the effects of a change in the entrance slit width will depend on both the solid angle over which radiation from the light source is available and the angle over which the components of the spectrometer itself are able to collect the radiation passing through the slit. This topic is of considerable importance and is discussed further in Section 9.2.4.

9.1.3 Requirements of Instruments Used for AAS and AFS

In principle a relatively simple monochromator should suffice for most analytical atomic absorption or fluorescence measurements. In practice, however, many commercial spectrophotometers are supplied with relatively expensive, high quality grating monochromators. This is partly because such instruments are convenient to use, allow the maximum precision in measurements and enable good sensitivity to be obtained under what might otherwise be considered adverse conditions. A high-quality monochromator may also prove extremely useful if the spectrometer is to be used for flame-emission spectrometry as a means of extending the facilities available to a particular laboratory. The versatility thus obtained could account for the specifications of some instruments. Some general features are, however, desirable in any dispersive devices to be used for routine applications. These include: (1) good stability in the wavelength setting mechanism to minimize drift from the peak intensity attainable at the resonance line, (2) good stability of the slit-width setting so that changes in intensity are not caused by drift, (3) an adjustable slit width allowing the optimum value to be chosen for each application, (4) a clear and accurate wavelength calibration allowing the required resonance line to be quickly and conveniently selected, preferably with an accuracy of 0·1 nm or better, (5) a rapid wavelength changing device to facilitate changes between lines hundreds of nanometres apart, (6) entrance and exit apertures situated to allow convenient placing of auxiliary equipment. Other features that may be useful include provision of a variable speed mechanical wavelength drive to allow spectra to be scanned, an exposed slit mechanism on which an image of the light source is easily focused, and interchangeable prisms or gratings to allow optimum performance to be obtained at any wavelength.

The specific requirements for AAS and AFS differ considerably, although the same device can often be applied to both techniques with satisfactory performance. Thus, for atomic fluorescence measurements it is important that the peak transmission be relatively high to give good sensitivity. With many fluorescence cells improved sensitivity would also be obtained if light were accepted over a wide solid angle. A high transmission is frequently not necessary for atomic absorption, but sufficient light must be transmitted to obtain adequate signal : noise ratios at the photomultiplier-amplifier system. A narrow spectral bandpass is generally unnecessary for atomic fluorescence work and would frequently prove a disadvantage as (for a given size and quality of instrument) it would be obtained at the expense of high transmission. An exception may occur when fluorescence measurements are made with a continuum source. Any scattering of incident radiation that occurred would then occupy a much wider wavelength range than the fluorescence radiation. Thus, the fluorescence signal : background intensity ratio may be improved by reducing the spectral bandwidth. A narrow spectral bandpass is not essential for many atomic absorption applications but may result in improved sensitivity in certain applications. For example a reduction in bandpass can sometimes avoid detection of non-absorbing or weakly absorbing lines at wavelengths close to that of the resonance line. It may also be possible to avoid the detection of intense background radiation emitted by certain flame cells; even with modulated systems this radiation reaches the detector and can increase the noise level on the absorption signal so that precision is lost. A narrow spectral bandpass is also necessary to obtain highest sensitivity when a continuum light source is used for atomic absorption measurements.

9.2 CONVENTIONAL MONOCHROMATORS

9.2.1 General Constructional Details

A brief description of monochromator construction is given here together with details of some components. The dispersive elements (prisms and gratings) are discussed in the following sections together with some of the most commonly used optical arrangements.

Most monochromators used in atomic absorption spectrometers are of compact design and an effort is usually made to avoid optical systems requiring a case of awkward dimensions; a simple rectangular box is by far the most popular shape for integration with the remainder of the instrument. A substantial base is required to maintain accurate alignment of optical components and a well-fitting cover is needed to minimize stray light, especially if very weak fluorescence signals are to be measured. The construction of the slit mechanism is important if the full potential of the dispersive element is to be obtained, and in a monochromator this applies equally to both of the slits required. A single fixed slit width is inadequate except for very simple instruments with a limited range of application. Several manufacturers provide a choice of slit widths at low cost by the provision of a set of interchangeable plates into which stainless steel slit jaws have been fitted. A slit mechanism with movable jaws is, however, more convenient and versatile and is fitted to most instruments. The edges of the slit jaws must be straight and sharp and it is essential that they remain accurately parallel as the width is changed. In some assemblies only one jaw is moved, but symmetrical movement of both jaws is preferable as the centre of the observed line then remains in the same position. Adjustment is usually effected with a micrometer screw and this should be accurately calibrated to allow the width to be adjusted and reset to within a few μm. Complete closure of the jaws can cause damage to their edges and the mechanism should incorporate a device to prevent this. Thus an auxiliary shutter is useful to allow "zero" light levels to be registered at the detector and read-out device. In some designs the slit jaws may be partly concealed by the mechanism or by a protective cover; for many applications it is useful if an image of the light source or even of a flame absorption cell can actually be focused onto the metal plates forming the slit jaws. Ideally, the entrance and exit slit-widths should be individually optimized; monochromators used in analytical AAS and AFS, however, often have a linked mechanism so that adjustments to both are made simultaneously. In some instruments the maximum slit-width is quite small and this may prove a disadvantage for fluorescence applications where a wide spectral bandpass is quite permissible but high transmission of radiation is essential; a maximum aperture of 2-3 mm is provided with many instruments and is suitable for most applications.

Although spectrographs are available in which the dispersive element is the only optical component (principally the concave grating, which is unsuitable for scanning monochromators, see Section 9.2.3), it is generally necessary to use several lenses or mirrors for collimating or focusing radiation within the instrument. The number of such components should be minimized because, apart from any considerations of expense, it is essential to avoid loss of radiation (by absorption, etc.), minimize stray light and maximize the contrast of the system. As a result, multi-component lenses such as those found in cameras are not used and, of course, the common coating techniques are generally impracticable if a wide wavelength range is needed. Instead of attempting to make each component perfect, it is more usual to carefully select the essential components so that their individual aberrations tend to cancel within the system.

At the present time, mirror systems are more widely used than lenses in the scanning monochromators needed for AAS and AFS. This arises because the usual disadvantage of mirrors in a spectrograph (the difficulty of providing minimum aberrations over the wide field angle needed to photograph a relatively wide region of the spectrum) no longer applies and in other respects mirrors have many advantages over lenses, particularly from the point of view of mass production at moderate cost. For example, chromatic aberrations are often a problem with lenses but cannot arise with front-surfaced mirrors. Similarly, high-quality mirror coatings are available over the whole visible and uv region, but even relatively expensive silica lenses may give considerable loss of light by absorption in wavelength regions of interest in AAS and AFS. Further losses also occur with lenses owing to surface reflections, material defects (e.g. bubbles) and other effects which are more easily avoided with mirrors. Finally, a large mirror may also be produced at substantially lower cost than an equivalent lens owing to the relative ease of mirror manufacture. The size of these components is a major factor in determining the transmission of the monochromator for a given spectral bandpass. In the past this would have been governed by the expense of the dispersive element but at the present time large replicate gratings (see Section 9.2.3) are available at quite reasonable cost. Thus the availability of large, low-cost mirrors is particularly useful for the production of large aperture* monochromators at reasonable cost. As a result it is economically feasible to use an instrument of aperture f/2 or f/3 in an atomic absorption spectrometer when f/8 or f/11 would be adequate. This not only improves the precision of measurements under adverse conditions (i.e. low-intensity light source) but also makes the spectrometer

* The aperture is usually expressed as an "f-number"—the ratio of focal length to effective diameter for an optical component.

much more useful for alternative application to atomic fluorescence or flame emission spectrometry.

As mentioned above, it is best to obtain a compact and convenient shape for the monochromator by the choice of a suitable optical system. This may not always be possible (particularly with prism instruments) and it may be necessary to add additional components simply to deflect the input or output beams; this is often the case where a "straight through" monochromator is required. It would also be possible to use such devices to scan the spectrum across the exit slit. In most cases, however, wavelength selection is achieved by rotation of the prism or grating. This is best carried out with a screw or gear system allowing accurate selection without backlash and reproducible operation over a long period. Wavelength calibration with dial or drum scales has been used extensively in the past; more convenient operation, however, is obtained with digital counters reading linearly in wavelength units. This is easily achieved with grating instruments (for which the dispersion does not vary greatly with wavelengths) and some prism instruments use ingenious drive mechanisms to obtain the same effect. Calibration in arbitrary units is inconvenient for routine applications where the wavelength setting is frequently changed, particularly when the calibration is not linear. Provision for fine adjustment of the wavelength setting is necessary if the instrument is to be easily set to the wavelength of a particular atomic line. Therefore, to allow rapid changes from one line to another the monochromator should be fitted with either coarse adjustment controls or a high-speed motor drive. A slow mechanical drive allowing scanning and recording of spectra is also often fitted. Several speeds are desirable to give correct matching of spectral bandpass and scan rate and the drive should be smooth to avoid vibration of the optical components.

Recently a number of modified monochromators have been reported with which it is possible to make intensity measurements at several discrete lines either while scanning rapidly or by automatic selection of lines.[5-9] These devices are useful for rapid, multi-element analysis either by AAS or AFS and may see further development. Automatic selection of lines is probably of greater interest as the rest of the spectrometer may remain unchanged (whereas rapid scanning requires that each measurement be made in an extremely small time interval). The system described by Cordos and Malmstadt[9] uses a conventional commercial monochromator and widely available digital and logic circuit components and may be programmed to change between a large number of preset wavelengths with an accuracy of 0·02 nm. Such a device should provide worthwhile time saving in many routine applications, quite apart from its application to multi-element analysis.

9.2.2 Prism Instruments

Most early spectroscopes used a prism as the dispersive element and these devices are still used today although, as mentioned earlier, many present-day commercial monochromators used for analytical AAS or AFS employ replicate gratings. Most applications have used a prism with flat surfaces and triangular cross section and only these are mentioned here. Many other types of prism, including some with curved surfaces, have, however, also been tested for use in specialized instruments.

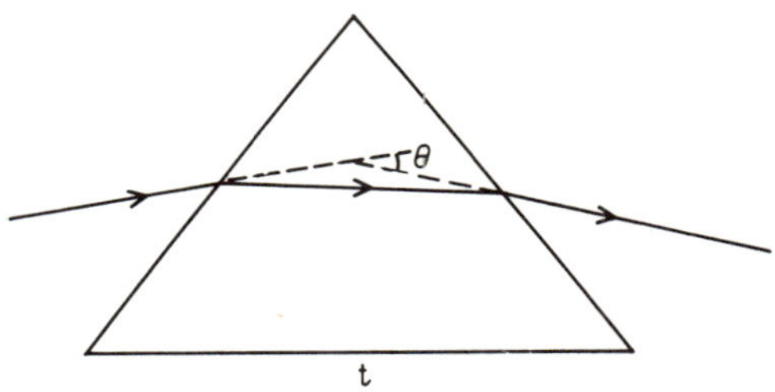

Fig. 9-2. Path of a ray through a section of a prism.

The passage of a ray of light through an isosceles prism is illustrated in Fig. 9-2. The ray is deflected as it enters and leaves the prism owing to refraction, its total deflection from the original path being known as the angle of deviation, θ. This angle assumes a minimum value when the ray passes symmetrically through the prism (i.e. enters and leaves at the same angle to the face), giving the position of minimum deviation. Under given conditions the value of θ depends on the refractive index, μ, of the prism material. Thus, as μ is a function of wavelength for a particular material, the prism acts as a dispersive element. For most spectroscopic applications, optical distortions are minimized by using the prism near to minimum deviation and also by arranging that the slit is parallel to the prism edge and using a parallel beam of light.

When a prism with a base of length t is completely illuminated by a light beam of width w it may be shown that the angular dispersion is given by

$$\frac{d\theta}{d\lambda} = \frac{t}{w}\frac{d\mu}{d\lambda} \tag{9-2}$$

and the resolving power is given by

$$\frac{\lambda}{\Delta\lambda} = t\frac{d\mu}{d\lambda} \tag{9-3}$$

where $d\mu/d\lambda$ is either found graphically or from an empirical formula such as that of Hartmann:

$$\mu = \mu_0 + C/(\lambda - \lambda_0) \tag{9-4}$$

where μ, C and λ_0 are experimentally determined constants for each material.

Hence

$$\frac{d\mu}{d\lambda} = -\frac{C}{(\lambda - \lambda_0)^2} \tag{9-5}$$

and it can be seen that both the dispersion and resolving power will decrease with increasing wavelength. It is also apparent that reducing the height of the prism (i.e. the effective value of w) will reduce the dispersion of the instrument but not its resolving power, which depends only on the length of the base. These simple rules apply only at minimum deviation and the situation is rather more complicated under other conditions. In practice it is found that the dispersion shows a rapid increase as θ is increased from its minimum value but the resolving power is reduced slightly.

Many different prism materials, including liquids, have been used successfully in the wavelength region 200 to 800 nm, but the choice for modern commercial instruments is not particularly wide. Obviously the material must have high transparency over the wavelength region of interest and this should be as wide as possible for a general purpose instrument. A high value of $d\mu/d\lambda$ is also essential to obtain good dispersion. The material should also be available in large pieces of optical quality, be easily polished, resistant to attack by air, water vapour, etc. and show a small dependence of refractive index on temperature. Optical quality glasses meet many of these requirements and are available at moderate cost in large pieces. They are widely used for instruments designed for use in the visible region and give high dispersion. Unfortunately, they are of little use for many AAS and AFS applications owing to their very low transmission in the important ultraviolet region. The useful limit lies in the range 350 to 400 nm for traditional crown-and-flint glasses and even special uv-transmitting glass cannot be used below *ca*. 300 nm. Applications in the region upwards from 200 nm generally require a quartz prism; large pieces of crystalline material are difficult to obtain, however,

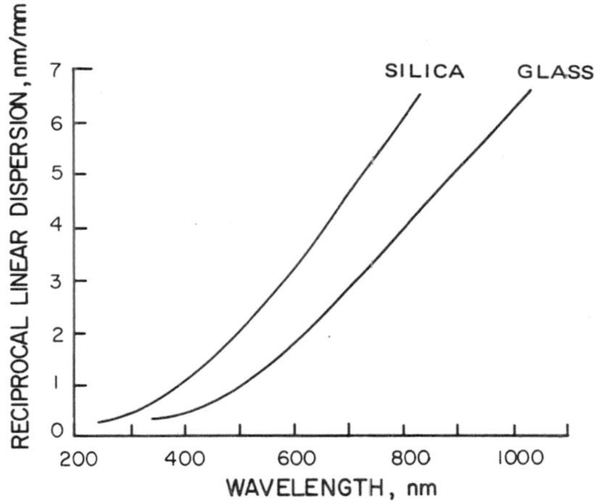

Fig. 9-3. Reciprocal linear dispersion obtained with glass and silica prisms in the same spectrometer.

and may therefore be expensive. As a result, many prism instruments supplied commercially have a fused silica prism. This material gives good performance in the ultraviolet but poor dispersion in the visible region owing to a rapidly decreasing value of $d\mu/d\lambda$. Some of the more expensive prism instruments are, therefore, supplied with interchangeable glass and silica prisms. The relative dispersions of these materials in a typical commercial instrument are illustrated in Fig. 9-3.

Some of the classical optical arrangements used in prism instruments are illustrated in Fig. 9-4. In practice, improved performance may be obtained by elaborating on these designs. For example, use of a prism made up of several components, or instruments with two or even three separate prisms, is quite common. Those arrangements in which the entrance and exit slits appear to coincide use a tilted mirror, an off-axis mirror, or a beam deflecting device.

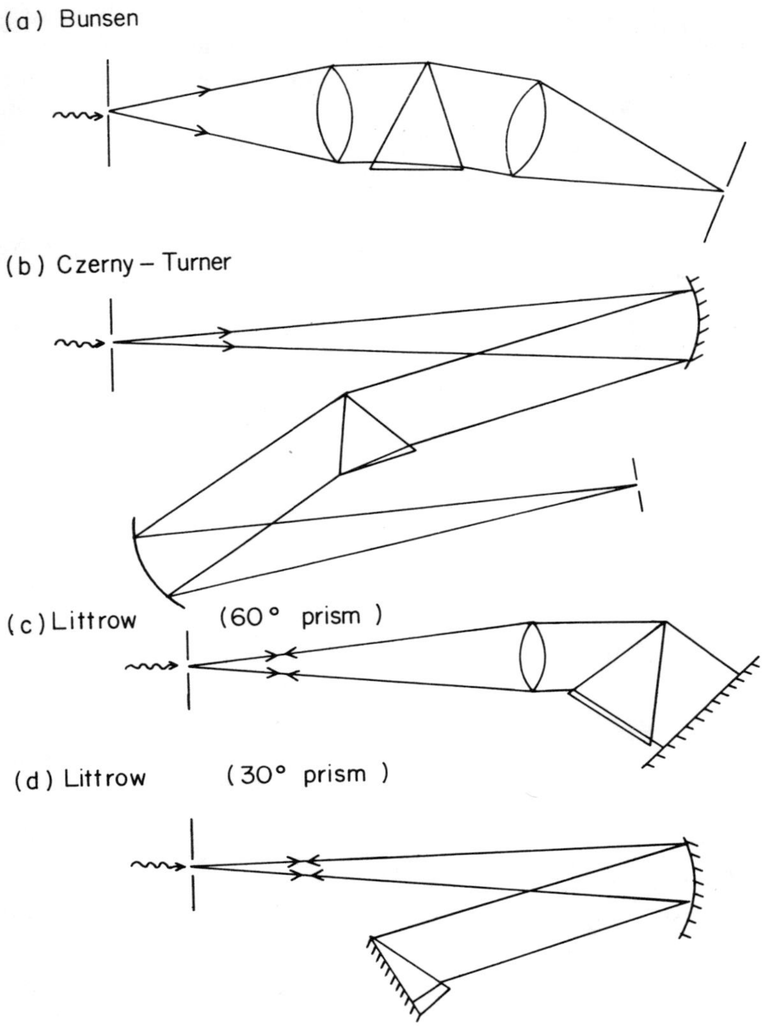

Fig. 9-4. Schematic layout of prism spectrometer systems.

9.2.3 Grating Instruments

As mentioned earlier, many commercial monochromators available at the present time use a diffraction grating as the dispersive element because of the availability of relatively high-quality replicate gratings at reasonable cost. (A replicate grating is simply a mass produced copy of a grating on which the spacings have been individually ruled; only the highest quality instruments are now fitted with original gratings.) Although a transmission grating would be the direct equivalent of the prism, most instruments today use front surface reflecting gratings which, like mirrors, are easily produced to cover a wide wavelength range with good performance. Concave gratings of this type ruled on concave mirrors combine the dispersive and focusing functions and thus reduce the number of components, loss of light and aberrations of the image. For this reason they are widely used for spectrographs. Plane gratings are, however, more suitable for all scanning monochromators as constant focusing conditions are maintained while the grating is rotated to scan the spectrum. Hence we shall be concerned here only with plane, reflection gratings.

A diffraction grating is simply a convenient practical method of obtaining a large number of very narrow, long apertures of the required dimensions. These apertures are of two types: those that transmit or reflect incident radiation and those that are opaque or give no reflection (depending on whether the grating is of the transmission or reflection type). When the grating is illuminated by parallel light and the transmitted or reflected radiation is brought to a focus two optical effects are obtained. The apertures produce a large number of light beams that have different path lengths but the same frequency and the same initial phase, i.e. that originate from coherent sources. Thus an interference pattern is obtained. At the same time, diffraction occurs at each of the narrow apertures (diffraction results from interference which arises from mutual interactions of different parts of the *same* wavefront so that it is observed at a single slit). As a result a combined interference and diffraction fringe pattern is seen with the characteristic intensity maxima and minima. When the grating is illuminated normally by light of wavelength λ, the principal maxima occur at angles, θ, given by the equation

$$\mu(a + b) \sin \theta = n\lambda \tag{9-6}$$

where a and b are the widths of the two types of spaces on the grating, μ is the refractive index and n is an integer known as the order of interference.

In spectroscopic applications, the grating is illuminated by light from a slit and an image of the entrance slit is seen as a sharp line at each maximum. The spectrum may be scanned by changing the value of θ. In practice this is complicated by several factors. In the first place it can be seen from equation (9-6) that if light of wavelength

λ_1, is observed at an angle θ, this same angle will also correspond to light of a different wavelength, λ_2, in the second order, and so on. Thus, the grating will produce a series of overlapping spectra, the wavelengths observed at a given angle being such that $\lambda_1 = 2\lambda_2 = 3\lambda_3 = \ldots n\lambda_n$. Obviously the wavelength range corresponding to one order (i.e. the range without overlap) will decrease with increasing values of n. To avoid difficulty in interpretation of spectra most monochromators applied to AAS or AFS therefore use low orders, frequently the first. Even then special precautions are needed to avoid confusion from higher orders over the complete visible and uv range (see below). A second factor requiring the use of low orders is the efficiency with which dispersion is obtained; the diffraction pattern of the fringes means that approximately $1/\pi^2 n^2$ of the incident radiation goes into each order. Thus much higher transmission is obtained by using low orders, although in fact the situation can be altered by using the blazing technique (see below).

On the other hand, both the dispersion and resolving power of the grating are improved by using higher orders. The angular dispersion is given by

$$\frac{d\theta}{d\lambda} = \frac{n}{\mu(a + b) \cos \theta} \qquad (9\text{-}7)$$

and the resolving power by

$$\frac{\lambda}{\Delta\lambda} = nN \qquad (9\text{-}8)$$

where N is the total number of spaces on the grating. Thus it can be seen that if the use of high orders is impracticable it is necessary to use a grating ruled with a large number of fine spaces in order to obtain high dispersion and good resolution. Nevertheless, there is obviously a finite limit to the combination of wide wavelength range and high resolution that is possible. The popularity of grating instruments is largely due to the success with which these features have been combined at reasonable cost. Reference to equation (9-7) shows that, for a given grating and order, the dispersion is not directly dependent on wavelength. By correct design of the optical system it is possible to use the grating to give a "normal" spectrum in which the linear dispersion is constant over the whole spectral range of interest. This allows the production of monochromators having linear wavelength scales. An additional advantage is that the distance from the zero-order image to the spectral line obtained for wavelength λ is directly proportional to λ. Hence the calibration may be easily checked by noting the reading obtained at zero order (when the incident radiation is reflected or transmitted unchanged). The problem of overlap of orders has received considerable attention as even in the first order it is relatively small (e.g. the second-order image of a line 250 nm will coincide with the first-order image of one at 500 nm). A common solution is to use auxiliary filters which transmit only the (narrow) wavelength range of immediate interest. This is, however, inconvenient and reduces the efficiency of the instrument. A much more acceptable solution is to replace the simple apertures mentioned earlier by apertures formed at a controlled angle or *blaze* to the surface of the grating. This changes the diffraction properties of the grating so that for any wavelength range it is possible to direct a very high proportion of the incident radiation into a chosen order. Thus the efficiency of the grating is improved and the "ghost" spectra of other orders can be considered as of negligible intensity for many applications. The highest blaze efficiency is obtained in the first order so that this is still generally preferred, although the use of higher orders is made much more convenient than with a simple grating. Very high blaze efficiency can be obtained only over a narrow wavelength range. Thus it is usual to accept a lower peak value (perhaps 60%) in return for good efficiency over a wide wavelength range. Typically, a grating blazed at 500 nm in the first order would retain about 50% of its peak efficiency at 250 or 750 nm. As a result, some monochromators are supplied with two interchangeable gratings blazed for the uv and visible regions respectively.

Although the resolving power (and hence spectral bandpass) of a grating depends in theory only on the number and separation of the apertures, its actual performance is affected considerably by the degree of perfection achieved in the production process. Thus, loss of resolution and efficiency and increased scattered light or ghost spectral lines will be observed with poor-quality gratings. Reflection gratings require the ruling of many thousands of fine lines on to a surface that has been polished to a perfect optical finish. This is most easily achieved by coating an optically polished glass or silica plate with a layer of aluminium by vacuum-evaporation; this then allows ruling in a soft material that could not otherwise be easily optically finished. By using specially designed ruling machines and a very fine diamond point it is possible to obtain groove densities of several thousand lines per millimetre over large areas. Even with modestly

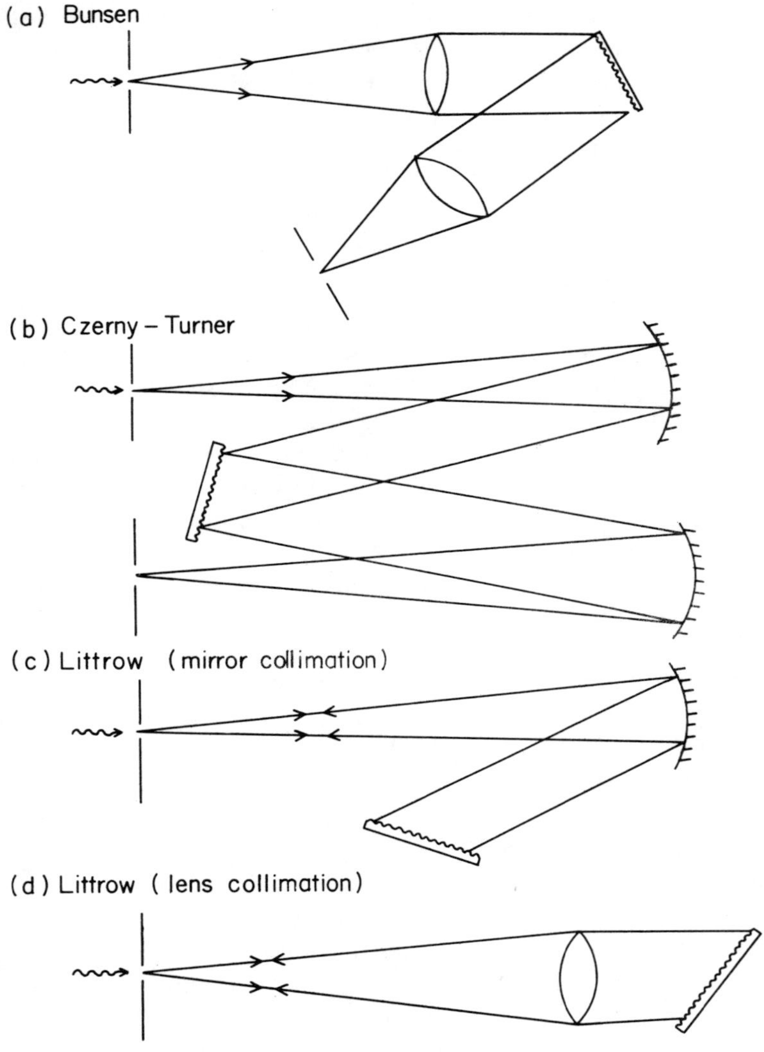

Fig. 9-5. Schematic layout of plane reflection grating spectrometric systems.

priced atomic absorption spectrometers it is possible to supply gratings of (typically) 50 x 50 mm size with more than 1000 lines/mm. Although these will be replicates, modern production techniques result in a quality almost as good as that of the original gratings.

Some of the optical arrangements that are commonly used for plane reflection gratings are illustrated in Fig. 9-5. They are equivalent to the prism systems described in Section 9.2.2 and similar comments apply. The Czerny-Turner arrangement is particularly popular as it uses mirrors and directly provides an instrument with conveniently placed entrance and exit slits within compact dimensions.

9.2.4 Illumination of the Monochromator

Even the largest monochromators have a relatively low transmission at any given wavelength. As a result it is important to ensure that the method of illumination gives maximum utilization of the available energy, particularly for fluorescence measurements where the sensitivity may depend directly on the amount of radiation collected. To achieve this several factors must be taken into account:

(1) The intensity of the source will vary inversely with the square of its distance from the monochromator.
(2) Efficient use will only be made of the collimation system of the monochromator if it is completely illuminated, i.e. if the source subtends a solid angle at the slit that is at least as large as that subtended by the condenser.
(3) If the source has a large surface area of uniform brightness, as much as possible of that area should be included within the solid angle of radiation that reaches the collimator.
(4) If the source is of extended depth, each cross-section (taken along the optical axis) should satisfy condition (3).
(5) When the other conditions are met, use of an auxiliary condensing lens or mirror cannot further increase the brightness of the spectrum. Indeed, loss of energy will always occur at each extra optical component.

It is therefore apparent that an increase in the intensity of a spectral line cannot be obtained by using a condensing lens to focus an image of the source onto the entrance slit *unless* the source itself is too small to fill the collimator with light or the source cannot be brought sufficiently close to the slit as a result of other instrumental requirements. An exception occurs with a long source when, in general, it is possible to utilize a greater depth of source by using a lens to focus the back surface of the source on the slit. Wherever a condensing lens must be used, maximum illumination will be obtained if the lens completely fills the solid angle that can be accepted by the collimator. In some cases the aperture of silica lenses supplied with atomic absorption spectrometers is too small to meet this requirement. It is also worth remembering that with a given lens and a given (minimum) distance between the source and the slit, moving the lens closer to the slit will produce a smaller, brighter image but this does *not* increase the brightness of the observed spectral line. This arises because the energy per unit solid angle entering the monochromator is unchanged and the additional energy simply misses the collimating element and is wasted. At first sight these rules may seem somewhat complicated but it is usually possible to visualize clearly the situation with a particular monochromator simply by drawing a scale diagram of the collimator and the entrance optics, taking into account both the horizontal (across the slit) and vertical (along the length of slit) planes.

In atomic absorption applications it is nearly always necessary to use auxiliary lenses or mirrors to obtain a satisfactory optical path through the atom cell. Even though in AAS minimization of energy

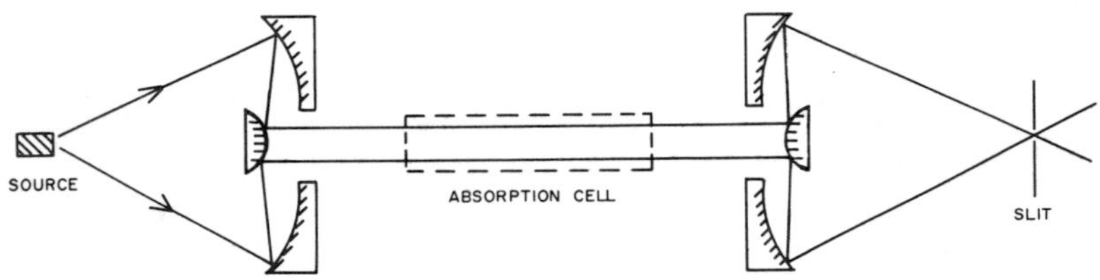

Fig. 9-6. Double inverted Cassegrain system for use with a narrow absorption cell.

loss in the optical system is not as important as in AFS, careful choice of the optical system may still give improved results. For example, the entire light beam may pass through the atom cell but only part of it fall on the collimator owing to incorrect use of a lens or mirror. The absorption signal observed for a given sample added to the atom cell would then be reduced. This often occurs when an extremely small image of the background light source is focused on the slit on the assumption that full utilization of the available energy will thus be achieved. Similarly, the precision of absorption measurements may be lowered if the intensity of the source radiation reaching the detector becomes too low. This often results from incorrect use of the optical components or simply because those components are unsuitable. For example, if an atomic absorption spectrometer is designed to use a narrow parallel beam of radiation, because the absorption cell is narrow (and all of the beam must pass through the cell), this might be obtained with two small condensing lenses, one to give a parallel light beam from the source, the other to focus this beam on the monchromator slit. Loss of efficiency then occurs at two points; the first lens is too small to collect all available radiation from the lamp and the second is too small to completely fill the solid angle accepted by the collimator. This could easily be avoided, for example, by using a mirror system such as that shown in Fig. 9-6. In some cases it may, of course, be more practicable to select an optical beam that meets the requirements of the monochromator and choose the dimensions of the absorption cell accordingly.

9.3 OTHER MONOCHROMATING DEVICES

9.3.1 Filters

Wavelength selection by means of a small optical filter transmitting a specific resonance line is in principle an attractive alternative to the monochromator devices described in Section 9.2. This is particularly true in atomic fluorescence spectroscopy where the transmission factor attainable with a filter is often much greater than that available with a conventional monochromator. In practice, however, the number of reported applications of optical filters is relatively small. For atomic absorption spectroscopy, the relatively poor resolution of even the best filters limits their application to simple instruments suitable for the determination of a few elements which exhibit simple atomic spectra. Typical element combinations for which simple AAS instruments of this type might be designed, for example, are sodium and potassium, mercury, thallium and lead, or calcium and magnesium. In atomic fluorescence spectroscopy, filters have been used in simple spectrometers and in multi-channel instruments to provide for wavelength selection in simultaneous multi-element analysis. Possibly the relatively small interest in filters to date results from the sophisticated design of most commercial atomic absorption spectrometers; the situation may change as the use of simple instruments dedicated to specific analytical applications becomes more widespread. It is worth remembering, however, that good-quality interference filters (especially those for the uv region) are still relatively expensive and if filters are required for a number of elements this approach may prove almost as costly as the use of a simple monochromator of adequate spectral bandpass and transmission characteristics.

The traditional optical filters, used for example in molecular devices which rely on the absorption of suitable dyes added to glass, plastic, gelatine or even used alone in solution. Commercial filters of this type are available with uv transmission capability, but they are usually most suitable for application to the visible region. The dyes or pigments used in the simplest versions (such as those intended for photography) have very wide transmission bands and are rarely suitable for spectroscopic applications. More satisfactory results are obtained by using combinations of "sharp" cut-off filters which are produced by several manufacturers. These are available in two series: long wavelength (transmitting red and yellow light) and short wavelength (transmitting blue and green light). By combining a filter from the long series with one from the short series it is possible to obtain a somewhat narrower bandpass with its peak at a preselected wavelength. The bandpass is, however, still wide for many applications; even with a peak transmission of only 15-20% the half-intensity width may be of the order of 50 nm. A further disadvantage is the possibility of change of spectral characteristics owing to the large amount of energy absorbed when the filters are irradiated with unwanted wavelengths at high intensity. Therefore, these filters are unsuitable for almost all AAS applications and the majority of applications to AFS. Improved bandwidth can be obtained by combining a number of filters but the transmission is then usually too low, particularly for fluorescence measurements. A detailed tabulation of transmission and stability data for commercially available filters is included in the "Handbook of Chemistry and Physics".*

Much better results are obtained using *interference filters.* These are simple dispersive devices utilizing the interference of light for spectral isolation. Unlike the diffraction grating, different path-lengths are obtained with a semi-reflecting surface to divide the amplitude of the light beam rather than by an actual division of the wavefront. A small piece of optically flat glass or silica is coated with a semi-reflecting layer of a material that has a high reflection coefficient but gives low absorption of light. This layer is covered by a non-reflecting spacing layer and finally by a second semi-reflecting layer. The reflecting surfaces of the two outside layers are face-to-face so that multiple reflections occur between them when a beam of light is passed through the filter. As a result the emergent beam is composed of many different path-lengths and an interference fringe

* Published annually by the Chemical Rubber Co., Cleveland, Ohio.

pattern is observed. The positions of the transmission maxima of this pattern are found from the equation

$$2\mu d \cos \theta = n\lambda \tag{9-9}$$

where μ is the refractive index of the spacing layer and d is its thickness. θ is the angle between the light beam and the normal to the filter surface. As the filter is usually used at 90° to the beam, equation (9-9) can be simplified to

$$2\mu d = n\lambda \tag{9-10}$$

Thus a filter can be made with a transmission peak at any required wavelength by depositing a spacing layer with the correct value of μd. This peak will be accompanied by other transmission maxima corresponding to different orders. For example, if μd = 150 nm, the filter would transmit at 300, 600 and 900 nm. Isolation of the required peak (e.g. 600 nm) does not present problems when the maxima are so widely spaced; in some cases use of a glass base ensures transmission only at a single visible wavelength, otherwise layers of absorbing dyes may be added to the filter to "block" the unwanted peaks.

Provided the filter has been correctly manufactured and is used at 90° to the incident radiation, the bandwidth at a given spacing (i.e. value of μd) depends only on the reflection coefficient of the reflecting layers. The relationship between the two is conveniently expressed in terms of the *finesse, N,* of the filter which is the ratio of the interval between successive peaks and the bandwidth. For example, a finesse of 50 for the filter mentioned above would mean a bandwidth of 600/50 = 12 nm. If the reflection coefficient is R* the finesse is found from

$$N = \frac{\pi R^{1/2}}{1 - R} \tag{9-11}$$

The peak transmission of the filter is also simply related to R, although the absorption coefficient, A, of the reflecting layers must also be taken into account. The fraction of incident radiation at the peak wavelength which is transmitted is given by

$$\tau_m = (1 - A - R/1 - R)^2 \tag{9-12}$$

Some numerical values of N and τ_m for different values of R and A are given in Table 9-1.

* i.e. $R \times 100$ gives the percentage of light reflected by the surface.

It will be appreciated that the reflecting layers of a high-quality interference filter must show simultaneously a high value of R and a low value of A. The former is achieved relatively easily in the visible region using the traditional coating materials of silver or sometimes aluminium, but low absorption values are more difficult to obtain reproducibly even when very thin layers of metal are used. Fortunately, the situation has improved considerably in recent years as a result of the use of multiple layers of dielectric materials such as ZnS, MgF_2, Sb_2O_3 and $PbCl_2$. By correct choice of the material and the number of layers used it is possible to obtain useful filters

TABLE 9-1

Variation of peak transmission with reflection and absorption coefficients of reflecting materials used in interference filters

A \ R	0·80	0·84	0·88	0·92	0·96	0·98
N	14·1	18·0	24·6	37·6	77·0	156
			τ_m			
0·18	0·01	—	—	—	—	—
0·14	0·09	0·02	—	—	—	—
0·10	0·25	0·14	0·03	—	—	—
0·06	0·49	0·39	0·25	0·06	—	—
0·02	0·81	0·77	0·69	0·56	0 25	—
0·01	0·90	0·88	0·84	0·77	0·56	0·25
0·005	0·95	0·94	0·92	0·88	0·77	0·56
0·001	0·99	0·99	0·98	0·98	0·95	0·90

throughout the visible and ultraviolet regions. At the present time reflecting layers combining a value of $R = 0{\cdot}98$ with $A \leqslant 0{\cdot}005$ are feasible, although such specifications are generally reserved for applications such as laser optics rather than for conventional commercial interference filters for routine use. It is also possible to reduce the bandwidth for a given peak transmission by using a wider separation (i.e. smaller wavelength interval between adjacent maxima) and relying on the blocking layers to prevent transmission at the extra harmonic peaks.

Interference filters are currently available from several manufacturers in a wide range of bandpass/transmission combinations. In the "narrow band" types of interest for AAS and AFS, typical filters for the visible region would be in the price range $30 to $300 with a peak transmission of about 65% for 6 to 8 nm bandpass or about

25% for 0·3 to 0·4 nm. The price will also depend on the degree of tolerance that the customer is prepared to accept in the specification of the peak transmission wavelength, as this factor is relatively difficult to control. This tolerance may be of the order of 2 nm in the cheaper filters with 6 to 8 nm bandpass, but the more expensive ultra-narrow versions can often be selected to ±0·1 nm. Although continual improvements are made in ultraviolet filters, which are now available for wavelengths down to about 200 nm, the specifications are generally somewhat poorer than in the visible region. For example, below 300 nm typical bandpass/transmission characteristics are 7-10 nm/1-5% or 23-30 nm/15-20%. As the resonance lines of many elements lie in the ultraviolet region this situation somewhat reduces the advantages of using interference filters for AAS and AFS applications.

The relatively high cost of these interference filters is also a disadvantage when wavelength selection for many elements is required. It may be possible to reduce this cost, particularly for fluorescence measurements where narrow bandpass is less important, by using a *variable interference filter*. These devices are manufactured by coating an interference filter onto a large disc (e.g. 15 cm diameter) in the form of a circular band with a continuously varying spacer thickness. The full width of the band is viewed through a slit about 1 mm wide so that as the disc is rotated a gradually changing peak wavelength is observed. Different filters are manufactured for both the visible and ultraviolet regions; a typical commercial device gives an average transmission of 40% for 20 to 25 nm bandpass, the region 450 to 700 nm being covered with a rotation of 270°.

Although interference filters have seen little use for wavelength selection in atomic absorption measurements, in atomic fluorescence spectroscopy several applications, particularly to multi-element analysis, have been reported. Mitchell and Johannson,[10] for example, utilized interference filters in a four-channel atomic fluorescence spectrometer. The filters utilized for transmission of the fluorescence radiation of each element were mounted on a rotating "filter-wheel" synchronized so that the appropriate filter passed the photomultiplier detector when the hollow-cathode lamp for that element was operated by a high-current pulse. Walsh[11] has described single and multi-channel atomic fluorescence spectrometers, the former using an interference filter and conventional monochromator, while the latter is based on a solar-blind photomultiplier (see Chapter 10).

9.3.2 Resonance Detectors

Early demonstrations[12] of the phenomenon of resonance fluorescence were made by irradiating an evacuated bulb containing a small amount of sodium metal with radiation from a sodium-discharge lamp. No radiation could be seen at 90° to the light beam until the bulb was warmed slightly to produce a low pressure of sodium vapour; under these conditions the characteristic *D*-lines are re-emitted as resonance fluorescence. In 1959, Russell and Walsh[13] suggested that this simple experiment could be used to permit the isolation of the resonance lines needed for atomic absorption measurements. All radiation from the background light source would be passed through the absorption cell and allowed to fall on a "resonance detector" containing vapour of the analyte element. A photomultiplier placed at 90° to this beam and positioned to receive fluorescence radiation generated in the detector would then respond only to changes in the analytical resonance line, just as if that line had been isolated using a conventional monochromator or filter. The advantages of this technique are immediately obvious: (1) the "bandpass" of the resonance detector is simply the width of its absorption line and under the conditions normally used the value is comparable with those obtainable only from dispersive instruments of extremely high resolution; (2) for many elements the "transmission" obtained with this narrow bandpass is much higher than could be achieved at this bandwidth by conventional techniques; and (3) unlike a conventional monochromator, it is unnecessary to select the peak wavelength manually and drift from this wavelength is impossible. In view of these advantages, considerable development of the technique has been carried out by Walsh and co-workers.[14-18] The main features are described here, although, as explained below, the advantages of application of the technique are really only realized for a few elements.

Resonance detectors are most efficient for atoms that exhibit simple spectra; a number of energy levels close to the ground state will give a complex resonance spectrum and result in poor absorption sensitivity. Thus the technique is best suited to the elements of groups I, II and III, for which the lines isolated are listed in Table 9-2. It can be seen that even in Group III several lines are detected for Al, Ga and In. The cloud of atoms in the resonance detector can be produced thermally for many metals simply by embedding a heating element in a small block of the appropriate metal (see Fig. (9-7), but in other cases it is necessary to use cathodic sputtering of

TABLE 9-2

Lines and groups of lines isolated by resonance detectors[18]

Element	Wavelength(s) of resonance radiation (nm)
Li	670·8
Na	589·0, 589·6
K	766·5, 769·9
Rb	780·0, 794·8
Cs	852·1, 894·4
Cu	324·7, 327·4
Ag	328·1, 338·3
Au	242·8, 267·6
Be	234·9
Mg	285·2
Ca	422·7
Sr	460·7
Ba	535·5
Zn	213·8
Cd	228·8
Hg	253·7
Al	308·2, 309·27, 309·28, 394·4, 396·2
Ga	287·4, 294·36, 294·42, 403·3, 417·2
In	303·9, 325·6, 325·9, 410·2, 451·1
Tl	276·8, 377·6

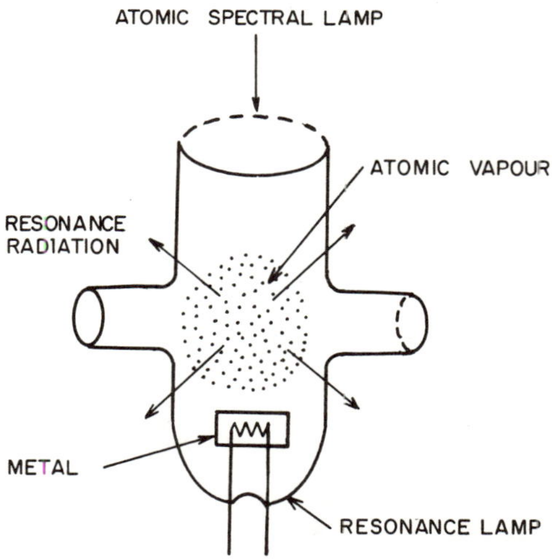

Fig. 9-7. Schematic diagram illustrating a resonance detector in which the metal atomic vapour is produced by electrical heating (from ref. 18).

the same type as in a hollow-cathode lamp (see Fig. 9-8). In practice, however, best results are obtained for most elements by using the sputtering method because the thermal types have a much longer warm-up time and require much more careful adjustment and regulation of the operating current. The current requirements of the sputtering types are similar to those of the hollow cathode lamp and the useful life is apparently as long, possibly in excess of 5000 hours.

Resonance detectors may be used for atomic absorption measurements with similar apparatus to that required when using conventional

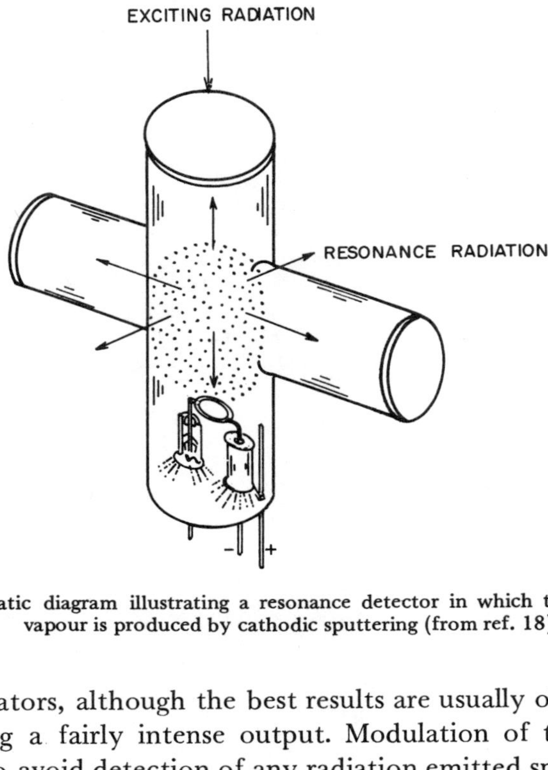

Fig. 9-8. Schematic diagram illustrating a resonance detector in which the metal atomic vapour is produced by cathodic sputtering (from ref. 18).

monochromators, although the best results are usually obtained with lamps having a fairly intense output. Modulation of the source is important to avoid detection of any radiation emitted spontaneously by the atoms in the resonance detector. In some respects these detectors would seem ideal for simple spectrometers, particularly of the multi-element type where several resonance detectors could view one absorption cell and automatic element selection would be easy (see Fig. 9-9). There are, however, several limitations, which will probably restrict their practical application. Perhaps the most important is the reduction in sensitivity compared with a conventional monochromator system. Optimum sensitivity will be obtained only with the elements of group II (a single absorption line) and the absorption sensitivity decreases steadily as the number of detected lines increases. Although the "transmission factor" is good by the standards of high-resolution instruments it may be fairly low compared with the monochromators used in ordinary spectrometers. Thus resonance detectors are mostly unsuitable for fluorescence measurements. The detectors also suffer problems normally

Fig. 9-9. Six-channel atomic absorption spectrophotometer incorporating resonance detectors as monochromators (from ref. 11).

associated only with spectral sources, i.e. limited life, varying performance, a warm-up time before each operation and possible electrical instability.

9.3.3 Selective Modulation

Selective modulation as a means of isolating spectral lines is closely linked to the principle of the resonance detector in that the formation of an atomic vapour of the element concerned is required.

However, instead of using the resonance radiation emitted by these atoms, selective modulation requires detection of the original light beam after it has passed through the absorbing cloud. During this passage only the resonance lines of the element in the cloud will be absorbed. Hence, if the cloud of absorbing atoms can be produced periodically it is possible to modulate the radiation at the resonance

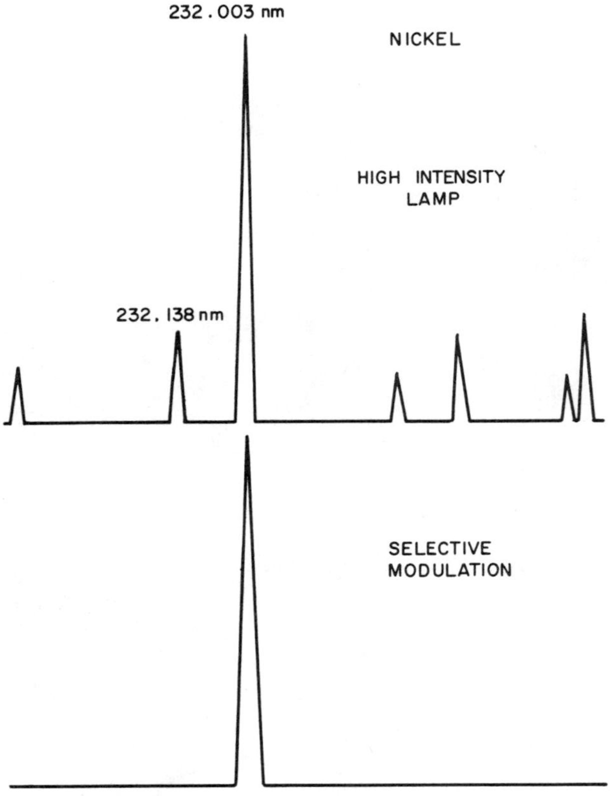

Fig. 9-10. Spectra showing isolation of the nickel resonance line at 232·003 nm by selective modulation. Lamp primary discharge: 39 mA. Lamp booster discharge: 400 mA. Modulator discharge: 20 mA.

lines without affecting that at any other wavelengths. A phase-sensitive detector modulated at the same frequency will, therefore, respond only to changes in intensity at the resonance lines. A demonstration of the effect of using selective modulation in this way to improve isolation of the nickel 232·0 nm resonance line is shown in Fig. 9-10. The application of this technique has been investigated by Walsh and his co-workers[18-20] and by Sebestyen.[21]

The cloud of absorbing atoms required for selective modulation is most conveniently generated by the sputtering technique as used in hollow-cathode lamps. Modulated resonance radiation is obtained by one of three methods:

(1) Using a pulsed hollow-cathode discharge as the light source but adjusting the pulse timing and duration so that the atoms produced also act as the selective modulation medium. Figure 9-11 illustrates the manner in which a modulated signal is

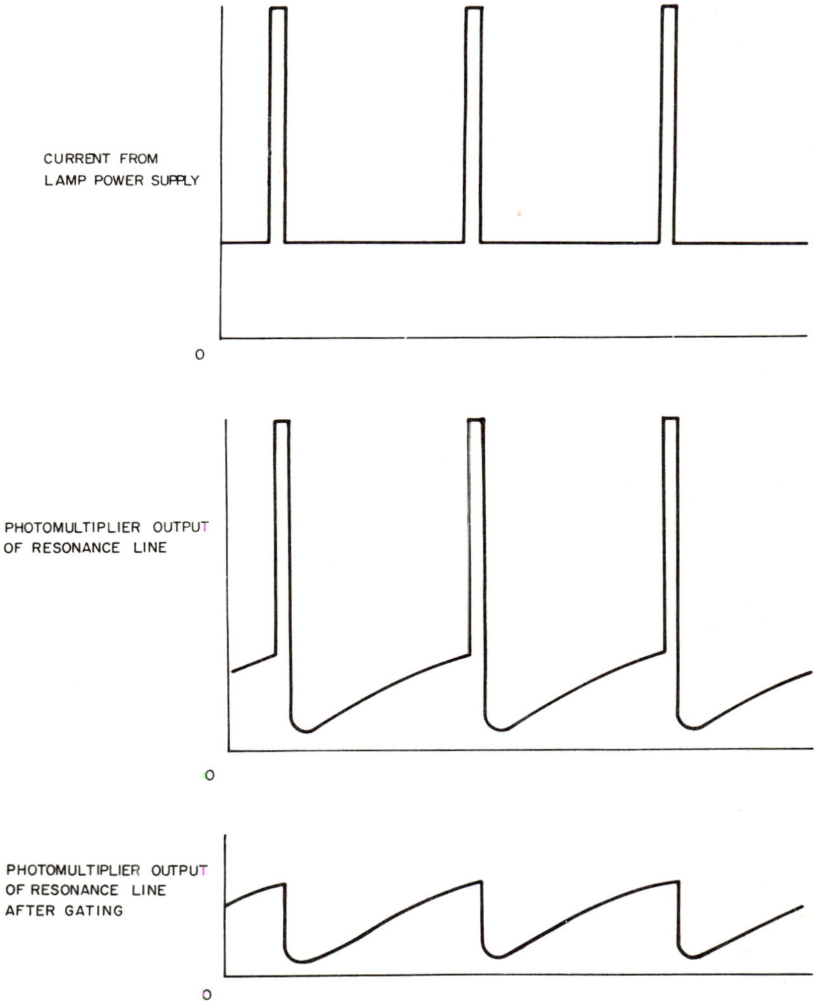

Fig. 9-11. Waveforms illustrating method of Lowe[20] for selective modulation of resonance lines by pulsed operation of a hollow-cathode lamp (from ref. 11).

obtained. The hollow cathode is operated at a steady direct current, and superimposed on this is a series of short high-current pulses. The increased sputtering due to each pulse causes a cloud of atoms of the cathode material to be formed inside and in front of the hollow cathode; by adjusting the pulse height and width it is possible to make the vapour cloud sufficiently dense to cause almost complete absorption of the resonance lines emitted by the lamp during its low dc operation. The absorption of the vapour cloud decreases between pulses and the output intensity of the resonance lines rises again to that of the dc level. The detection circuit is so

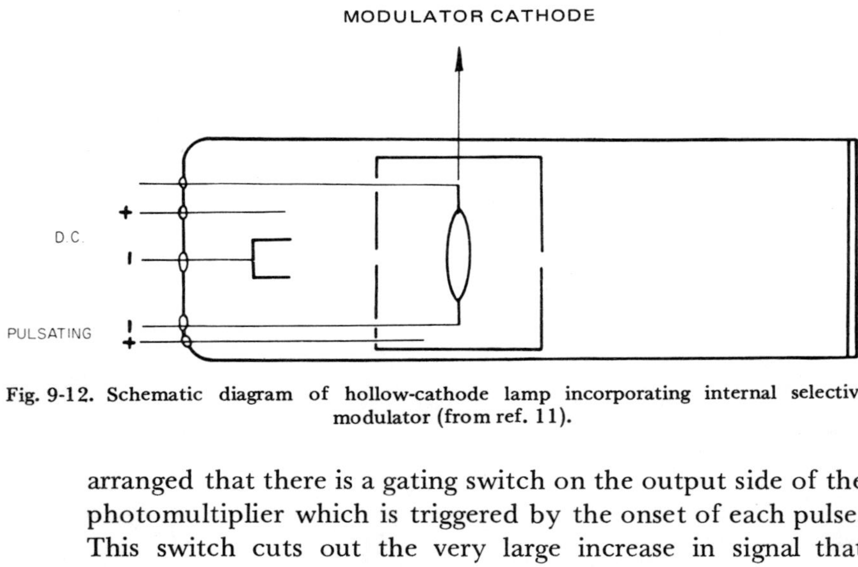

Fig. 9-12. Schematic diagram of hollow-cathode lamp incorporating internal selective modulator (from ref. 11).

arranged that there is a gating switch on the output side of the photomultiplier which is triggered by the onset of each pulse. This switch cuts out the very large increase in signal that would otherwise occur when the pulse is on. The net waveform obtained for the resonance wavelengths is thus ac in character.

(2) Using a conventional hollow-cathode discharge but with an auxiliary sputtering discharge mounted in front of the cathode to provide the cloud of cool atoms needed for selective modulation (see Fig. 9-12).

(3) Using a conventional hollow-cathode lamp with the selective modulator mounted in a separate cell in front of it (see Fig. 9-13). In almost all cases results are very poor unless a separate optical filter or a solar-blind photomultiplier is used to restrict detection of extraneous radiation and noise. The

sensitivity is generally lower than with a conventional monochromator and the range and linearity of calibration are also restricted. It may be possible, however, to obtain useful absorption results for many elements by combining selective modulation with a simple (wide bandpass) conventional monochromator.

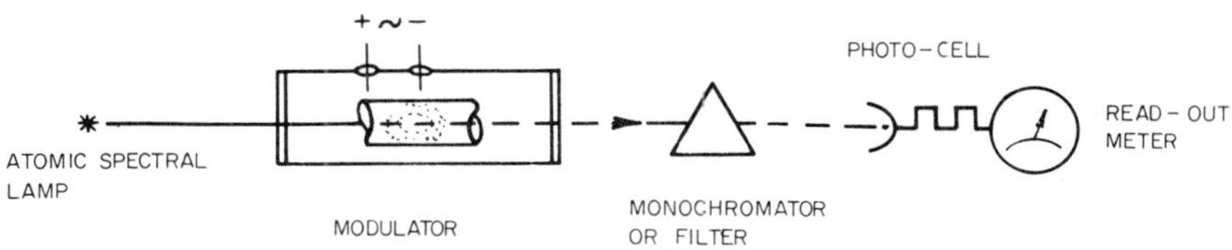

Fig. 9-13. Schematic diagram illustrating the apparatus for isolation of atomic resonance lines by selective modulation with a pulsating vapour produced by cathodic sputtering (from ref. 11).

REFERENCES

1. G. R. Harrison, R. C. Lord and J. R. Loofbourow, "Practical Spectroscopy", Prentice-Hall, Englewood Cliffs, N.J., 1948.
2. R. A. Sawyer, "Experimental Spectroscopy", Dover Publications, New York, 3rd ed., 1963.
3. R. S. Longhurst, "Geometrical and Physical Optics", Longmans, London, 2nd ed., 1967.
4. R. Mavrodineanu and R. C. Hughes, *Applied Optics,* 1968, **7,** 1281.
5. P. Buchhave and C. H. Church, ibid., 1968, **7,** 2200.
6. J. B. Dawson, D. J. Ellis and R. Milner, *Spectrochim. Acta,* 1968, **23B**, 695.
7. R. Roldan, *Rev. Sci. Instrum.,* 1969, **40,** 1388.
8. R. W. Strojek, G. A. Grover and T. Kuwana, *Anal. Chem.,* 1969, **41,** 481.
9. E. Cordos and H. V. Malmstadt, ibid., 1973, **45,** 425.
10. D. G. Mitchell and A. Johannson, *Spectrochim. Acta,* 1970, **25B,** 175.
11. A. Walsh, in "Atomic Absorption Spectroscopy", eds. R. M. Dagnall and G. F. Kirkbright, Butterworths, London, 1970.
12. A. C. G. Mitchell and M. W. Zemansky, "Resonance Radiation and Excited Atoms", University Press, Cambridge, 1934.
13. B. J. Russell and A. Walsh, *Spectrochim. Acta,* 1959, **15,** 883.
14. J. V. Sullivan and A. Walsh, ibid., 1966, **22,** 1843.
15. J. A. Bowman, *Anal. Chim. Acta,* 1967, **37,** 465.
16. B. S. Rawling and J. V. Sullivan, *Trans. Inst. Min. Met.,* 1967, **76,** C238.
17. P. L. Boar and J. V. Sullivan, *Fuel,* 1967, **46,** 230.
18. J. V. Sullivan and A. Walsh, *Applied Optics,* 1968, **7,** 1271.
19. J. A. Bowman, J. V. Sullivan and A. Walsh, *Spectrochim. Acta,* 1966, **22,** 205.
20. R. M. Lowe, ibid., 1969, **24B,** 191.
21. N. A. Sebestyen, ibid., 1970, **25B,** 261.

CHAPTER 10

Atomic Absorption and Fluorescence Instrumentation

The essential components of an atomic absorption or fluorescence spectrometer and their conventional arrangement were outlined in Chapter 1. In subsequent chapters we have described in detail atom cells and methods of sample introduction, light sources and monochromating devices. One major function of every spectrometer remains to be discussed here: the detection and processing of the analytical signal. Several other topics which have not been dealt with in detail elsewhere are also included in this chapter to complete the survey of existing practical instrumentation. Our principal concern here is with techniques used at present in commercial spectrometers, although one or two other recent developments are also described. No data are included on the specifications of currently available commercial spectrometers, principally because such data rapidly become obsolescent. For information on this topic the reader is referred to the extensive advertising literature available from all manufacturers, to the annual listing in *Annual Reports on Analytical Atomic Spectroscopy** and also the advertising pages of relevant journals (particularly *Analytical Chemistry*). A comprehensive survey of atomic absorption instrumentation, which is reasonably up to date at the time of writing, has also been published.[1]

10.1 RADIATION DETECTORS

The traditional method of recording and measuring a light signal uses the photographic technique. Although this method may sometimes prove useful for special projects involving AAS or AFS it is, in general, inconvenient for measuring fluorescence signals for analytical use and would be quite impracticable for routine atomic absorption analysis. Thus, electronic devices able to convert the light signal into an electrical signal for direct reading are now generally employed. Such devices must (a) be sufficiently compact for use in the optical systems of small, modern spectrometers, (b) give sufficient sensitivity to produce stable, noise-free signals at the low levels of radiation obtainable in many fluorescence measurements or when a monochromator is used with a narrow spectral bandpass, (c) have a reasonably uniform spectral response over the wavelength range of interest and (d) be simple to use and give reliable operation over long periods. Not all of these requirements are completely met by any one detector; simplicity and low cost must often be forefeited in return for increased sensitivity in most applications of interest here.

10.1.1 Photocells

Solid state photocells are the simplest of the photoelectric radiation detectors and are widely used for light-metering applications. They rely usually on the use of photo-conductive semi-conducting materials, but in fact several types were widely used long before the introduction of semi-conducting electronic components such as the transistor. The mechanism of their photoconductivity may be understood if it is recalled that in all semi-conductors the

* Published by Society for Analytical Chemistry, 9/10 Savile Row, London W.1.

electrons may be categorized according to the "band" of energy values in which they lie; when energy is supplied to the material by, for example, allowing light to fall upon it, some electrons in the valence band are excited into the conduction band. As the electrons lying in the conduction band are more mobile this process increases the conductivity of the material. Semi-conductors that show the photo-conductive effect include selenium, germanium, silicon, cadmium sulphide, cadmium selenide and lead sulphide. Their conductivity may be improved by introducing impurities of other elements; these can capture an electron in the conduction band and

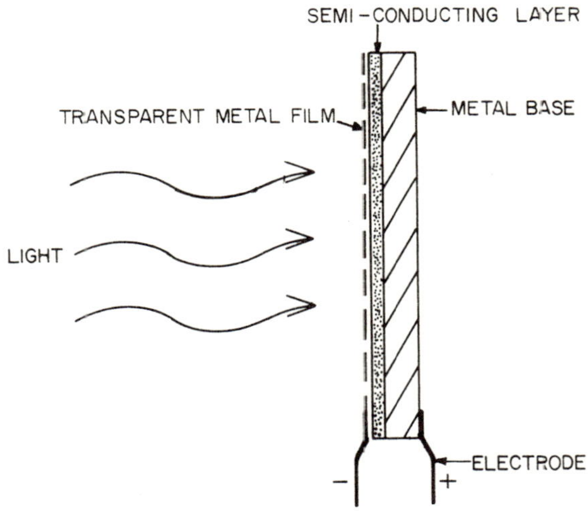

Fig.10-1. Photovoltaic light cell.

increase the time for which it remains there before returning to the valence band by recombination.

One means of utilizing the photoconductive effect to monitor light intensity is the *photovoltaic cell.* In this device a thin layer of semi-conducting material such as selenium or silicon is coated onto a metal base plate which is usually iron or aluminium (see Fig. 10-1). The semi-conductor is then overcoated with a very thin transparent layer of a precious metal such as gold, and electrodes are connected to each of the two metallic layers. A potential barrier arises at the junction between semi-conductor and metal in the absence of light and if the electrodes are connected to an external circuit no current will flow. When the junction is illuminated, however, the increase in conductivity causes the potential barrier to decrease as electrons are drawn across it into the transparent metal film, which attains a negative potential compared with the metal base. This voltage depends on the type of cell and the level of illumination but is commonly of the order of a few tens of millivolts and may reach a limiting value of *ca.* 0·5 V at very high illumination. The current obtained from the cell will depend on the external resistance of the circuit, as shown in Fig. 10-2, and is usually directly proportional to the incident radiation level over only a small range unless this resistance can be maintained at a very low value. The most commonly available photocells of this type show their peak

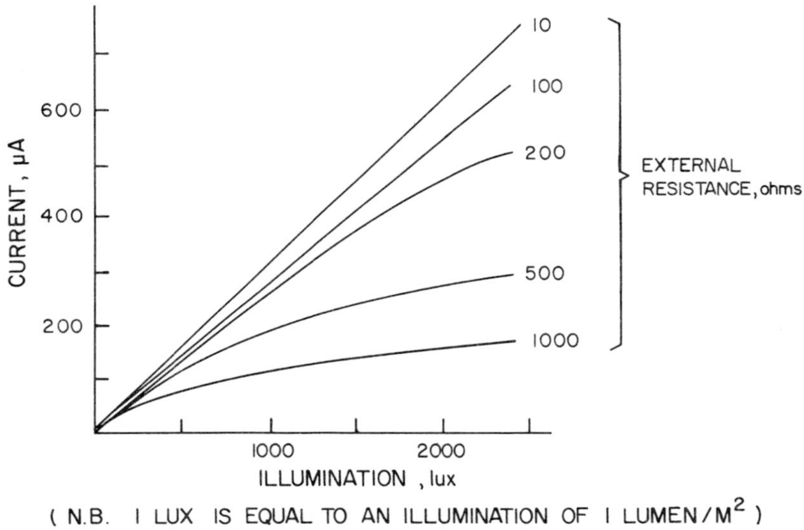

Fig. 10-2. Variation of current with illumination for a typical photovoltaic cell.

sensitivity in the visible region and usually high sensitivity can be obtained only by using a cell of large surface area (a typical cell may have a sensitivity of 0·5 mA/lumen). Thus the photovoltaic cell has seen little application to AAS or AFS in spite of its extreme simplicity and low cost.

The photoconductive effect may also be used in *photoresistance cells*. Semi-conductors such as cadmium sulphide or lead sulphide have a very high resistance in the dark (*ca.* 1–100 MΩ). As shown in Fig. 10-3, however, this decreases rapidly on exposure to light. Thus, light intensity may be measured simply by applying a fixed potential of a few volts across the cell and noting changes in the current. High sensitivity requires a relatively large sensitive area combined with closely spaced electrodes between which the resistance may be measured. This is commonly obtained by taking a thin, flat plate of the semi-conductor and depositing onto it the electrodes in the form of a thin metal film with the configuration shown in Fig. 10-4. The spectral response curves for cadmium sulphide and lead sulphide are shown in Fig. 10-5; lead sulphide is obviously only of use for infrared measurements while even cadmium sulphide shows its peak sensitivity at rather long wavelength for most AAS and AFS applications.

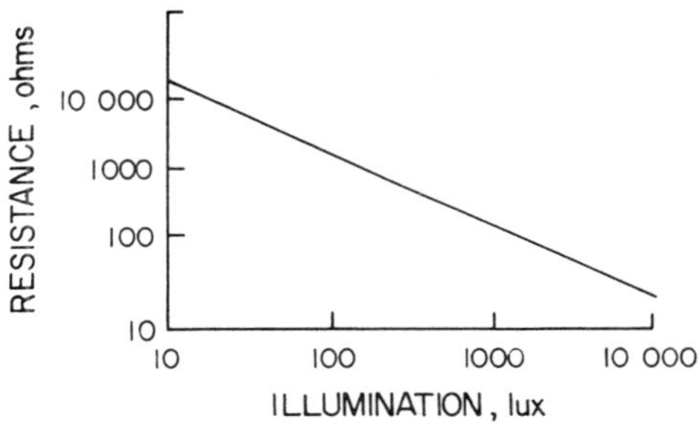

Fig. 10-3. Variation of cadmium sulphide resistance with illumination.

Fig. 10-4. Electrode configuration for cadmium sulphide photo-resistance cells.

Nevertheless, the sensitivity for a given size of cell is considerably better than with the photovoltaic type, a typical sensitivity being of the order of 1000 mA/lumen. Thus, cadmium sulphide cells may be of use for simple atomic absorption or fluorescence spectrometers. A possible disadvantage is the long response time of these cells which increases with decreasing illumination. Frequency response curves of a general purpose type are shown in Fig. 10-6, and it would, for example, be impossible to use a modulated system at the light levels available in usual spectrometric applications. Recent cells of this type

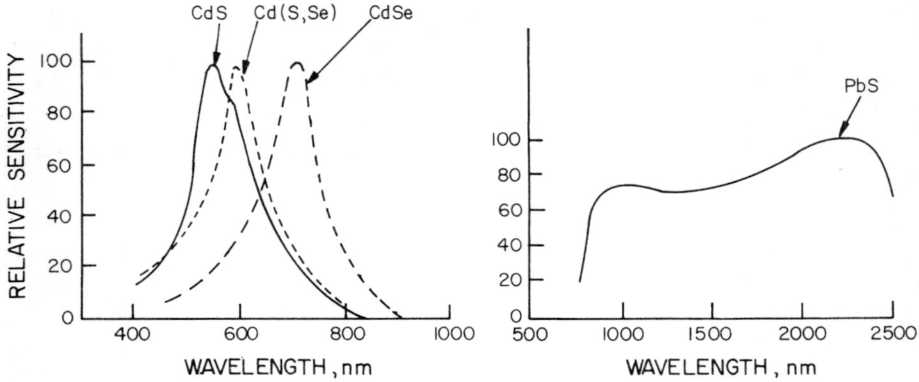

Fig. 10-5. Spectral response curves of cadmium sulphide/selenide and lead sulphide.

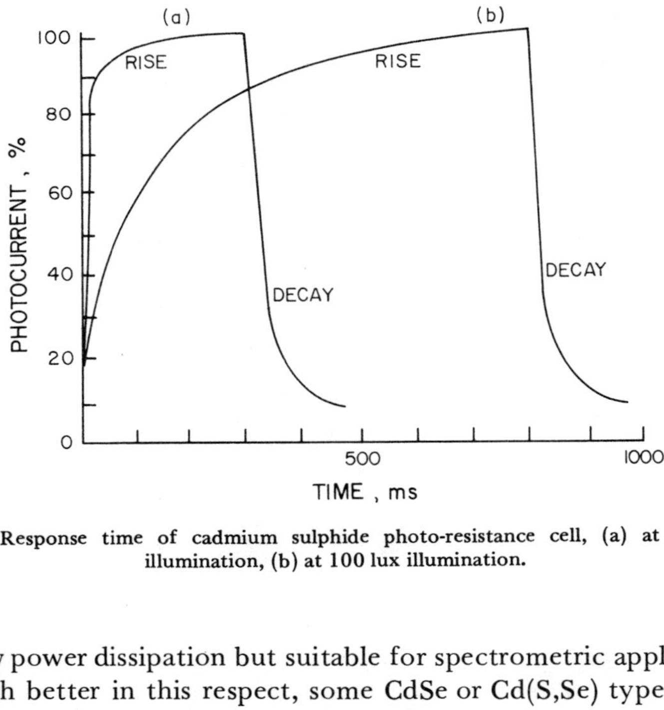

Fig. 10-6. Response time of cadmium sulphide photo-resistance cell, (a) at 1000 lux illumination, (b) at 100 lux illumination.

with low power dissipation but suitable for spectrometric applications are much better in this respect, some CdSe or Cd(S,Se) types having a response time of only a few milliseconds at 10 lux.

A modern development of the application of photoconductivity can be seen in *photo-diodes* and *photo-transistors*. In these potential barrier junctions are obtained using different types (p and n) of doped germanium or silicon crystals. A potential is applied to the cell and the current flowing depends on the level of illumination of the

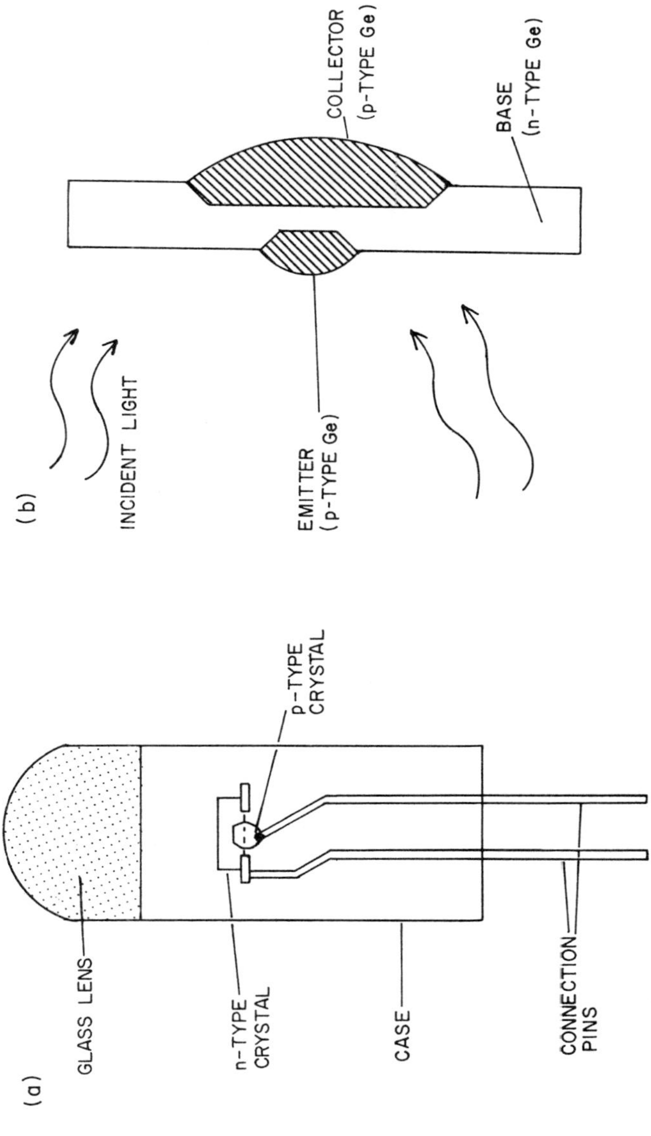

Fig. 10-7. Construction of (a) photo-diode and (b) photo-transistor.

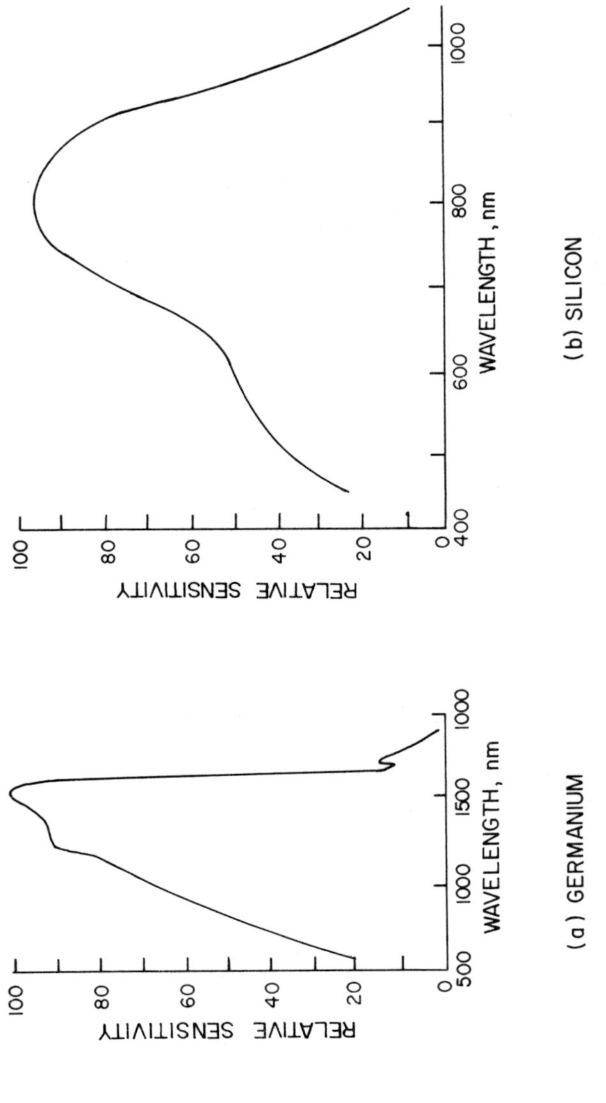

Fig. 10-8. Spectral response curves of photo-diodes and photo-transistors.

junction. Operation of the two types is similar except that the transistor incorporates an amplification factor of 400 or 500 and is therefore considerably more sensitive. Typical construction is shown in Fig. 10-7, which shows that with the transistor direct frontal illumination is not possible. The spectral response curves are shown in Fig. 10-8. It can be seen that germanium shows its peak sensitivity much further into the infrared than silicon and is thus probably less useful for application to atomic absorption or fluorescence measurements. Although the sensitivity is reasonably high (typically about 50 mA/lumen for the diode, 150 mA/lumen for the transistor) the extremely small size of these devices may limit the possibility of measurements at low light levels. Their small size may be turned to advantage in detectors used in simultaneous multi-element analysis in which a linear array of photodiodes replaces the photographic plate at the focal plane of a spectrograph. However, their frequency response is much better than that of cadmium sulphide cells (i.e. up to 50 kHz for diodes, 3 kHz for the transistors).

10.1.2 Photo-Emissive Cells

This type of detector depends upon the fact that when many metals are irradiated by photons of sufficient energy, electrons are emitted from the surface of the metal. The minimum energy required by an electron for escape is known as the work function, w, of the material, the energy of an electron emitted as the result of radiation of frequency ν being $(h\nu - w)$. Thus there is a threshold frequency, ν_0, for photoelectric emission given by $h\nu_0 = w$; radiation of lower frequency (longer wavelength) will not cause emission of electrons from that particular material. For many common metals the threshold lies well into the ultraviolet region, but for others, particularly the alkali metals, it occurs in the near ultraviolet and visible. Thus, these elements may be used to detect visible and ultraviolet radiation; it is necessary only to allow the light to fall onto a suitably shaped cathode made from an appropriate metal (or metals) and to collect the emitted electrons on an anode. Considerable effort has been applied to the development of photocathode materials and individual metals are now seldom used (except for certain specialized applications such as the detection of light in the far vacuum ultraviolet). The spectral response curves of several widely used materials are illustrated in Fig. 10-9. In many cases the short wavelength response is determined by the window material used for the photocell rather than by the cathode material itself.

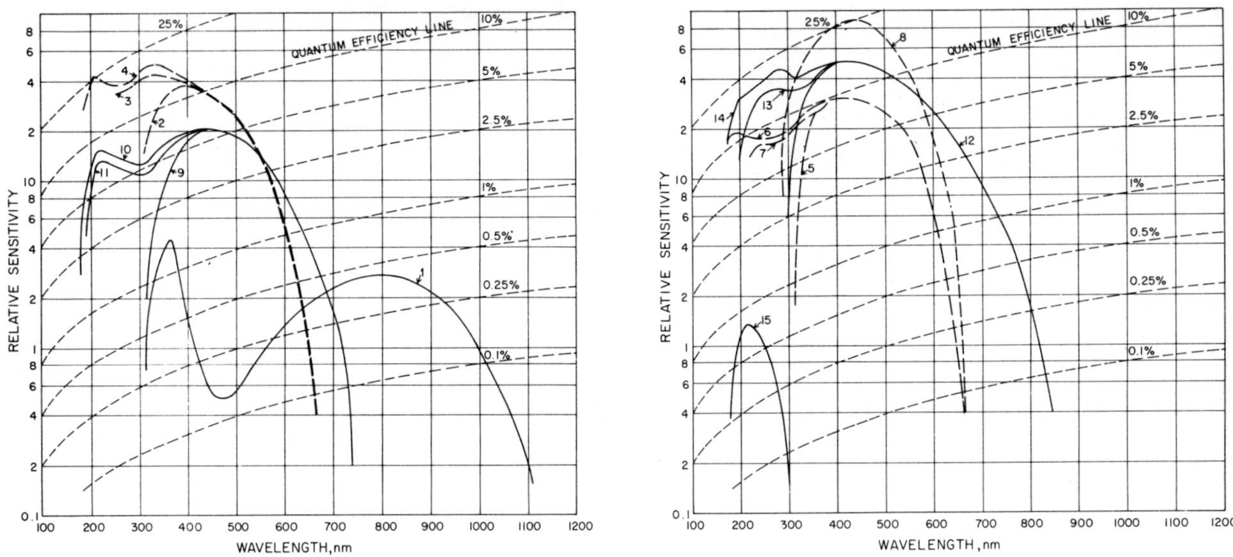

Curve	Cathode	Window*	Code	Curve	Cathode	Window*	Code
1	AgOCs	a	S1	9	AgBiOCs	a	S10
2	SbCs	a	S4	10	AgBiOCs	c	–
3	SbCs	b	S5	11	AgBiOCs	b	–
4	SbCs	c	S19	12	Multi-alkali (NaKCsSb)	a	S20
5	SbCs Semi-transparent cathode	a	S11	13	Multi-alkali (NaKCsSb)	b	–
6	SbCs Semi-transparent cathode	c	S13	14	Multi-alkali (NaKCsSb)	c	–
7	SbCs Semi-transparent cathode	b	S21	15	CsTe	c	–
8	Bialkali (KCsSb)	d	–				

*a = borosilicate glass, b = UV transmitting glass, c = silica, d = special glass

Fig. 10-9. Spectral response curves of photocells and photomultipliers (EMI, England).

Some combinations have been assigned "S" numbers which are given in the diagram and are used by many manufacturers, but this does not apply to the more recently developed cathode materials such as CsTe or "bialkali". Not all of the combinations shown in Fig. 10-9 are available in photocells, having been developed for the photomultiplier tubes described below but they are included here for convenience.

The photoelectrons emitted by the cathode are collected on an anode while applying a sufficiently strong potential between the two electrodes. To obtain efficient collection of electrons the electrodes are placed in an envelope which is evacuated or filled with a low pressure of an inert gas. At a particular wavelength the current flowing in the external circuit is directly proportional to the intensity of illumination. The voltage applied between the anode and cathode is needed to prevent emitted electrons accumulating in a cloud, or space charge, close to the cathode and repelling electrons back to it so that no photocurrent flows. In a vacuum cell the photocurrent increases with increasing applied potential until, at the saturation voltage, all emitted electrons are collected by the anode and the current reaches a maximum value for a particular light intensity (see Fig. 10-10(a)). In gas-filled cells (usually containing less than about 1 torr of a gas such as argon) the gas atoms are ionized by collision with electrons emitted by the cathode. The original and ionization electrons are then accelerated by the applied field and undergo further collisions with other gas atoms. Thus, for each electron leaving the cathode several electrons reach the anode and an amplification effect is obtained. As the kinetic energy of the electrons depends on the potential gradient, collisions producing ionization occur more frequently as the anode voltage is increased. Hence the amplification factor increases as the anode voltage is increased and the current-voltage curve is of the form shown in Fig. 10-10(b). The applied voltage must not be allowed to reach that at which a glow discharge would initiate within the cell and the amplification factor is usually limited to a maximum of between 5 to 10 times, depending on the type of cell.

In practice, the high-vacuum type of cell is often preferred to the gas-filled type in spite of the higher photocurrent obtained from the latter. This is partly because a vacuum cell operated above its saturation voltage is very stable whereas the sensitivity of gas-filled cells is considerably affected by small changes in gas pressure or by variations in the applied voltage. Furthermore, the amplification factor may not allow measurement of weaker light signals as it is accompanied by a large increase in the *dark current*. This is the current obtained from the cell when no light is falling on the cathode and is mainly due to thermionic emission of electrons. Thus it will increase with temperature and, as it too depends on the work function of the metal, is generally larger the longer the threshold wavelength of the cathode. A typical modern vacuum-type cell

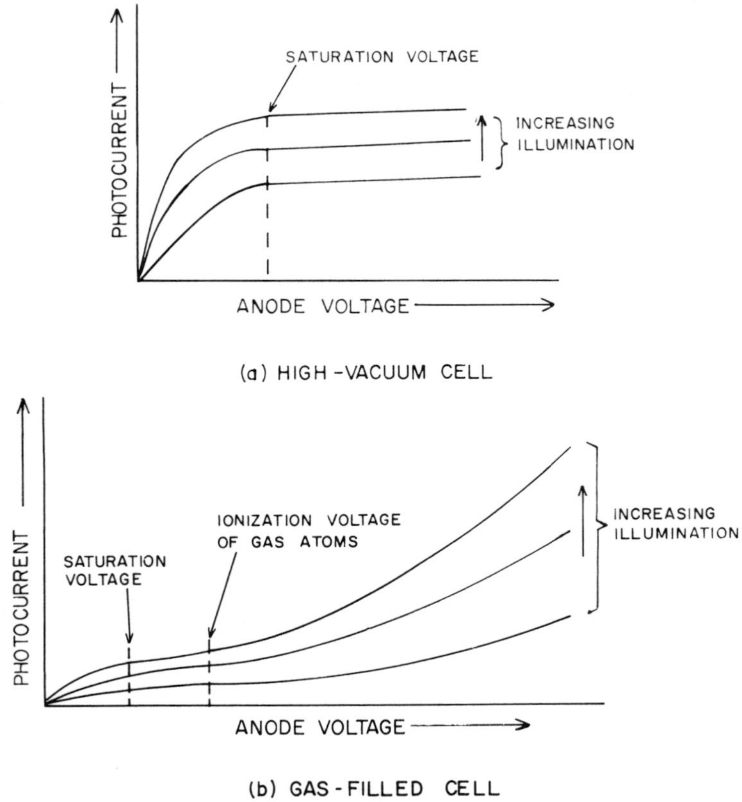

Fig. 10-10. Variation of photocurrent with applied potential for photo-emissive cells.

would have a dark current of the order of 1 nA (although values as low as 0·005 nA are available in cells designed for operation only at low light/current levels), whereas a value of 100 nA is more usual for the gas-filled type.

Although both the vacuum and gas-filled types are available with a variety of spectral response and sensitivity ranges, few of these are intended for spectrometric applications. Typical sensitivities for modern commercial types are 25-50 μA/lumen for the vacuum type and 150-300 μA/lumen for the gas filled cells. These values are usually obtained with anode voltages up to 250 V for the former but below about 100 V for the latter. The high-vacuum cells have an extremely rapid response, frequencies as high as 100 MHz being usable, but the cut-off frequency is considerably lower for the gas-filled type (usually 1–10 kHz). The main disadvantage of these cells is the relative small value for the maximum cathode current which may be drawn; for some of the more sensitive types it may be as low as 1 μA. Thus if very weak light signals are to be observed the cells must be used with stable, high-gain amplifiers. The much higher currents available from photomultiplier tubes have made this combination unattractive for most applications of interest in AAS or AFS.

10.1.3 Photomultiplier Tubes

These devices represent a logical development of the photo-emissive cell. Light is detected in the same manner but, as the name implies, the cell itself incorporates an amplification system so that a much higher output current is obtained. Amplification factors in excess of 10^6 are easily achieved and extensive development has ensured very small dark currents so that extremely low light levels may be detected and measured without difficulty. As a result the photomultiplier tube is used almost exclusively for the detection of radiation in the ultraviolet and visible regions in AAS and AFS.

The amplification of the photomultiplier is obtained by using the phenomenon of secondary electron emission. The principle of this can be understood by referring to the diagrammatic representation of a photomultiplier shown in Fig. 10-11. Photoelectrons emitted by the cathode are directed not to the anode but to the first of a number of target electrodes termed dynodes. A voltage is applied between the cathode and this dynode (just as between cathode and anode in the photo-emissive cell) to ensure efficient collection of electrons at the dynode. Each of the photoelectrons striking the dynode causes several secondary electrons to be released, producing an amplification effect. These electrons are directed towards a second dynode which is again held at a more positive potential than the preceding electrode so that the process is repeated. As several dynodes are used (commercial photomultipliers are available with as many as 14 dynodes although the use of 9 or 11 is more common) the number of electrons eventually reaching the anode as a result of each electron released by the cathode is very large. The actual gain observed will depend on the number of dynodes (or "stages"), the efficiency with which the design used is able to collect all of the electrons produced, and the voltages applied between the dynodes.

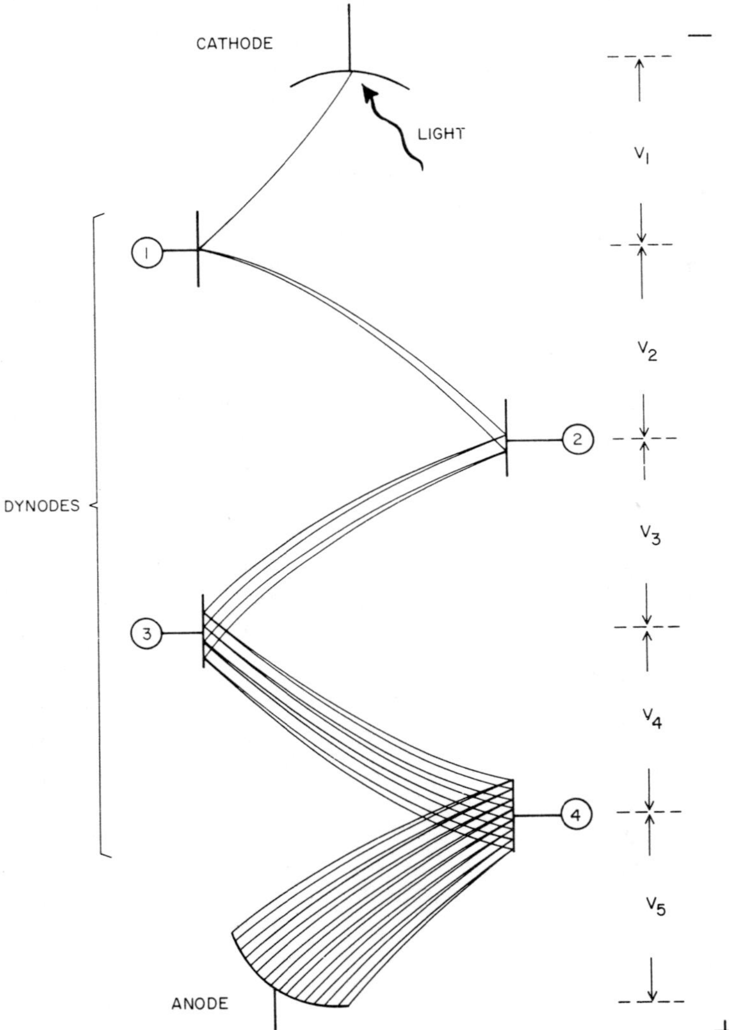

Fig. 10-11. Schematic representation of a photomultiplier tube.

The exact choice of dynode material is less critical than that of the photocathode material and SbCs is the most common, although BeCu and AgMg is also used.

The four most common designs of photomultiplier are shown in Fig. 10-12, each having certain advantages for particular applications.

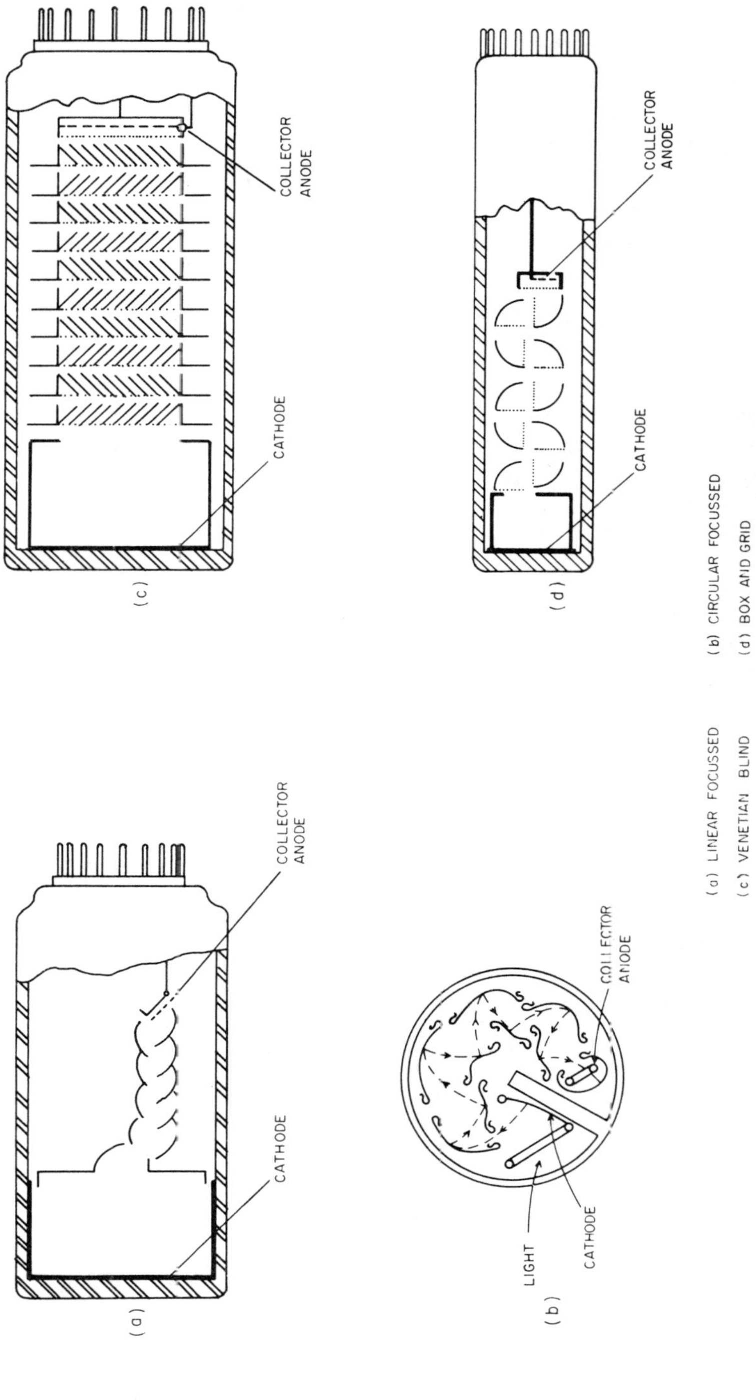

Fig. 10-12. Practical photomultiplier designs (EMI, England).

Thus, the venetian blind structure gives exceptionally low gain drift over long periods, the box and grid type is slightly less stable but more sensitive and more compact, the linear focused type has an extremely fast response and the circular focused type is compact and relatively inexpensive. In fact, the last-mentioned structure in side-window photomultipliers is probably the most widely used in both AAS and AFS for these reasons, even though it does not give the highest performance. The other three structures shown in Fig. 10-12 use semi-transparent cathodes, and in this case the spectral response curve of a particular cathode material may be modified (see Fig. 10-9). Not all end-window photomultipliers are necessarily of this type, however, and various modifications allowing reflecting cathodes have been used. The performance of a particular design (for a given cathode material and applied voltage) may also vary from batch to batch and between manufacturers.

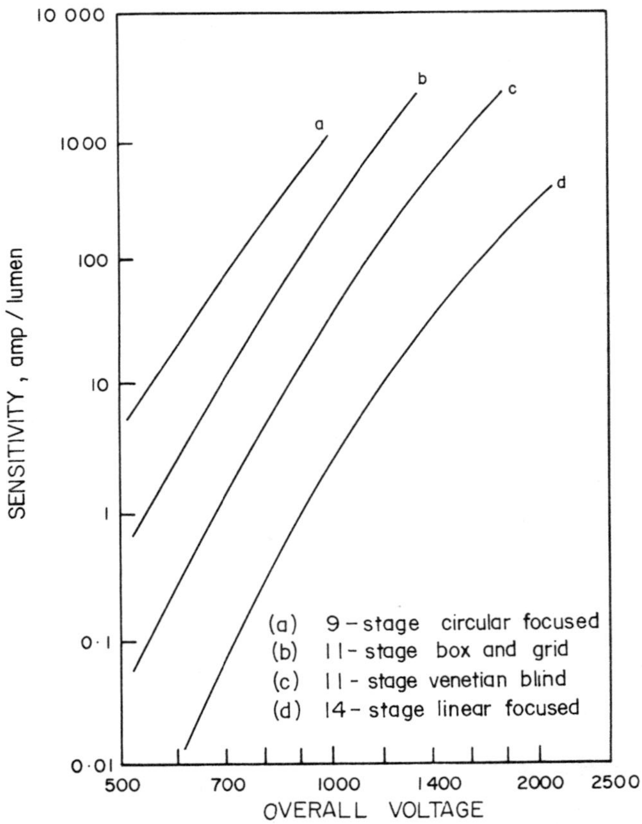

Fig. 10-13. Effect of applied voltage on photomultiplier sensitivity.

A particularly useful feature of the photomultiplier tube, apart from the large output currents at very low light levels, is the wide range over which the gain of a particular tube may be varied simply by altering the applied voltage. This effect arises because the higher the potential difference between two electrodes, the greater is the kinetic energy of the primary electron on impact and the larger the number of secondary electrons produced by each primary electron. The effect of applied voltage on sensitivity for several typical commercial photomultipliers is shown in Fig. 10-13. For most AAS and AFS applications a voltage source variable between about 500 and 1500 V is suitable, although potentials up to 2500 V may be

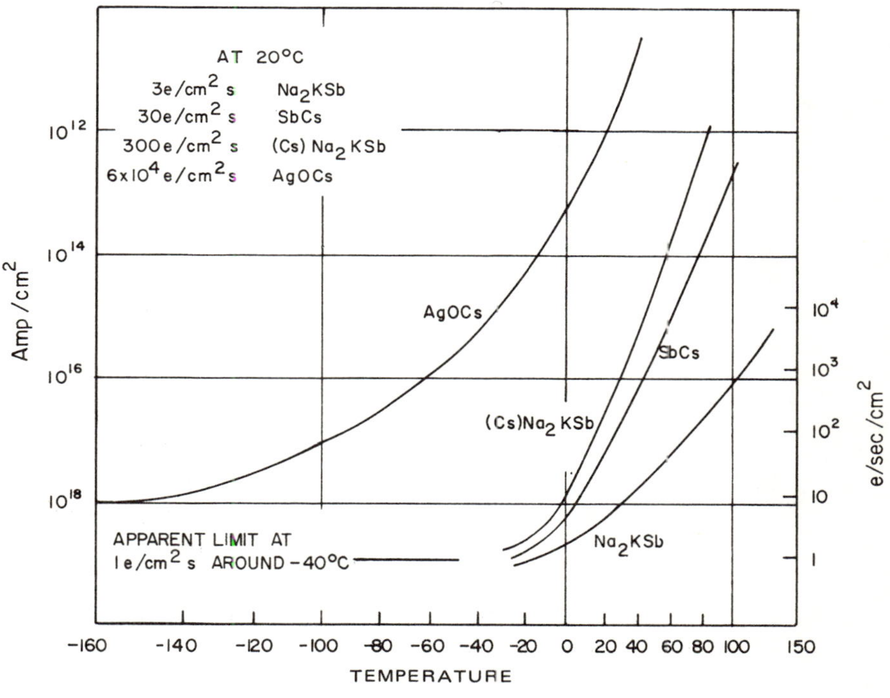

Fig. 10-14. Variation of photomultiplier cathode dark current with temperature for different cathode materials.

used with some photomultipliers. In general it is best to avoid very low or very high gain values as the relative effect of the dark current may be more serious under these conditions. The highest possible gain for a particular tube (obtained with the maximum safe voltage recommended by the manufacturer) is useful in most cases only if the dark current is reduced by cooling the photomultiplier to below room temperature. As can be seen from Fig. 10-14, maximum benefit is obtained at −40°C in most cases and at this temperature there is a very useful improvement compared with room temperature.

In view of the extremely rapid change of gain with applied voltage, it is essential to use a high-voltage power supply with high stability. As potentials up to 1500 V (at least) are desirable this unit will usually be more expensive than the photomultiplier itself and may contribute considerably to the overall cost of a spectrometer. Equally important is the method used to subdivide the applied voltage to supply the correct potentials between the dynodes and between the cathode or anode and their adjacent dynodes. In order to obtain optimum performance from the tube it should be protected from electrostatic fields by mounting it in a shield maintained at cathode potential; this requirement is obviously simplified by maintaining the cathode at earth potential and applying a high positive voltage at the anode. The output signal must then be measured via a coupling capacitor so that only pulsed or modulated light signals can be used with this system. Thus with steady light signals a high negative potential is applied to the cathode while the anode is earthed. In some spectrometers this condition is used while the tube is mounted in an earthed shield, in spite of the requirement mentioned above. In either case the applied voltage is divided using a chain of resistors (or dynode chain) as shown in Fig. 10-15. When the gain is to be varied by altering the applied voltage, it is important to maintain the cathode-to-first-dynode potential at a high value. This ensures optimum performance by using maximum gain for the first dynode, minimizes the effect of stray magnetic fields and produces the shortest response time for the tube. This is most easily achieved by replacing the first resistor with a Zener diode as shown in Fig. 10-15. Apart from this, a linear dynode chain is suitable for many applications. With pulsed operation, transient high currents may cause the build-up of an electron space charge and cause loss of linear response. This is avoided by applying a higher voltage to the last one or two dynodes. Some tubes are designed to avoid this problem at the anode and it is then necessary only to apply increased voltage at the penultimate dynode, as shown in Fig. 10-15(a).

As mentioned above, the overall sensitivity of the photomultiplier is extremely high. Maximum values of 100 to 200 A/lumen are common for the tubes used in AAS or AFS, but ratings up to 5000 A/lumen are available. It is important to remember, however, that the actual current that may be drawn from the anode is only a minute fraction of this value. An obvious limitation is the heat that can be dissipated by the components without damage. This typically limits the anode current to about 0·1 mA for compact tubes, and 0·5

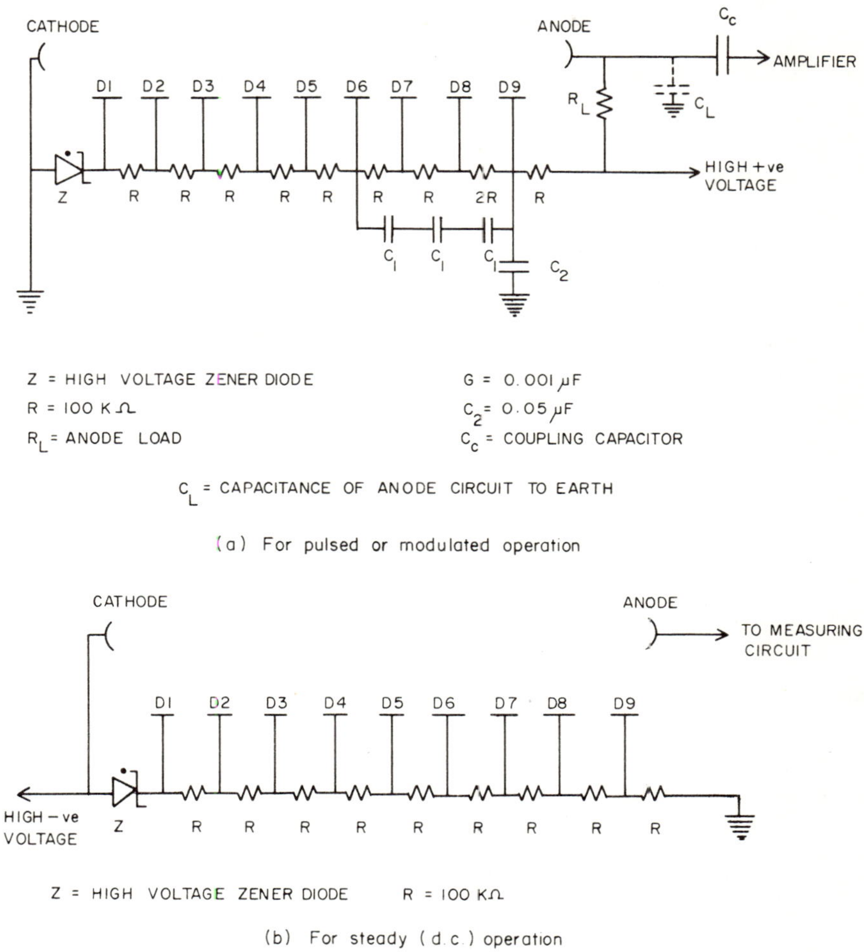

Fig. 10-15. Standard dynode chains for photomultiplier operation.

or 1 mA for the larger end-window types. In practice, the maximum current that can be drawn from the anode is limited to a much smaller value by a second consideration. If the anode current approaches the value of that flowing through the dynode chain, the voltage across the last resistor in the chain (and to a lesser extent the one before it) will vary with the anode current. This effect results in an increase in overall gain with increasing anode current so that the response of the photomultiplier becomes non-linear. To reduce the problem most manufacturers recommend that the anode current be kept below about 1/10th the value of the dynode chain current or even below 1/100th if ⩾1% linearity is required. The maximum value of the dynode current is itself limited, both by heat dissipation problems (especially as the resistors are usually mounted close to the photomultiplier) and by the maximum current available from high-voltage, stable power supply units. In practice, therefore, the maximum anode current is usually in the range 1–10 μA.

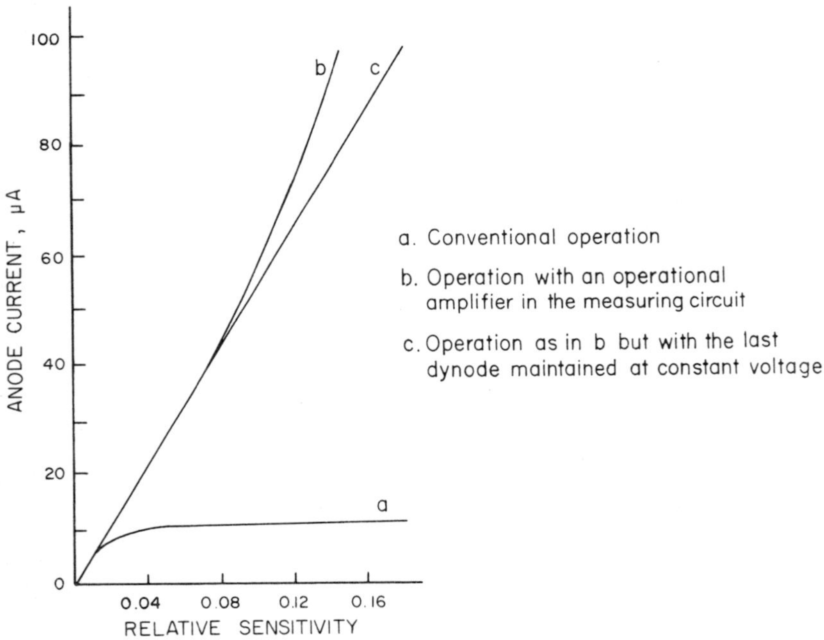

Fig. 10-16. Photomultiplier response curves at high anode currents.

These values may be increased by using higher value resistors for the last few stages of the dynode chain, the relative effect of change in anode current being reduced. Usually values of $2R$ or $3R$ are sufficient. The disadvantage of this method is that the overall voltage applied to the tube must be increased to obtain a given gain and care must be taken not to exceed the maximum rating of the tube. A more satisfactory solution is to stabilize the final dynode voltage. This is easily achieved by replacing the last resistor with a Zener diode or by using a voltage follower circuit. Use of a high anode

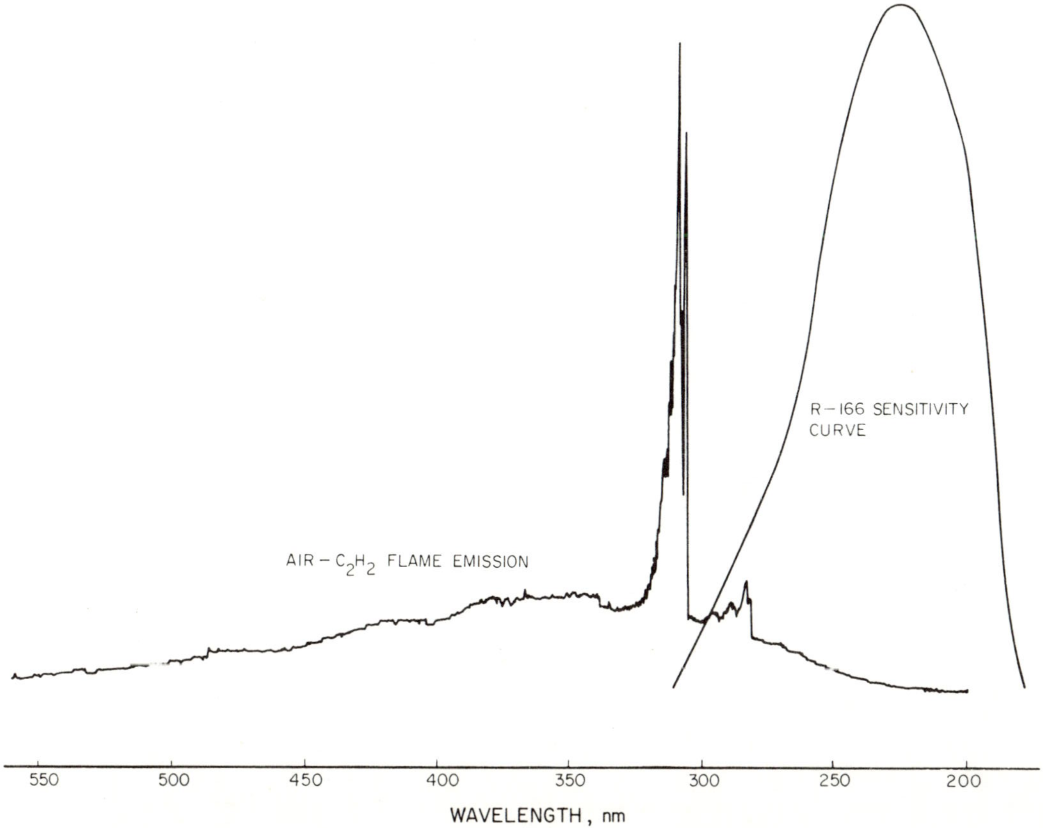

Fig. 10-17. Comparison of the background emission of an air-acetylene flame and the response of a solar-blind photomultiplier (HTV type R166).

current introduces an additional problem when the signal is measured. The voltage generated across the measuring resistor must be kept small relative to the applied voltage otherwise the effective voltage (and hence the gain) will change. Thus, if a high measuring resistor is used to allow small anode currents to be detected, the response will be non-linear as the signal increases. The problem can be avoided when operating the photomultiplier with the anode at earth potential by using an operational amplifier in the measuring circuit. This will maintain the anode at a constant "virtual earth" regardless of the photocurrent or the value of the measuring resistor. The effect of using a constant last dynode voltage together with this precaution is illustrated in Fig. 10-16.

The combination of high sensitivity and low permissible anode current obviously has the consequence that the photomultiplier must never be exposed directly to daylight when high voltage is being applied. Thus, care must be taken to avoid accidental exposure. For example if a filter is mounted directly in front of the photomultiplier housing it would be easy to remove the filter to exchange it for another without remembering to switch off the applied voltage. Although it is perfectly safe to expose a tube to daylight when the voltage is off, the most critical applications require that it be stored in the dark for several days before re-use. This arises because energy is trapped in the cathode and will, when the voltage is re-applied, appear as an excessive dark current unless time is allowed for it to dissipate. The problem of exposure to daylight is much less serious with tubes using photocathodes such as CsTe which has no response to visible light (see Fig. 10-9). These "solar-blind" photomultipliers are particularly useful for AAS or AFS applications where it is often necessary to measure only uv signals. In many circumstances (e.g. when using filters instead of a monochromator) these devices can give significantly reduced signal noise with flame atom cells. This is illustrated in Fig. 10-17, which compares the response characteristics of a solar-blind photomultiplier (HTV Type R166) with the background emission of an air-acetylene flame.

10.2 SIGNAL PROCESSING

In principle, the output current of the radiation detector could be measured directly using a micro-ammeter or recorded by applying the current across a load resistor and observing the voltage generated. In practice, however, with almost all AAS and AFS spectrometers, some form of electronic processing of the signal is undertaken to obtain that which is finally observed by the operator. The degree of complexity involved in this operation can vary greatly, but in all cases it is intended to increase the versatility and convenience of operation of the spectrometer. We shall describe the most widely used of the available techniques, emphasizing those aspects of interest to a potential user.

10.2.1 Analogue Techniques

Until recently almost all atomic absorption and fluorescence applications involved amplification of the relatively weak signals obtained from the photocell or photomultiplier with a relatively simple analogue circuit. Apart from providing a sufficiently large

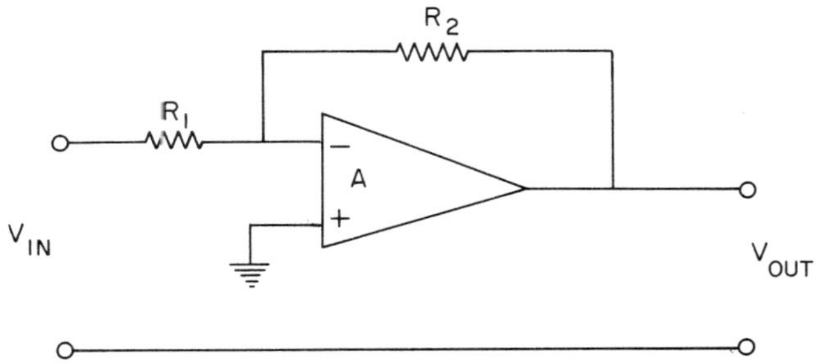

$$\text{EFFECTIVE GAIN} = \frac{V_{OUT}}{V_{IN}} = -\frac{R_2}{R_1}$$

Fig. 10-18. Feedback amplifier.

current or voltage to drive a meter or chart recorder, such amplifiers facilitate operations such as backing-off unwanted signals, scale expansion of small absorption signals and, in some cases, direct conversion of logarithmic absorption signals to a linear reading. The principal requirement for such amplifiers is that they should contribute the minimum additional noise to the signal and that they should be stable; i.e. their gain should remain constant. This is usually achieved by introducing the principle of feedback so that the gain of the amplifier no longer depends on the stability of the amplifier itself (i.e. of active components such as transistors or valves) but instead on the ratio (input : output) of two passive components such as resistors (see Fig. 10-18). Even so, the signal from a photocell illuminated by a steady source of radiation will be dc and the construction of drift-free dc amplifiers is notoriously difficult. For this reason a more convenient choice is the use of a *chopper amplifier* (see Fig. 10-19). The dc input is then filtered to remove any unwanted ac components and passed through an electronic "chopper" device. This generates an ac signal (usually a square-wave form) which may be amplified using an ac amplifier tuned to the frequency of the chopper. As well as removing all dc drift components, the narrow bandwidth of such an amplifier (compared to the alternative dc amplifier) reduces the noise level of the final signal. In fact, all methods used to reduce the noise levels of measuring systems are based on narrowing the bandwidth and this important point will be considered again below. The amplified ac signal is passed through a coupling capacitor to remove any dc components and then to a second filter unit. The dc signal obtained from this may be used to drive a conventional meter or potentiometric recorder.

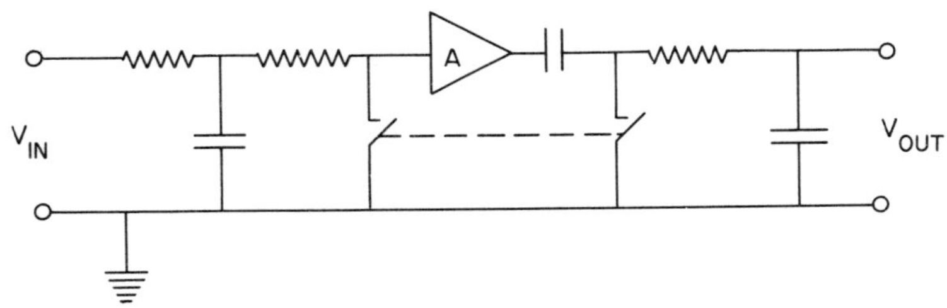

Fig. 10-19. Chopper amplifier.

In practice, simple chopper amplifiers are rarely used in absorption work and only occasionally for fluorescence measurements. This is because for optical applications considerable additional benefits may be obtained with little additional complexity by using the principle of *synchronous detection*. This is a logical extension of the chopper amplifier in which the output from the spectral light source itself is modulated so that the photodetector directly supplies an ac output. The use of modulated light sources with synchronous amplifiers is discussed further in Section 10.3.1, but briefly the technique has two main advantages for AAS and AFS. In the first place the tuned amplifier will respond only to the ac signal generated at the photodetector by the (modulated) light source so that any (dc) background emission and noise from the absorption or fluorescence cell should remain undetected. Secondly, the device used to modulate the light source can also generate a reference signal which is used to lock the amplifier exactly to the modulation frequency. It is thus possible to use an amplifier of exceptionally narrow bandwidth and to measure small signals in the presence of high noise levels.

With either type of amplifier there are several features that may be usefully incorporated into any spectrometer. For example, the gain (or perhaps attenuation) of the amplifier should be variable to allow variable scale expansion of small absorption signals. In certain cases, notably where the use of cool absorption cells results in very low signal noise levels or where very long measurement times are permissible, factors up to 100x may be usable. It is therefore useful if the controls of the instrument allow the degree of scale expansion to be easily ascertained or, at least, to be speedily reset to a predetermined value. An atomic absorption spectrometer having this facility of high, variable gain with calibrated controls is useful for

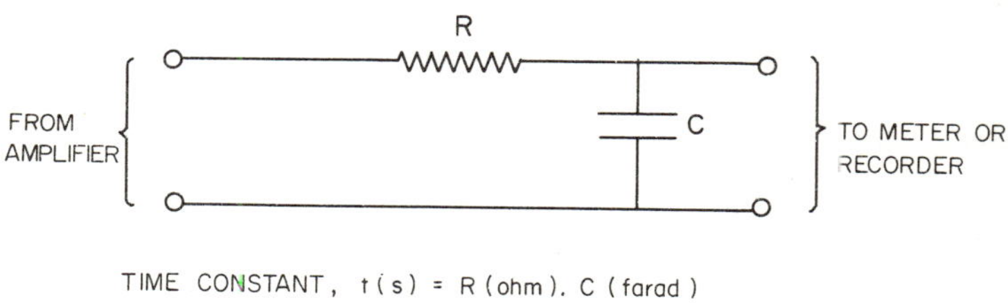

Fig. 10-20. Signal damping with a low pass RC circuit.

measurement of small atomic fluorescence signals. In order to make full use of such facilities for either AAS or AFS the noise reduction techniques described in Section 10.3 are particularly useful. Although the more complex techniques are rarely incorporated in commercial atomic absorption spectrometers, some form of variable bandwidth is desirable for most applications. It is then possible, with a single instrument, to make full use of the high gain amplifier or to measure less demanding signals rapidly. This facility is often referred to as *signal damping*. It may involve adjustable characteristics for the ac amplifier, including incorporation of filter circuits in combination with the amplifier, but the most common approach is to pass the final dc signal through a low pass RC circuit (see Fig. 10-20) before feeding it to the meter or recorder. The time constant of the measuring device is then easily varied by changing the resistance or capacitance of the circuit. When this facility is not incorporated in the spectrometer the same effect is achieved by connecting an appropriate capacitor across the meter or recorder input terminals.

Despite the trend towards digital presentation of measurements (see Section 10.2.3) some commercial spectrometers with analogue circuitry still rely on the simplest read-out device, a high-quality micro-ammeter. This should have a stable movement so that the pointer will not waver unnecessarily and a large, clear scale with a mirror to facilitate accurate readings. The most useful meters of this type are calibrated with both a linear 0 to 100 scale and a logarithmic absorbance scale; unless the former is present it will be inconvenient to use the instrument for emission or fluorescence work. If the meter reads absorbance values in linear fashion across the scale it should be possible to switch out the logarithmic section of the amplifier circuit so that linear readings of emission or fluorescence signals are possible. A suitable output from the meter circuit to permit use of a potentiometric chart recorder is also useful; this output may be taken either directly as a current (to be applied to a load resistor at the recorder input) or after conversion to a convenient voltage signal (e.g. 10 mV for full-scale meter deflection). Recorder output sockets which in operation disconnect the meter are best avoided as they offer no particular advantage and sometimes prove inconvenient.

10.2.2 Photon Counting

In the preceding section we have described the standard spectrometric practice in which a photocell is used as a transducer to convert light into a steady electrical signal, be it ac or dc, which may be processed by analogue techniques. The reader will recall, however, that emission of light from any source is a series of discrete events, energy being released in small packets known as quanta or photons (see Chapter 2). The energy of the photon determines the frequency of the emitted radiation, but the intensity depends on the number of photons reaching the observer. In the traditional technique these photons are integrated into a continuous signal, but it is also possible to use the rapid response of a transducer such as the photomultiplier tube in conjunction with modern electronic instrumentation to actually count the number of photons reaching the detector. Measurement of light intensity then becomes analogous to measurement of radioactivity using, for example, a Geiger counter. An obvious advantage of such a technique for some applications is that the signal may be directly processed by digital techniques without the need for any conversion system (such as a digital voltmeter). In fact, the inherent advantages of counting techniques make "photon counting" an attractive alternative to analogue methods even for applications where the signal is not finally required in digital form. These advantages include sensitivity to low light levels, improvement of signal-to-noise ratio, better precision of measurement (in a given time interval) and the ease of integrating signals accurately over long time intervals.

The photon counting technique was originally developed to improve measurement of low light levels[2-7] and this undoubtedly remains its most important application. Apart from the need to justify the increased cost of the instrumentation (see below) at light levels where conventional techniques give reasonable results, this situation arises because photon counting is, at present, unsuitable for measuring high light levels. The point at which it is necessary to revert to conventional current measurements is governed by the frequency response of the system. The count rate obviously increases with increasing light intensity and an error is introduced into the readings unless the system continues to resolve individual pulses as they arrive. As the emission rate of photons from the light source is random, the actual pulse rate may be greater or smaller than the average rate. Thus, the frequency response of the system must in practice be considerably faster than would be suggested by the maximum average counting rate required. In fact, it is found experimentally[7] that the photon counting system should have a frequency response at least 25 times greater than the maximum average counting rate to be measured in order to ensure less than 1% error at this rate. At present the frequency response is probably limited more by the electronic instrumentation than by the photomultiplier itself, which usually has a frequency response in the range of fifty to several hundred MHz (although it may be considerably slower for tubes with very large cathodes). With a typical tube and voltage used in AAS or AFS instrumentation a conventional current of about 0·01 μA would be equivalent to a count rate of 100 kHz and this is probably about the limit with instrumentation in present use. It may be possible, however, with more complex electronics to use a counting rate of up to 10 MHz, which with the same photomultiplier would be equivalent to a current of about 1 μA. This would then be typical of the light levels encountered in commercial atomic absorption spectrometers. Thus, at present, photon counting is of greatest interest for measurement of weak fluorescence signals, but it may soon be possible to apply its other advantages to high-precision atomic absorption measurements.

The apparatus required for a photon counting system is shown diagrammatically in Fig. 10-21. Most of the units are self-explanatory and would either be purchased individually or as a complete photon counting system. Thus, just as with the conventional use of the photomultiplier, it is useful to amplify the output before its measurement. Particular care must be taken in connecting the amplifier to the tube to avoid any loss of frequency response and it

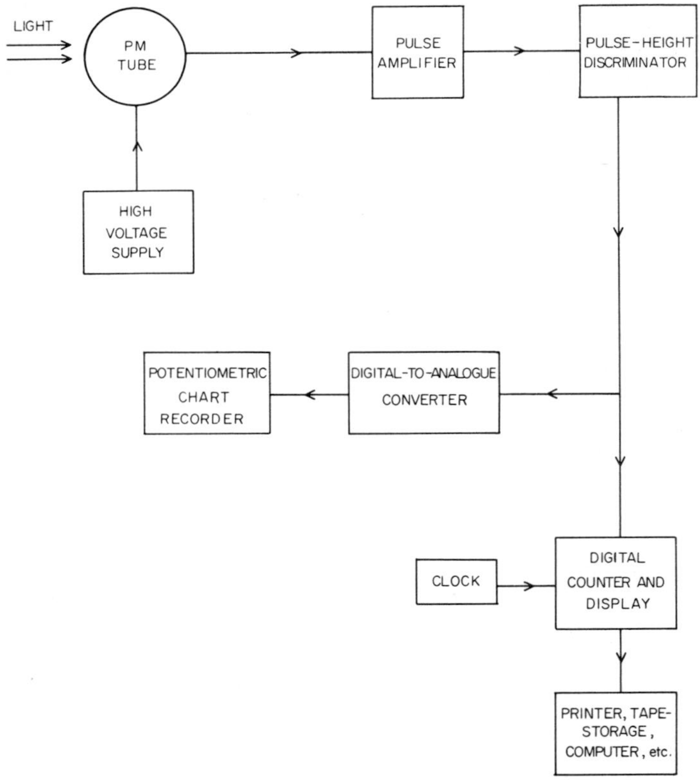

Fig. 10-21. Block diagram of a photon counting apparatus.

is useful if the system is able to reject any dc drift or low-frequency noise. The digital counter and printout is also self-explanatory and obviously requires a clock to control accurate counting. Similarly, a conventional chart recording is useful for applications such as scanning spectra and is easily obtained by directing the pulses to a digital-to-analogue converter. The one unit whose requirement is not obvious is the pulse-height discriminator; for low light-level measurements this is actually responsible to a considerable degree for the advantages of photon counting systems.

The pulse-height discriminator is an electronic instrument which is used to process the pulses received from the amplifier so that only those pulses lying between certain preselected values of the height (i.e. voltage) are sent to the counter and included in the measured value of the signal. Its use allows a considerable improvement in the signal-to-noise ratio obtainable from a photomultiplier tube operating at low light levels. This is possible because at these low light levels the significant noise arises not from the signal itself but from fluctuations in the photomultiplier dark current. The number

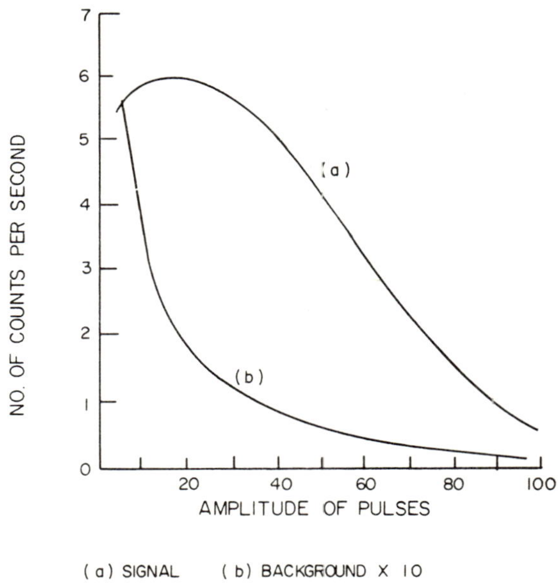

Fig. 10-22. Pulse distribution for a typical photomultiplier counting photons and background.

of counts obtained at discrete pulse heights (i.e. within a channel set to respond only to a small voltage interval) is plotted in Fig. 10-22 against the pulse height for a typical photomultiplier tube operated with (a) a light signal applied, and (b) in the dark. It can be seen that compared with pulses arising from photons (i.e. from the signal) the background count has a disproportionately large number of very small and very large pulses. Thus, it is possible to achieve a considerable improvement in signal-to-background ratio (which is equivalent to signal-to-noise ratio for very small signals) by counting only those pulses of intermediate height. This technique supplements the use of a cooled photomultiplier, as not all dark-current pulses are of thermal origin. In fact, photon counting is very efficient in this respect and when this technique is used little further improvement in signal-to-noise ratio may result when a cooled tube is employed.[8] Also, in practice, the largest reduction of background current is obtained by excluding the smallest pulses and successful results are often achieved using a discriminator in which only the smallest pulse height to be counted can be pre-selected. It should be noted that the pulse-height discriminator will also exclude some signal pulses so that the actual counting rate is reduced. The cut-off voltage of the

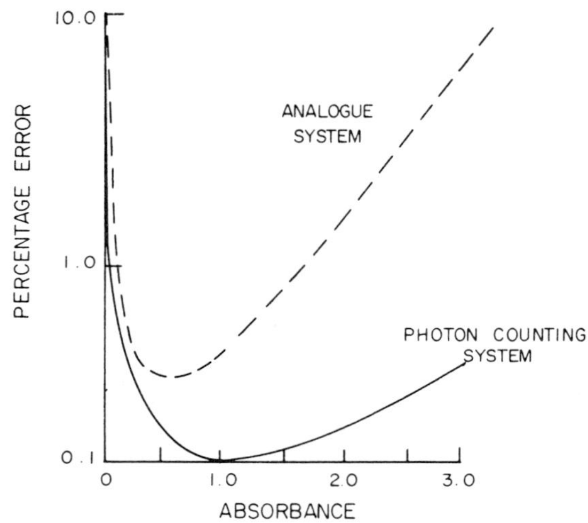

Fig. 10-23. Precision of absorbance measurements obtained by analogue and photon counting systems.[7]

discriminator must therefore be set in each individual case to give optimum results.

Provided it could be used at convenient light levels, photon counting would also result in a considerable improvement in the precision of atomic absorption measurements. Absorption measurements are readily made by dividing the counts (for a given time) of a signal and a reference channel (subtracting any background values if necessary) the absorbance being log (N_{ref}/N_{signal}). In a routine system used for many measurements these calculations could conveniently be carried out with a small computer. As with an analogue system used for measuring absorbance values, the percentage error under given conditions will vary with the absorbance value. The overall variation is, however, much more favourable with a counting system. This is illustrated in Fig. 10-23 which shows the calculated errors for absorbance values from 0 to 3·0 units measured with an analogue system, in which it is assumed a linear scale can be read with an accuracy of 0·1%, and with a photon counting system in which a reference beam count of 2×10^6 pulses is assumed. It can be seen that not only is the error of the counting system much smaller but it has a much broader minimum. In the important range from 0·25 to 2·5 absorbance units, the analogue error varies from about 0·3 to 4·6% whereas that of the counting system is less than 0·2%. The error of the counting system is also easily made smaller still by using higher counts. The maximum number that can be made in a convenient time is limited by the frequency response of the system as in most absorption measurements there is adequate light intensity (i.e. photons/second) available. For example, if the system can handle a counting rate of 200 kHz the errors mentioned above would apply to a ten-second count. Improving the frequency response to allow a 10 MHz rate would allow a ten times greater count (of 2×10^7 pulses) in only two seconds, or a fifty times greater count over the full ten seconds. Even ten seconds is relatively rapid by the usual standards of atomic absorption measurements, so that there remains scope for still further improvement in precision.

10.2.3 Digital Readout Systems

A completely digital system for AAS or AFS would require the use of the photon counting technique described above. This technique has, in fact, seen relatively little application to AFS and, as already discussed, is at present not generally suitable for the high light levels commonly used for AAS. Nevertheless, some of the advantages of digital techniques may be obtained without resorting to photon counting, and suitable systems have already been incorporated in several commercial spectrometers. In such systems the output of the photomultiplier is measured conventionally as a continuous current but is passed into an analogue-to-digital converter to obtain a digital signal. The advantage that is taken of having the signal in digital form may then vary greatly, depending on the application and on the cost and complexity of the spectrometer.

In the simplest case, digital conversion involves no more than simply replacing the conventional panel meter used to read the absorption or fluorescence signal with a small digital voltmeter. The advantages of such a conversion for AAS or AFS are debatable. A steady voltage is more easily read from a digital display than from a meter with a scale and pointer and is clearly visible at a considerable distance from the instrument. On the other hand, users of AAS or AFS seldom need to take their readings from the far side of the laboratory and, unfortunately, are often faced with rapidly fluctuating signals which can render difficult the reading (i.e. averaging) of an ever-changing digital display. Furthermore, many operations in both AAS and AFS require "peaking" of a signal, which is undoubtedly more easily achieved with an ordinary meter. In view of these considerations, the provision of a digital readout for the spectrometer is really justified only if it forms part of a more comprehensive digital system. This system may be present both before and after the digital display itself.

There are several relatively simple operations that may usefully be performed on the analytical signal before it is displayed for the operator. For example, in atomic absorption work it is possible to convert the decrease in transmittance of the absorption cell so that it may be displayed directly as an absorbance value. Similarly, in either absorption or fluorescence work, a facility that permits the correction of non-linear calibrations to provide a direct readout of concentration is often useful. When very small absorption or fluorescence signals must be measured, significant improvements in detection limits can be obtained by integrating the signal for a fixed period rather than by using "instantaneous" readings. These and similar operations can be performed using conventional analogue techniques but the circuitry is often complex and sometimes unreliable. Thus, most of these facilities would not be incorporated in an ordinary spectrometer. Once the signal has been converted to digital form, however, a wide range of quite complex operations may be performed economically by continuing to use digital techniques. In many cases, it is possible to purchase from electronic component manufacturers complete circuit modules performing a particular function and to incorporate them into the readout circuitry of the spectrometer. Commercial atomic absorption spectrometers are now available in which these techniques allow provision of features such as switchable transmittance/absorbance readout, variable period integration of signals (e.g. from 1 to 100 seconds), or direct concentration readout at reasonable cost. The adaptability of digital circuitry may sometimes make it useful even when a conventional meter readout is desired; the signal is simply converted to digital form and the reverse operation carried out after processing to regenerate an analogue signal. In many cases, even with digital readout, a digital-to-analogue converter will be incorporated to provide an analogue output for a potentiometric recorder.

The final link in the digital system is the utilization of the processed signal. In the majority of cases this simply involves the traditional manual recording of the signal by an operator. Some commercial spectrometers are, however, already available with digital (i.e. binary coded decimal, bcd) outputs which may be connected to a printing unit of the type used in computers. Analysis results may then be directly recorded in an easily readable digital form on paper, or even be punched on to paper tape or cards for future processing by a calculating machine or computer. If a keyboard is available on the printer the operator may add appropriate notes or comments to provide a complete record of the measurements. Furthermore, if the printer incorporates a memory facility, it may be programmed for a particular routine analysis so that the desired layout, headings, sample references, etc. are obtained automatically. Thus, the spectrometer is capable of rapid production of typed sample reports.

The provision of such a system is obviously expensive; the cost of an automatic printer may easily rival that of the spectrometer itself. As a result, any operation able to justify such an outlay may well benefit from the logical extension of these facilities; i.e. processing of the spectrometer output by a computer. Provision of a small purpose-built computer is possible, but expensive. In some cases where a large time-sharing computer is available it should be possible to link the spectrometer directly, the main item of expense being development of a suitable computer programme. Provided that the work load of the spectrometer is sufficiently great, the expense of a computer system is outweighed by its potential benefits. Not only is automatic printout of results with pre-determined notes possible, but also complicated operations such as statistical analysis or storage of results for subsequent tabulation in special format. Furthermore, the computer itself is able to perform the signal processing operations mentioned above so that the spectrometer no longer requires a large part of its digital circuitry. Finally, the computer may provide a feedback to the spectrometer to allow completely automatic operation in which the computer selects samples and standards, checks unsatisfactory results, etc. Techniques described in the next section for enhancement of signal : noise ratios may also be performed or assisted by the computer. The possibilities are endless; the only limiting factors are the expense and the opportunity for breakdown of the system.

10.3 ENHANCEMENT OF SIGNAL-TO-NOISE RATIOS

It should be apparent from many of the comments made in the foregoing pages that the problem of signal noise is of major importance in both atomic absorption and atomic fluorescence spectroscopy. At the practical level it is a major factor governing the detection limit of any determination and is often an important contributor to the level of precision that may be obtained. In many cases, significant improvement in signal-to-noise ratio is attainable simply by paying scrupulous attention to optimization of instrumental parameters as described elsewhere in this book. Nevertheless, analysis of the relative factors contributing to signal noise in AAS[9] and AFS[10] shows that for certain applications it is possible to obtain further improvements even with the most carefully optimized spectrometer. In this section are described several of the techniques available. Some are either already widely used in AAS or at least easily applicable; others require complex instrumentation or knowledge of data processing techniques. In these latter cases we shall merely attempt to describe available techniques and how they may be employed.

10.3.1 Modulation of Signals

As described in Section 10.2.1, the majority of atomic absorption or fluorescence spectrometers use a synchronous or tuned amplifier in conjunction with an amplitude-modulated light source. Such a system results in improved signal-to-noise ratios for two reasons. In the first place, an ac amplification system is inherently more stable with lower noise levels. The application of source-intensity modulation, first to AAS and later to AFS, was however based on the possibility of detecting the analytical signal without interference from light emitted by the atom cell itself. In many cases this results in a much greater improvement than would be obtained simply by using ac amplification. In fact, with several elements (particularly those having their resonance lines in the visible region) atomic absorption measurements would be most inconvenient were it not for the source modulation technique. Modulation of the light-source intensity is easily achieved by placing in front of it a rotating or vibrating sector (or "chopper"). Alternatively, direct modulation may be achieved by supplying the lamp with a modulated current of the desired frequency and waveform. As explained in Chapter 5, the latter technique is preferable whenever it can be used satisfactorily with the type of lamp in question. A rotating sector may be easier to arrange and is applicable to all sources (even a second flame), but it inevitably reduces the light intensity, may restrict the usable light beam, and can sometimes prove awkward to locate within a spectrometer. It is also worth noting that use of mechanical modulation requires a completely unmodulated (i.e. dc) signal from the source itself. For example, it would not be possible to obtain successful modulation at, say, 280 Hz if the source was operated either directly from the ac mains or from a supply showing a mains ripple; the source itself would then already be completely or partially modulated at 50 or 60 Hz.

To a certain extent the advantages of a modulated light source are obtained if the signal from the detector is processed with a *tuned amplifier*, i.e. one designed to respond only to the pre-selected modulation frequency. This device is simply a combination of a narrow band electronic filter with an amplifier and appropriate feedback between the two. Its disadvantage is that in order to allow for drift of the modulation frequency a relatively wide amplifier bandwidth (>*ca.* 1 Hz) must be used. As a result the amplifier may also amplify some noise, i.e. any signal not originating directly from the modulated source. Thus, there is only limited improvement in the signal-to-noise ratio obtained, say, for atomic absorption signals in the presence of strong emission of radiation by the atom cell. The problem is avoided by using a *synchronous* or *lock-in amplifier*, which eliminates the effects of frequency drift and may therefore use a much narrower bandpass. The improvement is achieved because such devices use a *reference waveform* which is either derived from the device modulating the light source or itself controls the modulation. Thus, a feedback system within the amplifier ensures that all frequencies other than that of the reference are rejected and small variations in the absolute frequency are unimportant. When the light source is modulated electrically the same waveform generator may be used to supply both the source power supply and the amplifier. With mechanical modulation an optical link is widely used to obtain a reference waveform. This usually involves placing a miniature light bulb and photocell on opposite sides of the chopper blade. Thus, the amplitude of signal emitted by the photocell is modulated in exactly the same way as the main light beam of the spectrometer.

In practice, use of a modulation system with reference waveform allows introduction of a further stage of noise rejection. The signal and the reference not only have identical frequencies but must also be exactly in phase. Thus, by using *phase-sensitive* circuitry in the lock-in amplifier it is possible to reject any components of the final signal that are not exactly in phase with the original waveform of the modulated light source. As a result the signal-to-noise ratio is further improved. The basic layout of the phase-sensitive detector is shown in Fig. 10-24. V_s and V_r are the amplitudes of the signal and reference waveforms, respectively, after they have been processed by a tuned amplifier. V_o is the amplitude of the output and is obtained

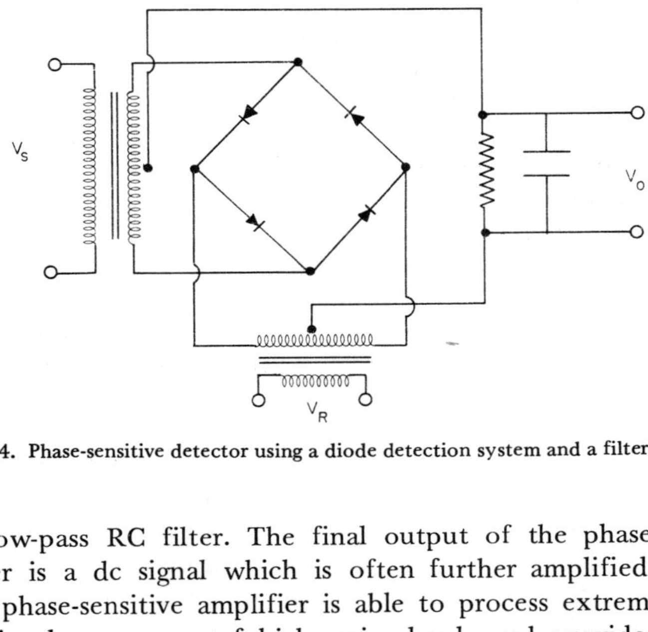

Fig. 10-24. Phase-sensitive detector using a diode detection system and a filtered output.

via a low-pass RC filter. The final output of the phase-sensitive amplifier is a dc signal which is often further amplified. Thus, a lock-in phase-sensitive amplifier is able to process extremely small signals in the presence of high noise levels and provide a stable, almost noise-free dc signal of useful amplitude for conventional meters or recorders.

The modulation technique described above, in which the intensity of the spectral light source is varied, has been applied almost universally to atomic absorption and fluorescence measurements. Modulation may, however, also be achieved in other ways. Several workers[11-14] have described a *sample modulation* technique. This usually involves use of a special nebulizer assembly able to modulate the supply of sample solution to a flame. Thus, a modulated absorption or fluorescence signal is obtained with dc source radiation. The system is most advantageous in flame emission spectrometry but advantages are also claimed for application to AAS and AFS. For example an absorption signal modulated in this way is unaffected by the presence of non-absorbed source lines in the bandpass of the monochromator, as these will not be amplified by the synchronous detection system. Similarly, the background absorption of the flame will not be modulated even at the resonance wavelength of the analyte element so that measurement will be unaffected by noise or drift of the flame background. To avoid any modulation of the background it is usual to arrange that the nebulizer operates alternately with sample and blank solutions rather than with simple intermittent sample aspiration. A constant check on the zero reading is therefore also obtained. Sample modulation also has some disadvantages. For example, the modulation frequency is usually rather low owing to mechanical difficulties and the apparatus is considerably more complex than the widely used pneumatic nebulizer. It is also easier to apply the technique to total consumption nebulizer-burner units or to diffusion flames than to the premixed flames with indirect nebulizers which are of widest application for AAS or AFS. Probably the technique is of greatest use when absorption measurements must be made in regions of high background absorption (e.g. below 220 nm for the air-acetylene flame) or when a continuum source is used. In the latter case, background absorption is always a more serious problem than with sharp line sources as a result of the wider effective absorption bandpass.

It is also possible to use *wavelength modulation* in which either the wavelength of the source or of the measurement is rapidly varied. Obviously, with a sharp line source the effect is exactly the same as conventional intensity modulation and no advantage is gained. Several workers have, however, applied the technique to atomic absorption measurements with a continuum source.[15-19] An obvious advantage of such a system is that background absorption of the flame near the resonance line is automatically eliminated. The system evolved by Snelleman[15] achieved this using a small oscillating mirror within the monochromator to vary the effective wavelength setting. A disadvantage of this method is that any flame emission at the measurement wavelength is also modulated. Elser and Winefordner[16] overcame the problem by combining a similar wavelength modulation method with conventional source modulation. Their experimental system is shown diagrammatically in Fig. 10-25. They were also able to improve the signal-to-noise ratio of measurements compared with conventional modulation of a continuum source. An alternative method of wavelength modulation which also avoids modulating flame emission has been described by Svoboda.[17,18] In this case the wavelength of a continuum source is modulated with an oscillating Fabry-Perot interferometer before passing the light beam through the flame. A conventional monochromator is placed after the flame to select and isolate the wavelength of measurement in the normal way. The interferometer

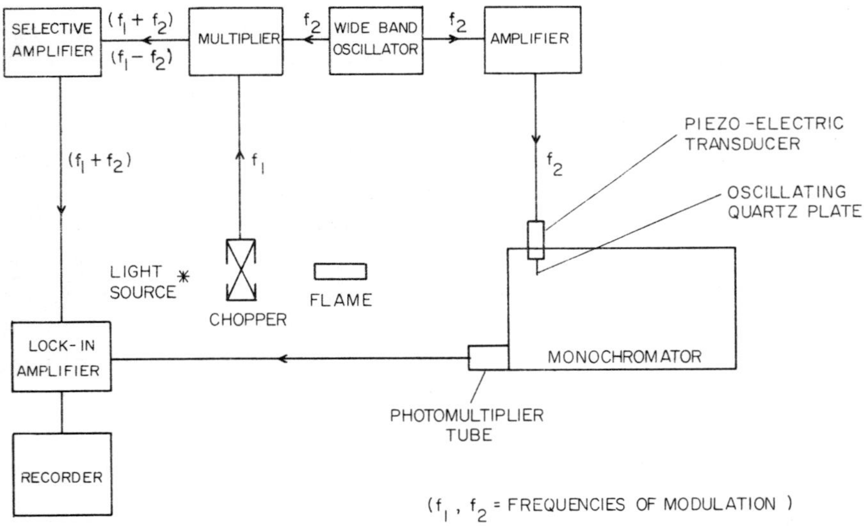

Fig. 10-25. Block diagram of experimental system for double-modulation optical scanning atomic absorption spectrometry.[10]

converts the continuum light source into a series of narrow wavelength bands (channels) each of which approximates to a sharp line spectral source. (Channel widths of a few milliangstroms are easily obtained.) The wavelengths at which these channels occur are then modulated by piezoelectrically changing the Fabry-Perot etalon plate separation. The intensity of a particular channel absorbed by the atomic species in the flame is then modulated as it passes through the wavelength of the absorption line and back (i.e. at twice the piezoelectric oscillation frequency). Even when the monochromator transmits not only the absorbed channel but nearby unabsorbed channels the atomic absorption may be detected by tuning to the appropriate frequency. Thus, the interferometer gives the continuum source the advantages of a sharp line source. Kirkbright and

Troccoli[19] have employed this type of system for the direct study of atomic line profiles for calcium in different flames by absorption against the channelled spectra from the continuum (xenon) source. The experimental spectrometers used by Svoboda *et al.* and Kirkbright and Troccoli were relatively complicated and difficult to adjust. Small Fabry-Perot interferometers of high quality are now widely used in laser work, however, and these may allow easier application of the technique.

10.3.2 Integration and Averaging of Signals

Most atomic absorption and fluorescence spectrometers allow improved measurements of very small signals by providing a *variable damping* facility. This simply increases the response time of the system so that the effect of random signal noise on a meter or recorder is evened out and small signals may be measured more precisely. The method is cheap (it usually involves merely switching a few capacitors of different values across the output terminals) and is reasonably effective. It is, however, by no means the most efficient way of extracting signals from noise and also has some disadvantages. For example, if a sudden noise "spike" occurs with a heavily damped system there will be a considerable time lapse before the reading returns to normal. Similarly, if heavy damping is used to obtain a noise-free steady signal there will be a long waiting period after introducing a sample before the actual reading is reached. A similar delay occurs at the end of the measurement before the zero reading is regained. Some manufacturers avoid this particular problem by using "automatic damping" which is only applied when the signal has reached a steady value. This is more than a time-saving device as heavy damping often cannot be used with small samples because all of the sample would be consumed before a reading could be taken.

Integration techniques provide a more efficient alternative to heavy damping, avoid some of the disadvantages, and are still reasonably simple. If the noise is random whereas the signal is not, true integration of a combined signal and noise for a set time should reduce the noise component to zero but greatly increase the size of the signal. Even in a practical case where the noise is not completely random the signal should increase more rapidly than the noise so that the signal-to-noise ratio is increased. The simplest analogue integrator is the RC circuit of Fig. 10-20 with a very long time constant. Thus the output would be a steadily increasing voltage which is characteristic of the input voltage and the time for which the capacitor has been charging. An integrated signal is obtained simply by noting the voltage at a fixed time after starting from zero. Harrison and Berry[20] have described the application to AAS of an improved analogue integrator using an operational amplifier and incorporating an electrical relay to set the integration time. Similar devices are available commercially.

Digital integration techniques are generally more versatile and also more accurate. Provided the digital circuitry is already in use, integration is obtained merely by a suitable choice of counting times. Thus, most atomic absorption spectrometers with digital readout (or a digital accessory unit) incorporate some form of integration as standard. Usually several pre-selected integration periods in the range 1 to 100 seconds are available. As the meter will "hold" one reading until the next is available to update it, correct choice of integration period will allow easier reading of even routine measurements; continuous changing of the meter reading while the operator attempts to note an average value is then avoided. The use of digital circuitry allows several additional features to be easily incorporated. For example a zero value can be stored and automatically subtracted from all integrated readings until it is updated by the operator. Improved precision is obtained by replacing integration over one long period with an average value of the integrated values obtained over several short periods.

The technique of *signal averaging* is extremely powerful for the extraction of signals from noise and may be considered quite separately to conventional integration. It involves the sampling of a repetitive signal and storage of successive samples in a register. When each sample is taken in a completely reproducible manner the signals are cumulative whereas random noise averages to zero. In the general case the increase in signal-to-noise ratio is proportional to the square root of the number of samples. Unlike the integrator, a signal averager is able to handle a signal which is made up of many components; for example, a spectrum which is scanned repeatedly. This is achieved by using a large number of channels, each of which samples successively at a pre-determined time after the sampling procedure has been initiated (triggered). Obviously, each sampling run must start at exactly the same point on the spectrum so that a synchronizing pulse would be generated by the spectrometer. Commercial signal averagers are available but are relatively expensive when many channels are required. Their application to AAS or AFS is limited as spectral scans are not usually required. It is worth noting that if the spectrometer provides a binary (digital) output, the technique may be carried out with a small computer suitably programmed. A single channel signal averager, often called a *boxcar integrator*, is more relevant to routine absorption or fluorescence measurements, with modulated sources. It has several advantages over straightforward demodulation (with a lock-in amplifier) and measurement of the resultant "average" dc signal, either directly or after integration. Thus, instead of filtering the output of a lock-in amplifier it is possible to sample a number of output pulses and to average them. Optimum results depend on the correct selection of a number of variables in the electronic system. Nevertheless, modern circuit techniques allow incorporation of such devices into spectrometers with the minimum of external controls. For example, a system may combine sophisticated averaging and integration techniques yet present the operator with merely a digital panel meter and a switch to select intervals of 1, 10 or 100 seconds between "readings".

10.3.3 Correlation Techniques

As with integration and signal averaging, correlation analysis relies on the periodic nature of a signal compared with the random nature of noise. It is, however, even more efficient and will extract a signal that is many times smaller than its accompanying noise. The instrumentation required to achieve this is complex, although a suitably programmed small computer is again adequate; to date only one application to AAS or AFS seems to have been reported.[21] The technique involves examining the correlation (i.e. interdependence) between the signal and a delayed version of itself (auto-correlation) or between the signal and a reference (cross-correlation). For each signal, this comparison is made at a number of different time intervals. Except near zero time interval, random noise can never show any correlation, whereas any periodic signal will generate another periodic waveform. Thus, the signal is readily extracted from the noise. The advantages of correlation analysis over the techniques mentioned above, apart from this greater efficiency, include direct background correction, elimination of drift, absence of synchronizing wave (for auto-correlation) and minimization of the effects of sudden noise impulses. This last advantage may be particularly useful; it is often difficult under such conditions to read the true level of the output from, say, a lock-in amplifier, particularly when a long time-constant is used. As shown in Fig. 10-26, similar pulses have little effect on the accuracy with which a

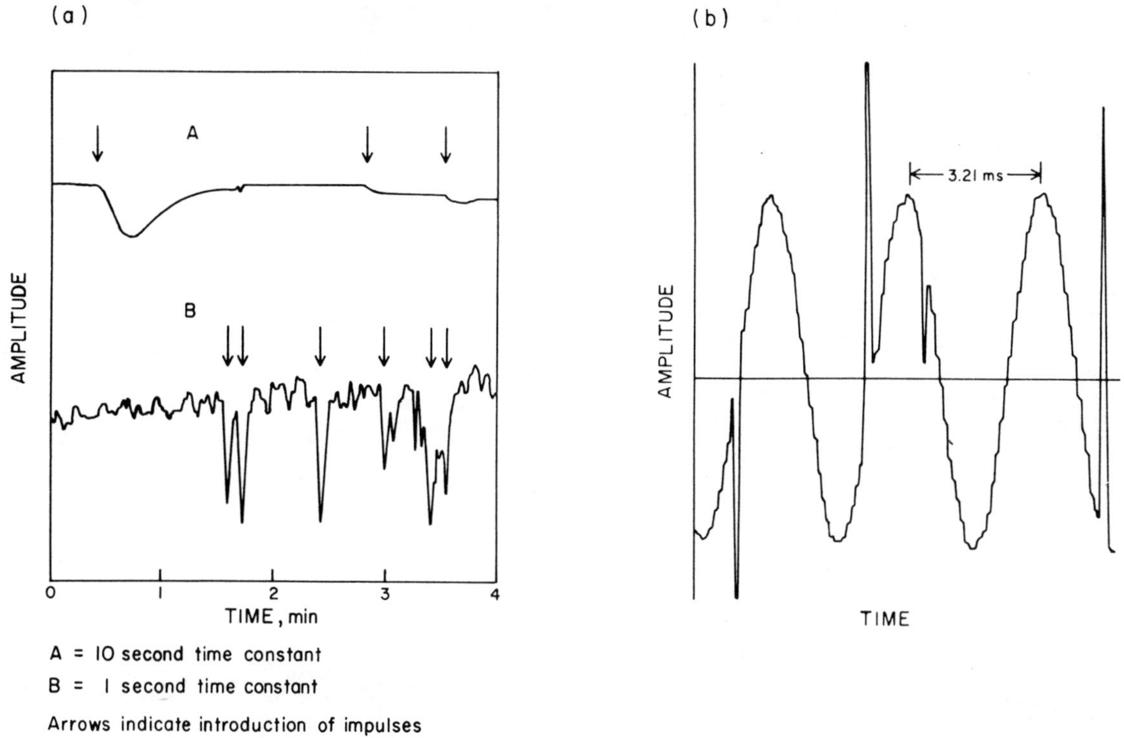

Fig. 10-26. Relative effects of impulse noise on (a) a chart recording from a lock-in amplifier and (b) a cross correlogram.[20] (In both cases the signal is the fluorescence obtained from a 100 ppm rhodium solution using a hollow-cathode lamp as source and an air-hydrogen flame.)

correlogram may be read (even if a noise peak coincided with a signal peak, the reading could merely be taken from an adjacent peak). In spite of its advantages and ease of application, the expense of the instrumentation required for correlation analysis is likely to hinder its routine application to analytical work in the near future. It is important, however, to remember the efficiency with which this and similar techniques can extract reliable data from signals that otherwise represent useless noise to the operator of a moderately priced atomic absorption spectrometer.

10.4 OPTICAL SYSTEMS

In this section we describe some of the variations that occur in the optical systems of commercial spectrometers. No attempt is made to explain fully their construction or to give comprehensive coverage, but it is hoped that the reader will obtain some guidance as to the usefulness of various features. For example, most double-beam spectrometers are claimed to have advantages over single-beam types, but at least one manufacturer lists the latter feature as an advantage over double-beam operation!

10.4.1 General Layout

Two distinct patterns are readily discernible amongst commercial atomic absorption spectrometers: the modular concept and the single-unit instrument. The former type in its classic form (see Fig. 1.2) offers maximum versatility with all optical components mounted on an accessible optical bench and ancillary units conveniently placed around them. Thus, the optical alignment is readily optimized for particular applications, components are easily changed when required and the system may often be applied in modes other than those envisaged by its designers. Such a system is ideal for research use and was quite widely available in early atomic absorption spectrometers. It is less attractive, however, in a spectrometer designed exclusively for routine analysis. The instrument requires careful alignment, components can be accidentally disturbed, lenses or mirrors may become dusty or degraded by sticky fingers, and more operator skill is required to ensure that the instrumental arrangement and operating parameters are optimized for a particular operation. Hence, routine trouble-free operation is more likely if all components are securely mounted in a single unit, optical alignment is preset by the manufacturer, and lenses, mirrors, etc., can only be disturbed by deliberate removal of a number of fixing bolts. A typical modern spectrometer based on this concept is shown in Fig. 10-27. It can be seen that even the lenses guiding the light beam through the flame have been recessed behind a metal shield to minimize the risk of soiling them. With the advent of AAS for the routine analysis of large numbers of samples this style of spectrometer has become increasingly popular and many commercial instruments now resemble the type shown in Fig. 10-27.

The trend towards a compact, easy to use spectrometer is understandable; the largest volume of sales is now for applications where this style has several advantages. The analytical chemist requiring specialized instrumentation for research purposes can probably obtain better results by purchasing individual optical, spectroscopic and electronic components from different manufacturers. The biggest problem in this respect may be the burner and nebulizer as suitable units have been developed largely for commercial atomic absorption spectrometers. Fortunately, several manufacturers supply their burners either separately or as "spares" so that no real difficulty need arise. Even the single-unit instruments show considerable variation between manufacturers as to ease of optical alignment and use, so that care is needed in the selection of a spectrometer. For example, hollow-cathode lamps usually need individual alignment owing to tolerance in manufacture. If elements are changed frequently the time required to change lamps may be an important consideration. Some instruments incorporate a movable turret or rack for several lamps, each of which may be aligned once and then moved into the correct position when required. The usefulness of this feature depends on the number of elements whose analysis is regularly required; obviously, a holder taking three lamps is inconvenient if samples are regularly analysed for elements needing six individual lamps. An alternative approach is to have a lamp-house for only a single lamp but to mount each lamp in a separate holder stored outside the spectrometer. Each lamp may then be aligned once (using adjustment screws provided on the holder) and simply be clipped into the spectrometer whenever required. This system is at its best in double-beam systems (see Section 10.4.2) where minimum warm-up of lamps is required; with single-beam spectrometers a turret holder conveniently allows lamps to be operated on "standby" and automatically switched to the main power supply when rotated into the operating position.

With either method of mounting the lamp correct alignment is

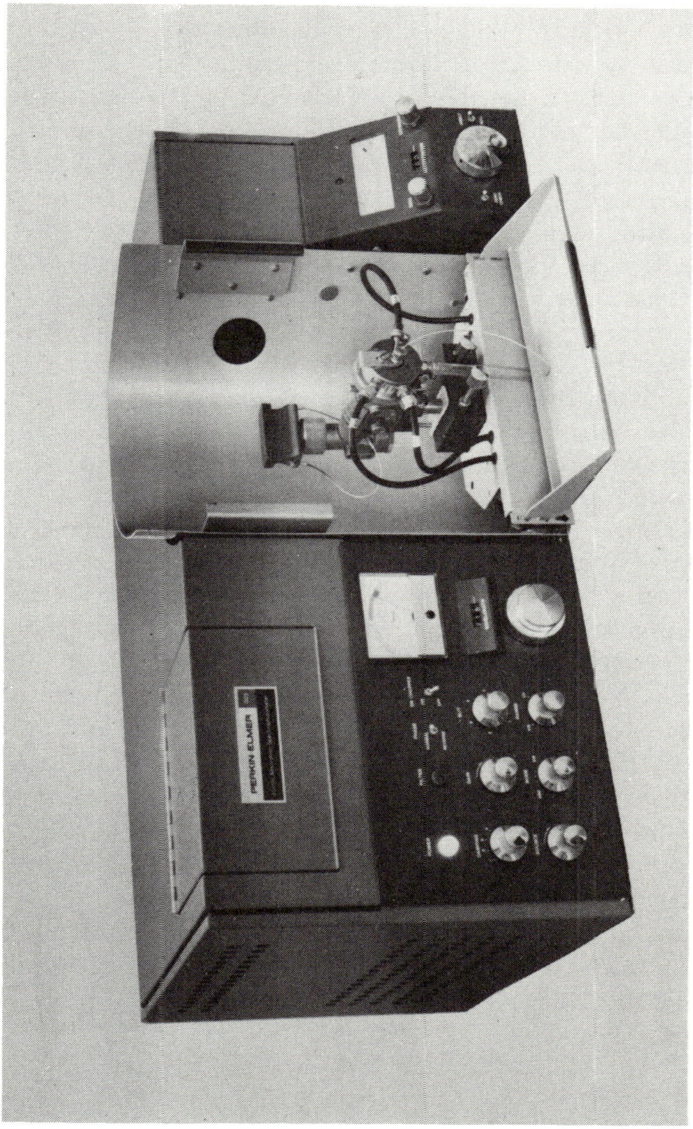

Fig. 10-27. A modern atomic absorption spectrometer constructed as a single unit.

required at some stage. The ease with which this is accomplished can vary considerably, depending on the design of the instrument. By careful positioning of the fixed optical components the manufacturer can ensure that correct alignment of the light beam with both the burner and monochromator is obtained by adjusting the lamp holder until maximum transmitted light is indicated by the readout meter. This is somewhat easier than attempting to follow the path of the light beam with pieces of white card! The adjustment of the lamp itself is probably best achieved with correctly placed screws of reasonably fine pitch but adequate length; considerable frustration results during lamp alignment with holders that are poorly designed and have inadequate provision for adjustment. Alignment is also facilitated if the optical arrangement has been carefully chosen by the manufacturer. Thus, the extent to which the path of the light beam depends on the exact position of the lamp will vary with the size and focal length of lenses and mirrors, their distance from the lamp, burner or monochromator, the size and location of beam-restricting apertures, etc.

It should not be forgotten that the flame itself is a major component of the optical system. Hence the burner should also be provided with accurate and reproducible adjustment for the vertical and lateral position of the flame. With careful design, optimization of this variable can also be made part of a simple sequence; once the light beam is correctly placed with respect to the remainder of the optical system the flame position may be set simply by adjusting the burner until maximum absorbance is observed for the element of interest. Again it is important that the available adjustment of the burner should be adequate for a wide range of conditions. This is best achieved if there is sufficient space in the burner compartment. A large burner compartment ensures ease of replacement or cleaning of the burner and allows any accessories to be fitted rapidly. Optical components, such as lenses or restricting apertures, may also then be at sufficient distance from the flame to prevent overheating and degradation.

10.4.2 Double-Beam Operation

Most early atomic absorption and fluorescence spectrometers were designed for single-beam operation, i.e. in these instruments the light passing through the absorption cell or originating from the fluorescence cell was the only radiation received at the detector. With these systems any fluctuation in the intensity of the light source appears as an analytical signal and unless the source is very stable frequent zero checks are required. Some of the problems of drift of the light source are avoided by double-beam operation in which the detector also receives a reference light beam which has originated from the same light source but has not passed through the atom cell. Many atomic absorption spectrometers with this facility are now available commercially.

The optical arrangement of an early double-beam system is shown in Fig. 10-28. Radiation from the source is divided into two beams by a silica beam splitter at position A. The height and direction of the beams are then varied by a number of mirrors until they are merged at position B in such a way that two identical vertically adjacent beams may be passed through the monochromator. After leaving the monochromator further mirrors are used to deflect each beam to its own photomultiplier tube. The signals from these two detectors are then amplified and balanced by the readout system so that absorbance values in the sample channel are always corrected for any changes in source intensity. This system has several disadvantages; the optics are complicated, two photomultipliers (and their power supplies) are required and most important of all, there is no compensation for variations in gain of the individual photomultipliers. Much more satisfactory results are obtained with an alternative system which is now widely used. As shown in Fig. 10-29, the light beam from the light source is divided by a rotating sector, the rear surface of which is reflecting. Thus, the light is alternately passed directly into the flame or reflected backwards to a mirror which deflects the reference beam through the rear of the spectrometer. The two beams are subsequently recombined by a half-silvered mirror and pass through the monochromator to the photomultiplier as a single beam. The signal from the photomultiplier is, however, processed by an amplifier able to separate it into the reference and sample channels using the same techniques as conventional lock-in amplifiers (the two channels merely have different phase angles). Both signals could be displayed separately by the readout device but it is more common to present them simply as an absorbance signal which is automatically corrected for variations in source intensity. Several spectrometers now simultaneously present the reference beam signal on a separate meter so that it may be used as a continuous check of source intensity no matter what absorption measurements are in progress in the sample beam.

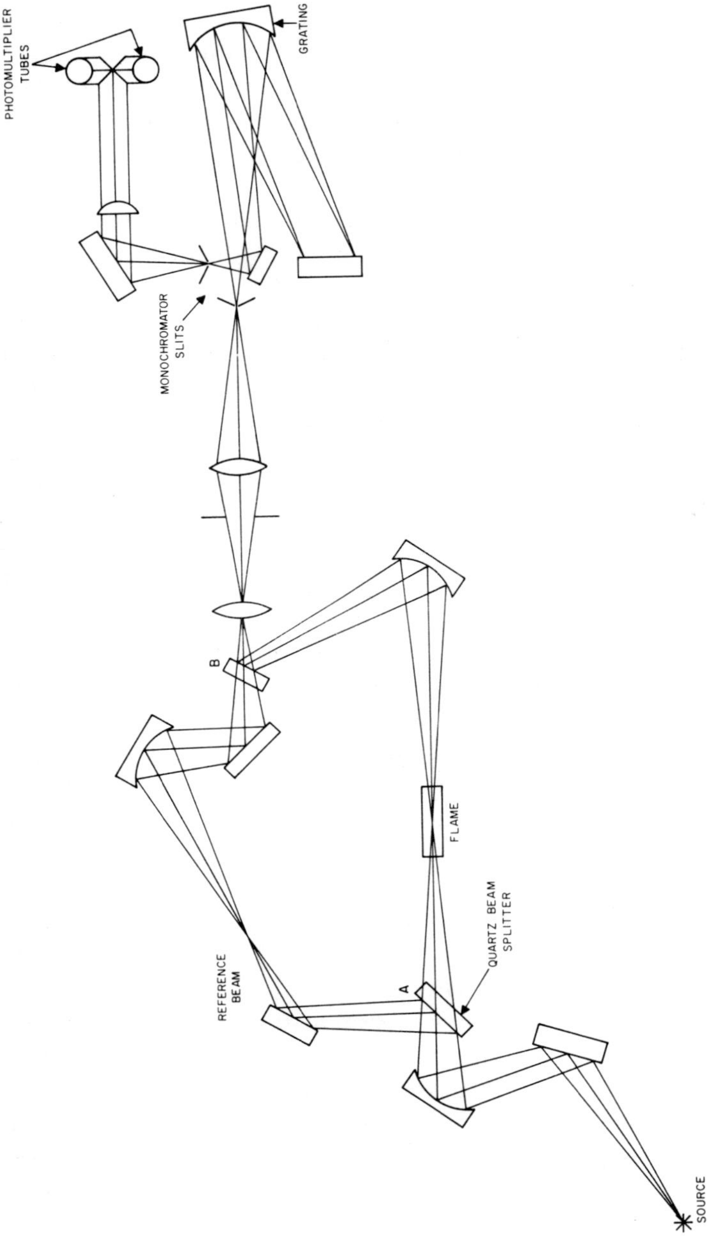

Fig. 10-28. A double-beam spectrometer using separate photomultiplier tubes for sample and reference beams.

It should be appreciated that double-beam systems only compensate for variations in source intensity; unlike solution absorption spectrophotometers the reference channel does not incorporate a "blank-cell" for correction of background absorbance. Thus, the value of double-beam operation with modern hollow-cathode lamps of high stability is sometimes questioned. On balance, there still appear to be advantages in incorporating double-beam optics in instruments designed for routine use with minimum attention. Even the best hollow-cathode lamps may show some drift of output, and all lamps require a warm-up period before a stable output is reached. Thus, with a single-beam instrument there is a tendency to run on standby

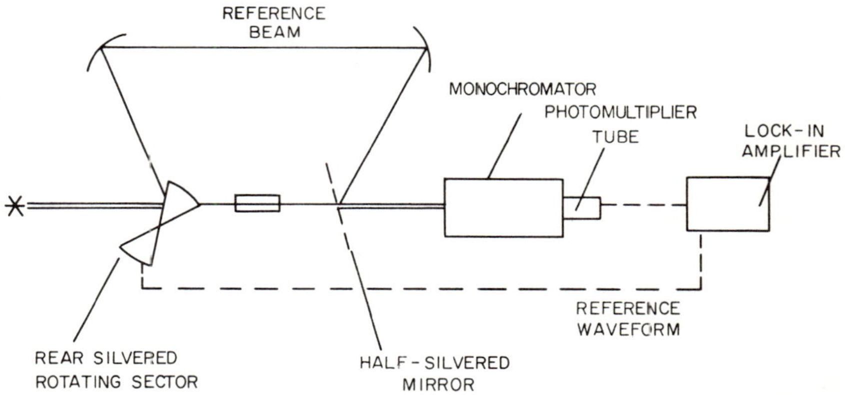

Fig. 10-29. A double-beam spectrometer using a single photomultiplier tube and electronic separation of reference and sample channels.

as many as possible of the lamps which might be needed during the day. This may reduce their useful life; with the double-beam facility, however, the lamps need be operated only when actually in use. For almost all elements double-beam measurements may be made within a minute or two of switching on the lamp and many manufacturers provide no facilities for warm-up. Hence, no additional power supply is required and over the life of a well-used spectrometer the convenience of double-beam operation may be obtained at very little extra expense. Double-beam operation may also be particularly advantageous for atomic fluorescence measurements, which are frequently made using the more intense electrodeless discharge lamps. As discussed in Chapter 5, these sources may often be less stable than hollow-cathode lamps. It is important to remember that with both absorption and fluorescence signals the system cannot compensate for any variations of the source except for the total intensity of the line. For example, a sudden increase in the intensity of a lamp may be accompanied by an increase in the spectral line width. This would result in a decrease in the measured peak absorbance of a sample but there is no way in which the double-beam system described above can correct this change.

10.4.3 Background Compensation Techniques

There are two situations in which absorption by species other than the analyte atoms may prove troublesome during atomic absorption measurements and which cannot be avoided by conventional double-beam operation. As described in Chapter 6, many flames show relatively strong natural absorption of radiation in certain spectral regions, particularly at short wavelengths. Sample solutions containing high concentrations of certain anions or compounds may also absorb over relatively wide spectral bandwidths owing to either scattering of radiation or to molecular absorption. This effect is discussed further in Section 12.1.4 and 12.3.1. In either case the spurious absorption signal can prove inconvenient and may cause errors, particularly if the operator is unaware of its existence. In view of this, several attempts have been made to obtain automatic background compensation and suitable accessories are available for some commercial spectrometers. Automatic background compensation is, of course, obtained by the sample modulation techniques and some of the wavelength modulation techniques described in Section 10.3.1. These have seen little general application, however,

In one commercial system[22] the background absorbance is measured using a deuterium or hydrogen discharge lamp acting as continuum source. The lamp is placed near to the hollow-cathode lamp of a double-beam spectrometer. The optics are adjusted so that the chopper, which normally passes light from the hollow cathode alternately through the reference and flame channels, now alternately passes light from either lamp through the flame. Absorption of the broad band of continuum radiation by sample atoms is negligible but both types of radiation are absorbed equally* as the result of background effects. Thus the double-beam circuitry of the readout section automatically compensates for changes in background absorption. It should be noted that conventional double-beam operation is no longer available to compensate for variations in

* As explained in Chapter 3, this is not strictly true under all conditions. Nevertheless, the system works well within the limitations imposed by other experimental factors.

the hollow-cathode lamp intensity. Thus the compensation accessory is made easily demountable and is only used when necessary.

A slightly different compensation method has been described by Woodriff *et al.*[23] again using a continuum source. These workers used a polarizing beam splitter to obtain horizontal polarization of one light beam and vertical polarization of the other. The two beams may then be combined for passage through the absorption cell and monochromator but separated optically using further polarizers before being detected by a photomultiplier. This system may allow easier alignment of the two beams throughout the length of a long absorption cell such as a furnace, but probably has little advantage over the other system.

Another approach to the problem is to make background corrections using a line from the conventional hollow-cathode lamp which lies close to the resonance line but which is not absorbed by the sample atom.[24,25] As the reference line should show similar intensity variations to the resonance line, this method would have the advantages of both background correction and conventional double-beam operation. Some problems arise with this technique, however. It is often difficult to find a suitable nearby non-absorbing line and there is no guarantee that its intensity variations will always follow those of the resonance line, particularly if a line emitted by the filler gas must be used. Also, although manual corrections are readily made, a fully automatic system using this principle would require quite a complex spectrometer. In fact, it only becomes fully practicable with the type of multichannel instrument described in Section 10.4.4.

10.4.4 Multichannel Techniques

A number of attempts have been made to use either AAS or AFS for multielement analysis[26-32] and several commercial spectrometers have been designed for the rapid simultaneous determination of several elements. In most cases these instruments have seen little application to practical problems. This may be attributed to operational problems such as the size and expense of spectrometers, difficulty of maintaining the complete system in optical alignment, poor sensitivity compared with conventional single-element instruments and the effort required by an operator to manage such a system unless it is almost completely automated. There is undoubtedly a place for absorption or fluorescence spectrometers able to handle the simultaneous analysis of, say, six elements in a single sample. It seems, however, that the full benefit of such a system is only realized when it is combined with complex sample handling and data processing techniques. Thus, the complete spectrometer becomes extremely expensive and will probably only be justified when designed specifically for specialized applications. A typical example may be clinical analysis in which a single laboratory will process

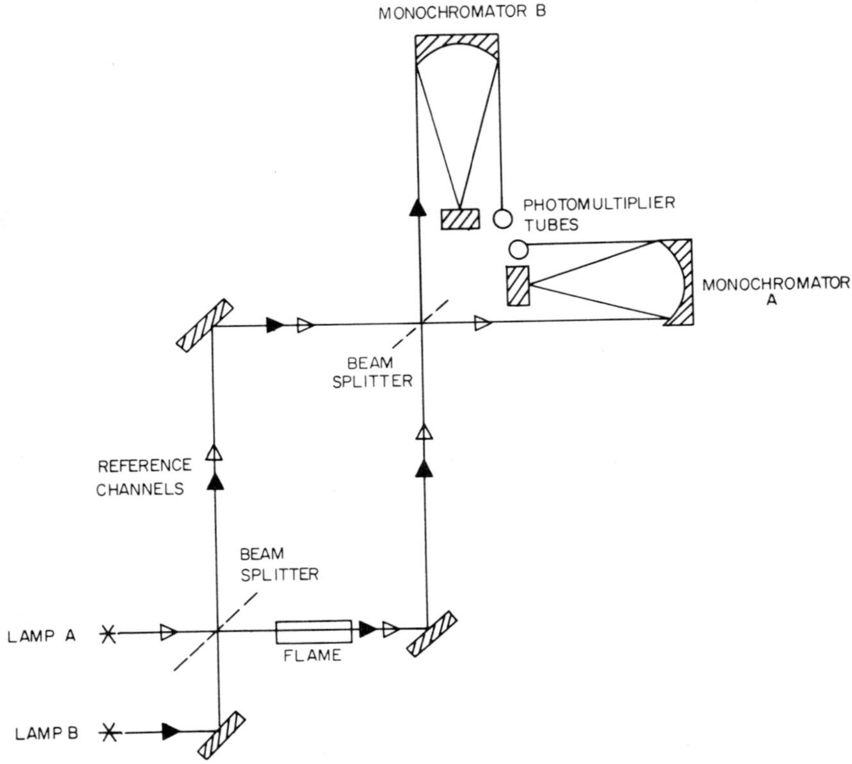

Fig. 10-30. A dual-channel, double-beam optical system.

many samples in each of which the determination of several elements within a relatively narrow range of concentrations is required.

Recently, several commercial examples have appeared of a new type of atomic absorption spectrometer which is expensive but still within the price range of a general purpose instrument. This is the double-beam, dual-channel spectrometer. As shown in Fig. 10-30 each channel has its own hollow-cathode source, monochromator and detector *and* is provided with a conventional double-beam reference channel. In some less expensive systems the second monochromator may be replaced with a tunable interference filter. Each channel also has its own readout unit so that the spectrometer may obviously be used for simultaneous two-element analysis. Of more interest for many applications, however, is the ability to program the electronics so that one channel may be used as a reference for the other. For example, if one hollow cathode is replaced by a continuum source the spectrometer provides background compensation and double-beam operation. Alternatively, both hollow cathodes may be retained but one channel set to a non-absorbed line so a different type of background correction is obtained. Similarly, one channel may be used to measure an internal standard element added to the sample and a readout obtained which is compensated for effects such as variations in nebulizer sensitivity. Such techniques also tend to minimize the effects of flame noise in certain spectral regions.

10.5 BURNER SYSTEMS

The design and construction of burners has already been described in Chapter 6 and that of nebulizers in Chapter 8; this brief section contains some comments on points of practical interest which could not be conveniently included elsewhere.

In most commercial spectrometers the position of the burner, other than fine adjustment, is selected by the designer and cannot be readily changed. A good design will ensure that the burner is easily accessible for cleaning or replacement but is well protected from draughts. At the same time protection must be provided for the operator against radiation and heat emitted by the flame and against the possibility of an explosion. The force of an explosion should be dissipated through a safety valve or bursting diaphragm to avoid any chance of disintegration of the burner or nebulizer. Probably the most useful safety feature is a removable cover across the front of the burner chamber, providing easy access when necessary while preventing accidental contact with the flame. The cover should be semi-transparent so that the operator can see the flame when adjusting gas flow or check that it is still burning. In some cases a photoelectric device is provided to switch off gases should the flame extinguish and this is particularly necessary for spectrometers designed for fully automatic operation. An extraction hood must usually be mounted over the spectrometer to remove the burnt gas. This is particularly important when using a high-temperature flame in a small laboratory or when samples are likely to produce toxic fumes. The ventilation provided is a useful safety factor against unlit gas jets. Care is needed in the choice of extractor. A low extraction rate with a large, high-temperature flame may lead to overheating of the ducting and fan. On the other hand, an excessively high extraction rate will cause draughts and may result in noisy signals.

Many atomic absorption spectrometers incorporate gas controls or offer them as an optional module but the purchaser must usually arrange suitable gas supplies. Air can be obtained from cylinders but frequent changes will be needed if, say, an air-acetylene flame is operated for long periods. It is usually more economical and more convenient to purchase a small compressor; this must be capable of supplying an adequate air flow at the relatively high pressure required by many pneumatic nebulizers while allowing sufficient margin for efficient operation of regulators. The compressor should be provided with a dust filter at its inlet and with a suitable filter to remove condensed water from the supply line. It is also advantageous to choose one of the "oil-free" types of compressor, particularly if the same regulation system is to be used with nitrous oxide. Apart from the possible use of town gas, any other gases required will be obtained from storage cylinders. Hydrogen, nitrogen, argon, etc., are stored in compressed form, propane and nitrous oxide as liquids and acetylene in special cylinders in which it is maintained as a solution in acetone. In most cases the cylinders may be used until they are empty but with acetylene a fresh cylinder should be installed when the pressure has dropped to 75-100 psi. This arises because considerable quantities of acetone distil over as the pressure falls and can significantly alter the flame characteristics. The problem can be avoided by installing a suitable filter (e.g. of activated carbon) in the acetylene line but the filter will require frequent replacement or regeneration and may prove more troublesome than simply changing the cylinder at more frequent intervals. Wherever possible all gas cylinders but particularly those containing flammable gases, should be kept in the open air. If this is impracticable, care must be taken to keep them well away from the burner and secured against accidental damage.

The storage pressure within the cylinder is relatively high, particularly with a full cylinder of compressed gas, and some form of reduction valve is required before the supply can be connected to the burner regulator unit. This is achieved almost universally with a pressure regulator which is fitted onto the outlet of the cylinder.

Such regulators generally provide a variable outlet pressure together with gauges to measure both this and the cylinder pressure. Pressure control is provided by a spring-loaded diaphragm whose position is set by a regulating screw. The diaphragm maintains a constant outlet pressure (relative to the atmosphere) until the cylinder pressure falls below the value selected. In practice, the pressure drop required is often extremely large and a single diaphragm is unable to maintain a sufficiently constant pressure for spectrometric applications. For this reason it is usually desirable to use a two-stage regulator. In this a non-adjustable diaphragm reduces the cylinder pressure to a convenient pre-selected value (e.g. 100 psi) while a second adjustable diaphragm controls the final reduction to the selected operating pressure. These regulators are available in a very wide range and care must be taken to select suitable models. Points to note are: suitability for the gas (e.g. corrosion resistance); correct thread for the gas-cylinder outlet; correct inlet pressure range (the fill presure of a cylinder varies considerably depending on the gas); and correct outlet pressure range for the application (e.g. a 0-200 psi range would not be correct if 5 psi is normally required). Regulators of a similar type but with relatively low inlet pressures are available for attachment to air compressors; they are essential for satisfactory results in most applications. Regulators used with nitrous oxide are often found to "ice-up". This occurs because of the large amount of heat extracted in vaporizing the liquid gas in the cylinder and it may cause variations in output pressure.

In order to obtain optimum results in AAS or AFS a relatively complex gas-control system must be installed between the cylinder regulator and the burner. Adequate systems are provided by some manufacturers but others are less satisfactory in this respect. It may, therefore, be desirable to modify the existing system or to assemble a purpose-built unit. The following features should be incorporated.

(1) The unit should be close to the burner.
(2) Each gas should be controlled by a pressure regulator similar to that used at the cylinder. This regulator provides a final step-down in pressure to the operating value. A gauge should be available to monitor the pressure of each gas.
(3) In some cases the flow rate of the gases can be controlled adequately using a fine needle valve on the outlet of the pressure regulator. The most satisfactory results are obtained, however, by using a flow controller. Simple diaphragm units

are widely available at relatively low cost. In the present application such a controller would be supplied with gas at a constant pressure by the pressure regulator and then supply a constant flow rate of gas regardless of any fluctuations in pressure downstream (such as that due to variations in the back-pressure of the flame). A regulating screw attached to the diaphragm allows the flow rate to be varied.

(4) The flow rate of each gas must be accurately measured so that flame conditions can be set reproducibly from day to day. Most flowmeters are pressure-sensitive so that they are best installed between the pressure regulator and the needle valve or flow controller. They may then be calibrated at the operating pressure.

(5) Electrical flow transducers are available but perfectly satisfactory results are obtained using a critical orifice and manometer or a rotameter. The latter type is fitted to many commercial units. It consists of a glass or plastic tube whose bore is tapered from one end to the other. The tube is mounted vertically with the narrow end at the bottom and a small spherical or conical "float" placed inside. The float is lifted by the gas flow and the position at which it reaches a balance is characteristic of the flow rate. Different flow rates are obtained by using tubes of different bore size and/or floats of different density. The tube is either calibrated directly or else marked with an arbitrary scale and supplied with calibration graphs for different gases. It is important to bear in mind that the readings will vary with the temperature, pressure and type of gas, so that standard conditions must be used in order to obtain accurate results. The calibration will also be invalid if the rotameter is not mounted vertically. More expensive types are supplied with conical floats which spin rapidly when the rotameter is supplied with gas. Most atomic absorption spectrometers are supplied with relatively inexpensive rotameters using spherical floats, both a glass and metal ball often being placed in one rotameter to widen its useful range.

(6) The method of operating a pneumatic nebulizer is also an important consideration in designing the gas-control unit. In most cases the oxidant gas is used for nebulization. Reproducible sample addition will only be obtained if the nebulizer is operated at constant pressure. Hence, if all the oxidant is

passed through the nebulizer it will not be possible to vary its flow rate. A better arrangement is to split the oxidant stream after the flow meter and to use a nebulizer that passes only a relatively small flow of gas. The rest of the oxidant may then be passed via a flow controller to the burner and its flow rate varied as required. The total flow reaching the burner will, of course, be indicated by the flow meter.

(7) The needle valves or flow controllers used to vary the gas flows should not be used to shut off the flow. Much more satisfactory operation is obtained with miniature "toggle" valves, preferably mounted on the high-pressure side of the unit. These allow the gas flows to be switched off without disturbing the flow setting and provide a simple means of rapid shut down.

(8) If the nitrous oxide-acetylene flame is used, it may be ignited and extinguished via an air-acetylene flame (see below). The control unit should therefore provide a simple means of change-over between the two oxidants. A three-way valve will suffice to change both the nebulizer and by-pass supplies, provided the same flow-control system is used for both gases. If different flow rates are required for the two gases a more complicated arrangement will be necessary. In any event, rapid change-over is necessary as an extremely sooty acetylene diffusion flame will burn while no oxidant is available.

All flames routinely used for AAS or AFS are perfectly safe provided the equipment is properly maintained and care is taken during operation, particularly when igniting or extinguishing the flame. Careful procedures are best ensured by using a standardized sequence of operations. Before attempting to light the flame the following checks should be carried out.

(1) Extractor switched on.

(2) All gas cylinders and compressor on and pressures correctly set.

(3) Nebulizer drain tube contains water (it is necessary to drain excess water from the nebulizer and at the same time provide a head of pressure to overcome the back-pressure of the flame on the burner. This is usually achieved simply by incorporating a trap with a head of several inches of water in the nebulizer drain tube).

(4) Correct flow rates preset for all gases.

When lighting the flame, the oxidant should be switched on first. As soon as the correct flow is reached the fuel gas may be switched on and the flame ignited almost immediately. To shut down, the fuel should be switched off and the oxidant left flowing for a short time to flush out any remaining fuel and to cool the burner. Owing to its somewhat higher burning velocity, most operators prefer not to use this system with the nitrous oxide-acetylene flame. Usually an air-acetylene flame is ignited as described, the fuel flow is increased to give a sooty, luminous flame, and the air is replaced by nitrous oxide. The fuel flow is then adjusted to give the required flame appearance. The reverse procedure is followed to extinguish the flame.

Fully automatic burner control units are now available which relieve the operator of the need to remember the operational sequence. A combination of electronic and servo systems is provided which switch on the gases in correct sequence, check their flows, ignite the flame and, if required, change to nitrous oxide. The flame is extinguished merely by pressing an "off" button to initiate the reverse sequence.

REFERENCES

1. C. Veillon, "Handbook of Commercial Scientific Instruments: Vol. 1, Atomic Absorption", Marcel Dekker, New York, 1972.
2. F. T. Areccki, E. Gatti and A. Sona, *Rev. Sci. Instrum.*, 1966, **37**, 942.
3. R. R. Alfano and N. Ockman, *J. Opt. Soc. Am.*, 1968, **58**, 90.
4. G. A. Morton, *Applied Optics*, 1968, **7**, 1.
5. F. Roblen, ibid., 1971, **10**, 777.
6. M. Jonas and Y. Alan, ibid., 1971, **10**, 2437.
7. H. V. Malmstadt, M. O. Franklin and G. Horlick, *Anal. Chem.*, 1972, **44**(8), 63A.
8. Y. D. Harker, J. D. Masso and D. F. Edwards, *Applied Optics*, 1969, **8**, 2563.
9. J. D. Winefordner and T. J. Vickers, *Anal. Chem.*, 1964, **36**, 1947.
10. J. D. Winefordner, M. O. Parsons, J. M. Mansfield and W. J. McCarthy, ibid., 1967, **39**, 436.
11. W. Lang, *Spectrochim. Acta*, 1967, **23A**, 471.
12. W. Trampisch and R. Herrmann, ibid., 1969, **24B**, 215.
13. K. Rüdiger, B. Gutsche, H. Kirchhof and R. Herrmann, *Analyst*, 1969, **94**, 204.
14. M. Marinkovic and T. J. Vickers, *Anal. Chem.*, 1970, **42**, 1613.
15. W. Snelleman, *Spectrochim. Acta*, 1968, **23**, 403.
16. R. C. Elser and J. D. Winefordner, *Anal. Chem.*, 1972, **44**, 698.
17. V. Svoboda, ibid., 1968, **40**, 1384.
18. G. J. Nitis, V. Svoboda and J. D. Winefordner, *Spectrochim. Acta*, 1972, **27B**, 345.
19. G. F. Kirkbright and O. E. Troccoli, *Spectrochim. Acta*, 1973, **28B**, 33.

20. W. W. Harrison and F. E. Berry, *Anal. Chim. Acta,* 1969, **47**, 415.
21. G. M. Hieftje, R. E. Bystroff and R. Lin, *Anal. Chem.,* 1973, **45**, 253.
22. H. L. Kahn, *Atomic Absorption Newsletter,* 1968, **7**, 40.
23. R. Woodriff, B. R. Culver and K. W. Olson, *Applied Spectroscopy,* 1970, **24**, 530.
24. A. C. Menzies, *Anal. Chem.,* 1960, **32**, 898.
25. J. W. Robinson, ibid., 1961, **33**, 1226.
26. L. R. P. Butler and A. Strasheim, *Spectrochim. Acta,* 1965, **21**, 1207.
27. A. Walsh, *Proc. 13th Colloq. Spec. Intern. Ottawa,* 1967, p. 257.
28. F. Brech, ibid., p. 294 and p. 1194.
29. R. Mavrodineanu and R. C. Hughes, *Applied Optics,* 1968, **7**, 1281.
30. A. Strasheim and H. G. C. Human, *Spectrochim. Acta,* 1968, **23B**, 265.
31. J. B. Dawson, D. J. Ellis and R. Milner, ibid., 1968, **23B**, 695.
32. D. G. Mitchell and A. Johansson, ibid., 1970, **25B**, 175.

CHAPTER 11

Practical Techniques of Atomic Absorption and Atomic Fluorescence Spectroscopy

In the previous chapters we have described the underlying theory of AAS and AFS and have discussed in some detail the individual components utilized in atomic absorption and fluorescence spectroscopy. This chapter is devoted to the consideration of some practical analytical aspects of AAS and AFS measurements. These include the selection of the spectral line used for analysis, the optimization of instrumental conditions, and the general technique of instrument operation. The attainable sensitivity, the preparation and measurement of samples and standards, and the evaluation of the precision and accuracy of the techniques are also discussed. Techniques for the preconcentration and separation of analyte elements are also described briefly.

11.1 SELECTION OF WAVELENGTH OF MEASUREMENT

11.1.1 Atomic Absorption Spectroscopy

It may be predicted from equation (3-31) (Section 3.4) that for a given element the atomic line providing the highest sensitivity will be the ground-state line having the highest absorption oscillator strength, f_{ji}. For many elements that exhibit simple line spectra the most suitable absorption line is easily established. Tables of *gf* values for the atomic lines of different elements may be consulted to provide a guide to the most sensitive absorption lines for a particular element.

The prediction of the most suitable line for AAS measurements is more difficult with elements exhibiting a complex spectrum; accurate values of *gf* may not be available for the ground-state lines of interest. Additionally it is necessary to consider *gf* values for atomic lines that originate from levels somewhat above the ground state, as even in relatively low-temperature atom cells the population of these states may be greater than that of the ground state for some elements. An example of this effect has already been described for aluminium (see Chapter 2.4.3), where the two levels of the ground state doublet ($3p\ ^2P_{1/2,3/2}$) are separated by only 0·014 eV. Thus, the 309·3 nm line ($g = 4$) provides greater sensitivity in hot flames than does the 308·2 nm ($g = 2$) ground-state line although both lines have similar oscillator strengths (*ca.* 0·22).

Even if the ground state of a particular atom is the most highly populated state it may not give rise to the most sensitive absorption lines. For the same atom a line whose energy level lies somewhat above the ground state (and may not be highly populated) may produce a higher absorption coefficient if it has a much higher transition probability than any of the ground-state lines. An example is found in the zirconium 360·12 nm line (transition between 0·154 and 3·600 eV) which gives greater absorption in a nitrous oxide-acetylene flame than does the ground-state line of highest *gf* value, i.e. 298·54 nm (transition between 0 and 4·155 eV).

A practical technique for the selection of the most suitable lines for some elements was described by Allan.[1,2] The radiation from a continuum source is directed through a flame containing a high concentration of an element, and its intensity at different absorption lines is measured photographically[1,2] or photoelectrically[3] using an emission spectrograph. For a particular flame this procedure then permits evaluation of the relative absorption coefficients at different lines for the element of interest; its use to establish the most suitable lines for AAS of manganese and iron,[1] cobalt and nickel,[2] molybdenum[4] and the lanthanide elements[3] has been described.

The notes given above are intended to help in the selection of the most sensitive absorption line for a particular element. However, there are several practical reasons that may lead to the choice of a line other than the most sensitive one. These practical factors can have serious effects in certain cases and they will be discussed in some detail.

(*a*) *Emission at non-absorbing lines of the analyte element*

When a non-resonance line is emitted from the spectral lamp by the analyte element at a wavelength close to the most sensitive line of the element concerned, this may be transmitted by the monochromator and lead to low sensitivity and non-linear calibration curves (see Section 12.3). A well-known example of this effect occurs in the determination of nickel by AAS. Although the use of the Ni 232·00 nm resonance line provides highest sensitivity, this line lies close to several other non-absorbing or poorly absorbing nickel atomic lines (e.g. 231·98 and 232·14 nm) and is also close to the nickel ion line at 231·6 nm which is emitted with appreciable intensity from conventional hollow-cathode lamp sources. Even when a narrow monochromator bandpass (<0·1 nm) is available it is difficult to isolate only the Ni 232·00 nm line. With the narrow slit-width required to produce this resolution at 232·00 nm the signal : noise ratio obtained may be lower. The use of a high-intensity lamp[5] or selective modulation[6] has been shown to provide increased sensitivity and linear calibration curves for nickel at 232·0 nm (see Fig. 11-1). However, with conventional hollow-cathode sources used with relatively broad spectral bandpass many workers have preferred to use the Ni 341·48 nm line. Although this line yields

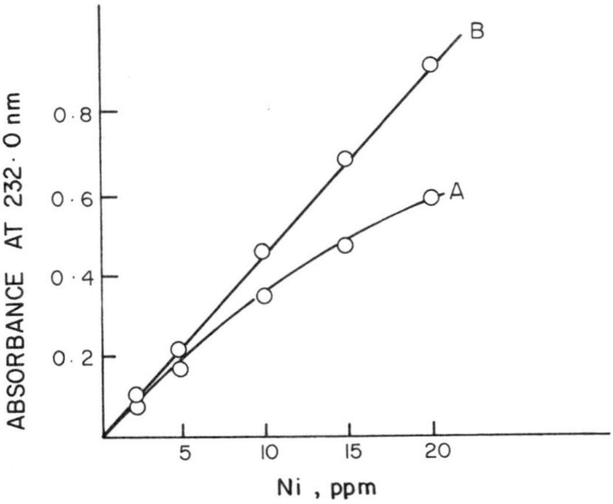

Fig. 11-1. Calibration curves for Ni at 232·0 nm with conventional (A) and high-intensity (B) hollow-cathode lamp sources (adapted from ref. 5).

a lower absorption coefficient it allows more linear calibration curves under these conditions and scale expansion of the absorption signal may readily be performed to help compensate for the lower sensitivity.

(*b*) *Emission at non-absorbing lines of the filler gas or cathode shell*

With early forms of hollow-cathode lamp low sensitivity and non-linearity in calibration curves was frequently encountered owing to transmission by the monochromator of lamp filler-gas lines or continuum background of wavelength similar to that of the analytical line. The incidence of this type of "spectral interference" has decreased considerably with improvement in the technology of hollow-cathode lamp manufacture. Slavin[7] has cited excellent examples of typical early difficulties with lead hollow-cathode sources. Early lead lamps, in which argon was used as filler gas, exhibited argon ion lines near the most sensitive Pb 217·0 nm line, a problem which, when identified, was readily eliminated by the use of neon filler gas.[8] A similar problem was also observed with continuum emission from hydrogen impurity which overlaid the Pb 283·3 nm resonance line.[7] When copper is employed as the metal of the support electrode for high-brightness tin hollow-cathode lamps the transmission of the Cu 224·7 nm line has been found to reduce the sensitivity obtained for the determination of tin at 224·6 nm.[9] Although the intensity of this copper line progressively decreases on prolonged operation of the lamp, the Sn 235·4 or 286·3 nm lines might be chosen for use under some conditions where a more linear calibration curve is required.

(*c*) *Background absorption in the atom cell*

Other practical problems that may lead to the choice of a line other than that of highest oscillator strength are associated with the existence of high flame background absorption and noise at the latter line. This is commonly observed at wavelengths below 200 nm, where the atmosphere, optical components and flame gases may absorb an appreciable fraction of the resonance-line radiation from the source. Thus, flame and atmospheric absorption hinder the use of the mercury line of greatest oscillator strength at 184·9 nm (f_{ji} = 1·2), and the line at 253·7 nm (f_{ji} = *ca.* 0·03) is invariably used for the determination of mercury in flame cells. In the same manner, the high flame background absorbance below 200 nm combined with the relatively low-intensity output from arsenic hollow-cathode lamps (which necessitates the use of high gain at the detector) usually results in superior detection limits at the As 193·7 line over those obtained at the As 189·0 nm line. Even at the As 193·7 nm line appreciable flame background absorption is observed in hydrocarbon-air or hydrogen-air flames, although the cool argon-hydrogen/entrained air flame[10,11] and the hot fuel-rich nitrous oxide-acetylene flame[12] have been reported to exhibit much higher transmission at this wavelength.

At wavelengths longer than 200 nm, molecular band emission or absorption from the flame-gas components may give rise to excessive signal noise and poor detection limits at otherwise sensitive absorption lines for some elements. In the determination of bismuth, for example, the Bi 222·83 nm and 306·77 nm lines produce comparable sensitivities in an air-acetylene flame. However, the Bi 306·77 nm line almost coincides with the strong OH 306·8 nm band head, which both absorbs part of the source radiation at 306·77 nm and is also emitted with appreciable intensity by the flame. Thus, use of this line can lead to excessive signal noise, and the Bi 222·83 nm line may be preferred. A second example of this effect of molecular band emission and absorption from flame gases occurs when the nitrous oxide-acetylene flame is used instead of the air-acetylene flame for the determination of chromium in order to eliminate chemical interference from other elements (notably iron). Whereas the Cr 357·87 or 359·35 nm lines provide highest sensitivity, less signal noise may be obtained at the somewhat less sensitive Cr 360·53 or 425·43 nm lines owing to the intense CN band head at 359·0 nm (degraded to shorter wavelengths) which is emitted by the fuel-rich nitrous oxide-acetylene flame. In practice the situation is made more complex when argon-filled chromium hollow-cathode lamps are employed owing to the emission of argon lines at 357·66, 358·23 and 360·65 nm which influence the sensitivity attainable.

(*d*) *Absorption by concomitant elements in the sample*

Absorption in the flame gases by molecules or atoms formed from the concomitant elements in the sample may also affect the choice of the analytical line to be used for a particular element. Spectral interference of this type (described in Section 12.3.1) due to molecular absorption by the sample may be corrected for by using a continuum background source. However, this facility is frequently not available and it may then be necessary to utilize an alternative absorption line of the analyte element. Similarly, correction with a continuum source cannot be applied in the few observed cases of spectral overlap interference caused by atomic absorption from concomitant elements and the use of an alternative absorption line is then necessary.

11.1.2 Atomic Fluorescence Spectrometry

In AFS, selection of the atomic line used for quantitative intensity measurements is governed by the nature of the fluorescence process for the element to be determined. For elements whose atoms show simple optical spectra the resonance fluorescence emission stimulated at the ground-state line of highest absorption oscillator strength has most frequently been employed for analytical AFS. This wavelength, of course, then corresponds to that usually employed for AAS for that element, and in AFS provides higher absolute signal intensities than direct-line or step-wise line fluorescence (see Section 4.3.2 and 4.3.3). In some cases, however, direct-line fluorescence may be more intense than the resonance fluorescence. Thus, for example, the Tl 535·0 nm direct-line fluorescence signal has been observed to be twice as intense as the resonance fluorescence emission at 377·6 nm.[13] Similarly, the Pb 405·8 nm direct-line fluorescence in a number of flames has been reported to be more intense than the resonance fluorescence emission at 283·3 nm.[14]

The relative intensities of different lines of an element emitted by the source used to excite atomic fluorescence, and the variation in photomultiplier response with wavelength, are also important factors that may influence the atomic line used for analysis by AFS. This latter effect of detector sensitivity, particularly, may enhance the apparent signal intensity obtained at non-resonance lines compared to that observed for the resonance line (see thallium and lead above). This results from the fact that the quanta of radiation emitted in the non-resonance fluorescence processes are usually of lower energy (longer wavelength) than the quanta absorbed and re-emitted in the resonance fluorescence process, and the decreased spectral sensitivity of many photomultiplier tubes in the ultraviolet region. An example of this effect encountered by several workers is found in the determination of selenium by AFS,[15] where higher intensity is observed for the Se 204·0 nm direct-line fluorescence than for the Se 196·1 nm resonance fluorescence owing to the rapid decrease in sensitivity with decreasing wavelength near 200 nm shown by the photomultiplier employed. As in AAS, when high flame background absorption or flame emission noise is registered at the wavelength of the resonance line, an alternative line of lower absolute intensity may yield higher signal : noise ratios. Thus in some instances it may be preferable to avoid the measurement of resonance fluorescence emission.

The existence of particulate material giving rise to scatter of radiation in the flame during sample introduction is difficult to differentiate from fluorescence when resonance fluorescence is measured. Under these conditions the detection of scattered radiation may be avoided by measurement of stepwise or direct-line fluorescence at longer wavelengths together with the use of a simple filter in the incident light path (to select the resonance excitation line and reject any emission from the source at the wavelength to be measured). A practical example of the application of non-resonance fluorescence measurement is found in the determination of bismuth in aluminium alloys.[16] Interference with the determination of bismuth occurs owing to scatter of source radiation when large amounts of the refractory oxide-forming matrix element (aluminium) are introduced into an air-acetylene flame. This is avoided by measurement of bismuth direct-line fluorescence at 302·5 nm with excitation at the Bi 206·2 nm resonance line.

The range of linearity of the analytical growth curve in AFS for a particular element may also be inadequate for some applications when resonance fluorescence is measured. Under conditions where re-absorption of the fluorescence emission occurs before it leaves the atom cell, curvature of the analytical growth curve towards the concentration axis occurs. The effect may frequently be eliminated by measurement of direct-line fluorescence; unless a very hot flame is employed the thermal population of the lower state of the direct-line emission will usually be too low to result in appreciable reabsorption of the emitted direct-line fluorescence.

11.2 OPTIMIZATION OF INSTRUMENTAL CONDITIONS

11.2.1 Spectral Bandpass

The wavelength range over which the spectrometer transmits incident radiation when a particular wavelength is selected has a substantial effect on both the sensitivity and selectivity obtained in analytical AAS and AFS. The manner in which this wavelength interval is controlled by the basic characteristics of the monochromator or filter employed is discussed in Chapter 9. In all commercially available instruments, and in most homemade spectrometers, the method of wavelength selection for either AAS or AFS will already have been chosen; hence in the determination of a particular element only the *spectral bandpass*, governed by the slit-width employed, can usually be optimized. When a monochromator is employed as the dispersive element, variation of the slit-width employed controls both the spectral bandpass and the radiant power received at the photomultiplier detector. Thus, optimization of this parameter usually involves a compromise between these two effects.

(*a*) *Optimization of spectral bandpass in atomic absorption spectroscopy*

In practical AAS procedures the monochromator slit-width is optimized to produce high-sensitivity, linear calibration curves and satisfactory signal : noise ratio with the resonance line of the element to be determined. In order to achieve good sensitivity (for 1% absorption) and adequate linear working range it is necessary to employ a sufficiently small spectral bandpass (slit-width) to ensure isolation of the resonance line of interest from nearby non-resonance lines, weakly absorbing lines, ion lines, or filler gas lines or continuum emission from the spectral source employed. The improvement in sensitivity and linear absorbance range obtained by decreasing the spectral bandpass for the determination of nickel at 232·0 nm with a conventional hollow-cathode lamp is shown in Fig. 11-2; the improvement results directly from the exclusion of unwanted radiation from the source. The use of a narrow slit-width also reduces the radiant power received at the photomultiplier detector owing to flame emission at the analyte resonance line or nearby wavelengths; flame emission noise should therefore be minimized. When the flame emission from the analyte, concomitant elements or the flame gases themselves takes the form of broad-band or continuum emission from molecular species or particles, the effect

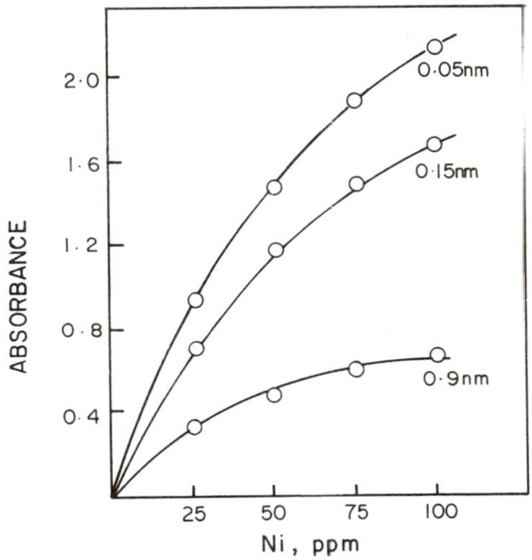

Fig. 11-2. Effect of spectral bandpass on calibration graphs for nickel in air-acetylene flame using conventional hollow-cathode lamp source.

of slit-width on the signal : noise ratio at the detector may be quite marked.*

In view of the above statements it might be expected that the use of a narrow slit-width in AAS would be advantageous under most conditions. Unfortunately, however, for monochromatic radiation from the analytical resonance line of the source the radiant power received by the photomultiplier is directly proportioned to the slit-width employed.† Thus, when the radiance of the analytical line from the source is low, use of a narrow slit-width may result in a poor signal : noise ratio on amplification of the power received at the detector. A common example of this effect in AAS occurs with the

* For a continuum the radiant power received at the detector is proportional to the square of the slit-width (see also Chapter 9).

† For a monochromatic source the radiant power received at the detector is equal to the product of the radiance of the source (watts cm^{-2} $ster^{-1}$) and the luminosity obtained for the monochromator (cm^2 ster). The luminosity may be shown to be proportional to the spectral bandpass (slit-width) for both prism and grating monochromators[17,18,19] (see also Chapter 9).

hollow-cathode sources of the volatile elements, arsenic and selenium, where much of the radiant power at their resonance lines below 200 nm is absorbed by the optical components and the flame gases. Use of high amplifier gain to compensate for the low light levels may then give rise to poor signal : noise ratios. On the other hand, if the slit-width is increased to circumvent this difficulty, lower sensitivity and restricted linear working range may result for these elements as the source often exhibits an inadequate signal : background intensity ratio at the resonance line employed.

A practical problem of using narrow slit-widths in AAS concerns the need to locate and maintain the monochromator setting at the wavelength of peak intensity from the spectral source. During an extended period of operation the wavelength setting may drift owing to thermal expansion of the components or vibration of the instrument. This results in the need for repetitive checking and resetting of the 100% transmission reading by fine adjustment of the wavelength, a procedure that may be inconvenient when automatic sample handling and read-out facilities are employed. The above considerations of spectral bandpass and optimization of slit-width are not applicable when a filter or non-dispersive system is employed for AAS. Thus when a resonance detector[20,21] is employed for wavelength selection in AAS, considerations of slit adjustment do not arise (see Section 9.3.2). With this device it is necessary to arrange adequate shielding to restrict the amount of stray radiation in the system, and the operating conditions of both the spectral source and resonance detector determine the radiant power which is received at the photomultiplier from the primary source. Similarly, no slit is employed to control the spectral bandpass and radiant power received at the detector when the selective modulation principle[22-24] is applied in AAS for selection of the resonance line (see Section 9.3.3). However, light stops, auxiliary interference filters and solar-blind photomultipliers may be used. With both of the above methods of wavelengths selection, problems of wavelength drift cannot arise; from this viewpoint they show particular promise for use in automated multi-channel spectrometers.

(*b*) *Optimization of spectral bandpass in atomic fluorescence spectrometry*

The radiant power of the fluorescence emitted from the atom cell in AFS is usually several orders of magnitude less than the power transmitted from the source by the atom cell in AAS. It is usual, therefore, to employ a dispersive element of high luminosity to maximize the power received at the photodector. With most dispersive elements of moderate size and cost, such as grating or prism monochromators or interference filters, this usually requires acceptance of a greater spectral bandpass (use of wide slit-width). In general terms, however, analytical AFS may be performed satisfactorily with a wider spectral bandpass than would be acceptable in AAS. This is possible because in AFS the atom cell itself acts as a monochromator (or resonance detector) to absorb resonance radiation from the spectral line source and subsequently re-emit it at the same or a different wavelength, while rejecting unwanted source radiation. The diagram in Fig. 11-3 illustrates the possible use of a wider spectral bandpass in AFS than in AAS under conditions where the source emits non-absorbing lines at wavelengths similar to that of the analyte resonance line. The same effect permits the successful use of continuum sources or broad line sources in AFS without recourse to the use of a high-resolution monochromator; unabsorbed radiation from the continuum is not received at the detector. At the same time, however, when a continuum source such as a xenon arc lamp is employed in conjunction with a wide spectral bandpass in the monochromator, other spectral interference may result. It is possible for the source simultaneously to stimulate fluorescence from more than one element at wavelengths which are relatively close together. Spectral interference may result under these conditions when the monochromator transmits both the fluorescence line of the analyte element and that of a concomitant element. In AFS work with a 450-watt xenon arc source and a grating monochromator of large spectral bandpass, Cresser and West have observed interference in this way by Ni 323·0 nm on Cu 327·4 nm, Cu 324·7 nm on Ag 328·1 nm, Ni 341·5 nm on Ag 338·3 nm, Co 419·1 nm on In 410·2 nm and Co 352·7 nm on Ni 352·4 nm.[25] When a continuum source is employed to stimulate resonance fluorescence, any particulate material from the sample present in the atom cell may scatter radiation over a wide wavelength range. The radiant power received at the detector from this "continuum" scatter is proportional to the square of the slit-width, so that its interference effect on the analytical signal is more serious than when scatter only at the resonance line wavelength is produced using a line source (where the radiant power is directly proportional to the slit-width). It should be remembered that the problem of scatter in AFS is not removed by source modulation.

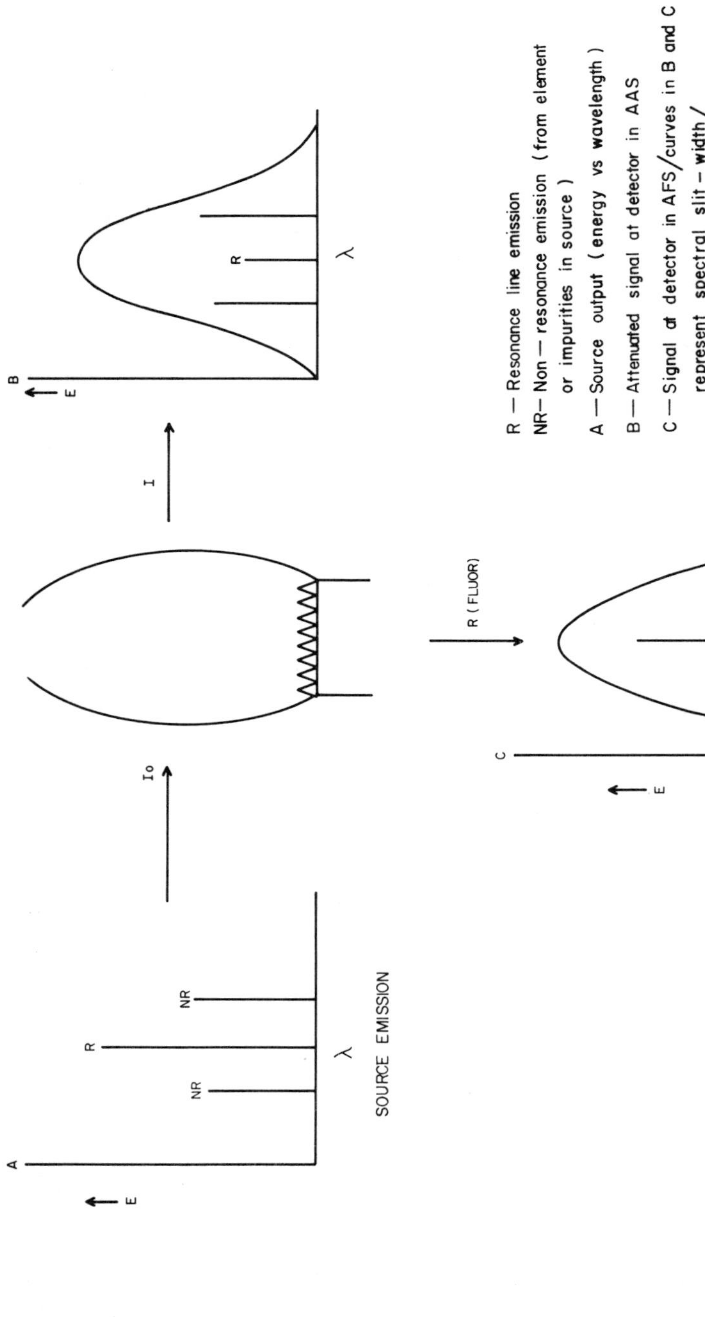

Fig. 11-3. Illustration of possible use of wider spectral bandpass in AFS than in AAS.

However, when a continuum source rather than a line source is used to excite resonance fluorescence, a correction can be made for scattered light by scanning the wavelength region on either side of the resonance line emission and subtracting the observed "continuum" background caused by scattered light.

Flame background emission and atomic or molecular emission from the sample introduced into the atom cell are also collected at the detector in AFS. When a wide spectral bandpass is employed the radiant power received from this emission may be large. This effect may then result in considerable signal from flame background which must be "backed off" when a dc source and detector system are employed. Even with a modulated source and detector, the high level of radiation reaching the detector when a wide spectral bandpass is used may have an adverse effect on the signal : noise ratio obtained at particular wavelengths. However, the effect of atomic or molecular emission from the sample is not usually a serious problem when a modulated light source is used because the useful fluorescence lines of many elements lie below 300 nm where thermal emission from the sample is usually of low intensity (except when very hot flames are employed).

In view of the less stringent requirements with respect to spectral bandpass, the use of interference filters and non-dispersive assemblies employing solar-blind photomultipliers (see Section 9.3) is more acceptable for AFS than for AAS. As described for AAS, problems of optimization of spectral bandpass and wavelength adjustment are then minimized; in commercially available instruments that employ this type of system the appropriate filters are preselected by the manufacturer. Even so, however, the attainable signal : noise ratio for the determination of an element by AFS at a particular wavelength still depends on the efficient collection of modulated fluorescence radiation and rejection (or minimizing) of the background emission from the atom cell.

11.2.2 Atom Cell Atmosphere and Illumination

It is usual practice, when an instrument is set up to undertake a particular determination by AAS or AFS, to vary the flow rate of fuel (for a flame) or inert gas (for a non-flame cell) to obtain the highest attainable concentration of free atoms of the analyte element. In conjunction with this experimental procedure the position of the atom cell in the beam of radiation from the source may be optimized to produce the highest sensitivity and signal : noise ratio. As described in Section 6.5, in flame cells the influence of fuel flow rate and height of observation on sensitivity are particularly marked in both AAS and AFS. When non-flame cells of the filament type (see Section 7.3) are employed, variation in height of observation again has a pronounced effect on signal strength in both techniques; the effect of variation in the flow rate of the inert shielding gas employed affects the residence time of the analyte atoms in the incident beam of radiation from the source. When non-flame cells of the furnace type (see Section 7.2) are employed with either technique, the incident radiation beam may fill the cell volume, so that variation of the position of the cell in the beam is unnecessary; however, the flow rate of inert support gas employed still affects the residence time of analyte atoms.

(*a*) *Atomic absorption spectroscopy*

In AAS the spectral source radiation is directed along the optical axis of the atom cell to provide maximum radiant power at the photomultiplier detector. With most modern instrumentation this is achieved by focusing the radiation from the source into the atom cell and refocusing the divergent transmitted beam with a lens of suitable aperture to permit efficient filling of the collimating mirror or lens of the monochromator.* With this arrangement it is possible to direct a narrow beam of radiation through the cell so that within its length the diameter of the cone of radiation is completely encompassed by the cell. When this condition is fulfilled it is then possible, by fine control of the burner height and its position in the horizontal plane, to ensure that as much of the beam as possible traverses the region of highest analyte atom concentration within the cell. As described in Section 6.5, the distribution of the analyte atom concentration is seldom homogeneous over the cell volume in flames and for many elements is extremely dependent on the fuel : oxidant concentration ratio in the gases supplied to the flame. The above experimental optimization of the burner height etc. is therefore usually performed using several different fuel flow rates to obtain the highest available sensitivity.

(*b*) *Atomic fluorescence spectroscopy*

The general theoretical principles governing the optical arrangement of the fluorescence cell have been discussed in Section 4.4 and

* L'Vov[26] has demonstrated that use of this optical arrangement does not result in the registration of greater background radiance from the atom cell than in earlier optical arrangements in which a parallel or convergent beam of radiation from the source was passed through the cell.

will not be repeated here. However, it is important to emphasize the importance of correct illumination of the cell to obtain the most efficient absorption of the incident radiation, of collecting the largest possible fraction of the re-emitted fluorescence., and of minimizing re-absorption of resonance fluorescence within the atom cell.[27] For a given instrumental assembly, where the type of source, atom cell and monochromator entrance optics are fixed, experimental optimization of the illumination conditions should be undertaken for each analyte element to be determined. This is particularly important for elements that are difficult to atomize, where the distribution of the atomic population in the atom cell may be far from uniform. The position of the atom cell in the incident radiation beam from the source may be adjusted to ensure that the analytical signal : noise ratio and the linear range of the fluorescence growth curve are sufficient for a particular application. It may be necessary to compromise between the two requirements, i.e. to accept a smaller absolute signal intensity in order to achieve a wide linear concentration range, or vice versa. The fuel flow rate (or flow rate of inert support gas for a non-flame cell) should then again be optimized to produce highest signal : noise ratio as in AAS.

11.2.3 Source Operating Power

The power at which the source is operated may have a marked effect on the sensitivity and the analytical calibration curves in both techniques. When a spectrometer is set up for the determination of a particular element by AAS or AFS, therefore, the source operating power should be optimized to produce the required analytical performance.

(*a*) *Atomic absorption spectroscopy*

As described in Chapter 5 the spectral line source employed in AAS must provide a narrow emission line profile, a stable output, a high radiant intensity at the resonance line of interest and an extended operating life-time. In the most commonly used sources, hollow-cathode lamps, vapour-discharge lamps and electrodeless discharge lamps, each of these factors may be affected by the source operating power. When sufficient radiant intensity is easily obtainable at the wavelength of the resonance line, the *minimum* operating power consistent with a stable output will generally give the highest analytical sensitivity. This will usually ensure adequate lamp lifetime for modern types of these sources. The effect of lamp operating power on the profiles of lines emitted by hollow-cathode and electrodeless discharge lamp sources was discussed in Section 5.2.5 (see Figs. 5-15 and 5-16). The practical effects of the substantial increase in half width due to self-absorption at high power are illustrated in Fig. 11-4 which shows the corresponding variation with lamp operating power of mercury absorbance using two different EDL sources and an air-propane flame. In this particular example the mercury vapour pressure, and hence self-absorption, at the resonance

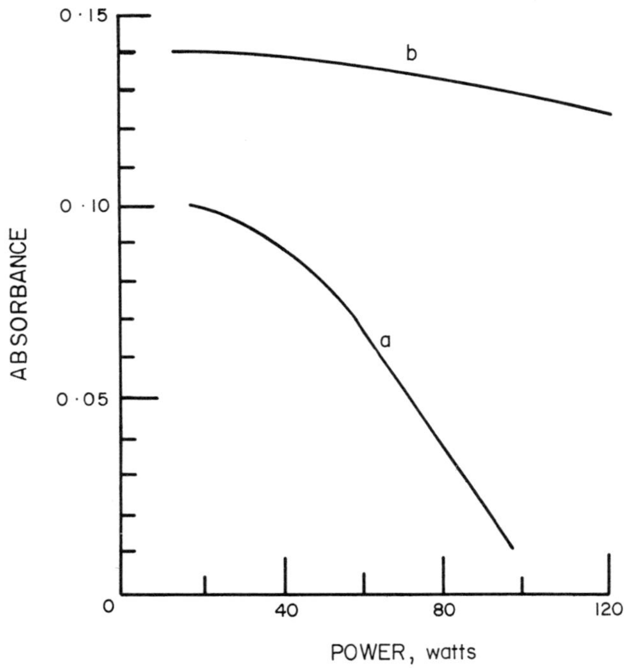

Fig. 11-4. Effect of microwave power on absorbance for Hg in air/propane flame with EDL source. (a) EDL source 40 mm in length; (b) EDL source 250 mm in length.

line at 253·65 nm, shows a much slower increase with power when a long EDL source is employed.[28]

It may be necessary to use high operating power for the hollow-cathode sources of some elements where the radiant intensity available at the resonance line is low. When this occurs with relatively volatile elements (e.g. arsenic and selenium) which have their strongest absorption lines at wavelengths at which background absorption in the atom cell is high, the lamp lifetime may be reduced substantially. Although electrodeless discharge lamp sources often provide high radiant intensity at the resonance line for just those elements whose volatility is a problem in hollow-cathode sources, the emission from these sources may suffer appreciably from self-absorption broadening at all but the lowest operating power.

Manufacturers of commercial instrumentation for AAS usually recommend the optimum operating power for each of the sources they supply. This recommendation should be adhered to where possible as it will normally represent a suitable compromise between the requirements for intensity, stability, line-width and operating lifetime.

(*b*) *Atomic fluorescence spectrometry*

As the intensity of the fluorescence radiation observed in AFS depends on the incident radiance of the source employed, the source operating power is an important parameter which may require careful adjustment. Unlike AAS, however, the resonance line-width requirement is of much less importance than that of obtaining maximum output intensity (consistent with a sufficiently long operating lifetime). Indeed, as explained in Chapter 4, the maximum transfer of energy to atoms in the fluorescence cell will be obtained when the excitation and absorption line profiles are similar. Even if the source line exceeds the width of the absorption line, unabsorbed radiation at the edges of the emission line is simply "wasted" and there are no other adverse effects. However, self-reversal of the source line, where the central most strongly absorbed part of the line profile is emitted from the lamp at reduced intensity compared to the wings, leads to reduced sensitivity and must be avoided.

When continuum sources such as the xenon arc are employed for AFS the power consumption is normally limited by the lamp type and supply requirements for a stable discharge. Even when variation of the power applied is possible, the maximum recommended current is usually required for optimum fluorescence measurements. When hollow-cathode lamp sources are used for AFS the optimum current will usually be the maximum value consistent with satisfactory lamp lifetime and absence of self-reversal of the resonance line (or even serious self-absorption, which in certain cases can actually result in a reduction of output radiant intensity on further increase of lamp power). When such sources are operated in the pulsed mode the average current through the source should be approximately equal to the current that provides satisfactory continuous operation of the same lamp. For electrodeless discharge lamp sources it may be necessary to optimize the operating power in conjunction with the position of the lamp in the tuning cavity and the temperature at which the lamp is operated. Metal-vapour-discharge lamp sources should be operated at currents that provide high spectral radiance without self-reversal of the metal resonance line.

11.2.4 Detector Response and Amplifier Characteristics

It was explained in Chapter 10 that the instrumental systems usually available for AAS and AFS allow adjustment of the signal observed for a given light level reaching the detector. This is achieved by varying the voltage applied to the dynode chain of the photomultiplier or by altering the gain setting of the electronic amplifier used to process the current from the detector. Most commercial atomic absorption spectrometers have one or both of these facilities available. It is then important that the instrument settings be optimized for each determination. When the photomultiplier operating voltage or the gain of the amplifier is increased, the dark current observed with no light reaching the detector and the "noise level"* of the signal increase. Over a certain optimum range of applied voltage to the photomultiplier (*ca.* 500 to 1000 V for typical tubes used in AAS and AFS) the ratio of the photocurrent to the dark current may remain virtually constant. It should generally be ensured that the photomultiplier is operating in this range for both techniques, so that adequate precision is achieved. The overall gain settings employed with a particular flame spectrometer for the determination of a given element will depend on the spectral radiance of the atomic line from the source and the spectral slit-width employed (in AAS), and on the intensity of the fluorescence emission signals from the atom cell and the spectral slit-width (in AFS). In most practical systems, particularly where flame cells are employed, it is usually the source and atom cell "noise" that govern the limit of detection for a particular element, rather than the detector shot-noise.

The instrument to be employed may also be equipped with a variable time constant for the output from the amplifier to the meter or chart recorder. As explained in Chapter 10.2, this allows short-term fluctuations in the signal to be damped out, the amount of damping usually being varied by increasing the capacitance applied across the output terminals. The damping must be reset for each determination; since increasing the damping also increases the

* So-called "shot-noise"; see Chapter 10.1.

response time of the instrument, it is usually best to utilize the minimum damping, allowing adequate precision for a particular measurement. This is particularly important when flame cells are used, as the longer time needed to obtain the final signal level with heavy damping requires that considerably more sample solution be nebulized. The sample should be nebulized for a period not less than *ca.* 10 times the effective response time of the instrument and a factor as high as 20 times may be preferable. Thus, when only small volumes of sample solution are available it is frequently necessary to work with short time constants (perhaps as short as 0·5–1 s) and accept a rapidly fluctuating signal and lower precision. For non-flame cells where a continuous analytical signal is not easily guaranteed for a long period of time, it is not possible to achieve high precision and accuracy by employing long time constants. L'Vov[28a] has demonstrated that for observation of transient absorption signals in a graphite furnace the highest signal : noise ratio is available when the response time of the detector matches the residence time of the sample atoms in the absorption cell.

11.2.5 Nebulizer Uptake Rate and Efficiency

In both AAS and AFS the sensitivity and detection limits obtainable for a particular element with a flame cell depend on the nebulizer performance characteristics. In addition, the drop-size distribution may also control the extent to which physical and chemical (condensed phase) interferences are encountered from other elements. With the simple pneumatic units used on most commercial spectrometers the sample solution uptake rate at the nebulizer depends on the pressure drop across the capillary outlet; this in turn depends on the operating pressure of the nebulizing gas. The sample uptake rate and volume of gas produced after the nebulizer jet, however, both increase almost proportionally when the gas operating pressure before the jet is varied. The concentration of aerosol in the gas, and thus the analyte concentration per unit volume of the flame gases, may therefore not vary substantially when the nebulizing gas pressure is varied with a nebulizer assembly having no other means of adjustment. Additionally, the range over which the nebulizer gas operating pressure may be varied is determined by the required volume and stability of the flame to be supported. Thus, under these conditions very little adjustment is possible and the optimum gas flow to the nebulizer may be similar for all determinations.

When indirect nebulizers are employed with premixed flames the use of an adjustable nebulizer may permit greater flexibility in optimization of operating conditions. The most common of these devices provide for variation in positioning of the sample uptake capillary within the concentric nebulizing gas outlet (see Section 8.2.1). The sample uptake rate, and the yield of solution reaching the flame, may then be varied without variation in the nebulizer gas flow-rate, so that the flame volume and stability remain unchanged. It has been shown[29] that when an adjustable nebulizer of this type is used the mean drop size increases with increasing uptake rate, while the *fraction* of the sample solution reaching the burner decreases. When an adjustable nebulizer is available, therefore, the optimization of the uptake rate to produce the greatest sensitivity and precision or freedom from chemical interferences is worthwhile for a particular analytical application.

Apart from this method of variation of the nebulizer efficiency with a particular sample solution via use of an adjustable nebulizer, the efficiency of the nebulizer may usually only be further "optimized" by changing the nature of the sample solution. The addition of a water-miscible organic solvent, such as acetone or ethanol, to the aqueous sample solution usually results in some enhancement of sensitivity in AAS and AFS despite the dilution effect of the added solvent. The improvement results from increased nebulizer efficiency due to smaller mean droplet size caused by the lower surface tension of the mixed solvent and more rapid solvent evaporation from the droplets in the spray chamber (with indirect nebulizers) or in the flame (with integral nebulizer-burners). In addition, there may be less cooling of the flame by the mixed solvent compared to that which occurs with a wholly aqueous solution, although this effect will depend on whether a lean or a rich flame is employed. With mixed organic solvent/water solutions a more favourable environment may be created in the flame for the formation of atoms of the analyte or concomitant elements.[30] The organic solvent/water mixtures used in this way frequently have a higher viscosity than have aqueous solutions; this lowers the sample uptake rate of the nebulizer. However, provided the percentage of organic solvent in the solution is carefully optimized, the depression of the analytical signal which this would cause can usually be outweighed by the effect of enhanced nebulizer efficiency. In fact the lower sample uptake rate under these conditions may be an advantage when the amount of sample solution available is limited.

The use of water miscible organic solvent mixtures to enhance AAS sensitivities in premixed air-acetylene flames was investigated in an early paper by Allan.[31] Typical enhancement factors in AAS obtained for copper in organic solvent/water mixtures are shown in Table 11-1. Allan concluded that a solution containing 20% acetone + 20% isobutanol was most satisfactory, and demonstrated that the enhancements observed in this mixture were the same for magnesium, zinc, iron, manganese and copper, as would be expected if the phenomenon responsible is improved sample transport to the flame.

TABLE 11-1

Enhancement of copper absorption in a premixed air-acetylene flame by use of water-miscible solvents[a]

Solvent	Relative sensitivity
0·1 M HCl	1·0
Methanol, 40%	1·7
Ethanol, 40%	1·7
Isopropyl alcohol, 40%	1·8
Acetic acid, 10%	1·0
Dioxan, 40%	1·5
Acetone, 40%	2·0
Acetone, 80%	3·5
Acetone, 20% + isobutanol, 20%	2·35

[a] From ref. 31 (J. E. Allan, *Spectrochim. Acta*, 1961, 17, 467).

11.3 ANALYTICAL SENSITIVITY

11.3.1 Atomic Absorption Spectroscopy

The *sensitivity* attainable for the determination of an element by AAS with given instrumentation is defined as the concentration of the element (usually as ppm, μg/ml or mg/l in aqueous solution) which produces a 1% absorption signal (0·0044 absorbance) under optimum experimental conditions. This value is sometimes also referred to as the "percentual concentration limit" or "percentual sensitivity".[32,33] The sensitivity value depends primarily on the absorption path-length and the efficiency of nebulization and atomization in the absorption cell. This sensitivity value defines the slope of a linear calibration graph of absorption against concentration and hence it is frequently useful when a knowledge of the

TABLE 11-2

Sensitivity and detection limit data for different elements by AAS[a]

Element	Wavelength, nm	Flame[b]	Sensitivity, ppm/1% absorption	Detection[c] limit, ppm
Ag	328·1	A–A	0·06	0·002
Al	309·3	A–N	0·9	0·03
As	193·7	Ar–H	0·25	0·1
Au	242·8	A–A	0·2	0·02
B	249·7	A–N	40	2·5
Ba	553·6	A–N	0·3[d]	0·02
Be	234·9	A–N	0·03	0·001
Bi	223·1	A–A	0·8	0·04
Ca	422·7	A–N	0·05[d]	0·001
Cd	228·8	A–A	0·25	0·001
Ce	520·0	A–N	—	—
Co	240·7	A–A	0·15	0·01
Cr	357·9	A–A	0·1	0·003
Cs	852·1	A–A	0·3	0·05
Cu	324·7	A–A	0·1	0·002
Dy	421·2	A–N	0·8[d]	0·2
Er	400·8	A–N	0·9[d]	0·1
Eu	459·4	A–N	0·6[d]	0·04
Fe	248·3	A–A	0·1	0·01
Ga	287·4	A–N	1·2	0·1
Gd	407·8	A–N	16[d]	4
Ge	265·1	A–N	2·5	1
Hf	286·6	A–N	15	8
Hg	253·7	A–A	10	0·5
Ho	410·4	A–N	1·5[d]	0·1
In	303·9	A–A	0·9	0·05
Ir	264·0	A–N	12	2
K	766·5	A–A	0·02	0·005
La	550·1	A–N	30[d]	2
Li	670·8	A–A	0·03	0·0006
Lu	335·9	A–N	6[d]	3
Mg	285·2	A–A	0·007	0·0001
Mn	279·5	A–A	0·05	0·002
Mo	313·3	A–N	0·6	0·03
Na	589·0	A–A	0·015[d]	0·002
Nb	334·4	A–N	20[d]	1
Nd	463·4	A–N	13[d]	2
Ni	232·0	A–A	0·1	0·01
Os	290·9	A–N	1·0	0·5
Pb	283·3	A–A	0·5	0·02
Pd	247·6	A–A	0·2	0·02
Pr	495·1	A–N	22[d]	10
Pt	265·9	A–A	2·5	0·1
Rb	780·0	A–A	0·05	0·005
Re	346·0	A–N	12	1

TABLE 11.2 (*continued*)

Element	Wavelength, nm	Flame[b]	Sensitivity, ppm/1% absorption	Detection[c] limit, ppm
Rh	353·5	A–A	0·5	0·03
Ru	349·9	A–A	0·5	0·3
Sb	217·6	A–A	0·6	0·1
Sc	391·2	A–N	0·65[d]	0·1
Se	196·0	Ar–H	0·25	0·1
Si	251·6	A–N	2	0·08
Sm	429·7	A–N	7[d]	2
Sn	224·6	Ar–H	0·6	0·02
Sr	460·7	A–A	0·15[d]	0·01
Ta	274·2	A–N	40	2
Tb	432·6	A–N	8[d]	3
Tc	261·6	A–A	3	—
Te	214·3	A–A	0·5	0·09
Ti	365·3	A–N	2	0·09
Tl	276·8	A–A	0·5	0·03
Tm	371·8	A–N	0·7[d]	0·2
U	358·5	A–N	50	30
V	318·4	A–N	1·5	0·06
W	400·9	A–N	20	3
Y	410·2	A–N	2[d]	0·3
Yb	398·8	A–N	0·2[d]	0·04
Zn	213·9	A–A	0·015	0·002
Zr	360·1	A–N	6	5
S	180·7	A–N	1	5[e]
P	178·3	A–N	5.4	21[f]
I	183·0	A–N	12	25[g]

[a] Perkin-Elmer, Model 403, in "Analytical Methods for Atomic Absorption Spectrophotometry", The Perkin-Elmer Corp., Norwalk, Connecticut, 1971.

[b] A–A = air-acetylene; A–N = acetylene-nitrous oxide; Ar–H = argon-hydrogen-entrained air.

[c] Defined as concentration giving deflection equal to twice the standard deviations of background (baseline) readings.

[d] In presence of excess potassium to suppress ionization.

[e] G. F. Kirkbright and P. J. Wilson, *Anal. Chem.*, 1974, in press.

[f] G. F. Kirkbright and M. Marshall, ibid., 1973, 45, 1610.

[g] G. F. Kirkbright, T. S. West and P. J. Wilson, *Atomic Absorption Newsletter*, 1972, 11, 53.

required analyte concentration range in sample and standard solutions is needed. The attainable 1% absorption sensitivity values for particular elements under optimum experimental conditions remain fairly constant between well-designed instruments of the same type. Some values obtained for different elements by AAS with a typical commercial instrument are shown in Table 11-2. For analytical work a disadvantage of using the 1% absorption sensitivity is that it yields no information concerning the attainable precision or the minimum concentration that may be determined.

Adequate analytical precision may be obtained with modern AAS instruments for many elements at solution concentrations that produce less than 1% absorption. The term *detection limit* is used to define the lowest detectable concentration of the element in the sample solution. This value is often also referred to as the "fluctuational concentration limit" or "fluctuational sensitivity".[32,33] It may be defined as that concentration of the element that produces an absorption signal equivalent to twice (or three times) the standard deviation in the noise fluctuation of the background (zero absorption) signal under the experimental conditions employed. Whether use is made of twice or three times the standard deviation in the background depends on the confidence level required (95 or 99%). Some workers also use a definition of the detection limit based on the analyte concentration required to produce an absorption signal : background noise ratio of 2; at this limit the relative standard deviation in the analysis is 50%.

The detection limit, in contrast to the 1% absorption sensitivity, thus depends on the instrumental stability and may vary considerably from instrument to instrument and from one type of sample matrix and solvent to another. The fluctuation in the signal may result from source, flame or detector noise. The contribution from fluctuation in intensity from the source depends on whether a single-beam or double-beam instrument is employed; appreciable elimination of source noise is obtained with double-beam spectrometers.[34] The source intensity arriving at the detector is subject to disturbance by the necessity to pass it through the absorption cell. When a flame cell is used the entire width of the light beam should be passed through the centre of the flame, to minimize absorption fluctuation caused by slight variations in shape of the secondary reaction zone. However, small fluctuations in flame background absorption in the interconal zone may still occur below 250 nm owing to slight changes in the fuel : oxidant ratio. In addition to this "absorption noise" from the atom cell the detector may receive appreciable emitted radiation at analyte wavelengths above *ca.* 250 nm from the flame background itself or from concomitant elements in the blank solution. As described elsewhere (Section 11.1.1) this gives rise to noise at the detector even when an a.c. source and detection system are employed. Provided that the radiation flux falling on the photomultiplier detector is sufficiently large, the detector noise may make a relatively small contribution to the overall background fluctuation level in AAS.

The detection limit defined above represents the relative value in concentration units and gives no indication of the response time of the instrument or the amount of sample solution required for a measurement. The background fluctuational level may be reduced by the use of a longer time constant (RC filtering) or by integration. As this increases the time for which the sample must be nebulized into the flame, however, no gain in absolute detection limit (weight of sample) is achieved. This absolute value depends on the relative detection limit and the minimum volume of sample required for the measurement. Typical relative detection limits obtained for different

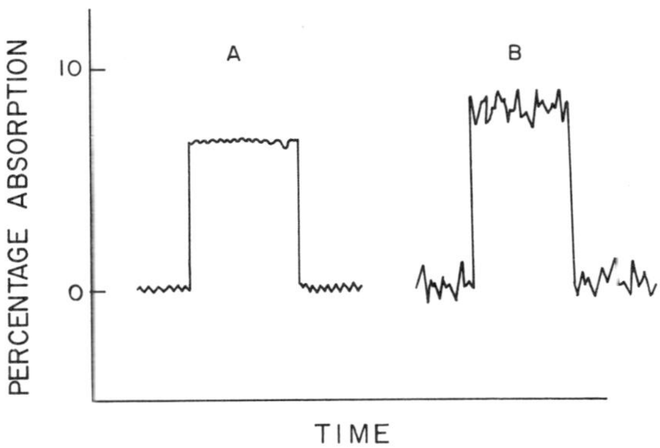

Fig. 11-5. Illustration of differing detection limits obtained for two elements A and B which exhibit similar 1% absorption sensitivities.

elements with a commercial double-beam AAS instrument are shown in Table 11-2.

Two elements A and B may sometimes show similar absorption sensitivities but quite different detection limits owing to differences in the magnitude of background noise from the source, flame or detector for each element. This is illustrated in Fig. 11-5; the detection limit for element A is considerably superior to that for element B although the 1% absorption sensitivity value for element B is slightly greater than that for element A. Thus, comparisons between different elements and different instruments may sometimes be confusing and a further factor is needed to convey an estimate of the attainable precision in the absorbance range used for calibration curves (e.g. 0·1 to 0·5 absorbance).

Smith and Feldman[35] have proposed the term "*performance index*" which is defined as the aqueous concentration and relative standard deviation which yields 0·44 absorbance measured at, or normalized to, a spectral noise bandwidth of 0·25 Hz (equivalent to a 1-second time constant). In many respects this term is of more use to the analyst than the 1% absorption sensitivity or detection limit defined above. The value of 0·44 absorbance is close to that for least photometric error (0·43 absorbance) demonstrated by Ringbom,[36] while at the same time the 1% absorption sensitivity is easily obtained by dividing the concentration term in the performance index by 100 (provided that a linear relationship exists between

TABLE 11-3

Values of performance index in AAS (from refs. 35 and 37)

	Performance index	
Element	% RSD	Concentration for 0·44 absorbance, ppm
Ca	0·2	6·0
Cr	0·3	5·0
Cu	0·3	2·0
Pb	0·5	10
Mg	0·44	0·8
Pt	0·30	40

absorbance and concentration). The detection limit may also be estimated from the relative standard deviation term in the index. Some typical values for performance index obtained with commercial instrumentation are shown in Table 11-3.[35,37]

11.3.2 Atomic Fluorescence Spectroscopy

The "sensitivity" of a procedure for the determination of an element by AFS is usually defined only by the relative *detection limit*, i.e. the concentration of the element in solution which produces a fluorescence signal equivalent to twice (or three times) the standard deviation in the background noise. Some workers also use a detection limit based on the concentration required to produce a signal equal to twice the maximum (peak-to-peak) noise level of the background. In AFS it is the background and noise from the atom cell and detector which principally limit the attainable detection limits. In contrast to AAS, near the detection limit in AFS the source may make a relatively minor contribution to the noise. For small signal intensities the noise level in the re-emitted fluorescence radiation (which originated from the source) is small compared to the flame flicker noise and noise at the detector due to the overall level of the flame background and shot-noise. In early fluorescence studies with intense unmodulated sources and dc amplifiers, low

TABLE 11-4

AFS detection limits for elements in flames

Element	Wavelength, nm	Flame*	Source†	Detection limit, ppm‡	Reference (see below)
Ag	328·1	Air/H_2	EDT	0·0001	*a*
Al	396·1	Argon sep. N_2O/C_2H_2	EDT	0·1	*b*
As		Ar—H_2—air	EDT	0·2	*c*
Au	267·6	Air/H_2	EDT	0·2	*a*
	242·8	O_2—H_2—Ar	H—HCl	0·005	*d*
Be	234·8	Sep. N_2O/C_2H_2	EDT	0·01	*e*
Bi	302·6	Ar—H_2—air	Iodine EDT	0·05	*f*
	306·8	N_2 sep. air/C_2H_2	EDT	0·04	*g*
Ca	422·7	Air/H_2	EDT	0·02	*a*
		H_2—air	450 W Xe	0·02	*h*
Cd	228·8	Air/H_2	EDT	0·000001	*a*
Co	240·7	Air/C_3H_8	EDT	0·005	*i*
Cr	357·9	Air/H_2	Demountable cooled HCL	1·0	*j*
Cu	324·7	Air/H_2	H—HCL	0·001	*k*
	327·4	Air/H_2	Hot HCL	0·001	*l*
Fe	248·3	Air/C_2H_2	EDT	0·009	*m*
		Ar/O_2/H_2	H—HCL	0·02	*n*
Ga	403·3 417·2	Air/H_2	H—HCl	0·5	*o*
Ge	265·1	Ar sep. N_2O/C_2H_2	EDT	0·1	*p*
Hg	253·7	O_2/H_2 org. solv	Hg pen light	0·002	*q*
In	410·5 451·1	Air/H_2	H—HCl	0·2	*o*
Mg	285·2	Air/propane	H—HCl	0·001	*r*
Mn	279·5	Air/C_2H_2	EDT	0·001	*s*
Mo	313·3	Ar. sep. N_2O/C_2H_2	EDT	0·46	*b*
Ni	232·0	Ar/O_2/H_2	H—HCl	0·003	*n*
		Air/H_2	H—HCl	0·003	*k*

TABLE 11-4 (*continued*)

Element	Wavelength, nm	Flame*	Source†	Detection limit, ppm‡	Reference (see below)
Pb	405·8	$Ar/O_2/H_2$	EDT	0·01	*t*
		Air/H_2	H—HCl	0·02	*o*
Pd	340·4	Air/H_2	H—HCL	2·0	*o*
Pt	265·9	Air/H_2	H—HCL	50	*o*
Sb	231·1	Air/C_3H_8	EDT	0·05	*u*
Se	196·0	Air/C_3H_8	EDT	0·25	*v*
Si	251·6	Ar sep. N_2O/C_2H_2	EDT	0·55	*w*
Sn	303·4	N_2 sep. $Ar/O_2/H_2$	EDT	0·1	*x*
Sr	460·7	Ar/H_2—air	EDT	0·03	*a*
Te	238·3 238·6	Air/C_3H_8	EDT	0·12	*v*
Tl	377·6	Ar/H_2—air	EDT	0·008	*a*
V	318·4	Ar sep. N_2O/C_2H_2	EDT	0·07	*b*
Zn	213·9	Air/H_2	EDT	0·00004	*a*

* Ar sep. N_2O/C_2H_2 = argon separated premixed nitrous oxide-acetylene flame; N_2 sep. air/C_2H_2 = nitrogen separated premixed air-acetylene flame.

† H—HCL = high intensity hollow-cathode lamp; EDT = electrodeless discharge tube; 450 W Xe = 450 watt xenon arc source.

‡ Detection limit usually defined as concentration resulting in S/N = 2:1.

[a] R. E. Zacha, M. P. Bratzel and J. D. Winefordner, *Anal. Chem.*, 1968, **40**, 1733.

[b] R. M. Dagnall, G. F. Kirkbright, T. S. West and R. Wood, *Anal. Chem.*, 1970, **42**, 1029.

[c] R. M. Dagnall, K. C. Thompson and T. S. West, *Talanta*, 1968, **15**, 677.

[d] J. Matousek and V. Sychra, *Proc. Intern. Conf. AAS, Sheffield*, 1969.

[e] D. N. Hingle, G. F. Kirkbright and T. S. West, *Analyst*, 1968, **93**, 522.

[f] R. M. Dagnall, K. C. Thompson and T. S. West, *Talanta*, 1967, **4**, 1467.

[g] R. S. Hobbs, G. F. Kirkbright and T. S. West, *Talanta*, 1971, **18**, 859.

[h] D. R. Demers and D. W. Ellis, *Anal. Chem.*, 1968, **40**, 860.

[i] B. Fleet, K. V. Liberty and T. S. West, *Anal. Chim. Acta*, 1969, **45**, 205.

[j] G. Rossi and N. Omenetto, *Talanta*, 1969, **16**, 263.

[k] D. C. Manning and P. Heneage, *Atomic Absorption Newsletter*, 1967, **6**, 124.

[l] J. I. Dinnin, *Anal. Chem.*, 1967, **39**, 1491.

[m] L. Ebdon, G. F. Kirkbright and T. S. West, *Anal. Chim. Acta*, 1969, **47**, 563.

[n] J. Matousek and V. Sychra, *Anal. Chem.*, 1969, **41**, 518.

[o] D. C. Manning and P. Heneage, *Atomic Absorption Newsletter*, 1968, **7**, 80.

[p] G. F. Kirkbright, R. M. Dagnall, T. S. West and R. Wood, *Analyst*, 1970, **95**, 425.

[q] T. J. Vickers and S. P. Merrick, *Talanta*, 1968, **15**, 873.

[r] T. S. West and X. K. Williams, *Anal. Chim. Acta*, 1968, **42**, 29.

[s] L. Ebdon, G. F. Kirkbright and T. S. West, *Talanta*, 1970, **17**, 965.

[t] R. F. Browner, R. M. Dagnall and T. S. West, *Anal. Chim. Acta*, 1970, **50**, 375.

[u] R. M. Dagnall, K. C. Thompson and T. S. West, *Talanta*, 1967, **14**, 1151.

[v] R. M. Dagnall, K. C. Thompson and T. S. West, *Talanta*, 1967, **14**, 557.

[w] R. M. Dagnall, G. F. Kirkbright, T. S. West and R. Wood, *Anal. Chim. Acta*, 1969, **47**, 407.

[x] R. F. Browner, R. M. Dagnall and T. S. West, *Anal. Chim. Acta*, 1969, **46**, 207.

background hydrogen flames or inert gas separated flames were used almost exclusively to permit good detection limits to be obtained. Even with the use of a modulated source and tuned amplifier, however, the use of such low background flames results in less unwanted radiation at the detector and hence a better signal : noise ratio.

TABLE 11-5

Detection limits obtained with a six-channel atomic fluorescence spectrometer[38]

Element		Wavelength, nm	Detection limit, ppm		Linear range, ppm
Aluminium	Al	396·2	0·07	(N_2O)	0–50
Antimony	Sb	231·1	0·05		0–45
Arsenic	As	197·2	0·07	(EDL)	0–50
Barium	Ba	553·5	0·2		0–40
Cadmium	Cd	228·8	0·002		0–8
Calcium	Ca	422·7	0·0003		0–30
Chromium	Cr	357·9	0·001		0–27
Cobalt	Co	240·7	0·002		0–25
Copper	Cu	324·8	0·0006		0–9
Gold	Au	242·8	0·009		0–10
Iron	Fe	248·3	0·002		0–19
Lead	Pb	405·8	0·01	(EDL)	0–160
Magnesium	Mg	285·2	0·0001		0–8
Manganese	Mn	279·5	0·0005		0–8
Mercury	Hg	253·7	0·4	(EDL)	0–50
Molybdenum	Mo	313·3	0·06		0–30
Nickel	Ni	232·0	0·003		0–21
Silver	Ag	328·1	0·002		0–9
Strontium	Sr	460·7	0·0008		0–14
Tin	Sn	224·6	0·2		0–50
Vanadium	V	318·5	0·25	(N_2O)	0–40
Zinc	Zn	213·9	0·001		0–6

1. All data obtained in simultaneous multielement mode of operation basis with aqueous solutions using a slightly fuel-rich air-C_2H_2 flame and pulsed hollow cathode light sources unless otherwise noted. (N_2O = nitrous oxide flame; EDL = electrodeless discharge lamp.)

2. Detection limits are defined as the concentration in solution of an element which can be detected with 95% certainty. This is that quantity of the element that gives a reading equal to twice the standard deviation of a series of at least ten determinations at or near blank level.

Table 11-4 gives some AFS detection limits taken from the literature. Whilst these values all apply to flame cells, a variety of instrumentation has been used by different workers to obtain these detection limits and various authors do not necessarily present their values according to the definition of detection limits given above. Table 11-4 also gives brief details of the source and flame type used to obtain each detection limit. In addition, developments in AFS instrumentation may render these values rapidly obsolete; they are recorded here in order to demonstrate the range and flexibility of the technique. Detection limit data for AFS with non-flame cells are given in Chapter 7. Table 11-5 shows the values of detection limit obtained for various elements when a commercial six-channel atomic fluorescence spectrometer is used.[38]

11.4 INSTRUMENT OPERATION

11.4.1 Operating Procedures

General instructions for the setting up of AAS instruments, and AFS accessories that may be used with these instruments, are invariably provided by the manufacturers. These instructions include advice on the positioning of the instrument in the laboratory to avoid draughts and bright sunlight. The use of an exhaust chimney is recommended for the removal of the burnt gases which frequently contain toxic materials from the samples nebulized when flame cells are employed.

Manufacturers' recommendations concerning the details and sequence of operations to be performed in an experimental method by AAS or AFS may vary but the following outline procedure is normally performed.

(a) The source and detector power is switched on, and the source power is set to the manufacturers' recommended value (or to an experimentally predetermined optimum value). With modern hollow-cathode lamps and photomultiplier-amplifier systems very little warm-up time is required. When electrodeless discharge lamps are used as sources in AFS some additional attention to the tuning of the lamp and microwave cavity assembly to give minimum reflected power may be necessary. Moreover, the time taken to achieve intense and stable emission may vary from element to element.

(b) The source position and optical system are adjusted so that the source radiation (in AAS) or fluorescence radiation (in AFS) is incident upon the monochromator slit;* the analytical line and spectral slit-width to be employed are then selected at the monochromator.

(c) The flame is ignited and the fuel : oxidant flow rate ratio adjusted to the recommended settings. The position of the incident

* This is readily carried out in AFS by temporarily placing a small mirror on the burner so as to deflect the source radiation through 90°.

light beam in the flame is usually optimized by variation of the height and horizontal position of the burner relative to the light beam. Information concerning the correct procedures for igniting and extinguishing the flame is invariably provided in the instruction manual with commercial instrumentation.

(d) The burner is allowed to warm up while solvent is nebulized. During this period the burner head, burner connections and the walls of the spray chamber reach their equilibrium temperatures. This procedure is more important with indirect nebulizers than with total consumption nebulizer-burner units.

(e) The detector response is set via adjustment of the amplifier gain or photomultiplier operating voltage. In AAS the response is adjusted to obtain a reading of 100% transmission (zero absorbance) at the meter or recorder. The zero transmission (infinite absorbance) setting is checked by obstructing the light path between the source and detector with a shutter. In AFS the sensitivity is set to an established optimum setting for the sample concentration range anticipated, and any small contribution from stray or scattered light is "backed-off" electrically so that the meter or recorder reads zero deflection.

(f) Pure solvent is nebulized into the flame and the 100% and 0% transmission settings (in AAS) and zero deflection setting (in AFS) are re-checked and adjusted if necessary. The instrument may then be used for analytical absorption or fluorescence measurements.

(g) The blank, sample and standard solutions are nebulized into the flame sequentially and the absorbance or fluorescence intensity reading is noted for each. Each solution should be aspirated for sufficient time to enable the analytical signal to reach a steady value that can be measured with confidence. The time required depends on the characteristics of the nebulizer-burner assembly and on the response time setting at the detector; it would normally be *ca*. 10-20 times the effective time constant of the system (see Section 11.2.4).

(h) The nebulizer and burner assembly may be washed by aspiration of pure solvent between each sample solution measured. This is particularly important when liquid samples of high dissolved solid content are dealt with in order to minimize clogging of the burner head or nebulizer capillary. The concentration of the element determined in the sample solution is most frequently found by relating its absorbance or fluorescence intensity to the absorption or fluorescence signals produced under the same experimental conditions by solutions of known concentration. These latter values may be plotted as a "calibration graph" (or curve) of analytical signal against concentration of the element in the standard solution; the sample solution concentration is then established from this graph. Alternatively, with direct concentration read-out facilities, the meter or recorder of the detector is calibrated directly in concentration units while the solutions of known concentration are aspirated.

11.4.2 Calibration Graphs

There has been some work directed towards the possibility of absolute analysis by AAS, without the need for empirical calibration; this is discussed briefly in Sections 3.5.3 and 11.4.3. At the present time, however, absolute analysis by AAS or AFS is possible only with experimental difficulty and under very restricted analytical circumstances. In practice, when either technique is used for routine analysis, a calibration curve of the signal strength against solution concentration is employed.

As mentioned above (11.4.1) the 100% transmission in AAS, or the zero fluorescence signal in AFS, is usually set while nebulizing pure solvent or a blank solution whose composition with respect to the concomitant species matches that of the samples. If the calibration graph is known to be linear over the desired concentration range, one standard solution should ideally be sufficient to enable the calibration graph to be constructed. It is usually preferable, however, to employ several standard solutions of different concentrations in order to obtain good precision. Even when the calibration graph is known to be non-linear the use of sufficient standard solutions to construct the graph may still permit quite accurate analysis.

Under conditions for least photometric error in AAS, i.e. measurement between 0·2 and 0·7 absorbance unit, the calibration curve might typically cover a 20- to 200-fold increase in concentration above the detection limit. This often results in some inconvenience, as this range above the detection limit may correspond to the use of solutions of relatively high dissolved matrix element content which may cause nebulization or atomization difficulties. Thus, calibration curves in AAS are normally constructed for use between somewhat lower absorbance values, e.g. 0·02 to 0·5. Apart from such considerations of the photometric accuracy, the accuracy of determinations when curved calibration graphs are employed is affected by the actual shape of the curve. The rate of change of analytical signal strength with concentration governs the reliability with which the concentrations of samples producing similar absorbance values may be differentiated. The slope of the curve may vary considerably over the concentration range of interest; depending on the shape of the graph (bending away from or towards the concentration axis) the certainty with which it is possible to distinguish between samples producing high absorbances may be good or poor compared to the situation at low absorbance values. Some modern instruments used in AAS provide for correction of read-out so that absorbance values observed appear to follow a linear calibration even though the actual uncorrected calibration data are non-linear.

Frequent checks of the calibration are recommended during operation in both AAS and AFS. Standard solutions should be nebulized with each set of sample solutions to ensure that inaccurate analyses cannot result from factors such as wavelength drift, change in fuel : oxidant flow rate ratio or variable nebulizer performance.

(*a*) *AAS calibration graphs*

The dependence of the analytical absorption signal on analyte concentration has already been discussed in Section 3.3. When absorbance is measured by using a sharp-line source, i.e. in the case of the majority of practical AAS analyses, the calibration graph of absorbance against concentration should be linear over a wide range of concentration. The source line is never ideally sharp, however, and this results in a curvature of the graph towards the concentration axis; the greater the line-width the greater is the curvature and the lower the concentration at which it becomes significant. When a continuum source is employed there is no simple relationship between absorbance and concentration and the percentage absorption must be measured. As in this case the percentage absorption is directly proportional to concentration at low concentrations, and to the square root of the concentration at higher values, it is best to plot the calibration graph as log $(\Delta I/I_0)$ *vs* log (concentration); the graph has a slope of 1 at low values and of $\frac{1}{2}$ at high values. Any tendency for the slope to decrease further at very high concentrations may be reduced by increasing the bandpass of the monochromator.

The above deviations from linearity predictable from the theoretical basis of the techniques are frequently accompanied by other deviations which may be described as interference effects. As these vary with the nature of the sample rather than the instrumentation they may often prove more troublesome. These types of interference will be discussed in detail in Chapter 12, but the principal effects may be repeated in summary. When a non-absorbing spectral line is emitted by the source at a wavelength close to that of the absorption line, a reduction in sensitivity will be observed if this line is transmitted by the monochromator. The effect will progressively become more serious as the absorbance at the analytical line increases; the calibration graph bends *towards* the concentration axis. When an atom is easily ionized, the number of atoms present in the absorption cell at low sample concentrations will be seriously depleted by the formation of ions; the percentage ionization decreases with increasing concentration of atoms in the cell; the calibration graph bends *away* from the concentration axis.

(b) AFS calibration graphs

As shown in Section 4.4 the theoretical dependence of atomic fluorescence signal strength on the analyte concentration is more complex than that in AAS; fortunately the usual applications of AFS for analytical work allow a fairly simple approach from the practical viewpoint. This is based on the assumption that AFS is most frequently applicable for the measurement of low analyte concentrations in the atom cell, or that the amount of absorption occurring in the atom cell is small. The graph of concentration against fluorescence signal is then linear for both continuum and sharp-line sources. When determinations are attempted at high concentrations (the exact value depends on the element concerned) the calibration graph bends towards the concentration axis. This is partly explained by the theoretical predictions given in Section 4.4 and partly by the increasingly serious re-absorption of the resonance fluorescence signals as the atom population in the cell increases. In addition to these deviations from linearity the same type of deviation due to ionization effects will cause concavity of the calibration graphs in AFS as in AAS.

11.4.3 Other Calibration Methods

Although most routine analytical determinations by AAS and AFS are carried out with calibration graphs as described above, a number of other simple techniques are available. These methods sometimes show advantages over the conventional calibration procedure for certain analyses and the most useful ones are described below.

(a) Standard additions method

This method may be applied within the range of concentrations corresponding to linear calibration graphs. The signal is measured for the sample alone and for the sample plus aliquots of a solution of the analyte element of known concentration. The main advantage of the method in AAS and AFS is that, in effect, exact matching of the sample and "standard" solutions (sample to which known amounts of analyte element are added) is achieved automatically. This is most useful in the analysis of samples where complex matrix element effects are observed on the analytical signal. An analysis may be carried out by this method by making only two measurements, i.e. for the sample alone and for the sample plus one aliquot of standard analyte solution, but three or four measurements (sample plus one, two and three aliquots of standard analyte solution) usually result in greater accuracy. The absorbances measured for these solutions should be small in order to ensure as far as possible that the assumption of a linear relationship between absorbance and concentration is valid. The standard additions method gives incorrect results in AAS and AFS when molecular band absorption or scattering from particulate material in the flame is encountered. When this method is employed, therefore, the use of a continuum source background corrector (in AAS) or non-absorbed line (AAS or AFS) to check for this potential source of error is recommended.

(b) Dilution method

This method is an adaptation of the standard additions method that may be used in preference under certain circumstances. A measurement of the analytical signal S_1, from a volume V_1 of a solution containing a known concentration C_1 of the analyte element is made. A volume V_2 of the unknown sample, containing a concentration C_2 of analyte, is then added and a second signal, S_2, is then recorded. The unknown concentration, C_2, is then given by:

$$C_2 = C_1 \frac{S_2(V_1 + V_2)(V_2 - S_1 V_1/V_2)}{S_1}$$

Provided that sufficient accuracy is obtained with only two measurements the method is rapid and easily applied. Smaller volumes of sample solution may be used than with the addition method as it is not necessary to record a signal for the sample solution alone. Also, with concentrated sample solutions, or at least analyte solutions that produce a large analytical signal, it is not necessary to perform an additional operation to obtain sufficient dilution to reach the linear part of the calibration. This method is subject to similar errors to the standard additions method when molecular band absorption or scattered radiation is present.

(*c*) *Absolute analysis by AAS*

A technique that would allow absolute analyses by AAS would offer several advantages over the conventional methods utilizing calibration against standards. The feasibility of absolute AAS methods has been discussed by Rann,[39] de Galan[40] and de Galan and Samaey[41] and is also dealt with briefly in Section 3.5.3 of this volume. Several reports of the measurement of free atom fraction and oscillator strengths by AAS have also been made, and these measurements involve experimental difficulties similar to those of absolute analysis.[42-45] There are two principal problems associated with absolute analysis by AAS. Firstly, it is necessary that an exact theoretical expression be used to relate the experimental absorption measurements to the number of absorbing atoms. Secondly, it is necessary to calibrate experimentally the apparatus employed, both in order to evaluate all of the parameters that must be substituted into the basic equation, and to relate the number of absorbing atoms to the actual concentration of the analyte sample. Although each of these presents difficulties, the most serious from the viewpoint of the analytical chemist is the requirement of an accurate expression over a sufficiently wide range of experimental conditions, and the selection of an absorption cell in which the number of absorbing atoms can be readily related to the sample concentration. The simplest expression is that for the integrated absorption coefficient given in Section 3.2.2. As explained, however, it is difficult to use this to establish a routine analytical technique A more convenient measurement would be that of the peak absorbance obtained with the sharpest possible source line under conditions of low optical density. Even under these limited circumstances, however, the measurement cannot be applied with equal success to all elements; the absorption line used may have a hyperfine structure which is sufficiently wide or well-separated in the source and/or atom cell to cause a serious error in the measurement of the absorbance. The problem has been considered in some detail by L'Vov;[46] he suggested that measurement in a high-pressure cell might offer one solution. Under these conditions the Lorentzian broadening would result in a much wider absorption line than that obtained with other absorption cells at atmospheric pressure. The hyperfine structure would then have an insignificant effect on the absorbance measurement for a much wider range of elements and the errors due to use of a broadened source line would also be reduced.

The ease with which the number of absorbing atoms may be related to the actual sample concentration also varies with the type of absorption cell used and the analyte element. It is often possible to measure fairly accurately the proportion of the sample that actually reaches a region of the absorption cell where the analyte atoms can give rise to a signal, although the experimental technique may prove more simple for a L'Vov furnace, for example, than for a conventional premixed flame and pneumatic nebulizer. The measurement of the fraction of the sample that actually forms atoms after reaching the region in which absorption can occur, however, is more difficult. The most promising solution might be the use of an atom cell in which the free atom fraction is unity for the element concerned. With the atom cells currently available, particularly high-temperature flames, this would mean that a useful range of elements would be amenable to absolute analysis. However, after those elements have been omitted for which the lack of a suitable sharp-line source or the effect of hyperfine structure cause additional problems the range of elements becomes much smaller. Also, the combined effects of inaccuracies in the various calibrations required and deviations from the theoretical assumptions which must be made are likely to severely restrict the attainable accuracy of the results.

11.4.4 Internal Standardization in AAS

The use of an internal standard to reduce errors in flame emission spectroscopy is well established. The procedure usually adopted is to add to each of the sample and standard solutions a known amount of an internal standard element and to monitor its line emission intensity and that of the analyte element for each solution. The measurement of the two signals may be accomplished simultaneously (with a two-channel spectrometer) or sequentially (with a single-channel spectrometer). The ratio of the signal intensities is then used as a measure of the analyte concentration rather than the analyte emission intensity alone. Provided that certain criteria concerning the suitability of the element used as internal standard are fulfilled, this technique may compensate for errors otherwise incurred via changes which can occur in the rate at which sample is supplied to the flame and in flame conditions. These include change in sample viscosity and surface tension, nebulizer performance, variation of fuel : oxidant concentration ratio or change in flame geometry due to disturbance by draughts etc. It is usual in flame emission spectroscopy to select the internal standard element so that its emission wavelength, intensity and dependence of degree of atomization on flame temperature, stoicheiometry and height of observation are similar to that of the analyte element.

In AAS the use of internal standardization permits similar compensation for errors caused by variation in the nebulizer and flame characteristics. As it is preferable to measure absorption signals for both sample and internal standard simultaneously, however, a

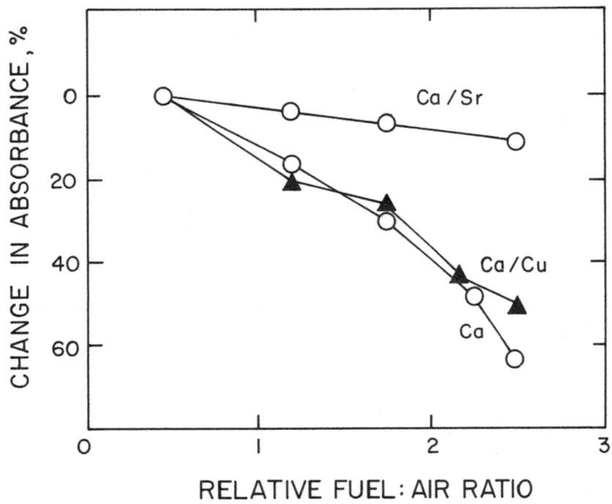

Fig. 11-6. Effect of strontium internal standard correction of effect of fuel : air rate on calcium absorbance. Use of copper internal standard does not result in appreciable correction (adapted from ref. 37).

two-channel spectrometer must be employed. With this type of instrument Feldman *et al.*[47,48] have studied the use of the technique in AAS. These workers have shown that, as in flame emission work, it is necessary to use an internal standard element whose atomization process in the flame cell employed is subject to disturbances by change in nebulizer and flame characteristics in a similar manner to that of the analyte element.[49] A good example is found in the examination of the relative suitability of copper and strontium as internal standard elements in the determination of calcium by AAS. The absorbance for copper is little affected by changes in the fuel : oxidant concentration ratio supplied to the flame; the critical dependence of the absorbance for strontium on the fuel : oxidant ratio, however, is similar to that observed for the analyte element (calcium). As shown in Fig. 11-6 the analytical signal in the calcium determination may be corrected for error produced by small random changes in the fuel : oxidant ratio by using strontium as internal standard and utilizing the Ca/Sr absorbance ratio; the use of copper as internal standard results in insignificant correction. Feldman has demonstrated the advantages of the technique in practical analyses by AAS. It is claimed that approximately a two-fold improvement in precision may be obtained by using internal standardization in the determination of aluminium in cement, copper in aluminium alloy and nickel in stainless steel.[50]

11.4.5 Scale Expansion and Zero Suppression in AAS

For AAS work with samples that produce low absorbance readings the scale expansion technique may be employed to allow the values to be read more accurately. An auxiliary source of potential is simply used to adjust the low end of the absorbance scale (e.g. 80 to 100% on the transmission scale) to cover the full scale at the meter or recorder read-out (zero to full-scale deflection). As the scale expansion is obviously applied to both the analytical signal and the noise it is necessary to have a favourable signal : noise ratio in order to benefit fully from this technique. David[51] first modified the single-beam atomic absorption instrument of Box and Walsh in this manner to provide up to five-fold scale expansion. With double-beam instrumentation for AAS, where appreciable compensation for source noise is accomplished, larger scale expansion factors may be realized. It is frequently sufficient, when working with low absorption values and scale expansion, to plot percentage absorption (rather than absorbance) directly against concentration; only slight deviation from linearity occurs. The ability to utilize greater sample dilution if scale expansion is employed may assist in the reduction of interference effects. This effect has been demonstrated by Belcher and Brooks[52] in the determination of strontium in coal ash.

In addition to its use in recording low absorption values, the technique of scale expansion may be employed in conjunction with zero suppression in order to increase the precision obtained in the measurement of high absorption values.[53,54] The high portion of the transmission range (low absorption) is suppressed by zero adjustment and the scale is expanded so that only a small intermediate part of the percentage absorption scale is presented across the entire scale of the meter or recorder. The sample solution, and standard solutions whose percentage absorption values closely bracket that of the sample, are then nebulized into the flame.[54] In a similar manner to simple scale expansion for low absorption values, it is again usually sufficient over the small range of absorption measured to then establish the concentration of the unknown sample by plotting percentage absorption (rather than absorbance) against concentration. Results with precision as good as 0·2% have been claimed for this technique in the determination of unknown samples. Figure 11-7 shows a typical example, where the range is

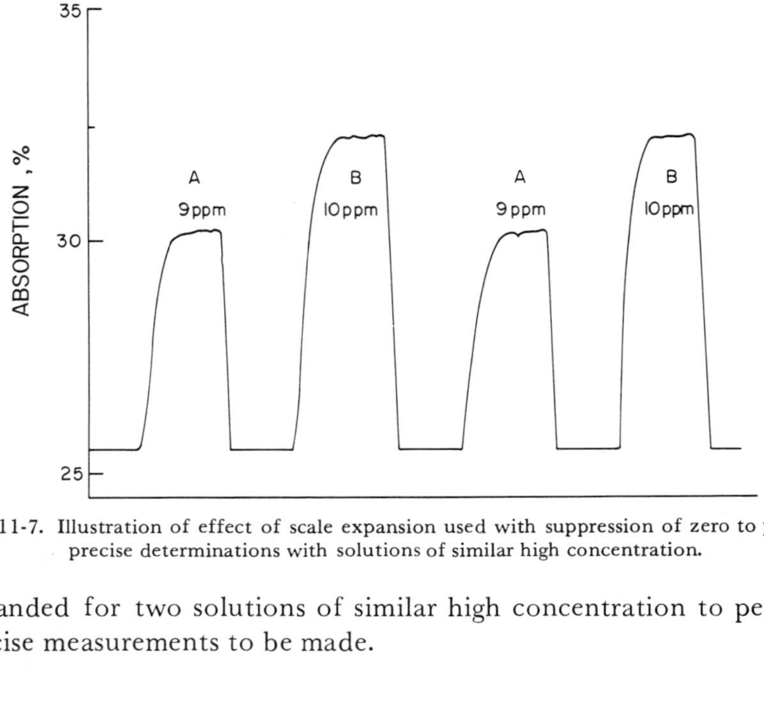

Fig. 11-7. Illustration of effect of scale expansion used with suppression of zero to permit precise determinations with solutions of similar high concentration.

expanded for two solutions of similar high concentration to permit precise measurements to be made.

11.5 PRECISION AND ACCURACY

Two types of error are recognized in any quantitative measurement. The first type is determinate or systematic error, which is attributable to a definite cause, even though the cause may not be known. Such errors are uni-directional. The magnitude may be constant from sample to sample, proportional to the size of the sample, or may vary in a complex fashion. Examples of this type of error in AAS or AFS are found in the chemical and physical interference effects on the analytical signal from concomitant elements present in samples.

The second type of error is indeterminate or *random* error brought about by the effects of uncontrolled variables. Real random errors are equally likely to cause high or low results; their existence thus causes a scatter of the results about the true value (in the absence of systematic error).

A measure of the random error is given by the *precision* of a measurement; this refers to the reproducibility obtained if the same measurement is made repeatedly. The *accuracy* of a measurement is determined by the systematic error (the deviation of the measurement from the "correct" value). Thus it is possible to carry out a precise analysis by AAS or AFS without obtaining an accurate result, but it is rare that an accurate analysis is obtained with an imprecise measurement.

11.5.1 Precision

The truly random error which is associated with a measurement is characterized by a "normal distribution" in which there is a symmetric distribution of the deviations from the true result. This situation is shown in Fig. 11.8; in the absence of systematic error the true value, A, occurs most frequently. A valid and useful estimate of the *precision* of a series of measurements is the *standard deviation*, s, of the usually small number of practical measurements (the sample),* where

$$s = \left(\frac{1}{n-1} \sum (x_i - \bar{x})^2 \right)^{1/2} \tag{11-1}$$

in which x_i is the result obtained from one measurement and $\bar{x}$ is the mean of n measurements. The summation term in this expression is inconvenient for calculating machines; if $\bar{x}$ is not known, rapid machine calculation can be achieved by using the relationship

$$\sum (x_i - \bar{x})^2 = \sum x_i^2 - \frac{1}{n} \left(\sum x_i \right)^2 \tag{11-2}$$

If $\bar{x}$ is known a still more convenient expression for calculation is

$$\sum (x_i - \bar{x})^2 = \sum x_i^2 - n(\bar{x})^2 \tag{11-3}$$

* The term *sample* refers to the small number of measurements which form part of the *population*. The term *population* refers to an infinitely large number of analytical measurements obtained from operations involving the same material. The standard deviation of a *sample*, s, is calculated using $(n - 1)$ results. The standard deviation of a population is symbolized as σ and is calculated using n results. The standard deviation of a sample is an estimate of σ.

The value of the standard deviation calculated from equation (11-3) is expressed in the same dimensions or units as the data being evaluated. It may also be expressed as the relative standard deviation, RSD, (or coefficient of variation) where

$$RSD = \frac{S}{\bar{x}} \times 100 \qquad (11\text{-}4)$$

As shown in Fig. 11-8, when only random error is considered a normal distribution curve may be drawn in which the relative

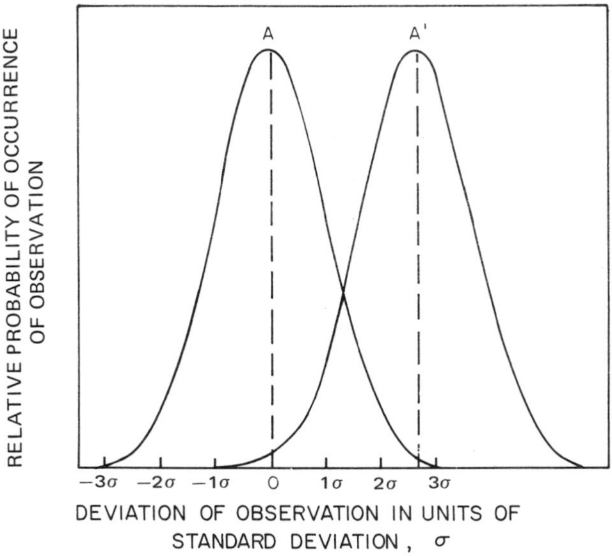

Fig. 11-8. Normal distribution curve for random error (true value A); effect of systematic error to shift curve to apparent true value A′.

frequency of occurrence of a measurement, x_i, or its error, $(x_i - A)$, is plotted against multiples of the standard deviation employed. Because of the symmetrical nature of the curve there is a negative error for every positive error of the same value. The relative frequency of measurements having a small error is very great. It is found that greater than 68% of all measurements lie within the range $A \pm \sigma$, while greater than 95% and 99% of all measurements lie within the ranges $A \pm 2\sigma$ and $A \pm 3\sigma$ respectively.

In order to obtain a reliable estimate of the reproducibility of a particular determination, via calculation of the standard deviation, as many results as possible (usually more than 10) should be obtained.

The overall attainable precision for a complete analytical procedure in AAS or AFS including sampling, sample pretreatment and measurement, is invariably different from that of the measurement step alone. For the purposes of the complete specification of an analytical method the overall standard deviation should be found by taking samples through the whole procedure. The standard deviation in the measurement step alone, obtained by repetitive measurements on a single prepared sample, gives an indication of the instrumental performance. An estimate of the detection limit is frequently obtained in this way, for example, by measurement of the concentration of the analyte which produces a relative standard deviation of 50%.

11.5.2 Photometric Error in Absorbance Measurement

In the technique of molecular absorption spectrophotometry in solution it is common practice to attempt to make analytical measurements between limiting absorbance values to minimize measurement error with non-linear absorbance scales. Ringbom[36] demonstrated that, when the instrumental noise in the system is constant, the variation in relative error in measurement of absorbance $\Delta A/A$ (or concentration $\Delta C/C$) with the absorbance value may be shown by differentiation of the absorbance expression to follow the curve shown in Fig. 11-9. The value of $\Delta A/A$ can thus be seen to be minimized at absorbance values between *ca.* 0·2 and 0·7 while the actual minimum on the curve occurs at about 40% transmission, i.e. $A = ca.\ 0{\cdot}4$.

In AAS, however, these limiting values for absorbance measurement are seldom adopted. High absorbance values may frequently be obtained only when the matrix element concentration in solutions of the analyte is too great to allow trouble-free routine nebulization; in addition, analytical curves in AAS are often non-linear at high absorbance values. At the low absorbance end of the scale, absorbance values considerably less than 0·2 are routinely measurable with good precision with modern instrumentation using scale expansion, signal integration or a long response time (damping). As mentioned elsewhere (Section 11.4.2), therefore, AAS calibration graphs are frequently constructed for use between *ca.* 0·02 and 0·5 absorbance unit.

With many AAS instruments the reading scale may be linear in absorbance values. In these instruments the photomultiplier output is transformed logarithmically before amplification. The photometric error then becomes more difficult to analyse because that part of the error due to the detector is proportional to the radiant power of the transmitted beam, whereas the errors in the amplification and recorder are proportional to the logarithm of the transmitted radiant power. When the linearity of the detector is good compared to that of the amplifier and recorder, the overall error will be a nearly constant fraction of an absorbance unit. The relative error is then constant across the scale and the error in determination of the analyte concentration in the sample solution decreases with increasing absorbance values (see Fig. 11-9).

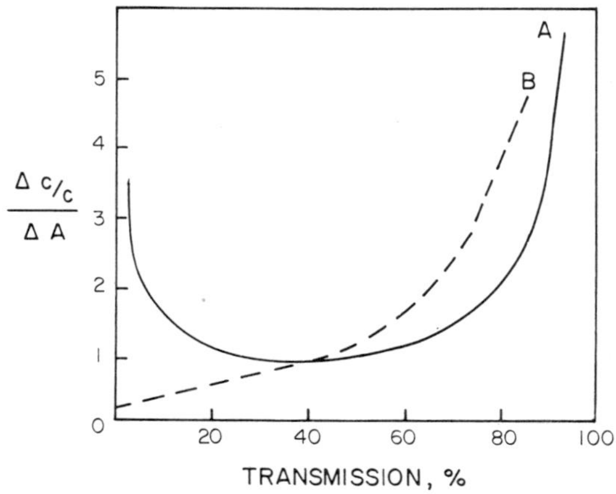

Fig. 11-9. Relative photometric error variation with percentage transmission. A, Linear transmission scale; B, linear absorbance scale.

11.5.3 Accuracy

The effect of the presence of systematic error in an analysis is to shift the most frequently observed value in the normal distribution curve (e.g. to A' in Fig. 11-8). Systematic error may be introduced at each stage of a complete analytical procedure. Loss of analyte element (or gain by contamination) may occur during sample pretreatment, or uni-directional changes in the analytical signal may be induced by the chemical or physical interference effects of concomitant materials.

Evaluation of the accuracy of an AAS or AFS procedure may be made using standard analysed samples (such as those available from the National Bureau of Standards, U.S.A., or the Bureau of Analysed Samples Ltd. in England). An alternative method is by reference of the results to those given by a well-established referee method employing a classical absolute technique or another instrumental method. The "recovery" of analyte element may also be checked by adding it in known amounts to samples and conducting the analysis in the normal fashion. In a technique for estimating inaccuracy which does not require standard additions of analyte, Youden[55] proposed the analysis of two samples of different analyte concentrations. Equal weights of each are then taken to make a third "sample" whose concentration will be the average of the two initial values. This sample is then analysed and the difference between the calculated and measured results provides an indication of the systematic error in the determinations.

11.6 PRETREATMENT OF SAMPLES

AAS and AFS are widely applicable to the analysis of many different types of solid and liquid samples where the matrix may be organic or inorganic in nature. The type of sample and the quantity available will determine the choice of the sampling procedure and the way in which the samples are prepared for analysis. Well-established sampling procedures exist for bulk samples which ensure that a homogenous sample is taken for analysis. These procedures are reviewed thoroughly in standard texts[56-58] concerned with quantitative analysis and will not be dealt with here.

In the technique of *average sampling* care must be taken that the composition of the sample is accurately representative of the average composition of the material. This type of sampling is the most common. The material to be analysed is usually a heterogeneous mineral mixture, raw material or unfinished or final product. The method of average sampling used depends on the degree of inhomogeneity and on the physical state and chemical properties of the sample. When very large amounts of material (e.g. bulk ore) are sampled, a representative raw sample is first taken. Laboratory samples may then be prepared from this by further grinding and homogenization.

In *discrete sampling* the sampling is confined to small parts of the material. For large amounts of material discrete sampling is usually easily performed. This method is used if heterogeneous parts of an otherwise homogeneous material are to be examined, e.g. oxide or sulphide inclusions in alloy castings. For such sampling from small stratified samples it may be difficult to ensure that each individual sample is representative of the material in a particular stratum.

Sampling difficulties may also be encountered when only minute amounts of solid or liquid samples of a biomedical or forensic origin are available. The usual procedure under these conditions is to obtain as many *raw* samples as possible and analyse these directly.

11.6.1 Dissolution of Samples

Once laboratory samples have been obtained care should be taken to avoid loss of the analyte element, or its gain by contamination, during any initial washing, drying or grinding of the material or on storage. The type of sample to be analysed obviously dictates the manner in which it is treated to render it suitable for analysis by AAS or AFS. The principal operation encountered for solid samples, however, is dissolution, with or without preliminary wet or dry ashing. Owing to the fact that the sample must frequently be treated with relatively large volumes of reagents during the ashing and/or dissolution procedure, care must be taken not to contaminate it with reagents containing either the analyte element or possible interfering substances. When this is known to occur, and is unavoidable, it is then essential to employ a procedure blank. If sufficient contamination occurs in this way to necessitate taking blanks through the whole procedure, the analytical precision attainable in the determination is usually substantially decreased.

Water which has been distilled in quartz apparatus or purified by ion-exchange should always be employed for dilution of sample solutions. Many acids and other reagents are available commercially in "analytical reagent" grades. It is still often advisable, however, to assume that even these reagents contain traces of the analyte element or an interferent as impurity, or that they may become contaminated from the vessels used for storage or dissolution; a procedure blank should still be undertaken. Frequently the magnitude of this blank may be decreased by pretreatment of the acids and other reagents by solvent extraction or ion-exchange procedures which remove unwanted impurities.

The material chosen for the apparatus used for the dissolution and subsequent storage of samples and standards is also frequently important, particularly in trace analysis. Adsorption of ions onto glassware from dilute standard and sample solutions is a well-known cause of inaccuracy. Dilute solutions should therefore not be stored for long periods but rather prepared fresh from more concentrated stock solutions every few days. In addition to loss of analyte by adsorption in this way, contamination of solutions can occur when previously adsorbed ions on the surface of glass containers are released on subsequent use of the vessels for sample or standard solution preparation. The apparatus itself may be attacked by the sample solutions to release contaminant species. The most common form of contamination in this way is from silicate leached from glassware. Mitchell[59] has reported the study of contamination from apparatus made from different materials. Minczewski[60] recommends the use of the following materials in order of preference: fluorinated polymers, polyethylene, quartz, platinum and glass.

The most common dissolution procedure for solid samples with an inorganic matrix is acid dissolution. The acids employed most frequently for metals, alloys and ores are HCl, H_2SO_4, HNO_3, H_3PO_4 and $HClO_4$. Phosphoric acid (H_3PO_4) is used in admixture with H_2SO_4 for some types of metallurgical sample; its possible chemical interference effect on the AAS or AFS of the analyte must be carefully checked before its use, however, particularly when large volumes are used. Hydrofluoric acid is often used in a mixture with a second acid for those samples where the formation of a stable fluoride complex assists the dissolution and aids the retention of the sample in solution.

Fusion procedures are sometimes employed for the dissolution of samples which are not readily attacked by acids. Fusion methods, however, are usually slow and it is often difficult to obtain fusion mixtures which are sufficiently free from contamination from traces of inorganic ions.

The most commonly encountered matrices that require fusion treatment for dissolution are those in which a high silicate content is expected, or for finished or part-finished refractory materials which have undergone heat treatment during processing.

11.6.2 Ashing Techniques

Both dry and wet ashing techniques are in widespread use for the destruction of organic matter in samples prior to inorganic trace analysis by AAS or other instrumental techniques. Each procedure will be briefly considered here for its suitability with different analyte elements.

(*a*) *Dry ashing*

In this technique the sample is heated in air in a suitable container at a temperature which may vary between *ca.* 400 and 800°C. The heating is usually performed by direct application of a gas flame to the base of the container. Solid samples must usually be dried in a drying oven prior to ashing and aqueous solutions are evaporated in small portions in the crucible used for the ashing procedure. When ashing of a very large amount of sample is necessary, it is advisable to weigh the sample in a weighing bottle and to ash it in portions after transference to the crucible; the weight of sample ashed is then determined by a final weighing of the bottle.

The dry ashing technique is simple, and large samples can be treated in this way. The procedure is relatively free from difficulties with contamination, although the introduction of impurities from the atmosphere must be avoided.[61] For some samples difficulty may be encountered with the retention of analyte materials by the walls of the crucible or with the subsequent dissolution of the ash (e.g., formation of acid-insoluble silicates). The most serious disadvantage of the dry ashing technique, however, occurs when the element to be determined is volatile. Thus, for example, in samples to be analysed for lead, mercury, cadmium, arsenic or selenium, dry ashing leads to partial or complete loss of the analyte element. The elements bismuth, chromium, copper, iron, nickel, vanadium and zinc may also be lost by volatilization as the metals, chlorides or organometallic compounds under certain conditions.

Some elements may be retained by the crucible employed and prove difficult to remove. Thus, Gorsuch[62] found significant retention of lead in silica crucibles above 550°C. In the presence of organic matter low recoveries of copper on dry ashing have been attributed to the retention of copper on the silica crucible employed.[50] This may be due to reduction of copper oxide and diffusion into the crucible walls of the copper metal formed. Similar difficulties have been reported for recovery of cobalt on dry ashing.[63] Thiers, Williams and Yoe,[64] however, recovered trace amounts of cobalt from blood on dry ashing for eighteen hours at 450°C. Thiers[65] recommends the following procedure for the routine analysis of biological samples with dry ashing: A liquid or solid sample in a platinum dish is heated on a hot-plate in filtered air for a period between 15 minutes and several hours, increasing the temperature gradually, until the sample has a charred appearance. The dish is then transferred to a muffle furnace and the temperature is raised from 250° to 450°C within an hour. This temperature is then maintained until the ashing is completed. If any carbon remains in the residue after prolonged ashing it may be decomposed with hot 12 M HNO_3. The ash obtained is dissolved in a small volume of hot 6 M

HCl; if silica is present it should be volatilized away by evaporation with sulphuric and hydrofluoric acids.

Dry ashing is often carried out with the aid of additional oxidant. The sample may be moistened with a salt solution or a few drops of acid before combustion, or pure magnesium nitrate or acetate salt may be added as an ashing base. These procedures serve to speed the process and minimize losses by volatilization of some analyte elements, although the danger of sample contamination from trace impurities in the ashing aid used is increased. Dry combustion procedures in which an oxygen bomb[66] or oxygen flask[67] is used are frequently also employed. A particular advantage of the use of the oxygen flask technique may be the elimination of errors otherwise encountered from loss or contamination on sample transfer from the ashing vessel to the other apparatus used for storage of the sample solution. If the flask used in the technique is suitably shaped it may itself be employed as the vessel from which nebulization into the flame is effected. Belisle, Green and Winter[68] have described a flow-through oxygen flask technique which may be used with larger samples than the conventional oxygen flask method.

Gleit and co-workers[69-71] have described a low-temperature dry ashing technique which depends on radio-frequency excitation of oxygen. The active oxygen is passed over the organic material to be ashed and oxidation ensues at sample temperatures which may be maintained below 100°C. Particular advantages of this technique are the ability to retain the morphology of the sample and relative freedom from losses due to sample volatility. It is also claimed that the introduction of contamination from the container and the atmosphere are reduced. The quantitative recovery of seventeen elements after complete destruction of organic matrix material has been demonstrated using this technique. In a typical demonstration using radiotracer techniques it has been reported that selenium in alfalfa was recovered with 99% efficiency. In other work, however, incomplete recovery of the elements arsenic, selenium and mercury has been reported.[72] Two commercial assemblies utilizing this low-temperature ashing principle have been produced. This technique should find extensive application if rapid and efficient oxidation can be ensured for a wide range of matrices.

(*b*) *Wet ashing*

Organic matter may be destroyed by oxidation at lower temperatures than those used in dry ashing procedures when the technique of wet ashing is employed. The oxidizing reagents most commonly used are nitric acid, sulphuric acid and perchloric acid, either used singly or in admixture. At the lower temperatures employed the danger of loss of analyte by volatilization is decreased, although difficulty may still be encountered from loss of elements such as mercury, selenium and arsenic. The contamination risk from the presence of impurities in the digestion reagents is obviously greater than in the dry ashing procedure and the technique also requires more care in operation. From the standpoint of recovery of the analyte element, however, the wet ashing technique is usually preferable.

The most generally useful acid mixture appears to be a combination of HNO_3, H_2SO_4 and $HClO_4$; the volume ratios in which these are mixed varies according to the application and to the preference of different workers, but a mixture commonly used is $HNO_3 : H_2SO_4 : HClO_4 = 3 : 1 : 1$. While sufficient sulphuric acid is present there is little risk of explosion from the perchloric acid in this mixture and the solution may be evaporated until sulphur trioxide fumes appear. When sulphuric acid is omitted from the digestion mixture much more care must be taken to avoid explosion and the sample must not be taken to dryness; additional nitric acid is frequently added after the solution has evaporated until fumes of perchloric acid form.

Wet ashing procedures using acid mixtures containing molybdenum (VI)[73-75] or vanadium(V)[76] as a catalyst are finding increasing application. The incorporation of these catalysts is particularly useful in wet ashing where the analyte is arsenic or selenium. The procedure is rapid, only a few minutes digestion frequently being required, and appears to be very effective in minimizing loss with volatile analyte elements. Simon and co-workers[75] have investigated a variety of methods for pretreatment of urine samples before analysis and recommend this catalytic wet-digestion technique.

Johnson[77] has described the effective use of 50% H_2O_2 in conjunction with H_2SO_4 for the wet oxidation of organic matter. Other workers,[78] however, have reported loss of several elements during this procedure.

In a study of both wet and dry ashing techniques for the elements, lead, mercury, zinc, iron, strontium, chromium, antimony, molybdenum, silver, cadmium, copper, cobalt, selenium and arsenic, Gorsuch[79] found that a wet ashing procedure using HNO_3 and $HClO_4$ was generally the most satisfactory for effective recovery of these elements (except mercury).

The problems associated with destruction of organic material in samples prior to the determination of mercury have received considerable attention. Oxidation with a mixture of nitric and sulphuric acids has been employed by many workers; the temperature of the digestion mixture is maintained just below the boiling point of sulphuric acid. Wet ashing with mixtures containing perchloric acid appears to be unsuitable as loss of volatile mercury halides occurs. An oxidation procedure using 50% hydrogen peroxide has been described by Polley and Miller,[80] and other workers have employed nitric acid in the presence of potassium permanganate.[81] A variety of procedures have been described for the recovery of mercury by reduction to the element in solution using stannous chloride, hydroxylamine hydrochloride, ascorbic acid or other reductants. In some of these procedures organic matter is destroyed by prior oxidation in the same solution using strong oxidants such as potassium permanganate. The mercury may then be swept out from the solution in a current of nitrogen or argon and determined by the cold-vapour technique (see Chapter 13.4.34).

11.7 PRE-CONCENTRATION AND SEPARATION TECHNIQUES

11.7.1 Solvent Extraction

The "organic solvent effect" of increased sample transport to the flame may be usefully exploited by the solvent extraction of the analyte element into a water-immiscible solvent. This technique may also accomplish pre-concentration of the analyte element (by extraction from a large volume of aqueous solution into a small volume of organic solvent) when its concentration in the original sample solution is too low to permit its determination by AAS or AFS. The inherent spectral selectivity of AAS or AFS usually permits the use of a relatively non-selective solvent extraction system for the analyte element (in contrast to molecular absorption spectrometry in solution). However, when large amounts of matrix element or other species which give rise to chemical or physical interference are present in the aqueous sample, a degree of selectivity in the solvent extraction procedure may be useful for the elimination of these effects.

Both chelate and ion-association solvent extraction systems have been successfully employed in AAS by many workers. In most commonly employed chelate extraction procedures the pH of the aqueous sample solution is adjusted to a value suitable for formation of the complex to the extracted, and the solution is shaken with the organic solvent containing the organic chelating agent. After the phases have been allowed to separate, the organic solvent containing the chelate of the analyte element is nebulized directly into the flame for absorption or fluorescence measurements to be made.

The organic solvent chosen should not be appreciably soluble in water (to avoid problems with emulsification and slow separation) and should provide a substantial "organic solvent effect". It must also be directly combustible in the flame without the production of a smoky flame or toxic products. Aromatic solvents such as benzene or toluene generally give rise to the production of a smoky flame which may affect the analytical precision, while chlorinated solvents such as chloroform and carbon tetrachloride give rise to unpleasant and toxic combustion products. Allan[31] has reported that esters and ketones are generally satisfactory for use with the air/acetylene flame, since they produce stable flames, are completely burnt and do not contribute to the background absorption of the flame. Methyl isobutyl ketone (MIBK) has been employed extensively for solvent extraction followed by AAS (see, for example, refs. 31, 82-87) but other solvents have also commonly been employed, such as ethyl acetate,[31,83,88,89] isobutyl acetate,[90] amyl acetate,[31] propyl acetate,[31] methyl-n-amyl ketone,[91,92] methyl isopentyl ketone[93] and butanol.[90,94]

As mentioned above, unless the solvent extraction step must effect the removal of chemical or physical interferences as well as the pre-concentration of the analyte element, the most generally useful extraction methods are those that are non-selective and not critically dependent for their efficiency on the pH of the sample solution. A single set of conditions for solvent extraction (i.e. sample pH, reagent concentration) can then be applied to the pre-concentration of a range of elements. The chelating reagent that has been used most frequently in this way in AAS is ammonium pyrrolidine dithiocarbamate (APDC).[95] This reagent, which is generally used dissolved in water for direct extraction with methyl isobutyl ketone (MIBK), reacts with about 30 metal ions to form extractable metal chelates. The pH ranges over which the APDC complexes of different metal ions are formed are shown in Table 11-6. It is evident that some degree of selectivity can be introduced into solvent extraction procedures even with this "blanket" reagent by careful control of the pH of the sample solution. The APDC reagent is usually used as a 1% or 2% aqueous solution; its stability in aqueous solution depends on the pH, the rate of decomposition increasing in acid medium. It is usual to prepare the reagent solution fresh daily. Some of the typical applications of APDC extraction to AAS are given in Table 11-7.

8-Hydroxyquinoline (oxine) has also been employed for both selective and non-selective pre-concentration procedures prior to AAS determinations. The influence of pH and reagent concentration on the extraction characteristics of many metal ions is well-known from use of this reagent for gravimetry and molecular spectrophotometry in solution. In addition, the application of auxiliary masking agents, such as EDTA, cyanide or fluoride, to improve the

TABLE 11-6

pH range of formation of the complexes of various metal ions with ammonium pyrrolidine dithiocarbamate[95] (APDC)

pH range	Elements forming APDC complexes
2	W
2-4	Nb, U
2-6	As, Cr, Mo, V, Te
2-8	Sn
2-9	Sb, Se
2-14	Ag, Au, Bi, Cd, Co, Cu, Fe, Hg, Ir, Mn, Ni, Os, Pb, Pd, Pt, Ru, Rh, Tl, Zn

selectivity of the extraction of particular elements has received much attention. Morrison and Freiser[96] and Stary[97] have listed conditions for the extraction of different metal ions with oxine, but the detailed extraction equilibria have not been studied for many elements. Table 11-8 shows extraction conditions for the most common elements whose pre-concentration is frequently required for AAS and which may easily be accomplished quantitatively with oxine, while Table 11-7 contains references to some published applications of oxine extraction prior to AAS determinations. 8-Hydroxyquinoline is frequently employed as a solution in chloroform for direct extraction of metal ions from aqueous solution. As chloroform may form toxic combustion products on nebulization, however, the extraction of the 8-hydroxyquinoline complexes into solvents such as ethyl acetate,[83] butanol[94] or MIBK is generally preferred.

TABLE 11-7

Some typical AAS applications employing pre-concentration or separation by solvent extraction

Organic reagent	Analyte	Solvent	Material	Reference
APDC	Cu	MIBK	Fertilizer, plants and soil	119
	Zn	MIBK	Fertilizer, plants and soil	120
	Fe, Mn, Ni	MIBK	KCl	121
	Cd, Tl, Hg	MIBK	Blood, urine	122
	Co	MIBK	Cements	123
	Co	MIBK	Plant digests	125
	Ni, Pb, Bi, Hg	MIBK	Urine	130
8-Hydroxyquinoline	Mg	MIBK	Brine, natural waters	124
	Mo	n-BuOH	Nb and Ta	94
	Cu	Ethyl acetate	Nb and Ta	88
	Mn, Co, Cu, Ni	$CHCl_3$	Brine	132
8-Hydroxyquinaldine	Mn	$CHCl_3$	Aluminous materials	133
TTA (thenoyl-trifluoroacetone)	Fe	MIBK	Urine	126
Na-DDTC (sodium diethyl dithiocarbamate)	Cd, Tl, Hg	MIBK	Blood, urine and tissue	127
	Cu	MIBK	Serum	128
Cupferron	V	MIBK	Water	131
Bathophenanthroline	Fe	MIBK	Serum	129
Diphenylthiocarbazone	Cr	MIBK	Brine	134
Toluene-3,4-dithiol	Mo	MIBK	Brine	134

Other chelating agents which may be successfully employed for pre-concentration by solvent extraction are diphenylthiocarbazone, 2-thenoyltrifluoroacetone, and cupferron (ammonium salt of *N*-nitrosophenyl hydroxylamine). Some applications of these reagents to AAS are shown in Table 11-7.

A second useful type of solvent extraction system is that based on the formation of extractable ion-association complexes between metal ions and simple inorganic ligands in aqueous solution. The ligands commonly encountered are the halides, thiocyanates and

TABLE 11-8

Practical conditions for solvent extraction of metal ions as their 8-hydroxyquinoline complexes

Metal ion	Soln. pH	Reagent conc. and solvent	Ref.
Al	4·5–11	0·1 M in $CHCl_3$	98, 99, 100
Bi	2·5–11	0·1 M in $CHCl_3$	100
Ca	>10·7	0·5 M in $CHCl_3$	100-102
	10–11	2% in MIBK	52
Cd	5·5–9·5	0·1 M in $CHCl_3$	100
		+ alcohols	103
Co(II)	4·5–10·5	0·1 M in $CHCl_3$	100
Cr(III)	6–8	Boil + excess oxine extract + $CHCl_3$	104
		0·1 M in $CHCl_3$	98
Cu	2–12	0·1% in ethyl acetate	88
Fe(III)	2–12	0·1 M in $CHCl_3$	98, 100
Hg(II)	>3	0·1 M in $CHCl_3$	100
		0·1 M in $CHCl_3$	100
Mg	10·7–13·6	0·1% in $CHCl_3$	105, 106
		+ butylamine	107, 108
		oxine in MIBK	
Mn(II)	6·5–10	0·1 M in $CHCl_3$	98, 100
Mo(VI)	1·0–5·5	0·01 M in $CHCl_3$	100
	4·5	1% aq. oxine, BuOH + butylamine	94
Nb(V)	6–9 (2% tartrate)	4% in $CHCl_3$	109
Ni	4–10	0·01 M in $CHCl_3$	100
	4·5–9·5	0·07 M in $CHCl_3$	98
Pb	6–10	0·1 M in $CHCl_3$	100
Pd	0–10	0·01 M in $CHCl_3$	100
Sn(IV)	2·5–5·5	0·07 M in $CHCl_3$	98
Sr	>11·5	0·5 M in $CHCl_3$	100
Th	4–10	0·1 M in $CHCl_3$	110, 100
		oxine in MIBK	110
Ti	2·5–9·0	0·1 M in $CHCl_3$	100
Tl(III)	3·5–11·5	0·01 M in $CHCl_3$	100
U(VI)	5–9	0·1 M in $CHCl_3$	111, 112, 100
		oxine in MIBK	112, 113, 114
V(V)	2–6	0·1 M in $CHCl_3$	100
		also with other solvents	66, 67
			115, 116, 117
W(VI)	2·5–3·5	0·14 M in $CHCl_3$	100
Zn	4–5	0·1 M in $CHCl_3$	100
		oxine in MIBK	118
Zr(IV)	1·5–4·0	0·1 M in $CHCl_3$	100

organic phosphate esters (such as tributyl phosphate). Ion association system extraction methods usually require a strongly acidic aqueous medium for efficient extraction, and the organic solvents employed are most frequently relatively polar (esters, ketones, lower alcohols and ethers) and suitable for nebulization into flames commonly employed in AAS and AFS.

Ion association extraction procedures are generally capable of extracting higher absolute weights of metal ions in moderate volumes of solution than are organic chelate extraction procedures, and large amounts of relatively expensive reagent are not required. In addition to the use of solvent extraction for extraction of the analyte element(s) away from the bulk matrix components in the aqueous solution, therefore, with these procedures it is frequently possible to undertake the extraction of the bulk matrix element(s) to leave the analyte element in the original aqueous solution. This general approach may sometimes be chosen when pre-concentration of the analyte is unnecessary and adequate AAS or AFS sensitivity is obtained in aqueous solution. Bulk matrix element extraction may then be employed to eliminate interference by chemical or physical effects in the nebulizer or flame employed. Table 11-9 gives details

TABLE 11-9

Ion association systems for different ions shown to provide efficient extraction

Ion	Aqueous medium	Organic solvent
As(III)	11 M HCl	Benzene
Au(III)	10% HCl	Ethyl acetate
Au(III)	3 M HBr	Ethyl ether
Au(III)	7 M HI	Ethyl ether
Cd	7 M HI	Ethyl ether
	1·5 N H_2SO_4 + 1·5 M KI	Ethyl ether
Fe(III)	1 M HCl	0·1 M TOPO in cyclohexane
Fe(III)	4 M HBr	Ethyl ether
Fe(III)	6 M HCl	Ethyl ether
Fe(III)	8 M HCl	Iso-propyl ether
Ga(III)	7 M HCl	Iso-propyl ether
	3 M HCl	Tributyl phosphate
Ga(III)	7 M NH_4CNS + 0·5 M HCl	Ethyl ether
Ga(III)	5 M HBr	Ethyl ether
Ge(IV)	11 M HCl	Benzene
Hg(II)	7 M HI	Ethyl ether
In(III)	4 M HBr	Ethyl ether
In(III)	1·5 M KI + 1·5 N H_2SO_4	Ethyl ether
Mo(VI)	1 M HCl	0·1 M TOPO in cyclohexane
Mo(VI)	5 M HCl	Amyl acetate

TABLE 11-9 (*continued*)

Ion	Aqueous medium	Organic solvent
Mo(V)	1 M NH_4CNS/0·5 M HCl	Ethyl ether
Nb(V)	11 M HCl	Di-isopropyl ketone
Nb(V)	0·6 M DBPA/1 M HNO_3	Di-n-butyl ether
Nb(V)	9 M HF + 6 M H_2SO_4	Di-isopropyl ketone
Nb(V)	10 M HF + 6 M H_2SO_4 + 2·2 M NH_4F	MIBK
Pb(II)	Excess KI/5% HCl	MIBK
Ta(V)	10 M HF + 6 M H_2SO_4 + 2·2 M NH_4F	MIBK
Sb(III)	7 M HI	Ethyl ether
Sb(V)	5 M HBr	Ethyl ether
Sb(V)	6·5-8·5 M HCl	Isopropyl ether
Sc(III)	8 M HCl	Tributyl phosphate
Sn(II)	4·6 M HF	Ethyl ether
Sn(II)	7 M HI	Ethyl ether
Sn(II)	1·5 M KI + 1·5 N H_2SO_4	Ethyl ether
Sn(IV)	3 M NH_4CNS + 0·5 M HCl	Ethyl ether
Sn(IV)	1·2-4·6 M HF	Ethyl ether
Sn(IV)	0·6 M DBPA + 1 M HNO_3	Di-n-butyl ether
Tl(I)	0·5 M HI	Ethyl ether
Tl(III)	6 M HCl	Ethyl ether
Tl(III)	>1 M HBr	Ethyl ether
Tl(III)	0·5 M HI	Ethyl ether
Ti(IV)	7 M HCl	0·1 M TOPO in cyclohexane
U(VI)	1 M HCl	0·1 M TOPO in cyclohexane
V(V)	pH 1·5-2·0	0·6 M TOPO in kerosene
Zn(II)	1 M NH_4CNS + 0·5 M HCl	Ethyl ether
Zr(IV)	1 M HCl	0·1 M TOPO in cyclohexane
Zr(IV)	0·06 M DBPA + 1 M HNO_3	Di-n-butyl ether

TOPO = tri-n-octylphosphine oxide; DBPA = di-n-butylphosphoric acid; MIBK = methyl iso-butyl ketone.

of some simple ion-association procedures for different elements where the method has been shown to permit virtually quantitative extraction of the species concerned. In addition to those shown in Table 11-9, other elements may be extracted under similar conditions with varying efficiencies. Morrison and Freiser[96] have presented a more extensive compilation of data concerned with efficiency of extraction of different elements with halogen acids and other ion-association systems.

11.7.2 Ion Exchange

Ion-exchange procedures are applicable during pre-treatment of AAS and AFS sample solutions either to pre-concentrate the analyte element from dilute sample solutions or to remove from the sample solution other ions that may exert a chemical or physical interference effect.

The ion-exchangers most commonly employed are the high molecular-weight organic resins containing acidic or basic functional groups. Sulphonated styrene/divinylbenzene copolymer resins, such as Amberlite IR-120 and Dowex-50, are the most generally encountered types of cation exchanger. Styrene/divinylbenzene copolymers into which quaternary ammonium groups are introduced, such as Amberlite IRA-400 and Dowex-1, are commonly employed as anion exchangers. All of these resins may be employed for either batch or column techniques. The sample solution is passed onto the resin and retained in contact with it for sufficient time to ensure equilibration; non-exchanged ions may then be washed from the resin, while those ions retained must be stripped out by treatment with a suitable eluting agent which reverses the ion-exchange process.

As AAS and AFS are relatively free from interference by moderate amounts of foreign ions, it is not usually necessary for the exchanger selectively to retain only the analyte when ion-exchange techniques are used for pre-concentration of sample solutions. Retention of a group of cations or anions that includes the analyte element is usually sufficient and these ions may then be stripped from the resin with a strong acid or base. Batch operation, in which the sample solution is simply shaken with the ion-exchange resin, is frequently satisfactory for pre-concentration procedures. After filtration the analyte and other exchanged ions are stripped from the resin by treating it with the minimum necessary volume of acid or base (in order to achieve the highest possible concentration factor). In order to ensure quantitative exchange from the initial solution the pH and equilibration time used for a given resin must be carefully controlled; the same factors must also be controlled in the stripping procedure to ensure complete removal of the analyte element from the ion exchanger. In some cases the analyte ion may be difficult to remove from the resin after ion exchange. This can lead to low recoveries and contamination of subsequent samples if the same batch of ion exchanger is employed. Under these conditions it may be preferable to ash the resin to recover the analyte element.

As mentioned above, ion-exchange procedures may be employed prior to AAS or AFS determination in order to remove matrix elements from the sample solution. In this type of application relatively large amounts of ion exchanger are often required and the

TABLE 11-10

Some selected ion-exchange procedures used for the pre-concentration or separation of inorganic species prior to determination by instrumental techniques

Analyte element(s) concentrated or separated	Matrix or accompanying elements	Ion-exchange process	Reference
Si	Water	Cl-anion exch. as SiF_6^{2-}; elution with boric acid	135
Cu	Water	H-form cation exch.; elution + 5 M HCl or hot 1 M Na_2SO_3	136 137
Au, Bi, Cd	Sea water	Cl-anion exch.; elution + 0·25 M HNO_3; ash resin	138
Cs, Rb	Sea water	NH_4-form cation exchange	139
Sn	H_2O_2	H-form cation exch.; elution + 5 M HCl	140
Co	Ni	Cl-form anion exch. from HCl soln.; elute + HNO_3/HCl	141
Cu, Cd, Co, Zn and Fe	Al and Mg	Cl-form anion exch.; elution + HNO_3/HCl	142
Pt	Chlorate cell liquor	Cl-form anion exch. from HCl medium; ash resin	143
U	Uranium leach liquors	SO_4-form anion exch.	144
Rare earths	Zirconium	H-form cation exch. HF soln.; elution + 6 M HCl	145
Cu	Milk	H-form cation exch.; elution + 6% HCl	146
Cu	Mineral oils	H-form cation exch. from soln. diluted + 2-propanol; elution + 10% H_2SO_4	147
Si	Pu	Absorption of Pu on H-form cation exch. from 0·2 N HNO_3/0·05 N NH_2OH	148
Mn	U	Absorption of U on Cl-form anion exch. from 9 M HCl	149
B	Al alloys	Absorption of Al on cation exch. from pH 5 acetate soln.	150
Fluoride	Water	Acetate anion exch.; elution + 0·2 M acetic acid/0·005 M Be soln.	151

capacity of the resin to retain large amounts of the concomitant species is an important consideration. No concentration of the analyte element is obtained and, indeed, some dilution is usually inevitable as the sample solution containing the unadsorbed analyte must be completely washed from the resin.

Many descriptions of the use of ion-exchange procedures for the pre-concentration and separation of traces of inorganic ions are found in the literature. In many of these studies instrumental techniques other than AAS or AFS are used for the determination of the analyte after ion-exchange pretreatment. The practical procedures detailed may, however, be directly applicable even when the final determination is by flame spectroscopy. Table 11-10 lists some selected examples from the literature; the reader is referred to two excellent treatises devoted solely to ion exchange processes[152,153] for further detailed treatment of their analytical application.

11.7.3 Other Techniques

Precipitation techniques have found considerable application for separation and pre-concentration in trace analysis but the direct precipitation of traces of an analyte element is seldom employed (owing to the difficulty of ensuring quantitative deposition and recovery of small amounts of material). Precipitation of bulk amounts of interfering matrix elements may, however, be achieved provided that care is taken to eliminate errors due to co-precipitation of the required analyte. The phenomenon of co-precipitation may be turned to advantage for pre-concentration in trace analysis by the use of collecting or scavenging agents. In these procedures a collector precipitate is formed and co-precipitates traces of the element to be determined. This precipitate may then be filtered off and dissolved in a small volume of solution to perform the analysis. Inorganic precipitates of sparingly soluble hydroxides, sulphides or carbonates are frequently employed in this fashion. The element chosen to act as the carrier ion for precipitation should have chemical properties similar to those of the ion to be co-precipitated, and must be added to the sample solution in sufficient quantity to provide enough precipitate for filtration without special apparatus. Furthermore, it must not cause interference in the subsequent determination of the analyte by AAS or AFS. The most frequently used carrier ions are manganese, iron and aluminium, but organic precipitating agents such as 8-hydroxyquinoline and thionalide have also been employed. An advantage with the use of organic collectors, compared to inorganic reagents, is that these are usually easily purified to remove traces of inorganic ions which would otherwise cause contamination. Examples of this type of application are found in the separation and pre-concentration by Dehm *et al.*[154] of seven heavy metal impurities in inorganic salts using 8-hydroxyquinoline and thionalide with aluminium carrier, and the effective recovery of trace metals from sea water using thio-acetamide with tin and indium carriers.[155] The use of organic precipitants has been reviewed recently[156] and the application of co-precipitation techniques for analysis of sea-water has been discussed by Joyner and co-workers.[157]

Other well-established techniques for separation and pre-concentration in trace analysis include volatilization, column or paper chromatography, fire (or dry) assay (particularly for the noble elements) and electrochemical deposition. At the time of writing, however, with the possible exception of fire assay these techniques have not found extensive application prior to determinations by flame spectroscopy.

REFERENCES

1. J. E. Allan, *Spectrochim. Acta,* 1954, **10,** 800.
2. J. E. Allen, *Nature,* 1960, **187,** 1110.
3. V. G. Mossotti and V. A. Fassel, *Spectrochim. Acta,* 1964, **20,** 1117.
4. D. J. David, *Analyst,* 1961, **86,** 730.
5. J. V. Sullivan and A. Walsh, *Spectrochim. Acta,* 1965, **21,** 721.
6. J. A. Bowman, J. V. Sullivan and A. Walsh, *Spectrochim. Acta,* 1966, **22,** 205.
7. W. Slavin, "Atomic Absorption Spectroscopy", Interscience, New York, 1968, p. 120.
8. W. Slavin and D. C. Manning, *Appl. Spectrosc.,* 1965, **19,** 65.
9. L. Capacho-Delgado and D. C. Manning, *Spectrochim. Acta,* 1966, **22,** 1505.
10. H. L. Kahn and J. E. Schallis, *Atomic Absorption Newsletter,* 1968, **7,** 5.
11. T. C. Rains, in "Flame Emission and Atomic Absorption Spectrometry", eds. J. A. Dean and T. C. Rains, Marcel Dekker, New York, 1969, p. 363.
12. G. F. Kirkbright and L. Ranson, *Anal. Chem.,* 1971, **43,** 1238.
13. W. J. McCarthy, M. L. Parsons and J. D. Winefordner, *Spectrochim. Acta,* 1967, **23B,** 25.
14. R. F. Browner, R. M. Dagnall and T. S. West, *Anal. Chim. Acta,* 1970, **50,** 375.
15. R. M. Dagnall, K. C. Thompson and T. S. West, *Talanta,* 1967, **14,** 557.
16. R. S. Hobbs, G. F. Kirkbright and T. S. West, *Talanta,* 1971, **18,** 859.
17. G. Hansen, *Optik,* 1946, **1,** 227; **1,** 269.
18. P. Jacquinot, *J. Opt. Soc. Amer.,* 1954, **44,** 761; 1955, **45,** 996.
19. J. W. Perry, *J. Opt. Soc. Amer.,* 1955, **45,** 995.
20. B. J. Russell and A. Walsh, *Spectrochim. Acta,* 1959, **15,** 883.
21. J. V. Sullivan and A. Walsh, *Applied Optics,* 1968, **7,** 1271.

22. J. V. Sullivan and A. Walsh, *Spectrochim. Acta,* 1965, **21,** 721.
23. J. A. Bowman, J. V. Sullivan and A. Walsh, *Spectrochim. Acta,* 1966, **22,** 205.
24. R. M. Lowe, *Spectrochim. Acta,* 1969, **24B,** 191.
25. M. S. Cresser and T. S. West, *Spectrochim. Acta,* 1970, **25B,** 61.
26. B. V. L'Vov, "Atomic Absorption Spectroscopy", Adam Hilger, London, 1970.
27. C. Th. J. Alkemade, in "Atomic Absorption Spectroscopy", eds. R. M. Dagnall and G. F. Kirkbright, Butterworths, London, 1971.
28. M. S. Cresser, G. F. Kirkbright, M. Sargent and T. S. West, unpublished work.
28a. B. V. L'Vov, in "Atomic Absorption Spectroscopy", eds. R. M. Dagnall and G. F. Kirkbright, Butterworths, London, 1971.
29. D. A. Davies, R. Venn and J. B. Willis, *J. Sci. Ins.,* 1965, **42,** 816.
30. A. Strasheim, F. W. E. Strelow and E. Norval, *J. S. Afric. Chem. Inst.,* 1961, **20,** 25.
31. J. E. Allan, *Spectrochim. Acta,* 1961, **17,** 467.
32. J. Ramirez-Munoz, *Talanta,* 1966, **13,** 87.
33. J. Ramirez-Munoz and W. F. Ulrich, *Flame Notes, Beckman,* 1966, **1,** 33.
34. H. L. Kahn and W. Slavin, *Appl. Optics,* 1963, **2,** 931.
35. S. B. Smith Jnr. and F. J. Feldman, *8th Nat. Meeting Soc. Appl. Spectrosc.,* Anaheim, Calif. U.S.A., 1969.
36. A. Z. Ringbom, *Z. Anal. Chem.,* 1934, **115,** 332.
37. G. D. Christian and F. J. Feldman, "Atomic Absorption Spectroscopy: Applications in Agriculture, Biology and Medicine' , Wiley-Interscience, New York, 1970, p. 124.
38. AFS—6 Spectrometer, Technicon Instrument Corp., Tarrytown, New York, 1971.
39. C. S. Rann, *Spectrochim. Acta,* 1968, **23B,** 827.
40. L. de Galan, *Spectrochim. Acta,* 1969, **24B,** 629.
41. L. de Galan and G. F. Samaey, *Anal. Chim. Acta,* 1970, **50,** 39.
42. B. V. L'Vov, *Spectrochim. Acta,* 1961, **17,** 761.
43. S. R. Koirtyohann and E. E. Pickett, Paper presented at XIIIth Colloq. Spectroscop. Intern., Ottawa, 1967.
44. L. de Galan and J. D. Winefordner, *J. Quant. Spec. Radiat. Transfer,* 1967, **7,** 251.
45. L. de Galan and G. F. Samaey, *Spectrochim. Acta,* 1970, **25B,** 245.
46. B. V. L'Vov, *Industrial Laboratory,* 1963, **28,** 987.
47. S. B. Smith Jnr., J. A. Blasi and F. J. Feldman, *Anal. Chem.,* 1968, **40,** 1525.
48. F. J. Feldman, J. A. Blasi and S. B. Smith Jnr., *Anal. Chem.,* 1969, **41,** 1095.
49. F. J. Feldman and S. B. Smith Jnr., *15th Spectrosc. Symp., Canada,* Toronto, 1968.
50. G. D. Christian and F. J. Feldman, "Atomic Absorption Spectroscopy: Applications in Agriculture, Biology and Medicine", Wiley-Interscience, New York, 1970, pp. 129, 192.
51. D. J. David, *Analyst,* 1961, **86,** 730.
52. C. B. Belcher and K. A. Brooks, *Anal. Chim. Acta,* 1963, **29,** 202.
53. W. Lang and R. Herrman, *Mikrochim. Acta,* 1963, 1053.
54. H. L. Kahn, *Atomic Absorption Newsletter,* 1967, **6,** 51.
55. W. J. Youden, *Ind. Eng. Chem., Anal. Ed.,* 1947, **19,** 943.
56. L. Erdey, "Gravimetric Analysis, Vol. 1", Pergamon Press, Oxford, 1963, p. 13.

57. N. H. Furman, "Scott's Standard Methods of Analysis", Van Nostrand, New York, 1939.
58. Willard and Diehl, "Advanced Quantitative Analysis", Van Nostrand, New York, 1945.
59. R. L. Mitchell, *J. Sci. Food Agric.*, 1960, **11**, 553.
60. J. Minczewski, in "Trace Characterisation", eds. W. W. Meinke and B. F. Scribner, Nat. Bureau of Standards Monograph 100, 1965.
61. N. E. Thiers, in "Trace Analysis", eds. J. H. Yoe and H. J. Koch, John Wiley, New York, 1957.
62. T. T. Gorsuch, *Analyst*, 1959, **84**, 135.
63. P. Ambler and N. Griggs, *Ind. Eng. Chem., Anal. Ed.*, 1941, **13**, 102.
64. R. E. Thiers, J. F. Williams and J. H. Yoe, *Anal. Chem.*, 1955, **27**, 1725.
65. R. E. Thiers, in "Methods of Biochemical Analysis, Vol. 5", ed. D. Glick, Interscience, New York, 1957.
66. R. A. Anduze, *Anal. Chem.*, 1957, **29**, 90.
67. W. Schöniger, *Mikrochim. Acta*, 1955, 123; 1956, 869.
68. J. Belisle, C. D. Green and L. D. Winter, *Anal. Chem.*, 1968, **40**, 1006.
69. C. E. Gleit and W. D. Holland, *Anal. Chem.*, 1962, **34**, 1454.
70. C. E. Gleit and W. D. Holland, *Anal. Chim. Acta*, 1965, **32**, 346.
71. C. E. Gleit, *Anal. Chem.*, 1965, **37**, 314.
72. C. E. Mulford, *Atomic Absorption Newsletter*, 1966, **5**, 135.
73. D. W. Bolin and O. E. Stamberg, *Ind. Eng. Chem., Anal. Ed.*, 1964, **16**, 345.
74. L. M. Cummins, T. L. Martin, G. W. Maag and D. C. Maag, *Anal. Chem.*, 1964, **36**, 382.
75. R. K. Simon, G. D. Christian and W. C. Purdy, *Amer. J. Clin. Path*, 1968, **49**, 733.
76. J. S. McNultey, *Ind. Eng. Chem., Anal. Ed.*, 1947, **19**, 809.
77. W. C. Johnson, *Analyst*, 1967, **92**, 403.
78. J. L. Down and T. T. Gorsuch, 1967, *Analyst*, **92**, 398.
79. T. T. Gorsuch, *Analyst*, 1959, **84**, 135.
80. D. Polley and V. L. Miller, *Anal. Chem.*, 1955, **27**, 1162.
81. A. C. Rolfe, F. R. Russell and N. T. Wilkinson, *Analyst*, 1955, **80**, 523.
82. H. Goto, Y. Kakita and T. Furukawa, *J. Chem. Soc. Japan*, 1958, **79**, 1513.
83. J. E. Allan, *Analyst*, 1961, **86**, 530.
84. E. Berman, *Atomic Absorption Newsletter*, 1964, **3**, 111.
85. E. A. Boling, *Spectrochim. Acta*, 1966, **22**, 425.
86. D. C. Burrell, *Atomic Absorption Newsletter*, 1965, **4**, 309.
87. D. Trent and W. Slavin, *Atomic Absorption Newsletter*, 1964, **3**, 118.
88. G. F. Kirkbright, M. K. Peters and T. S. West, *Analyst*, 1966, **91**, 411.
89. R. J. Magee and A. K. Rahman, *Talanta*, 1965, **12**, 409.
90. G. F. Kirkbright, A. M. Smith and T. S. West, *Analyst*, 1967, **92**, 411.
91. J. B. Willis, *Anal. Chem.*. 1962, **34**, 614.
92. L. R. P. Butler and P. M. Mathews, *Anal. Chim. Acta*, 1966, **36**, 319.
93. M. E. Morgan, *Atomic Absorption Newsletter*, 1964, **21**, 1.
94. G. F. Kirkbright, M. K. Peters and T. S. West, *Analyst*, 1966, **91**, 705.
95. H. Malissa and E. Schoffman, *Mikrochim. Acta*, 1965, 187.
96. G. H. Morrison and H. Freiser, "Solvent Extraction in Analytical Chemistry", John Wiley, New York, 1957.
97. J. Stary, "The Solvent Extraction of Metal Chelates", Pergamon Press, Oxford, 1964.
98. C. H. R. Gentry and L. G. Sherrington, *Analyst*, 1946, **71**, 432; 1950, **75**, 17.

99. J. P. Riley, *Anal. Chim. Acta,* 1958, **19,** 413.
100. J. Stary, *Anal. Chim. Acta,* 1963, **28,** 132.
101. F. Umland and K. U. Meckenstock, *Z. Anal. Chem.,* 1959, **165,** 161.
102. H. Goto and E. Sudo, *Japan Analyst* (*Bunseki Kagaku*), 1961, **10,** 171.
103. F. Umland and W. Hoffmann, *Z. Anal. Chem.,* 1954, **168,** 268.
104. A. J. Blair and D. A. Pantony, *Anal. Chim. Acta,* 1956, **14,** 545.
105. F. Umland and W. Hoffmann, *Angew. Chemie,* 1956, **68,** 704; 1957, **69,** 396.
106. F. Umland and W. Hoffmann, *Anal. Chim. Acta,* 1957, **17,** 234.
107. Y. Kekita and H. Goto, *Sci. Rep. Res. Inst., Tohoku Univ.,* 1960, **A12,** 334.
108. H. Goto and E. Sudo, *Sci. Rep. Res. Inst., Tohoku Univ.,* 1959, **A11,** 355.
109. I. P. Alimarin and I. M. Gibalo, *Vestnik Moskov. Univ., Ser. Mat., Mekh., Astr., Fiz., Khim.* 1956, **11,** No. 2, 185; 1958, **14,** No. 2, 185.
110. D. Dyrssen, *Svensk kem. Tidskr.,* 1953, **65,** 43.
111. E. P. Bullwinkel and P. Noble, *J. Amer. Chem. Soc.,* 1958, 2955.
112. D. Dyrssen and V. Dahlberg, *Acta Chem. Scand.,* 1953, **7,** 1186.
113. R. F. Clayton, W. H. Hardwick, M. Moreton-Smith and R. Todd, *Analyst,* 1958, **83,** 13.
114. W. E. Clifford, E. P. Bullwinkel, L. A. McClaine and P. Noble, *J. Am. Chem. Soc.,* 1958, **80,** 2959.
115. A. W. Ashbrook and K. Conn, *Chemist-Analyst,* 1961, **50,** 47.
116. N. Naito and K. Sugawara, *Bull. Chem. Soc., Japan,* 1957, **30,** 799.
117. E. Sudo, *J. Chem. Soc. Japan, Pure Chem. Sect.,* 1951, **72,** 817; *C.A.,* 1952, **46,** 6034.
118. Y. Kakita and H. Goto, *Sci. Rep. Res. Inst., Tohoku Univ.,* 1960, **A12,** 334.
119. J. E. Allan, *Spectrochim. Acta,* 1961, **17,** 459.
120. J. E. Allan, *Analyst,* 1961, **86,** 530.
121. S. Sprague and W. Slavin, *Atomic Absorption Newsletter,* 1964, May.
122. E. Berman, *Atomic Absorption Newsletter,* 1967, **6,** 57.
123. M. Montagut-Buscas, J. Obiols and E. Rodriguez, *Atomic Absorption Newsletter,* 1967, **6,** 61.
124. M. Suzuki, M. Yanagisawa and T. Takeuchi, *Talanta,* 1965, **12,** 989.
125. J. Stupar, *Mikrochim. Acta,* 1966, 722.
126. G. Devoto, *Russ. Med. Sarda,* 1968, **71,** 357.
127. E. Berman, *Atomic Absorption Newsletter,* 1967, **6,** 57.
128. E. Berman, *Atomic Absorption Newsletter,* 1965, **4,** 296.
129. A. Zettner, L. C. Sylvia and L. Capacho-Delgado, *Am. J. Clin. Pathol.,* 1966, **45,** 533.
130. J. B. Willis, *Anal. Chem.,* 1962, **34,** 614.
131. S. L. Sachdev, J. W. Robinson and P. W. West, *Anal. Chim. Acta,* 1967, **37,** 12.
132. R. E. Mansell and H. W. Emmel, *Atomic Absorption Newsletter,* 1965, **4,** 365.
133. R. C. Calkins, *Appl. Spect.,* 1966, **20,** 146.
134. B. Delaughter, *Atomic Absorption Newsletter,* 1965, **4,** 273.
135. R. Wickbold, *Z. Anal. Chem.,* 1959, **171,** 81.
136. B. Tuck and E. M. Osborn, *Analyst,* 1960, **85,** 105.
137. N. S. McNutt and R. H. Maier, *Anal. Chem.,* 1962, **34,** 276.
138. R. R. Brooks, *Analyst,* 1960, **85,** 745.
139. A. A. Smales and L. Salmon, *Analyst,* 1955, **80,** 37.
140. G. F. Reynolds, *Analyst,* 1957, **82,** 46.

141. A. Mizuike, Y. Iida and S. Hirano, *J. Chem. Soc. Japan, Ind. Chem. Sect.*, 1958, **61**, 1459.
142. *Idem*, ibid., 1959, **62**, 1491.
143. Y. Iida and A. Mizuike, *Japan Analyst*, 1964, **13**, 68.
144. J. N. Van Niekerk, J. F. Dewet and F. T. Wybenga, *Anal. Chem.*, 1961, **33**, 213.
145. H. J. Hettel and V. A. Fassel, *Anal. Chem.*, 1955, **27**, 1311.
146. H. A. Cranston and J. B. Thompson, *Ind. Eng. Chem., Anal. Ed.*, 1946, **18**, 323.
147. H. Buckwald and L. G. Wood, *Anal. Chem.*, 1953, **25**, 664.
148. C. E. Pietri and A. W. Wenzel, *Anal. Chem.*, 1963, **35**, 209.
149. F. Nakashima, *Anal. Chim. Acta*, 1964, **30**, 167.
150. G. F. Kirkbright, R. M. Dagnall, L. Ranson and T. S. West, unpublished work.
151. F. S. Kelso, J. M. Matthews and H. P. Kramer, *Anal Chem.*, 1964, **36**, 577.
152. O. Samuelson, "Ion Exchanges in Analytical Chemistry", John Wiley, New York, 1953.
153. J. Inczedy, "Analytical Applications of Ion Exchangers", Pergamon Press, New York, 1966.
154. R. L. Dehm, W. G. Dunn and E. R. Loder, *Anal. Chem.*, 1961, **33**, 607.
155. E. C. Mallory Jnr., in "Trace Inorganics in Water—Advances in Chemistry Series No. 73", ed. R. B. Gould, Am. Chem. Soc., Washington, D.C., 1968.
156. G. V. Miasoyedova, *Zh. Anal. Khim.*, 1966, **21**, 598.
157. T. Joyner, M. L. Healy, D. Chakravarti and T. Koyanagi, *Env. Sci. and Tech.*, 1967, **1**, 417.

CHAPTER 12

Interferences

The techniques of analytical atomic absorption and fluorescence spectroscopy are subject to the same types of interference as those encountered in flame emission spectroscopy. When concomitant elements effect an alteration in those physical or chemical properties or processes that control the final population of neutral ground state atoms of the sample in the absorption cell, interference with the analytical signal results. The principal types of interference encountered thus result from physical effects and from chemical effects in the condensed phase or vapour phase. Each of these types of interference, and the occurrence of spectral interference, will be considered separately.

12.1 PHYSICAL INTERFERENCES

In AAS and AFS the term physical interference is usually employed to denote any influence of concomitant materials present in the sample solution with the analyte element on one or more of the physical processes involved during the nebulization and atomization, or on one of the physical properties of the sample solution. The principal interferences encountered in both techniques result from the effect of solution viscosity on sample aspiration, and the effect of the surface tension, vapour pressure and temperature of the sample solution on the nebulization process and solvent evaporation. Solute evaporation effects and the scattering of incident radiation from the source may also result in physical interference. Many examples of the occurrence of physical interferences are described in the literature; extensive documentation of these effects will not be attempted here. The nature of the interferences caused by effects on each of the physical processes and properties mentioned above will be considered separately and illustrated, where applicable, by reference to some commonly encountered examples.

12.1.1 Aspiration

When an interference with the observed atomic absorption or fluorescence signal may be traced to an effect on the sample aspiration rate at a pneumatic nebulizer, this is most frequently caused by variation in viscosity of the sample solution. The sample solution is drawn up the capillary in indirect and direct nebulizers by the pressure drop created by the high-velocity gas stream at the outlet of the sample capillary (see Chapter 8.2.1(a)). The hydrostatic head (due to the column of liquid lifted) and the viscosity of the sample solution oppose the aspiration. The pressure due to the hydrostatic head is usually greatly exceeded by the aspiration pressure, however, so that the aspiration rate (sample uptake rate, ml min^{-1}) is not significantly affected by small changes in the hydrostatic head or density of the sample liquid. The force of capillarity also opposes the aspiration, but this effect is also very small, and any variation caused by different surface tensions in the sample liquids is unimportant in relation to the aspiration pressure.

When the sample liquid leaves the capillary under conditions of laminar flow, and with low velocity, the sample flow-rate theoretically is proportional to the pressure drop along the capillary and the fourth power of the capillary diameter and inversely proportional to the viscosity of the liquid. The dependence of the flow-rate on the fourth power of the capillary diameter has been demonstrated experimentally for the nebulizer of a flame-photometer,[1] and Winefordner and Latz[2] have shown that the flow-rate closely obeys the

Poiseuille equation for viscous flow for low but not for high flow-rates. The variation in nebulizer uptake rate caused by change in the viscosity of sample solutions does not necessarily produce a proportional change in the atomic absorption or fluorescence observed in the flame. Willis[3] has shown that with the indirect nebulizers employed with premixed flames, variation in sample uptake rate may be compensated for by changes in nebulization efficiency and drop-size distribution which occur simultaneously; the change in atomic concentration of the analyte element which is produced in the flame may then be quite small. The interference effect caused by variation in sample viscosity on the flow-rate for total consumption nebulizer-burners, however, may be serious. Because of the large amount of solution entering the flame, a considerable fraction of which exists as large droplets, substantial temperature changes may occur when the sample uptake rate varies. The residence time of the sample droplets in hot flames supported at these burners is also short; if a large number of large droplets enter the flame when the uptake rate is increased the desolvation process may be less efficient. Owing to this temperature effect and the effect on desolvation, when the sample uptake rate is increased substantially, decreased absorption or fluorescence signals may be obtained with this type of burner.

When an interference effect in AAS or AFS can be directly attributed to varying viscosities in the sample solutions it may be overcome by preparation of the standard solutions used to construct the calibration curve at similar viscosity in the same solvent. Alternatively, the samples may be diluted to minimize the variation in viscosity or the method of standard addition to the sample solution may be employed.

12.1.2 Nebulization

The physical processes that occur during nebulization have a large influence on the sensitivity and selectivity of flame methods of analysis (see Chapter 8). Owing to their complexity, however, and the difficulty of designing experiments to alter one parameter affecting the nebulization while maintaining the other parameters unchanged, nebulization is probably the least well-characterized physical process involved in analytical flame spectroscopy. The drop-size distribution and nebulizer efficiency depend on the nature of the nebulizing gas and sample solvent in a manner which may vary from one nebulizer type and geometry to another. The principal effects for indirect nebulizers have been described in Chapter 8.2.1. The following paragraphs summarize these effects from the viewpoint of changes in physical properties of sample solutions. As described in Chapter 8.2, the effect of the surface tension, density and viscosity of the sample solution on the drop-size distribution at a concentric nebulizer may be demonstrated from the empirical equation established by Nukiyama and Tanasawa.[4] The equation may be employed to show the general manner in which the mean droplet diameter is controlled by these physical properties of the solution nebulized. As pointed out earlier, with the nebulizers usually encountered on flame spectrometers the second term of the empirical equation is significant and the viscosity becomes an important factor. The mean droplet diameter then decreases with increase in the velocity of the air at the orifice and with decrease in surface tension and viscosity of the sample solution. As the mean droplet diameter affects the nebulizer efficiency, efficiency of transport of solution through connecting tubing to the burner, desolvation and solute vaporization in the flame, changes in the viscosity and surface tension in the sample solution can result in interference with the analytical signal observed. This property is used to advantage in the use of organic solvents of low viscosity and surface tension and at high vapour pressure to enhance sensitivity in flame methods of analysis (see Chapter 11.7).

Solvent evaporation from droplets in the nebulizer chamber affects the drop-size distribution of the sample reaching the burner.[1,5-7] The extent of solvent evaporation depends on the drop-size and residence time in the nebulizer, the molecular weight of the vapour, the diffusion coefficient and saturation vapour pressure of the solvent in air and the temperature. The smaller droplets may lose their solvent completely by evaporation during their residence time in the nebulizer. As the evaporation proceeds, however, the decreasing drop-size and increasing dissolved solid concentration change the effective saturation vapour pressure for the vapour. When a concomitant material causes a substantial change in the drop-size distribution through its effect on the solution viscosity or surface tension, or in the dissolved solid concentration, the change in saturation vapour pressure which occurs may affect the rate of evaporation. This change is probably small, however, and overshadowed by the effect on the analytical signal of the gross changes in drop-size distribution and nebulization efficiency caused by change in viscosity or surface tension. As described in Chapter 8.2, the extent of evaporation can be increased substantially by preheating the gas used for nebulization[8] or by using a heated nebulizer chamber.[9] After removal of most of the solvent in a heated nebulizer chamber the gases may be cooled to condense and collect the solvent.[10] The fine droplets and dry particles may then pass into the flame and a substantial increase in the fraction of the sample transported is obtained.

The analytical absorbance or fluorescence signals observed may also be affected by the temperature of the solution aspirated. The viscosity, surface tension and vapour pressure of the solution nebulized are temperature-dependent. Change in temperature of the sample solution may therefore affect the uptake rate and drop-size distribution, although a proportional change in analytical signal may not occur. With indirect nebulizer systems it is generally observed that when sample solution temperatures differ only slightly (e.g. $\pm 3^\circ C$) the errors are insignificant.

12.1.3 Solvent and Solute Vaporization in Flames

Although the smaller droplets may lose their solvent by evaporation in the nebulizer chamber of indirect nebulizers, the desolvation process is usually completed only when the aerosol enters the flame. The rate of the desolvation process in the flame is controlled by droplet size, boiling point and specific heat of vaporization of the solvent, thermal conductivity of the gas and the flame temperature. The process is rapid with well-designed indirect nebulizers and the premixed flames normally employed in AAS. Interference with the analytical signal due to the radiation scattering and loss of atomic concentration which would accompany incomplete desolvation is therefore seldom encountered. The situation is quite different with integral nebulizer-burner units, however, where incomplete desolvation occurs owing to the large droplets introduced which may have only a short residence time in the flame. Parsons and Winefordner[11] reported only 50-90% evaporation for water in such a flame in the part of the flame normally used for analysis.

After complete desolvation of the aerosol droplets, the dry particles which remain must be vaporized and dissociated to provide the atomic population for analysis. The size and other physical properties of the particle determine the rate of vaporization. When the composition of the dry particles varies in the presence of concomitant elements these physical properties may vary and give rise to disturbance of the analytical signal. Thus, for example, when many large particles are formed from a sample solution which has a high dissolved solid content, the rate of vaporization may be slower than that of the smaller particles produced from pure aqueous solutions of the analyte element. Other interferences, frequently classified as chemical interferences, result from the different physical properties of the dry particles of compounds formed between analyte and concomitant elements in the flame compared to corresponding properties of the particles formed for the analyte element in the absence of concomitant elements. The physical properties of most concern in these cases are the melting and boiling points of the compounds. Further discussion of these effects is presented in Chapter 6.

12.1.4 High Dissolved Solid Content Samples and Light Scattering

When a sample solution contains a high concentration of other dissolved solids a reduction in the analytical atomic absorption signal may be observed compared to that obtained with a pure aqueous solution of the same analyte concentration. The sample uptake rate and nebulizer efficiency may be lower owing to the increased viscosity of the solutions, and incrustation of the nebulizer and burner head may lead to high signal noise levels and drift. Physical occlusion of analyte in the large particles formed from such solutions on desolvation may also retard the vaporization process in the flame.

In addition to the interference effects mentioned above in AAS and AFS for samples of high dissolved solid content, with such solutions problems from scattered radiation may also become significant. Willis[12] first proposed that the scattering of the incident radiation by particles in the flame might explain the small unspecific absorption often obtained in the presence of high concentrations of concomitant elements in AAS. For small particles, whose diameter is much smaller than the wavelength of the radiation measured, Rayleigh scattering (proportional to the fourth power of the frequency) should occur. Willis observed a wavelength-dependence of this type in the nebulization of concentrated (5%) solutions of simple salts into a low-temperature flame. Where the particle diameter is larger than the wavelength of the radiation measured, the Mie[13] scattering theory predicts that only a slight dependence of scattering on wavelength should occur. This type of behaviour was observed by Gidley[14] when 10% solutions of the salts of elements that form refractory oxides (Ti, Zr and Hf) were nebulized.

Koirtyohann and Pickett[15] calculated the mean size of solid particles formed on desolvation with a burner for premixed flames; using the Mie theory they concluded that the experimentally observed non-specific absorption could not be accounted for wholly on the basis of light scattering. The same authors[16] demonstrated that absorption by gaseous molecular species caused at least part of the observed non-specific absorption. For example, the molecular absorption

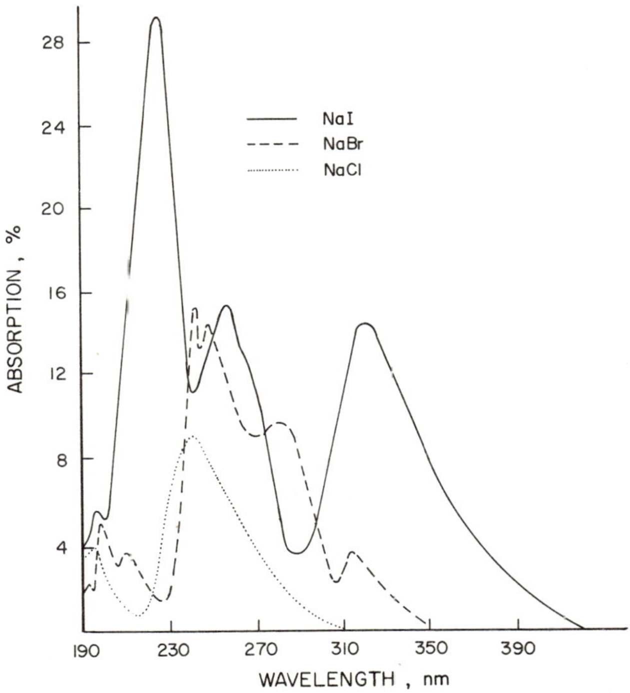

Fig. 12-1. Absorption spectra of alkali metal halides obtained with a xenon arc source and a cool nitrogen-hydrogen diffusion flame (10 cm path length) with 1% sample solution.

spectra of CaOH and SrOH have been observed[16,17] and may give rise to non-specific background absorption effects. Non-specific background absorption, either by scattering or molecular absorption, may be more serious in cool flames than in hot flames. Figure 12-1 illustrates typical absorption spectra from alkali metal halide salts produced in a cool nitrogen-hydrogen diffusion flame. The severity of the background absorption is seen to increase as the melting and boiling points and dissociation energies of the halides increase.[18] Interference by non-specific absorption may be corrected for by measurement of the absorbance at wavelengths near to the resonance line using a non-absorbing line from the source and subtracting this from the absorbance at the resonance line. Alternatively, the absorbance at wavelengths in the neighbourhood of the resonance line may be measured using a continuum source[19,20] and the value obtained subtracted similarly (see also 12.3.1).

The phenomenon of light scattering by particles in the flame is more important in atomic fluorescence spectroscopy than in atomic absorption spectroscopy. When resonance fluorescence signals are measured, the signal which results from scattering of the primary source radiation by particles in the flame cannot be distinguished from the atomic resonance fluorescence signal. Veillon and coworkers,[21] who employed a continuum source for AFS, corrected for scattered radiation by using a scanning monochromator. When stepwise or direct-line fluorescence signals are employed for analysis, scattered radiation from the source (if it emits at the wavelength measured) may frequently be eliminated by the use of a simple filter in the incident light beam from the source to select only that line used for irradiation. When the source does not emit at the measured line this is not necessary. Problems of scattered radiation are more serious with integral nebulizer-burner systems than when direct nebulizers and premixed flames are employed. Ellis and Demers,[22] however, have employed an integral nebulizer-burner with a hydrogen-entrained air flame and report fewer scattering problems with this flame than with the hydrogen-oxygen flame at a similar burner. With premixed air-hydrocarbon and nitrous oxide-acetylene flames very few scattering phenomena have been observed.[23,24] Scattering has been observed to cause serious interferences in these flames only with samples containing very high concentrations of matrix elements. In the determination of minor element components in aluminium alloys by AFS it has been shown possible to "back-off" the small scatter signal obtained when dilute aluminium solutions are nebulized into an air-acetylene flame by incorporating a similar amount of aluminium into the blank and standards as that present in the sample solutions. Little loss in analytical precision was observed.[25]

12.2 CHEMICAL INTERFERENCES

In AAS and AFS any reaction of concomitant materials present with the analyte element in the sample solution that occurs in the solid or vapour phase to affect the population of free atoms of analyte element formed in the atom cell may be termed a chemical interference. Such reactions in the solid phase are frequently also referred to as *condensed* phase interferences.

12.2.1 Chemical Interferences in the Solid Phase

Non-specific interferences in the solid phase, where no compounds are formed between the analyte and concomitant material in the flame, may occur for any analyte element and are physical rather than chemical in nature. An example of this type of non-specific solid phase interference is the relatively small depressive effect on AAS signals for analyte elements which is caused by occlusion of analyte within solid particles of salts of the matrix element in the flame. Specific interferences in the solid phase, however, are usually due to a reaction of the analyte element or one of its compounds with concomitant material to form a different compound with different thermochemical properties. This type of *chemical interference* is frequently specific to one analyte element or a small group of elements, e.g., the interference of phosphate with the alkaline earth elements in AAS in air-hydrocarbon flames due to compound formation. The literature is replete with references to interference effects of a "chemical" nature. The exact conditions under which these were observed are often difficult to establish and the results may not be reproducible from one laboratory to another. This is often the result of the critical dependence of the observed effect on the flame composition, height of observation and drop-size distribution from the nebulizer employed.

It is important to realize that it is the *rate* of evaporation of the particles of compounds formed between the analyte and concomitant elements compared to the evaporation rate of the analyte particles alone that determines the severity of the chemical interference effect. As described in Chapter 6, it is not essential for the flame temperature to be higher than the melting or boiling points of the compound formed in the dry particle for vaporization to proceed to completion. The flame temperature and composition, however, do influence the extent of formation and *rate* of evaporation of solid compounds in the flame, and in high-temperature flames the incidence of chemical interference is lower than in cool flames. Owing to this effect of compound formation on the rate of release of vapour of the analyte, the severity of any chemical interference may also depend on the size of the particles formed (and therefore on nebulizer performance—see also Chapter 8.2.2) and the height of observation of absorption or fluorescence in the flame (i.e. the residence time in the flame).

Perhaps the best-known chemical interference effect in analytical flame spectroscopy is the depression of calcium atomic concentration in the presence of phosphate. The effect has been observed and studied, by both emission and absorption spectroscopy, by many workers. Sulphuric acid also has a similar effect on the determination of calcium. Figure 12-2 shows the typical manner in which the calcium signal, observed in emission or absorption at the 422·7 nm

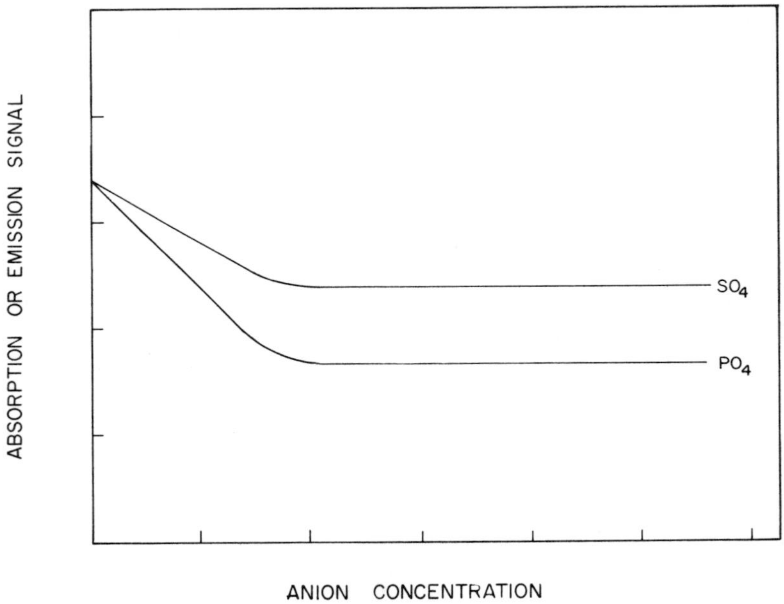

Fig. 12-2. Typical results for suppression by phosphate or sulphate of calcium 422·7 nm emission or absorption in air-acetylene flame.

line or in emission at the CaOH bond at 554 nm, varies with the concentration of phosphate or sulphate. Similar results may be obtained with premixed hydrocarbon-air flames and with oxy-hydrogen or oxy-acetylene flames supported at integral nebulizer-burner units. The interference observed is virtually linear with the anion concentration up to a certain atomic concentration ratio of P/Ca, after which there is no further suppression with increasing anion concentration. The P/Ca concentration ratio above which the suppression is constant is not dependent on the calcium concentration, height of observation in the flame or flame temperature. The relative severity of the interference, however, is affected by these factors. Values of the P/Ca ratio at the knee of the curve are reported in the literature to be between 0·3 and 1·1. The severity of interference is usually greater with turbulent flames burning at integral nebulizer-burner units than for premixed flames with indirect nebulizer. The experimental observations may be explained by the postulation that a relatively non-volatile compound is formed between calcium and phosphorus in the flame. The particles of this compound, which probably contains only calcium and phosphorus in association with some oxygen, persist in the flame and evaporate only slowly. The limited volatility of the particles accounts for the observation that the severity of the interference effect observed is less when the calcium atomic absorption signal is measured higher in the flame (after a longer residence time) or in hotter flames (more rapid evaporation). As the evaporation rate also depends on the particle size, the observation of less serious interference with premixed flames and indirect nebulizers than with integral nebulizer-burner units might be accounted for by the smaller drop-size and particle size obtained with the former. It has also been observed that the relative depression of the calcium signal by phosphorus is smaller at lower calcium concentrations in the sample solution; this effect leads to curvature of the calibration graph for calcium towards the concentration axis when phosphorus is present. This effect can also be explained as being due to the relatively greater evaporation rate of the smaller particles of Ca—P—O compound formed at low concentrations than that of the larger particles formed with more concentrated sample solutions.

As well as chemical interference from concomitant anions on the determination of cations, which is quite prevalent in hydrogen-air and hydrocarbon-air flames used in AAS and AFS, mutual chemical interference between cations may also be encountered. A well-known example lies in the interference of aluminium salts in the determination of the alkaline earth elements and magnesium in hydrocarbon-air flames. The magnitude of the interference observed depends on the anion or acid which accompanies the aluminium. Dickson and Johnson[26] found that the degree of depression of calcium atomic absorption by aluminium varied with the anion present in the order $PO_4^{3-} > SO_4^{2-} > Cl^- > ClO_4^-$. It is most probable that a non-volatile Al—Ca—O compound is formed, and the extent of its formation depends on the manner in which the aluminium salts with different anions decompose in the flame.

Many other examples of chemical interference for various analyte and concomitant elements have been observed. Gilbert[27] has listed elements that may cause interference with various analyte elements by the formation of mixed oxide species of low volatility. The list includes B, Be, Cr, Fe, Mo, Si, Ti, U, V, W and the rare-earth elements.

The occurrence of specific chemical interference between concomitant element and analyte, rather than a non-specific physical interference in the solid phase, may frequently be recognized from the shape of the graph of the analytical signal against the concentration of the concomitant element. Where compound formation is prevalent, such graphs normally show a pronounced break or "knee" similar to that described for calcium and phosphate. The position of the break on the concentration axis usually remains independent of height of observation, although the depressive effect may be less severe when observations are made high in the flame. Specific chemical interference and a non-specific solid-phase interference may occur at the same time. Thus, for example, in addition to the specific effect of aluminium on calcium described above, due to formation of a low-volatility Al—Ca—O compound, in the presence of a very high concentration of aluminium considerable amounts of aluminium oxide are also formed. In this way the Al—Ca—O compound is bound within relatively large particles of aluminium oxide and the release of calcium atoms to the flame is further impeded. Under these conditions the shape of the graph of the analytical signal against the interfering element concentration is modified and a sharp break may no longer be observed. A useful indication of the occurrence of specific chemical interference rather than a vapour-phase interference can be obtained by experiments conducted with separate nebulizer units for the analyte and concomitant elements. When a condensed phase interference exists, separate introduction into the flame in this way of the species involved in compound formation should yield no interference effect. If, on the other hand, the interference is due to a reaction in the vapour phase, the interference effect should still be observable. Experiments of this type have been undertaken by several workers[28-31] to confirm the occurrence of condensed-phase interference of phosphate and other anions with calcium and magnesium in flame emission and absorption spectroscopy.

12.2.2 Elimination of Solid-Phase Interferences

The depressive effect of concomitant elements caused by chemical interference in the solid phase may frequently be eliminated, or at least minimized, by judicious choice of operating conditions during the analysis. Where the effect itself may not be eliminated, errors due to the interference may be prevented by suitable preparation of standards and samples.

It is important first to optimize the nebulizer and flame conditions to minimize chemical interference. This is achieved by attention to the following factors.

(*i*) *Particle size.* In order to achieve a high rate of particle evaporation the particles should be as small as possible (see also Chapter 8). Careful adjustment of sample uptake rate, use of an organic solvent, and the insertion of barrier surfaces within the nebulizer chamber to assist the formation of as high a proportion of small droplets as possible, are thus advisable. Large droplet and particle size accounts for much of the greater severity of condensed-phase interference observed with integral nebulizer-burner units compared to premixed flames with indirect nebulizers. When there is no lack of sensitivity for the detection of the analyte by absorption or fluorescence in the samples to be analysed, dilution of the sample solutions assists the formation of small particles and minimization of interference.

(*ii*) *Height of observation.* When atomic absorption or fluorescence measurements are made as high as possible in the flame commensurate with good stability and low noise levels, the chemical interference effects are minimized. The low-volatility compound formed between the analyte and concomitant elements will then have the longest possible residence time in the flame to release the analyte. It has been noted by various workers that the interference effect of phosphate on calcium is greatly diminished when the calcium emission or absorption is viewed high in the particular flames used.[28,29,32] As shown in Fig. 12-3, Harrison and Wadelin demonstrated the elimination of the interference of aluminium on magnesium absorption at 285·2 nm when observations are made sufficiently high above the burner plate in an air-acetylene flame burning at a triple slot (Boling) burner.[32]

(*iii*) *Flame temperature and composition.* The rate of vaporization of the compound formed in the flame will depend on the flame temperature and composition. The most serious interferences in the solid phase are usually observed in cooler flames (i.e. in air-hydrogen, argon-hydrogen-entrained air and even air-hydrocarbon flames). Even in air-hydrocarbon flames, however, control of the fuel : oxidant concentration ratio may result in less serious chemical interference.

In contrast to the requirement of a fuel-rich flame to promote the dissociation of molecules in the vapour phase and provide high sensitivity and freedom from vapour-phase interference, a lean flame may sometimes be required for the minimization of chemical interferences in the solid phase. This probably results from the more rapid vaporization of certain species in the somewhat hotter lean flames available. This is illustrated in Fig. 12-4 for the interference of aluminium with magnesium investigated by Harrison and Wadelin.[32]

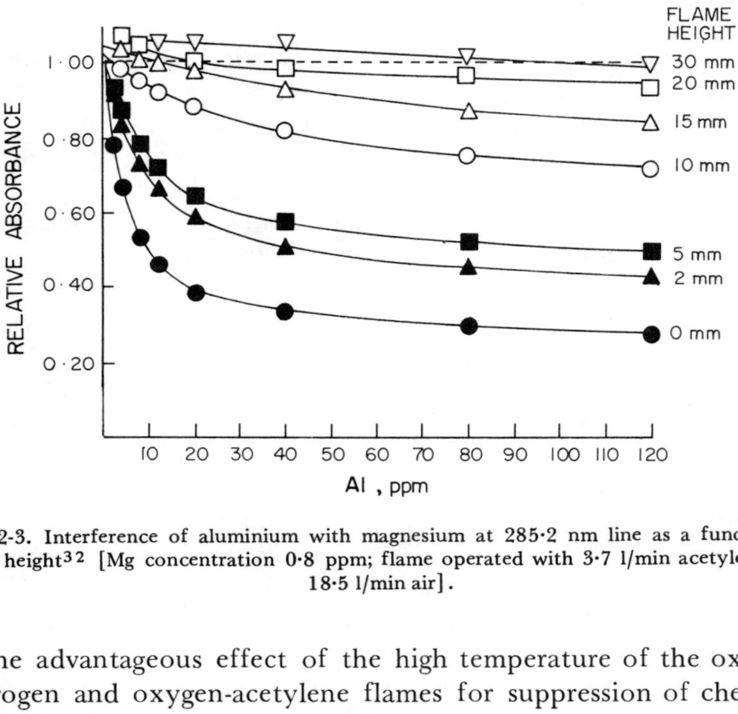

Fig. 12-3. Interference of aluminium with magnesium at 285·2 nm line as a function of flame height[32] [Mg concentration 0·8 ppm; flame operated with 3·7 l/min acetylene and 18·5 l/min air].

The advantageous effect of the high temperature of the oxygen-hydrogen and oxygen-acetylene flames for suppression of chemical interference effects may be difficult to realize in practice owing to the adverse effects on sample residence time and particle size that accompany their use with integral nebulizer-burner units. The use of hot premixed nitrous oxide-acetylene and oxygen-acetylene flames, however, has been shown to possess great advantages from the viewpoint of elimination of chemical interferences.[33,34,35] In a consideration of the chemical or solute vaporization interferences for the alkaline earth-anion systems, Fassel and Becker[35] have demonstrated that even at a 200 ppm calcium concentration no significant depression of the calcium emission or absorption signal occurs up to P/Ca molar ratios of 100 for premixed oxygen-nitrogen-acetylene, nitrous oxide-acetylene or oxygen-acetylene flames when an efficient indirect nebulizer is used (Fig. 12-5). At molar ratios greater than 100, the dissolved solid content of the solution significantly affects the physical properties of the sample solutions to reduce the nebulization efficiency. Figure 12-5 also shows for comparison the interference effects reported for calcium on phosphate when an air-acetylene flame and indirect nebulizer are used and the CaO

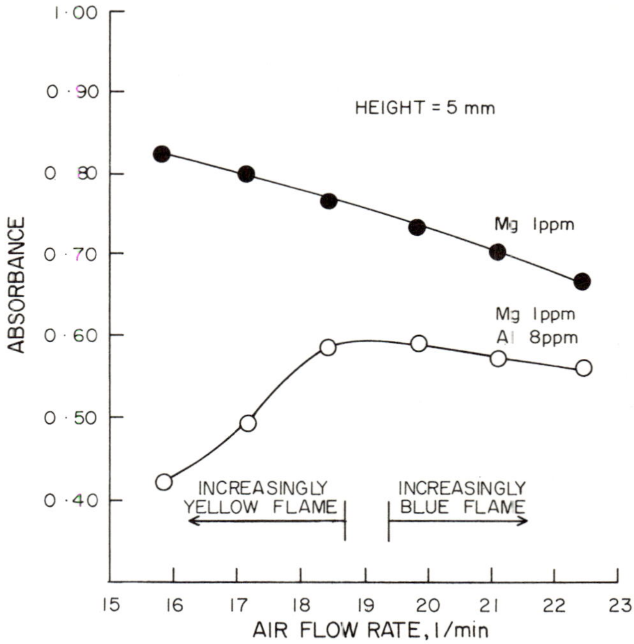

Fig. 12-4. Dependence of magnesium sensitivity and aluminium-magnesium interference on flame composition at 285·2 nm line[32] [Flame operated with 3·7 l/min acetylene].

emission is measured at 620 nm.[28] The elimination of the interference of phosphate with calcium does not appear to depend upon the flame stoicheiometry or height of observation in the hot nitrous oxide-acetylene flame. The same workers have also observed no significant interference from phosphate in the determination of the elements magnesium, barium and strontium in oxygen-nitrogen-acetylene and nitrous oxide-acetylene flames, or from sulphate or borate in the determination of calcium. Although these observations were made in emission, there is no reason to expect that appreciably different results would be obtained by absorption or fluorescence measurements. Harrison and Wadelin[32] have also confirmed the elimination of the interferences of aluminium, titanium and zirconium in the AAS determination of magnesium, due to spinel formation, that occurs in air-hydrocarbon flames when a hotter nitrous oxide-acetylene flame is employed. Again the interference is dependent on the flame stoicheiometry and becomes less important as the flame is made more fuel-lean (hotter); indeed, an actual enhancement in magnesium absorbance was observed by these workers in the presence of aluminium, titanium or zirconium when a rich flame was employed. The use of hot flames to eliminate

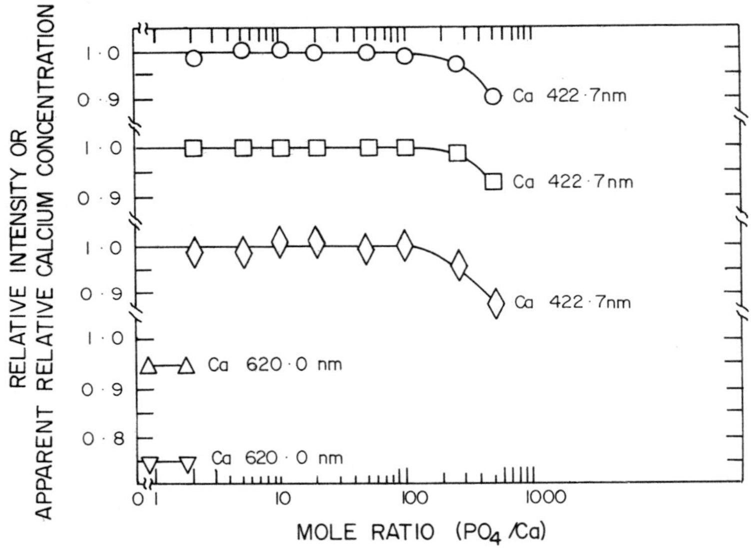

Fig. 12-5. Comparison of phosphate interference with calcium[33]

○ $O_2-C_2H_2$ slot burner
□ $N_2O-C_2H_2$ slot burner
◇ $N_2O-C_2H_2$ premix burner
} 200 ppm Ca in a 1:1 ethanol/water mixture. Measurements 20 mm above burner

△ Air–C_2H_2 premix burner, 112 mm above inner cones
▽ Air–C_2H_2 premix burner, 32 mm above inner cones
} 116 ppm Ca in water

chemical interferences may give rise to somewhat reduced sensitivity owing to ionization of the analyte atoms for some elements. Suppression of ionization, however, is a relatively simple matter, and the use of these hot flames should be the recommended method for the avoidance of chemical effects from concomitant elements.

When it is not possible, through optimization of the nebulizer and flame conditions, to eliminate a particular chemical interference the use of a releasing or a protective agent, or suitable addition of the interfering element to the standard solutions used to prepare the calibration graph, may achieve a satisfactory result.

12.2.3 Use of Releasing and Protective Agents

The depression in AAS or AFS signal caused by decrease in the analyte vaporization rate in the presence of an interfering element may frequently be minimized by the use of a releasing or a protective agent. The substances employed for this purpose are also sometimes termed "buffers". The general mode of action of *releasing* agents appears to rest on the formation of a compound between the interferent and the releasing agent which is more thermally stable than that formed between the analyte and the interferent. A competition governed by mass-action considerations therefore exists between the analyte and the releasing agent for reaction with the interfering species, and in the presence of sufficient of the particular additive the analyte is "released" for measurement. *Protective* agents, on the other hand, appear to minimize solute vaporization interferences by several different possible mechanisms. An organic protective agent may form a chelate complex with either interferent or analyte and prevent their chemical interaction so that each is vaporized separately. When no complex is formed between the analyte or interferent and the protective agent, the protective agent may enhance the analyte vaporization rate by formation of particles of large surface area which vaporize rapidly.

(*i*) *Releasing agents.* Some of the most commonly employed releasing agents are cations with which the interfering species combine in preference to the analyte element. Thus, in the determination of magnesium or strontium, calcium may be employed as a releasing agent for the interference from phosphate, silicate or aluminium. Similarly, strontium may be used to prevent the interference of phosphate in the calcium determination. Usually a considerable excess of the releasing agent must be employed. It then consumes the interfering species by formation of a compound of low volatility and allows the analyte to vaporize without interference by compound formation. David[36] utilized magnesium as a releasing agent to suppress the interference of phosphate, sulphate, aluminium and silicate in the determination of calcium in plant materials by AAS. Platte and Marcy[37] were able to suppress the interference of silica in the determination of iron and manganese by AAS by the addition of calcium as releasing agent. Figure 12-6, adapted from their work, illustrates the effect of this suppression. More recently, strontium has been employed as a releasing agent for suppression of interferences in the multichannel determination of Ca, Cu, Mg, Mn and Zn in soil extracts by AFS.[38] Strontium addition has also been utilized to release the interference of aluminium in the determination of magnesium in aluminium alloys by AFS.[39]* Table 12-1 illustrates the observed effect of strontium to release this interference, and also indicates the slight loss in precision incurred when excessive amounts of this releasing agent are employed. This most probably results from a degradation of the signal : noise ratio due to stray light radiated from the SrOH band emission in the flame. Lanthanum is particularly suitable for use as a releasing agent in the determination of

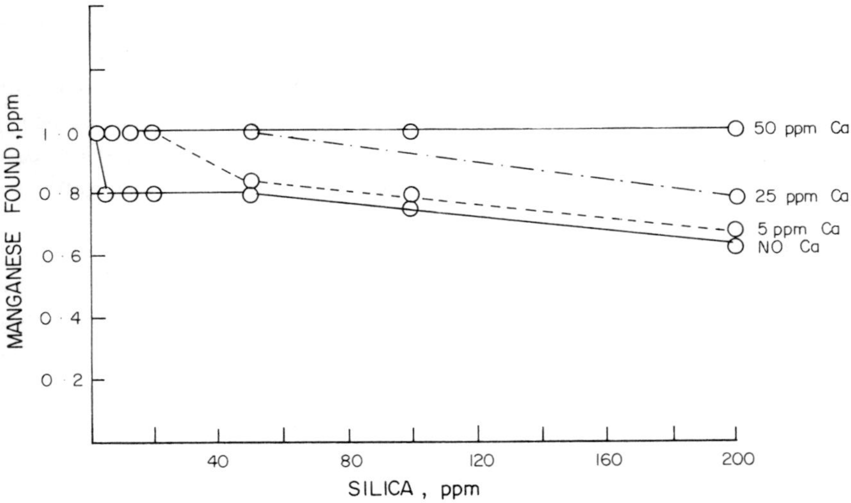

Fig. 12-6. Effect of addition of calcium on suppression of silica interference in determination of manganese (adapted from ref. 37).

the alkaline earth and other elements as it does not exhibit background emission or absorption at their resonance lines in hydrocarbon-air flames. Other rare earths have been employed to suppress interference of aluminium, phosphate and sulphate in the determination of strontium.[41] It has also been demonstrated that, in the presence of 8-hydroxyquinoline[42] or EDTA[43] a considerably lower concentration of lanthanum may be used to suppress chemical interferences from aluminium. Similarly, the use of less lanthanum is required to release interference of aluminium in the determination of magnesium when glycerol is present.[44]

* Mansell[40] has demonstrated from a consideration of the free energies of formation of the species involved that, as $MgAl_2O_4$ is the least stable of the alkaline earth aluminates, the elements Ca, Sr, Be and La will result in the release of Mg and simultaneous formation of the aluminate of the releasing element.

Copper may be used in the determination of platinum by AAS to suppress the interference from many elements, including Sb, W, Co and Ni.[45] This observation has been confirmed by other workers[46] but the precise nature of the releasing action remains unclear.

Further examples of the application of releasing agents to suppression of chemical interferences will be found in Chapter 13.

(*ii*) *Protective agents.* In the determination of the alkaline earth elements by AAS the addition of protective agents that form chelate complexes with the analyte or interferent elements may result in the elimination of solute vaporization interference. The two most extensively studied protective agents used in this way are 8-hydroxyquinoline and EDTA. 8-Hydroxyquinoline has been used to suppress the

TABLE 12-1

Effect of various concentrations of strontium releasing agent on magnesium atomic fluorescence signal obtained in the presence of aluminium

Solution (concentration, μg/ml)	Signal	Relative standard deviation, %
1 Mg	1·00	1·17
1 Mg + 100 Al	0·84	1·34
1 Mg + 100 Al + 100 Sr	0·97	1·40
1 Mg + 100 Al + 250 Sr	1·00	2·32
1 Mg + 100 Al + 500 Sr	1·01	2·98
1 Mg + 100 Al + 750 Sr	1·00	3·51
1 Mg + 100 Al + 1000 Sr	1·01	3·52
1 Mg + 100 Al + 1500 Sr	1·01	4·82
1 Mg + 100 Al + 3000 Sr	1·01	4·84

interference of aluminium in the determination of Ca and Mg[47,48] (see Fig. 12-7), and it has also been recommended for elimination of the interference of Co, Ni and Cu in the AAS determination of iron.[49] Addition of EDTA to analyte solutions has been shown by several groups of workers to eliminate the interference of Se, Te, B, Al, Si, phosphate and sulphate in the determination of magnesium and calcium.[50,51] The mode of action of this type of protective agent presumably depends on the formation of relatively more volatile *compounds* between the analyte or interferent and the protector to increase the vaporization rate.

The addition of protective reagents that are not normally considered to form compounds with the analyte or interferent element may also be effective in the suppression of interference. The use of the sugars, dextrose and sucrose, for example, has been shown to be successful in suppression of the interference of phosphate in the determination of calcium and strontium.[52,53,54] Ethylene glycol, glycerol and mannitol have also been used for this purpose.[53,55] Ammonium chloride appears to function as a protective agent in a similar fashion, and the addition of relatively high concentrations of NH_4Cl to sample solutions has been successful in minimizing the interference of, for example, Al in the determination of Na and Cr[56] and Fe in the determination of Cr.[57] Mostyn and Cunningham also

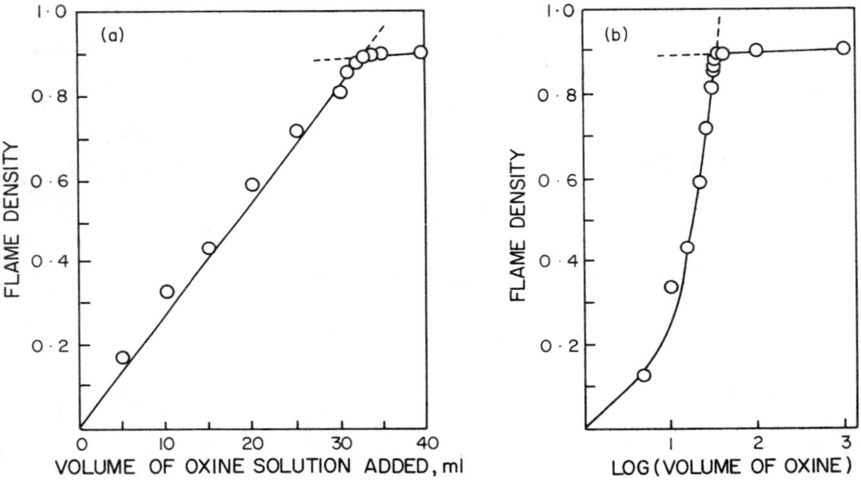

Fig. 12-7. Suppression of aluminium interference in the AAS determination of magnesium using 8-hydroxyquinoline.[47]

recommended the use of NH_4Cl for the removal of interferences in the AAS determination of molybdenum in the air-acetylene flame,[58] and David[59] employed its addition as a protector for the determination of Na, K, Ca and Mg in soils. No full explanation of the manner in which this type of protective agent functions has been proposed; the vaporization efficiency may be improved by the presence of these protective agents which decompose or sublime at relatively low temperatures.

12.2.4 Chemical Interferences in the Vapour Phase

Suppression or enhancement of the absorption or fluorescence signal may be caused by the effect of concomitant elements on the dissociation and ionization equilibria that control the concentration of atoms of the analyte element in the flame after vaporization of dissolved solid in a sample. Excitation interference, in which concomitant elements affect the analytical signal by disturbance of the equilibria between the electronically excited and unexcited states of the analyte atom, are also considered as vapour-phase interferences but may be shown to be insignificant in analytical flame spectroscopy when conditions of thermal equilibrium prevail within the part of the flame employed.[60] Vapour-phase interference, as distinct from condensed-phase interference effects, is expected to occur even when the analyte and concomitant elements are introduced into the flame via separate nebulizers.

(*a*) *Dissociation equilibria*

An account of the most important dissociation equilibria occurring between analyte atoms and the bulk constituents of the flame gases or other species added to the flame with the analyte has been given in Chapter 6.

When sample solutions of low dissolved-solid content are nebulized there is little tendency for the analyte element to form stable gaseous compounds with the other species (halides, etc.) introduced simultaneously into the flame from the sample. Under these conditions, when an analyte element forms a stable molecule in the gaseous phase, this is usually an oxide or hydroxide formed with atomic oxygen or hydroxyl species from the flame gases. In the flame the partial pressure of the flame species involved ([O] or [OH]) is usually several orders of magnitude greater than the partial pressure of the analyte element; the concentration of the flame species may therefore be regarded as being unaffected by the small amount consumed to form the molecular compound with the analyte element. The degree of dissociation attained, and thus the analytical sensitivity in AAS or AFS, then depends on the equilibrium constant for the dissociation of the molecule. The degree of dissociation may therefore remain substantially constant provided that the temperature and flame-gas composition are maintained unaltered (by control of gas flow-rates). When the partial pressure of the flame-gas species involved is high and effectively constant, being stabilized also by its relation to the dissociation equilibria of the bulk constituent gases, little vapour-phase interference with the analyte partial pressure caused by disturbance of the dissociation equilibria of its compound can be caused by the introduction of a concomitant element which also forms a gaseous compound in the flame.

When the partial pressure of the flame-gas partner (e.g. atomic oxygen) is low, however, and those of the analyte and concomitant elements are similar, depletion of the flame species may occur by gaseous compound formation (e.g. oxide) with the concomitant element. This depletion then gives rise to a greater degree of dissociation of the gaseous compound (oxide) formed by the analyte element. This effect results in an interference with the analytical signal, which increases. This type of enhancement of absorbance in the presence of concomitant elements that form thermally stable oxides has been widely observed in AAS with fuel-rich nitrous oxide-acetylene flames. Robinson and co-workers,[61,62] for example, observed an enhancement of vanadium absorption by the addition of aluminium or titanium which could not be explained by an effect on the ionization equilibria. Similarly, titanium effects an enhancement of aluminium absorption and vice versa[62,63] and yttrium has been reported to enhance sensitivity in the detection of the rare earth elements.[64]

It is important to distinguish between the interference effects due to shifting of the dissociation equilibria and those caused by disturbance of the ionization equilibria in these flames. Martin[65] has observed enhancement for the absorbance of uranium in the nitrous oxide-acetylene flame in the presence of various cations, and the interference of aluminium and tungsten in the determination of vanadium and molybdenum by AFS in this flame has been reported.[66]

It is well established that molecules of the halides of the alkali and alkaline earth metals may be formed in the vapour phase in some flames and give rise to suppression of the observed analytical signal.[67,68] The formation of halides is most pronounced in cool flames and when high concentrations of halide or halogen acid are present in the sample solution. Mandelstam[67] explained this behaviour from a consideration of the dissociation equilibria for various alkali-metal halides. In the air-acetylene flame the dissociation of halides is complete even when relatively high concentrations of halide ion are introduced into the flame with the analyte element. In cooler flames, such as air-coal gas or air-propane, however, incomplete dissociation of alkali-metal halides may be obtained even at fairly low concentrations. When the concentration of halogen acid or halide ion in the sample solution is increased, the dissociation may be suppressed and depression of the analytical signal observed. This explanation may hold for the cooler flames, but does not explain the depressive effect of halogen acids observed in air-acetylene flames. In this hotter flame even considerable increase in halogen concentration should not cause significant metal halide formation. To account for the experimental observations of this depression it is necessary to postulate that the ionization equilibrium for the alkali metal is disturbed by the addition of the strongly electronegative halogen so that the degree of ionization of the metal is increased. For further discussion of this effect see ref. 69.

Vapour-phase interference from halides and acids is not usually encountered for easily atomized elements other than the alkali metals. Slight depression of the atomic absorption signals sometimes encountered for elements such as Ag, Zn, Cd and Bi in the presence of acids, can usually be accounted for by their effect on the nebulization of the sample. Little information is available concerning the extent to which true vapour-phase interactions are involved in the interference effects observed for different anions in the AAS of elements which form stable oxides. Many of these effects, however, may also be largely due to the effect of the anion on the nebulization process and the rate of vaporization of the solute particles in the flame.

(*b*) *Ionization equilibria*

The effect of the ionization of species added to flames on the population of free atoms of easily ionized analyte elements has been described in Chapter 6.5.4. The equilibria outlined there may be invoked to explain the interference effects from concomitant elements due to their ionization. The mutual interference effects of the alkali metals in the determination of these elements are probably the most widely encountered ionization interference effects in flame spectroscopy. As a result of their low ionization potentials, quite pronounced ionization effects are observed in the determination of the alkali metals in the presence of each other by AAS.* The enhanced absorption observed in the determination of sodium, for example, in the presence of potassium is explained by the suppression of the degree of ionization of sodium owing to the higher electron concentration in the flame when potassium is added to the flame and undergoes ionization. Few other analyte elements suffer serious ionization interference in hydrocarbon-air flames; these

* In a premixed air-acetylene flame at 2500 K, aspiration of 10^{-3} M solutions of each of the alkali metals produces a degree of ionization of *ca.* 0·95 for Cs, 0·75 for Rb, 0·6 for K, 0·2 for Na and 0·1 for Li.

flames are not hot enough to give rise to significant ionization of elements other than the alkali metals and alkaline earths. In hotter flames, such as oxy-acetylene and nitrous oxide-acetylene, however, some other elements are appreciably ionized. Thus in addition to K, Na, Li, Rb, Cs, Ca, Ba, and Sr, elements with ionization potentials between 6 and 7·5 eV are also ionized to some extent. Ionization effects may therefore result in mutual interference in AAS between elements such as Al, V, Ti and some rare earths in these flames. Such interelement effects due to interaction of the ionization equilibria for

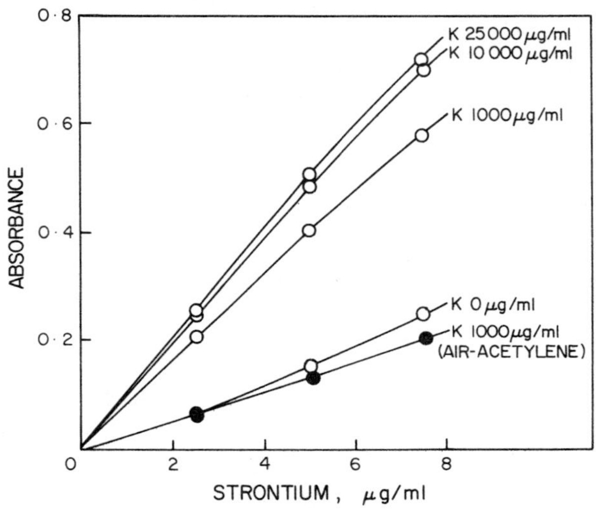

Fig. 12-8. Effect of potassium concentration on the calibration curve for strontium.[70] ●, Air-C_2H_2 flame; ○, N_2O–C_2H_2 flame.

these elements must be carefully distinguished from those due to competitive dissociation equilibria (see Section 12.2.4(a)).

The most commonly used methods of overcoming ionization interference effects in AAS and AFS are to make the concentration of the concomitant element the same in standards and the analytical sample, or to suppress the analyte ionization by the addition of a high concentration of an easily ionized element to the standards and samples to "swamp" the effect by providing a very high electron concentration in the flame. Figure 12-8 shows a typical example of the effect of addition of potassium (ionization potential 4·3 eV) to suppress strontium ionization on the AAS calibration graphs for strontium at 460·7 nm in air-acetylene and nitrous oxide-acetylene flames.[70] As a general rule, the lower the ionization potential of the element added, the lower the required concentration for effective suppression of the ionization of the analyte element. The required concentration should be determined experimentally. The addition of excessively high concentrations of ionization suppressor to samples and standards may lead to degradation of the attainable signal : noise ratio owing to the high dissolved-solid content of the solutions or stray radiation reaching the detector because of the intense emission of the suppressor element. In some cases the concomitant elements themselves may be present at sufficiently high concentration in a sample to suppress the analyte ionization completely. The suppression of magnesium ionization in this way in its determination in silicate rocks by the concomitant elements present has been reported.[71]

12.3 SPECTRAL INTERFERENCES

The term "spectral interference" is commonly used in flame emission spectroscopy to denote any effect on the analyte emission signal from emission by concomitant elements in the same wavelength region. In order to achieve minimal spectral interference in flame emission spectroscopy a monochromator of high resolution is required. The selective nature of the absorption process involved in AAS and AFS, however, and the use of modulated light sources and ac detection systems, results generally in somewhat greater freedom from spectral interference from concomitant elements in these techniques than in flame emission spectroscopy. There are, however, some interference effects that occur in these techniques as a result of the emission, absorption or scattering properties of the concomitant elements and these may be termed spectral interferences.

12.3.1 Spectral Interferences in AAS

Three principal types of interference with AAS signals may be considered as spectral interferences. These are spectral "overlap" and radiation scatter interferences, flame emission interference and interference from unwanted source radiation.

(*a*) *Spectral overlap and radiation scatter interference*

When the wavelength of one of the resonance lines of a concomitant element is almost coincident with the resonance line of the analyte element monitored for its determination by AAS, inaccurate (high) results may be obtained in the determination. This effect corresponds to a true spectral interference by the atomic absorption process. The severity of the interference effect depends on the wavelength separation and the line-widths of the source line and the absorption lines in the flame; when a sharp line from a hollow-cathode lamp is employed, the separation of the line-widths of the source line of the analyte element and absorption line of the concomitant element must be less than *ca.* 0·01 nm for a substantial effect to be observed. Specific examples of this type of interference that have been observed in practice are given in Table 12-2. Figure 12-9 shows the results obtained by Fassel and co-workers[72] for the spectral interference of the vanadium absorption line at 308·211 nm in the determination of aluminium by AAS at the 308·215 nm line.

TABLE 12-2

Spectral line interferences observed in AAS

Analyte and source emission line, nm	Interferent and flame absorption line, nm	Wavelength separation, nm	Reference
Cu 324·754	Eu 324·753	0·001	72
Fe 271·902	Pt 271·904	0·002	72
Si 250·689	V 250·690	0·001	72
Al 308·215	V 308·211	0·004	72
Hg 253·652	Co 253·649	0·003	73
Mn 403·307	Ga 403·298	0·009	74

This type of interference is usually overcome by selection of an alternative absorbing line of the analyte element for its determination by AAS when the interfering concomitant element is suspected or known to be present in the sample.

A second type of true spectral interference may occur when the concomitant element forms molecules in the vapour phase; the absorption spectra of these molecular species may then overlap the analyte resonance line emission from the source. Koirtyohann and Pickett[19] observed molecular absorption by alkali halides in an oxy-hydrogen flame passed into a long tube. These workers have also observed absorption spectra which closely resembled the emission spectra for the alkaline-earth oxide and hydroxide molecules. The most serious interference of this type observed was by CaOH absorption in an air-acetylene flame in the vicinity of the barium 553·55 nm resonance line. The absorption from a solution containing

1% calcium was about that expected for 75 ppm barium. Capacho-Delgado and Sprague[17] also observed this effect for calcium and barium; these workers also demonstrated that the interference effect disappears in a nitrous oxide-acetylene flame owing to less extensive formation of calcium oxide. Similar, but less serious, interferences of the same type have been observed with the absorption signals for lithium at 760·8 nm due to SrO bands, sodium 589·0 nm due to CaO and chromium 357·9 nm due to MgOH. This type of interference, which results from high background absorption in the presence of concomitant elements, may be corrected for by measurement of the

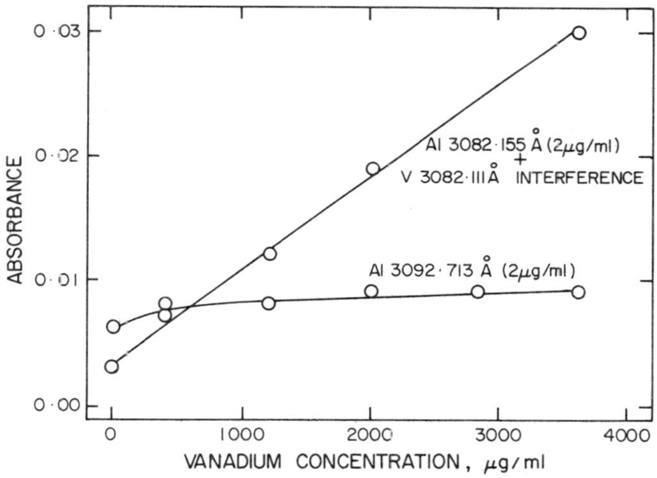

Fig. 12-9. Spectral interference of vanadium on aluminium absorbance at 308·215 nm.[73]

absorption at wavelengths near the resonance line using a continuum source, and subtraction of this absorbance from that at the wavelength of the resonance line.

Solid unevaporated particles may be present in the flame gas in the presence of high concentrations of dissolved solids, particularly when the matrix element forms a refractory oxide and a low-temperature flame is employed. A small fraction of the incident radiation from the source may then be scattered by the particles to produce an apparent high value of absorbance for the analyte element. The severity of interference by scattering in this way depends on the flame type, wavelength of measurement and nature of the sample. Billings[75] and Koirtyohann and Pickett[16] have studied this effect of scattering in detail. The latter authors have reported that it is not possible to account for the non-specific absorption effects observed simply on the basis of light scattering by particles, and that molecular absorption probably accounts for most of the light losses observed. The use of a continuum source to correct for both effects is recommended.[16]

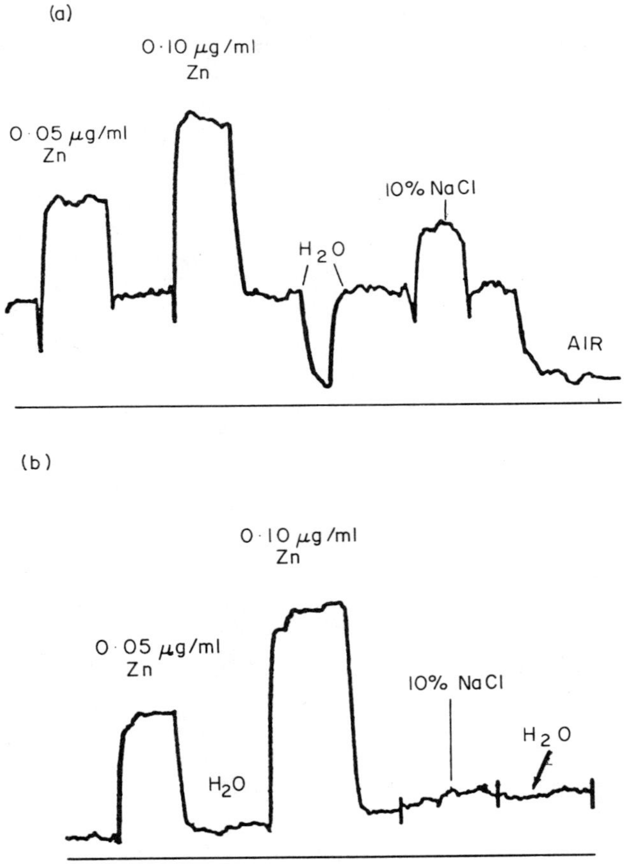

Fig. 12-10. Determination of zinc in sodium chloride using a Perkin-Elmer 303 spectrometer at 3x scale expansion.[20] (a) Standard instrument; (b) with deuterium compensation accessory.

Kahn[20] has described the design and application of such a system using a deuterium arc source and a commercial double-beam atomic absorption spectrophotometer. Radiation from the hollow-cathode lamp and from the deuterium source is passed alternately through the flame. Non-specific background absorption is nominally equal for the beams from both sources, while the element being determined absorbs a significant fraction of only the hollow-cathode radiation. When the ratio of the beam intensities is measured, therefore, the effect of non-specific background absorption is eliminated. Figure 12-10 illustrates the effect of this device in elimination of background absorption in the determination of zinc in 10% sodium chloride solutions.[20]

Light scattering is also considered briefly in Section 12.1.4 on physical interferences.

(*b*) *Flame emission interference*

Analyses with early atomic absorption instrumentation, which employed dc source and detector operation, were prone to serious errors owing to detection and amplification of the flame emission at the resonance line wavelength from the analyte element or atomic or molecular emission in the same wavelength region from concomitant elements. Predictably, this effect was most serious for those elements whose resonance lines used for AAS lie in the visible region of the spectrum. Analyses with modern instrumentation, which invariably employs ac source and detector operation, are very much less subject to this type of interference. When the flame emission is intense at the wavelength to which the monochromator is set, a large dc signal is received at the photomultiplier. Although this should not be amplified it may be registered as noise; in extreme cases photomultiplier saturation may result. The effect of flame emission in this way is frequently observed for several elements in the nitrous oxide-acetylene flame owing to its high background emission over a wide range of wavelengths in both the visible and ultraviolet regions of the spectrum. If a sample solution contains a high concentration of an element which produces molecules in the vapour phase whose band emission is in the same wavelength region as the analyte line from the hollow cathode, loss of analytical precision may result from the increase in the signal noise level which is observed. This effect is minimized by operation at narrow spectral slit-width and high source operating power. Recently, a somewhat different type of flame emission interference in AAS has been observed. In the determination of calcium in a nitrous oxide-acetylene flame spectral interference has been observed due to intense emission from the flame reflected from the window of the hollow-cathode lamp.[76] As a mechanical chopper was employed to modulate the source output, this reflected flame emission was in phase with the hollow-cathode radiation received and amplified at the photomultiplier.

(c) Unwanted source radiation

The wavelength of an atomic line from the filler gas used in the hollow-cathode, a line from an impurity in the cathode material, or a non-absorbing line of the analyte element itself may lie close to the wavelength of the analyte resonance line from the source. This unwanted radiation is not absorbed by the analyte atoms in the atom cell but may be transmitted to the detector and amplified at the detector system. Instead of $\log I_0/I$ being measured as absorbance, therefore, where I_0 and I are the incident and transmitted intensities of the resonance line, $\log(I_0 + i)/(I + i)$ is recorded, where i is the unabsorbed radiation intensity. The overall sensitivity is lowered and non-linear calibration graphs are obtained. This effect is not a true spectral "interference" as it occurs even in the absence of the concomitant element. The effect is minimized by operation at narrow spectral band-pass.

12.3.2 Spectral Interferences in AFS

At the present stage of development of the technique of AFS no single particular type of instrumental assembly has found widespread usage; both line and continuum sources have been recommended and dc and ac source and detector systems are both employed. It is to be expected, however, that ac systems will prevail over dc systems except in certain special cases (e.g. with some non-flame cells). Problems encountered from flame background emission noise should be no greater in AFS than in AAS, except that smaller signal strengths require detection against this background. In contrast to AAS, in the absence of scatter, unwanted source radiation does not reach the detector in AFS, so that the type of effect described in 12.3.1 is seldom encountered.

Spectral interference effects in AFS analogous to those described in Section 12.3.1 may, however, occur when a high-intensity hollow-cathode lamp or other sharp-line source is employed; the frequency of occurrence of spectral line overlap interference should be no greater in resonance fluorescence measurements than in AAS. The effect of overlap of the analyte emission line from the source and concomitant element absorption lines in the flame will give rise to an inaccurate AFS signal for the analyte element. This may occur both when resonance and non-resonance fluorescence is measured; a somewhat greater effect may be observed in the case of resonance fluorescence, however, as the radiation emitted as fluorescence at the same wavelength may also be absorbed by atoms of the concomitant element before it leaves the flame. The effect of spectral overlap interference of this type on the signal which registers as atomic fluorescence at the detector is governed by the fluorescence characteristics of the elements concerned; an enhancement or a suppression may occur depending on whether the concomitant element re-emits the absorbed radiation as resonance or non-resonance fluorescence and on the quantum efficiency of this fluorescence process compared to that for the fluorescence of the analyte element. The effect of spectral overlap of cadmium 228·802 nm on arsenic 228·812 nm and iodine 206·163 nm on bismuth 206·170 nm[77] and gallium 403·298 nm on manganese 403·307 nm has been observed in AFS.[78] Spectral interference by molecular absorption similar to that observed in AAS should also be encountered in AFS. The effect of non-specific absorption in AFS is to attenuate the source intensity available to the analyte atom or to re-absorb the fluorescence emission; both effects then cause a suppression of the observed signal. No interferences in AFS have yet been clearly assigned to this effect, but it is possible that some interference effects observed in cool flames in the determination of those elements whose absorption and emission in AFS occurs at short wavelengths (less than *ca.* 230 nm) in the presence of high concentrations of alkali metal salts may be explained in this way.

When a continuum source, such as a xenon arc, or a line source operated under conditions of high self-absorption, is employed for AFS it is possible for the source simultaneously to stimulate fluorescence from more than one element at wavelengths that are relatively close together. Spectral interference may result under these conditions if the analysing monochromator is allowed to transmit both the fluorescence line of the analyte element and that of the concomitant element. As mentioned in Section 11.2.1, the spectral interference in this way by nickel 323·3 nm with copper 324·7 nm, copper 324·7 nm with silver 328·1 nm, nickel 341·5 nm with silver 338·3 nm, cobalt 419·1 nm with indium 410·2 nm, and cobalt 352·7 nm with nickel 352·4 have been recorded in work with a 450-watt xenon arc source and a grating monochromator used with a wide spectral bandpass.[79]

The effect of scattered light from particulate material in the flame is more important in AFS than in AAS. In measurements of resonance fluorescence, scatter of incident radiation from the source by particles produced from concomitant elements in the flame may give rise to high results. The measurement of direct-line or stepwise fluorescence, on the other hand, permits any problems from scattered light to be eliminated. When resonance fluorescence measurements are made with a continuum source, a correction for scattered light may be made by scanning the wavelength region on either side of the resonance line emission and subtracting the observed "continuum" background caused by scattered light. The problems that are encountered owing to scattered light in AFS have not been found to be as extensive as was originally expected; with premixed flames and efficient nebulizer units, even relatively high concentrations of refractory oxide-forming elements may be tolerated, e.g. in an air-acetylene flame in the determination of elements such as iron,[80,81] manganese,[82] nickel and cobalt.[81] Studies of the use of premixed nitrous oxide-acetylene flames in AFS also reveal relatively little interference by scattering of radiation, see for example ref. 66.

REFERENCES

1. R. E. Bernstein, *S. African J. Med. Sci.*, 1955, **20**, 57.
2. J. D. Winefordner and H. W. Latz, *Anal. Chem.*, 1961, 33, 1727.
3. J. B. Willis, *Spectrochim. Acta,* 1967, **23A**, 811.
4. S. Nukiyama and Y. Tanasawa, Translation of Reports 1-6, *Trans. Soc. Mech. Eng. Japan,* **4, 5, 6**, 1938, by E. Hope, Def. Res. Board, Dept. of Nat. Def., Ottawa, Canada.
5. C. Th. J. Alkemade, Ph.D. Thesis, University of Utrecht, Holland, 1954.
6. R. D. Caton, Jnr. and R. W. Bremner, *Anal. Chem.*, 1954, **26**, 805.
7. P. Porter and G. Wyld, *Anal. Chem.*, 1955, **27**, 733.
8. R. A. G. Rawson, *Analyst,* 1966, **91**, 630.
9. C. A. Dubbs, *Anal. Chem.*, 1952, **24**, 1654.
10. A. Hell, W. F. Ulrich, N. Shifrin and J. Ramirez-Munoz, *Appl. Optics,* 1968, **7**, 1317.
11. M. P. Parsons and J. D. Winefordner, *Anal. Chem.*, 1966, **38**, 1593.
12. J. B. Willis, "Methods of Biochemical Analysis", ed. D. Glick, Vol. XI, Wiley, New York, 1963.
13. G. Mie, *Ann. Physik,* 1908, **25**, 377.
14. J. A. Gidley in "Limits of Detection in Spectrochemical Analysis", Hilger and Watts Ltd., London, 1964, p. 25.
15. S. R. Koirtyohann and E. E. Pickett, *Anal. Chem.*, 1966, **38**, 1087.
16. S. R. Koirtyohann and E. E. Pickett, *Anal. Chem.*, 1966, **38**, 585.
17. L. Capacho-Delgado and S. Sprague, *Atomic Absorption Newsletter,* 1965, **4**, 363.
18. G. F. Kirkbright and S. J. Wood, unpublished work.
19. S. R. Koirtyohann and E. E. Pickett, *Anal. Chem.*, 1965, **37**, 601.
20. H. L. Kahn, *Atomic Absorption Newsletter,* 1968, **7**, 40.

21. C. Veillon, J. M. Mansfield, M. L. Parsons and J. D. Winefordner, *Anal. Chem.*, 1966, **38**, 204.
22. D. W. Ellis and D. R. Demers, *Anal. Chim. Acta*, 1966, **38**, 1945.
23. R. M. Dagnall, P. Young and T. S. West, *Talanta*, 1966, **13**, 803.
24. D. N. Hingle, G. F. Kirkbright and T. S. West, *Analyst*, 1968, **93**, 522.
25. R. M. Dagnall, G. F. Kirkbright, T. S. West and R. Wood, *Analyst*, 1972, **97**, 245.
26. R. E. Dickson and C. M. Johnson, *Appl. Spectroscopy*, 1966, **20**, 214.
27. P. T. Gilbert Jnr., "Analysis Instrumentation", *Proc. Xth Nat. Anal. Inst. Symp., San Francisco, 1964*, Plenum Press, New York, 1964.
28. C. Th. J. Alkemade and M. H. Vorhuis, *Z. Anal. Chem.*, 1958, **163**, 91.
29. S. Fukishima, *Mikrochim. Acta*, 1959, 596.
30. W. Schuchknecht and H. Schinkel, *Z. Anal. Chem.*, 1958, **162**, 266.
31. W. T. Elwell and J. F. Gidley, in "Atomic Absorption Spectrophotometry", 2nd edition, Pergamon Press, London, 1966.
32. W. W. Harrison and W. H. Wadelin, *Anal. Chem.*, 1969, **41**, 374.
33. M. D. Amos and J. B. Willis, *Spectrochim. Acta*, 1966, **22**, 1325.
34. D. C. Manning and L. Capacho-Delgado, *Anal. Chim. Acta*, 1966, **36**, 312.
35. V. A. Fassel and D. A. Becker, *Anal. Chem.*, 1969, **41**, 1522.
36. D. J. David, *Analyst*, 1959, **84**, 536.
37. J. A. Platte and V. M. Marcy, *Atomic Absorption Newsletter*, 1965, **4**, 289.
38. R. M. Dagnall, G. F. Kirkbright, T. S. West and R. Wood, *Anal. Chem.*, 1971, **43**, 1765.
39. R. M. Dagnall, G. F. Kirkbright, T. S. West and R. Wood, *Analyst*, 1972, **97**, 245.
40. R. E. Mansell, *Appl. Spectroscopy*, 1968, **22**, 790.
41. J. I. Dinnin, *Anal. Chem.*, 1969, **32**, 1475.
42. I. Rubeska and B. Moldan, *Anal. Chim. Acta*, 1967, **37**, 421.
43. P. B. Adams and W. O. Passmore, *Anal. Chem.*, 1966, **38**, 630.
44. O. Menis, ed., NBS Technical Note 402, 1966.
45. A. Strasheim and G. J. Wessles, *Appl. Spectroscopy*, 1963, **17**, 65.
46. W. Slavin, "Atomic Absorption Spectroscopy", Interscience, New York, 1968, p. 142
47. F. J. Wallace, *Analyst*, 1963, **88**, 259.
48. M. Yanagisawa, M. Suzuki and T. Takeuchi, *Talanta*, 1967, **14**, 933.
49. J. M. Ottaway, D. T. Coker, W. B. Rowston and D. R. Bhattarai, *Analyst*, 1970, **95**, 567.
50. A. C. West and W. D. Cooke, *Anal. Chem.*, 1960, **32**, 1471.
51. T. V. Ramakrishna, J. W. Robinson and P. W. West, *Anal. Chim. Acta*, 1966, **36**, 57.
52. M. J. Pro and A. P. Mathers, *J.A.O.A.C.*, 1954, **37**, 945.
53. C. A. Baker and F. W. J. Garton, U.K. At. Energ. Auth., AERE-R-3490, 1961.
54. T. C. Rains, H. E. Zittel and M. Ferguson, *Talanta*, 1963, **10**, 367.
55. T. C. Rains, H. E. Zittel and M. Ferguson, *Anal. Chem.*, 1962, **34**, 778.
56. L. Barnes, Jnr., *Anal. Chem.*, 1966, **38**, 1085.
57. A. Giammarise, *Atomic Absorption Newsletter*, 1966, **5**, 113.
58. R. A. Mostyn and A. F. Cunningham, *Anal. Chem.*, 1966, **38**, 121.
59. D. J. David, *Analyst*, 1960, **85**, 495.
60. C. Th. J. Alkemade, *Appl. Optics*, 1968, **7**, 1261.
61. S. L. Sachdev, J. W. Robinson and P. W. West, *Anal. Chim. Acta*, 1967, **37**, 12.
62. T. V. Ramakrishna, P. W. West and J. W. Robinson, *Anal. Chim. Acta*, 1967, **39**, 81.

63. J. B. Headridge and D. P. Hubbard, *Anal. Chim. Acta,* 1967, **37,** 151.
64. R. J. Jaworowski, R. P. Weberling and D. J. Bracco, *Anal. Chim. Acta,* 1967, **37,** 284.
65. M. J. Martin, *Analyst,* 1971, **96,** 843.
66. R. M. Dagnall, G. F. Kirkbright, T. S. West and R. Wood, *Anal. Chem.,* 1970, **42,** 1029.
67. S. L. Mandelstam, *Compt. Rend. Acad. Sci. U.R.S.S.,* 1939, **22,** 403.
68. T. F. Borovik-Romanova, *Zh. Analit. Khim.,* 1961, **16,** 664.
69. I. Rubeska in "Flame Emission and Atomic Absorption Spectrometry", Vol. I, eds. J. A. Dean and T. C. Rains, Marcel Dekker, New York, 1969.
70. J. A. Bowman and J. B. Willis, *Anal. Chem.,* 1967, **39,** 1210.
71. R. W. Nesbitt, *Anal. Chim. Acta,* 1966, **35,** 413.
72. V. A. Fassel, J. O. Rasmuson and T. C. Cowley, *Spectrochim. Acta,* 1968, **23B,** 579.
73. D. C. Manning and F. Fernandez, *Atomic Absorption Newsletter,* 1968, **7,** 24.
74. J. E. Allan, *Spectrochim. Acta,* 1969, **24B,** 13.
75. G. K. Billings, *Atomic Absorption Newsletter,* 1965, **4,** 357.
76. M. Marinkovic, P. J. Slevin and T. J. Vickers, *Appl. Spectroscopy,* 1971, **25,** 372.
77. R. M. Dagnall, K. C. Thompson and T. S. West, *Talanta,* 1967, **14,** 1467.
78. L. Ebdon, G. F. Kirkbright and T. S. West, unpublished work.
79. M. S. Cresser and T. S. West, *Spectrochim. Acta,* 1970, **25B,** 61.
80. L. Ebdon, G. F. Kirkbright and T. S. West, *Anal. Chim. Acta,* 1969, **47,** 563.
81. J. Matousek and V. Sychra, *Anal. Chem.,* 1969, **41,** 518.
82. L. Ebdon, G. F. Kirkbright and T. S. West, *Talanta,* 1970, **17,** 965.

CHAPTER 13

Analytical AAS and AFS Characteristics of the Elements and Applications Data

13.1 INTRODUCTION

At the time of writing AAS and AFS have been applied to the determination of 71 and 35 elements respectively. As described elsewhere in this volume, the sensitivity achievable in both techniques is determined principally by the oscillator strength of the absorption line employed, the fluorescence yield (for AFS), the degree of atomization in the atom cell and the efficiency of the process by which the sample is introduced into the cell. The selectivity attainable depends greatly on the corresponding nebulization and atomization characteristics of the concomitant elements present with the analyte elements in samples. This chapter contains tabular material relating to the AAS and AFS characteristics of each element and the sensitivities attainable by both techniques. Notes on important experimental and instrumental operating conditions are included for each element and the applications of AAS and AFS to its determination in samples of industrial, geological, biological and environmental importance are reviewed. The applications notes that are given for each element are not intended to present a completely comprehensive coverage of the literature; any attempt to do this would be of little value in view of the rapid rate of expansion of the AAS literature and the availability of periodicals such as *Annual Reports on Analytical Atomic Spectroscopy*.* The notes are intended to reflect the state of the applications literature at the time of writing and to present brief methodology and key references to enable the reader more easily to locate those publications of interest for a particular type of sample material to be examined.

13.2 TERMINOLOGY AND ABBREVIATIONS

The tables, which are concerned with sensitivity and applications data for each element, contain abbreviations for the different types of source, flame, organic solvent, etc. employed under various conditions. A list of these abbreviations follows.

Sources

HCL	hollow-cathode lamp (shielded or unshielded)
HIL	high-intensity hollow-cathode lamp (with secondary discharge, i.e. Sullivan and Walsh or Lowe type)
VDL	metal vapour discharge lamp
EDL	microwave or RF excited electrodeless-discharge lamp
150 W Xe arc	150 watt high-pressure xenon arc lamp
450 W Xe arc	450 watt high-pressure xenon arc lamp

Premixed flames

A-town-gas	air-town gas flame
A-P	air-propane flame

* Published by the Society for Analytical Chemistry, London.

A-B	air-butane flame
A-H	air-hydrogen flame
A-coal-gas	air-coal gas flame
A-LPG	air-liquid petroleum gas flame
A-P-B	air-propane-butane flame
A-A	air-acetylene flame
N-A	nitrous oxide-acetylene flame
O_2-A	oxygen-acetylene flame
Ar-O_2-H_2	argon-oxygen-hydrogen flame
N_2-O_2-H_2	nitrogen-oxygen-hydrogen flame
N_2-O_2-A	nitrogen-oxygen-acetylene flame
Ar sep. A-A	argon-separated air-acetylene flame
N_2 sep. A-A	nitrogen-separated air-acetylene flame
Ar sep. N-A	argon-separated nitrous oxide-acetylene flame
N_2 sep. N-A	nitrogen-separated nitrous oxide-acetylene flame

Unpremixed flames

O_2/H_2	oxygen-hydrogen flame
O_2/A	oxygen-acetylene flame
Air/H_2	air-hydrogen flame
Ar/H_2	argon-hydrogen-entrained air flame
N_2/H_2	nitrogen-hydrogen-entrained air flame

Organic solvents and other reagents

MIBK	methyl iso-butyl ketone (hexone)
APDC	ammonium pyrrolidine dithiocarbamate
DDTC	diethyl dithiocarbamate (Na or NH_4 salt)
TCA	trichloroacetic acid
TTA	thenoyltrifluoroacetone
TBP	tributylphosphate

13.3 GENERAL NOTES

1. ND refers to the use of a non-dispersive system (i.e. no monochromator) or use of simple interference filter.
2. *f* refers to the oscillator strength (and excludes the statistical weight, *g*).
3. AAS sensitivity values (for 1% absorption) refer to standard burner arrangements where possible, e.g. 50 mm slot for premixed nitrous oxide-acetylene and 100 mm slot for premixed air-acetylene.

4. All sensitivity and detection limit data refer to aqueous solutions unless otherwise stated.
5. The values given for dissociation energy refer to the monoxide species unless otherwise stated.
6. The values for the atomization efficiency (β values) are taken from L. de Galan and J. D. Winefordner, *J. Quant. Spectrosc. Rad. Transfer,* 1967, 7, 251, for the air-acetylene flame, and J. B. Willis, *Spectrochim. Acta,* 1970, **25B**, 487, for the nitrous oxide-acetylene flame.

13.4 ELEMENTS AND APPLICATIONS DATA

13.4.1 Aluminium

Al, atomic weight 26·98

Technique	Line, nm	*f*	Source	Flame	Sensitivity, ppm/1% abs.	Detection limit, ppm	Reference (see below)
AAS	309·27 309·28	0·23	HCL	N-A	0·7	0·02	1, 2
	308·2	0·22	HCL	N-A	1	—	1
	396·15	0·15	HCL	N-A	0·9	—	1
	394·40	0·15	HCL	N-A	1·4	—	1
	237·34	—	HCL	N-A	2·3	—	1
AFS	396·15	0·15	EDL	Ar sep. N-A	—	0·1	3

Ionization potential: 5·98 eV.
Dissociation energy, AlO: 5·0 eV.
Atomization efficiency: A-A, $< 10^{-5}$; N-A, 0·42

Notes on analytical conditions

1. A fuel-rich nitrous oxide-acetylene flame is required. The absorbance obtained depends critically on the height of observation in the flame and on the flame stoicheiometry.

2. With the 309·2 nm lines the sensitivity depends on the spectral band-width employed; sensitivity increases as the spectral slit-width decreases. Flame emission may also give rise to poor detection limits when wide spectral band-pass is used.

3. Early workers used the oxy-acetylene,[4,5] oxy-hydrogen[5,6] and oxygen-nitrogen-acetylene[7] flames for AAS determination of Al, usually with Al in an organic solvent such as ethanol or in MIBK after extraction as its cupferron complex.[5,6]

4. Aluminium is ionized to the extent of *ca.* 14% in nitrous oxide-acetylene flames at low concentrations.[8] In a general study of interferences on Al absorption, Urbain and Varlot[9] suggested interference observed from HCl and acetic acid was due to influence on Al ionization; other cations generally also caused enhancement by suppression of ionization, a conclusion reached in early work on AAS of Al in oxy-acetylene flame by Manning.[10] The addition of KCl to sample and standard solutions is recommended to suppress ionization. Other workers have studied interference effects on Al absorption in the nitrous oxide-acetylene flame. Marks and Welcher,[11] for example, investigated the effects of the presence of Mn, Ni, Cr, Fe, Ti, V, Al and Co on determination of Al; most effects could be controlled by correct selection of flame conditions and the use of a large excess of a matrix diluent such as KCl to minimize differences in the degree of atomization of Al caused by the presence of other elements. Other workers have studied the mutual interferences observed between Al, Si and Fe in the presence of pyrophosphate and NaCl.[12] Hofer[13] has investigated and utilized the enhancement effect of NH_4F on Al absorption in the air-acetylene flame; this effect is stated not to be observed in the nitrous oxide-acetylene flame. The presence of 100-fold excesses by weight of a range of 18 cations and 4 anions was found to be without effect on the atomic fluorescence signal produced by 20 ppm of Al in a nitrogen-separated nitrous oxide-acetylene flame.[3]

Applications notes and key references

The most important applications of the determination of Al by AAS to date lie in analysis of inorganic matrices such as minerals and metallurgical products. A large volume of published work is concerned with Al determination in siliceous materials such as rocks and minerals,[14,15,16] coal ash,[15] sulphide ores[17] and bauxite,[18] usually after sample preparation via fusion with lithium metaborate and dissolution of the melt in HNO_3 or HCl. Direct HF dissolution in open[19] or closed[20] vessels has also been adopted for analysis of silicate rock[21] and bauxite samples. The excess HF is removed by volatilization or complex formation with H_3BO_3 before final dilution of the sample for nebulization. Similar acid dissolution procedures using HF or HF + HCl have been described for the determination of Al in cement[22,23,24] and slags and sinters.

Aluminium has been determined by AAS in a variety of samples of metallurgical interest.[25,26] Endo *et al.*[27] have analysed iron ore samples for Al after sample dissolution in HCl + HNO_3 and fusion of any insoluble residue with $K_2S_2O_7$; the melt is extracted with water and the solutions combined for nebulization. Other workers[28] have used a similar dissolution and fusion procedure (with $Na_2CO_3 + Na_2B_4O_7$) for sample preparation in determination of Al in low and high alloy steels, ores and slags by AAS. Thulbourne and Scholes,[29] in the determination of Al in iron and steel-making slags, used perchloric acid and reported that the effect of oxides that suppress Al ionization is stabilized by the addition of 0·05% iron in acid solution. Direct acid dissolution has been employed for the AAS determination of Al in uranium alloys,[30] brass and bronze,[31] tin-lead solders (with addition of HBF_4),[32] vanadium metal[33] and high-temperature cobalt-base alloys.[34]

Other samples in which the AAS determination of Al has been reported include wool,[35] soil extracts,[36] uranium compounds,[37] red fuming HNO_3,[38] sea-water,[39] baking powder[40] and lubricating oils.[41]

References

1. M. D. Amos and J. B. Willis, *Spectrochim. Acta,* 1966, **22,** 1325.
2. S. Slavin, W. B. Barnett and Herbert L. Kahn, *Atomic Absorption Newsletter,* 1972, **11,** 37.
3. R. M. Dagnall, G. F. Kirkbright, T. S. West and R. Wood, *Anal. Chem.,* 1970, **42,** 1029.
4. W. Slavin and D. C. Manning, *Anal. Chem.,* 1963, **35,** 253.
5. F. B. Dowling, C. L. Chakrabarti and G. R. Lyles, *Anal. Chim. Acta,* 1963, **28,** 392.
6. C. L. Chakrabarti, G. R. Lyles and F. B. Dowling, *Anal. Chim. Acta,* 1963, **29,** 489.
7. M. D. Amos and P. E. Thomas, *Anal. Chim. Acta,* 1965, **32,** 139.
8. J. B. Willis, *Applied Optics,* 1968, **7,** 1295.
9. H. Urbain and M. Varlot, *Meth. Phys. Analyse,* 1970, **6,** 373.
10. D. C. Manning, *Atomic Absorption Newsletter,* 1964, **3,** 84.
11. J. Y. Marks and G. G. Welcher, *Anal. Chem.,* 1970, **42,** 1033.
12. A. P. Ferris, W. B. Jenson and R. C. Shapland, *Analyst,* 1970, **95,** 574.
13. A. Hofer, *Z. Anal. Chem.,* 1971, **253,** 206.
14. J. C. Van Loon and C. M. Parissis, *Analyst,* 1969, **94,** 1057.
15. John W. Yule and Glenda A. Swanson, *Atomic Absorption Newsletter,* 1969, **8,** 30.
16. P. L. Boar and L. K. Ingram, *Analyst,* 1970, **95,** 124.
17. F. J. Langmyhr and P. E. Paus, *Anal. Chim. Acta,* 1970, **50,** 515.
18. Sverre. H. Omang, *Anal. Chim. Acta,* 1969, **46,** 225.
19. J. C. Van Loon, *Atomic Absorption Newsletter,* 1968, **7,** 3.
20. B. Bernas, *Anal. Chem.,* 1968, **40,** 1682.
21. F. J. Langmyhr and P. E. Paus, *Anal. Chim. Acta,* 1968, **43,** 397.
22. *Idem,* ibid., 1968, **43,** 508.
23. J. T. H. Roos and W. J. Price, *Analyst,* 1969, **94,** 89.
24. T. P. Lees, *Chemy. Ind.,* 1969, **36,** 1249.

25. J. Reid, J. M. Galloway, J. Macdonald and B. B. Bach, *Metallurgia,* 1970, **81**, 243.
26. J. E. Sheridan, *Spectrovision,* 1971, **25**, 10.
27. Yoshihide Endo, Toshihiko Hata and Yuki Nakahara, *Japan Analyst,* 1968, **17**, 679.
28. P. Koenig, K. H. Schmitz and E. Thiemann, *Z. Anal. Chem.*, 1969, **244**, 232; *Atomic Absorption Newsletter,* 1970, **9**, 103.
29. C. Thulbourne and P. H. Scholes, BISRA Open Report MG/D/676/70, 1970.
30. G. Baudin, J. Normand and J. Fijalkowski, *Spectrochim. Acta,* 1968, **23B**, 587.
31. P. Johns and W. J. Price, *Metallurgia,* 1970, **81**, 75.
32. J. Y. Hwang and Linda M. Sandonato, *Anal. Chem.*, 1970, **42**, 244.
33. R. K. Hansen, R. Z. Bachman, J. W. O'Laughlin and Charles V. Banks, *Anal. Chim. Acta,* 1969, **46**, 217.
34. G. G. Welcher and Owen H. Kriege, *Atomic Absorption Newsletter,* 1970, **9**, 61.
35. F. R. Hartley and A. S. Inglis, *Analyst,* 1967, **92**, 622.
36. L. C. Blakemore, *N. Z. Journ, Agric. Res.*, 1968, **11**, 515.
37. C. R. Walker and O. A. Vita, *Anal. Chim. Acta,* 1968, **43**, 27.
38. R. G. Lundy and W. F. Watje, *Atomic Absorption Newsletter,* 1969, **8**, 124.
39. Derek W. Spencer and P. L. Sachs, *Atomic Absorption Newsletter,* 1969, **8**, 65.
40. W. Holak, *J. Ass. Off. Analyt. Chem.*, 1970, **53**, 877.
41. D. R. Jackson, C. Salama and R. Dunn, *Can. Spectrosc.*, 1970, **15**, 17.

13.4.2 Antimony

Sb, atomic weight 121·75

Technique	Line, nm	f	Source	Flame	Sensitivity, ppm/1% abs.	Detection limit, ppm	Reference (see below)
AAS	206·84	0·1	HCL	A-A	1·2	—	1
	217·59	0·04	HCL	A-A	1·4	—	1
	231·15	0·03	HCL	A-A	2·0	—	1
	231·15	0·03	HCL	A-coal-gas	1·5	—	2
AFS	231·15		EDL	A-P	—	0·05	3
	231·15		EDL	A-H	—	0·4	4
	217·59		HIL	N_2 sep. A-A	—	0·04	5

Ionization potential: 8·64 eV.
Dissociation energy SbO: 3·2 eV.

Notes on analytical conditions

1. Mostyn and Cunningham[1] found that the 206·84 nm line gives rise to considerably greater noise levels than 231·15 and 217·59 nm; the use of the 231·15 nm line is preferred by most workers for the determination of low concentrations of Sb in solution by AAS.

2. Highest AAS sensitivity is observed in the air-acetylene flame

just above the primary zone and when a slightly fuel-rich flame is used.

3. The most complete study of interferences on Sb determination by AAS appears to be that of Mostyn and Cunningham.[1] These workers examined the effect on Sb absorption of the presence of each of 16 different metal ions; only Cu interfered at this level by absorption of radiation at the Sb 217·59 nm line. No interference was found at the Sb 231·15 nm line. Various acids were found to decrease the Sb absorption, however, and at 1% concentration in solution after dissolution of lead alloys, lead caused some background absorption at the Sb 217·59 nm line. It is therefore recommended that the standards be matched with respect to major components of the sample matrix and reagents used to prepare the samples.

4. Slavin and Sattur[6] have reported that the high results obtained in the determination of Sb in Pb alloys by AAS at 217·59 nm when a wide (0·7 nm) spectral bandwidth is employed are due to overlap of the Sb 217·025 nm line from the source, which may be transmitted by the monochromator at 217·59 nm, with the Pb 216·999 nm absorption line in an air-acetylene flame. These workers recommended that the Sb 206·84 or 231·15 nm line be used when Sb is to be determined in the presence of lead.

Applications notes and references

Most applications of AAS to Sb determinations have been made in inorganic matrices. Mostyn and Cunningham[1] determined Sb in Pb, Sn, Cu and Al alloys after dissolution of samples in dilute HNO_3, or HCl, while Yanagisawa *et al.*[7] determined Sb in crude Cu, Cu and Al alloys and Sn by extracting Sb as its chloro-complex from 6 N HCl into MIBK for direct nebulization into an air-acetylene flame. Other workers have used a similar extraction procedure in the AAS determination of Sb in uranium and nickel,[8] and Burke[9] has determined Sb in nickel by using MnO_2 as a scavenging agent. Hwang and Sandonato[10] determined Sb in tin-lead solders by AAS after dissolution of samples in concentrated HBF_4 : 70% HNO_3 : H_2O (2 : 3 : 5 v/v). In a study of the elemental composition of pistol bullets by AAS, Brunelle, Hoffman and Snow[11] determined Sb after dissolution of samples in 3 N HNO_3; they recommended that the standard solutions should be matched for Pb content with the samples in order to compensate for enhanced signals caused by their high Pb content. Meranger and Somers[12] have described the determination of Sb in TiO_2 by AAS in air-acetylene at 217·6 nm after fusion of samples with $KHSO_4$ and glucose and extraction of Sb from HCl medium into MIBK.

References

1. R. A. Mostyn and A. F. Cunningham, *Anal. Chem.*, 1967, **39**, 433.
2. B. M. Gatehouse and J. B. Willis, *Spectrochim. Acta*, 1961, **17**, 710.
3. R. M. Dagnall, K. C. Thompson and T. S. West, *Talanta*, 1967, **14**, 1151.
4. K. E. Zacha, M. P. Bratzel, J. D. Winefordner and J. M. Mansfield, *Anal. Chem.*, 1968, **40**, 1733.
5. P. L. Larkins, *Spectrochim. Acta*, 1971, **26B**, 477.
6. S. Slavin and T. W. Sattur, *Atomic Absorption Newsletter*, 1968, **7**, 99.
7. Masaaki Yanagisawa, Masami Suzuki and Tsugio Takeuchi, *Anal. Chim. Acta*, 1969, **47**, 121.
8. C. R. Walker, O. A. Vita and R. W. Sparks, *Anal. Chim. Acta*, 1969, **47**, 1.
9. K. E. Burke, *Anal. Chem.*, 1970, **42**, 1536.
10. J. Y. Hwang and Linda M. Sandonato, *Anal. Chem.*, 1970, **42**, 744.
11. R. L. Brunelle, C. M. Hoffman and K. B. Snow, *J. Ass. Off. Analyt. Chem.*, 1970, **53**, 470.
12. J. C. Meranger and E. Somers, *Analyst*, 1968, **93**, 799.

13.4.3 Arsenic

As, atomic weight 74·92

Technique	Line, nm	*f*	Source	Flame	Sensitivity, ppm/1% abs.	Detection limit, ppm	Reference (see below)
AAS	188·99	—	HCL	A-H	1	—	1
	193·70	0·095	HCL	A-H	0·9	—	2
	193·70	0·095	HCL	A-A	1·5	—	2
	197·20	0·07	HCL	A-H	1·2	—	2
	197·20	0·07	HCL	A-A	2·1	—	2
	188·99	—	EDL	Ar/H_2	—	0·1	3
	193·70	0·095	EDL	Ar/H_2	—	0·1	3
	197·20	0·07	EDL	Ar/H_2	—	0·2	3
	193·70	0·095	EDL	A-A	2	0·5	4
	193·70	0·095	EDL	N_2/H_2	0·75	0·1	4
	193·70	0·095	HCL	A-A	1	0·25	5
	193·70	0·095	HCL	Ar/H_2	0·2	0·05	5
	193·70	0·095	EDL	N_2 sep. N-A	2·0	1·0	6
AFS	235·0	—	EDL	N_2/H_2	—	0·2	7
	235·0	—	EDL	A-A	—	1·0	7
	193·70	—	HIL	A-H	—	1·0	8

Ionization potential: 9·8 eV.
Dissociation energy, AsO: 4·9 eV.

Notes on analytical conditions

1. The As 193·70 line is preferred by most workers as it is less strongly absorbed by the flame gases than the 188·99 nm line and

gives better sensitivity than the 197·20 nm line. As with selenium, use of the cool argon-hydrogen-entrained air flame yields better sensitivity than that obtainable in the air-acetylene flame and improved detection limits also result from the greater transparency of this flame at 193·70 nm. Use of the cooler flame may give rise to more severe chemical interference effects, however. Although poorer sensitivity and detection limits are obtained, the use of the nitrogen-separated nitrous oxide-acetylene flame has been shown to be almost completely transparent at 193·70 nm and to permit determination of As with considerable freedom from interference effects.[6]

2. An oxidizing lean mixture is usually preferred for As when an air-acetylene flame is employed. Interference in the air-acetylene flame has been reported from Mg, Al, Ca, V, Cr(VI), Fe, Co, Ni, Cu and Mo when present at 100-fold excess in the determination of As at 193·70 nm;[9] Johns[10] found relatively few interferences, however, in this flame for a wide range of elements at the 20-fold excess level. This author[10] observed serious interference for these elements under the same conditions when the argon-hydrogen flame was employed, however, and recommended that this flame be used only for analysis of pure solutions.

Applications notes and references

Menis and Rains[11] have determined As by AAS in cast iron and high-purity selenium. These workers developed a procedure in which As was extracted into chloroform with diethyl ammonium diethyldithiocarbamate and then stripped from this complex by addition of copper(II) in 2 M HCl to return As to aqueous medium before nebulization into an argon-hydrogen flame at a total consumption nebulizer-burner. Hill[12] has described the AAS determination of As in iron ores, carbon and high-alloy steels in a method that necessitates separation of As by a distillation procedure before its determination using an argon-hydrogen flame. Gandrud and Marshall[13] determined As in copper using a nitrous oxide-acetylene flame, while Hwang and Sandonato[14] determined As in tin-lead solders in an air-acetylene flame after sample dissolution in $HBF_4 : HNO_3 : H_2O$ (2 : 3 : 5 v/v). Thompson, Morgan and Purdue have determined As in atmospheric particulate matter,[15] and Leaton[16] has described the AAS determination of As in pharmaceutical preparations.

Several indirect methods for the determination of traces of As by AAS have been developed.[17,18,19] In each case these are based on formation and extraction of molybdoarsenic acid prior to determination of the molybdenum content of the complex at 313·2 nm in a reducing air-acetylene or nitrous oxide-acetylene flame. The method developed by Devoto[18] was applied to the determination of As in urine.

Holak[20] has described a technique for enhancement of AAS sensitivity for As in which arsine is generated from the sample and collected in a liquid nitrogen trap; the trap is then allowed to warm and the arsine is swept to the flame for absorption measurements. This type of device has recently become available commercially as an accessory for AAS instruments and offers improved sensitivity and selectivity for As trace analysis.

References

1. C. K. Coogan, J. D. Morrison, A. Walsh and J. K. Wilmhurst, *Nature*, 1963, **200**, 319.
2. M. D. Amos, *The Element*, No. 16, p.1
3. O. Menis and T. C. Rains, ACS Nat. Meeting, Chicago, 1967; *Chem. and Eng. News*, 1967, **45**, 47.
4. R. M. Dagnall, K. C. Thompson and T. S. West, *Atomic Absorption Newsletter*, 1967, **6**, 117.
5. H. L. Kahn and J. E. Schallis, *Atomic Absorption Newsletter*, 1968, **7**, 5.
6. G. F. Kirkbright and Leslie Ranson, *Anal. Chem.*, 1971, **43**, 1238.
7. R. M. Dagnall, K. C. Thompson and T. S. West, *Talanta*, 1968, **15**, 677.
8. D. C. Manning and P. Heneage, *Atomic Absorption Newsletter*, 1967, **6**, 124.
9. G. F. Kirkbright, M. Sargent and T. S. West, *Atomic Absorption Newsletter*, 1969, **8**, 34.
10. P. Johns, *Spectrovision*, 1970 (24), 6.
11. O. Menis and T. C. Rains, *Anal. Chem.*, 1969, **41**, 1952.
12. W. T. Hill, *Spec. Tech. Publ., Am. Soc., Test. Mat.*, 1969, No. **443**, 83.
13. Bruce Gandrud and J. C. Marshall, *Appl. Spectrosc.*, 1970, **24**, 367.
14. J. Y. Hwang and Linda M. Sandonato, *Anal. Chem.*, 1970, **42**, 744.
15. R. J. Thompson, G. B. Morgan and L. J. Purdue, *Atomic Absorption Newsletter*, 1970, **9**, 53.
16. J. R. Leaton, *J. Ass. Off. Analyt. Chem.*, 1970, **53**, 237.
17. R. S. Danchik and D. F. Boltz, *Anal. Letters*, 1968, **1**, 901.
18. G. Devoto, *Boll. Soc. Ital. Biol. Sper.*, 1968, **44**, 425.
19. T. V. Ramakrishna, J. W. Robinson and P. W. West, *Anal. Chim. Acta*, 1969, **45**, 43.
20. W. Holak, *Anal. Chem.*, 1969, **41**, 1712.

13.4.4 Barium

Ba, atomic weight 137·34

Technique	Line, nm	f	Source	Flame	Sensitivity, ppm/1% abs.	Detection limit, ppm	Reference (see below)
AAS	553·56	1·4	HCL	A-A	8	—	1
	553·56	1·4	HCL	N-A	0·4†	—	2
	553·56	1·4	HCL	N-A	0·3†	0·02	3, 4
	305·11	—	HCL	N-A	4·8	—	3
	455·40	Ba ion line	HCL	N-A	2·0	1·0	—
AFS	553·55	1·4	Ruby-laser pumped-dye laser	—	—	<1 ppm	5

Ionization potential: 5·21 eV.
Dissociation energy, BaO: 5·85 eV.
Atomization efficiency: A-A, 0·0011; N-A, 0·3.
† In presence of alkali metal to suppress ionization.

Notes on analytical conditions

1. Even under fuel-rich conditions the air-acetylene flame does not provide for efficient atomization of Ba. In the hot fuel-rich nitrous oxide-acetylene flame higher sensitivity is obtained as long as ionization is suppressed. Ba is ionized to the extent of *ca.* 90% in this flame and a large excess of potassium ionization buffer is required to suppress this. The Ba ion line at 455·4 nm may be used for AAS in this flame when ionization suppression is not employed.

2. As with other alkaline earth elements, interference from aluminium, phosphate and silicate is encountered for Ba determination by AAS in the air-acetylene flame. These effects may be minimized using lanthanum releasing agent. Koirtyohann and Pickett[6] observed spectral interference from Ca on Ba absorption owing to overlap of the Ba 553·55 nm line by CaOH molecular band absorption. Both chemical interference from Al, Si and PO_4 and the latter spectral interference may be minimized using the nitrous oxide-acetylene flame.

Applications notes and key references

Few data are available in the literature concerning the determination of Ba in samples of metallurgical or geological origin. Barium in rock samples may be determined in a similar fashion to calcium using a nitrous oxide-acetylene flame and either HF dissolution or carbonate or $LiBO_2$ fusion techniques for sample decomposition. Lees[7] has determined Ba in cement by AAS, and Adams and Passmore[8] and Jones[9] have examined glass samples for Ba. The latter worker added strontium to the glass and ceramic frit sample solutions to prevent interference from Al and phosphate on the Ba absorption signal. Sedykh[10] has determined Ba in glass samples by a method in which HF containing $HClO_4$ or HNO_3 is used for sample dissolution and lanthanum is added to eliminate matrix effects encountered in the air-acetylene flame employed for AAS.

The AAS determination of Ba additives in lubricating oils has been described by several workers.[11-15] As the concentration of additive is frequently relatively high the analysis is straightforward, particularly when the nitrous oxide-acetylene flame is employed. The oil may be diluted with xylene, *n*-heptane or white spirit, and calibration standards are produced from barium cyclohexane butyrate. Alkali metal ionization buffer may be added to the organic solvent medium in the form of potassium cyclohexane butyrate, potassium alkyl sulphonate or sodium naphthenate.

Barium in biological materials has been determined by Tompsett;[16] in the procedure described the sample was digested with H_2SO_4 in the presence of an oxidant, the digest was diluted, centrifuged and the $BaSO_4$ precipitate dissolved in ammoniacal EDTA for AAS determination. Miro and Sutton[17] subjected animal feed, cereals, milk, shellfish, excreta, soft tissue and bone samples to dry ashing at 475-500°C and converted the residual elements to nitrates. After dissolution in 5 M HNO_3/MeOH mixture the sample solutions were subjected to ion-exchange chromatography; ^{133}Ba radiotracer experiments showed 85-95% recovery of barium in the eluate fraction and a coefficient of variation of 5% for AAS determination of Ba down to 3·4 μg Ba per gram of sample was obtained. Strasheim and co-workers[18] have also described an ion-exchange chromatographic separation of barium from maize meal and bread flour samples prior to its determination by AAS.

In an early paper Billings and Harriss[19] described the examination of marine waters by AAS, and more recently Edelbeck and West[20] have described a solvent extraction procedure for preconcentration of Ba from water samples before its determination by AAS. In this method 200 ml samples were buffered to pH 10·6, potassium ionization buffer was added, and the solution was extracted with hexafluoroacetylacetone in iso-amyl acetate for nebulization into the nitrous oxide-acetylene flame. Total air-borne Ba has been determined by collection on filter paper, reduction to Ba sulphide and dissolution of the ashed residue in dilute HNO_3 for AAS.[21] Thompson, Morgan and Purdue[22] have determined Ba in atmospheric particulate matter by sample collection on a glass-fibre filter, ashing at low temperature in an oxygen plasma and extracting the ash with HCl/HNO_3 mixture to provide a solution suitable, after centrifuging off SiO_2, for AAS.

References

1. B. M. Gatehouse and J. B. Willis, *Spectrochim. Acta,* 1961, **17**, 710.
2. M. D. Amos and J. B. Willis, *Spectrochim. Acta,* 1966, **22**, 1325.
3. "Analytical Methods for Atomic Absorption Spectrophotometry", Perkin-Elmer Corp., Norwalk, Conn., U.S.A., 1971.
4. H. L. Kahn, *Atomic Absorption Newsletter,* 1967, **6**, 96.
5. M. B. Denton and H. V. Malmstadt, *Appl. Phys. Letters,* 1971, **18**, 485.
6. S. R. Koirtyohann and E. E. Pickett, *Anal. Chem.,* 1966, **38**, 585.
7. T. P. Lees, *Chemy. Ind.,* 1969, **36**, 1249.
8. P. B. Adams and W. O. Passmore, *Anal. Chem.,* 1966, **38**, 630.
9. A. H. Jones, *Anal. Chem.,* 1965, **37**, 1761.
10. E. H. Sedykh, *Steklo Keram.,* 1971 (4), 46.
11. S. Sprague and W. Slavin, *Atomic Absorption Newsletter,* 1963, **2**, 4.
12. R. A. Mostyn and A. F. Cunningham, *J. Inst. Petrol.,* 1967, **53**, 101.
13. E. E. Peterson and H. L. Kahn, *Atomic Absorption Newsletter,* 1970, **9**, 71.
14. D. Pickles and C. C. Washbrook, *Proc. Soc. Anal. Chem.,* 1970, **7**, 13.
15. M. Kashiki and S. Oshima, *Anal. Chim. Acta,* 1971, **55**, 436.
16. S. L. Tompsett, *Proc. Ass. Clin. Biochem.,* 1968, **5**, 125.
17. M. Miro and D. C. Sutton, *Appl. Spectrosc.,* 1970, **24**, 220.
18. A. Strasheim, F. W. E. Strelow and E. Norval, *J. S. Afr. Chem. Inst.,* 1967, **20**, 25.
19. G. K. Billings and R. C. Harriss, *Texas J. Sci.,* 1965, **17**, 129.
20. L. Edelbeck and P. W. West, *Anal. Chim. Acta,* 1970, **52**, 447.
21. M. Magyar, J. L. Bove, B. Nathanson, S. Siebenberg and E. F. Ferrand, *Envir. Sci. Technol.* 1971, **5**, 358.
22. R. J. Thompson, G. B. Morgan and L. J. Purdue, *Atomic Absorption Newsletter,* 1970, **9**, 53.

13.4.5 Beryllium

Be, atomic weight 9·012

Technique	Line, nm	f	Source	Flame	Sensitivity, ppm/1% abs.	Detection limit, ppm	Reference (see below)
AAS	234·86	0·24	HCL	N-A	0·02	—	1
	234·86	0·24	HCL	N-A	0·03	0·001	2
	234·86	0·24	HCL	O_2/A	0·2	—	3
AFS	234·86	0·24	EDL	Sep. N-A	—	0·01	4
	234·86	0·24	HIL	N-A	—	0·5	5
	234·86	0·24	EDL	N-A	—	0·04	6
ND	234·86	0·24	HIL	N_2 sep. N-A	—	0·008	7

Ionization potential: 9·32 eV.
Dissociation energy, BeO: 4·6 eV.

Notes on analytical conditions

1. A fuel-rich nitrous oxide-acetylene flame is required for atomization of Be. Amos and Willis[1] also reported Be absorption in an oxygen-nitrogen-acetylene flame, and Slavin and Manning[3] utilized a very rich oxy-acetylene flame, at a total consumption nebulizer burner unit and nebulized Be in ethanol solutions.

2. A broad spectral band-pass (up to 1 nm) may be employed with the Be 234·86 nm line in AAS; the absorption signal, however, is markedly dependent upon the acetylene flow-rate and height of observation in the flame.

3. The determination of Be by AAS and AFS is relatively free from interferences. The depressive effect on Be absorption of large amounts of aluminium was noted by Amos and Willis[1] and by Ramakrishna *et al.*[8] Fleet and co-workers[9] recorded decreased sensitivity in the presence of large amounts of silicon and magnesium. Thomas and Pickering[10] have discussed the effect of a variety of additives on the absorption signal obtained for Be. The presence of high concentrations of Al, Ca and Mo has been reported to cause interference with the Be atomic fluorescence signal observed at 234·86 nm in the nitrous oxide-acetylene flame.[4]

4. Kirkbright, Sargent and West[11] observed improved detection limits for Be determination by AAS using a nitrogen-separated nitrous oxide-acetylene flame. Bratzel, Dagnall and Winefordner[12] obtained a detection limit of 2 μg/ml of Be using 2 μl samples in

conjunction with a platinum loop atomization device for AFS studies with an EDL source.

Applications notes and key references

Few applications of the determination of Be by AAS or AFS have been reported. Peterson[13] has described the determination of Be in aluminium alloys by AAS. Samples were dissolved in 50% HCl, H_2O_2 was added dropwise to complete the dissolution and the solution was filtered, diluted and nebulized into a nitrous oxide-acetylene flame. Thomas and Chumnong[14] have determined Be in beryl by an AAS procedure in which the dried, crushed sample is fused with KOH, the melt dissolved in dilute HNO_3 and H_2O_2 added before dilution and nebulization into the nitrous oxide flame. Standards were prepared to contain HNO_3, KOH, 150 ppm Si and 50 ppm of Al to match the sample solutions approximately and compensate for interference effects.

Bokowski[15] has described an AAS procedure for the determination of Be in air, swabs and biological samples; preconcentration was affected by a modified acetylacetone extraction method which permitted detection of Be in urine down to 2 μg per litre. H_3PO_4, Al, Si and SO_4 could be tolerated to concentrations of 0·1, 1, 1 and 80 mg/ml respectively. The method of Ramakrishna *et al.*[8] for the determination of Be by AAS was described as applicable to air and water analysis. Hicks and co-workers[16] have investigated the effect of organic solvents present in waste waters on the AAS determination of Be.

References

1. M. D. Amos and J. B. Willis, *Spectrochim. Acta*, 1966, **22**, 1325.
2. "Analytical Methods for Atomic Absorption Spectrophotometry", Perkin-Elmer Corp., Norwalk, Conn., U.S.A., 1971.
3. W. Slavin and D. C. Manning, *Anal. Chem.*, 1963, **35**, 253.
4. D. N. Hingle, G. F. Kirkbright and T. S. West, *Analyst*, 1968, **93**, 522.
5. J. W. Robinson and C. J. Hsu, *Anal. Chim. Acta*, 1968, **43**, 109.
6. M. P. Bratzel, R. M. Dagnall and J. D. Winefordner, *Anal. Chem.*, 1969, **41**, 1527.
7. P. L. Larkins and J. B. Willis, *Spectrochim. Acta*, 1971, **26B**, 477.
8. T. V. Ramakrishna, P. W. West and J. W. Robinson, *Anal. Chim. Acta*, 1967, **39**, 81.
9. B. Fleet, K. V. Liberty and T. S. West, *Talanta*, 1970, **17**, 203.
10. P. E. Thomas and W. F. Pickering, *Talanta*, 1971, **18**, 127.
11. G. F. Kirkbright, M. Sargent and T. S. West, *Talanta*, 1969, **16**, 1467.
12. M. P. Bratzel Jnr., R. M. Dagnall and J. D. Winefordner, *Appl. Spectrosc.*, 1970, **24**, 518.
13. E. A. Peterson, *Atomic Absorption Newsletter*, 1969, **8**, 53.

13.4.6 Bismuth

Bi, atomic weight 208·98

Technique	Line, nm	f	Source	Flame	Sensitivity, ppm/1% abs.	Detection limit, ppm	Reference (see below)
AAS	222·83	0·002	HCL	A-A	1·5	–	1
	223·06	0·012	HCL	A-A	0·7	–	1
	306·77	0·25	HCL	A-A	2·1	–	1
	227·66	–	HCL	A-A	9·5	–	1
	206·17	–	HCL	A-A	5·5	–	1
	202·12	–	HCL	A-A	50	–	1
	211·03	–	HCL	A-A	18	–	1
	306·77	0·25	HCL	A-coal gas	2	–	2
	306·77	0·25	HCL	O_2/A	2	1	3
AFS	302·46	–	EDL(I_2)	N_2/H_2	–	0·05	4
	306·77	–	EDL	Ar/H_2	–	0·7	5
	306·77	–	EDL	N_2 sep. A-A	–	0·04	6
	302·46	–	EDL(I_2)	N_2 sep. A-A	–	0·09	6
	206·17	–	EDL(I_2)	A-A	–	0·4	7

Ionization potential: 7·29 eV.
Dissociation energy, BiO: 4·0 eV.

14. P. E. Thomas and H. Chumnong, *Resonance Lines*, 1969, **1**, 5.
15. D. L. Bokowski, *Atomic Absorption Newsletter*, 1967, **6**, 97.
16. J. E. Hicks, R. T. McPherson and J. W. Salyer, *Anal. Chim. Acta*, 1972, **61**, 441.

Notes on analytical conditions

1. The 306·77 nm line coincides with the band head of the strongest OH band in the flame; difficulty may therefore arise from noise due to absorption of source radiation at this wavelength; the Bi 223·06 nm is preferred for most AAS applications

2. Greater sensitivity for Bi AAS may be obtained using an air-coal gas or air-propane flame rather than air-acetylene; interference effects may become more marked in these cooler flames, however.

3. The Bi 206·17 nm line may be used for excitation of direct-line fluorescence of Bi at 302·46 nm using the I 206·163 nm line from an iodine EDL source.[4] Flame separation with nitrogen assists the observation of Bi fluorescence at 306·77 without interference from OH radical radiation and is also beneficial even when a modulated iodine EDL source is used to stimulate Bi direct-line fluorescence at 302·46 nm.[6]

4. No interference has been observed in AAS for Bi in the presence of large amounts of Pb or Sn,[8] or in Bi AFS using the direct-line fluorescence at 302·46 nm from 100-fold excesses of As, Cd, Fe, K, Na, Sb, Zn, Cl, NO_3, PO_4; interference from SO_4, Al or Mg was overcome by raising the temperature of the diffusion flame used.[4]

5. Amos, Bennett and Brodie observed improved detection limits for Bi by both emission and absorption techniques when the air-acetylene flame was separated by nitrogen or argon.[9]

Applications notes and references

In early work Willis[10] reported the determination of Bi in urine by extraction with ammonium tetramethylene dithiocarbamate into MIBK and nebulization of the organic extract into an air-coal gas flame for AAS. Kinser[11] determined Bi in animal body tissues after sample dissolution in HNO_3, subsequent dry ashing and dissolution of the residue in dilute HCl; standard solutions containing approximately similar amounts of inorganic material to those present in the ashed samples were employed to compensate for light scattering.

A number of workers have investigated the determination of Bi in rock, ore and soil samples. Ward and Nakagawa[12] have determined Bi directly by AAS in altered rocks after dissolution in concentrated HNO_3, and Ward and co-workers[13] and Huster[14] have also described AAS methods for this and other elements in soils and rocks. Billings and Ragland[15] have reported the determination of Bi (and also Mo, Co, Ni, Cr, Cu and Fe) in oil-field brines by methods based on extraction with APDC into MIBK before nebulization. Ion-exchange separation of Bi prior to its AAS determination in galena and pure lead has been recommended,[16] and Burke[17] has utilized preconcentration on MnO_2 before determination of Bi in nickel oxide samples. Other workers have described the determination of Bi in iron ores,[18] ferrous alloys,[19] aluminium alloys[20] and tin-lead solders.[21] Bismuth has been determined by AFS in aluminium alloy samples without preconcentration or separation from the matrix element utilizing the Bi direct-line fluorescence at 302·46 nm,[6] and in nickel-based alloys after preconcentration by solvent extraction.[22]

References

1. D. C. Manning, J. Vollmer and F. Fernandez, *Atomic Absorption Newsletter*, 1967, **6**, 17.
2. B. M. Gatehouse and J. B. Willis, *Spectrochim. Acta*, 1961, **17**, 710.
3. D. Marshall and W. G. Schrenk, *Appl. Spectrosc.*, 1967, **21**, 246.
4. R. M. Dagnall, K. C. Thompson and T. S. West, *Talanta*, 1967, **14**, 1467.

5. K. E. Zacha, M. P. Bratzel, J. D. Winefordner and J. M. Mansfield, *Anal. Chem.*, 1968, **40**, 1733.
6. R. S. Hobbs, G. F. Kirkbright and T. S. West, *Talanta*, 1971, **18**, 859.
7. K. C. Thompson and P. C. Wildy, *Analyst*, 1970, **95**, 776.
8. T. W. Sattur, *Atomic Absorption Newsletter*, 1966, **5**, 37.
9. M. D. Amos, P. A. Bennett and K. G. Brodie, *Resonance Lines*, 1970, **2**, 3.
10. J. B. Willis, *Anal. Chem.*, 1962, **34**, 614.
11. R. E. Kinser, *Amer. Ind. Hyg. Assoc. J.*, 1966, **27**, 260.
12. F. N. Ward and H. M. Nakagawa, *Prof. Pap. U.S. Geol. Surv.*, 575D, 1967, D239-241.
13. F. N. Ward, H. M. Nakagawa, T. F. Harms and G. H. Van Sickle, *Bull. U.S. Geol. Surv.*, No. 1289, 1969.
14. J. Husler, *Atomic Absorption Newsletter*, 1970, **9**, 31.
15. G. K. Billings and P. C. Ragland, *Canad. Spectrosc.*, 1969, **14**, 8.
16. Z. Sulcek, Povondra and V. Kratochvil, *Coll. Czech. Chem. Comm.*, 1969, **34**, 3711.
17. K. E. Burke, *Anal. Chem.*, 1970, **42**, 1536.
18. Yoshihide Endo, Toshihiko Hata and Yuki Nakahara, *Japan Analyst*, 1969, **18**, 833.
19. J. B. Headridge and J. Richardson, *Analyst*, 1970, **95**, 930.
20. S. E. Hermon and R. J. Rennie, *Metallurgia*, 1970, **82**, 201.
21. J. Y. Hwang and Linda M. Sandonato, *Anal. Chem.*, 1970, **42**, 744.
22. R. M. Dagnall, M. R. G. Taylor and T. S. West, *Laboratory Practice*, 1971, **20**, 209.

13.4.7 Boron

B, atomic weight 10·8

Technique	Line, nm	*f*	Source	Flame	Sensitivity, ppm/1% abs.	Detection limit, ppm	Reference (see below)
AAS	249·68	0·32	HCL	N-A	100	–	1
	249·77	0·33	HCL	N-A	50	–	1
				N-A	40	2·5	2
AFS: not reported							

Ionization potential: 8·3 eV.
Dissociation energy: BO = 7·95 eV.

Notes on analytical conditions

1. A fuel-rich nitrous oxide-acetylene flame is required for atomization. In early work Amos and Willis obtained absorption in the oxygen-nitrogen-acetylene flame.[1] Sanders[3] observed that argon-filled hollow-cathode lamps of boron were brighter but gave lower AAS sensitivity than neon-filled lamps. A reduction in the size of the hollow-cathode was also reported to give increased intensity and moderate improvement in sensitivity. Some instrumentation does not

permit resolution of the boron 249·68 and 249·77 nm lines; working under these conditions Manning[4] obtained a detection limit for boron of *ca.* 6 ppm.

2. Using separate hollow-cathode lamp sources for ^{11}B and ^{10}B and containing a boron sample in a hollow-cathode discharge cell, Goleb[5] has investigated the possibility of the determination of $^{11}B : {}^{10}B$ isotope abundance ratio. This was unsuccessful, however, as even with the low pressure cooled discharge cell used the half-widths of the spectral lines appear to be greater than the isotope displacement (0·001 nm for both 249·68 and 249·77 nm lines).

Applications notes and key references

The principal applications of the AAS determination of boron have been to fertilizers, plants, biological materials and food. Melton and co-workers[6,7] have described the determination of B in plants and fertilizers; after extraction of fertilizer with hot water (or dissolution of residue after dry-ashing of plant material) the boron is extracted into a 20% solution of 2-ethylhexane-1,3-diol in MIBK. The organic phase is then nebulized into a nitrous oxide-acetylene flame. Weger, Hossner and Ferrara[8] have utilized a similar extraction procedure using a mixture of 2-ethylhexane-1,3-diol and chloroform (1 : 19) for the determination of total and acid-soluble boron in fertilizers. Boron in foods was determined by Holak[9] after digestion with HNO_3/H_2SO_4 and extraction with 2-ethylhexane-1,3-diol in MIBK. Bader and Brandenburger[10] have reported determination of B in biological materials; tissue samples (10 g) were dried in the presence of 1% Li_2CO_3 under an infrared lamp at *ca.* 200°C before dry-ashing at 600°C, extraction of the ash with 6 N HCl, filtration and dilution to 10 ml for nebulization into the nitrous oxide-acetylene flame.

The determination of boron in ferroboron,[11] KCl[12] and polymers[13] has also been described.

References

1. M. D. Amos and J. B. Willis, *Spectrochim. Acta,* 1966, **22,** 1325.
2. "Perkin-Elmer AAS Methods Handbook, 1971", Perkin-Elmer Corp., Norwalk, Conn., U.S.A.
3. J. Sanders, *Resonance Lines,* 1969, **1,** 7.
4. D. C. Manning, *Atomic Absorption Newsletter,* 1966, **5,** 127; 1967, **6,** 35.
5. J. Goleb, *Anal. Chem. Acta,* 1966, **36,** 130.
6. J. R. Melton, W. L. Hoover, P. A. Howard and J. L. Ayers, *J. Ass. Off. Analyt. Chem.,* 1970, **53,** 682.

7. J. R. Melton, W. L. Hoover and P. A. Howard, *J. Ass. Off. Analyt. Chem.*, 1969, **52**, 950.
8. Steve J. Weger Jnr., Lloyd R. Hossner and Louis W. Ferrara, *J. Agric. Ford. Chem.*, 1969, **17**, 1276.
9. W. Holak, *J. Ass. Off. Analyt. Chem.*, 1971, **54**, 1138.
10. H. Bader and H. Bradenburger, *Atomic Absorption Newsletter*, 1968, **7**, 1-3.
11. D. C. Smith, J. R. Johnson and G. C. Soth, *Appl. Spectrosc.*, 1970, **24**, 576.
12. R. Harris, *Atomic Absorption Newsletter*, 1969, **8**, 42.
13. M. Olivier, *Z. Anal. Chem.*, 1969, **248**, 145.

13.4.8 Cadmium

Cd, atomic weight 112·40

Technique	Line, nm	f	Source	Flame	Sensitivity, ppm/1% abs.	Detection limit, ppm	Reference (see below)
AAS	228·80	1·2	HCL	A-coal gas	0·03	–	1
	326·11	0·002			20	–	1
	228·80	1·2	HCL	A-A	0·025	0·001	2
	228·80	1·2	HCL	Ar/H_2	0·01	0·001	3
AFS	228·80	1·2	EDL	O_2/H_2	–	0·000001	4
	228·80	1·2	VDL	Air/H_2	–	0·005	5
	228·80	1·2	VDL	N_2 sep. A-A	–	0·0005	6
	326·11	0·002	VDL	N_2 sep.	–	1·0	6
ND	228·80	1·2	VDL	A-A	–	0·0002	7
	228·80	1·2	VDL	A-P	–	0·0001	8
	228·80	1·2	VDL	O_2/H_2	–	0·0002	9

Ionization potential: 6·11 eV.
Dissociation energy CdO: 3·8 eV.
Atomization efficiency: A-A, 0·50.

Notes on analytical conditions

1. Cadmium HCL sources are usually operated at low current to avoid decreased sensitivity due to self absorption which may otherwise be experienced. The spectral band-width employed is usually not critical.

2. The AAS determination of Cd is subject to very few chemical interferences when an air-acetylene flame is used. The presence of large amounts of silicate may depress the Cd absorbance. Dagnall *et al.*[8] studied the effects on the AFS signal for Cd at 228·80 nm produced by 100-fold excesses of 41 cations and 18 anions; in no case was a deviation greater than ±5% produced in the signal. Bratzel and Winefordner[10] have described the effect of different acid (HCl) concentrations on cadmium atomic fluorescence; these workers

concluded that the acid type and concentration in samples and standards should be matched. Bratzel *et al.*[11,12] have also examined the determination of Cd by AFS in various premixed and unpremixed flames and the effect of flame-type on signal-to-noise ratios and interference effects. According to this work hydrogen-supported flames are preferred for Cd unless chemical interference is expected, when the nitrous oxide-acetylene flame is preferable.

3. The ease of thermal atomization of Cd renders it particularly suitable for detection in non-flame atom cells; the reader is referred to Chapter 7 for further information.

4. Cd is readily preconcentrated by solvent extraction with APDC into MIBK. Sachdev and West[13] have also reported its extraction as the dithizone complex from NH_4 tartrate solution at pH 7·5 into ethyl propionate.

Applications and key references

The determination of Cd in ores and alloys has been reported by a number of workers. Poluektov and Vitkun[14] used a propane-butane-air flame for the AAS determination of Cd in metallic zinc and zinc ores, and other workers have described its AAS determination in lead and tin-alloys[15] and tin-lead solders[16] after acid dissolution of samples. Headridge and Smith[17] determined Cd in zinc-based alloys using an induction heated furnace. Taskaeva and Vainshtein[18] applied AAS to the determination of Cd in mineral raw materials containing up to 90% Si, 40% Ca, P, F or C, 3% of K, Al or Fe, 4% of Na or 2% of Mg. Yamada and co-workers[19] utilized the absorption tube technique for Cd in silicates after its extraction with APDC into MIBK. Cadmium in uranium compounds has been determined using the residual aqueous solution after extraction of the uranium into kerosene with tributylphosphate.[20,21] Gates and Black[22] have described the determination of Cd in CdZn arsenide by AAS after removal of arsenic by distillation.

Willis[23,24] determined Cd in urine by AAS directly on undiluted samples. In order to correct for light scatter at 228·80 nm under these conditions Wall and Rhodes[25] proposed the use of the Pb 220·35 nm line from a Pb hollow-cathode source. Other workers have preferred to dilute the sample[26] or to employ solvent extraction for separation of Cd from biological samples before nebulization into air-acetylene or air-hydrogen flames. Berman[27] employed APDC extraction of Cd into MIBK after wet ashing of blood, urine and tissue samples, and Lehnert and co-workers[28] used a similar procedure for serum and urine after digestion with HNO_3, H_2SO_4 and $HClO_4$. Klaus[29] has used AFS for determination of Cd in biological materials after wet digestion. Cd in atmospheric particulate matter has been determined by flame AAS after collection using filters,[30] and other workers have determined this element in plating wastes[31] and river water[31,32,33] after preconcentration by extraction with 2-mercaptobenzothiazole,[31] dithizone[32] or its ternary complex with 1,10-phenanthroline and perchlorate.[33]

Other materials in which Cd has been determined by AAS include wines,[34] fruit juices and soft drinks,[35] foods[36,37] and sodium chloride.[37]

References

1. B. M. Gatehouse and J. B. Willis, *Spectrochim. Acta,* 1961, **17,** 710.
2. "Analytical Methods for Atomic Absorption Spectrophotometry", Perkin-Elmer Corp., Norwalk, Connecticut, 1971.
3. H. L. Kahn and J. E. Schallis, *Atomic Absorption Newsletter,* 1968, **7,** 5.
4. K. E. Zacha, M. P. Bratzel Jnr., J. D. Winefordner and J. M. Mansfield Jnr., *Anal. Chem.,* 1968, **40,** 1733.
5. N. Omenetto and G. Rossi, *Anal. Chim. Acta,* 1968, **40,** 195.
6. R. S. Hobbs, G. F. Kirkbright, M. Sargent and T. S. West, *Talanta,* 1968, **15,** 997.
7. P. L. Larkins, *Spectrochim. Acta,* 1971, **26B,** 477.
8. R. M. Dagnall, T. S. West and P. Young, *Talanta,* 1966, **13,** 803.
9. J. M. Mansfield, J. D. Winefordner and C. Veillon, *Anal. Chem.,* 1965, **37,** 1049.
10. M. P. Bratzel Jnr. and J. D. Winefordner, *Anal. Chim. Acta,* 1967, **39,** 394.
11. M. P. Bratzel, R. M. Dagnall and J. D. Winefordner, *Anal. Chem.,* 1969, **41,** 713.
12. *Idem,* ibid., 1969, **41,** 1527.
13. S. L. Sachdev and P. W. West, *Anal. Chim. Acta,* 1969, **44,** 301.
14. N. S. Poluektor and R. A. Vitkun, *Zh. Anal. Khim.,* 1962, **17,** 935.
15. J. L. Jimenez Seco, Aurora Gomez Coedo and Maria T. Dorado, *Revta Metal,* 1969, **5,** 602.
16. J. Y. Hwang and Linda M. Sandonato, *Anal. Chem.,* 1970, **42,** 744.
17. J. B. Headridge and D. Risson Smith, *Talanta,* 1971, **18,** 247.
18. T. P. Taskaeva and E. E. Vainshtein, *Zh. Anal. Khim.,* 1967, **22,** 852.
19. J. Yamada, C. Iida and K. Yamasaki, *Japan Analyst,* 1969, **18,** 1088.
20. J. Stupar, B. Podobnik and J. Korosin, *Croat. Chim. Acta,* 1965, **37,** 141.
21. C. R. Walker and O. A. Vita, *Anal. Chim. Acta,* 1968, **43,** 27.
22. O. R. Gates and R. W. Black, *Rep. U.S. Atom. Energ. Comm.,* NRL-7146, 1970.
23. J. B. Willis, *Nature,* 1961, **191,** 381.
24. J. B. Willis, *Anal. Chem.,* 1962, **34,** 614.
25. H. Wall and C. Rhodes, *Canad. Spectrosc.,* 1968, **13,** 51.
26. P. Pulido, K. Fuwa and B. L. Vallee, *Analyt. Biochem.,* 1966, **14,** 393.
27. E. Berman, *Atomic Absorption Newsletter,* 1967, **6,** 57.
28. G. Lehnert, K. H. Schaller and T. Haas, *Z. Klin. Chem. Klin. Biochem.,* 1968, **6,** 174.
29. R. Klaus, *Z. Klin. Chem.,* 1966, **4,** 299.

30. R. J. Thompson, G. B. Morgan and L. J. Purdue, *Atomic Absorption Newsletter,* 1970, **9,** 53.
31. J. L. Robinson, R. G. Barnekow and P. F. Lott, *Atomic Absorption Newsletter,* 1969, **8,** 60.
32. Makoto Kanke, Yasuhisa Hayashi, Takahiro Kumumaru and Yuroku Yamamoto, *Nippon Kagaku Zasshi,* 1971, **92,** 983.
33. Y. Yamamoto, T. Kumumaru, Y. Hayashi and Y. Octani, *Nippon Kagaku Zasshi,* 1971, **92,** 345.
34. J. C. Meranger and E. Somers, *J. Ass. Off. Analyt. Chem.,* 1968, **51,** 922.
35. J. C. Meranger, *Bull. Envir. Contam. Toxicol.,* 1970, **5,** 271.
36. *Idem,* ibid., 1968, **3,** 360.
37. R. J. Gajan, J. H. Gould, J. O. Watts and J. A. Fiorino, *J. Ass. Off. Analyt. Chem.,* 1973, **56,** 876.
38. S. Skurnik-Sarig, A. Glasner, M. Zidon and D. Weiss, *Talanta,* 1969, **16,** 1488.

13.4.9 Caesium

Cs, atomic weight 132·905

Technique	Line, nm	*f*	Source	Flame	Sensitivity, ppm/1% abs.	Detection limit, ppm	Reference (see below)
AAS	852·11	0·8	VDL	A-coal gas	0·15	—	1
	455·54	—	VDL	A-coal gas	20	—	1
	852·11	0·8	VDL	A-A	0·3[a]	0·05	2
	852·11	0·8	VDL	A-H	0·2	—	3
AFS: not reported							

[a] In the presence of large excess KCl to suppress ionization.
Ionization potential: 3·87 eV.

Notes on analytical conditions

1. Caesium vapour discharge lamp sources should be operated at the lowest current consistent with good stability and receipt of sufficient energy at the detector; this is necessary to minimize line-broadening and reversal due to self-absorption. A relatively broad spectral band-pass (2–4 nm) may be employed at the Cs 852·11 nm line to gain energy when the detector sensitivity at this wavelength is low.

2. Caesium is ionized to the extent of *ca.* 70% at analytical concentrations in the air-acetylene flame. As the ionization potential of Cs is lower than that of sodium or potassium a very large excess of these ions must be added to sample and standard solutions to suppress Cs ionization and increase sensitivity. The ionization in the cooler air-coal gas and air-hydrogen flames is less severe.

3. Interference with Cs absorption may be observed in the presence of high concentrations of mineral acids.

Applications notes and key references

In comparison with radiochemical and flame emission spectrometric methods for the determination of Cs very few applications of AAS to the determination of this element in "real" sample materials have been reported. Strelow and co-workers[4] have described a procedure by which Cs may be separated from other elements by cation exchange before its determination in silicates by AAS. Mountjoy and Wahlberg[5] have determined between 1 and 10 ppm Cs in geological samples after its separation as insoluble Cs phosphomolybdate. Sanui and Pace[6] have investigated interferences in the determination of Cs in biological materials by AAS. These workers examined premixed air-acetylene and air-propane flames and preferred the latter; La and Rb or K were added to standards and samples to suppress chemical and ionization interferences, respectively.

References

1. B. M. Gatehouse and J. B. Willis, *Spectrochim. Acta,* 1961, **17**, 710.
2. "Analytical Methods for Atomic Absorption Spectrophotometry", Perkin-Elmer Corp., Norwalk, Conn., U.S.A., 1971.
3. J. E. Schallis and H. L. Kahn, *Atomic Absorption Newsletter*, 1968, **7**, 75.
4. F. W. E. Strelow, C. J. Liebenberg and F. von S. Toerien, *Anal. Chim. Acta,* 1968, **43**, 465.
5. W. Mountjoy and J. S. Wahlberg, *Prof. Papers U.S. Geol. Surv.*, 600B, 1968, B119.
6. H. Sanui and N. Pace, *Analyt. Biochem.*, 1968, **25**, 330.

13.4.10 Calcium

Ca, atomic weight 40·08

Technique	Line, nm	f	Source	Flame	Sensitivity, ppm/1% abs.	Detection limit, ppm	Reference (see below)
AAS	422·67	1·5	HCL	A-A	0·07	0·001	1
	422·67	1·5	HCL	N-A	0·05	—	1, 2
	239·86	0·04	HCL	A-A	20	—	3
AFS	422·67	1·5	EDL	Air/H_2	—	0·02	4
	422·67	1·5	450 W Xe arc	Air/H_2	—	0·02	5
	422·67	1·5	HCL	A-H	—	0·2	6
	422·67	1·5	N_2 pumped dye laser	A-H	—	0·005	7
ND	422·67	1·5	pulsed HCL	A-A	—	0·003	8

Ionization potential: 6·11 eV.
Dissociation energy, CaO: 5·0 eV.
Atomization efficiency: A-A, 0·14; N-A, 1·4.

Notes on analytical conditions

1. The air-coal gas, air-propane and air-hydrogen flames may be used for AAS; considerably more troublesome chemical interferences are observed in these cooler flames, however.

2. Chemical interference (suppression) from phosphate, silicate, aluminium, titanium, zirconium and other species that are difficult to atomize have been reported for Ca determination in the air-acetylene flame. The severity of interference is dependent upon flame stoicheiometry and height of observation; interferences are least when absorbance is measured high in the flame (see also Chapter 12.2). This latter procedure entails loss of sensitivity, however, and it may be preferable to measure Ca absorbance just above the primary zone of the premixed flame and utilize La, Sr or EDTA as releasing agent for chemical interferences.[9-15] The most common chemical interferences are eliminated when the nitrous oxide-acetylene flame is employed;[2,17] with this flame KCl should be added to sample and standard solutions to suppress Ca ionization.

3. For measurement of Ca absorption with decreased sensitivity the Ca 239·86 nm line may be used. Alternatively, Thompson[18] has recommended utilization of the Ge 422·657 nm line which overlaps the Ca 422·673 nm line; this line provides for linear calibration up

to 1500 ppm in the air-acetylene flame and 1000 ppm in the nitrous oxide-acetylene flame (1% absorption is obtained for 8·3 and 6·0 ppm Ca, respectively).

4. Enhancement of sensitivity in the determination of Ca by AAS may be obtained using long-path absorption tubes[19] and non-flame atomization devices such as hollow-cathode spluttering cells,[20,21] and the carbon rod,[22] carbon tube[23] and Ta ribbon filament.[24]

5. Calcium may be extracted from alkaline medium (pH 11-12) into MIBK as its 8-hydroxyquinoline complex. Szego and Calvert[25] have described a separation procedure for Ca based on solvent extraction with a naphthenic acid-cyclohexylamine-MIBK mixture (50 : 3 : 47 v/v).

Applications notes and key references

Atomic absorption spectrometry is the preferred technique for the determination of calcium in a wide variety of materials and permits rapid, accurate and relatively trouble-free analyses. Table 13-1 lists some of the published AAS and AFS applications work for calcium.

There is no attempt at completely comprehensive coverage; any such coverage would be quickly outdated owing to the continuing rapid expansion of the literature concerned with the applications of AAS to calcium determinations. Sufficient data and references are given however, to enable the reader to select publications pertinent to his own interests.

TABLE 13-1

Some applications of the determination of calcium by AAS and AFS

Sample type	Technique; sample pretreatment	Reference
Steel	AAS; add Na and Sr to samples and standards to suppress interferences; N-A flame	26
Iron ores	AAS; A-A flame	27
Ferrosilicon	AAS; dissolution with HF/HNO_3	28
Mg alloys	AAS; dissolve in HCl/HNO_3, add La; A-A flame	29
Al alloys	AAS; dissolve in 50% HCl, add La to release interferences	30
U compounds	AAS; remove U with tributylphosphate extraction; N-A flame	31
Mineral slags and wool	AAS; dissolve in $HF/HCl/HNO_3$; add H_3BO_3 to complex fluoride	32
Chrome refractories	AAS; fuse with $Na_2CO_3/Na_2B_4O_7$, dissolve in 2 N H_2SO_4, add $LaCl_3$	15
Chrome refractories	AAS; dissolve in H_3PO_4, heat to 310°C, dilute; N-A flame	33
Slags and sinters	AAS; decompose with HF or HF/HCl; use NaCl to suppress ionization; N-A flame	34

TABLE 13-1 (*continued*)

Sample type	Technique; sample pretreatment	Reference
Slags	AAS	35
Coal ash	AAS; dissolve in HF; add La; A-A flame	36
Coal ash	AAS; fuse with Li borate, dissolve in HCl, add $LaCl_3$	37
Coal ash	AAS; dissolve in $HClO_4$/HF, evaporate, dilute add La	38
Coal ash and silicate rocks	AAS; fused with $LiBO_2$, dissolve in dilute HNO_3 containing tartaric acid. Add KCl to suppress ionization; N-A flame	39
Clinkers, siliceous limestone, cement	AAS; dissolve in HCl in PTFE vessel; add HF, H_3BO_3 and $LaCl_3$	40
Silicate rocks	AAS; dissolve in HF/H_2SO_4 or fuse with Na_2CO_3	41
Silicate and carbonate minerals	AAS; dissolve in HF/HNO_3, use La to suppress interferences	42
Silicate rocks	AAS; HF/HNO_3 dissolution, add $HClO_4$, fume, dissolve in HCl, add La; A-A flame	43
Silicates	AAS; fuse with $LiBO_2$; dissolve melt in HCl; N-A flame	44
Silicates	AAS; fuse with $LiBO_2$. Comprehensive scheme	45
Silicates	AAS; fuse with Li_2CO_3/H_3BO_3; dissolve melt in HCl, add $LaCl_3$; A-A flame	46
Silicate minerals and rocks	AAS, HF decomposition; N-A flame	47
Silicates	AAS; decompose with HF in sealed vessel; add H_3BO_3, dilute; N-A flame	48
Silicates	AAS; fuse with Li borate/Sr borate mixture; dissolve in citric acid. A-A flame	49
Bauxite	AAS; HF dissolution in sealed vessel; N-A flame	50
Cement	AAS; dissolve in HCl, filter, dilute, add La; N-A flame	
Cement	AAS; dissolve in HCl, add La in sample and standards; use scale expansion and zero suppression with similar standards to gain precision and accuracy	52
Cement	AAS; dissolve in 4 M HCl, evaporate, dissolve in HCl, dilute, add Sr to suppress interference from Al, Si, P; A-A flame	53
Cement	AAS; disolve in HCl, add HF, then H_3BO_3 and $LaCl_3$	54
Cement	AAS	55
Glass	AAS; dissolve in $HClO_4$/HF, fume to remove Si, dissolve in $HClO_4$, dilute, add EDTA + La buffer	56
Glass	AAS; HCl/HF dissolution; MeOH added for high B samples. Evaporate to dryness twice, dissolve in HCl. Use Sr releasing agent	57

TABLE 13-1 (*continued*)

Sample type	Technique; sample pretreatment	Reference
Lubricating oils	AAS; dilute with white spirit	58
Lubricating oils	AAS; dilute with xylene, add K cyclohexane butyrate to suppress ionization; N-A flame	59
Lubricating oils	AAS; dilute with white spirit, add K alkyl sulphonate; N-A flame	60
Lubricating oils	AAS; dilute with MIBK, add La nitrate and iodine; N-A flame	61, 62
Oils	AAS; dilute with cyclohexanone/butanol/ methylated spirit (5:3:2 v/v). Compatible solvent for oils and aqueous Ca standards; A-A flame	63
Oils	AAS; add Na petroleum sulphonate or Triton X emulsifier, dilute with H_2O, add KCl/HNO_3, dilute; N-A flame	64
Serum	AAS; dilute with water (1:20) or precipitate protein with TCA, add $SrCl_2$ to suppress interference. Add Na and K to standards	65
Serum, urine	AAS; dilute with water or EDTA (serum) or 1% Sr or La soln. (urine)	66, 67
Serum	AAS; add TCA; dilute with $SrCl_2$. Add Sr and Na to standards	68
Serum, cerebrospinal fluid	AAS; automated system	69
Serum, urine	AAS; precipitate protein with TCA, dilute with $LaCl_3$	70
Serum, urine	AAS; dilute (1:50) with $LaCl_3$ solution	71
Serum	AAS; carbon rod atomization	72
Serum, urine	AAS; dilute with NaCl/KCl, add $LaCl_3$; N-A flame	73
Serum, urine diet, faeces	AAS	74
Serum	AAS; dilute (1:50) with $HCl/LaCl_3/SrCl_2$ solution; Sr internal standard	75
Tissue, urine	AAS; coprecipitate with Sr oxalate, dissolve in HCl	76
Blood	AAS; ash in active oxygen; dissolve in $HCl/LaCl_3$/ CsCl	77
Plasma	AAS; dilute with solution of $SrCl_2$	78
Tissues	AAS; extract with acetic acid/TCA; add Na or K to suppress ionization; N-A flame	79
Plasma, uterine fluid	AAS; dilute with $LaCl_3$ solution	80
Cerebrospinal fluid	AAS; dilute solution (1:20) with water; A-A flame	81
Phospholipids	AAS; dissolve in isopentyl acetate	82
Finger nails	AAS; digest with $HNO_3/HClO_4$, dilute	83
Blood, ghost-cell membranes	AAS; dilute with 50% ethanol, add 1 M HCl and dilute NaCl	84
Cellular membranes	AAS; digest with $HNO_3/HClO_4$ or ash at low temperature in active oxygen; add La	85
Urinary calculi	AAS; dissolve in HCl	86

TABLE 13-1 (*continued*)

Sample type	Technique; sample pretreatment	Reference
Saliva	AAS; dilute with 1% Na_2 EDTA solution	87
Animal feeds, faeces, urine	AAS	88
Brain tissue	AAS; dry ash, dissolve in HCl; use La to suppress interferences	89
Vegetable oils	AAS; dissolve in MIBK, extract into 2 M HNO_3, add $LaCl_3$; A-A flame	90
Sheep tissues	AAS	91
Fats	AAS; dissolve in iso-amyl acetate/methanol mixture (85:15 v/v)	92
Plant material	AAS; digest with acid. Use Na/K/Mg in acid standards	93
Plant ash	AAS; remove phosphate by ion-exchange; A-coal gas flame	94
Plant material	AAS; dry ash, add Sr to solutions	95
Herbage	AAS; automatic; N-A flame	96
Plant tissue	AAS; dry ash or extract with $(NH_4)_2$ EDTA solution	97
Sugar beet leaves	AAS; digest with $HNO_3/HClO_4/H_2SO_4$ in presence of Na molybdate; A-A flame	98
Corn and tobacco leaves	AAS	99
Grass, straw, manure	AAS; extract with 1 M HCl for 24 hours	100
Fertilizers	AAS; dissolve in HCl; sample and standards contain 0·1 M HCl	101
Soils	AAS; extract with NH_4Cl, add Sr	102
Soils	AAS; decompose with HF/HCl	103
Soil extracts	AFS; extract with NH_4Cl, acetic acid or EDTA; A-A flame	104
Soil extracts	AAS; extract with 0·5 M acetic acid, add Sr to suppress phosphate interference	105
Milk	AAS; dry ashing procedure	106
Colostrum, milk	AAS	107
Beer	AAS; dilute 1:100 with water	108
Fruit juices	AAS; add HCl, centrifuge, dilute supernate with dilute HCl, add $LaCl_3$; A-A flame	109
Waters	AAS; add $LaCl_3$ in HCl; use similar concentrations of each in standards	110
Industrial waters	AAS; acidify with HCl, autoclave, add La to suppress interferences	111
Waters	AAS	112
Sea-water	AAS	113
Atmospheric particulate matter	AAS; collect on filter, low-temperature ashing in active oxygen	114
Paint, vinyl additives	AAS	115
Synthetic fibres	AAS; dry ash, dissolve ash in HNO_3	116

References

1. "Analytical Methods for Atomic Absorption Spectrophotometry", Perkin-Elmer Corp., Norwalk, Conn., U.S.A., 1971.
2. M. D. Amos and J. B. Willis, *Spectrochim. Acta,* 1966, **22,** 1325.
3. B. M. Gatehouse and J. B. Willis, *Spectrochim. Acta,* 1961, **17,** 710.
4. K. E. Zacha, M. P. Bratzel Jnr., J. D. Winefordner and J. M. Mansfield Jnr., *Anal. Chem.,* 1968, **40,** 1733.
5. D. R. Demers and D. W. Ellis, *Anal. Chem.,* 1968, **40,** 860.
6. D. C. Manning and P. Heneage, *Atomic Absorption Newsletter,* 1968, **7,** 80.
7. L. M. Fraser and J. D. Winefordner, *Anal. Chem.,* 1971, **43,** 1693; 1972, **44,** 1444.
8. D. R. Demers and D. G. Mitchell, "Advances in Automated Analysis", Vol. II, Thurman Associates, Miami, p. 507, 1971.
9. J. B. Willis, *Nature,* **1969, 186,** 249.
10. W. Slavin, S. Sprague and D. C. Manning, *Atomic Absorption Newsletter,* 1963, **2,** 49.
11. J. H. Gibson, W. E. L. Grossman and W. D. Cooke, *Anal. Chem.,* 1963, **35,** 266.
12. T. V. Ramakrishna, J. W. Robinson and P. W. West, *Anal. Chim. Acta,* 1966, **36,** 57.
13. C. D. Christian and F. J. Feldman, *Anal. Chim. Acta,* 1968, **40,** 173.
14. R. E. Dickson and C. M. Johnson, *Appl. Spectrosc.,* 1966, **20,** 214.
15. N. Sidiropoulos, *Analyst,* 1969, **94,** 389.
16. R. M. Dagnall, G. F. Kirkbright, T. S. West and R. Wood, *Anal. Chem.,* 1971, **43,** 1765.
17. D. C. Manning and L. Capacho-Delgado, *Anal. Chim. Acta,* 1966, **36,** 312.
18. K. C. Thompson, *Analyst,* 1970, **95,** 1043.
19. S. R. Koirtyohann and E. E. Pickett, *Anal. Chem.,* 1965, **37,** 601.
20. J. A. Goleb and J. K. Brody, *Anal. Chim. Acta,* 1963, **28,** 457.
21. B. W. Gandrud and R. K. Skogerboe, *Appl. Spectrosc.,* 1971, **25,** 243.
22. M. D. Amos, P. A. Bennett, K. G. Brodie, P. M. Y. Lung and J. P. Matousek, *Anal. Chem.,* 1971, **43,** 211.
23. B. V. L'Vov, *Spectrochim. Acta,* 1969, **24B,** 53.
24. H. M. Donega and T. E. Burgess, *Anal. Chem.,* 1970, **42,** 1421.
25. L. Szega and J. M. Calvert, *Anal. Chim. Acta,* 1968, **42,** 1.
26. M. L. Taylor and C. B. Belcher, *Anal. Chim. Acta,* 1969, **45,** 219.
27. Y. Endo, T. Hata and Y. Nakahara, *Japan Analyst,* 1968, **17,** 679.
28. F. J. Langmyhr and P. E. Paus, *Ana. Chim. Acta,* 1969, **45,** 173.
29. R. E. Mansell, H. W. Emmel and E. L. McLaughlin, *Appl. Spectrosc.,* 1966, **20,** 231.
30. C. F. Bell, *Atomic Absorption Newsletter,* 1966, **5,** 73.
31. C. R. Walker and O. A. Vita, *Anal. Chim. Acta,* 1968, **43,** 27.
32. J. E. Sheridan, *Spectrovision,* 1971, **25,** 10.
33. R. P. Lucas and B. C. Ruprecht, *Anal. Chem.,* 1971, **43,** 1013.
34. J. Reid, J. M. Galloway, J. MacDonald and B. B. Bach, *Metallurgia,* 1970, **81,** 243.
35. H. Goto, S. Ikeda and I. Atsuya, *Japan Analyst,* 1964, **13,** 111.
36. C. B. Belcher and K. A. Brooks, *Anal. Chim. Acta,* 1963, **29,** 202.
37. R. B. Muter and C. F. Cockerell, *Appl. Spectrosc.,* 1969, **23,** 493.
38. E. L. Obermiller and R. W. Freedman, *Fuel,* 1965, **44,** 199.
39. P. L. Boar and L. K. Ingram, *Analyst,* 1970, **95,** 124.

40. F. J. Langmyhr and P. E. Paus, *Anal. Chim. Acta,* 1969, **44,** 445.
41. D. J. Trent and W. Slavin, *Atomic Absorption Newsletter,* 1964, **3,** 17, 118.
42. G. K. Billings and J. A. S. Adams, *Atomic Absorption Newsletter,* 1964, **3,** 65.
43. C. B. Belt Jnr., *Anal. Chem.,* 1967, **39,** 676.
44. J. W. Yule and G. A. Swanson, *Atomic Absorption Newsletter,* 1969, **8,** 30.
45. J. C. Van Loon and C. M. Parissis, *Analyst,* 1969, **94,** 1057.
46. S. H. Omang, *Anal. Chim. Acta,* 1969, **46,** 225.
47. F. J. Langmyhr and P. E. Paus, *Anal. Chim. Acta,* 1969, **47,** 371.
48. B. Bernas, *Anal. Chem.,* 1968, **40,** 1682.
49. K. Govindaraju, *Appl. Spectrosc.,* 1970, **24,** 81.
50. F. J. Langmyhr and P. E. Paus, *Anal. Chim. Acta,* 1968, **43,** 508.
51. L. Capacho-Delgado and D. C. Manning, *Atomic Absorption Newsletter,* 1967, **92,** 553.
52. S. Sprague, *Atomic Absorption Newsletter,* 1963, **2,** 43.
53. T. Takeuchi and M. Suzuki, *Talanta,* 1964, **11,** 1391.
54. J. T. H. Roos and W. J. Price, *Analyst,* 1969, **94,** 89.
55. F. J. Feldman, J. A. Blasi and S. B. Smith Jnr., *Anal. Chem.,* 1969, **41,** 1095.
56. P. B. Adams and W. O. Passmore, *Anal. Chem.,* 1966, **38,** 630.
57. A. H. Jones, *Anal. Chem.,* 1965, **37,** 1761.
58. D. W. Kohlenberger, *Atomic Absorption Newsletter,* 1969, **8,** 108.
59. G. E. Peterson and H. L. Kahn, *Atomic Absorption Newsletter,* 1970, **9,** 71.
60. D. Pickles and C. C. Washbrook, *Proc. Soc. Anal. Chem.,* 1970, **7,** 13.
61. M. Kashiki and S. Oshima, *Anal. Chim. Acta,* 1971, **55,** 436.
62. M. Kashiki, S. Yamazoe and S. Oshima, *Anal. Chim. Acta,* 1971, **54,** 533.
63. S. T. Holding and J. W. Noar, *Analyst,* 1970, **95,** 1041.
64. L. C. Westwood, *Dev. Appl. Spectrosc.,* 1971, **9,** 235.
65. J. B. Willis, *Nature,* 1960, **186,** 249.
66. J. B. Willis, *Spectrochim. Acta,* 1960, **16,** 259.
67. J. B. Willis, *Anal. Chem.,* 1961, **33,** 556.
68. F. W. Sundermann and J. E. Carrol, *Am. J. Clin. Pathol.,* 1965, **43,** 302.
69. B. Klein, J. H. Kaufman and M. Oklander, *Clin. Chem.,* 1967, **13,** 797.
70. A. M. Briscoe and C. Rogan, *J. Appl. Physiol.,* 1965, **20,** 453.
71. D. F. Trudeau and E. F. Freier, *Clin. Chem.,* 1967, **13,** 101.
72. M. D. Amos, K. G. Brodie, J. P. Matousek and B. Stevens, *Pittsburgh Conf., on Anal. Chem. and Appl. Spectry.,* Cleveland, Ohio, 1971.
73. M. L. Girard and F. Rousselet, *Ann. Pharm. Fr.,* 1967, **25,** 271.
74. E. G. Gimblet, A. F. Marney and R. W. Bosnes, *Clin. Chem.,* 1967, 13, 204.
75. J. Pybus, F. J. Feldman and G. N. Bowers, *Clin. Chem.,* 1970, **16,** 998.
76. H. E. Parker, *Atomic Absorption Newsletter,* 1963, **2,** 23.
77. N. Sanui, *Anal. Biochem.,* 1971, **42,** 21.
78. D. J. Bell, J. L. Braidwood, J. G. Campbell, J. Culbert, J. Filshie and P. E. Lake, *Comp. Biochem. Physiol.,* 1966, **19,** 133.
79. F. T. W. Varghese, A. Lipton and G. J. Huxham, *Lab. Prac.,* 1969, **18,** 419.
80. S. E. Solomsh, *Br. Poult. Sci.,* 1971, **12,** 165.
81. C. F. Decker, A. Aras and L. E. Decker, *Anal. Biochem.,* 1964, **8,** 344.
82. B. Montford and S. C. Cribbs, *Talanta,* 1969, **16,** 1079.
83. W. W. Harrison and A. B. Tyree, *Clin. Chim. Acta,* 1971, **31,** 63.

84. C. Long and B. Movat, *Biochem. J.*, 1971, **123**, 829.
85. H. Sanui and N. Pace, *Appl. Spectrosc.*, 1966, **20**, 135.
86. A. Hodgkinson, *J. Clin. Pathol.*, 1971, **24**, 147.
87. E. Newbrun, *Nature*, 1961, **192**, 1182.
88. M. A. Carces and J. L. Evans, *J. Animal Sci.*, 1971, **32**, 789.
89. N. J. McBroom, Y. J. Lancaster and K. Weiss, *Analyt. Biochem.*, 1971, **42**, 178.
90. R. A. Foss and D. M. Houston, *Atomic Absorption Newsletter*, 1969, **8**, 82.
91. J. F. Standish and C. B. Ammerman, *J. Animal Sci.*, 1971, **33**, 481.
92. R. Guillaumin and N. Drouhin, *Rev. Fr. Corps. Gras*, 1965, **12**, 735.
93. D. J. David, *Analyst*, 1959, **84**, 536.
94. W. H. Hinson, *Spectrochim. Acta*, 1962, **18**, 427.
95. W. L. Berry and C. M. Johnson, *Appl. Spectrosc.*, 1966, **20**, 209.
96. N. T. Faithfull, *Lab. Prac.*, 1971, **20**, 41.
97. J. H. Baker and T. Greweling, *J. Agr. Food Chem.*, 1967, **15**, 340.
98. M. Harrison and C. Andre, *Appl. Spectrosc.*, 1969, **23**, 354.
99. A. Wallace, E. Frohlich and O. R. Lunt, *Nature*, 1966, **209**, 634.
100. B. J. Hale, *J. Agr. Sci.*, 1947, **37**, 236.
101. C. H. McBride, *J. Ass. Off. Analyt. Chem.*, 1967, **50**, 401.
102. D. J. David, *Analyst*, 1960, **85**, 495.
103. S. Pawluk, *Atomic Absorption Newsletter*, 1967, **6**, 53.
104. R. M. Dagnall, G. F. Kirkbright, T. S. West and R. Wood, *Anal. Chem.*, 1971, **43**, 1765.
105. W. S. G. MacPhee and D. F. Ball, *J. Sci. Food Agr.*, 1967, **18**, 376.
106. G. K. Murphy and U. Rhea, *J. Dairy Sci.*, 1967, **50**, 313.
107. R. F. Elliot, G. W. Vander Novt, R. L. Gilbreath and H. Fisher, *J. Animal Sci.*, 1971, **32**, 1128.
108. S. W. Frey, *Atomic Absorption Newsletter*, 1964, **3**, 127.
109. J. T. H. Roos and W. J. Price, *J. Sci. Food Agr.*, 1970, **21**, 51.
110. M. J. Fishman and S. C. Downs, *U.S. Geol. Surv. Water-Supply Papers*, 1540-C, 1966.
111. J. A. Platte and V. M. Marcy, *Atomic Absorption Newsletter*, 1965, **4**, 289.
112. L. R. P. Butler and D. W. Brink, *S. Afr. Ind. Chem.*, 1963, **17**, 152.
113. G. K. Billings and R. C. Harriss, *Texas J. Sci.*, 1965, **17**, 129.
114. R. J. Thompson, G. B. Morgan and L. J. Purdue, *Atomic Absorption Newsletter*, 1970, **9**, 53.
115. N. G. Eider, *Appl. Spectrosc.*, 1971, **25**, 313.
116. W. Slavin, *Atomic Absorption Newsletter*, 1965, **4**, 192.

Spectrochim, 1966, **22**, 1325.

13.4.11 Cerium

Ce, atomic weight 140·12

Technique	Line, nm	f	Source	Flame	Sensitivity, ppm/1% abs.	Detection limit, ppm	Reference (see below)
AAS	520·0	–	HCL	N-A	30	–	1
	569·7	–	HCL	N-A	39	–	1
AFS: not reported							

Ionization potential: 5·6 eV.
Dissociation energy, CeO: 8·3 eV[2].

Notes on analytical conditions

1. With a triple-pass optical system, fuel-rich oxy-acetylene flame and xenon arc continuum source Mossotti and Fassel[3] did not detect absorption for Ce in ethanolic solution. Amos and Willis[4] also failed to detect absorption at a number of Ce lines. The cerium(I) absorption spectrum has not been completely determined and the choice of the most suitable line is difficult. The complexity of the cerium emission spectrum necessitates the use of narrow spectral band-pass if this element is to be detected. Additionally, the high dissociation energy of CeO may lead to a low value for the Ce free atom fraction, even in hot flames. The value for the Ce ionization potential also suggests that Ce would be appreciably ionized in the nitrous oxide-acetylene flame.

2. Johnson and co-workers[5] have described an indirect method for the determination of Ce by AAS via formation of molybdocerophosphoric acid.

Applications notes and key references

See Lanthanum.

References

1. "Varian-Techtron Hollow Cathode Lamp Data book", Varian-Techtron, Walton-on-Thames, U.K.
2. P. N. Walsh, D. F. Dever and D. White, *J. Phys. Chem.*, 1961, **65**, 1410.
3. V. G. Mossotti and V. A. Fassel, *Spectrochim. Acta*, 1964, **20**, 1117.
4. M. D. Amos and J. B. Willis, *Spectrochim. Acta*, 1966, **22**, 1325.
5. H. N. Johnson, G. F. Kirkbright and R. J. Whitehouse, *Anal. Chem.*, 1973, **45**, 1603.

13.4.12 Chromium

Cr, atomic weight 51·996

Technique	Line, nm	f	Source	Flame	Sensitivity, ppm/1% abs.	Detection limit, ppm	Reference (see below)
AAS	357·87	0·34	HCL	A-A	0·1	0·003	1, 2
	357·87	0·34	HCL	N-A	0·25	—	1
	359·35	0·27	HCL	A-A	0·15	—	1
	360·53	0·19	HCL	A-A	0·22	—	1
	425·44	0·10	HCL	A-A	0·3	—	1
	427·48	—	HCL	A-A	0·38	—	1
AFS	359·35	0·27	EDL	Air/H_2	—	10	3
	357·87	—	150 W Xe arc	A-H	—	10	4
	359·35	—	HCL	A-H	—	1	5
	359·35	—	Hg VDL	A-H	—	5	6
	359·35	—	Dye laser	A-A	—	0·02	7
ND	357·87	—	Pulsed HCL	N_2 sep. A-A	—	0·02	7a

Ionization potential: 6·76 eV.
Dissociation energy, CrO: 4·2 eV.
Atomization efficiency: A-A, 0·064.

Notes on analytical conditions

1. A narrow spectral band-pass is preferable in order to obtain high sensitivity in AAS for Cr. Kinson and co-workers[8] observed interference from the argon 357·66 and 360·65 nm lines on chromium determination at 357·87 and 360·53 nm when an argon-filled HCL source was employed; with large spectral band-pass, interference from the argon 358·23 nm line may also be experienced when the Cr 357·87 nm line is employed.

2. The AAS sensitivity for Cr in the air-acetylene flame is critically dependent upon the flame stoicheiometry. Fuel-rich flames produce highest sensitivity, but interference is observed from Fe, Ni and other elements in these fuel-rich flames. These effects are minimized in a leaner flame and it may therefore be necessary to sacrifice sensitivity to gain freedom from interference. Interferences in the rich flame, however, may be suppressed by the addition of ammonium chloride[9,10] or ammonium bifluoride[10a] to sample and standard solutions. Although lower sensitivity is obtained in the nitrous oxide-acetylene flame, fewer chemical interferences are encountered.

3. As well as the air-acetylene and nitrous oxide-acetylene flames, the air-hydrogen flame has been shown to provide high sensitivity for Cr determination by AAS[11] and AFS.[3-6].

4. Chromium(VI) may be extracted into MIBK from 3 M HCl solution,[11] or with diphenyl carbazone[12] or APDC from pH 0·5 or pH 3-7 respectively into MIBK.

Applications notes and key references

Kinson, Hodges and Belcher[8] first published a method for the determination of Cr in low-alloy iron and steel by AAS; these workers used a slightly fuel-rich air-acetylene flame and dissolved 1 g samples in 15% H_3PO_4/15% H_2SO_4 and oxidized with HNO_3 before evaporation and redissolution of the residue in water. Other workers have reported similar procedures for alloy samples using the air-acetylene[13-16] or nitrous oxide-acetylene flames,[17,18] and sample dissolution in mineral acid. The presence of H_3PO_4 in the acid solvent employed serves to prevent precipitation of tungsten as tungstic acid. Endo and co-workers[19] have described the AAS determination of Cr in iron ores. Other materials in which Cr has been determined by straightforward AAS procedures include aluminium salts[20] and alloys,[21,22] nickel and nickel alloys,[23] silver,[24] zirconium alloys,[25] uranium compounds[26] and plutonium nitrate solutions[27] and ferro-silicon.[28] Billings and Ragland[29] have described the determination of Cr in geological samples. Elrod and Ezell[30] have reported a procedure for AAS of Cr in aluminous samples in which samples are decomposed with 50% H_2SO_4 or H_2SO_4/HNO_3/HCl and $K_2S_2O_8$ is added to the solutions to prevent depression of Cr absorption by Al, Fe and Ti in the air-acetylene flame employed. Chromium in siliceous materials may be determined by AAS after sample decomposition by heating with 40% HF in a PTFE lined bomb.[31] This method has been applied to the determination of Cr in bauxite. Chromium in chrome ores and chrome-magnesite refractories may be determined in a nitrous oxide-acetylene flame;[32] samples are dissolved in H_3PO_4.

Slavin and co-workers[33] employed AAS for the determination of Cr in urine, plasma and whole blood without preconcentration; wet digestion with HNO_3 was used for whole blood samples. Pierce and Cholak[34] utilized a dry ashing procedure for blood and urine samples, and Feldman *et al.*[35] found that although both wet and dry ashing procedures gave good recoveries the wet ashing procedure was more rapid. In the method developed by the latter workers Cr was oxidized to Cr(VI) with $KMnO_4$ and extracted into MIBK from HCl solution. Devoto[36] utilized H_2O_2 to oxidize Cr from blood and urine samples which had been wet-ashed with HNO_3 and extracted

Cr(VI) from pH 1 medium with a 10% solution of tributylphosphate in MIBK. In an early AAS paper Williams, David and Iisma[37] determined Cr in faeces of sheep; samples were digested with phosphoric acid–manganous sulphate–potassium bromate to destroy organic matter. More recently Arthur[38] has also determined Cr in animal feed and excreta by AAS. Premi and Cornfield[39] extracted Cr from plant materials and organic residues by boiling with 6 M HCl and examined the extracts by AAS using the air-acetylene flame. Meranger and Somers have determined traces of Cr in wines,[40] fruit juices[41] and sea-foods[42] by direct AAS procedures; ethanol in wines and CO_2 in carbonated beverages is removed by evaporation while sea-food samples were dry ashed and the residue dissolved in HNO_3 before Cr determination. Davidson and Secrest[42a] have recently described the determination of Cr in biological materials using a graphite furnace atomizer.

The determination of Cr in lubricating oils by AAS has been reported by a number of workers. Sprague and Slavin[43,44] have described a rapid AAS technique for determination of Cr and other metals in jet engine oils; samples were diluted ten-fold with MIBK and nebulized directly in the flame. Burrows and co-workers[45] reported a similar procedure for determination of Cr in railway engine oils, and Means and Ratcliffe[46] employed four-fold dilution of jet-engine lubricants with *p*-xylene prior to nebulization into the air-acetylene flame. Other work has confirmed the findings of these workers[47-49] and it is now generally agreed that the presence of metal particles in used lubricating oil samples causes little difficulty when hot flames are employed.

Miller, Fraser and Winefordner[50] have described the determination of Cr and other metals in jet-engine lubricating oils using a spectrometer combining flame emission and atomic fluorescence; the oils were diluted 1 : 4 with CCl_4 and nebulized into an argon-hydrogen-entrained air flame. Gardels, Demers and Mitchell[51] have demonstrated the simultaneous determination of Cr and five other metals in oils with a six-channel atomic fluorescence instrument. These workers diluted lubricating oil standards 1 : 5 with MIBK or xylene and employed a stoicheiometric air-acetylene flame for atomization. The results compared favourably with those obtained by AAS.

Chromium in industrial and fresh waters has been determined directly by AAS by Platte and Marcy[52] and Midgett and Fishman.[53] Burelli and Ecrement[54] have described a method for Cr determination in water utilizing MIBK extraction; the method was applied to measurement of water flow using the addition of $Na_2Cr_2O_7$ as a tracer. Delaughter[12] determined Cr in brines using extraction with diphenylthiocarbazone.

Yiu-Kee Chau *et al.*[55] have determined Cr by formation of the Cr-acetylacetonate complex by boiling at pH 6-7 with acetylacetone and extracting the complex into MIBK. The latter workers determined Cr in 5 litre samples of sea-water using pre-concentration by coprecipitation on ferric hydroxide.

References

1. "Analytical Methods for Atomic Absorption Spectrophotometry", Perkin-Elmer Corp., Norwalk, Conn., U.S.A., 1971.
2. W. Slavin, "Atomic Absorption Spectroscopy", Wiley-Interscience, New York, 1968.
3. K. E. Zacha, M. P. Bratzel, J. D. Winefordner and J. M. Mansfield Jnr., *Anal. Chem.*, 1968, **40**, 1733.
4. D. C. Manning and P. Heneage, *Atomic Absorption Newsletter*, 1968, **7**, 80.
5. G. Rossi and N. Omenetto, *Talanta*, 1969, **16**, 263.
6. N. Omenetto and G. Rossi, *Anal. Chim. Acta*, 1968, **40**, 195.
7. L. M. Fraser and J. D. Winefordner, *Anal. Chem.*, 1971, **43**, 1963; 1972, **44**, 1444.

7a. D. Demers and D. G. Mitchell in "Advances in Automated Analyses", Vol. II, p. 507, Thurman Associates, Miami, 1971.

8. K. Kinson, R. J. Hodges and C. B. Belcher, *Anal. Chim. Acta*, 1963, **29**, 134.
9. A. Giammarise, *Atomic Absorption Newsletter*, 1966, **5**, 113.
10. L. Barnes Jnr., *Anal. Chem.*, 1966, **38**, 1083.

10a. A. Purushottam, P. P. Naidu and S. S. Lal, *Talanta*, 1973, **20**, 631.

11. F. J. Feldman and W. C. Purdy, *Anal. Chim. Acta*, 1965, **33**, 273.
12. B. Delaughter, *Atomic Absorption Newsletter*, 1965, **4**, 273.
13. M. Beyer, *Atomic Absorption Newsletter*, 1965, **4**, 212.
14. A. Gomez Coedo and J. L. Jimenez-Seco, *Revta Metal*, 1968, **4**, 316.
15. D. M. Knight and M. K. Pyzyna, *Atomic Absorption Newsletter*, 1969, **8**, 129.
16. W. J. Price and P. A. Cooke, *Spectrovision*, 1967, **18**, 2.
17. F. J. Feldman, J. A. Blasi and S. B. Smith Jnr., *Anal Chem.*, 1969, **41**, 1095.
18. D. R. Thomerson and W. J. Price, *Analyst*, 1971, **96**, 321.
19. Y. Endo, T. Hata and Y. Nakahara, *Japan Analyst*, 1969, **18**, 833.
20. G. B. Marshall and T. S. West, *Analyst*, 1970, **95**, 343.
21. G. F. Bell, *Atomic Absorption Newsletter*, 1966, **5**, 73.
22. J. L. Jimenez-Seco and Aurora Gomez Coedo, *Revta Metal*, 1969, **5**, 290.
23. R. Dyck, *Atomic Absorption Newsletter*, 1965, **4**, 170.
24. J. M. Viebock, *Anal. Letters*, 1970, **3**, 373.
25. J. H. Schlewitz and M. G. Shields, *Atomic Absorption Newsletter*, 1971, **10**, 39.
26. C. R. Walker and O. A. Vita, *Anal. Chim. Acta*, 1968, **43**, 27.
27. M. Ganivet and A. Benhaman, *Anal. Chim. Acta*, 1969, **47**, 81.
28. F. J. Langmyhr and P. E. Paus, *Anal. Chim. Acta*, 1969, **45**, 173.
29. G. K. Billings and P. C. Ragland, *Can. Spectrosc.*, 1969, **14**, 8.

30. B. B. Elrod and J. B. Ezell, *Atomic Absorption Newsletter*, 1969, **8**, 40.
31. F. J. Langmyhr and P. E. Paus, *Anal. Chim. Acta*, 1968, **43**, 508.
32. R. P. Lucas and B. C. Ruprecht, *Anal. Chem.*, 1971, **43**, 1013.
33. W. Slavin, S. Sprague, F. Rieders and V. Cordova, *Atomic Absorption Newsletter*, 1964, **4**, 7.
34. J. O. Pierce and J. Cholak, *Arch. Environ. Health*, 1966, **13**, 208.
35. F. J. Feldman, E. C. Knoblock and W. C. Purdy, *Anal. Chim. Acta*, 1967, **38**, 489.
36. G. Devoto, *Bull. Soc. Ital. Biol. Sper.*, 1968, **44**, 1251.
37. C. H. Williams, D. J. David and O. Iisma, *J. Agric. Sci.*, 1962, **59**, 381.
38. D. Arthur, *Can. Spectrosc.*, 1970, **15**, 134.
39. P. R. Premi and A. H. Cornfield, *Spectrovision*, 1968, **19**, 15.
40. J. C. Meranger and E. Somers, *J. Ass. Off. Analyt. Chem.*, 1968, **51**, 922.
41. J. C. Meranger, *Bull. Envir. Contam. Toxicol.*, 1970, **5**, 271.
42. J. C. Meranger and E. Somers, *Bull. Envir. Contam. Toxicol.*, 1968, **3**, 360.
42a. I. W. F. Davidson and W. L. Secrest, *Anal. Chem.*, 1972, **44**, 1801.
43. S. Sprague and W. Slavin, *Atomic Absorption Newsletter*, 1965, **4**, 367.
44. S. Slavin and W. Slavin, *Atomic Absorption Newsletter*, 1966, **5**, 106.
45. J. A. Burrows, J. C. Heerdt and J. B. Willis, *Anal. Chem.*, 1965, **37**, 579.
46. E. A. Means and D. Ratcliffe, *Atomic Absorption Newsletter*, 1965, **4**, 174.
47. D. R. Jackson, *Can. Spectrosc.*, 1970, **15**, 17.
48. H. L. Kahn, G. E. Peterson and D. C. Manning, *Atomic Absorption Newsletter*, 1970, **9**, 79.
49. J. H. Taylor, T. T. Bartels and N. L. Crump, *Anal. Chem.*, 1971, **43**, 1780.
50. R. L. Miller, L. M. Fraser and J. D. Winefordner, *Appl. Spectrosc.*, 1971, **25**, 477.
51. M. Gardels, D. Demers and D. G. Mitchell, "Advances in Automated Analysis", Vol. II, p. 513, Thurman Associates, Miami, 1971.
52. J. A. Platte and V. M. Marcy, *Atomic Absorption Newsletter*, 1965, **4**, 289.
53. M. R. Midgett and N. J. Fishman, *Atomic Absorption Newsletter*, 1967, **6**, 128.
54. F. P. Burelli and F. Ecrement, *Chim. Analyt.*, 1969, **51**, 561.
55. Yiu-Kee Chau, Soo-San Sim and Yu-Hou Wing, *Anal. Chim. Acta*, 1968, **43**, 13.

13.4.13 Cobalt

Co, atomic weight 58·93

Technique	Line, nm	f	Source	Flame	Sensitivity, ppm/1% abs.	Detection limit, ppm	Reference (see below)
AAS	240·72	0·22	HCL	A-A	0·15	0·01	1, 2
	242·49	0·19	HCL	A-A	0·18	—	2
	241·16	—	HCL	A-A	0·27	—	—
	252·14	0·19	HCL	A-A	0·30	—	—
	243·58	—	HCL	A-A	0·45	—	—
	341·26	—	HCL	A-A	3·0	—	—
	352·68	—	HCL	A-A	2·2	—	—
	347·40	—	HCL	A-A	7·5	—	—
	301·76	—	HCL	A-A	16	—	—
AFS	240·72	0·22	HIL	A-H	—	0·01	3
	240·72	0·22	EDL	A-P	—	0·005	4
	240·72	0·22	HIL	A-H	—	0·04	5
	240·72	0·22	EDL	Ar/H_2	—	0·1	6
	240·72	0·22	HCL	A-H	—	0·3	7
ND	240·72	0·22	HIL	N_2 sep. A-A	—	0·0015	8

Ionization potential: 7·86 eV.
Dissociation energy, CoO: *ca.* 3·7 eV.
Atomization efficiency: A-A 0·052.

Notes on analytical conditions

1. A narrow spectral band-pass is necessary to achieve the highest sensitivity and linearity in Co AAS owing to the emission of non-absorbing lines from the HCL source near to the Co 240·72 nm line. Harrison[9] reported that significant variation of absorption signal was obtained with HCL operating current at 240·72 nm and ascribed this to change in the relative intensity of this line and a nearby non-absorbed line; Slavin,[10] however, attributed this variation to line broadening from the source at high current. The use of a high intensity HCL source for Co, as described by Sullivan and Walsh[11] gives rise to higher sensitivity and extended linear range.

2. A fuel-lean air-acetylene flame is preferred for cobalt; under these conditions few interferences have been reported. In a study of Co absorption in various flames Sachdev *et al.* observed interference from other cations and anions only at very high concentrations, and in the air-acetylene flame McPherson[13] demonstrated that 2000 ppm of Ni, Cr and W, 1000 ppm of Mo and Cu and 500 ppm of Si caused no interference on the Co absorption signal. Enhancement of Co absorption in the presence of very large amounts

of iron (>5000 ppm) may be observed when the solution nebulized contains perchloric acid. Little information is available concerning interferences in the determination of Co by AFS.

3. Cobalt may be extracted into MIBK with APDC between pH 2 and 4.[14]

Applications notes and key references

Sprague and Slavin[15] determined Co in iron and steel by an AAS procedure in which samples were dissolved in HCl/HNO_3 and filtered before nebulization into an air-acetylene flame; standard solutions containing a similar amount of iron and acid were employed. McPherson, Price and Scaife[16] also reported a method for the determination of low concentrations of Co in steels and nickel by AAS. Thomerson and Price have recommended HCl/HNO_3 mixture, with subsequent additions of $HClO_4$ for oxidation, for dissolution of steel samples; in the comprehensive scheme for steel analysis used by these workers Co is determined at 240·72 nm in a lean air-acetylene flame. Norris and West[17] dissolved samples in HCl/HNO_3 and determined 10 to 200 ppm Co in steel directly by AFS using an argon-separated air-acetylene flame and EDL source. The determination of Co in reduced ores and oxides has been reported by Kinson and co-workers.[18] Samples were treated with bromine-methanol (1 : 19), filtered, and HCl added to the filtrate; this was titrated with hydroxylamine hydrochloride and then treated with H_2O_2 and boiled before nebulization. Viebock[19] determined Co in silver by precipitation of the silver as silver chloride, filtration, and nebulization of the filtrate into an air-acetylene flame for AAS. Donaldson and co-workers[20] have determined Co in high-purity Nb, Ta, Mo and W metals by AAS in a procedure in which Co is extracted into chloroform as its diantipyrinylmethane complex and then back-extracted into ammonia buffer. Levine[21] has described the determination of Co in tungsten carbide by AAS in an air-acetylene flame after sample dissolution in HF, and addition of HNO_3 and then H_3BO_3 to complex fluoride. Co in alumina has been determined by dissolution in HCl in a pressure vessel at 270°C and solvent-extraction with 8-hydroxyquinoline.[22] Kapetan[23] has determined Co in gold plating solutions by direct AAS at 240·7 nm using the air-acetylene flame. Ward and co-workers[24] determined Co in rocks and soils after sample dissolution in hot HNO_3 and preconcentration by solvent extraction. Marine sediment fractions have been analysed for Co and other metals by Angino.[25] In the procedure reported the organic fraction is extracted with sodium hypochlorite solution and the residue is then extracted with dilute HCl; both solutions may then be nebulized into the flame. Burrell[26] extracted the asphaltic fraction of marine sediments with chloroform and diluted this with methanol (3 : 2 v/v) for nebulization into the air-acetylene flame. Cobalt in oil-field brines has been determined by AAS after preconcentration with a chelating ion-exchange resin.[27] The determination of Co in KCl brines via extraction into MIBK with APDC has been discussed by Sprague and Slavin,[28] and Burrell[29] determined Co in natural waters after co-precipitation by addition of ferric chloride, and extraction of iron from the dissolved precipitate into isopropylether, and the extraction of Co with APDC into MIBK for nebulization into the flame for AAS.

Cobalt at relatively high concentrations (>0·1 ppm) in urine may be determined directly by AAS without sample pretreatment,[30] or may be extracted from acid medium with APDC into MIBK when higher sensitivity is required. Plant digests have been examined for Co by AAS by Stupar;[31] cobalt was extracted with APDC into MIBK, and the chelate was then destroyed and the resulting aqueous solution nebulized. Steyn and Brandt[32] utilized an AAS method for the determination of Co in citrus leaves, and more recently other workers have analysed sugar-beet leaves for Co.[33] In the latter procedure the samples were heated with $H_2SO_4/HClO_4/HNO_3$ mixture to which sodium molybdate had been added and the solutions were diluted with methanol before nebulization. Jago, Wilson and Lee[34] have described a procedure for Co in plant and animal tissues using a similar digestion medium and preconcentration of Co for AAS by extraction with 1-nitroso-2-naphthol into MIBK. Extractable Co in soils has been examined by Ure and Mitchell[35] by AAS using acetic acid extractant; down to 0·05 ppm Co in soils was determinable by the procedure described.

Other sample types in which Co has been determined by AAS without significant complications include beer,[36] wine,[37] fruit juices and soft drinks,[38] pharmaceuticals[39] and paint and vinyl additives.[40]

References

1. J. E. Allan, *Nature*, 1960, **187**, 1110.
2. "Analytical Methods for Atomic Absorption Spectrophotometry", Perkin-Elmer Corp., Norwalk, Conn., U.S.A., 1971.
3. J. Matousek and V. Sychra, *Anal. Chem.*, 1969, **41**, 518.
4. B. Fleet, K. V. Liberty and T. S. West, *Anal. Chim. Acta*, 1969, **45**, 205.

5. D. C. Manning and P. Heneage, *Atomic Absorption Newsletter,* 1968, **7,** 80.
6. K. E. Zacha, M. P. Bratzel Jnr., J. D. Winefordner and J. M. Mansfield Jnr., *Anal. Chem.,* 1968, **40,** 1733.
7. G. Rossi and N. Omenetto, *Talanta,* 1969, **16,** 263.
8. P. L. Larkins, *Spectrochim. Acta,* 1971, **26B,** 477.
9. W. W. Harrison, *Anal. Chem.,* 1965, **37,** 1168.
10. W. Slavin, "Atomic Absorption Spectroscopy", Interscience, New York, 1968, p. 100.
11. J. V. Sullivan and A. Walsh, *Spectrochim. Acta,* 1965, **21,** 721.
12. S. L. Sachdev, J. W. Robinson and P. W. West, *Anal. Chim. Acta,* 1967, **38,** 491.
13. G. L. McPherson, *Atomic Absorption Newsletter,* 1965, **4,** 186.
14. S. Sprague and W. Slavin, *Atomic Absorption Newsletter,* 1964, **3,** 37.
15. S. Sprague and W. Slavin, *Atomic Absorption Newsletter,* 1964, **3,** 72.
16. G. L. McPherson, J. W. Price and P. H. Scaife, *Nature,* 1963, **199,** 371.
17. J. D. Norris and T. S. West, *Anal. Chim. Acta,* 1971, **55,** 359.
18. K. Kinson, J. E. Dickeson and C. B. Belcher, *Anal. Chim. Acta,* 1968, **41,** 107.
19. J. M. Viebock, *Anal. Letters,* 1970, **3,** 373.
20. E. M. Donaldson, D. J. Charette and V. H. E. Rolko, *Talanta,* 1969, **16,** 1305.
21. S. L. Levine, *Atomic Absorption Newsletter,* 1969, **8,** 58.
22. B. Fleet, K. V. Liberty and T. S. West, *Analyst,* 1968, **93,** 701.
23. J. P. Kapetan, *Spec. Tech. Publ. Am. Soc. Test, Mater.,* 1969, **443,** 78.
24. F. N. Ward, H. M. Nakagawa, T. F. Harms and G. H. Van Sickle, *Bull. U.S. Geol. Surv.,* 1969, **1289,** 45.
25. E. E. Angino, in W. Slavin, "Atomic Absorption Spectroscopy", Interscience, New York, 1968, p. 246.
26. D. C. Burrell, *Atomic Absorption Newsletter,* 1965, **4,** 328.
27. O. K. Galle, *Appl. Spectrosc.,* 1971, **25,** 664.
28. S. Sprague and W. Slavin, *Atomic Absorption Newsletter,* 1964, **3,** 37.
29. D. C. Burrell, *Atomic Absorption Newsletter,* 1965, **4,** 309.
30. J. B. Willis, *Anal. Chem.,* 1962, **34,** 614.
31. J. Stupar, *Mikrochim. Acta,* 1966, 722.
32. W. J. A. Steyn and P. J. Brandt, *J. S. Afr. Chem. Inst.,* 1965, **18,** 92.
33. M. Harrison and C. Andre, *Appl. Spectrosc.,* 1969, **23,** 354.
34. J. Jago, P. E. Wilson and B. M. Lee, *Analyst,* 1971, **96,** 349.
35. A. M. Ure and R. L. Mitchell, *Spectrochim. Acta,* 1967, **23B,** 79.
36. S. W. Frey, W. G. DeWitt and B. R. Bellomy, *Am. Soc. Brew. Chem., Proc.,* 1966, 172.
37. J. C. Meranger and E. Somers, *J. Ass. Off. Analyt. Chem.,* 1968, **51,** 922.
38. J. C. Meranger, *Bull. Envir. Contam. Toxicol.,* 1970, **5,** 271.
39. J. R. Leaton, *J. Ass. Off. Analyt. Chem.,* 1970, **53,** 237.
40. N. G. Eider, *Appl. Spectrosc.,* 1971, **25,** 313.

13.4.14 Copper

Cu, atomic weight 63·54

Technique	Line, nm	f	Source	Flame	Sensitivity, ppm/1% abs.	Detection limit, ppm	Reference (see below)
AAS	324·75	0·74	HCL	A-coal gas	0·1	—	1
	327·40	0·38	HCL	A-coal gas	0·2	—	1
	222·57	0·004	HCL	A-coal gas	2	—	1
	217·89	0·01	HCL	A-coal gas	0·4	—	1
	324·75	0·74	HCL	A-A	0·1	0·002	2
	324·75	0·74	HCL	A-A	0·1	—	3
AFS	324·75	0·74	HIL	A-H	—	0·001	4
	324·75	0·74	HIL	Ar/H_2	—	0·003	5
	327·40	0·38	Hot HCL	A-H	—	0·001	6
	324·75	0·74	Hg arc	A-H	—	0·1	7

Ionization potential: 7·72 eV.
Dissociation energy, CuO: 4·9 eV.
Atomization efficiency: A-A, 0·98; N-A, 1·00.

Notes on analytical conditions

1. Copper is easily determined by AAS; the Cu 324·75 nm line is usually employed. The Cu 249·2 nm line may be useful for analysis of Cu as a major constituent (*ca.* 10 ppm/1% absorption).

2. The sensitivity attainable depends on the HCL operating current and decreases at higher currents.[1,3] Frank, Schrenk and Meloan succeeded in the use of an Fe hollow-cathode lamp for the AAS determination of copper.[8]

3. Thomerson[9] has described a differential spectrophotometric method for the determination of Cu by AAS; this has been shown to provide for good precision in the determination of large amounts of Cu in alloys.

4. With a modulated source and detector system few interferences are encountered in Cu AAS in the air-acetylene flame. Several workers have observed interference in this and other flames in the presence of mineral acids,[10,11] however, and the presence of very large amounts of elements such as Fe, Cr, Nb may depress the Cu absorption signal observed. Smith, Elser and Winefordner[5] examined 31 species for interference on the AF signal for Cu using an unmodulated system; only Al, Si and PO_4^{3-} were observed to interfere.

Applications notes and key references

As AAS offers a sensitive and selective method for Cu it is one of the elements most frequently determined by this technique. A very wide variety of matrices have been studied and Table 13-2 gives only a general guide to the applications literature; no attempt is made to give comprehensive coverage to the literature for Cu as its routine AAS determination is now commonplace.

TABLE 13-2

Some applications of the determination of copper by AAS and AFS

Sample type	Technique; sample pretreatment	Reference
Silver	AAS; dissolve in HNO_3, extract Ag with soln. of 66% tri-iso-octyl phosphorothioate in CCl_4, det. Cu in aqueous phase	12
Nickel and cobalt salts	AAS; pH1 extract into CCl_4 with dithizone	13
U compounds	AAS	14
W metal	AAS; dissolve in HNO_3/HF; use A-A flame	15
Niobium and tantalum	AAS; dissolve in HF/HNO_3; extract + 8-hydroxy-quinoline into ethyl acetate at pH 4·5; A-P or A-A flame	16
Aluminium alloys	AAS; dissolve in HCl, treat with H_2O_2; A-A flame	17
Aluminium alloys	AFS; dissolve in HCl, evaporate + HF, dissolve in HCl/HNO_3; A-A flame	18
Tin and lead based alloys	AAS, dissolve in 45% HF: 15N HNO_3:H_2O + 50% tartaric acid (2:3:4:8); A-A flame	19
Tin and lead solders	AAS; dissolve in HBF_4:HNO_3:H_2O (2:3:5 v/v); A-A flame	20
Pistol bullets	AAS; dissolve in 3N HNO_3; prepare standards to contain 1% lead to match samples	21
Ferrous alloys	AAS	22
Tool steel	AAS; dissolve 0·5 g samples in H_2SO_4/H_3PO_4/$HClO_4$/H_2O	23
Ferrosilicon	AAS	24
Steel	AAS; sample dissolved in 4·5% H_3PO_4/4·5% H_2SO_4	25
Nickel plating solutions	AAS; warm, dilute and mix solution before nebulization	26
Nickel and chromium electroplating solutions	AAS; A-A flame	27
Gold plating solutions	AAS; A-A flame, 1-30 ppm Cu without dilution	28
Iron ores	AAS; dissolve sample in HCl/HNO_3; use A-A flame	29
Silicates	AAS; absorption tube technique	30, 31, 32

TABLE 13-2 (*continued*)

Sample type	Technique; sample pretreatment	Reference
Sulphide minerals and ores	AAS	33
Serum	AAS	34
Serum and urine	AAS; serum nebulized directly after dilution or protein precipitated with trichloroacetic acid; urine subjected to extraction procedure	35
Serum and urine	AAS; dilute serum with 0·1 N HCl (1:19). For urine add 8·7 ml concentrated HCl/litre	36
Serum	AAS; dilute (1:10) + 1% HCl. Take 0·2 ml sample for nebulization; A-B flame	37
Urine	AAS; urine at pH 6 extracted + NaDDTC into MIBK	38
Urine	AAS; 2 ml sample + conc. H_2SO_4. Nebulize. 3-slot burner	39
Serum	AAS: 2 ml sample + 2 ml 20% trichloroacetic acid, heat to 90°C, cool, centrifuge, spray supernate	40
Hair and brain	AAS; sample + $KClO/HClO_4$. Heat. Dissolve digest in water. Determine Cu in hair directly; for Cu in brain extract soln. + APDC into MIBK	41
Hair	AAS; wet digestion + $HNO_3/HClO_4$	42
Finger nails	AAS; wet digestion + $HNO_3/HClO_4$	43
Brain and muscle	AAS; wet digestion with acids	44
Butter and butter oil	AAS; extract + HNO_3; 3383 Å line used to correct for scatter	45
Butter	AAS; extract + warm 2 M HNO_3; fats removed + light petroleum. A-B flame + water-cooled burner	46
Milk	AAS; graphite furnace of Massman type; 25 μl samples	47
Foods	AAS; dry or wet ashing; extraction + APDC into MIBK; A-A flame	48
Food, drinks	AAS; dry ash and dissolved in HCl, or wet ash + 50% H_2O_2/H_2SO_4	49
Meat and meat products	AAS; dry ash at 500°C + $Mg(NO_3)_2$; wet ash + HNO_3	50
Marine shrimp	AAS; dry ash, dissolve residue in HNO_3 (1:1), dilute, A-A flame	51
Fruit juices and carbonated beverages	AAS	52
Tea	AAS; wet or dry ash; A-A flame	53
Wines	AAS; dry ash	54
Wines	AAS; ethanol removed by evaporation; solution nebulized directly	55
Beer and wine	AAS; 25 ml sample extracted + APDC into MIBK. Cu standards prepared in decarbonated beer of low Cu content; A-P flame	56
Sugar beet leaves	AAS; digest + $HNO_3/HClO_4/H_2SO_4$/molybdate; dilute digest with water or methanol; A-A flame	57
Plant materials	AAS; extract + boiling 6N HCl	58

TABLE 13-2 (*continued*)

Sample type	Technique; sample pretreatment	Reference
Leaf	AAS	59
Soils	AAS; dissolve in $HNO_3/HCl/HClO_4$	60
Soil extracts	AAS; extraction with EDTA; mechanically separated A-A flame	61
Soil extracts	AAS	62
Soil extracts	AFS; extract + ammonium acetate, acetic acid or EDTA; N_2 separated A-A flame, pulsed HCl excitation	63
Plating wastes and river water	AAS; pH 10, add NaCl, extract + 2% 2-mercapto-benzothiazole into butyl acetate	64
Sea-water	AAS; extract + APDC into ethyl acetate	65
Sea-water	AAS; 500 ml sample extracted + APDC into MIBK from solution at pH 3	66
Sea-water	AFS; solvent ext. + Na DDTC; N_2 sep. A-A flame, pulsed HCL excitation	67
Water	AAS	68
Water	AFS; A-A flame, pulsed HCL excitation. Solvent extraction	69
Hydrocarbon fuels	AFS; direct determination in kerosine using $N_2/O_2/H_2$ or kerosine flame	70
Lubricating oils	AAS; sample diluted + twice its volume of MIBK; A-A flame	71
Lubricating oils	AAS; carbon filament atomization	72
Lubricating oils	AFS; oils nebulized directly (total consumption nebulizer-burner) $H_2/Ar/air$ flame or diluted + MIBK	73
Petroleum products	AAS; direct injection	74
Raw rubber	AAS; ash 10 g sample; dissolved residue in HCl; A-A flame	75
Atmospheric particulate matter	AAS; collect sample on organic membrane filter	76
Atmospheric particulate matter	AAS	77
Tobacco smoke	AAS; tobacco smoke condensate in MIBK, after equilibrating with aqueous Cu solution	78

References

1. J. E. Allen, *Spectrochim. Acta*, 1961, **17**, 459.
2. "Analytical Methods for Atomic Absorption Spectrophotometry", Perkin-Elmer Corp., U.S.A., 1971.
3. B. M. Gatehouse and J. B. Willis, *Spectrochim. Acta*, 1961, **17**, 710.
4. D. C. Manning and P. Heneage, *Atomic Absorption Newsletter*, 1967, **6**, 124.

5. R. Smith, R. C. Elser and J. D. Winefordner, *Anal. Chim. Acta,* 1969, **48,** 35.
6. J. I. Dinnin, *Anal. Chem.,* 1967, **39,** 1491.
7. N. Omenetto and G. Rossi, *Anal. Chim. Acta,* 1968, **40,** 195.
8. C. W. Frank, W. G. Schrenk and C. E. Meloan, *Anal. Chem.,* 1966, **38,** 1005.
9. D. R. Thomerson, *Spectrovision,* 1971, (25), 12.
10. T. V. Ramakrishna, J. W. Robinson and P. W. West, *Anal. Chim. Acta,* 1967, **37,** 20.
11. Toshihisa Maruta, Masami Suzuki and Tsugio Takeuchi, *Anal. Chim. Acta,* 1970, **51,** 381.
12. G. Mathieu and S. Guiot, *Anal. Chim. Acta,* 1970, **52,** 335.
13. G. H. Jamro and R. W. Frei, *Mikrochim. Acta,* 1970, 429.
14. C. R. Walker and O. A. Vita, *Anal. Chim. Acta,* 1968, **43,** 27.
15. J. W. Husler, *Atomic Absorption Newsletter,* 1969, **8,** 1.
16. G. F. Kirkbright, M. K. Peters and T. S. West, *Analyst,* 1966, **91,** 411.
17. J. L. Jimenez Seco and Aurora Gomez Coedo, *Revta Metal,* 1969, **5,** 290.
18. R. M. Dagnall, G. F. Kirkbright, T. S. West and R. Wood, *Analyst,* 1972, **97,** 245.
19. J. L. Jimenez Seco, Aurora Gomez Coedo and Maria T. Dorado, *Revta Metal,* 1969, **5,** 602.
20. J. Y. Hwang and Linda M. Sandonato, *Anal. Chem.,* 1970, **42,** 744.
21. R. L. Brunelle, C. M. Hoffman and K. B. Snow, *J. Ass. Off. Analyt. Chem.,* 1970, **53,** 470.
22. J. L. Jimenez Seco and Aurora Gomez Coedo, *Revta Metal,* 1970, **6,** 64.
23. D. M. Knight and M. K. Pyzyna, *Atomic Absorption Newsletter,* 1969, **8,** 129.
24. F. J. Langmyhr and P. E. Paus, *Anal. Chim. Acta,* 1969, **45,** 173.
25. K. Kinson and C. B. Belcher, *Anal. Chim. Acta,* 1964, **31,** 180.
26. R. G. Shafto, *Atomic Absorption Newsletter,* 1964, **3,** 115.
27. Janet E. Parker, *Spectrovision,* 1970, (24), 10.
28. J. P. Kapetan, *Spec. Tech. Publ. Amer. Soc. Test. Mater.,* No. 443, 1969.
29. Yoshihide Endo, Toshihiko Hata and Yuki Nakahara, *Japan Analyst,* 1968, **17,** 679.
30. Masato Nagura and Chuzo Iida, *Japan Analyst,* 1968, **17,** 1513.
31. C. Iida and Kazuo Yamasaki, *Anal. Letters,* 1970, **3,** 251.
32. M. Miksovsky and B. Moldan, *Chemicke Zvesti,* 1970, **24,** 128.
33. F. J. Langmyhr and P. E. Paus, *Anal. Chim. Acta,* 1970, **50,** 515.
34. R. Herrman and W. Lang, *Z. Klin. Chem.,* 1963, **1,** 182.
35. M. M. Parker, F. L. Humoller and D. J. Mahler, *Clin. Chem.,* 1967, **13,** 40.
36. J. B. Dawson, D. J. Ellis and H. Newton-John, *Clin. Chim. Acta,* 1968, **21,** 33.
37. M. L. Girard and F. Rousselet, *Ann. Pharm. Fr.,* 1967, **25,** 353.
38. K. M. Pirke and D. Stamm, *Z. Klin. Chem. Klin. Biochem.,* 1970, **8,** 449.
39. Harvey Spector, Sheldon Glusman, Peter Jatlow and David Seligman, *Clin. Chim. Acta,* 1971, **31,** 5.
40. A. D. Olson and W. B. Hamlin, *Atomic Absorption Newsletter,* 1968, 7, 69.
41. E. T. Backer, *Clin. Chim Acta,* 1969, **24,** 233.
42. W. W. Harrison, J. P. Yurachek and C. A. Benzon, *Clin. Chim. Acta,* 1969, **23,** 83.
43. W. W. Harrison and A. B. Tyree, *Clin. Chim. Acta,* 1971, **31,** 63.
44. B. H. Lehmann, J. D. L. Hansen and P. J. Warren, *Brit. J. Nutr.,* 1971, **26,** 197.
45. J. B. Willis, *Aust. J. Dairy Technol.,* 1964, **19,** 70.

46. M. L. Girard, F. Rousselet and J. Adda, *Chim. Analyt.*, 1965, **47**, 627.
47. D. C. Manning and F. Fernandez, *Atomic Absorption Newsletter,* 1970, **9**, 65.
48. A. G. Cameron and D. R. Hackett, *J. Sci. Fd. Agric.*, 1970, **21**, 535.
49. Analytical Methods Committee of Society for Analytical Chemistry, *Analyst,* 1971, **96**, 741.
50. E. F. Dalton and A. J. Malanoski, *J. Ass. Off. Analyt. Chem.*, 1969, **52**, 1035.
51. G. A. Knauer, *Analyst,* 1970, **95**, 476.
52. J. C. Meranger, *Bull. Envir. Contam. Toxicol.*, 1970, **5**, 271.
53. K. E. Burke and C. H. Albright, *J. Ass. Off. Analyt. Chem.*, 1970, **53**, 531.
54. P. B. Zeeman and L. R. P. Butler, *Appl. Spectrosc.*, 1962, **16**, 120.
55. J. C. Meranger and E. Somers, *J. Ass. Off. Analyt. Chem.*, 1968, **51**, 922.
56. J. P. Weiner and L. Taylor, *J. Inst. Brew.*, 1969, **75**, 195.
57. Merle Harrison and Charles Andre, *Appl. Spectrosc.*, 1969, **23**, 354.
58. P. R. Premi and A. H. Cornfield, *Spectrovision,* 1968, (19), 15.
59. E. G. Bradfield and D. Spincer, *J. Sci. Fd. Agric.*, 1965, **16**, 33.
60. J. E. Allen, *Spectrochim. Acta,* 1961, **17**, 459.
61. A. M. Ure and M. L. Barrow, *Anal. Chim. Acta,* 1970, **52**, 247.
62. S. F. du Plessis, *S. Afr. J. Agric. Sci.*, 1966, **9**, 1033.
63. R. M. Dagnall, G. F. Kirkbright, T. S. West and R. Wood, *Anal. Chem.*, 1971, **43**, 1765.
64. J. L. Robinson, Russell G. Barnekow and P. F. Lott, *Atomic Absorption Newsletter,* 1969, **8**, 60.
65. R. J. Magee and A. K. Matior Rahman, *Talanta,* 1965, **12**, 409.
66. R. R. Brooks, B. J. Presley and I. R. Kaplan, *Anal. Chim. Acta,* 1967, **38**, 321.
67. M. Jones, G. F. Kirkbright, L. Ranson and T. S. West, *Anal. Chim. Acta,* 1973, **63**, 210.
68. L. R. P. Butler and D. W. Brink, *S. Afr. Ind. Chem.*, 1963, **17**, 152.
69. D. Demers, M. Gardels and D. Mitchell, *Proc. Technicon Intern. Symp.*, New York, 1970.
70. D. H. Cotton and D. R. Jenkins, *Spectrochim. Acta,* 1970, **25B**, 283.
71. D. R. Jenkins, C. Salama and R. Dunn, *Can. Spectrosc.*, 1970, **15**, 17.
72. J. F. Alder and T. S. West, *Proc. 3rd SAC Conf.*, Durham, U.K., 1971.
73. R. Smith, C. M. Stafford and J. D. Winefordner, *Can. Spectrosc.*, 1969, **14**, 38.
74. R. Goux, *Methy. Phys. Analyse,* 1970, **6**, 118.
75. Michal Gulez and Danuta Raczynska, *Chimia Analit.*, 1970, **15**, 95.
76. J. Y. Hwang and F. J. Feldman, *Appl. Spectrosc.*, 1970, **24**, 371.
77. R. J. Thompson, G. B. Morgan and L. J. Purdue, *Atomic Absorption Newsletter,* 1970, **9**, 53.
78. L. W. Michael, E. E. Mendenam and H. G. Petering, *Environ. Sci. Technol.*, 1971, **5**, 249.

13.4.15 Dysprosium

Dy, atomic weight 162·50

Technique	Line, nm	*f*	Source	Flame	Sensitivity, ppm/1% abs.	Detection limit, ppm	Reference (see below)
AAS	421·17	–	Xe arc	O_2/A	–	0·5	1
	421·17	–	HCL	N-A	1·7	–	2
	421·17	–	HCL	N-A	0·9	–	3
	419·49	–	HCL	N-A	3·4	–	2
	419·49	–	HCL	N-A	1·5	–	3
	418·68	–	HCL	N-A	1·1	–	3
	416·80	–	HCL	N-A	6·2	–	3
	404·60	–	HCL	N-A	1·0	–	3
AFS: not reported							

Ionization potential: 6·2 eV.

Notes on analytical conditions

1. Amos and Willis[2] used a slightly fuel-rich flame for atomization of Dy and observed increase of sensitivity with decrease of source operating current. Manning[3] recorded absorption at several Dy ion lines in a nitrous oxide flame (Dy(II) 353·17 nm gave 1% absorption for 1·7 ppm Dy) and added excess potassium to sample solutions to achieve the sensitivities listed above for the strongest atom absorption lines. Jaworowski and co-workers[4] obtained a sensitivity of 4 ppm Dy/1% absorption at the Dy 407·79 nm line with 80% ethanolic sample solutions.

2. Enhancement of Dy absorption in the presence of H_2SO_4 and H_3PO_4 has been reported.[2]

Applications notes and key references

See Lanthanum.

References

1. V. G. Mossotti and V. A. Fassel, *Spectrochim. Acta,* 1964, **20,** 1117.
2. M. D. Amos and J. B. Willis, *Spectrochim. Acta,* 1966, **22,** 1325.
3. D. C. Manning, *Atomic Absorption Newsletter,* 1966, **5,** 127.
4. R. J. Jaworowski, R. P. Weberling and D. J. Bracco, *Anal. Chim. Acta,* 1967, **37,** 284.

13.4.16 Erbium

Er, atomic weight 167·26

Technique	Line, nm	*f*	Source	Flame	Sensitivity, ppm/1% abs.	Detection limit, ppm	Reference (see below)
AAS	400·80	—	Xe arc	O_2/A	—	1	1
	400·80	—	HCL	N-A	1·4	—	2
	400·80	—	HCL	N-A	1·5	—	3
	386·28	—	HCL	N-A	2·3	—	2
	386·28	—	HCL	N-A	2·3	—	3
	389·27	—	HCL	N-A	4·5	—	2
	389·27	—	HCL	N-A	4·0	—	3
	415·11	—	HCL	N-A	2·3	—	3
AFS: not reported							

Ionization potential: 6·08 eV.

Notes on analytical conditions

1. In addition to the above Manning[3] also recorded sensitivities for seven other less sensitive Er lines in the nitrous oxide-acetylene flame for sample solutions containing 100 ppm potassium as KCl; without the use of potassium to suppress ionization sensitivity values at six Er(II) ion lines were also obtained. The most sensitive of these was the Er(II) line at 337·28 nm for which 1% absorption was obtained with 1·2 ppm Er. Amos and Willis[2] used a slightly fuel-rich flame and a spectral band-pass of 0·17 nm for their study of Er absorption.

2. Little information is available concerning chemical interferences in the determination of Er by AAS.

Applications notes and key references

See Lanthanum.

References

1. V. G. Mossotti and V. A. Fassel, *Spectrochim. Acta,* 1964, **20,** 1117.
2. M. D. Amos and J. B. Willis, *Spectrochim. Acta,* 1966, **22,** 1325.
3. D. C. Manning, *Atomic Absorption Newsletter,* 1966, **5,** 127.

13.4.17 *Europium*

Eu, atomic weight 151·96

Technique	Line, nm	f	Source	Flame	Sensitivity, ppm/1% abs.	Detection limit, ppm	Reference (see below)
AAS	462·72	—	Xe arc	O_2/A	—	0·4	1
	462·72	—	HCL	N-A	3·0	—	2
	466·19	—	HCL	N-A	3·0	—	2
	459·40	—	HCL	N-A	1·8	—	2
	459·40	—	HCL	N-A	0·6	—	3
AFS: not reported							

Ionization potential: 5·67 eV.

Notes on analytical conditions

1. Less sensitive Eu lines at 311·14, 321·06 and 321·28 nm were also investigated by Manning[4,5] using a nitrous oxide-acetylene flame. Skogerboe and Woodriff[6] used an oxy-hydrogen flame supplied with Eu as a source for the determination of this element in ethanolic sample solutions nebulized into a fuel-rich oxy-acetylene flame; 1% absorption was obtained for solutions containing 2·3 ppm Eu.

2. At the 100 ppm level in sample solutions Eu has been calculated to be 75% ionized in the nitrous oxide-acetylene flame.[5] This accounts for the relatively good sensitivity obtained at the Eu 337·28, 326·48 and 349·91 nm ion lines (1·2, 2·3 and 3·6 ppm/1% absorption respectively) in the absence of an ionization suppressor such as sodium or potassium. Kinnunen and Lindsjo[7] observed depression of Eu absorption in the presence of Si, Al or fluoride in the nitrous oxide-acetylene flame.

Applications notes and key references

See Lanthanum.

References

1. V. G. Mossotti and V. A. Fassel, *Spectrochim. Acta,* 1964, **20,** 1117.
2. M. D. Amos and J. B. Willis, *Spectrochim. Acta,* 1966, **22,** 1325.
3. "Analytical Methods for Atomic Absorption Spectrophotometry", Perkin-Elmer Corp., Norwalk, Conn., U.S.A. 1971.
4. D. C. Manning, *Atomic Absorption Newsletter,* 1966, **5,** 127.
5. *Idem,* ibid., 1966, **5,** 63.
6. R. K. Skogerboe and R. A. Woodriff, *Anal. Chem.,* 1963, **35,** 1977.
7. J. Kinnunen and O. Lindsjo, *Chemist-Analyst,* 1967, **56,** 25.

13.4.18 Gadolinium

Gd, atomic weight 157·25

Technique	Line, nm	f	Source	Flame	Sensitivity, ppm/1% abs.	Detection limit, ppm	Reference (see below)
AAS	407·87	—	Xe arc	O_2/A	—	6	1
	407·87	—	HCL	N-A	42	—	2
	378·31	—	HCL	N-A	46	—	2
	368·41	—	HCL	N-A	38	—	2
	405·82	—	HCL	N-A	51	—	2
	407·87	—	HCL	N-A	16	4	3
AFS: not reported							

Ionization potential: 6·16 eV.
Dissociation energy, GdO: 5·9 eV.

Notes on analytical conditions

1. Amos and Willis[2] employed a slightly fuel-rich flame for AAS of Gd with a spectral slit-width of *ca.* 0·1 nm. Manning[4] obtained higher sensitivities than those of Amos and Willis shown above when sample solutions contained 1000 ppm of potassium as KCl to suppress Gd ionization. The Gd 378·31 nm line may suffer from interference from emission in the fuel-rich nitrous oxide-acetylene flame.

2. Little information is available concerning chemical interferences in the determination of Gd by AAS.

Applications notes and key references

See Lanthanum.

References

1. V. G. Mossotti and V. A. Fassel, *Spectrochim. Acta,* 1964, **20,** 1117.
2. M. D. Amos and J. B. Willis, *Spectrochim. Acta,* 1966, **22,** 1325.
3. "Analytical Methods for Atomic Absorption Spectrophotometry", Perkin-Elmer Corp., Norwalk, Conn., U.S.A., 1971.

13.4.19 Gallium

Ga, atomic weight 69·72

Technique	Line, nm	*f*	Source	Flame	Sensitivity, ppm/1% abs.	Detection limit, ppm	Reference (see below)
AAS	287·42	0·32	HCL	A-A	2·3	0·07	1
	287·42	0·32	HCL	A-A	3	—	2
	294·42/36	0·29	HCL	A-A	2·4	—	1 (see note 1)
	294·42/36	0·29	HCL	A-A	3	—	2
	417·12	0·14	HCL	A-A	3·7	—	1
	417·12	0·14	HCL	A-A	5	—	2
	403·30	0·13	HCL	A-A	6·2	—	1
	403·30	0·13	HCL	A-A	7	—	2
	250·0	—	HCL	A-A	22	—	1
	254·0	—	HCL	A-A	28	—	1
AFS	417·12	0·14	EDL	O_2/H_2	—	10	3
	417·12	0·14	EDL	Air/H_2	—	1	4
	403·30	0·13	HIL	A-H	—	0·5	5
	403·30	0·13	HIL	A-H	—	0·02	6

Ionization potential: 6·0 eV.
Dissociation energy, GaO: 2·5 eV.
Atomization efficiency: A-A, 0·16.

Notes on analytical conditions

1. Both Ga 294·36 and 294·42 nm lines are transmitted unless very narrow spectral band-pass is available; significant curvature of calibration graphs towards the concentration axis may be observed at this wavelength.

2. The HCL source used by Mulford[1] contained a niobium cathode containing metallic gallium; the design of this HCL permits operation at high current (when the Ga is molten) to yield high emission intensities without loss of sensitivity.

3. The air-acetylene flame has been used virtually exclusively for AAS of Ga. Most workers report few problems from interferences with this flame. Boltz and co-workers[7] tested interference from 43 ions on determination of Ga at 294·4 nm in dilute $HClO_4$; 31 ions were reported not to interfere and tolerance limits for the interfering species were presented. The air-hydrogen flame has been used most frequently for AFS of Ga; no significant amount of data on interferences in AFS is available for this flame, however.

4. Manning and Chabot[8] have drawn attention to the reduction in sensitivity for Ga which may be observed in AAS in the air-acetylene

flame when the acetylene cylinder pressure falls as it becomes depleted.

5. The determination of Ga by AAS or AFS using non-flame atomization has attracted considerable attention (see Chapter 7 and refs. 9, 10, 11).

6. Gallium may be extracted into MIBK–pentanol (4 : 1) as its 8-hydroxyquinoline complex for determination by AAS;[7] Mulford[1] extracted gallium from aqueous solution at pH 4 with ammonium pyrrolidinedithiocarbamate into MIBK.

Applications notes and key references

Very few applications of the determination of Ga by AAS have been published. However, using a continuum source (H_2-lamp) and photographic registration of the spectrum Tarasevich *et al.* have reported the determination of gallium in semi-conductor alloys.[12]

References

1. C. E. Mulford, *Atomic Absorption Newsletter,* 1966, **5**, 28.
2. B. M. Gatehouse and J. B. Willis, *Spectrochim. Acta,* 1961, **17**, 710.
3. J. M. Mansfield, J. D. Winefordner and C. Veillon, *Anal. Chem.,* 1965, **37,** 1049.
4. K. E. Zacha, M. P. Bratzel, J. D. Winefordner and J. M. Mansfield, *Anal. Chem.,* 1968, **40,** 1733.
5. D. C. Manning and P. Heneage, *Atomic Absorption Newsletter,* 1968, **7,** 80.
6. P. Benetti, N. Omenetto and G. Rossi, *Appl. Spectry.,* 1971, **25,** 57.
7. H. K. L. Gupta, F. J. Amore and D. F. Boltz, *Atomic Absorption Newsletter,* 1968, **7,** 107.
8. D. C. Manning and H. Chabot, *Atomic Absorption Newsletter,* 1968, **7,** 94.
9. M. P. Bratzel, R. M. Dagnall and J. D. Winefordner, *Anal. Chim. Acta,* 1969, **48,** 197.
10. R. G. Anderson, I. S. Maines and T. S. West, *Anal. Chim. Acta,* 1970, **51,** 355.
11. G. H. Morrison and Y. Talmi, *Anal. Chem.,* 1970, **42,** 809.
12. N. I. Tarasevich, G. V. Kozyreva, N. P. Ivanov and O. N. Mironova, *Vest. Mosk. Univ. Ser. Khim.,* 1968, (5), 78; *Anal. Abs.,* 1970, **18,** 757.

13.4.20 Germanium

Ge, atomic weight 72·59

Technique	Line, nm	f	Source	Flame	Sensitivity, ppm/1% abs.	Detection limit, ppm	Reference (see below)
AAS	265·16	0·84	HCL	N-A	2·2	1·0	1
	259·25	0·37	HCL	N-A	4·3	—	
	270·96	0·43	HCL	N-A	5·2	—	
AFS	265·16	0·84	EDL	Ar sep. N-A	—	0·1	2

Ionization potential: 7·38 eV.
Dissociation energy, GeO: 6·7 eV.

Notes on analytical conditions

1. A fuel-rich hot flame is required. Early work has been reported in AAS using oxy-acetylene[3,4] and air-acetylene flames.[3] Detection of atomic fluorescence from Ge was also reported in the oxygen-nitrogen-acetylene flame[5] (detection limit 15 ppm).

2. No specific solute vaporization interferences have been reported in AAS. No interferences were observed in AFS determination from 18 cations and 4 anions present at 1000 ppm in determining 10 ppm Ge.[2]

3. Indirect AAS determination via germanomolybdic acid formation has also been reported (see Chapter 14.2).

Applications notes and key references

There are few reported applications of the determination of Ge by AAS or AFS. Yanagisawa, Suzuki and Takeuchi[6] determined Ge in synthetic fibres at concentrations of 0·01 to 0·02%; after dry ashing the residue was dissolved in 7·5 M HCl and Ge extracted into MIBK for determination at 265·16 nm in the N-A flame. Fe, Sb, As, Sn, Au were reported not to interfere.

Use of chloride and HCl in sample preparation for Ge analysis should be avoided as $GeCl_4$ is easily lost by volatilization.

References

1. D. C. Manning, *Atomic Absorption Newsletter*, 1967, **6**, 35.
2. R. M. Dagnall, G. F. Kirkbright, T. S. West and R. Wood, *Analyst*, 1970, **95**, 425.
3. M. D. Amos and J. B. Willis, *Spectrochim. Acta*, 1966, **22**, 1325.
4. W. Slavin, *Appl. Spectry.*, 1966, **20**, 281.

5. R. M. Dagnall, K. C. Thompson and T. S. West, *Anal. Chim. Acta*, 1968, **41**, 551.
6. M. Yanagisawa, M. Suzuki and T. Takeuchi, *Anal. Chim. Acta*, 1969, **46**, 152.

13.4.21 Gold

Au, atomic weight 196·966

Technique	Line, nm	*f*	Source	Flame	Sensitivity, ppm/1% abs.	Detection limit, ppm	Reference (see below)
AAS	242·79	0·3	HCL	A-coal gas	0·3	—	1
	267·59	0·19	HCL	A-coal gas	1·3	—	1
	242·79	0·3	HCL	A-A	0·2	—	2
	267·59	0·19	HCL	A-A	0·4	—	2
AFS	267·59	0·19	EDL	O_2/H_2	—	0·2	3
	242·79	0·3	HIL	O_2-Ar-H_2	—	0·005	4
	242·79	0·3	HCL	N_2 sep. A-A	—	0·005	5

Ionization potential: 9·22 eV.
Atomization efficiency: A-A, 0·63; N-A, 0·71.

Notes on analytical conditions

1. The 242·79 nm line may be used with a spectral band-pass as great as 2 nm without much loss of sensitivity; the sensitivity is not significantly affected by the current employed when modern shielded sources are used.

2. A hot fuel-lean flame is usually preferable for gold determinations.

3. Strasheim and Wessels[6] studied interferences in AAS of Au and reported enhancements from Pt and Pd at high concentration in the detection of 5 ppm of Au; slight depression of absorbance was observed from HNO_3 and HCl and severe depression from H_2SO_4 was reported at high Au concentrations. Other workers[7] have found large amounts of lead and tellurium to be tolerable and no effect has been observed from Ni, Co, Pb or Fe at 5, 10, 100 and 100-fold excesses, respectively. The presence of cyanide gives rise to suppression of Au absorbance; $H_2SO_4/HClO_4$ digestion may be used to eliminate cyanide before nebulization. Matousek and Sychra[4] eliminated interferences in determination of Au by AFS in $O_2/H_2/Ar$ flame by addition of lanthanum as releasing agent.

4. Most solvent extraction procedures commonly employed for the separation and preconcentration of Au depend on extraction as

the anionic chloro or bromo complex from concentrated HCl or HBr into MIBK; extraction with APDC into MIBK may also be used.

Applications notes and key references

Most applications of the determination of gold by AAS have been in geochemistry and metallurgy. Butler *et al.*[8] determined gold in cyanide solutions by an automatic procedure involving addition of HCl, oxidation with $KMnO_4$ and extraction with MIBK. By this technique a 100-fold increase in sensitivity may be obtained in the determination of gold.[9] Skewes[10] used an air-acetylene flame and nebulized mill cyanide solutions directly for Au analysis; in this method EDTA was used at pH 11 to avoid precipitation of calcium and good agreement with the results obtained by the fire-assay method was claimed. Schuler[11] determined Au in barren cyanide solution by coprecipitation with tellurium, filtering the precipitate, ashing at 900°C, evaporating with *aqua regia* and dissolving the residue in HCl for nebulization. Gold in cyanide solutions has also been determined by Groenewald[12] by a method in which AuCN is extracted directly into di-isobutylketone containing 1% trialkyl-methyl ammonium chloride. Where the fire-assay method is employed the Au content may be determined by AAS. Van Loon[13] has reported a method in which the bead is dissolved in HNO_3, HCl and 1% lanthanum solution is added, and the solution is diluted with 6 N HCl before nebulization. In another recent method the bead is dissolved in a suitable acid, the solution is evaporated to dryness and the residue is dissolved in 2% KCN for nebulization; down to 0·08 ppm Au in the sample was determined by this technique.[14]

The determination of Au in ores and rocks has been studied by several workers. The sample is usually dissolved in *aqua regia* (HF may be added), the solution evaporated to dryness and the residue dissolved in water or HCl and the solution examined directly,[15,16] after extraction of Au into MIBK,[17,18,19] and with or without coprecipitation with tellurium.[19,20] Huffman and co-workers[21] have examined both the CN-atomic absorption and fire-assay-AAS methods for determination of Au in geological materials. Gold has also been determined in picogramme amounts in rocks, ores and metallurgical products using a carbon rod atomizer.[22] Pollock and Andersen[23] have described the determination of Au in copper after its extraction as the Ferroin-$AuBr_4$ complex into chloroform.

Gold may be determined in serum and urine after dilution with sodium dodecylsulphate or water, respectively.[24] Uchida *et al.*[25]

have determined Au in serum, urine and tissue after dry ashing, dissolution and extraction with MIBK for nebulization.

Other samples in which Au has been determined by AAS include plant liquors and electrolytes,[26] natural water,[27] tin-lead solders[28] and polymers.[29]

References

1. B. M. Gatehouse and J. B. Willis, *Spectrochim. Acta,* 1961, **17,** 710.
2. D. C. Manning and J. Vollmer, *Atomic Absorption Newsletter,* 1967, **6,** 38.
3. K. E. Zacha, M. P. Bratzel, J. D. Winefordner and J. M. Mansfield, *Anal. Chem.,* 1968, **40,** 1733.
4. J. Matousek and V. Sychra, *Anal. Chim. Acta,* 1970, **49,** 175.
5. P. L. Larkins, *Spectrochim. Acta,* 1971, **26B,** 447.
6. A. Strasheim and G. J. Wessels, *Appl. Spectrosc.,* 1963, **17,** 65.
7. T. Toyoguchi and H. Shimizu, *Bunseki Kaguka,* 1967, **16,** 565.
8. L. R. P. Butler, A. Strasheim, F. W. E. Strelow, P. Mathews and E. C. Feast, *Proc. XIIth Colloq. Spect. Intern.,* Exeter, 1965.
9. F. W. E. Strelow, E. C. Feast, P. M. Mathews, C. J. C. Bothma and C. R. Van Zyl, *Anal. Chem.,* 1966, **38,** 115.
10. H. R. Skewes, *Proc. Australas, Inst. Min. Metall.,* 1964, (211), 217.
11. V. C. O. Schuler, *J. S. Afr. Inst. Min. Metall.,* 1967, **67,** 362.
12. T. Groenewald, *Anal. Chem.,* 1968, **40,** 863.
13. J. C. Van Loon, *Z. Anal. Chem.,* 1969, **246,** 122.
14. Silve Kallemann and Everett W. Hobart, *Talanta,* 1970, **17,** 845.
15. G. G. Carlson and J. C. Van Loon, *Atomic Absorption Newsletter,* 1970, **9,** 90.
16. W. Schmidt, *Flame Notes,* 1969, **4,** 9.
17. S. L. Law and T. E. Green, *Tech. Prog. Rep. U.S. Bur. Mines,* 1968, No. 7.
18. N. Ichinose, *Japan Analyst,* 1971, **20,** 660.
19. M. A. Hildon and G. R. Sully, *Anal. Chim. Acta,* 1971, **54,** 245.
20. *Japan Analyst,* 1967, **16,** 565.
21. C. Huffman, J. D. Mensik and L. B. Riley, *Circ. U.S. Geol. Surv.,* 1967, No. 544.
22. M. P. Bratzel and C. L. Chakrabarti, *18th Spect. Symp. Canada,* Toronto, October 1971.
23. E. N. Pollock and S. I. Andersen, *Anal. Chim. Acta,* 1968, **41,** 441.
24. A. Lorber, R. L. Cohen, C. C. Chang and H. E. Anderson, *Arthritis Rheumat.,* 1968, **11,** 170.
25. T. Uchida, C. Iida, N. Yasahara and M. Nakagawa, *Analyt. Letters,* 1971, **4,** 555.
26. T. P. Mikhailova and V. A. Rezepina, *Analyst,* 1970, **95,** 769.
27. A. Zlatkis, W. Bruening and Ernst Bayer, *Anal. Chem.,* 1969, **41,** 1692.
28. J. Y. Hwang and Linda M. Sandonato, *Anal. Chem.,* 1970, **42,** 744.
29. M. Olivier, *Z. Anal. Chem.,* 1969, **248,** 145.

13.4.22 Hafnium

Hf, atomic weight 178·49

Technique	Line, nm	f	Source	Flame	Sensitivity, ppm/1% abs.	Detection limit, ppm	Reference (see below)
AAS	307·29	0·02	HCL	N-A	14	–	1
	307·29	0·02	HCL	O_2-N_2-A	25	–	1
	286·64	–	HCL	N-A	15	8	2
	286·64	–	HCL	O_2-N_2-A	30	–	1
	289·83	–	HCL	O_2-N_2-A	70	–	1
	296·49	–	HCL	O_2-N_2-A	80	–	1
	368·22	–	HCL	O_2-N_2-A	80	–	1
AFS: not reported							

Ionization potential: 6·8 eV.

Notes on analytical conditions

1. Other less sensitive lines examined for AAS of Hf by Amos and Willis[1] were 295·07, 302·05, 377·76, 290·48 and 294·08 nm. A fuel-rich flame is required.

2. The presence of HF in samples results in greatest sensitivity.[1,3] Enhancement of the Hf absorption is observed in the presence of Fe.

3. Although Hf atomic fluorescence was reported by Zacha *et al.*[4] it seems probable that the signal observed in an air-hydrogen flame for this element resulted from scatter of the incident radiation.[5]

Applications notes and key references

Although Hf has been included in several reports concerned with interference effects in the nitrous oxide-acetylene flame,[6,7,8] no significant volume of applications data has been published for this element. Bond and Canterford[9] have reported the determination of Hf in rocks; samples were first fused with LiF/H_3BO_3. The presence of 1000 ppm of Li enhances the Hf absorption.

References

1. M. D. Amos and J. B. Willis, *Spectrochim. Acta*, 1966, **22**, 1325.
2. "Analytical Methods for Atomic Absorption Spectrophotometry", Perkin-Elmer Corp., Norwalk, Conn., U.S.A., 1971.
3. A. M. Bond, *Anal. Chem.*, 1970, **42**, 932.
4. K. E. Zacha, M. P. Bratzel, J. D. Winefordner and J. M. Mansfield, *Anal. Chem.*, 1968, **40**, 1733.
5. J. D. Winefordner, *Anal. Chem.*, 1970, **42**, 206R.
6. V. S. Sastri, C. L. Chakrabarti and D. E. Willis, *Talanta*, 1970, **16**, 1093.

7. P. E. Thomas and W. F. Pickering, *Talanta,* 1971, **18,** 127.
8. H. A. Von Dershau and H. Prugger, *Z. Anal. Chem.*, 1969, **247,** 8.
9. A. M. Bond and Caterford, *Anal. Chem.*, 1971, **43,** 134.

13.4.23 Holmium

Ho, atomic weight 164·93

Technique	Line, nm	*f*	Source	Flame	Sensitivity, ppm/1% abs.	Detection limit, ppm	Reference (see below)
AAS	416·30	—	Xe arc	O_2/A	—	2	1
	416·30	—	HCL	N-A	3·5	—	2
	416·30	—	HCL	N-A	2·4	—	3
	410·38	—	HCL	N-A	2·2	—	2
	410·38	—	HCL	N-A	1·4	—	3
	405·39	—	HCL	N-A	2·8	—	2
	405·39	—	HCL	N-A	1·9	—	3
AFS: not reported							

Notes on analytical conditions

1. The values reported by Manning[3] were obtained in the presence of excess potassium to suppress ionization; other less sensitive Ho lines examined were 404·08, 410·11, 410·86, 412·72 and 417·32 nm. In the absence of potassium to suppress ionization the Ho(II) ion line at 345·60 nm gave a sensitivity of 16 ppm Ho/1% absorption.

2. Little information is available concerning chemical interferences in the determination of Ho by AAS.

Applications notes and key references

See Lanthanum.

References

1. V. G. Mossotti and V. A. Fassel, *Spectrochim. Acta,* 1964, **20,** 1117.
2. M. D. Amos and J. B. Willis, *Spectrochim. Acta,* 1966, **22,** 1325.
3. D. C. Manning, *Atomic Absorption Newsletter,* 1966, **5,** 127.

13.4.24 *Indium*

In, atomic weight 114·82

Technique	Line, nm	f	Source	Flame	Sensitivity, ppm/1% abs.	Detection limit, ppm	Reference (see below)
AAS	303·97	0·36	HCL	A-A	0·9	0·05	1
	325·61	0·37	HCL	A-A	0·9	—	1
	410·5	0·14	HCL	A-A	2·6	—	1
	451·13	—	HCL	A-A	2·8	—	1
	256·02	—	HCL	A-A	11	—	1
	275·39	—	HCL	A-A	26	—	1
AFS	451·13	—	EDL	Ar/H_2	—	0·1	2
	410·5	0·14	HIL	A-H	—	0·2	3
				A-H	—	0·015	4

Ionization potential: 5·78 eV.
Dissociation energy, InO: 1·1 eV.
Atomization efficiency: A-A, 0·67.

Notes on analytical conditions

1. A narrow spectral band-pass is required when the In 325·61 nm line is used in order to avoid transmission of the weakly absorbing In 325·86 nm line.

2. Hollow-cathode lamps using indium metal contained in a titanium cup cathode have been recommended for AAS.

3. A relatively lean (oxidizing) air-acetylene flame may be employed.

4. Few interferences have been encountered in the determination of In by AAS in the air-acetylene flame; large excess of Mg and Zn, however, may cause suppression of the absorption signal.[1]

5. Indium may be extracted with ammonium pyrrolidine dithiocarbamate into MIBK between pH 2 and 10;[5] extraction with 8-hydroxyquinoline[6] or cupferron,[7] or from hydrobromic acid solution (see Chapter 11.7.1), may also be useful.

6. As noted for Ga, Manning and Chabot[8] demonstrated that loss of sensitivity for In may be observed in the presence of acetone vapour in the flame owing to its use for dissolution of acetylene in gas cylinders.

Applications notes and key references

Few applications of the determination of In by AAS have been reported. The determination of In in Cu-In-Te alloys using an EDL source has been described,[9] and its determination in semi-conductor alloys using a continuum source and photographic registration of absorption has also been reported.[10] Spitzer and Tesik[11,12] determined In in pyrites, calcined pyrites and works extracts from chloridizing roasts (of Cu ore). These workers described the preconcentration of indium by co-precipitation as the hydroxide with iron(III) and extraction as the iodide complex into cyclohexanone before nebulization into an air-acetylene flame.

References

1. C. E. Mulford, *Atomic Absorption Newsletter,* 1966, **5,** 28.
2. K. E. Zacha, M. P. Bratzel, J. D. Winefordner and J. M. Mansfield, *Anal. Chem.,* 1968, **40,** 1733.
3. D. C. Manning and P. Heneage, *Atomic Absorption Newsletter,* 1968, **7,** 80.
4. P. Benetti, N. Omenetto and G. Rossi, *Appl. Spectrosc.,* 1971, **25,** 57.
5. C. E. Mulford, *Atomic Absorption Newsletter,* 1966, **5,** 88.
6. J. Stary, *Anal. Chim. Acta,* 1963, **28,** 132.
7. J. Stary and J. Smizanska, *Anal. Chim. Acta,* 1963, **29,** 546.
8. D. C. Manning and H. Chabot, *Atomic Absorption Newsletter,* 1968, **7,** 94.
9. N. P. Ivanov and G. V. Kozyreva, *Vest. Mosk. Univ.,* Ser. II, 1967, **22,** 102.
10. N. I. Tarasevich, G. V. Kozyreva, N. P. Ivanov and O. N. Mironova, *Vest. Mosk. Univ. ser. Khim.,* 1968, (5), 78; *Anal. Abs.,* 1970, **18,** 757.
11. H. Spitzer and G. Tesik, *Z. Anal. Chem.,* 1968, **241,** 218.
12. H. Spitzer and G. Tesik, *Erzmetall,* 1969, **22,** 383.

13.4.25 Iodine

I, atomic weight 126·90

Technique	Line, nm	*f*	Source	Flame	Sensitivity, ppm/1% abs.	Detection limit, ppm	Reference (see below)
AAS	183·0	0·016	EDL	N_2 sep. N-A	12	25	1
	183·0	0·016	EDL	L'Vov furnace	$3{\cdot}3 \times 10^{-11}$ g	–	2
	206·16	–	EDL	L'Vov furnace	–	2×10^{-9} g	2
	206·16	–	EDL	A-A	1300	600	3
Indirect AAS: see Chapter 14							
AFS: not reported							

Ionization potential: 10·45 eV.

Notes on analytical conditions

1. In order to utilize the I 183·0 nm line for direct AAS measurements the light path between the EDL source and nitrogen-separated nitrous oxide-acetylene flame, and between the flame and

monochromator, must be purged with nitrogen or argon. A nitrogen-purged monochromator should also be employed.

2. The flame transmission (fuel-rich) with a 50 mm slot burner is *ca.* 75% of that of nitrogen alone at the 183·0 nm line.

3. Similar sensitivities are obtained for iodine in the nitrous oxide-acetylene flame when sample solutions containing iodide, iodate or periodate are nebulized. The high atomization efficiency obtained in this flame also results in freedom from interference in the presence of high concentration of Cu, Co, Cr, K, Na, Ni, Mg, Mo, Zn, Al, Hg, fluoride, chloride, bromide, nitrate, phosphate and sulphate.[1]

4. Enhanced sensitivity is obtained (a) when iodide is introduced into the flame in MIBK or ethanol (*ca.* 5 ppm/1% absorption), or (b) by the use of the Leipert amplification reaction in which iodide is oxidized to iodate and then treated with excess iodide in acid medium. The reaction $IO_3^- + 5I^- + 6H^+ \rightarrow 3I_2 + 3H_2O$ then provides 6-fold amplification; the liberated iodine must then be separated from the excess iodide by extraction into MIBK and nebulized directly into the flame. An effective overall sensitivity of 0·32 ppm iodide (for 1% absorption) may be obtained in this manner.[4]

5. The determination of iodine by AAS at 183·0 nm using the L'Vov furnace requires the use of an argon-purged system. In the particular studies of L'Vov,[2] a vacuum monochromator was utilized, although a nitrogen or argon-filled instrument would also prove satisfactory.

6. Several indirect AAS methods have been proposed for the determination of iodine. These are outlined in Chapter 14.

Applications notes and key references

Although the direct determination of iodine by AAS has now been shown to be feasible using both flame and non-flame cells, no applications of these methods have yet been reported.

References

1. G. F. Kirkbright, T. S. West and P. J. Wilson, *Atomic Absorption Newsletter*, 1972, **11**, 53.
2. B. V. L'Vov, in "Atomic Absorption Spectroscopy", eds. R. M. Dagnall and G. F. Kirkbright, Butterworths, London, 1970, p. 28.
3. K. C. Thompson, *Spectrosc. Letters*, 1970, 3, 59.
4. G. F. Kirkbright, T. S. West and P. J. Wilson, *Atomic Absorption Newsletter*, 1972, **11**, 113.

13.4.26 Iridium

Ir, atomic weight 192·22

Technique	Line, nm	*f*	Source	Flame	Sensitivity, ppm/1% abs.	Detection limit, ppm	Reference (see below)
AAS	208·88	–	HCL	A-A	7·7	–	1
	263·94/97		HCL	A-A	13	2	1, 2
	266·48	–	HCL	A-A	15	–	1
	284·97	–	HCL	A-A	18	–	1
	237·28	–	HCL	A-A	20	–	1
	254·40	–	HCL	A-A	34	–	1
AFS	254·40	–	150 W Xe arc	A-H	–	100	3
	ND 254·40	–	HIL	N_2 sep. A-A	–	4	4

Ionization potential: 9·3 eV.

Notes on analytical conditions

1. Mulford[5] briefly reported Ir atomic absorption in the air-acetylene flame and obtained a sensitivity of 34 ppm Ir/1% absorption at 284·97 nm with a detection limit of 4 ppm. Manning and Fernandez[1] conducted a more extensive study of Ir absorption; some of the most sensitive absorption lines investigated by these workers are listed above. The effect of nearby Ir non-absorbing lines and emission from the Cu cathode shell necessitated the use of narrow spectral band-pass. Although 1% absorption sensitivity is best at 208·88 nm, the 263·94/97 lines gave best signal-to-noise ratio and were recommended for routine analysis.

2. A number of workers have reported interferences from other platinum and base metals on the determination of Ir in the air-acetylene flame.[6-9] Fuhrman[6] used La addition to eliminate interference from Na and K, while Van Loon[7] and Grimaldi and Schnepfe[8] reported the use of Cu + Na for suppression of interferences from a range of metal ions. Luecke and Zielke[10] employed the nitrous oxide-acetylene flame for Ir AAS using a preliminary sample chlorination technique.

Applications notes and key references

In the method described by Van Loon[7] for AAS determination of Ir in precious metal concentrates, SiO_2 is removed by heating with HF and the sample is then sintered with Na_2O_2 for 30 minutes and fused. The melt is extracted with dilute HCl, concentrated HCl is added, the solution is evaporated to small volume and HNO_3 is added. The solution is then treated again with HCl to destroy nitrates, copper solution is added to give a Cu concentration greater than 500 ppm on dilution to 100 ml, and the sample is then nebulized into an air-acetylene flame for AAS at 284·97 nm. Grimaldi and Schnepfe[8] used AAS to determine Ir in mafic rocks after its preconcentration by the fire assay technique with a gold bead; the method is applicable to the determination of Ir in rock samples containing down to 2·5 ppm Ir.

References

1. D. C. Manning and F. Fernandez, *Atomic Absorption Newsletter,* 1967, **6,** 15.
2. "Analytical Methods for Atomic Absorption Spectroscopy", Perkin-Elmer Corp., Norwalk, Conn., U.S.A., 1971.
3. D. C. Manning and P. Heneage, *Atomic Absorption Newsletter,* 1968, **7,** 80.
4. P. L. Larkins, *Spectrochim. Acta,* 1971, **26B,** 477.
5. C. E. Mulford, *Atomic Absorption Newsletter,* 1966, **5,** 63.
6. D. L. Fuhrman, *Atomic Absorption Newsletter,* 1969, **8,** 105.
7. J. C. Van Loon, *Atomic Absorption Newsletter,* 1969, **8,** 6.
8. F. S. Grimaldi and M. M. Schnepfe, *Talanta,* 1970, **17,** 617.
9. A. Janssen and F. Umland, *Z. Anal. Chem.,* 1970, **251,** 101.
10. W. Luecke and H. J. Zielke, *Z. Anal. Chem.,* 1971, **253,** 20.

13.4.27 Iron

Fe, atomic weight 55·847

Technique	Line, nm	f	Source	Flame	Sensitivity, ppm/1% abs.	Detection limit, ppm	Reference (see below)
AAS	248·33	0·34	HCL	A-A	0·1	0·01	1, 2
	248·81	–	HCL	A-A	0·2	–	1
	252·28	0·30	HCL	A-A	0·2	–	1
	271·90	0·15	HCL	A-A	0·3	–	1
	302·05/6	0·08	HCL	A-A	0·4	–	1
	296·69	0·06	HCL	A-A	1·1	–	1
	371·99	0·04	HCL	A-A	1·0	–	1
	385·99	0·035	HCL	A-A	1·9	–	1
	344·06	0·055	HCL	A-A	2·6	–	1
	382·44	0·007	HCL	A-A	8·7	–	3
	367·99	0·005	HCL	A-A	8·9	–	3
AFS	248·33	0·34	HIL	$Ar-O_2-H_2$	–	0·02	4
	248·33	0·34	EDL	Air/H_2	–	0·25	5
	248·33	0·34	HCL	A-H	–	0·4	6
	248·33	0·34	150 W Xe arc	A-H	–	1·0	6
	248·33	0·34	HCL	A-H	–	2	7
	248·33	0·34	HgVDL	A-H	–	1	8
	248·33	0·34	EDL	A-A	–	0·009	9
ND	248·33	0·34	HIL	N_2 sep. A-A	–	0·003	10
ND	248·33	0·34	HIL pulsed HCL	N_2 sep. A-A	–	0·1	11

Ionization potential: 7·87 eV.
Dissociation energy, FeO: 4·0 eV.
Atomization efficiency: A-A, 0·66.

Notes on analytical conditions

1. Several other less frequently used absorption lines of Fe are available which provide equal or lower sensitivity than those listed above. Narrow spectral band-pass is required when the most sensitive Fe 248·33 nm line is used in order to prevent transmission of nearby weakly absorbing lines from the source.

2. A fuel-lean stoicheiometric air-acetylene flame is preferred for AAS to minimize interference effects. When the correct flame stoicheiometry is chosen only silicon interferes significantly in the air-acetylene flame in AAS,[1,12] and this may be suppressed by the addition of $CaCl_2$ to sample solutions.[1] Other workers have confirmed this freedom from interferences in the air-acetylene flame.[13,14] Various ions have been found to interfere with Fe

absorption when cooler air-coal-gas or air-propane flames are employed.[15,16]

3. In the determination of Fe by AFS in the air-acetylene flame Ebdon, Kirkbright and West[9] found no interference in the presence of 1000 ppm of Co, Cr(VI), Mo, Cd, Al, Zn, Hg, Pb, Ta or Ti. Significant interferences were observed from Cu, V, Mn, Th, Cd and phosphate at this concentration.

4. The use of T-piece adaptors and absorption tube devices has been recommended for AAS of Fe by a number of workers in order to increase sensitivity.[13,17,18,19] In their work with organic liquid fuel flames, Bailey and Rankin[20] reported a 1% absorption sensitivity of 0·16 ppm Fe at 248·33 nm in an air-hexane flame.

5. Iron may be preconcentrated by solvent extraction, ion-exchange or co-precipitation prior to its determination by AAS (see Chapter 11.7).

Applications notes and key references

AAS and AFS provide sensitive and relatively selective methods for the determination of Fe in a wide variety of sample types. Table 13-3 shows some of the published applications of both techniques to Fe determination.

TABLE 13-3

Applications of AAS and AFS to the determination of iron

Sample type	Technique; sample treatment	Reference
Aluminium alloys	AAS	21
Zinc alloys	AAS; dissolve in HCl, oxidize + HNO_3	22
Aluminium alloys	AFS: dissolve in HCl, evaporate, add HF, evaporate to dryness; dissolve residue in HCl/HNO_3 (3:1), dilute, nebulize, A-A flame	23
Silver	AAS; dissolve in HNO_3, extract Ag + 66% soln. of tris-iso-octyl phosphorothioate in CCl_4	24
Silver	AAS; dissolve, remove Ag by precipitation as AgCl	25
Gold	AAS	26
Copper-base alloys	AAS	27
Cobalt	AAS; preconcentrate by solvent extraction or ion-exchange	28
Cobalt-base alloys	AAS	29
Zirconium alloys	AAS; dissolve in HF/HNO_3, add H_3BO_3, dilute, nebulize into A-A flame	30
Uranium, nickel	AAS	31
Uranium	AAS; extract U with TBP in carbon tetrachloride; nebulize residual aqueous phase	32

TABLE 13-3 (*continued*)

Sample type	Technique; sample treatment	Reference
Uranium alloys	AAS; direct determination	33
Tin-lead solders	AAS; 1% solution of sample after acid dissolution	34
High-temperature alloys	AAS; measurement at 371·9 nm	35
Tungsten carbide	AAS; 1 g sample + 4 ml H_3PO_4 + 3 ml HCl, heat, add 4 ml HNO_3, dilute to 50 ml. A-A flame, prepare standards to contain sodium tungstate	36
Ni-plating solutions	AAS	37
Coal ash	AAS; dissolve in $HClO_4$/HF, evaporate, dilute	38
Coal ash and silicate rocks	AAS; fuse + $LiBO_2$	39
Siliceous materials	AAS; dissolve with H_2SO_4/HF or fuse with Na_2CO_3	40
Silicate minerals	AAS; dissolve in HNO_3/HF	41
Silicate rocks bauxite, limestone, cement, clinkers	AAS; sample decomposed + HF; A-A flame	42
Silicates	AAS; decomposition + HF in sealed vessel. A-A flame	43
Silicates and carbonates	AAS; fuse + $LiBO_2$; dissolve melt in HCl. Use La releasing agent; N-A flame	44
Silicates	AAS; $LiBO_2$ fusion	45
Chrome refractories	AAS; add H_3PO_4, heat to 310°C, dilute; N-A flame	46
Cement	AAS; comprehensive scheme	47, 48
Marine sediments	AAS; determination of Fe in asphaltic fraction after extraction with $CHCl_3$; 3:2 methanol: $CHCl_3$ solvent nebulized	49
Serum	AAS; extract Fe-bathophenanthroline complex into MIBK	50
Serum	AAS; dilute + $LaCl_3$. Use $LaCl_3$ in standards	51
Serum, liver, milk	AAS; deproteinize serum and milk + TCA; acid digestion for liver	52
Plasma	AAS; dilute 1:1 with water; add egg albumen to standards	53
Blood	AAS; dilute whole blood or toluene hemolyzate + water, or wet ash	54
Blood	AAS	55
Plasma	AAS; carbon rod atomizer, 0·5 μl samples	56, 57
Urine	AAS; direct aspiration, wide-slot burner	58
Urine	AAS; wet ash + H_2SO_4, extract with TTA into MIBK	59
Brain, hair	AAS; wet digest with acid	60, 61
Amoebae, tissue	AAS; digest + chromic acid, dilute with water; A-A flame	62
Plant materials	AAS; digest + $HClO_4$/H_2SO_4/HNO_3. Filter, dilute + water. A-A flame. Standards contain 1% H_2SO_4	63

TABLE 13-3 (*continued*)

Sample type	Technique; sample treatment	Reference
Plant materials	AAS; wet ash + $HClO_4/HNO_3$. Standards contain 10% $HClO_4$	1
Feeds	AAS; dry ash or digest with HNO_3	64
Soil extracts	AAS; oxalate extraction	65
Soil	AAS	66
Soils	AAS; extract + 1% Na_2EDTA	1
Fertilizers	AAS; dissolve in HCl, evaporate to dryness. Repeat HCl evaporation, filter, dilute	67
Milk	AAS; dry ash. Use method of standard additions	68
Milk	AAS; boil with HCl, evaporate, dissolve in HCl, extract + APDC into MIBK; A-A flame	69
Soft drinks	AAS; direct determination; A-A flame	70
Fruit juices, drinks	AAS; treat + HCl, centrifuge, aspirate supernatant fluid	71
Beer	AAS; remove CO_2 by filtration or agitation	72
Wine and beer	AAS; direct determination after decarbonation, or extract with APDC into MIBK	73
High-purity water	AFS; direct nebulization. Air + H_2 flame, non-dispersive system	74
Industrial waters	AAS; add $CaCl_2$ to eliminate Si interference	12
Natural waters	AAS; extract with DDTC into MIBK	75
Brines	AAS; preconcentrate with chelating ion-exchange resin; elute with NH_4Cl at pH 4·5; A-A flame	76
Sea-water	AAS; extract with DDTC into MIBK	77
KCL brines	AAS; extract with APDC into MIBK	78
Vegetable and animal oils	AAS; dilute 1:3 with n-heptane for nebulization	70
Oil feed stocks	AAS; dilute sample with n-heptane	79
Lubricating oils	AAS; dilute oil 1:4 with *p*-xylene; A-A flame	80
Oils	AAS; Co octoate internal standard; N-A flame	81
Jet-engine oils	AAS; dilute samples 1:10 with MIBK	82
Jet-engine oils	AAS; nebulize directly or dilute with MIBK; Air/H_2 or Ar/H_2 flame	83
Engine oils	AAS; dilute samples with MIBK	84
Jet-engine oils	AFS; dilute 1:4 with CCl_4; Ar/H_2 flame	85
Fuel oils	AAS; direct determination or acid extraction	86
Hydrocarbon fuels	AFS; kerosine organic fuel flame; 450 W Xe arc source	87
Diesel oils	AAS	88
Red fuming HNO_3	AAS	89
Atmospheric particulate matter	AAS; collect on membrane filter, ash, dissolve in HNO_3/HCl	90
Sea and river water	AAS; extract with 8-hydroxyquinoline from pH3–6 into MIBK; A-A flame.	91

References

1. J. E. Allan, *Spectrochim. Acta,* 1969, **15,** 800.
2. "Analytical Methods for Atomic Absorption Spectrophotometry", Perkin-Elmer Corp., Norwalk, Conn., U.S.A., 1971.
3. W. Slavin, *Atomic Absorption Spectroscopy,* Interscience, New York, 1968, p. 116.
4. J. Matousek and V. Sychra, *Anal. Chem.,* 1969, **41,** 518.
5. K. E. Zacha, M. P. Bratzel Jnr., J. D. Winefordner and J. M. Mansfield Jnr., *Anal. Chem.,* 1968, **40,** 1733.
6. D. C. Manning and P. Heneage, *Atomic Absorption Newsletter,* 1968, **7,** 80.
7. G. Rossi and N. Omenetto, *Talanta,* 1969, **16,** 263.
8. G. Rossi and N. Omenetto, *Anal. Chim. Acta,* 1968, **40,** 195.
9. L. Ebdon, G. F. Kirkbright and T. S. West, *Anal. Chim. Acta,* 1969, **47,** 563.
10. P. L. Larkins, *Spectrochim. Acta,* 1971, **26B,** 477.
11. D. G. Mitchell and A. Johannson, *Spectrochim. Acta,* 1970, **25B,** 175.
12. J. A. Platte and W. M. Marcy, *Atomic Absorption Newsletter,* 1965, **4,** 289.
13. S. L. Sachdev, J. W. Robinson and P. W. West, *Anal. Chim. Acta,* 1967, **38,** 491.
14. A. Zettner and L. Mansbach, *Am. J. Clin. Pathol.,* 1965, **44,** 517.
15. J. Vlastnik, *Hutn. Listy,* 1967, **22,** 201.
16. W. T. Elwell and J. A. F. Gidley, "Atomic Absorption Spectrophotometry", 1st edition, Pergamon Press, Oxford, 1961.
17. J. P. Mislan and S. J. Mark, At. Energy of Canada Ltd., Chalk River, AEICL-2228, 1965.
18. J. Stupar, *Mikrochim. Acta,* 1966, 722.
19. C. Iida and M. Nagura, *Spectrosc., Letters,* 1970, **3,** 63.
20. B. W. Bailey and J. M. Rankin, *Spectrosc. Letters,* 1969, **2,** 233.
21. G. F. Bell, *Atomic Absorption Newsletter,* 1966, **5,** 73.
22. S. B. Smith, J. A. Blasi and F. J. Feldman, *Anal. Chem.,* 1968, **40,** 1525.
23. R. M. Dagnall, G. F. Kirkbright, T. S. West and R. Wood, *Analyst,* 1972, **97,** 245.
24. G. Mathieu and S. Ginot, *Anal. Chim. Acta,* 1970, **52,** 335.
25. J. M. Viebock, *Anal. Letters,* 1970, **3,** 373.
26. V. C. O. Schuler, A. V. Jansen and G. S. James, *J. S. Afr. Inst. Min. Metall.,* 1962, **62,** 807.
27. T. W. Sattur, *Atomic Absorption Newsletter,* 1966, **5,** 37.
28. C. Mathien, J. P. Tardif and T. V. Danh, *Anal. Chim. Acta,* 1971, **53,** 109.
29. G. G. Welcher and O. H. Kriege, *Atomic Absorption Newsletter,* 1970, **9,** 61.
30. J. H. Schlewitz and M. G. Shields, *Atomic Absorption Newsletter,* 1971, **10,** 39.
31. R. M. Walker, O. A. Vita and R. W. Sparks *Anal. Chim. Acta,* 1969, **47,** 1.
32. M. L. Jursik, *Atomic Absorption Newsletter,* 1967, **6,** 21.
33. G. Baudin, J. Normand and J. Fijalkowski, *Spectrochim. Acta,* 1968, **23,** 587.
34. J. Y. Hwang and L. M. Sandonato, *Anal. Chem.,* 1970, **42,** 744.
35. A. F. Cunningham, *Atomic Absorption Newsletter,* 1969, **8,** 70.
36. C. B. Belcher, *Anal. Chim. Acta,* 1963, **29,** 340.
37. R. G. Shafto, *Atomic Absorption Newsletter,* 1964, **3,** 115.
38. E. L. Obermiller and R. W. Freedman, *Fuel,* 1965, **44,** 199.

39. P. L. Boar, *Analyst,* 1970, **95**, 124.
40. D. J. Trent and W. Slavin, *Atomic Absorption Newsletter,* 1964, **3**, 17; 1964, **3**, 118.
41. G. K. Billings and J. A. S. Adams, *Atomic Absorption Newsletter,* 1964, **3**, 65.
42. F. J. Langmyhr and P. E. Paus, *Anal. Chim. Acta,* 1968, **43**, 508; 1969, **44**, 445; 1969, **45**, 176; 1968, **43**, 397.
43. B. Bernas, *Anal. Chem.,* 1968, **40**, 1682.
44. J. W. Yule and G. A. Swanson, *Atomic Absorption Newsletter,* 1969, **8**, 30.
45. J. C. Van Loon and C. M. Parissis, *Analyst,* 1969, **94**, 1057.
46. R. P. Lucas and B. C. Ruprecht, *Anal. Chem.,* 1971, **43**, 1013.
47. J. T. H. Roos and W. J. Price, *Analyst,* 1969, **94**, 89.
48. L. Capacho-Delgado and D. C. Manning, *Analyst,* 1967, **92**, 553.
49. D. C. Burrell, *Atomic Absorption Newsletter,* 1965, **4**, 328.
50. A. Zettner, L. C. Sylvia and L. Capacho-Delgado, *Am. J. Clin. Pathol.,* 1966, **45**, 533.
51. W. Schmidt, *Z. Anal. Chem.,* 1968, **243**, 198.
52. D. G. Spruile, V. W. Hays and G. L. Cromwell, *J. Animal Sci.,* 1971, **33**, 376.
53. E. C. Zaino, *Atomic Absorption Newsletter,* 1967, **6**, 43.
54. A. Zettner and A. H. Mensch, *Am. J. Clin. Pathol.,* 1967, **48**, 225.
55. O. W. Van Assendelft, W. G. Zijlstra, A. Buursma, E. J. Van Kampen and W. Hoeck, *Clin. Chim. Acta,* 1968, **22**, 281.
56. D. P. Sandoz, J. P. Matousek and B. J. Stevens, *Proc. 10th Nat. SAS Meeting,* St. Louis, Mo., U.S.A., October 1971.
57. J. P. Matousek and B. J. Stevens, *Clin. Chem.,* 1971, **17**, 363.
58. A. Zettner and L. M. Mansbach, *Am. J. Clin. Pathol.,* 1965, **44**, 517.
59. G. Devoto, *Russ. Med. Sarda,* 1968, **71**, 357.
60. W. W. Harrison, G. Martin and M. D. Brown, *Clin. Chim. Acta,* 1968, **21**, 55.
61. W. W. Harrison, J. P. Yurachek and C. A. Benson, *Clin. Chim. Acta,* 1969, **23**, 83.
62. S. Christensen, *Atomic Absorption Newsletter,* 1970, **9**, 126.
63. D. J. David, "Modern Methods of Plant Analysis", Vol. 5, eds. Paech and Tracy, Springer Verlag, Berlin, 1962; *Atomic Absorption Newsletter,* 1962, **1**, 45.
64. M. Heckman, *J. Ass. Off. Analyt. Chem.,* 1968, **51**, 776.
65. L. C. Blakemore, *N. Z. J. Agric. Res.,* 1968, **11**, 515.
66. S. Pawluk, *Atomic Absorption Newsletter,* 1967, **6**, 53.
67. C. H. McBride, *J. Ass. Off. Analyt. Chem.,* 1965, **48**, 406.
68. G. G. Roussos and B. H. Morrow, *Appl. Spectrosc.,* 1968, **22**, 769.
69. P. Aufila and V. Aufila, *Suamen Kemi,* 1971, **44**, 161.
70. W. Slavin, *Atomic Absorption Newsletter,* 1965, **4**, 330.
71. W. J. Price and J. T. H. Roos, *J. Sci. Food Agric.,* 1969, **20**, 437.
72. S. W. Frey, *Atomic Absorption Newsletter,* 1964, **3**, 127.
73. J. P. Weiner and L. Taylor, *J. Inst. Brewing,* 1969, **75**, 195.
74. G. B. Marshall and A. C. Smith, *Proc. 3rd CISAFA,* Paris, 1971.
75. J. Nix and T. Goodwin, *Atomic Absorption Newsletter,* 1970, **9**, 119.
76. O. K. Galle, *Appl. Spectrosc.,* 1971, **25**, 664.
77. T. Joyner and J. S. Finley, *Atomic Absorption Newsletter,* 1966, **5**, 4.
78. S. Sprague and W. Slavin, *Atomic Absorption Newsletter,* 1964, **3**, 37.
79. R. C. Barras, *Jarrell-Ash Newsletter,* **13**, June 1962.
80. E. A. Means and D. Ratcliffe, *Atomic Absorption Newsletter,* 1965, **4**, 174.

81. G. S. Golden, *Appl. Spectrosc.*, 1971, **25**, 668.
82. S. Sprague and W. Slavin, *Atomic Absorption Newsletter*, 1965, **4**, 367; 1966, **5**, 106.
83. R. Smith, C. M. Stafford and J. D. Winefordner, *Can. Spectrosc.*, 1969, **14**, 2.
84. J. A. Burrows, J. C. Heerdt and J. B. Willis, *Anal. Chem.*, 1965, **37**, 579.
85. R. L. Miller, L. M. Fraser and J. D. Winefordner, *Appl. Spectrosc.*, 1971, **25**, 477.
86. J. Scott and F. C. A. Killer, *Proc. Soc., Anal. Chem.*, 1970, **7**, 18.
87. D. H. Cotton and D. R. Jenkins, *Spectrochim. Acta*, 1970, **25B**, 283.
88. D. R. Jackson, C. Salama and R. Dunn, *Can. Spectrosc.*, 1970, **15**, 3.
89. R. G. Lundy, *Atomic Absorption Newsletter*, 1969, **8**, 124.
90. R. J. Thompson, G. B. Morgan, and L. J. Purdue, *Atomic Absorption Newsletter*, 1970, **9**, 53.
91. Kazuo Hiiro, Takashi Tanaka and Toshihiko Sawada, *Japan Analyst*, 1972, **21**, 635.

13.4.28 Lanthanum

La, atomic weight 138·905

Technique	Line, nm	f	Source	Flame	Sensitivity, ppm/1% abs.	Detection limit, ppm	Reference (see below)
AAS	357·44	0·12	HCL	N-A	110	–	1
	357·44	0·12	HCL	N-A	120	–	2
	364·95	–	HCL	N-A	140	–	2
	392·76	–	HCL	N-A	150	–	2
	403·72	–	HCL	N-A	200	–	2
	407·92	–	HCL	N-A	200	–	2
	418·73	–	HCL	N-A	50	–	2
	494·98	–	HCL	N-A	53	–	2
	550·13	0·15	HCL	N-A	35	–	2, 3
AFS: not reported							

Ionization potential: 5·61 eV.
Dissociation energy, LaO: 8·4 eV.

Notes on analytical conditions

1. A hot fuel-rich flame is necessary for AAS of La. Mossotti and Fassel[4] first reported La absorption in a fuel-rich oxy-acetylene flame at a total consumption nebulizer-burner using a xenon arc continuum source. Although Amos and Willis[1] used the La 357·44 nm line, its use results in high noise levels due to intense emission from CN at this wavelength in the nitrous oxide-acetylene flame.

2. KCl added to solutions suppresses La ionization in the nitrous oxide-acetylene flame. Little information is available concerning chemical interferences in the determination of La by AAS.

Applications notes and key references

Although several workers have reported general studies of the atomic absorption characteristics of the lanthanide elements,[1,2,4,5] relatively few applications reports have been published. Kinnunen and Lindsjo[6] studied the determination of the rare-earth elements in phosphate rock by AAS. These workers prepared calibration curves in a solution containing the other rare-earth elements at similar concentrations to those expected in samples; the rare-earths were precipitated as their oxalates, converted to their oxides by ignition and dissolved in HCl/NaCl solution for nebulization. Van Loon, Galbraith and Aarden[7] have described the determination of Eu, Tb, Dy, Ho, Er, Tm, Yb and Lu in minerals by AAS in the nitrous oxide-acetylene flame. Samples were decomposed in HF or by alkaline carbonate fusion, depending on sample type; the rare-earths were precipitated as their hydroxides, dissolved in HNO_3 and diluted with lanthanum buffer solution. McMahon and Fricioni[8] have described the determination of cerium, lanthanum and other rare-earth elements in steel by atomic absorption and flame emission methods. The determination of rare-earths in Mg-rare-earth alloys by AAS after sample dissolution in HCl has also been reported.[9] Bek and Sychra[10] have discussed the determination of lanthanum by flame spectroscopy.

References

1. M. D. Amos and J. B. Willis, *Spectrochim. Acta*, 1966, **22**, 1325.
2. D. C. Manning, *Atomic Absorption Newsletter*, 1966, **5**, 127.
3. "Analytical Methods for Atomic Absorption Spectrophotometry", Perkin-Elmer Corp., Norwalk, Conn., U.S.A., 1971.
4. V. G. Mossotti and V. A. Fassel, *Spectrochim. Acta*, 1964, **20**, 1117.
5. R. J. Jaworowski, R. P. Weberling and D. J. Bracco, *Anal. Chim. Acta*, 1967, **37**, 284.
6. J. Kinnunen and O. Lindsjo, *Chemist-Analyst*, 1967, **56**, 25.
7. J. C. Van Loon, J. H. Galbraith and H. M. Aarden, *Analyst*, 1971, **96**, 47.
8. M. McMahon and R. B. Fricioni, *Pittsburgh Conf. on Anal. Chem. and Appl. Spectrosc.*, Cleveland, U.S.A., 1971.
9. S. G. Melamed, A. M. Saltykova and L. F. Tchelidze, *Zavod. Lab.*, 1971, **37**, 166.
10. F. Bek and V. Sychra, *Proc. Colloq. Spectrosc. Intern.*, XVI, Heidelberg, Germany, 1971.

13.4.29 Lead

Pb, atomic weight 207·21

Technique	Line, nm	*f*	Source	Flame	Sensitivity, ppm/1% abs.	Detection limit, ppm	Reference (see below)
AAS	283·31	0·21	HCL	A-A	0·5	0·02	1
	217·00	0·39	HCL	A-A	0·2	—	—
AFS	405·8		EDL	$Ar-O_2-H_2$	—	0·01	2
			HCL	A-H	—	0·02	3
	405·8		HCL	$Ar-O_2-H_2$	—	0·02	4

Ionization potential: 7·4 eV.
Dissociation energy, PbO: 4·1 eV.
Atomization efficiency: A-A, 0·44.

Notes on analytical conditions

The 217·00 nm line yields higher sensitivity than the 283·31 nm line;[5] if argon is used rather than neon as filler gas for HCL sources the Ar(II) line at 217·14 may interfere; additionally, the flame background absorption in the air-acetylene flame is greater at the 217·00 nm line and may give rise to poorer detection limit unless an inert gas-separated flame is employed.[6] No significant chemical interferences have been reported for the determination of lead by AAS. In AFS in the argon-oxygen-hydrogen flame Browner *et al.*[2] observed no effect at 100-fold excess from 30 cations and anions; only aluminium was found to interfere. Sychra and Matousek[4] reported interference in AFS from 0·1 M HF and molybdate as well as from aluminium in this flame. The interferences were reduced or eliminated by addition of EDTA.

Applications notes and key references

AAS has been used extensively for the determination of small amounts of lead in a wide variety of matrices. Of particular interest is the determination of lead in blood and urine; this has been widely undertaken by conventional AAS with the air-acetylene flame after preconcentration of lead by solvent extraction, for example with ammonium pyrrolidine dithiocarbamate or Na diethyldithiocarbamate[14] into methyl iso-butyl ketone. Blood samples may be dry-ashed or simply treated with trichloroacetic acid to precipitate protein before solvent extraction. Lead in urine has been preconcentrated by coprecipitation with bismuth[15] or thorium[16] before nebulization into flames for AAS determination. A number of workers have described rapid methods for the direct determination of lead in small volumes of blood and urine using boat-in-flame and non-flame devices. Kahn and Sebestyen,[17] for example, have determined lead in 0·02 ml samples of blood and 0·2 ml samples of urine using a "sampling-boat" introduced into the flame of a commercial AAS instrument, while Delves[18] has determined lead in blood by vaporization of the lead from a nickel crucible into a nickel absorption tube positioned in an air-acetylene flame. In this technique a 10 μl whole blood sample is evaporated in a nickel crucible at 140°C for 30 seconds, 20μl of 100 vol. H_2O_2 is added and the sample is again evaporated. The crucible is then introduced into a premixed air-acetylene flame below the entrance of a nickel absorption tube. The lead vaporized is determined by AAS at 283·3 nm; the sensitivity claimed is 0·1 ng Pb and a coefficient of variation for 3 ng Pb of 4% has been reported. Cernik and Sayers[19] have reported a similar method using the Delves cup technique but in which no chemical sample pretreatment is required; in this method the blood sample is collected on filter paper, dried, and a 9 mm disc of paper containing the sample is punched out and placed in the nickel crucible. The carbon filament non-flame device has also been used for the direct determination of traces of Pb in 2 μl samples of blood and urine by atomic fluorescence spectroscopy.[20] In this method the direct-line fluorescence of lead at 405·8 nm is measured; detection limits for lead in whole blood and non-diluted urine of 0·005 μg/ml were reported.

The determination by AAS of traces of lead in beer[21] and wines,[21-24] fruit juices,[25] sea-foods,[26] canned food,[27] fish flour[28] and other plant and animal products[29] has been described. Dry-ashing of solid samples is most frequently employed, followed by dissolution of the residue in HNO_3 before nebulization into an air-acetylene flame. The determination of lead in air and air-borne particulates has been undertaken by a number of workers. Burnham *et al.*[30] collected samples by passing air through fibre-glass filters for 24 hours, igniting the filter at 500°C and extracting the lead with HNO_3. These workers obtained results in the range 0·1-3·2 μg Pb per cubic metre for 38 samples; the results agreed well with those obtained by polarography. In a second method the filter containing the lead sample was ashed at 25-250°C in the presence of active oxygen and a standard additions method was employed for the determination.[31] Thilliez[32] has reported continuous monitoring of the atmosphere over the range 1-500 μg Pb per cubic metre using a

Fuwa-Vallee burner. Loftin, Christian and Robinson[33] have described a technique for the continuous determination of lead in which the air is passed over white-hot carbon rods to reduce lead to the atomic state and then passed to a heated chamber for determination of Pb at 283·3 nm by AAS; a lower limit of sensitivity of 30 pg was reported. The direct determination of lead in air-borne particulates using non-flame AAS with a modified graphite sampling cup has been described by Matousek and Brodie.[33a]

In early work Robinson[34] demonstrated the determination of lead in petrol in the range 0-70 ppm after 10-fold dilution of samples with 2,2,4-trimethyl-pentane. Many workers have since routinely determined lead in petrol, either by simple dilution and nebulization into a flame,[35-38] or by extraction into iodine monochloride solution and determination of lead in this solution compared with aqueous lead standards.[39] This latter procedure eliminates problems from the different responses that may otherwise be obtained for lead tetramethyl and lead tetraethyl. Lead in lubricating oils may be determined after dilution with MIBK,[40] iso-propanol/toluene[41] or iodine/MIBK[42] or by formation of an emulsion with aqueous solution.[43] The use of a carbon rod atomizer for determination of lead in lube oils by AAS has also been described.[44] Winefordner and coworkers[45] have examined the use of both turbulent and premixed flames for the determination of lead in lubricating oils by AFS and compared the results with those obtained by emission spectrography and AAS; no particular advantage was claimed for use of AFS in this application.

The use of AAS for the determination of lead in minerals, ores, metals and alloys is widespread. These determinations may usually be made without complication after the sample has been brought into solution by fusion or acid dissolution. The reader is referred to the following selected references for details of such determinations: silicates (rock),[46-49] Fe ores,[50] soils,[51] ferrosilicon,[52] steel,[53,54] copper alloys,[53-56] uranium,[57] gold,[58] and nickel[59] plating solutions.

References

1. "Analytical Methods for Atomic Absorption Spectrophotometry", Perkin-Elmer Corp., Norwalk, Conn., 1971.
2. R. F. Browner, R. D. Dagnall and T. S. West, *Anal. Chim., Acta,* 1970, **50,** 375.
3. D. C. Manning and P. Heneage, *Atomic Absorption Newsletter,* 1967, **6,** 124.
4. V. Sychra and J. Matousek, *Talanta,* 1970, **17,** 363.
5. D. C. Manning and J. Vollmer, *Atomic Absorption Newsletter,* 1967, **6,** 38.
6. G. F. Kirkbright and L. Ranson, unpublished work.

7. J. B. Willis, *Anal. Chem.*, 1962, **34**, 614.
8. J. O. Pierce and J. Cholak, *Arch. Envir. Health*, 1966, **13**, 208.
9. D. W. Hessel, *Atomic Absorption Newsletter*, 1968, **7**, 55.
10. G. Devoto, *Boll. Soc. Ital. Biol. Sper.*, 1968, **44**, 421.
11. S. Selander, K. Cramer, B. Boerjesson and G. Mandorf, *Brit. J. Ind. Med.*, 1968, **25**, 209.
12. R. O. Farrelly and J. Pybus, *Clin. Chem.*, 1969, **15**, 566.
13. P. P. Donovan and D. T. Feeley, *Analyst*, 1969, **94**, 879.
14. E. Berman, V. Valavanis and A. Dubin, *Clin. Chem.*, 1968, **14**, 239.
15. L. Kopito and S. Shwachman, *J. Lab. Clin. Med.*, 1967, **70**, 326.
16. N. Zurlo, A. M. Griffini and G. Colombo, *Anal. Chim. Acta*, 1969, **47**, 203.
17. H. L. Kahn and J. S. Sebestyen, *Atomic Absorption Newsletter*, 1970, **9**, 33.
18. H. T. Delves, *Analyst*, 1970, **95**, 431.
19. A. A. Cernik and M. H. P. Sayers, *Brit. J. Ind. Medicine*, 1971, **28**, 392.
20. M. D. Amos, P. A. Bennett, K. G. Brodie, P. W. Y. Lung and J. P. Matousek, *Anal. Chem.*, 1971, **43**, 211.
21. J. P. Weiner and L. Taylor, *J. Inst. Brew.*, 1969, **75**, 195.
22. P. B. Zeeman and L. R. P. Butler, *Appl. Spectrosc.*, 1962, **16**, 120.
23. J. C. Meranger and E. Somers, *J. Ass. Off. Analyt. Chem.*, 1968, **51**, 922.
24. D. Roosels and J. V. Vanderkeel, *Atomic Absorption Newsletter*, 1968, **7**, 9.
25. J. C. Meranger, *Bull. Envir. Contam. Toxicol.*, 1970, **5**, 271.
26. J. C. Meranger and E. Somers, *Bull. Envir. Contam. Toxicol.*, 1968, **3**, 360.
27. G. Bionda, R. Ciurlo and L. Biino, *Ricerce Scient.*, 1967, **37**, 665.
28. A. Strasheim, E. Norval and L. R. P. Butler, *J. S. Afric. Chem. Inst.*, 1964, **17**, 55.
29. William L. Hoover, John C. Reagar and Julianna C. Garner, *J. Ass. Off. Analyt. Chem.*, 1969, **52**, 708.
30. Carole D. Burnham, Carl E. Moore, Eugene Kanabrocki and Dan M. Hattori, *Envir. Sci. Technol.*, 1969, **3**, 472.
31. C. D. Burnham, C. E. Moore, T. Kowalski and J. Krasniewski, *Appl. Spectrosc.*, 1970, **24**, 411.
32. G. Thilliez, *Anal. Chem.*, 1967, **39**, 427.
33. H. P. Loftin, C. M. Christian and J. W. Robinson, *Spectroscopy Letters*, 1970, **3**, 161.

33a. J. P. Matousek and K. G. Brodie, *Anal. Chem.*, 1973, **45**, 1606.

34. J. W. Robinson, *Anal. Chim. Acta*, 1961, **24**, 451.
35. R. M. Dagnall and T. S. West, *Talanta*, 1964, **11**, 1553.
36. M. Kashiki, S. Yamazoe and S. Oshima, *Anal. Chim. Acta*, 1971, **53**, 95.
37. D. J. Trent, *Atomic Absorption Newsletter*, 1965, **4**, 348.
38. R. A. Mostyn and A. F. Cunningham, *J. Inst. Petrol*, 1967, **53**, 101.
39. P. Johns, *Spectrovision*, 1971, **26**, 15.
40. D. R. Jackson, C. Salama and R. Dunn, *Canadian Spectrosc.*, 1970, **15**, 17.
41. R. Goux, *Meth. Phys. Analyse*, 1970, **6**, 118.
42. M. Kashiki, S. Yamazoe and S. Oshima, *Anal. Chim. Acta*, 1961, **54**, 533.
43. L. C. Westwood, *Developments Appl. Spectrosc.*, 1971, **9**, 235.
44. K. G. Brodie and J. P. Matousek, paper presented at Symposium on Analytical Chemistry, Adelaide, Australia, May 1971.
45. R. Smith, C. M. Stafford and J. D. Winefordner, *Canadian Spectrosc.*, 1969, **14**, 38.
46. Chuzo Iida, Tamotsu Tanaka and Kazuo Yamasaki, *Japan Analyst*, 1966, **15**, 1100.

47. S. Moldan, I. Rubeska and M. Miksovsky, *Anal. Chim. Acta,* 1970, **50,** 342.
48. Masato Nagura and Chuzo Iida, *Japan Analyst,* 1968, **17,** 1513; 1969, **18,** 492.
49. A. S. Bazhov, *Zh. Analit. Khim.,* 1968, **23,** 1640.
50. Yoshihide Endo, Toshihika Hata and Yuki Nakahara, *Japan Analyst,* 1969, **18,** 833.
51. R. Emmermann and W. Luecke, *Z. Anal. Chem.,* 1969, **248,** 325.
52. F. J. Langmyhr and P. E. Paus, *Anal. Chim. Acta,* 1969, **45,** 173.
53. W. T. Elwell and J. A. F. Gidley, *Anal. Chim. Acta,* 1961, **24,** 71.
54. R. M. Dagnall, T. S. West and P. Young, *Anal. Chem.,* 1966, **38,** 358.
55. P. Johns and W. J. Price, *Metallurgia,* 1970, **81,** 75.
56. J. M. Hall and C. Woodward, *Spectroscopy Letters,* 1969, **2,** 113.
57. C. R. Walker and O. A. Vita, *Anal. Chim. Acta,* 1968, **43,** 27.
58. J. P. Kapetan, *Spec. Tech. Publ. Amer. Soc. Test. Mat.,* No. 443, 1969, 78.
59. R. G. Shafto, *Atomic Absorption Newsletter,* 1964, **3,** 115.

13.4.30 Lithium

Li, atomic weight 6·94

Technique	Line, nm	f	Source	Flame	Sensitivity, ppm/1% abs.	Detection limit, ppm	Reference (see below)
AAS	670·78	0·7	HCL	A-coal gas	0·03	–	1
	323·26	0·03	HCL	A-coal gas	15	–	1
	670·78	0·7	HCL	A-A	0·03	–	2
AFS: no analytical studies reported.							

Ionization potential: 5·39 eV.
Atomization efficiency: A-A 0·2; N-A, 0·44.

Notes on analytical conditions

1. Lithium hollow-cathode lamp sources are generally satisfactory for AAS and show a high signal : background intensity ratio at the Li resonance lines. A relatively broad spectral band-pass may therefore be employed.

2. In the air-acetylene flame Li may be determined without trouble from chemical interferences. Excess KCl should be added to samples and standards to suppress ionization of lithium. The interference of Sr observed by several workers,[3,4,5] has been attributed to molecular absorption at the Li 670·78 nm line by SrOH species in the flame.[4]

3. The relatively large isotope displacement at the $^2P_{3/2} - {}^2S_{1/2}$ and $^2P_{1/2} - {}^2S_{1/2}$ lines of the Li 670·8 nm doublet for the ^{6}Li and ^{7}Li isotopes permits the direct determination of the isotopic

abundance ratio by AAS using a flame[6,7] or hollow cathode[8] as the atom cell. This is discussed further in Chapter 14.3.

Applications notes and key references

Relatively little work has been published concerning the determination of Li by AAS in metallurgical, geochemical and biological samples. Flame emission spectroscopy is capable of extremely high sensitivity for the determination of Li, however, and some workers have preferred to retain the emission technique rather than to adopt absorption procedures.

TABLE 13-4

Some applications of the determination of lithium by AAS

Sample type	Technique; sample pretreatment	Reference
Li-aluminium alloys (isotope)	AAS; dissolve in HCl, use acetone extraction for Li, evaporate, dissolve residue in HCl; A-A flame	9
Silicates	AAS; fuse sample with sodium borate, dissolve melt in HNO_3; A-A flame	10
Cement	AAS; dissolve in HCl, evaporate, redissolve in 50% HCl, filter, add Ca to standards	11
Serum	AAS; dilute 1:10 with water. Match Na and K in standards and samples	12
Serum	AAS; dilute 1:20 with 0·1 N HCl	13
Serum	AAS; dilute with Sterox SE solution. Match Mg, Ca, Na and K in sample and standards	14
Serum, urine	AAS; dilute 1:10 (serum) and 1:100 (urine) with water. Standards prepared to contain NaCl	15
Serum	AAS; deproteinize with ethanol, dilute with water	16
Serum	AAS; dilute 1:10 with water; A-A flame	17
Plants	AAS; wet ash with $HNO_3/H_2SO_4/HClO_4$	18
Sea-water	AAS; match major element concentrations in samples and standards	19
Waters	AAS; aspirate sample directly	3
Water	AAS	20

References

1. B. M. Gatehouse and J. B. Willis, *Spectrochim. Acta*, 1961, **17**, 710.
2. "Analytical Methods for Atomic Absorption Spectrophotometry", Perkin-Elmer Corp., Norwalk, Conn., U.S.A., 1971.
3. M. J. Fishman and S. C. Downs, *U.S. Geol. Surv. Water Supply Papers*, 1540-C, 1966.
4. S. R. Koirtyohann and E. E. Pickett, *Anal. Chem.*, 1966, **38**, 385.
5. A. N. Zaidel and E. P. Korennoi, *Opt. i. Spektrosk.*, 1961, **10**, 570.
6. A. N. Zaidel and E. P. Korennoi, *Opt. i. Spektrosk.*, 1961, **10**, 299.

7. D. C. Manning and W. Slavin, *Atomic Absorption Newsletter*, 1962, **1**, 39.
8. J. A. Goleb and Y. Yokohama, *Anal. Chim. Acta*, 1964, **30**, 213.
9. J. A. Wheat, *Appl. Spectrosc.*, 1971, **25**, 328.
10. J. V. O'Gorman and N. H. Suhr, *Analyst*, 1967, **96**, 335.
11. L. Capacho-Delgado and D. C. Manning, *Analyst*, 1967, **92**, 553.
12. J. A. Bowman, *Anal. Chim. Acta*, 1967, **37**, 465.
13. V. Lehmann, *Clin. Chim. Acta*, 1968, **20**, 523.
14. A. E. Woods, R. D. Crowder, J. T. Coates and J. J. Wittrig, *Atomic Absorption Newsletter*, 1968, **7**, 85.
15. A. Zettner, K. Rafferty and H. J. Jarecki, *Atomic Absorption Newsletter*, 1968, **7**, 32.
16. B. G. Blijenberg and B. Leijnse, *Clin. Chim. Acta*, 1968, **19**, 97.
17. J. Pybus and G. N. Bowers Jnr., *Clin. Chem.*, 1970, **16**, 139.
18. F. Z. Patassy, *Plant Soil*, 1965, **22**, 395.
19. E. E. Angino and G. K. Billings, *Geochim. Cosmochim. Acta*, 1966, **30**, 153.
20. S. D. Soman, V. K. Panday and K. T. Joseph, *Am. Ind. Hyg. Assoc. J.*, 1969, **30**, 527.

13.4.31 Lutecium

Lu, atomic weight 174·97

Technique	Line, nm	*f*	Source	Flame	Sensitivity, ppm/1% abs.	Detection limit, ppm	Reference (see below)
AAS	331·21	—	Xe arc	O_2/A	—	50	1
	331·21	—	HCL	N-A	21	—	2
	355·96	—	HCL	N-A	12	—	2
	355·96	—	HCL	N-A	6	3	3
AFS: not reported							

Ionization potential: 6·15 eV.
Dissociation energy, LuO: 5·3 eV.

Notes on analytical conditions

1. Amos and Willis[2] also reported Lu absorption at the less sensitive 356·71, 356·78, 337·65, 328·17, 327·90, 317·14, 308·15 and 298·93 nm lines in a slightly fuel-rich nitrous oxide-acetylene flame.

2. A large excess (>1000 ppm) of potassium (as KCl) should be added to Lu sample and standard solutions to suppress ionization.

3. Little information is available concerning chemical interferences in the determination of Lu by AAS.

Applications notes and key references

See Lanthanum.

References

1. V. G. Mossotti and V. A. Fassel, *Spectrochim. Acta*, 1964, **20**, 1117.
2. M. D. Amos and J. B. Willis, *Spectrochim. Acta*, 1966, **22**, 1325.
3. "Analytical Methods for Atomic Absorption Spectrophotometry", Perkin-Elmer Corp., Norwalk, Conn., U.S.A., 1971.

13.4.32 Magnesium

Mg, atomic weight 24·305

Technique	Line, nm	*f*	Source	Flame	Sensitivity, ppm/1% abs.	Detection limit, ppm	Reference (see below)
AAS	285·21	1·2	HCL	A-A	0·007	0·0001	1
	285·21	1·2	HCL	N-A		–	2
	202·58	–	HCL	A-A	0·3	–	3
	279·55	Mg ion line	HCL		5	–	4
AFS	285·21	1·2	EDL	Ar/H_2	–	0·008	5
	285·21	1·2	EDL	Ar/H_2	–	0·01	6
	285·21	1·2	HIL	A-P	–	0·001	7
	285·21	1·2	HCL	A-H	–	0·01	8
	285·21	1·2	HCL	A-H	–	0·03	9
	285·21	1·2	450 W Xe arc	Air/H_2	–	0·04	10
ND	285·21	1·2	HIL	N_2 sep. A-A	–	0·0002	11
ND	285·21	1·2	HIL	A-H	–	0·00005	12
ND	285·21	1·2	Pulsed HCL	A-H	–	0·001	13

Ionization potential: 7·64 eV.
Dissociation energy, MgO: 4·3 eV.
Atomization efficiency: A-A, 0·59.

Notes on analytical conditions

1. Much early work with Mg was undertaken using the premixed air-propane or air-coal gas flames;[14-19] although equal to slightly better sensitivity may be attained in these flames than in the air-acetylene flame, chemical interference effects are more troublesome.

2. The use of the Mg(II) 279·55 nm line for absorption measurements is subject to interferences by alkali metal ions, which depress the Mg(II) ion population.

3. When a Mg-Al cathode material is used HCL sources are usually operated at moderate current to achieve good ratios of line signal to background. When the cathode material is Mg alone it is necessary to

avoid use of high operating current in order to minimize line broadening and obtain a long operating life-time. Absorption measurements with the Mg 285·21 nm line, however, can be made without difficulty using a 1–2 nm spectral band-pass.

4. Chemical interference from Al and Si is observed in the air-acetylene flame; lanthanum and strontium are effective as releasing agents for these suppression effects. Phosphate does not interfere with the Mg absorption signal provided that a small mean droplet diameter from the nebulizer is obtained.[20,21] The use of the nitrous oxide-acetylene flame is recommended for elimination of chemical interference effects in the Mg determination.

5. A considerable volume of work has been published concerned with enhancement of AAS sensitivity for Mg using devices such as heated air supply to the nebulizer,[22] long path absorption tubes,[23-25] and double nebulizers.[26] Mg has also been determined with high absolute sensitivity by AAS and AFS using various non-flame cell devices[27-31] (see also Chapter 7).

Applications notes and key references

The high sensitivity attainable in the determination of Mg by AAS has resulted in its widespread application to the determination of Mg traces in a variety of sample types. Table 13-5 lists some of the studies that have been published; although not comprehensive the data given should permit the reader to select those publications pertinent to his own interests.

TABLE 13-5

Some applications of the determination of magnesium by AAS and AFS

Sample type	Technique; sample pretreatment	Reference
Cast iron	AAS; dissolve in 50% HCl, add HNO_3, evaporate, bake at 200°C; dissolve in HCl, dilute, add $SrCl_2$	32
Cast iron	AAS; Sr used as releasing agent	33
Steel	AAS; Na and Sr used for interference suppression; N-A flame	34
Fe ores	AAS; dissolve in HCl, add HNO_3, evaporate to dryness, dissolve in HCl, filter. Fuse residue with $K_2S_2O_7$, extract melt + H_2O. Combine solutions. Add Sr to suppress interference; A-A flame	35
Chrome ores and chrome-magnesite	AAS; dissolve in H_3PO_4, heat to 310°C, dilute; N-A flame	36
Ni alloys	AAS; match acid and Ni concentration in sample and standards	17

TABLE 13-5 (*continued*)

Sample type	Technique; sample pretreatment	Reference
Ni and Ni alloys	AAS; dissolve in 50% HCl. Prepare standards from Ni carbonyl	37
Al alloys	AAS; oxine used to suppress interferences.	16
Al alloys	AAS; dissolve in HCl, add H_2O_2. Use Sr releasing agent	15
Al alloys	AAS; 1% La added as releasing agent to samples and standards; A-A flame	38
Al alloys	AAS; dissolve in NaOH	39, 40
Al alloys	AAS; dissolve in 50% HCl, add H_2O_2 if Cu is present, filter off or volatilize Si (HF/HNO_3), add La releasing agent; A-A flame	41
Al alloys	AAS; extract Mg with 8-hydroxyquinaldine into MIBK, nebulize into A-A flame	42
Al alloys	AFS; dissolve in HCl; N_2 sep. A-A flame	43
Zn alloys	AAS; coprecipitate Mg with strontium phosphate; dissolve precipitate in HCl/acetic acid	44
Zn alloys	AAS; dissolve in HCl, oxidize with HNO_3; A-A flame	45
Uranium	AAS; dissolve in HCl. Match U and acid content in samples and standards; A-A flame	46
Uranium compounds	AAS; extract bulk of U with tributyl phosphate; nebulize aqueous soln.; N-A flame	47
Tellurium	AAS; dissolve in HCl/HNO_3; A-A flame	48
Silicate rocks	AAS; HF/H_2SO_4 dissolution of Na_2CO_3 fusion; A-A flame	49
Silicate and carbonate rocks	AAS; dissolve in HF/HNO_3. 1% La added to suppress interferences	50
Silicate rocks	AAS	18
Silicate and carbonate rocks	AAS; dissolve in $HF/HClO_4$. Add oxine to suppress Al interference; A-A flame	51
Silicates	AAS; N-A flame used to avoid interferences observed in A-A flame	52
Silicates	AAS; decompose with HF at 110°C in sealed vessel; N-A flame	53
Silicates and carbonates	AAS; fuse with $LiBO_2$ in Pt crucible, dissolve in HCl. Add 1% La; N-A flame	54
Silicates	AAS; fuse with Li_2CO_3 and H_3BO_3. Dissolve in HCl, add $LaCl_3$	55
Silicates	AAS; fuse with Li_2CO_3, H_3BO_3, $SrCO_3$ and Co_3O_4. Dissolve in citric acid or $SrCO_3$/citric acid mixture	56
Silicate rocks and coal ash	AAS; fuse with $LiBO_2$. Dissolve in dilute HNO_3/tartaric acid. Add KCl; N-A flame	57
Coal ash	AAS; dissolve in HF. Use La to suppress interferences	58
Coal ash	AAS; dissolve in $HF/HClO_4$	59
Slags and sinters	AAS; samples dissolved at 120°C in sealed vessel with HF/HCl, or with HF alone at 100°C. Add H_3BO_3; N-A flame	60
Cement	AAS; dissolve in HCl; N-A flame	61
Cement	AAS; dissolve in HCl. Use method of standard additions	62

TABLE 13-5 (*continued*)

Sample type	Technique; sample pretreatment	Reference
Cement	AAS; dissolve in HCl in PTFE vessel. Add HF, H_3BO_3 and $LaCl_3$	63, 64
Glass	AAS; $HF/HClO_4$ dissolution. EDTA + La used for interference suppression	65
Glass	AAS; HCl/HF dissolution; MeOH added for high B content samples. Evaporate to dryness twice, dissolve in HCl. Use Sr to suppress interferences	66
Mineral slag and wool	AAS; dissolve at 135°C with HF in sealed vessel, add H_3BO_3; N-A flame	67
Jet engine oils	AAS; dilute 1:10 with MIBK. NBS metallorganic standard	68, 69
Jet engine oils	AAS; dilute 1:4 with xylene. A-A flame. Use metal naphthenate standard	70
Railway engine oils	AAS; dilute with MIBK	71
Jet engine oils	AFS; air-hydrogen or Ar/H_2-entrained air flame	6
Jet engine oils	AFS; dilute 1:4 with CCl_4. Ar/H_2-entrained air flame	72
Lubricating oils	AFS; dilute with MIBK. A-A flame	73
Lubricating oils	AAS; carbon rod atomization	74
Serum	AAS; dilute 20-fold with 1% EDTA or $SrCl_2$	74
Serum	AAS; dilute 1:50 with water O_2/H_2 flame in long path tube	77
Plasma	AAS	78, 79
Serum	AAS; deproteinize with TCA; add $SrCl_2$ to sample and standards	80
Serum	AAS; direct method after dilution	81, 82, 83, 84, 85, 86
Cerebrospinal fluid	AAS; dilute with water	87, 88
Serum, cerebrospinal fluid	AAS; dilute 1:20 with acid 8-hydroxyquinoline solution	89
Serum, urine	AAS; dilute 1:100 with dilute HCl; A-P flame	90, 91, 92
Serum, plasma, cerebrospinal fluid	AAS; dilute with strontium chloride solution (0·25%)	93
Plasma, uterine fluid	AAS; dilute with $LaCl_3$	94
Serum, urine, faeces	AAS	95
Feeds, faeces, urine	AAS; digest with acid	96
Urine	AAS; dilute 1:20 with water	97
Urine	AAS; 100 μl sample diluted with $LaCl_3/CsCl$ solution	98
Serum, urine	AAS; add $SrCl_2/HClO_4$ to dilute solution; A-A flame	99
Serum, urine	AAS; ionized Mg adsorbed on cellophane coated with Congo Red; elute with NaCl for AAS	100

TABLE 13-5 (*continued*)

Sample type	Technique; sample pretreatment	Reference
Urinary calculi	AAS; dissolve in HCl	101
Cellular membranes	AAS; digest with $HClO_4/HNO_3$ or dry ash in active oxygen. Add La as suppressor	102
Brain tissue	AAS; digest with HNO_3 and then $HClO_4$; A-A flame	103
Brain tissue	AAS; dry ash, dissolve in HCl, add $LaCl_3$	104
Plasma, bone, tissues	AAS; dilute with $SrCl_2$ (plasma); wet ash with HNO_3 and dilute with $SrCl_2$	105
Saliva	AAS	106
Plasma, muscle, bone	AAS; dry ash bone and muscle at 600°C, dissolve in HCl, add $LaCl_3$. Dilute plasma with water	107
Rat heart tissue	AAS; extract cations with acetic acid/ trichloro-acetic acid. Centrifuge, wash, spray solution. Add Li or K to suppress ionization in N-A flame	108
Feeds, tissues	AAS	109, 110
Human hair, fingernails	AAS; wash, digest with $HNO_3/HClO_4$, dilute	111
Plants	AAS; wet ash with $H_2SO_4/HClO_4/HNO_3$	112
Plants	AAS; dry ash; add Sr to samples and standards.	113
Plant tissues	AAS; extract with 0·1 M diammonium EDTA	114
Plant materials	AAS; Kjeldahl digest. La addition not necessary	115
Chloroplasts	AAS; dry ash, dissolve in HCl, evaporate, dissolve residue in water and dilute with dilute H_2SO_4	116
Plants, soil, water	AAS; N-A flame	117
Plants and soil extracts	AAS	118
Soil extracts	AAS; extract with 1 M NH_4Cl; add $SrCl_2$	119
Soil extracts	AAS; use NH_4 acetate extraction	120
Soil extracts	AAS; extract with NH_4 nitrate. Add Sr to suppress Al and Si interference	121
Soil extracts	AFS; extract with NH_4 acetate, acetic acid or EDTA. Add Sr and Na to control interferences. A-A flame	122
Fertilizers	AAS; dissolve sample in HCl. Sample and standards contain 0·1 M HCl	123
Soils	AAS; dissolve in HF/HCl	124
Animal feeds	AAS	125, 126
Vegetable oils and fats	AAS; dissolve in $LaCl_3$/MeOH/isoamyl acetate	127
Milk	AAS; dry ash	128
Milk	AAS; deproteinize with TCA, filter, dilute, add La	129
Marine shrimps	AAS; dry, char at 200°C, ash at 550°C. Dissolve in HNO_3 (1:1), dilute; A-A flame	130
Fruit juices	AAS; add HCl, centrifuge, dilute	131
Beer, wort, yeast	AAS	132
Wine and beer	AAS; decarbonate if necessary, aspirate directly or after dilution	133
Fresh waters and brines.	AAS; add La in HCl. Match standards and samples for La and HCl	134

TABLE 13-5 (*continued*)

Sample type	Technique; sample pretreatment	Reference
Industrial waters	AAS; acidify with HCl. Autoclave at 120°C, add 1% La	135
Waters	AAS	136
Sea-water	AAS; aspirate directly or after dilution	137
Synthetic fibres	AAS; dry ash, dissolve in HNO_3	138
Wool, cotton and synthetic polymers	AAS; dissolve polymer or wet ash with H_2SO_4/H_2O_2	139
Paint, vinyl additives	AAS	140

References

1. "Analytical Methods for Atomic Absorption Spectrophotometry", Perkin-Elmer Corp., Norwalk, Conn., U.S.A.
2. M. D. Amos and J. B. Willis, *Spectrochim. Acta.* 1966, **22**, 1325.
3. G. Peterson, *Atomic Absorption Newsletter,* 1966, **5**, 117.
4. B. M. Gatehouse and J. B. Willis, *Spectrochim. Acta,* 1961, **17**, 710.
5. K. E. Zacha, M. P. Bratzel Jnr., J. D. Winefordner and J. M. Mansfield Jnr., *Anal. Chem.,* 1968, **40**, 1733.
6. R. Smith, C. M. Stafford and J. D. Winefordner, *Can. Spectrosc.,* 1969, **14**, 2.
7. T. S. West and X. K. Williams, *Anal. Chim. Acta,* 1968, **42**, 29.
8. D. C. Manning and P. Heneage, *Atomic Absorption Newsletter,* 1967, **6**, 124.
9. G. Rossi and N. Omenetto, *Talanta,* 1969, **16**, 263.
10. D. R. Demers and D. W. Ellis, *Anal. Chem.,* 1968, **40**, 860.
11. P. L. Larkins, *Spectrochim. Acta,* 1971, **26B**, 477.
12. P. L. Larkins and J. B. Willis, *Spectrochim. Acta,* 1971, **26B**, 491.
13. D. G. Mitchell and A. Johansson, *Spectrochim. Acta,* 1970, **25B**, 175.
14. A. C. Menzies, *Anal. Chem.,* 1960, **32**, 898.
15. W. T. Elwell and J. A. F. Gidley, "Atomic Absorption Spectrophotometry", Pergamon Press, London. 1st edition 1961, 2nd edition 1966.
16. F. J. Wallace, *Analyst,* 1963, **88**, 259.
17. T. R. Andrew and P. N. R. Nicols, *Analyst,* 1962, **87**, 259.
18. R. J. Firman, *Spectrochim. Acta,* 1965, **21**, 341.
19. D. J. Halls and A. Townshend, *Anal. Chim. Acta,* 1966, **36**, 278.
20. C. H. McBride, *J. Ass. Off. Analyt. Chem.,* 1965, **48**, 406.
21. W. Slavin, *Atomic Absorption Newsletter,* 1967, **6**, 9.
22. R. A. G. Rawson, *Anal. Chim. Acta,* 1966, **91**, 630.
23. I. Fuwa and B. L. Vallee, *Anal. Chem.,* 1963, **35**, 942.
24. S. R. Koirtyohann and E. E. Pickett, *Anal. Chem.,* 1965, **37**, 601.
25. C. Iida and M. Nagura, *Spectrosc. Lett.,* 1970, **3**, 63.
26. M. Tsuhako and E. Sekido, *Japan Analyst,* 1970, **19**, 525, 531.
27. T. S. West and X. K. Williams, *Anal. Chim. Acta,* 1969, **45**, 27.
28. R. G. Anderson, I. S. Maines and T. S. West, *Anal. Chim. Acta,* 1970, **51**, 355.

29. M. D. Amos, P. A. Bennett, K. G. Brodie, P. W. Y. Lung and J. P. Matousek, *Anal. Chem.*, 1971, **43**, 211.
30. M. P. Bratzel, R. M. Dagnall and J. D. Winefordner, *Appl. Spectrosc.*, 1970, **24**, 518.
31. D. Welz and E. Wiedeking. *Anal. Chem.*, 1970, **252**, 111.
32. C. B. Belcher and H. M. Bray, *Anal. Chim. Acta*, 1962, **26**, 322.
33. H. Goto, S. Ikeda and I. Atsuya, *Japan Analyst*, 1964, **13**, 111.
34. M. L. Taylor and C. B. Belcher, *Anal. Chim. Acta*, 1969, **45**, 219.
35. Y. Endo, T. Hata and Y. Nakahara, *Japan Analyst*, 1968, **17**, 679.
36. R. P. Lucas and B. C. Ruprecht, *Anal. Chem.*, 1971, **43**, 1013.
37. R. Dyck, *Atomic Absorption Newsletter*, 1965, **4**, 170.
38. R. E. Mansell, H. W. Emmel and E. L. McLaughlin, *Appl. Spectrosc.*, 1966, **20**, 231.
39. L. Wilson, *Anal. Chim. Acta*, 1968, **40**, 503.
40. L. Wilson, *Anal. Chim. Acta*, 1964, **30**, 377.
41. G. F. Bell, *Atomic Absorption Newsletter*, 1966, **5**, 73.
42. M. Suzuki, M. Yanagisawa and T. Takeuchi, *Talanta*, 1965, **12**, 989.
43. R. M. Dagnall, G. F. Kirkbright, T. S. West and R. Wood, *Analyst*, 1972, **97**, 245.
44. T. Miura, *Japan Analyst*, 1965, **14**, 310.
45. S. B. Smith, J. A. Blasi and F. J. Feldman, *Anal. Chem.*, 1968, **40**, 1525.
46. J. R. Humphrey, *Anal. Chem.*, 1965, **37**, 1604.
47. C. R. Walker and O. A. Vita, *Anal. Chim. Acta*, 1968, **43**, 27.
48. I. G. Yudelevich, V. L. Kustas and G. V. Poleva, *Zh. Analit. Khim.*, 1971, **26**, 1378.
49. D. J. Trent and W. Slavin, *Atomic Absorption Newsletter*, 1964, **3**, 17, 118.
50. G. K. Billings and J. A. S. Adams, *Atomic Absorption Newsletter*, 1964, **3**, 65.
51. I. Rubeska and B. Moldan, *Acta Chim. Lung.*, 1965, **44**, 367.
52. R. W. Nesbitt, *Anal. Chim. Acta*, 1966, **35**, 413.
53. B. Bernas, *Anal. Chem.*, 1968, **40**, 1682.
54. J. W. Yule and G. A. Swanson, *Atomic Absorption Newsletter*, 1969, **8**, 30.
55. S. H. Omang, *Anal. Chim. Acta*, 1969, **46**, 225.
56. K. Gavindaraju, *Appl. Spectrosc.*, 1970, **24**, 81.
57. P. L. Boar and L. K. Ingram, *Analyst*, 1970, **95**, 124.
58. C. B. Belcher and K. A. Brooks, *Anal. Chim. Acta*, 1963, **29**, 202.
59. E. L. Obermiller and R. W. Freedman, *Fuel*, 1965, **44**, 199.
60. J. Reid, J. M. Galloway, J. MacDonald and B. B. Bach, *Metallurgia*, 1970, **81**, 243.
61. L. Capacho-Delgado and D. C. Manning, *Analyst*, 1967, **92**, 553.
62. T. Takeuchi and M. Suzuki, *Talanta*, 1964, **11**, 1391.
63. J. T. H. Roos and W. J. Price, *Analyst*, 1969, **94**, 89.
64. F. J. Langmyhr and P. E. Paus, *Anal. Chim. Acta*, 1969, **44**, 445.
65. P. B. Adams and W. O. Passmore, *Anal. Chem.*, 1966, **38**, 630.
66. A. H. Jones, *Anal. Chem.*, 1965, **37**, 1761.
67. J. E. Sheridan, *Spectrovision*, 1971, **25**, 10.
68. S. Sprague and W. Slavin, *Atomic Absorption Newsletter*, 1965, **4**, 367.
69. S. Slavin and W. Slavin, *Atomic Absorption Newsletter*, 1966, **5**, 106.
70. E. A. Means and D. Ratcliffe, *Atomic Absorption Newsletter*, 1965, **4**, 174.
71. J. A. Burrows, J. C. Heerdt and J. B. Willis, *Anal. Chem.*, 1965, **37**, 579.

72. R. L. Miller, L. M. Fraser and J. D. Winefordner, *Appl. Spectrosc.*, 1971, **25**, 477.
73. D. G. Mitchell, D. Demers and M. Gardels, "Advances in Automated Analysis", Vol. II, Thurman Associates, Miami, 1971.
74. K. G. Brodie and J. P. Matousek, *Anal. Chem.*, 1971, **43**, 1557.
75. J. B. Willis, *Nature*, 1959, **184**, 186.
76. J. B. Willis, *Spectrochim. Acta*, 1960, **16**, 273.
77. W. E. C. Wacker, C. Iida and K. Fuwa, *Nature*, 1964, **202**, 659.
78. F. W. Heaton, *Nature*, 1965, **207**, 1292.
79. K. C. Scotts, U. R. Bohman and A. L. Lesperance, *J. Animal Sci.*, 1971, **32**, 354.
80. F. W. Sunderman and J. E. Carrol, *Amer. J. Clin. Pathol.*, 1965, **43**, 302.
81. F. W. Sunderman Jnr. and J. E. Carrol, *Clin. Pathol. Serum Electrolytes*, 1966, 199.
82. K. H. Hyatt, L. Levy, M. Nichaman and M. Oscherwicz, *Appl Spectrosc.*, 1966, **20**, 142.
83. J. L. Hansen and E. F. Freier, *Amer. J. Med. Technol.*, 1967, **33**, 158.
84. R. Gray and E. L. Pruden, *Amer. J. Med. Technol.*, 1967, **33**, 349.
85. L. W. Fleming and W. K. Stewart, *Clin. Chim. Acta*, 1966, **14**, 131.
86. W. K. Stewart, F. Hutchinson and L. W. Fleming, *J. Lab. Clin. Med.*, 1963, **61**, 858.
87. C. F. Decker, A. Aras and L. E. Decker, *Anal. Biochem.*, 1964, **8**, 344.
88. K. Paschen and C. Fuchs, *Clin. Chim. Acta*, 1971, **35**, 401
89. C. Iida, K. Fuwa and W. E. Wacker, *Anal. Biochem.*, 1967, **18**, 18.
90. D. B. Horn and A. L. Latner, *Clin. Chim. Acta*, 1963, **8**, 974.
91. C. G. Thin and P. A. Thomson, *J. Clin. Pathol.*, 1967, **20**, 280.
92. J. B. Dawson and F. W. Heaton, *Biochem. J.*, 1961, **80**, 99.
93. M. L. Girard and F. Rousselet, *Ann. Pharm. Fr.*, 1967, **25**, 271.
94. S. E. Solomsh, *Br. Poultry Sci.*, 1971, **12**, 165.
95. E. G. Gimblet, A. F. Marney and R. W. Bonsnes, *Clin. Chem.*, 1967, **13**, 204.
96. M. A. Garces and J. L. Evans, *J. Animal Sci.*, 1971, **32**, 789.
97. J. B. Willis, *Anal. Chem.*, 1961, **33**, 556.
98. B. W. Grunbaum and N. Pace, *Microchem. J.*, 1970, **15**, 666.
99. J. Pybus, *Clin. Chim. Acta*, 1969, **23**, 309.
100. F. W. Heaton, *Clin. Chim. Acta*, 1967, **15**, 139.
101. A. Hodgkinson, *J. Clin. Pathol.*, 1971, **24**, 147.
102. H. Sanui and N. Pace, *Appl. Spectrosc.*, 1966, **20**, 135.
103. T. L. Chang, T. A. Gover and W. W. Harrison, *Anal. Chim. Acta*, 1966, **34**, 17.
104. M. J. McBroom, Y. J. Lancaster and K. Weiss, *Anal. Biochem*, 1971, **42**, 178.
105. F. D. Griffith, H. E. Parker and J. C. Rogler, *J. Nutr.*, 1964, **83**, 15.
106. A. D. Care, *Nature*, 1963, **199**, 818.
107. B. J. Hunt, *Clin. Chem.*, 1969, **15**, 979.
108. F. T. N. Varghese, A. Lipton and G. J. Huxham, *Lab. Pract.*, 1969, **18**, 419.
109. R. M. Forbes, *J. Nutr.*, 1971, **101**, 35.
110. F. L. Wang, R. Wang, E. A. Khairallah and R. Schwartz, *J. Nutr.*, 1971, **101**, 1201.
111. W. W. Harrison, J. P. Yurachek and C. A. Benson, *Clin. Chim. Acta*, 1969, **23**, 83; 1971, **31**, 63.
112. D. J. David, *Analyst*, 1958, **83**, 655.

113. W. L. Berry and C. M. Johnson, *Appl. Spectrosc.*, 1966, **20**, 209.
114. J. H. Baker and T. Greweling, *J. Agr. Food Chem.*, 1967, **15**, 340.
115. R. E. Collier, *Hilger J.*, 1967, **10**, 52.
116. R. Herrman and W. Neu, *Experientia*, 1965, **21**, 436.
117. V. K. Panday and A. K. Ganguly, *Anal. Chim. Acta*, 1970, **52**, 417.
118. E. G. Bradfield and D. Spincer, *J. Sci. Food Agric.*, 1965, **16**, 33.
119. D. J. David, *Analyst*, 1960, **85**, 495.
120. J. Lacy, *Analyst*, 1965, **90**, 65,
121. T. R. Williams, B. Wilkinson, G. A. Wadsworth, D. H. Barter and W. J. Beer, *J. Sci. Food Agric.*, 1966, **17**, 344.
122. R. M. Dagnall, G. F. Kirkbright, T. S. West and R. Wood, *Anal. Chem.*, 1971, **43**, 1765.
123. C. H. McBride, *J. Ass. Off. Analyt. Chem.*, 1965, **48**, 406; 1967, **50**, 401.
124. S. Pawluk, *Atomic Absorption Newsletter*, 1967, **6**, 53.
125. H. E. Parker, *Atomic Absorption Newsletter*, 1966, **2**, 23.
126. D. C. Cresswell and C. C. Brook, *J. Animal. Sci.*, 1971, **33**, 366.
127. R. Guillaumin, *Atomic Absorption Newsletter*, 1966, **5**, 19.
128. G. K. Murphy and U. Rhea, *J. Dairy Sci.*, 1967, **50**, 313.
129. J. B. Brooks, G. A. Luster and D. G. Easterly, *Atomic Absorption Newsletter*, 1970, **9**, 93.
130. G. A. Knaver, *Analyst*, 1970, **95**, 476.
131. J. T. H. Roos and W. J. Price, *J. Sci. Food Agric.*, 1970, **21**, 51.
132. S. W. Frey, W. G. Dewitt and B. R. Bellomy, *Am. Soc. Brew. Chem., Proc.*, 1966, 172.
133. J. P. Weiner and L. Taylor, *J. Inst. Brew.*, 1969, **75**, 195.
134. M. J. Fishman and S. C. Downs, *U.S. Geol. Surv. Water Supply Papers*, 1540-C (1966).
135. J. A. Platte and V. M. Marcy, *Atomic Absorption Newsletter*, 1965, **4**, 289.
136. L. R. P. Butler and D. Brink, *S. Afr. Ind. Chemist*, 1963, **17**, 152.
137. G. K. Billings and R. C. Harriss, *Texas J. Sci.*, 1965, **17**, 129.
138. W. Slavin, *Atomic Absorption Newsletter*, 1965, **4**, 192.
139. M. Olivier, *Z. Anal. Chem.*, 1969, **248**, 145.
140. N. G. Eider, *Appl. Spectrosc.*, 1971, **25**, 313.

13.4.33 Manganese

Mn, atomic weight 54·938

Technique	Line, nm	f	Source	Flame	Sensitivity, ppm/1% abs.	Detection limit, ppm	Reference (see below)
AAS	279·48	0·58	HCL	A-A	0·05	0·002	1
	279·83	—	HCL	A-A	0·09	—	2
	280·11	—	HCL	A-A	0·14	—	2
	403·08	—	HCL	A-A	0·85	—	2
	321·70	—	HCL	A-A	100	—	—
AFS	279·48	0·58	EDL	Ar/H_2	—	0·006	3
	279·48	0·58	HCL	A-H	—	0·05	4
	279·48	0·58	150W Xe arc	A-H	—	0·5	4
	279·48	0·58	HCL	A-H	—	0·05	5
	279/280	0·58	EDL	N_2 sep. A-A	—	0·001	6
ND	279·48	0·58	HCL	N_2 sep. A-A	—	0·01	7
ND	279/280	0·58	Pulsed HCL	N_2 sep. A-A	—	0·008	8

Ionization potential: 7·43 eV.
Dissociation energy, MnO: 4·0 eV.
Atomization efficiency: A-A, 0·45; N-A, 0·76.

Notes on analytical conditions

1. A spectral half-band width of 0·2 nm or less is required to isolate the Mn 279·48 nm line to obtain maximum sensitivity and linear calibration range. The absorption signal obtained is not markedly dependent on the lamp current employed in AAS.

2. Fuel-lean or stoicheiometric air-acetylene flames are most commonly preferred for Mn AAS and AFS. In AAS the absorption signal is not critically dependent on the height of observation in the flame.

3. Frank, Schrenk and Meloan[9] used the 279·47 nm Fe line from an iron HCL for determination of Mn at the 279·48 nm absorption line. Several groups of workers have reported the use of flames other than air-acetylene for the determination of Mn by AAS; air-coal gas,[10] air-hydrogen,[11] oxy-hydrogen,[12] oxy-acetylene[12] and nitrous oxide-acetylene[12] flames have been investigated. The sensitivity attainable for Mn in AAS may be enhanced by the use of T-piece adaptors[12] or long-path absorption tubes.[13,14] Bailey and Rankin[15] obtained a sensitivity of 0·06 ppm Mn/1% absorption using an air-hexane liquid fuel flame. Mn has also been widely

determined by both AAS and AFS using non-flame cells of the graphite furnace or rod type; L'Vov[16] reported 1% absorption for 2×10^{-13} g Mn in a graphite furnace, and Ebdon, Kirkbright and West obtained 1% absorption for 5×10^{-11} g Mn and an AFS detection limit of 6×10^{-13} g Mn using a carbon rod atomizer[17] (see also Chapter 7). Weak atomic fluorescence signals from Mn ions formed by thermal ionization of Mn atoms in an air-acetylene flame have been recorded.[6]

4. Mn determination by AAS in the air-acetylene flame has been shown to be virtually free from interferences.[2,12] Silicon, however, suppresses Mn absorption; Kinson and Belcher[13] were able to minimize this effect by sampling only a restricted volume of the flame, and Platte and Marcy[19] suppressed Si interference by addition of $CaCl_2$ to sample solutions. Sprague and Slavin[20] observed interference from molybdenum, silicon and tungsten on Mn absorption when a nearly empty cylinder of acetylene was used; on replacement with a full cylinder the interference from molybdenum and tungsten was not observed. Molybdenum absorbance has been shown by Manning and Chabot[21] to depend on the acetone content of acetylene used to support the flame. Ebdon, Kirkbright and West[6] investigated the effect of large amounts of each of 25 foreign ions on Mn AFS; only Si, Th, V and Mg were found to give scattering interference in the air-acetylene flame employed.

Applications notes and key references

Table 13-6 lists some of the published work concerned with the determination of Mn in samples of metallurgical, geochemical and biological origin.

TABLE 13.6
Some applications of the determination of manganese by AAS and AFS

Sample type	Technique, sample pretreatment	Reference
Ferrous alloys	AAS; dissolve in HCl, add HNO_3, filter	22
Cast irons and steels	AAS; dissolve in HNO_3/HCl, evaporate, add $HClO_4$, filter off SiO_2, aspirate ethanol/H_2O solution; A-A flame	23
Iron and steel	AAS	18,20
Aluminium and ferro-alloys	AAS; acid dissolution; A-A flame	24
Ferro-silicon	AAS; HF decomposition	25
Ferro-manganese	AAS; N-A flame	26
Steel	AAS; remote sampler (arc) and N-A flame	27

TABLE 13-6 (*continued*)

Sample type	Technique, sample pretreatment	Reference
Iron ores	AAS; dissolve in HNO_3/HCl; A-A flame	28
Nickel and nickel alloys	AAS; dissolve in HCl, dilute. Standard solutions prepared to contain nickel	29
Al alloys	AAS; dissolve in HCl, dilute; A-A flame	30
Al alloys	AFS; dissolve in HCl, remove Si by HF treatment, dissolve residue in HNO_3/HCl; N_2 sep. A-A flame	31
U compounds	AAS; hydrolyse UF_6, dissolve in HNO_3, evaporate, dissolve in dilute HNO_3; N-A flame	32
Co-base alloys	AAS; A-A flame	33
Cu-base alloys	AAS; A-A flame	34, 35
Brasses and bronzes	AAS; direct method; N-A flame	36
Cement	AAS; dissolve in HCl; A-A flame, comprehensive scheme	37
Cement	AAS; dissolve in 4 M HCl; A-A flame	38
Silicate rocks	AAS; dissolve in 40% HF; A-A flame	39
Siliceous materials	AAS; dissolve with H_2SO_4/HF or fuse with Na_2CO_3	40
Silicates	AAS; fuse with $LiBO_2$, dissolve in HNO_3	41
Glasses	AAS; HF dissolution	42
Bauxite	AAS; dissolve in HF; A-A flame	43
Aluminous materials	AAS; fuse with Na_2CO_3, dissolve in HCl; separate by extraction with 8-hydroxyquinaldine into $CHCl_3$; A-A flame	44
Blood serum	AAS; acid digestion; A-A flame	45
Urine	AAS; extract with APDC into MIBK	46
Liver and muscle tissue	AAS; dry ash at 450°C	47
Brain muscle	AAS; digest with acid	48
Foetal tissues	AAS; acid digestion	49
Skin	AAS; dry ash at 460°C; dissolve residue in HCl; A-A flame	50
Serum	AAS; graphite furnace	51
Phospholipids	AAS; dissolve in iso-pentyl acetate	52
Soils and plants	AAS; wet ash + $HClO_4/HNO_3$ (and HF for soils). Standards contain $HClO_4$	2
Plants, tree leaves	AAS; acid digestion or dry ash	53, 54
Fungicides, plants	AAS; HCl or HNO_3/HCl digestion	56
Soils	AAS; digest with $HF/HClO_4$ or extract Mn with NH_4 acetate	57
Soil extracts	AAS; extract with 1% EDTA for 2 hours	2
Soil extracts	AAS; extract with NH_4 acetate, pH 7; A-A flame	58
Soil extracts	AFS; extract with acetic acid, EDTA or NH_4 acetate; N_2 sep. A-A flame	59
Soil extracts	AAS; extract with EDTA; mechanically separated A-A flame	60

TABLE 13-6 (*continued*)

Sample type	Technique, sample pretreatment	Reference
Fertilizers	AAS; dissolve sample in HCl, boil nearly to dryness, add HCl, filter	61
Vegetable and animal oils	AAS; dilute 1 : 3 with n-heptane	62
Milk	AAS; boil with HCl, extract with APDC into MIBK; A-A flame	63
Water	AAS; extract with DDTC into MIBK, pH 3·6; A-A flame	64
Water	AAS; direct method	65
Water	AAS; acidify with HCl, heat at 250°F for 1 h; use $CaCl_2$ to suppress Si interference	19
Sea-water	AAS; extract with 8-hydroxyquinoline into MIBK from pH 8; A-A flame	66
Sea-water	AAS; extract with DDTC into MIBK	67
River and spring waters	AAS; make 0·1 N in HNO_3; A-A flame	68
River water	AAS; graphite furnace	51
KCL brines	AAS; extract with APDC into MIBK	69
Oil-field brines	AAS; preconcentrate with chelating ion-exchange resin; dilute with 1 M HCl	70
Atmospheric samples	AAS; collect on organic membrane	71
Atmospheric particulate matter	AAS; collect on membrane filter, ash, dissolve in HNO_3/HCl	72

References

1. "Analytical Methods for Atomic Absorption Spectrophotometry", Perkin-Elmer Corp, Norwalk, Conn., U.S.A., 1971.
2. J. E. Allan, *Spectrochim. Acta,* 1959, **15**, 800.
3. K. E. Zacha, M. P. Bratzel Jnr., J. D. Winefordner and J. M. Mansfield Jnr., *Anal. Chem.,* 1968, **40**, 1733.
4. D. C. Manning and P. Heneage, *Atomic Absorption Newsletter,* 1968, **7**, 80.
5. G. Rossi and N. Omenetto, *Talanta,* 1969, **16**, 263.
6. L. Ebdon, G. F. Kirkbright and T. S. West, *Talanta,* 1970, **17**, 965.
7. P. L. Larkins, *Spectrochim. Acta,* 1971, **26B**, 477.
8. D. R. Demers and D. G. Mitchell, "Advances in Automated Analysis", Technicon International Congress, New York, 1970, Vol. II, Thurman Associates, Miami, Florida, 1971.
9. C. W. Frank, W. G. Schrenk and C. E. Meloan, *Anal. Chem.,* 1966, **38**, 1005.
10. B. M. Gatehouse and J. B. Willis, *Spectrochim. Acta,* 1961, **17**, 710.
11. F. J. Feldman, R. E. Bosshart and G. D. Christian, *Anal. Chem.,* 1967, **39**, 1175.
12. S. L. Sachdev, J. W. Robinson and P. W. West, *Anal. Chim. Acta,* 1967, **38**, 491.
13. I. Rubeska and B. Moldan, *Appl. Optics,* 1968, **7**, 1341.
14. C. Iida and M. Nagura, *Spectrosc. Letters,* 1970, **3**, 63.

15. B. W. Bailey and J. M. Rankin, *Spectrosc. Letters,* 1969, **2**, 233.
16. B. V. L'Vov, *Spectrochim. Acta,* 1969, **24B**, 53.
17. L. Ebdon, G. F. Kirkbright and T. S. West, *Anal. Chim. Acta,* 1972, **61**, 15.
18. C. B. Belcher and K. Kinson, *Anal. Chim. Acta,* 1964, **30**, 483.
19. J. A. Platte and V. M. Marcy, *Atomic Absorption Newsletter,* 1965, **4**, 289.
20. S. Sprague and W. Slavin, *Atomic Absorption Newsletter,* 1964, **3**, 72.
21. D. C. Manning and H. Chabot, *Atomic Absorption Newsletter,* 1968, **7**, 94.
22. M. Beyer, *Atomic Absorption Newsletter,* 1965, **4**, 212.
23. D. P. Hubbard and H. H. Monks, *Anal. Chim. Acta,* 1969, **47**, 197.
24. I. Atsuya, *Japan Analyst,* 1966, **15**, 672.
25. F. J. Langmyhr and P. E. Paus, *Anal. Chim. Acta,* 1969, **45**, 173.
26. D. C. Smith, J. R. Johnson and G. C. Soth, *Appl. Spectrosc.,* 1970, **24**, 576.
27. R. K. Winge, V. A. Fassel and R. N. Kniseley, *Appl. Spectrosc.,* 1971, **25**, 636.
28. Y. Endo and T. Hata, *Japan Analyst,* 1968, **17**, 679.
29. R. Dyck, *Atomic Absorption Newsletter,* 1965, **4**, 170.
30. G. F. Bell, *Atomic Absorption Newsletter,* 1966, **5**, 73.
31. R. M. Dagnall, G. F. Kirkbright, T. S. West and R. Wood, *Analyst,* 1972, **97**, 245.
32. C. R. Walker and O. A. Vita, *Anal. Chim. Acta,* 1968, **43**, 27.
33. G. G. Welcher and O. H. Kriege, *Atomic Absorption Newsletter,* 1970, **9**, 61.
34. W. T. Elwell and J. A. F. Gidley, "Atomic Absorption Spectrophotometry", Pergamon Press, London, 2nd ed., 1966.
35. T. W. Sattur, *Atomic Absorption Newsletter,* 1966, **5**, 37.
36. P. Johns and W. J. Price, *Metallurgia,* 1970, **81**, 75.
37. J. T. H. Roos and W. J. Price, *Analyst,* 1969, **94**, 89.
38. T. Takeuchi and M. Suzuki, *Talanta,* 1964, **11**, 1391.
39. F. J. Langmyhr and P. E. Paus, *Anal. Chim. Acta,* 1968, **43**, 397; 1969, **47**, 371.
40. D. J. Trent and W. Slavin, *Atomic Absorption Newsletter,* 1964, **3**, 17, 118.
41. J. C. Van Loon and C. M. Parissis, *Analyst,* 1969, **94**, 1057.
42. A. H. Jones, *Anal. Chem.,* 1965, **37**, 1761.
43. F. J. Langmyhr and P. E. Paus, *Anal. Chim. Acta,* 1969, **43**, 508.
44. R. C. Calkins, *Appl. Spectrosc.,* 1966, **20**, 146.
45. R. H. Thompson and W. J. Blanchflower, *Lab. Pract.,* 1971, **20**, 859.
46. W. Slavin, "Atomic Absorption Spectroscopy", Interscience, New York, 1968, p. 204.
47. D. B. Cheek, G. K. Powell, R. Reba and M. Feldman, *Bull. John Hopkins Hosp.,* 1966, **118**, 338.
48. B. H. Lehmann, J. D. L. Hansen and P. J. Warren, *Brit. J. Nutrition,* 1971, **26**, 197.
49. C. T. Gamble, S. L. Hausard, B. R. Moss, D. J. Davis and E. R. Lidvall, *J. Animal. Sci.,* 1971, **32**, 84.
50. J. M. Hsu and W. L. Anthony, *J. Nutr.,* 1971, **101**, 525.
51. H. L. Kahn, *American Laboratory,* 1971, August, p. 35.
52. B. Montford and S. C. Cribbs, *Talanta,* 1969, **16**, 1079.
53. D. J. David, *Atomic Absorption Newsletter,* 1967, **1**, 45.
54. J. R. Buchanan and T. T. Muraoka, *Atomic Absorption Newsletter,* 1964, **3**, 79.
55. J. H. Baker and T. Greweling, *J. Agric. Food Chem.,* 1967, **15**, 340.
56. B. J. Gudzinowicz and V. J. Luciano, *J. Ass. Off. Analyt. Chem.,* 1966, **49**, 1.

57. L. R. Hossner and L. W. Ferrara, *Atomic Absorption Newsletter,* 1967, **6,** 71.
58. M. Nadirshaw and A. H. Cornfield, *Analyst,* 1968, **93,** 475.
59. R. M. Dagnall, G. F. Kirkbright, T. S. West and Roger Wood, *Anal. Chem.,* 1971, **43,** 1765.
60. A. M. Ure and M. L. Berrow, *Anal. Chim. Acta,* 1970, **52,** 247.
61. C. H. McBride, *J. Ass. Off. Analyt. Chem.,* 1965, **48,** 406.
62. W. Slavin, *Atomic Absorption Newsletter,* 1965, **4,** 330.
63. P. Aufila and V. Aufila, *Suamen Kemist.,* 1971, **44,** 161.
64. J. Nix and T. Goodwin, *Atomic Absorption Newsletter,* 1970, **9,** 119.
65. M. J. Fishman and S. C. Downs, *V.S. Geol. Surv. Water-Supply Papers,* 1540-C, U.S. Gov. Printing Office, Washington, D.C., 1966.
66. M. J. Orren, *J. S. Afr. Chem. Inst.,* 1971, **24,** 96.
67. T. Joyner and J. S. Finely, *Atomic Absorption Newsletter,* 1966, **5,** 4.
68. R. Nakagawa and Y. Ohyagi, *Nippon Kagaku Zasshi,* 1971, **92,** 620.
69. S. Sprague and W. Slavin, *Atomic Absorption Newsletter,* 1964, **3,** 37.
70. O. K. Galle, *Appl. Spectrosc.,* 1971, **25,** 664.
71. J. Y. Hwang and F. J. Feldman, *Appl. Spectrosc.,* 1970, **24,** 371.
72. R. J. Thompson, G. B. Morgan and L. J. Purdue, *Atomic Absorption Newsletter,* 1970, **9,** 53.

13.4.34 Mercury

Hg, atomic weight 200·59

Technique	Line, nm	*f*	Source	Flame	Sensitivity, ppm/1% abs.	Detection limit, ppm	Reference (see below)
AAS	253·65	0·03	HCL	A-A	10	0·5	1
	253·65	0·03	HCL	Ar/H_2	7·5	—	2
	253·65	0·03	EDL	N_2 sep. N-A	1·0	2·0[a]	3
	184·96	*ca.* 1	EDL	N_2 sep. N-A	0·05	0·02[a]	3
	253·65	0·03	HCL or VDL	Non-flame cold vapour techniques		Often < 1 ng (see below)	
AFS	253·65	0·03	EDL	Ar/H_2	—	0·1	4
	253·65	0·03	EDL	A-H	—	0·12	5
ND	253·65	0·03	VDL	A-town gas	—	0·00025	6
ND	253·65	0·03	VDL	N_2 sep. A-A	—	0·07	7
	253·65 (+ excitation at 184·96 + 253·65)	0·03	EDL	N_2 sep. A-A	—	0·02	8

[a] Hg(I) solutions nebulized.

Ionization potential: 10·43 eV.

Notes on analytical conditions

1. Low-pressure mercury VDL sources may be used for Hg determination by AAS provided that low lamp currents are used to avoid loss of sensitivity due to self-absorption at the Hg 253·65 nm line. Mercury HCL sources must also be operated at low current for this reason and to achieve long operating lifetime.

2. The Hg 184·96 nm line is potentially capable of providing higher sensitivity than the spin-forbidden 253·65 nm line. The atmosphere and most flames absorb very strongly at this wavelength, however; use of the nitrogen-shielded nitrous oxide-acetylene flame and an inert gas purged optical path have been shown to permit operation at 184·96 nm with high sensitivity and freedom from interferences in AAS.[3] Similarly in AFS the provision of a purged optical path between an EDL source and the flame provides for excitation at both 253·65 and 184·96 nm and enhanced fluorescence emission (resonance + stepwise line, respectively) at 253·65 nm.[8]

3. Several workers have observed that the response obtained for Hg(I) is greater than that for Hg(II) solutions in both AAS and AFS with flames used in conjunction with indirect nebulizer units.[9,10,5,11] This may possibly be accounted for by disproportionation of Hg(I) by the reaction $Hg_2^{2+} \rightarrow Hg^{2+} + Hg^0$ in the nebulizer, whereby free mercury vapour exists in the chamber and enters the flame when Hg(I) is used. This effect may be made even more pronounced by addition of reducing agents such as $SnCl_2$ or ascorbic acid to the Hg solution to be nebulized.

Cold vapour techniques and other methods of sensitivity enhancement

Elemental mercury exhibits an appreciable vapour pressure at room temperature; the vapour is monatomic and it is therefore unnecessary to employ a flame for mercury atomization. The determination of Hg by AAS in this way was undertaken by Woodson[12] appreciably before the work of Walsh on flame AAS. In early methods for the determination of Hg by the cold vapour technique the element was collected on a pad of cadmium sulphide and volatilized into a silica cell for absorption measurements at 253·65 nm;[13,14] broad-band absorption interference from organic compounds was compensated for using a continuum background source.[15] In later work Ulfvarson[16] and Brandenburger and Bader[17] described the collection of Hg by amalgamation onto gold or copper prior to measurement of the absorption by the cold vapour technique; the latter workers described both dynamic[17] and static[18] techniques for the determination of nanogramme amounts of mercury. Poluektov and co-workers[11] used stannous chloride to reduce mercury in sample solutions to elemental form and then passed air through the solution to sweep the vapour into an absorption cell. Organic material may be destroyed by oxidation (usually with $KMnO_4$) before reduction of mercury for absorption measurement.[19-22] Digestion with $KMnO_4/H_2SO_4$ may release both inorganic and organically bound mercury.[21] It has been shown,[23] however, that in dilute (2 N) sulphuric acid medium, when oxidative digestion is not employed, stannous chloride only reduces inorganic mercury. In alkaline medium, however, in the presence of cupric ions both inorganic and organic mercury are reduced by stannous chloride. A simple procedure is therefore possible for the determination of mercury in both forms.[23] This type of method has found widespread application to the AAS determination of mercury in biological samples (see Table 13-7) and many devices that may be employed with commercially available instrumentation for AAS have been described.

Other methods that have been employed to increase sensitivity for the determination of mercury by AAS or AFS include solvent extraction into MIBK with APDC,[24] with dithizone[25] into $CHCl_3$ or using an indirect method based on the extraction into 1,2-dichloroethane of the anionic mercury bromo-complex with Zn-dipyridyl and estimation of the Zn concentration by AAS.[26] The application of a platinum loop atomizer for the determination of nanogramme amounts of Hg by AFS has also been described.[27] Long path flame adaptors may also provide for increased absorption sensitivity for the determination of mercury.[28,29,30]

Applications notes and key references

Relatively little work has been published on the determination of mercury in samples of metallurgical or geochemical interest. Owing to its toxicity, however, considerable attention has been paid to the determination of mercury in biological fluids and in water, air, and plant and animal tissues. The literature concerned with these determinations has become voluminous since the development of the cold vapour technique for routine analytical purposes; the technique provides for rapid and precise analysis with very high sensitivity. Table 13-7 lists some of the publications concerned with the determination of mercury by AAS.

TABLE 13-7

Some applications of the determination of mercury by AAS

Sample type	Technique, sample pretreatment	Reference
Hg isotope analysis	AAS; ^{202}Hg EDL source used for 253·65 nm resonance line	31
Coals, ash	AAS; reduce with $SnCl_2$, collect vapour on Ag, vaporize Hg from Ag in an induction furnace atomizer	32
Rocks	AAS; heat rock sample to 500°C, vapour cell technique	33
Geological materials	AAS; dissolve in HF/HNO_3 in sealed vessel; add H_3BO_3, reduce with $SnCl_2$; cold vapour technique	34
Rocks and soils	AAS; cold vapour technique	19
Soils	AAS	35
Urine	AAS; digest with $KMnO_4/H_2SO_4$, cold vapour technique	36
Urine	AAS; extract with APDC into MIBK; nebulize	23
Urine, blood	AAS; adjust to pH 2·8 (urine), add TCA, extract with APDC into MIBK; A-A flame	37
Urine	AAS; burn sample in O_2/H_2 flame, condense out water vapour and salts from flame exhaust gases, measure absorption in gas stream	38
Biological samples	AAS; automated technique, absorption cell	39
Blood, urine	AAS; deposit on Pt, heat to release Hg vapour; cold vapour technique	40
Urine	AAS; digest with $KMnO_4/H_2SO_4$, destroy excess $KMnO_4$ with $NH_2OH\ .\ HCl$; reduce with $SnCl_2$; absorption cell	20
Tissues	AAS; $KMnO_4/H_2/SO_4$ digestion, reduce with $SnCl_2$; cold vapour technique	22
Urine	AAS; extract with APDC into MIBK; boat technique	41
Urine	AAS; solvent extraction. Use long path flame absorption tube	27
Biological samples, water	AAS; oxidize with HNO_3, reduce to element with $SnCl_2$, absorb vapour on charcoal; use sampling boat with charcoal sample	42
Biological samples	AAS; digest with HNO_3/H_2SO_4 and vanadium pentoxide	43
Wheat, fish	AAS; oxygen flask combusion, cold vapour technique	44
Foods	AAS; GLC column eluant analysed by cold vapour method	45
Fish tissue	AAS; digest with $H_2SO_4/KMnO_4$ at 70°C for 2 hours. Add $NH_2OH\ .\ HCl$, then $SnCl_2$, pass air to sweep Hg into cold vapour absorption cell	46

TABLE 13-7 (*continued*)

Sample type	Technique, sample pretreatment	Reference
Wheat, cereals	AAS; decompose in oxygen bomb, reduce with $SnCl_2$; cold vapour technique	47
Fish	AAS; acid digestion reduction to element; cold vapour technique	48, 49, 50
Fish, urine, water	AAS; reduce to element; use cold vapour technique	51
Fish, foods	AAS; acid digest, reduce, cold vapour technique	52
River water	AAS; reduce inorganic Hg with $SnCl_2$ in $2NH_2SO_4$, organic Hg with $SnCl_2$ in 1 N NaOH in presence of Cu^{2+}; cold vapour technique	23
Water	AAS; acidify with HNO_3, dilute; collect on Ag wire, volatilize; absorption cell	53
Water	AAS; acid digestion, reduce to element, cold vapour technique	21
Water and sediment samples	AAS; absorption cell	54
Eye drops	AAS; extract with APDC into MIBK	55
Pharmaceuticals	AAS; boil with HNO_3, dilute, filter; A-A flame	56
Paper, wood pulp	AAS; digest with HCl/HNO_3, reduce to element; cold vapour technique	57
Polymers	AAS; flame method	58
Air	AAS; direct measurement	59
Air	AAS; direct measurement; induction heated silica cell containing carbon rods	60

References

1. "Analytical Methods for Atomic Absorption Spectrophotometry", Perkin-Elmer Corp, Norwalk, Connecticut, 1971.
2. H. L. Kahn and J. E. Schallis, *Atomic Absorption Newsletter,* 1968, **7,** 5.
3. G. F. Kirkbright, T. S. West and P. J. Wilson, *Analyst,* 1973, **98,** 49.
4. K. E. Zacha, M. P. Bratzel Jnr., J. D. Winefordner and J. M. Mansfield, *Anal. Chem.,* 1968, **40,** 1733.
5. R. F. Browner, R. M. Dagnall and T. S. West, *Talanta,* 1969, **16,** 75.
6. P. D. Warr, *Talanta,* 1970, **17,** 543.
7. P. L. Larkins, *Spectrochim. Acta,* 1971, **26B,** 477.
8. G. F. Kirkbright, T. S. West and P. J. Wilson, *Anal. Chem. Acta,* 1973, **66,** 130.
9. D. N. Hingle, G. F. Kirkbright and T. S. West, *Analyst,* 1967, **92,** 759.
10. Handbook, "Techtron AA4 Flame Spectrophotometer", Techtron Pty. Ltd., Australia.
11. N. S. Poluektov, R. A. Vitkun and Yu. V. Zelyukova, *Zh. Anal. Khim.,* 1964, **19,** 937.
12. T. T. Woodson, *Rev. Sci. Instr.,* 1939, **10,** 308.
13. A. E. Ballard and C. D. W. Thornton, *Ind. Eg. Chem., Anal. Ed.,* 1941, **13,** 893.
14. C. W. Zuelke and A. E. Ballard, *Anal. Chem.,* 1950, **22,** 953.
15. A. E. Ballard, D. W. Stewart, W. O. Kamm and C. W. Zuelke, *Anal. Chem.,* 1954, **26,** 921.

16. U. Ulfvarson, *Acta Chem. Scand.*, 1967, **21**, 641.
17. H. Brandenburger and H. Bader, *Helv. Chim. Acta*, 1967, **50**, 1409.
18. H. Brandenburger and H. Bader, *Atomic Absorption Newsletter*, 1968, **7**, 53.
19. W. R. Hatch and W. L. Ott, *Anal. Chem.*, 1968, **40**, 2085.
20. G. Lindstedt, *Analyst*, 1970, **95**, 264.
21. S. H. Omang, *Anal. Chim. Acta*, 1971, **53**, 415.
22. R. Braun and A. P. Husbands, *Spectrovision*, 1971, **26**, 2.
23. Y. Umezaki and K. Iwamoto, *Japan Analyst*, 1971, **20**, 173.
24. J. B. Willis, *Anal. Chem.*, 1962, **34**, 614.
25. T. J. Vickers and S. P. Merrick, *Talanta*, 1968, **15**, 873.
26. Y. Yamamoto, T. Kumamaru, Y. Hayashi and Y. Otani, *Anal. Letters*, 1968, **1**, 955.
27. M. P. Bratzel Jnr., R. M. Dagnall and J. D. Winefordner, *Anal. Chem. Acta*, 1969, **48**, 197.
28. S. R. Koirtyohann and C. Feldman, in "Developments in Applied Spectroscopy", Vol. III, eds. J. E. Forrette and E. Lanterman, Plenum Press, New York, 1964.
29. B. Moldan, in "Atomic Absorption Spectroscopy", eds. R. M. Dagnall and G. F. Kirkbright, Butterworths, London, 1970.
30. C. Iida and M. Nagura, *Spectrosc. Letters*, 1970, **3**, 63.
31. K. R. Osborn and H. E. Gunning, *J. Opt. Soc. Am.*, 1955, **45**, 552.
32. M. R. Booth and J. Brown, *Proc. 18th Spectrosc., Symposium of Canada*, Toronto, Ontario, October, 1971.
33. J. A. Goleb, *Appl. Spectrosc.*, 1971, **25**, 522.
34. S. H. Omang and P. E. Paus, *Anal. Chim. Acta*, 1971, **56**, 393.
35. W. W. Vaughn and J. H. McCarthy Jnr., *U.S. Geol. Surv. Prof. Papers*, 501-D, 1964, p. 123.
36. J. L. Monkman, P. A. Moffet and T. F. Doherty, *Ind. Hyg. Found., Amer., Quart.*, 1956, **17**, 418.
37. E. Berman, *Atomic Absorption Newsletter*, 1967, **6**, 57.
38. O. Lindstrom, *Anal. Chem.*, 1959, **31**, 461.
39. G. Lindstedt and I. Skare, *Analyst*, 1971, **96**, 223.
40. G. Thilliez, *Chim. Anal.*, 1968, **50**, 226.
41. B. B. Mesman and B. S. Smith, *Atomic Absorption Newsletter*, 1970, **9**, 81.
42. A. E. Moffitt Jnr. and R. E. Kupel, *Atomic Absorption Newsletter*, 1970, **9**, 113.
43. M. Malaiyandi and J. P. Barrette, *Anal. Letters*, 1970, 3, 579.
44. E. G. Pappas and L. A. Rosenberg, *J. Ass. Off. Analyt. Chem.*, 1966, **49**, 782, 792.
45. H. Hey, *Z. Anal. Chem.*, 1971, **5**, 256.
46. J. F. Uthe, F. A. J. Armstrong and M. P. Stainton, *Fish. Res. Board. Can.*, 1970, **27**, 805.
47. R. Tkachuk, *Proc. Intern. Symp. on Indent. and Measurement of Environmental Pollutants*, Ottawa, Canada, June 1971.
48. M. J. Jeffus, J. S. Elkins and C. T. Kenner, *J. Ass. Off. Analyt. Chem.*, 1970, **53**, 1172.
49. R. K. Munns and D. C. Holland, *J. Ass. Off. Analyt. Chem.*, 1971, **54**, 202.
50. W. L. Hoover, J. R. Melton and P. A. Howard, *J. Ass. Off. Analyt. Chem.*, 1971, **54**, 860.
51. J. Y. Hwang, P. A. Ullucci and A. L. Malenfant, *Can. Spectrosc.*, 1971, **16**, 100.
52. V. A. Thorpe, *J. Ass. Off. Analyt. Chem.*, 1971, **54**, 206.

53. M. J. Fishman, *Anal. Chem.*, 1970, **42**, 1462.
54. G. W. Kalb, *Atomic Absorption Newsletter*, 1970, **9**, 84.
55. W. H. Harper, *Proc. Soc. Anal. Chem.*, 1970, **7**, 104.
56. J. R. Leaton, *J. Ass. Off. Analyt. Chem.*, 1970, **53**, 237.
57. D. C. Lee and C. W. Laufman, *Anal. Chem.*, 1971, **43**, 1127.
58. M. Olivier, *Z. Anal. Chem.*, 1969, **248**, 145.
59. C. M. Christian and J. W. Robinson, *Anal. Chim. Acta*, 1971, **56**, 466.
60. J. W. Robinson and Y. E. Araktingi, *Anal. Chim. Acta*, 1973, **63**, 29.

13.4.35 Molybdenum

Mo, atomic weight 95·94

Technique	Line, nm	*f*	Source	Flame	Sensitivity, ppm/1% abs.	Detection limit, ppm	Reference (see below)
AAS	313·26	0·20	HCL	A-A	0·8	—	1, 2
	313·26	0·20	..	N-A	0·4	—	3
	317·03	0·12	..	A-A	1·1	—	1
	379·83	0·13	..	A-A	1·3	—	1
	319·40	—	..	A-A	1·4	—	1
	386·41	—	..	A-A	1·7	—	1
	390·30	—	..	A-A	2·4	—	1
	315·82	—	..	A-A	2·8	—	1
	320·88	—	..	A-A	5·9	—	1
AFS	313·26	0·20	EDL	Ar sep. N-A	—	0·46	4

Ionization potential: 7·10 eV.
Dissociation energy MoO: 5·0 eV.

Notes on analytical conditions

1. A reducing fuel-rich air-acetylene flame[5] or nitrous oxide-acetylene flame[6] is required for the atomization of Mo for AAS. The absorption signal is quite dependent upon the height of observation in the flame; sensitivity may be enhanced by using a three-slot burner[7] or by inert gas separation of the flame.[8]

2. In the air-acetylene flame several workers have observed a number of interferences on Mo absorption; the suppression observed in the presence of Ca, Sr, Mn, Fe and sulphate was reported by David[5] to be eliminated by addition of a large excess of Al. Mostyn and Cunningham preferred the use of 2% NH_4Cl to suppress interferences in this flame. In early work with the nitrous oxide-acetylene flame, Kirkbright, Smith and West[6] found a depression of Mo absorption in the presence of large amounts of iron. Ramakrishna, Robinson and West[9] also observed interference from calcium and some other ions in this flame and have recommended

the addition of Al or Na_2SO_4 to control these effects. Other workers have also discussed interferences in the determination of Mo by AAS.[10,11] Manning and Chabot[12] found reduced sensitivity in Mo AAS in the presence of acetone vapour from depleted acetylene cylinders. The determination of Mo by AFS using an argon-separated nitrous oxide-acetylene flame has been reported to suffer interference from several ions present at a 100-fold excess.[4]

3. Molybdenum may be extracted into MIBK with APDC at pH 2,[13] with toluene-3,4-dithiol[14] or with 8-hydroxyquinoline[13,15] prior to its determination by AAS.

Applications notes and key references

A number of papers concerned with the determination of Mo in ferrous materials have been published. In the first report of the AAS of Mo, David[5] applied the method to stainless steel samples using an air-acetylene flame and aluminium as a releasing agent. Mostyn and Cunningham[1] determined Mo in ferrous alloys using the Mo 379·83 nm line and an air-acetylene flame with NH_4Cl as releasing agent. Kirkbright, Smith and West[6] determined Mo in alloy steels in a premixed nitrous oxide-acetylene flame; no releasing agent was employed but iron was added to the solutions of the standards in order to compensate for its depression of Mo absorption. Endo *et al.*[16] and Thomerson and Price[17] have employed a similar method for the AAS determination of Mo in steels. Knight and Pyzyna[18] also found it necessary to match the compositions of sample and standard solutions in the analysis of tool steel samples. Pollock[19] has described the determination of Mo in slags using $AlCl_3$ releasing agent. Kirkbright, Peters and West[15] determined between 0·006 and 0·06% Mo in niobium or tantalum in a procedure in which Mo(VI)-8-hydroxyquinolate was extracted from a fluoride/EDTA medium at pH 4·5 into n-butanol and the organic phase was nebulized into a nitrous oxide-acetylene flame for AAS. The interference of vanadium(V) and chromium(VI) was suppressed by addition of K_2SO_3 to the sample solution before extraction. Grison[20] has examined the AAS determination of Mo in samples from the re-processing of irradiated U fuel and Walker and co-workers[21] have determined Mo in U or Ni by AAS. Welcher and Kriege[22] have determined Mo in high-temperature cobalt-based alloys in a nitrous oxide-acetylene flame.

Butler and Mathews[13] determined down to 1 ppm Mo in silicate rocks after preconcentration by solvent extraction with APDC or

8-hydroxyquinoline. Molybdenum in mixtures containing MoS_2 has been determined by fusing the sample with NaOH and baking at 650°C for 30 min; the residue was dissolved in dilute H_2SO_4, filtered and nebulized into an air-acetylene flame.[23] Hutchinson[24] determined Mo in geological materials by AAS. After acid dissolution of samples, Mo was extracted with benzoin-α-oxime into $CHCl_3$, solvent evaporated off, the residue warmed with $HNO_3/HClO_4$, evaporated to dryness and redissolved in $HClO_4/NH_4Cl$ solution for nebulization into a luminous air-acetylene flame. Mostyn and Cunningham[25] also determined Mo in lubricating greases after ashing and extracting the ash with ammonia. Billings and Ragland[26] concentrated Mo from oil-field brines by extraction with APDC into MIBK before its AAS determination in a nitrous oxide-acetylene flame.

Blood and urine have been analysed for Mo after dry ashing and dissolution of the residue in 10% HCl,[27] and plant samples after dry ashing and extraction with APDC.[13] Molybdenum in dry-ashed citrus leaves has also been determined by AAS after formation and extraction of its APDC complex.[28] David,[5] and more recently Koirtyohann and Hamilton,[29] have determined Mo by AAS in fertilizer samples. Roussos and Morrow[30] proposed a direct nebulization method for the AAS determination of Mo in milk xanthine oxidase fractions and demonstrated that this gave similar results to those obtained after preliminary dry-ashing and HCl extraction. The direct determination of Mo in water samples by AAS has been studied by Chau and Lum-Shue-Chan[31] and by Farnsworth.[32] Delaughter[14] determined Mo in brines after extraction with toluene-3,4-dithiol. The determination of Mo by AFS appears to be relatively insensitive and has not found practical application at the time of writing.

References

1. R. A. Mostyn and A. F. Cunningham, *Anal. Chem.*, 1966, **38**, 121.
2. "Analytical Methods in Atomic Absorption Spectrophotometry", Perkin-Elmer Corp., Norwalk, Conn., U.S.A., 1971.
3. M. D. Amos and J. B. Willis, *Spectrochim. Acta*, 1966, **22**, 1325.
4. R. M. Dagnall, G. F. Kirkbright, T. S. West and R. Wood, *Anal. Chem.*, 1970, **42**, 1029.
5. D. J. David, *Analyst*, 1961, **86**, 730.
6. G. F. Kirkbright, A. M. Smith and T. S. West, *Analyst*, 1966, **91**, 700.
7. S. Sprague and W. Slavin, *Atomic Absorption Newsletter*, 1965, **4**, 293.
8. G. F. Kirkbright, M. Sargent and T. S. West, *Talanta*, 1969, **16**, 1467.

9. T. V. Ramakrishna, P. W. West and J. W. Robinson, *Anal. Chim. Acta,* 1969, **44**, 437.
10. H. Rampson, *Chim. Analyt.*, 1969, **51**, 627.
11. J. D. Kerbyson and C. Ratzkowski, *Can. Spectrosc.*, 1970, **15**, 43.
12. D. C. Manning and H. Chabot, *Atomic Absorption Newsletter,* 1968, **7**, 94.
13. L. R. P. Butler and P. M. Mathews, *Anal. Chim. Acta,* 1966, **36**, 319.
14. B. Delaughter, *Atomic Absorption Newsletter,* 1965, **4**, 273.
15. G. F. Kirkbright, M. K. Peters and T. W. West, *Analyst,* 1966, **91**, 705.
16. Y. Endo, T. Hata and Y. Nakahara, *Japan Analyst,* 1969, **18**, 878.
17. D. R. Thomerson and W. J. Price, *Analyst,* 1971, **96**, 321.
18. D. M. Knight and M. K. Pyzyna, *Atomic Absorption Newsletter,* 1969, **8**, 129.
19. E. N. Pollock, *Atomic Absorption Newsletter,* 1970, **9**, 47.
20. J. Grison, *Chim. Analyt.*, 1969, **51**, 605.
21. C. R. Walker, O. A. Vita and R. W. Sparks, *Anal. Chim. Acta,* 1969, **47**, 1.
22. G. G. Welcher and O. H. Kriege, *Atomic Absorption Newsletter,* 1970, **9**, 61.
23. R. J. Julietti and J. A. E. Wilkinson, *Analyst,* 1968, **93**, 797.
24. D. Hutchinson, *Analyst,* 1972, **97**, 118.
25. R. A. Mostyn and A. F. Cunningham, *J. Inst. Petrol.*, 1967, **53**, 101.
26. G. K. Billings and P. C. Ragland, *Can. Spectrosc.*, 1969, **14**, 8.
27. J. O. Pierce and J. Cholak, *Arch. Envir. Health,* 1966, **13**, 208.
28. W. J. A. Steyn and P. J. Brandt, *J. S. Afr. Chem. Inst.*, 1965, **18**, 92.
29. S. R. Koirtyohann and M. Hamilton, *J. Ass. Off. Analyt. Chem.*, 1971, **54**, 787.
30. G. G. Roussos and B. H. Morrow, *Appl. Spectrosc.*, 1968, **22**, 769.
31. Y. K. Chau and K. Lum-Shue-Chan, *Anal. Chim. Acta,* 1969, **48**, 205.
32. C. G. Farnsworth, *Water Sewage Wks.*, 1972, **119**, 52.

13.4.36 Neodymium

Nd, atomic weight 144·24

Technique	Line, nm	*f*	Source	Flame	Sensitivity, ppm/1% abs.	Detection limit, ppm	Reference (see below)
AAS	492·45	–	Xe arc	O_2/A	–	40	1
	463·42	–	HCL	N-A	35	–	2
	471·90	–	HCL	N-A	73	–	2
	489·69	–	HCL	N-A	48	–	2
	463·42	–	HCL	N-A	13	2	3
AFS: not reported							

Ionization potential: 5·45 eV.
Dissociation energy, NdO: 7·4 eV.

Notes on analytical conditions

1. A hot fuel-rich flame is required for atomization of Nd. Mossotti and Fassel[1] used an arrangement of three oxy-acetylene flames supported at total consumption nebulizer-burner units and a xenon arc continuum source for detection of Nd.

2. Nd is appreciably ionized in the hot flames required for its atomization; highest sensitivity is therefore obtained when potassium is added to samples and standards to suppress ionization. Jaworowski and co-workers[4] investigated the effect of organic solvents on Nd absorption in oxy-acetylene and nitrous oxide-acetylene flame using Nd solution in 80% methanol or ethanol.

3. Kinnunen and Lindsjo[5] observed suppression of Nd absorption in the nitrous oxide flame in the presence of Si, Al and fluoride; cerium, however, was reported to cause slight enhancement of absorption.

Applications notes and key references

See Lanthanum.

References

1. V. G. Mossotti and V. A. Fassel, *Spectrochim. Acta,* 1964, **20,** 1117.
2. M. D. Amos and J. B. Willis, *Spectrochim. Acta,* 1966, **22,** 1325.
3. "Analytical Methods for Atomic Absorption Spectrophotometry", Perkin-Elmer Corp., Norwalk, Conn., U.S.A., 1971.
4. R. J. Jaworowski, R. P. Weberling and D. J. Bracco, *Anal. Chim. Acta,* 1967, **37,** 284.
5. J. Kinnunen and O. Lindsjo, *Chemist-Analyst,* 1967, **56,** 25.

13.4.37 Nickel

Ni, atomic weight 58·71

Technique	Line, nm	f	Source	Flame	Sensitivity, ppm/1% abs.	Detection limit, ppm	Reference (see below)
AAS	232·00	0·095	HCL	A-A	0·12	0·01	1, 2
	231·09	–	HCL	A-A	0·2	–	–
	234·55	0·05	HCL	A-A	0·5	–	–
	341·48	0·3	HCL	A-A	0·5	–	–
	352·45	0·12	HCL	A-A	2·5	–	–
	346·17	0·16	HCL	A-A	1·0	–	–
	234·75	–	HCL	A-A	3·5	–	–
AFS	232·00	0·095	HIL	$Ar/O_2/H_2$	–	0·003	3
	232·00	0·095	HIL	A-H	–	0·003	4
	232·00	0·095	EDL	Air/H_2	–	0·04	5
	232·00	0·095	HgVDL	A-H	–	1·0	6
ND	232·00	0·095	HIL	N_2 sep. A-A	–	0·002	7
ND	232·00	0·095	Pulsed HCL	N_2 sep. A-A	–	0·07	8
	232·00	0·095	Hot HCL	Air/H_2	–	0·1	9
	361·0/ 352·4	0·095	N_2 pumped dye laser	A-A	–	0·05	9a

Ionization potential: 7·63 eV.
Dissociation energy, NiO: 4·3 eV.

Notes on analytical conditions

1. The spectrum from HCL sources exhibits many Ni lines; some of the most sensitive absorption lines are listed above. Other lines which may be used for AAS include Ni 339·10, 233·75, 336·96, 323·29 and 247·68 which give 1% absorption sensitivities of approximately 5, 2, 2·5, 4·0 and 35 ppm in a 10 cm air-acetylene flame. To obtain highest sensitivity with the Ni 232·00 nm line a narrow spectral band-width must be used to minimize the effect of transmission of the nearby weakly absorbing and non-absorbing lines (see Fig. 11-2); with most spectrometers the Ni 232·00 nm line cannot be resolved from the Ni 231·98 nm non-absorbing line and curvature of the calibration graph is always observed. Greater sensitivity and improved linearity may be achieved using resonance detector or selective modulation techniques[10,11] in AAS, and the use of the less sensitive Ni 341·48 nm line may be preferred with conventional instrumentation when greater linearity is essential. Frank, Schrenk

and Meloan[12] utilized the Fe 352·42 nm line for Ni AAS at 352·45 nm. When a multi-element lamp containing Ni and Fe is employed interference in the determination of nickel at 232·00 nm may result when Ni is determined in a Fe matrix; use of the Ni 352·45 or 341·48 nm line is then recommended.[1]

2. Although an air-acetylene flame is most commonly employed, air-coal gas,[13] air-propane[14] and oxy-acetylene[15] flames have been employed for Ni absorption work. The air-acetylene flame gives best freedom from interferences when fuel-lean, particularly when Fe and Cr are present. Several workers have studied interferences in AAS for Ni[16,17,18] and observed generally that the determination is relatively free of chemical interference effects.

3. Nickel may be preconcentrated by extraction into MIBK with APDC from weakly acidic solutions (pH 2–5), by ion-exchange[19] or by coprecipitation with iron at pH 9.[20]

Applications and key references

Table 13-8 lists some of the published work concerned with the determination of nickel by AAS and AFS in different sample types.

TABLE 13-8

Some applications of the determination of nickel by AAS and AFS

Sample type	Technique; sample pretreatment	Reference
Steel	AAS; dissolve in H_3PO_4/H_2SO_4	21
Steel	AAS; dissolve in HCl/HNO_3, add $HClO_4$, evaporate, fume, cool, dissolve residue in water	22
Steel	AFS; dissolve in HCl/HNO_3. Use argon-separated A-A flame	23
Iron ores	AAS	24
Reduced ores and oxides	AAS; sample treated with Br_2/MeOH, cool, filter, wash + MeOH, add 1 : 1 HCl to filtrate, titrate + NH_2OH . HCl, add 3% H_2O_2. Boil, cool, dilute nebulize	25
Cobalt	AAS	26
Uranium	AAS; remove bulk of U by extraction with tributyl-phosphate into CCl_4; nebulize aqueous solution	27
Silver	AAS	28
Cu and Cu alloys	AAS	29
Cu and brass	AAS; dissolve in HCl/HNO_3	30
Al alloys	AFS; dissolve in HCl, evap. + HF, dissolve in HCl/HNO_3; N_2 sep. A-A flame	31

TABLE 13-8 (*continued*)

Sample type	Technique; sample pretreatment	Reference
Al alloys	AAS; dissolve in 50% HCl, add 30% H_2O_2 if Cu present, filter off silica or volatilize + HF/HNO_3, dissolve residue in HCl	32
Zr alloys	AAS: dissolve in HF/HNO_3, add H_3BO_3 to complex fluoride; A-A flame	33
Sn-Pb solders	AAS	34
Pu nitrate	AAS	35
Silicate and carbonate rocks	AAS; dissolve in HF/HNO_3, add 1% La to control interferences	36
Silicates	AAS; absorption tube technique	37
Marine sediments	AAS	38
Rocks and soils	AAS	39
Glass	AAS	40
Urine	AAS; direct or extract 50 ml samples + APDC into MIBK	41
Urine	AAS; direct nebulization of sample	42
Urine, protein	AAS; wet ash + HNO_3/H_2SO_4, extract + dimethylglyoxime into $CHCl_3$, back extract into 0·5 M HCl, nebulize	43
Blood, plasma	AAS; extract + APDC into iso-butyl acetate at pH 2·5	44
Serum, blood	AAS; concentrated by solvent extraction	45
Animal tissue	AAS; acid digestion, solvent extraction	46
Serum	AAS; no pretreatment. Graphite tube furnace	47
Tea	AAS; wet digest + $HNO_3/HClO_4$	48
Vegetable and mineral oils	AAS; samples diluted 1 : 3 with n-heptane	49
Gas oils	AAS; dilute 1 + 5 with xylene	50, 51
Gas oils	AFS; direct determination, separated A-A flame	52
Gasoline	AAS; burn directly	53
Petroleum fractions	AFS; direct determination in n-heptane and xylene	54
Mineral oil	AAS; graphite tube atomizer, dissolve oil in xylene, filter	55
Lubricating oils	AFS; H_2/argon-air or H_2/air flames with total consumption nebulizer burner unit	56, 57
Lubricating oils	AAS; dilute 1 : 10 + MIBK. Use NBS organo-metallic standard	58, 59, 60
Lubricating oils	AAS; carbon rod furnace	61
Fuming HNO_3	AAS	62
Industrial waters	AAS	63
Natural waters	AAS; ppt. metals with $FeCl_3$, dissolve in HCl, ext. + APDC into MIBK	64
Natural waters	AAS; extract + DDTC into MIBK; A-A flame	65
Brines	AAS; extract + NaDDTC into MIBK	66

TABLE 13-8 (*continued*)

Sample type	Technique; sample pretreatment	Reference
Brines	AAS; chelating resin preconcentration; elute + NH_4Cl at pH 4·5; A-A flame	67
KCl brines	AAS; extract + APDC into MIBK	68
Fruit juices and soft drinks	AAS	69
Wine	AAS	70
Beer	AAS	71

References

1. "Analytical Methods for Atomic Absorption Spectrophotometry", Perkin-Elmer Corp., Norwalk, Conn., 1971.
2. J. E. Allan, *Nature,* 1960, **187,** 1110.
3. J. Matousek and V. Sychra, *Anal. Chem.,* 1969, **41,** 518.
4. D. C. Manning and P. Heneage, *Atomic Absorption Newsletter,* 1967, **6,** 124.
5. K. E. Zacha, M. P. Bratzel Jnr., J. D. Winefordner and J. M. Mansfield Jnr., *Anal. Chem.,* 1968, **40,** 1733.
6. G. Rossi and N. Omenetto, *Talanta,* 1969, **16,** 263.
7. P. L. Larkins, *Spectrochim. Acta,* 1971, **26B,** 477.
8. R. M. Dagnall, G. F. Kirkbright, T. S. West and R. Wood, *Analyst,* 1972, **97,** 245.
9. J. I. Dinnin, *Anal. Chem.,* 1967, **39,** 1491.

9a. L. M. Fraser and J. D. Winefordner, *Anal. Chem.,* 1972, **44,** 1444.

10. B. J. Russell and A. Walsh, *Spectrochim. Acta,* 1959, **15,** 883.
11. J. V. Sullivan and A. Walsh, *Spectrochim. Acta,* 1965, **21,** 721.
12. C. W. Frank, W. G. Schrenk and C. E. Meloan, *Anal. Chem.,* 1966, **38,** 1005.
13. J. B. Willis, *Anal. Chem.,* 1962, **34,** 614.
14. I. P. Ivanov and B. M. Talalaev, *Zav. Lab.,* 1966, **32,** 1481.
15. S. L. Sachdev, J. W. Robinson and P. W. West, *Anal. Chim. Acta,* 1967, **38,** 491.
16. K. Kinson and C. B. Belcher, *Anal. Chim. Acta,* 1964, **30,** 64.
17. F. W. Sunderman, *Amer. J. Clin. Pathol.,* 1965, **44,** 182.
18. J. A. Platte and V. M. Marcy, *Atomic Absorption Newsletter,* 1965, **4,** 289.
19. D. C. Belcher, *Anal. Chem.,* 1965, **37,** 1054.
20. D. C. Burrell, *Atomic Absorption Newsletter,* 1965, **4,** 328.
21. K. Kinson and C. B. Belcher, *Anal. Chim. Acta,* 1964, **30,** 64.
22. W. J. Price and P. A. Cooke, *Spectrovision,* 1967, **18,** 2.
23. J. D. Norris and T. S. West, *Anal. Chim. Acta,* 1971, **55,** 359.
24. T. Hata, Y. Endo and Y. Nakahara, *Japan Analyst,* 1969, **18,** 838.
25. K. Kinson, J. E. Dickeson and C. B. Belcher, *Anal. Chim. Acta,* 1968, **41,** 107.
26. G. Uny, M. Brule and J. Spitz, *Anal. Chim. Acta,* 1969, **44,** 29.
27. M. L. Jursik, *Atomic Absorption Newsletter,* 1967, **6,** 21.
28. J. M. Viebock, *Anal. Letters,* 1970, **3,** 373.
29. T. W. Sattur, *Atomic Absorption Newsletter,* 1966, **5,** 37.
30. P. Johns and W. J. Price, *Metallurgia,* 1970, **81,** 75.

31. R. M. Dagnall, G. F. Kirkbright, T. S. West and R. Wood, *Analyst,* 1972, **97,** 245.
32. G. F. Bell, *Atomic Absorption Newsletter,* 1966, **5,** 73.
33. J. H. Schlewitz and M. G. Shields, *Atomic Absorption Newsletter,* 1971, **10,** 39.
34. J. Y. Hwang and L. M. Sondanato, *Anal. Chem.,* 1970, **42,** 747.
35. M. Ganivet and A. Benhaman, *Anal. Chim. Acta,* 1969, **47,** 81.
36. G. K. Billings, P. C. Ragland and R. C. Harriss, *Texas J. Sci.,* 1966, **18,** 277.
37. C. Iida and K. Yamasaki, *Anal. Letters,* 1970, **3,** 251.
38. D. C. Burrell, *Atomic Absorption Newsletter,* 1965, **4,** 328.
39. F. N. Ward, H. M. Nakagawa, T. F. Harms and G. H. Van Sickle, *Bull. U.S. Geol. Surv.,* 1969, **45,** 1289.
40. A. H. Jones, *Anal. Chem.,* 1965, **37,** 1761.
41. J. B. Willis, *Anal. Chem.,* 1962, **34,** 614.
42. S. Sprague and W. Slavin, *Atomic Absorption Newsletter,* 1964, **3,** 160.
43. F. W. Sunderman, *Am. J. Clin. Pathol.,* 1965, **44,** 182.
44. K. H. Schaller, A. Kuehner and G. Lehnert, *Blut,* 1968, **17,** 155.
45. S. Nomoto and F. W. Sunderman, *Clin. Chem.,* 1970, **16,** 477.
46. G. D. O'Dell, W. J. Miller, S. L. Moore, W. A. King, J. C. Ellers and H. Jurecek, *J. Anim. Sci.,* 1971, **32,** 769.
47. H. L. Kahn, *American Laboratory,* 1971, August, p. 35.
48. K. E. Burke and C. H. Albright, *J. Ass. Off. Analyt. Chem.,* 1970, **53,** 531; 1971, **54,** 658.
49. W. Slavin, *Atomic Absorption Newsletter,* 1965, **4,** 330.
50. J. D. Kerber, *App. Spectrosc.,* 1966, **20,** 212.
51. D. J. Trent and W. Slavin, *Atomic Absorption Newsletter,* 1964, **3,** 131.
52. V. Sychra and J. P. Matousek, *Anal. Chim. Acta,* 1970, **52,** 376.
53. E. J. Moore, O. I. Milner and J. R. Glass, *Microchem. J.,* 1966, **10,** 148.
54. V. Sychra, J. Matousek and S. Masek, *Chem. Listy,* 1969, **63,** 177.
55. S. H. Omang, *Anal. Chim. Acta,* 1971, **56,** 470.
56. R. Smith, C. M. Stafford and J. D. Winefordner, *Can. Spectrosc.,* 1969, **14,** 38.
57. R. L. Miller, L. M. Fraser and J. D. Winefordner, *Appl. Spectrosc.,* 1971, **25,** 477.
58. S. Sprague and W. Slavin, *Atomic Absorption Newsletter,* 1965, **4,** 367; 2966, **5,** 106.
59. J. A. Burrows, J. C. Heerdt and J. B. Willis, *Anal Chem.,* 1965, **37,** 579.
60. E. A. Means and D. Ratcliffe, *Atomic Absorption Newsletter,* 1965, **4,** 174.
61. M. D. Amos, K. G. Brodie and J. P. Matousek, *Proc. Pittsburgh Conf. on Anal. Chem. and Applied Spectrosc.,* Cleveland, 1971.
62. R. G. Lundy and W. F. Watje, *Atomic Absorption Newsletter,* 1969, **8,** 124.
63. J. A. Platte and V. M. Marcy, *Atomic Absorption Newsletter,* 1965, **4,** 289.
64. D. C. Burrell, *Atomic Absorption Newsletter,* 1965, **4,** 309.
65. J. Nix and T. Goodwin, *Atomic Absorption Newsletter,* 1970, **9,** 119.
66. B. Delaughter, *Atomic Absorption Newsletter,* 1965, **4,** 273.
67. O. K. Galle, *Appl. Spectrosc.,* 1971, **25,** 664.
68. S. Sprague and W. Slavin, *Atomic Absorption Newsletter,* 1964, **3,** 37.
69. J. C. Meranger, *Bull. Envir. Contam. Toxicol.,* 1970, **5,** 271.
70. J. C. Meranger and E. Somers, *J. Ass. Off. Analyt. Chem.,* 1968, **51,** 922.
71. S. W. Frey, W. G. DeWitt and B. R. Bellomy, *Am. Soc. Brew. Chem., Proc.,* 1966, 172.

13.4.38 *Niobium*

Nb, atomic weight 92·906

Technique	Line, nm	f	Source	Flame	Sensitivity, ppm/1% abs.	Detection limit, ppm	Reference (see below)
AAS	334·37	—	HCL	N-A	20	1	1, 2
	358·03	—	HCL	N-A	22	—	1, 2
	334·91	—	HCL	N-A	24	—	2, 3
	407·97	—	HCL	N-A	28	—	2, 3
	334·91	—	HCL	$N_2/O_2/A$	27	—	3
	358·03	—	HCL	$N_2/O_2/A$	27	—	3
AFS: not reported							

Ionization potential: 6·88 eV.
Dissociation energy, NbO: 4·0 eV.

Notes on analytical conditions

1. Nb absorption is observed only at very high concentration when the air-acetylene flame is used.[4] Nb absorption was reported in a fuel-rich oxy-acetylene flame by Fassel and Mossotti[5] using the Nb 405·89, 407·97 and 410·09 nm lines. Amos and Willis[3] examined Nb absorption at a large number of lines using a 60% oxygen-nitrogen-acetylene flame. Headridge and Hubbard[6] have reported a sensitivity of 35 ppm at 358·03 nm for Nb in HF solutions containing 25% ethanol using a nitrous oxide-acetylene flame.

2. Nb is appreciably ionized in the nitrous oxide-acetylene flame and potassium should be added to samples and standards to suppress this effect. In a similar fashion to Ti, Zr and Ta, Nb absorption is enhanced in the presence of fluoride and iron. This chemical interference occurs in the vapour phase (see Chapter 12.2.4) and may result from formation of a volatile fluoro-complex that dissociates directly to provide Nb atoms, rather than via the oxide, and by competition by Fe for the available oxygen in the flame gases to minimize stable oxide formation by Nb.

3. Nb may be determined indirectly by AAS with high sensitivity via formation of molybdoniobophosphoric acid and its extraction into n-butanol for AAS determination of Mo.[7]

Applications notes and key references

The principal applications of AAS to the determination of Nb have been in metallurgy. Schiller[8] determined Nb in low alloy steels by AAS at 334·37 nm in a nitrous oxide-acetylene flame. Large samples (60 g for 0·01% Nb) were dissolved in 3 N H_2SO_4 and filtered; the residue containing much of the Nb was further treated with $HF/HNO_3/H_2O$ (2 : 2 : 1 v/v) to dissolve it, saturated aqueous KOH added and the solution diluted for nebulization. Down to 0·05% Nb in steel has been determined by AAS in a nitrous oxide-acetylene flame after direct dissolution in $HCl/HNO_3/HF$ mixture prior to nebulization.[10] Martin[9] has employed a similar method for determination of Nb in steels; in her method, however, the residue left after dissolution in 3 N H_2SO_4 was fused with ammonium sulphate/H_2SO_4 and the melt dissolved in tartaric acid. Smith and co-workers[11] determined Nb in ferroniobium by AAS after dissolving samples in HF or HF/HNO_3 (when silicon was present as SiO_2) or HF/HCl (when elemental silicon was present). Welcher and Kriege[12] determined Nb in high-temperature Co-base alloys after extraction into MIBK from 10 M HF/6 M HCl solution.

References

1. D. C. Manning, *Atomic Absorption Newsletter,* 1967, **6,** 35.
2. "Analytical Methods for Atomic Absorption Spectrophotometry", Perkin-Elmer Corp., Norwalk, Conn., U.S.A., 1971.
3. M. D. Amos and J. B. Willis, *Spectrochim. Acta,* 1966, **22,** 1325.
4. B. M. Gatehouse and J. B. Willis, *Spectrochim. Acta,* 1961, **17,** 710.
5. V. A. Fassel and V. G. Mossotti, *Anal. Chem.,* 1963, **35,** 252.
6. J. B. Headridge and D. P. Hubbard, *Anal. Chim. Acta,* 1967, **37,** 151.
7. G. F. Kirkbright, A. M. Smith and T. S. West, *Analyst,* 1968, **93,** 292.
8. R. Schiller, *Atomic Absorption Newsletter,* 1970, **9,** 111.
9. M. J. Martin, *Analyst,* 1972, **97,** 394.
10. D. R. Thomerson, *Spectrovision,* 1971, **26,** 13.
11. D. C. Smith, J. R. Johnson and G. G. Soth, *Appl. Spectrosc.,* 1970, **24,** 576.
12. G. G. Welcher and O. H. Kriege, *Atomic Absorption Newsletter,* 1970, **9,** 61.

13.4.39 Osmium

Os, atomic weight 190·2

Technique	Line, nm	*f*	Source	Flame	Sensitivity, ppm/1% abs.	Detection limit, ppm	Reference (see below)
AAS	290·90	—	HCL	A-A	3	—	1
	290·90	—	HCL	N-A	1	—	1
	305·9	—	HCL	A-A	5	—	1
	305·9	—	HCL	N-A	2	—	1
	290·90	—	HCL	A-P-B	17	—	2
	283·8	—	HCL	A-P-B	19	—	2
	305·9	—	HCL	A-P-B	20	—	2
	323·2	—	HCL	A-P-B	26	—	2
	326·2	—	HCL	A-P-B	38	—	2
AFS: not reported							

Ionization potential: 8·73 eV.

Notes on analytical conditions

1. Few data are available for the AAS determination of Os; the fuel-rich nitrous oxide-acetylene flame provides the best reported sensitivity. Osolinski and Knight[3] obtained greatest sensitivity in this flame at 290·90 nm and reported that the absorption did not depend on the valency of the Os in the sample solution. Little information is available concerning interferences in the AAS determination, although Sen Gupta[4] recommended the use of Cd and Cu addition to minimize interelement effects observed in the A-A flame.

2. The determination of Os by AFS has not been reported. This probably results from the lack of sufficiently intense spectral sources and the requirement for use of a hot flame at wavelengths where high flame background is encountered (i.e. between 280 and 330 nm).

Applications notes and key references

Osolinski and Knight[3] determined Os in biological products by AAS in a nitrous oxide-acetylene flame. Osmium was determined in acid thiourea complexes after their destruction with 30% hydrogen peroxide. Sen Gupta[4] determined Os in osmiridium, native platinum and sperrylite ($PtAs_2$) by AAS in the presence of Cd and Cu spectroscopic buffer.

References

1. J. B. Willis, in W. Slavin "Atomic Absorption Spectroscopy", Interscience, New York, 1968, p. 138.
2. D. F. Makarov, Yu. N. Kukushkin and T. A. Eroschevich, *Zhur. Analit. Khim.*, 1969, **24**, 1436.

3. T. W. Osolinski and N. H. Knight, *Conf. Assoc. Anal. Chemists,* Detroit, 1967; *Chem. Eng. News,* Oct. 16, 1967.
4. J. G. Sen Gupta, *Anal. Chim. Acta,* 1972, 58, 23.

13.4.40 Palladium

Pd, atomic weight 106·4

Technique	Line, nm	*f*	Source	Flame	Sensitivity, ppm/1% abs.	Detection limit, ppm	Reference (see below)
AAS	247·64	0·1	HCL	A-A	0·2	0·02	1
	244·79	0·074	HCL	A-A	0·3	—	2
	276·31	0·071	HCL	A-A	1·0	—	2
	244·79	0·074	HIL	A-A	0·5	0·03	3
AFS	244·79	0·074	HIL	$Ar/O_2/H_2$	—	0·04	3
	340·46	—	HCL	A-H		2	4
ND	244·79	0·074	HIL	N_2 sep. A-A	—	2	5
ND	244·79	0·074	HIL	A-H	—	0·06	6

Ionization potential: 8·33 eV.

Notes on analytical conditions

1. A narrow spectral band-pass should be used with the Pd 244·79 nm line to avoid transmission of the less sensitive Pd lines at 244·5 and 245·0 nm. Even with relatively narrow band-pass, however, emission from a Pd ion line from the source at 244·62 nm may be transmitted and give rises to decreased sensitivity and non-linear calibration graphs. The Pd 247·64 nm line, however, may be used with a spectral band-pass of 0·2-0·3 nm.

2. Although recent workers have preferred to use the fuel lean air-acetylene flame for AAS of Pd, in early work Lockyer and Hames,[7] Strasheim and Wessels[8] and Erinc and Magee[9] employed cooler air-coal gas and air-propane flames. Even with these cool flames these workers observed few interference effects, except for the depression of absorption caused by high concentrations of HCl, HNO_3 and H_2SO_4.[8,9] Sychra *et al.*[3] investigated the effect of foreign ions on the determination of Pd by AFS and observed no effects provided that $SrCl_2$ or EDTA was added to sample solutions to release chemical interferences in the argon-oxygen-hydrogen flame. Addition of Cu to suppress interferences observed in the determination of the platinum metals was used by Strasheim and Wessels,[8] and has since been employed by other workers for suppression of interferences from other elements on the AAS of Pd

in the air-acetylene flame.[10,11] Addition of vanadium,[12] cadmium,[10] Cd and Cu[13] or lanthanum[14] has also been reported to achieve a similar effect.

Applications notes and key references

The principal applications of the AAS determination of Pd have been in its assay in fire-assay beads. Schnepfe and Grimaldi[10] determined Pd and Pt in ores by fire-assay in a gold bead. These workers used an air-acetylene flame and suppressed interferences using a 0·5% solution of cadmium and copper. No interference was observed from platinum group or other metals present in samples. Van Loon[14] determined Pd, Pt and Au in a silver assay bead by heating the bead with HNO_3, removing nitrates by heating with HCl, and adding La and making the solution 6 N in HCl before nebulization. Kallman and Hobart[15] determined Pd in ores and concentrates by a procedure in which the precious metal bead was dissolved in acid, the solution evaporated to dryness and the residue dissolved in 2% KCN solution for AAS. Fishkova[16] has reported a similar combined fire-assay-AAS procedure for the determination of Pd using an air-propane flame. Takeuchi and Yanagisawa[17] determined Pd in silver by precipitation of AgCl and extraction of Pd from alkaline medium into MIBK using NaDDTC. Palladium has been determined by AAS in uranium alloys after acid dissolution;[18] in this method mutual interferences between the platinum metals are prevented by the high concentration of uranium present. Harrington[19] has shown that colloidal Pd in solution may be determined directly by AAS in an air-acetylene flame; identical results are obtained for a Pd^{2+} solution and a solution containing the same concentration of colloidal Pd with an average particle size of 6·5 nm. Coudert and Vergnaud[20] have described an interesting technique for the determination of metals in powders in which the samples are fed at constant rate into the central channel of the burner where it is then carried up by the current of gas. The sample (*ca.* 10 mg) is first diluted 100-times with $CaCO_3$ of uniform particle size. Palladium was determined on active carbon, alumina and a molecular sieve in this way; rectilinear calibration graphs were obtained for samples containing up to 5% Pd.

References

1. "Analytical Methods for Atomic Absorption Spectrophotometry", Perkin-Elmer Corp., Norwalk, Conn., U.S.A., 1971.
2. J. E. Allan, *Spectrochim. Acta*, 1962, **18**, 259.

3. V. Sychra, P. J. Slevin, J. Matousek and F. Bek, *Anal. Chim. Acta,* 1970, **52,** 259.
4. D. C. Manning and P. Heneage, *Atomic Absorption Newsletter,* 1968, **7,** 80.
5. P. L. Larkins, *Spectrochim. Acta,* 1971, **26B,** 477.
6. P. L. Larkins and J. B. Willis, *Spectrochim. Acta,* 1971, **26B,** 491.
7. R. Lockyer and G. E. Hames, *Analyst,* 1959, **84,** 385.
8. A. Strasheim and G. J. Wessels, *Appl. Spectry.,* 1963, **17,** 65.
9. G. Erinc and R. J. Magee, *Anal. Chim. Acta,* 1964, **31,** 197.
10. M. M. Schnepfe and F. S. Grimaldi, *Talanta,* 1969, **16,** 591.
11. A. Janssen and F. Umland, *Z. Anal. Chem.,* 1970, **251,** 101.
12. R. C. Mullett, D. C. G. Pearton, E. J. Ring and T. W. Steele, *Talanta,* 1972, **19,** 181.
13. J. G. Sen Gupta, *Anal. Chim. Acta,* 1972, **58,** 23.
14. J. C. Van Loon, *Z. Anal. Chem.,* 1969, **246,** 122.
15. S. Kallman and E. W. Hobart, *Talanta,* 1970, **17,** 845.
16. N. L. Fishkova, *Zav. Lab.,* 1970, **36,** 1461.
17. T. Takeuchi and M. Yanagisawa, *Bunseki Kagaku,* 1966, **15,** 1059.
18. J. M. Scarborough, *Anal. Chem.,* 1969, **41,** 250.
19. D. E. Harrington, *Atomic Absorption Newsletter,* 1970, **9,** 106.
20. M. A. Coudert and J. M. Vergnaud, *Anal. Chem.,* 1970, **42,** 1303.

13.4.41 Phosphorus

P, atomic weight 30·97

Technique	Line, nm	f	Source	Flame	Sensitivity ppm/1% abs.	Detection limit, ppm	Reference (see below)
AAS	177·50	0·154	EDL	L'Vov furnace	$3{\cdot}4 \times 10^{-12}$ g	–	1
	178·29	0·102	EDL	L'Vov furnace	$4{\cdot}4 \times 10^{-12}$ g	–	1
	178·77	0·051	EDL	L'Vov furnace	$7{\cdot}3 \times 10^{-12}$ g	–	1
	213·62 213·55	–	EDL	L'Vov furnace	$2{\cdot}2 \times 10^{-10}$ g	2×10^{-10} g	1
	177·50	0·154	EDL	N_2 sep.	4·8	29	2
	178·29	0·102	EDL	N-A	5.4	21	2
	178·77	0·051	EDL	N-A	8·8	37	2
AFS: not reported							

Ionization potential: 10·49 eV.

Notes on analytical conditions

1. In order to permit direct AAS measurements at the P 177·50, 178·29 and 178·77 nm lines L'Vov used an argon-purged graphite cuvette and a vacuum monochromator.

2. The transmission of the P 177·50, 178·29 and 178·77 nm lines by the slightly fuel-rich separated nitrous oxide-acetylene flame (70 mm path length) has been observed to be 28, 56 and 64%

respectively of that of nitrogen alone at these wavelengths. A nitrogen purged path between source and flame, and flame and monochromator, must be provided.

3. No significant chemical or physical interferences were observed on the phosphorus absorption signal at 178·29 nm in the presence of 40-fold weight excesses of Al, Ba, Ca, Co, Cr, Cu, K, Li, Mg, Mo, Na, Ni, Zn, chloride, bromide, fluoride, nitrate, sulphate, borate, carbonate and EDTA.[2] Iodide caused enhancement of absorbance at 178·29 nm but not at 177·50 and 178·77 nm; this may be attributable to spectral overlap interference by the iodine 178·276 nm absorption line.

4. A number of indirect AAS methods for the determination of phosphorus have been reported (see Chapter 14).

Applications notes and key references

The direct technique for the determination of phosphorus by AAS in the separated nitrous oxide-acetylene flame has been applied to its estimation in milk powder, beef extract and yeast extract.[2] These samples were diluted with water and the total phosphorus content determined by nebulization into the flame; the method of standard additions was employed in order to eliminate the need to match the nebulization characteristics of the sample and standard solutions. The results obtained agreed well with those from determination by the molecular absorptiometric technique based on the formation and measurement of molybdenum blue. The AAS procedure is considerably more rapid than the solution absorptiometric technique for these water-soluble samples as it is not necessary to effect preliminary wet-ashing to destroy organic matter.

The indirect AAS determination of phosphorus via formation and extraction of molybdophosphoric acid and determination of the associated molybdenum has found several applications. These are detailed in Chapter 14.

References

1. B. V. L'Vov, "Atomic Absorption Spectroscopy", eds. R. M. Dagnall and G. F. Kirkbright, Butterworths, London, 1970, p. 28.
2. G. F. Kirkbright and Maurice Marshall, *Anal. Chem.*, 1973, **45**, 1610.

13.4.42 Platinum

Pt, atomic weight 195·09

Technique	Line, nm	*f*	Source	Flame	Sensitivity, ppm/1% abs.	Detection limit, ppm	Reference (see below)
AAS	265·94	0·12	HCL	A-A	2·2	0·1	1
	306·47	–	HCL	A-A	4·6	–	1
	217·47	–	HCL	A-A	3·3	–	1
	262·80	–	HCL	A-A	5·3	–	1
AFS	265·94	0·12	HCL	A-H	–	50	2
ND	265·94	–	HIL	N_2 sep. A-A	–	0·3	3
ND	265·94	–	HIL	A-H	–	0·15	4

Ionization potential: 9·0 eV.

Notes on analytical conditions

1. The Pt 265·94 and 217·47 nm lines were employed in early work with air-coal gas[5] and air-propane-butane[6] flames as well as in the air-acetylene flame.[7,8] It is not clear from the early studies which line and flame is preferable for Pt AAS; Slavin[1] has reported the results of a study of the useful Pt lines in a hot air-acetylene flame and given sensitivity and noise data for 19 Pt lines other than the most sensitive lines listed above.

2. Strasheim and Wessels[6] observed interference from other noble metals in the AAS of Pt and recommended the addition of a high concentration of Cu to solutions to overcome these effects. Schnepfe and Grimaldi[9] have utilized a 0·5% Cd + Cu solution and Van Loon[10] has used 1% La solution as additive to suppress these interferences on Pt AAS. Janssen and Umland[11] observed that the use of 1% Na and Cu solutions effectively prevented interelement interferences in the determination of Pt in the air-acetylene flame. Pitts and Beamish[12] found that of several water-miscible organic solvents examined, acetone gave the best enhancement in Pt absorption in the air-acetylene flame. Interference effects were similar with acetone-HCl medium (4 : 1) to those in wholly aqueous solution, but in the nitrous oxide-acetylene flame only Rh, Ir and Ru interfered; these were masked by 0·2% $LaCl_3$. These workers also extracted Pt into ethyl acetate as $(PtCl_2\ (SnCl_3)_2)^{2-}$; Rh, Ir and Pd are also partially extracted but do not interfere.

3. Iida and Nagura[13] have described the detection of Pt by AAS using an absorption tube device and ring burner, and

Brandenburger[14] has determined Pt by evaporation from a filament after its electrolytic deposition. Donega and Burgess[15] detected 300 ng of Pt by AAS using atomization from an electrically heated tungsten or tantalum boat in an inert atmosphere.

Applications notes and key references

The application of the determination of Pt by AAS has been largely restricted to geochemistry and metallurgy. Swider[16] extracted Pt from ores with *aqua regia* at 95-100°C and evaporated the solution to dryness. The residue was dissolved in HCl, diluted 1:3 with water, filtered, and the Pt was separated by double extraction into MIBK in the presence of KI. The Pt was then determined by AAS in an air-acetylene flame; standards were prepared from Pt solutions extracted similarly. Donnelly[17] determined Pt in ores and mineral concentrates using the fire assay technique with Pb as the collector. The determination of Pt in basic rocks has been reported by Simonsen.[18] After dissolution in HNO_3/HF the Pt was extracted into MIBK with dithizone for AAS using an air-acetylene flame. Pitts, Van Loon and Beamish[19] have reported the applicability to the determination of Pt in precious metal concentrates of a procedure using $LaCl_3$ to suppress inter-element effects on the absorption signal.

References

1. W. Slavin "Atomic Absorption Spectroscopy", Interscience, New York, 1968.
2. D. C. Manning and P. Heneage, *Atomic Absorption Newsletter*, 1968, **7**, 80.
3. P. L. Larkins, *Spectrochim. Acta*, 1971, **26B**, 477.
4. P. L. Larkins and J. B. Willis, *Spectrochim. Acta*, 1971, **26B**, 491.
5. R. Lockyer and G. E. Hames, *Analyst*, 1959, **84**, 385.
6. A. Strasheim and G. J. Wessels, *Appl. Spectry.*, 1963, **17**, 65.
7. J. E. Allen, *Spectrochim. Acta*, 1962, **18**, 259.
8. B. M. Gatehouse and J. B. Willis, *Spectrochim. Acta*, 1961, **17**, 710.
9. M. M. Schnepfe and F. S. Grimaldi, *Talanta*, 1969, **16**, 591.
10. J. C. Van Loon, *Z. Anal. Chem.*, 1969, **246**, 122.
11. A. Janssen and F. Umland, *Z. Anal. Chem.*, 1970, **251**, 101.
12. A. E. Pitts and F. E. Beamish, *Anal. Chim. Acta*, 1970, **52**, 405.
13. C. Iida and M. Nagura, *Spectrosc., Letters*, 1970, **3**, 63.
14. H. Brandenburger, *Chimia*, 1968, **22**, 449.
15. H. M. Donega and T. E. Burgess, *Anal. Chem.*, 1970, **42**, 1521.
16. R. T. Swider, *Atomic Absorption Newsletter*, 1968, **7**, 111.
17. T. H. Donnelly, *Proc. Australas. Inst. Min. Metall.*, 1970, **236**, 61.
18. A. Simonsen, *Anal. Chim. Acta*, 1970, **49**, 368.
19. A. E. Pitts, J. C. Van Loon and F. E. Beamish, *Anal. Chim. Acta*, 1970, **50**, 181, 195.

13.4.43 Potassium

K, atomic weight 39·10

Technique	Line, nm	*f*	Source	Flame	Sensitivity, ppm/1% abs.	Detection limit, ppm	Reference (see below)
AAS	766·49	0·69	VDL	A-coal gas	0·03	—	1
	766·49	0·69	HCL	A-A	0·02	0·005	2
	769·90	0·69	HCL	A-A	0·05	—	2
	404·41/72	0·11	HCL	A-A	8·8	—	3
	404·41/72	0·11	HCL	A-H	8·0	—	3
	404·41/72	0·11	VDL	A-coal gas	5	—	1
AFS: no analytical studies reported							

Ionization potential: 4·34 eV.
Atomization efficiency: A-A, 0·25.

Notes on analytical conditions

1. Both hollow-cathode and vapour-discharge lamp sources may be employed for AAS of potassium; VDL sources should be operated at low current to minimize line broadening by self-absorption and loss of sensitivity.

2. Potassium is appreciably ionized at low concentrations in the air-acetylene flame; excess Na or Cs should be added to both samples and standards to suppress this effect. The degree of ionization is lower in the cooler air-coal gas and air-hydrogen flames.

3. The AAS determination of potassium is relatively free from chemical interferences when the air-acetylene flame is employed. Several workers, however, have observed suppression of K absorption in the presence of high concentrations of mineral acids; this is compensated for by matching the samples and standards with respect to the acid concerned.

Applications notes and key references

As for sodium, the flame emission spectrometric determination of potassium is capable of high sensitivity and selectivity; many workers have preferred to use this technique rather than AAS for its determination. Potassium is frequently determined at the same time as sodium in samples; applications of the AAS determination of potassium are listed together with those of sodium in Table 13-9.

References

1. B. M. Gatehouse and J. B. Willis, *Spectrochim. Acta*, 1961, **17**, 710.
2. "Analytical Methods for Atomic Absorption Spectrophotometry", Perkin-Elmer Corp., Norwalk, Conn., U.S.A., 1971.
3. J. E. Schallis and H. L. Kahn, *Atomic Absorption Newsletter*, 1968, **7**, 75.

13.4.44 Praseodymium

Pr, atomic weight 140·91

Technique	Line, nm	*f*	Source	Flame	Sensitivity, ppm/1% abs.	Detection limit, ppm	Reference (see below)
AAS	495·14	—	Xe arc	O_2/A	—	60	1
	495·14	—	HCL	N-A	72	—	2
	495·14	—	HCL	N-A	13	—	3
	491·40	—	HCL	N-A	19	—	3
	504·55	—	HCL	N-A	42	—	3
	513·34	—	HCL	N-A	23	—	3
	513·34	—	HCL	N-A	92	—	2

Ionization potential: 5·48 eV.
Dissociation energy; PrO: 7·7 eV.

Notes on analytical conditions

1. Amos and Willis[2] used a fuel-rich nitrous oxide-acetylene flame for AAS of Pr.

2. Praseodymium is appreciably ionized in hot flames and excess potassium should be added to both samples and standards. Little information is available referring specifically to chemical interferences in the determination of Pr by AAS.

Application notes and key references

See Lanthanum.

References

1. V. G. Mossotti and V. A. Fassel, *Spectrochim. Acta*, 1964, **20**, 1117.
2. M. D. Amos and J. B. Willis, *Spectrochim. Acta*, 1966, **22**, 1325.
3. D. C. Manning, *Atomic Absorption Newsletter*, 1966, **5**, 127.

13.4.45 Rhenium

Re, atomic weight 186·2

Technique	Line, nm	*f*	Source	Flame	Sensitivity, ppm/1% abs.	Detection limit, ppm	Reference (see below)
AAS	345·19	0·06	HCL	N-A	33	–	1
	346·05	0·2	HCL	N-A	12	2	1, 2, 3
	346·47	0·13	HCL	N-A	20	–	1
	346·05	0·2	HCL	N-A	1·4	–	4
AFS: not reported							

Ionization potential: 7·87 eV.

Notes on analytical conditions

1. Fassel and Mossotti[5] first determined Re by AAS using a rich oxy-acetylene flame, total consumption nebulizer-burner unit and continuum source; a detection limit of 25 ppm Re was obtained in ethanolic medium. With an HCL source and fuel-rich oxy-acetylene flame, Schrenk, Lehman and Neufeld[6] obtained a detection limit of 1–2 ppm Re at 346·05 nm.

2. A fuel-rich nitrous oxide-acetylene flame is preferred for AAS; the sensitivity attainable is markedly dependent upon the flame stoicheiometry and height of observation in the flame.

3. In the fuel-rich oxy-acetylene flame Schrenk and co-workers[6] observed interference from Pb, K, Mo, Fe, Ca, Mn, and Al on the Re absorption.

4. Flame-emission spectrometry may provide good sensitivity for Re when a hot flame is used. Kniseley and co-workers[7] reported enhanced emission sensitivity using a premixed oxy-acetylene flame rather than a turbulent flame, and Smith and Lawson[8] used a nitrogen-separated nitrous oxide-acetylene flame for emission work with Re.

5, Preconcentration by solvent extraction permits improvement in AAS sensitivity for Re. Re may be extracted into MIBK with APDC between pH 1 and 9.

Applications notes and key references

Few applications of the determination of Re by AAS have been reported. Rhenium may be determined in Re/Mo and Re/W alloys by AAS in a nitrous oxide-acetylene flame at 346·05 nm after dissolution of Mo alloys in ammonia/H_2O_2 and tungsten alloys in ammonia/citric acid/H_2O_2.[9] Tomkins and Frank[3] determined Re in organorhenium compounds by AAS in the nitrous oxide-acetylene flame in a solvent consisting of 57% ethanol : 43% water.

References

1. M. D. Amos and J. B. Willis, *Spectrochim. Acta,* 1966, **22,** 1325.
2. "Analytical Methods for Atomic Absorption Spectrophotometry", Perkin-Elmer Corp., Norwalk, Conn., U.S.A., 1971.
3. D. F. Tomkins and C. W. Frank, *Appl. Spectrosc.,* 1971, **25,** 539.
4. J. Ramirez-Munoz and M. E. Roth, *Flame Notes,* 1967, **2,** 18.
5. V. A. Fassel and V. G. Mossotti, *Anal. Chem.,* 1963, **35,** 252.
6. W. G. Schrenk, D. A. Lehman and L. Neufeld, *Appl. Spectrosc.,* 1966, **20,** 389.
7. R. N. Kniseley, A. P. D'Silva and V. A. Fassel, *Anal. Chem.,* 1963, **35,** 910.
8. R. Smith and A. E. Lawson, *Analyst,* 1971, **96,** 631.
9. G. M. Neumann, *Talanta,* 1971, **18,** 955.

13.4.46 Rhodium

Rh, atomic weight 102·90

Technique	Line, nm	*f*	Source	Flame	Sensitivity, ppm/1% abs.	Detection limit, ppm	Reference (see below)
AAS	343·49	0·73	HCL	A-A	0·35	0·03	1
	369·24	0·58	HCL	A-A	0·6	—	1
	339·68	0·53	HCL	A-A	0·7	—	1
	350·25	0·47	HCL	A-A	1·35	—	1
	365·80	0·82	HCL	A-A	1·75	—	1
	370·09	0·72	HCL	A-A	2·9	—	1
	350·73	0·47	HCL	A-A	8	—	1
AFS	369·24	0·58	HCL	A-H	—	0	2
	369·24	0·58	150 W Xe arc	A-H	—	10	2

Ionization potential: 7·45 eV.

Notes on analytical conditions

1. The early studies of Rh AAS were undertaken with air-coal gas and air-propane-butane flames using the 343·49 nm line.[3,4] Gatehouse and Willis[5] employed an air-acetylene flame and reported that sensitivity at this line was strongly dependent on the HCL current and flame stoicheiometry. Heneage[1] observed that a very lean air-acetylene flame was required for highest sensitivity. Deily[6] also employed a lean flame in AAS of Rh with organic solvents.

2. Although Lockyer and Hames[3] did not observe interferences from Fe and other platinum metals on Rh absorption, Strasheim and

Wessels[4] observed interference from noble metals and mineral acids. These effects could not be eliminated by the use of a high concentration of Cu in the solution. Kallman and Hobart[7] found that the presence of various anions had little effect on Rh absorption in a fuel-lean air-acetylene flame but depressed absorption in a fuel-rich flame. Various cations (as their chlorides in 10% HCl solution of Rh) had little effect, except for La^{3+} and UO_2^{2+} and $K_2Cr_2O_7$, which enhanced absorption significantly. The alkali metals, Al and Zn affected enhancement of the absorption and the use of 3% $KHSO_4$ or $NaHSO_4$ eliminated any depressant effects from cations and from H_3PO_4. Atwell and Hebert[8] studied Rh AAS in a nitrous oxide-acetylene flame and found no interferences at less than 3000 ppm of Cu, Ni, W, Fe, Cr, Pb, Pt, Pd, Ti and Zn, and also when less than 10% of HCl, HNO_3, H_2SO_4 and H_3PO_4 were present. Only Ir and Ru were found to interfere seriously but the interference was suppressed using a 0·5% solution of Zn.

Application notes and key references

Schnepfe and Grimaldi[9] determined Rh by AAS after pre-concentration by coprecipitation with Te or by the fire-assay method with a gold bead. Interferences were reduced by adding lanthanum to the final solution; as little as 0·07 ppm of Rh was determinable in a 3 g sample. Scarborough[10] has reported the determination of Rh in U alloys using an air-acetylene flame and 3-slot burner, and Hofer[11] described the AAS determination of Rh in Pt concentrates by a procedure in which $NaHSO_4$ and Zn were added to the sample solution (containing 0·01-0·1% Rh and up to 5% Pt) before nebulization into the flame. Sen Gupta[12] determined Rh and other noble and base metals in osmiridium, native Pt and sperrylite by AAS using a 0·5% Cd + 0·5% Cu solution as a spectroscopic buffer for the noble metals and 1% La or Sr addition for the base metals.

References

1. P. Heneage, *Atomic Absorption Newsletter,* 1966, **5**, 64.
2. D. C. Manning and P. Heneage, *Atomic Absorption Newsletter,* 1968, **7**, 80.
3. R. Lockyer and G. E. Hames, *Analyst,* 1969, **84**, 385.
4. A. Strasheim and G. J. Wessels, *Appl. Spectrosc.,* 1963, **17**, 65.
5. B. M. Gatehouse and J. B. Willis, *Spectrochim. Acta,* 1961, **17**, 710.
6. J. R. Deily, *Atomic Absorption Newsletter,* 1967, **6**, 66.
7. S. Kallman and E. W. Hobart, *Anal. Chim. Acta,* 1970, **51**, 20.
8. M. G. Atwell and J. Y. Hebert, *Appl. Spectrosc.,* 1969, **23**, 480.
9. M. M. Schnepfe and F. S. Grimaldi, *Talanta,* 1969, **16**, 1461.
10. J. M. Scarborough, *Anal. Chem.,* 1969, **41**, 250.

11. A. Hofer, *Z. Anal. Chem.*, 1972, **258**, 285.
12. J. G. Sen Gupta, *Anal. Chim. Acta*, 1972, **58**, 23.

13.4.47 Rubidium

Rb, atomic weight 85·47

Technique	Line, nm	*f*	Source	Flame	Sensitivity, ppm/1% abs.	Detection limit, ppm	Reference (see below)
AAS	780·02	0·8	VDL	A-coal gas	0·1	–	1
	420·18	–	VDL	A-coal gas	10	–	1
	780·02	0·8	VDL	A-A	0·05	0·005	2
	794·76	–	VDL	A-A	0·1	–	2
	420·18	–	VDL	A-A	6·0	–	2
	780·02	0·8	VDL	A-LPG	0·12	–	3
AFS: not reported							

Ionization potential: 4·2 eV.
Atomization efficiency: A-A, 0·16.

Notes on analytical conditions

1. Rb vapour-discharge lamp sources have been shown to exhibit much higher intensity and excellent signal : background intensity ratio at the Rb 780·02 nm line compared to the corresponding Rb hollow-cathode lamp sources.[3] Rubidium VDL sources should be operated at low current to minimize self-absorption.

2. The addition of excess KCl to standards and samples is required to suppress Rb ionization in the air-propane and air-acetylene flames. The use of the low temperature air-LPG flame results in a lower degree of ionization.

3. Interferences from mineral acid and from Al have been reported in the determination of Rb by AAS; Al interference may be suppressed by addition of excess Ca, and mineral acid interference should be compensated for by matching standards and samples.[4] In the same work no interference with Rb absorption was observed in the presence of high concentration of Ba, Cr, Cu, Ca, Fe, Mg, Mn, Li, Na, Si, Sr and Ti.

Applications notes and key references

Several groups of workers have determined Rb in geological samples by AAS. Billings and Adams[5] dissolved silicate and carbonate rock samples in HF/HNO_3 mixture and used lanthanum to suppress interferences in the air-acetylene flame; other similar methods requiring use of releasing agents and matching of samples and standards have also been described.[6,7] Billings and Harriss[8] have reported the direct determination of Rb in sea-water samples by AAS.

Very little work has been published concerned with the AAS determination of Rb in biological materials. In an early paper Rb in plants and soil extracts was determined by Stupar;[9] plant samples were dry ashed and matching of sample and standard solutions was necessary. The AAS determination of Rb in whole blood, serum, plasma and urine has been examined by Sutter, Platman and Fieve.[10]

References

1. B. M. Gatehouse and J. B. Willis, *Spectrochim. Acta,* 1961, **17,** 710.
2. "Analytical Methods for Atomic Absorption Spectrophotometry", Perkin-Elmer Corp., Norwalk, Conn., U.S.A., 1971.
3. W. Slavin, D. J. Trent and S. Sprague, *Atomic Absorption Newsletter,* 1965, **4,** 180.
4. A. S. Bazhov, Y. A. Lazarev and P. A. Koka, *Zavod. Lab.,* 1967, **33,** 173.
5. G. K. Billings and J. A. S. Adams, *Atomic Absorption Newsletter,* 1964, **3,** 65.
6. M. Vosten and S. Deutsch, *Earth Planet Sci. Letters,* 1967, **2,** 449.
7. M. Guillaume and I. Roelandts, *Proc. 3rd Int. Conf. Atom. Abs. and Fluor.,* Paris, 1971.
8. G. K. Billings and R. C. Harriss, *Texas J. Sci.,* 1965, **17,** 129.
9. J. Stupar, *Zh. Anal. Khim.,* 1964, **19,** 401.
10. E. Sutter, S. R. Platman and R. R. Fieve, *Clin. Chem.,* 1970, **16,** 602.

13.4.48 Ruthenium

Ru, atomic weight 101·07

Technique	Line, nm	*f*	Source	Flame	Sensitivity, ppm/1% abs.	Detection limit, ppm	Reference (see below)
AAS	349·89	0·1	HCL	A-A	0·30	–	1
	372·80	0·09	(See note 1)	A-A	0·25	–	2
AFS	372·80	0·09	150 W Xe arc	A-H	–	100	3
ND	372·80	0·09	HCL	N_2 sep. A-A	–	80	4

Ionization potential: 7·34 eV.

Notes on analytical conditions

1. Allan[2] estimated sensitivities of 0·25 ppm/1% absorption at the Ru 349·89 and 372·80 nm lines using a continuum source and photographic detection, and reported maximum sensitivity in a fuel-rich air-acetylene flame.

2. Scarborough[5] reported depression of Ru absorption in the air-acetylene flame in the presence of Mo and enhancement when Pt or Rh are present; in the determination of Mo, Ru, Pd and Rh in U alloys, however, the high concentration of U present in the solution (1-15 g/l) prevents mutual interferences between these ions. Ottaway and co-workers[6,7] have recommended the addition of Cu or Cd as their sulphates to suppress interferences in the determination of Ru in the air-acetylene flame.

3. Ru may be extracted with APDC into MIBK between pH 1 and 10.

Applications notes and key references

Schwab and co-workers[8] determined Ru in radioactive materials by AAS in an air-acetylene flame using La addition to suppress interferences. As mentioned above, Scarborough[5] has determined Ru in U alloys by AAS using an air-acetylene flame.

References

1. W. Slavin, "Atomic Absorption Spectroscopy", Interscience, New York, 1968, p. 154.
2. J. E. Allan, *Spectrochim. Acta,* 1962, **18,** 259.
3. D. C. Manning and P. Heneage, *Atomic Absorption Newsletter,* 1968, 7, 80.
4. P. L. Larkins, *Spectrochim. Acta,* 1971, **26B,** 477.
5. J. M. Scarborough, *Anal. Chem.,* 1969, **41,** 250.
6. W. B. Rowston and J. M. Ottaway, *Anal. Letters,* 1970, 3, 411.
7. W. B. Rowston, J. M. Ottaway and M. Stulik, *Proc. Soc., Anal. Chem.,* 1970, 7, 148.
8. M. R. Schwab and N. H. Hembree, *Atomic Absorption Newsletter,* 1971, **10,** 15.

13.4.49 Samarium

Sm, atomic weight 150·4

Technique	Line, nm	*f*	Source	Flame	Sensitivity, ppm/1% abs.	Detection limit, ppm	Reference (see below)
AAS	429·67	–	Xe arc	O_2/A	–	10	1
	429·67	–	HCL	N-A	21	–	2
	429·67	–	HCL	N-A	7	2	3
	476·03	–	HCL	N-A	29	–	2
	476·03	–	HCL	N-A	45	–	4
	520·06	–	HCL	N-A	25	–	4
	528·29	–	HCL	N-A	50	–	4
	511·72	–	HCL	N-A	55	–	4
AFS: not reported							

Ionization potential: 5·6 eV.
Dissociation energy, SmO: 6·1 eV.

Notes on analytical conditions

1. Manning[4] examined Sm absorption at 13 lines; in addition to the most sensitive lines listed above, absorption was also observed at the Sm 478·31, 488·37/9, 534·13, 458·16, 527·14, 472·84, 540·37 and 532·06 nm lines.

2. Ionization of Sm in the fuel-rich nitrous oxide-acetylene flame is suppressed by addition of excess of an alkali metal salt, usually potassium chloride. Jaworowski *et al.*[5] recommended nebulization of Sm in 80% alcoholic solutions in order to enhance sensitivity in AAS. Little detailed information is available concerned with chemical interferences in the determination of Sm by AAS.

Applications notes and key references

See Lanthanum.

References

1. V. G. Mossotti and V. A. Fassel, *Spectrochim. Acta,* 1964, **20,** 1117.
2. M. D. Amos and J. B. Willis, *Spectrochim. Acta,* 1966, **22,** 1325.
3. "Analytical Methods for Atomic Absorption Spectrophotometry", Perkin-Elmer Corp., Norwalk, Conn., U.S.A., 1971.
4. D. C. Manning, *Atomic Absorption Newsletter,* 1966, **5,** 127.
5. R. J. Jaworowski, R. P. Weberling and D. J. Bracco, *Anal. Chim. Acta,* 1967, **37,** 284.

13.4.50 Scandium

Sc, atomic weight 44·956

Technique	Line, nm	*f*	Source	Flame	Sensitivity, ppm/1% abs.	Detection limit, ppm	Reference (see below)
AAS	391·18	—	HCL	N-A	0·8	—	1
	391·18	—	HCL	N-A	0·65	0·1	2
	390·75	—	HCL	N-A	1·1	—	1
	390·75	—	HCL	N-A	0·7	—	3
	402·37	—	HCL	N-A	1·2	—	1
	402·37	—	HCL	N-A	0·9	—	3
	402·04	—	HCL	N-A	1·7	—	1
	402·04	—	HCL	N-A	1·2	—	3
	405·45	—	HCL	N-A	1·7	—	3
	326·99	—	HCL	N-A	2·8	—	1
	326·99	—	HCL	N-A	2·0	—	3
AFS: not reported							

Ionization potential: 6·54 eV.
Dissociation energy, ScO: 7·0 eV.

Notes on analytical conditions

1. Other less sensitive Sc lines investigated by Manning were 327·36 and 408·24 nm. A hot fuel-rich flame is required to permit detection of Sc absorption. Fassel and Mossotti[4] described the first analytical detection of Sc by AAS using fuel-rich oxy-acetylene flames supported at a total consumption nebulizer-burner unit and a xenon arc continuum source.

2. Scandium ionization in the nitrous oxide-acetylene flame may be controlled by the addition of excess K^+ to sample and standard solutions. Yiu-Kee Chau[5] investigated interferences on the Sc absorption in a fuel-rich nitrous oxide flame; the presence of 25-fold excess of U, Fe, Ni, Co, Cu, Mn, Zn, Mo, Pb, Cr, Na, Ca or Mg had no effect on the absorption produced by 4 ppm Sc after extraction with 8-hydroxyquinoline into butanol. Kriege and Welcher[6] observed interference from fluoride and sulphate in the determination of Sc by AAS.

Applications notes and key references

Riley and Taylor[7] determined Sc in sea-water by AAS after preconcentration on chelating resin and elution with dilute mineral acid. Yiu-Kee Chau and Pui-Yuen Wong[8] have also described the determination of Sc in sea-water by AAS in a method in which Sc is coprecipitated with ferric iron at pH 8–9, the precipitate is dissolved in concentrated HCl, passed through Deacidite FF ion-exchange resin, and Sc extracted from the eluate into butanol as its 8-hydroxyquinoline complex.

References

1. M. D. Amos and J. B. Willis, *Spectrochim. Acta,* 1966, **22,** 1325.
2. "Analytical Methods for Atomic Absorption Spectrophotometry", Perkin-Elmer Corp., Norwalk, Conn. U.S.A., 1971.
3. D. C. Manning, *Atomic Absorption Newsletter,* 1966, **5,** 127.
4. V. A. Fassel and V. G. Mossotti, *Anal. Chem.,* 1963, **35,** 252.
5. Yiu-Kee Chau, *Talanta,* 1968, **15,** 421.
6. O. H. Kriege and C. C. Welcher, *Talanta,* 1968, **15,** 781.
7. J. P. Riley and D. Taylor, *Anal. Chim. Acta,* 1968, **40,** 479.
8. Yui-Kee Chau and Pui-Yuen Wong, *Talanta,* 1968, **15,** 781.

13.4.51 Selenium

Se, atomic weight 78·96

Technique	Line, nm	*f*	Source	Flame	Sensitivity, ppm/1% abs.	Detection limit, ppm	Reference (see below)
AAS	196·03	0·12	HCL	A-A	3	1	1
	196·03	0·12	EDL	A-A	1	0·25	2
	196·03		EDL	N_2/H_2	0·5	0·1	2
	196·03	0·12	HCL	A-A	1	—	3
	196·03	0·12	HCL	A-A	1	0·5	4
	196·03	. .	. .	Ar/H_2	0·27	0·05	4
	196·03	. .	HCL	N-A	5	1·8	5
AFS	204·00	—	EDL	A-P	—	0·25	6
	196·03	—	EDL	Ar/H_2	—	0·4	7
	196·03	0·12	EDL	A-A	—	0·2	8

Ionization potential: 9·75 eV.
Dissociation energy, SeO: 3·5 eV.

Notes on analytical conditions

1. The other absorbing lines of the triplet ground state of selenium lie at 203·99 and 206·28 nm. These are seldom used for AAS, however, owing to the inefficient population of the $^3P_1$ and $^3P_0$ states in commonly used flames compared to that of the $^3P_2$ state which gives rise to the 196·03 nm line.

2. The Se 196·03 nm line lies in a region of high flame background absorbance in air-hydrocarbon flames. The argon-hydrogen and nitrogen-hydrogen flames are more transparent in this region and also provide better sensitivity (for 1% absorption). In practical analysis with these cool flames, however, care should be taken to

check for vaporization interferences from other elements present in the sample. Although lower sensitivity is obtained in the nitrous oxide-acetylene flame considerable freedom from interelement effects is observed with this flame.[5]

3. With some hollow-cathode lamp sources the loss of energy reaching the detector at 196·0 nm caused by flame background absorption may give rise to the need to use high gain settings and wide spectral band-pass at the monochromator; this may result in excessive signal noise. Se hollow-cathode lamp sources should not be operated at high power, however, as this may result in shorter lamp life due to the high volatility of Se. In order to permit more energy to reach the detector at 196·03 nm for AAS of Se, several workers have employed EDL sources[2,3,9] rather than hollow-cathode lamps. Inert gas separation of the air-acetylene[10] or nitrous oxide-acetylene[5] flames has also been shown to achieve a similar effect.

4. Johns[11] found remarkable freedom from interference from many metal ions at the 1000 ppm level in the signal obtained for Se in the argon/hydrogen flame; somewhat more marked effects (enhancements) were observed in the air-acetylene flame. Chakrabarti[12] found no interference in the air-acetylene flame from 1000-fold excesses (over a 10^{-3} M Se soln.) of Cd, Zn, Sn(IV), Cl^- or PO_4^{3-} in 10% HNO_3; many other cations and anions were found to interfere, however, under similar conditions. Kirkbright and co-workers[5,10] have also examined interference effects in inert gas-shielded air-acetylene and nitrous oxide-acetylene flames. Several ions (e.g. Mg, Al, Ca, Ti, Ni, Mo, Ta, Co, Fe and Cr) were observed to depress Se absorption at 100-fold weight excess over Se in the air-acetylene flame but the nitrous oxide-acetylene flame did not suffer from these effects.

Applications notes and key references

In the determination of Se in matrices where organic matter is to be destroyed care should be taken to avoid loss of Se by volatilization. Dry-ashing procedures are generally unsuitable, although the successful use of an active oxygen RF ashing device has been described by Mulford.[13] Wet digestion with a mixture of HNO_3, H_2SO_4 and $HClO_4$ may be successful, particularly if Mo(IV) is added to catalyse the digestion and oxidizing conditions are maintained. Siren[14] has determined Se in animal retina using AAS, while Rann and Hambly[3] determined Se in wheat after combustion by the Parr bomb technique. These workers also determined Se in galena;

samples were ignited with Na_2O_2 and Se determined by AAS in an aqueous extract of the melt. Nakahara and co-workers[15] have reported the determination of Se in sulphur, and Sprague *et al.*[16] examined its determination in various copper-containing solutions.

References

1. W. Slavin and S. Sprague, *Atomic Absorption Newsletter,* 1964, **18,** 1.
2. R. M. Dagnall, K. C. Thompson and T. S. West, *Atomic Absorption Newsletter,* 1967, **6,** 117.
3. C. S. Rann and A. N. Hambly, *Anal. Chim. Acta.* 1965, **32,** 346.
4. H. L. Kahn and Jane E. Schallis, *Atomic Absorption Newsletter,* 1968, **7,** 5.
5. G. F. Kirkbright and Leslie Ranson, *Anal. Chem.,* 1971, **43,** 1238.
6. R. M. Dagnall, K. C. Thompson and T. S. West, *Talanta,* 1967, **14,** 55; 1968, **15,** 677.
7. K. E. Zacha, M. P. Bratzel, J. D. Winefordner and J. M. Mansfield, *Anal. Chem.,* 1968, **40,** 1733.
8. M. S. Cresser and T. S. West, *Spectrosc. Letters,* 1969, **2,** 9.
9. A. S. Bazhov, *Zavod. Lab.* (English Translation), 1967, **33,** 1096.
10. G. F. Kirkbright, M. Sargent and T. S. West, *Atomic Absorption Newsletter,* 1969, **8,** 34.
11. P. Johns, *Spectrovision,* 1970, (24) 6.
12. C. L. Chakrabarti, *Anal. Chim. Acta,* 1968, **42,** 379.
13. C. E. Mulford, *Atomic Absorption Newsletter,* 1966, **5,** 135.
14. M. J. Siren, *Sci. Tools,* 1964, **11,** 37.
15. Taketoshi Nakahara, Makoto Munemori and Soichiro Musha, *Anal. Chem. Acta,* 1970, **50,** 51.
16. S. Sprague, D. C. Manning and W. Slavin, *Atomic Absorption Newsletter,* 1964, **3,** 27.

13.4.52 Silicon

Si, atomic weight 28·09

Technique	Line, nm	*f*	Source	Flame	Sensitivity, ppm/1% abs.	Detection limit, ppm	Reference (see below)
AAS	251·61	0·26	HCL	N-A	1·2	0·08	1
	250·69	0·20	HCL	N-A	3·1	—	
	251·43	0·54	HCL	N-A	3·8	—	
AFS	251·61	0·26	EDL	Ar sep. N-A	—	0·5	2

Ionization potential: 8·15 eV.
Dissociation energy, SiO: 8·0 eV.

Notes on analytical conditions

1. A fuel-rich nitrous oxide-acetylene flame is preferred for high sensitivity. Slavin *et al.*[3] observed that the sensitivity attainable does not depend critically on the size of the light beam in the flame.

2. Narrow spectral band-pass is required in AAS to isolate 251·61 nm line from the nearby less sensitive 251·43 and 251·92 nm Si lines.

3. Early work on AAS for Si at 251·61 nm by Amos and Willis[4] produced 1% sensitivity of 2·5 ppm in N-A flame and 10 ppm in a nitrogen-oxygen-acetylene flame. Hell, Ramirez-Munoz and Shifrin[5] reported a sensitivity of 0·3 ppm in a preheated N-A flame with a detection limit of 0·05 ppm Si.

4. Enhancement of Si absorbance in the N-A flame has been reported by several workers, e.g. Bowman and Willis[6] in the analysis of bauxite and Roos and Price[7] in the analysis of cement. No interferences at 100-fold weight excess over 500 ppm Si in AFS were observed for a wide range of elements (ref. 2).

Applications notes and key references

AAS has found extensive application in the analysis of Si and has been selected as the preferred technique for its determination in various matrices. Silicon has been widely determined in rocks, minerals and coal ash by nebulization of solutions obtained after fusion of powdered samples with $LiBO_2$ at *ca.* 900°C and dissolution of the melt in HNO_3, HCl or HF.[8-12] For determination of Si in cement[13,14,15] bauxite[16] and sulphide minerals[17] direct dissolution in acid has been employed successfully, although fusion procedures are also used. Slags and sinters have been analysed for silicon after dissolution in HF/HCl,[18] mineral slags or wools have been dissolved in HF prior to Si determination[19] and a number of workers have described its determination in cast iron,[14,20] steel,[20,21] iron ores and ferrosilicon.[22] AFS has been applied to the determination of Si in low alloy steels.[23] Price and Roos[14] and Campbell[24] have determined Si in Al alloys by AAS. Other materials analysed for Si by AAS include used lubricating oils,[25] polymers,[26] and soil.[27]

References

1. J. S. Cartwright, C. Sebens and D. C. Manning, *Atomic Absorption Newsletter,* 1966, **5**, 91.
2. R. M. Dagnall, G. F. Kirkbright, T. S. West and R. Wood, *Anal. Chim. Acta,* 1969, **47**, 407.
3. W. Slavin, A. Venghiattis and D. C. Manning, *Atomic Absorption Newsletter,* 1966, **5**, 84.
4. M. D. Amos and J. B. Willis, *Spectrochim. Acta,* 1966, **22**, 1325.
5. A. Hell, J. Ramirez-Munoz and N. Shifrin, *Pittsburgh Conf., Anal. Chem. and Appl. Spectry.,* 1966.
6. J. A. Bowman and J. B. Willis, *Anal. Chem.,* 1967, **39**, 1210.
7. J. T. H. Roos and W. J. Price, *Analyst,* 1969, **94**, 89.
8. J. W. Yule and G. A. Swanson, *Atomic Absorption Newsletter,* 1969, **8**, 30.

9. P. L. Boar and L. K. Ingram, *Analyst,* 1970, **95**, 124.
10. Sydney Abbey, *Geol. Surv. Pap. Can.*, No. 72, 1970.
11. Sverre H. Omang, *Anal. Chim., Acta,* 1969, **46**, 225.
12. J. C. Van Loon and C. M. Parissis, *Analyst,* 1969, **94**, 1057.
13. J. T. H. Roos and W. J. Price, *Analyst,* 1969, **94**, 89.
14. W. J. Price and J. T. H. Roos, *Analyst,* 1968, **93**, 709.
15. F. J. Langmyhr and P. E. Paus, *Anal. Chim. Acta,* 1969, **44**, 445.
16. F. J. Langmyhr and P. E. Paus, *Anal. Chim. Acta,* 1968, **43**, 508.
17. F. J. Langmyhr and P. E. Paus, *Anal. Chim. Acta,* 1970, **50**, 515.
18. J. Reid, J. M. Galloway, J. MacDonald and B. B. Bach, *Metallurgia,* 1970, **81**, 243.
19. J. E. Sheridan, *Spectrovision,* 1971, (25), 10.
20. J. J. McAuliffe, *Atomic Absorption Newsletter,* 1967, **6**, 69.
21. D. M. Knight and M. K. Pyzyna, *Atomic Absorption Newsletter,* 1969, **8**, 129.
22. F. J. Langmyhr and P. E. Paus, *Anal. Chim. Acta,* 1969, **45**, 157, 173.
23. G. F. Kirkbright, A. P. Rao and T. S. West, *Anal. Lett.,* 1969, **2**, 465.
24. D. E. Campbell, *Anal. Chim. Acta,* 1969, **46**, 31.
25. D. R. Jackson, C. Salama and R. Dunn, *Can. Spectrosc.,* 1970, **15**, 17.
26. M. Olivier, *Z. Anal. Chem.,* 1969, **248**, 145.
27. Y. La Flamme, *Atomic Absorption Newsletter,* 1968, **7**, 101.

13.4.53 Silver

Ag, atomic weight 107·87

Technique	Line, nm	*f*	Source	Flame	Sensitivity, ppm/1% abs.	Detection limit ppm	Reference (see below)
AAS	328·07	0·51	HCL	A-coal gas	0·05	–	1
	328·07	0·51	HCL	A-A	0·08	0·005	2
	338·29	0·25	HCL	A-coal gas	0·15	–	1
AFS	328·07	0·51	EDL	Air/H_2	–	0·0001	3
	328·07	..	HIL	A-P	–	0·002	4
	328·07	..	HCL	A-H	–	0·003	5
ND	328·07	..	HCL pulsed	A-H	–	0·002	6

Ionization potential: 7·57 eV.
Dissociation energy AgO: 1·4 eV.
Atomization efficiency: A-A, 0·66.

Notes on analytical conditions

1. With some HCL sources emission from a Cu 327·4 nm line from the lamp may cause loss of sensitivity unless a relatively narrow spectral band-pass is used. With early HCL sources the sensitivity was reported to be markedly dependent on the lamp current; this effect is minimized with shielded HCL sources.

2. In early work Lockyer and Hames[7] observed no interference from Au, Pt, Rh, Pd, Pb and Fe on Ag absorbance. Depression of absorption in the presence of large amounts of Al has been reported,[8] while Belcher *et al.*[9] examined a large number of cations and anions at 1000-fold excess over Ag and found interference only from Th, permanganate, iodate and tungstate. In AFS using an air-propane flame West and Williams[10] observed few interferences, even at 1000-fold excess; elements forming stable oxides, such as aluminium, produced depression of the fluorescence by chemical interference effects.

3. Silver may be extracted with APDC into MIBK. Other workers have used extraction as the dibutylamine-Ag-salicylic acid complex into MIBK.[9,10]

Applications and key references

Silver assay by AAS is in widespread use. Mineral assay for silver employing dissolution with HNO_3 and H_2O_2 was described by Fixman and Boughton.[11] It is necessary to prevent precipitation of Ag as AgCl in these methods. When a hydrochloric acid concentration of *ca.* 50% (by volume) is maintained silver salts will not precipitate;[12] alternatively in sulphide materials the Ag may be held in solution by addition of Hg(II)[13] or diethylenetriamine.[14] Silver in rocks has also been determined by AAS by digestion with HNO_3 followed by extraction with tri-iso-octylthiophosphate into MIBK;[15] a similar determination using a carbon rod atomizer has also been described by Bratzel and Chakrabarti.[16] In a combined fire-assay/AAS procedure for Ag Moloughney and Graham[17] dissolved the tin assay button in HCl, precipitated Ag with tin powder and redissolved the precipitate for nebulization into an air-acetylene flame. Other metallurgical products in which Ag has been determined by AAS include non-ferrous alloys,[18] copper matte,[19] aluminium alloys[8] and steel.[20] Hickey[21] has described the determination of Ag in fine Ag bullion by a modification of the standard Gay-Lussac method; 1 g of sample is dissolved in HNO_3 and exactly 100 ml of 0·54% NaCl solution is added. Ag is determined at 328·1 nm in the supernatant solution by nebulization into an air-coal gas flame.

Stupar[22] has described a method for the determination of Ag in plants after digestion and solvent extraction, and Emmerman and Luecke[23] determined down to 0·2 ppb Ag in soils using a technique in which the soil was heated in a Ta boat and the supernatant vapour was examined by AAS. The determination of Ag in water by AAS after preconcentration by ion-exchange procedures has been investigated.[24,25] With 700 ml samples of snow-water from AgI seeded snowfall, Warburton[25] determined as little as 11 ng Ag/ml of sample after cation exchange, elution of Ag with NH_4CNS in acetone, dilution with MIBK and AAS determination in an air-acetylene flame. Other workers have reported AAS procedures for Ag in water utilizing preconcentration by extraction with dithizone[26,26a] or with tri-iso-octylthiophosphate.[27] Ag has been determined directly in air by AAS[28] at 328·1 nm in a method in which the air sample containing Ag provides primary air to the atom cell. Silver in oils has been determined by AAS by dilution of the sample with MIBK on nebulization into an air-acetylene flame,[29] and Smith, Stafford and Winefordner investigated the application of AFS for Ag determination in oils.[30] Silver in lubricating oils has recently been determined by AAS with carbon filament atomization.[31] Miller, Fraser and Winefordner[32] have described the use of a combined AFS-flame-emission spectrometer for the determination of Ag and other metals in jet-engine oils.

Schmidt[33] has determined Ag in photosensitive paper by immersing the sample (100 sq. cm) in 12% Na thiosulphate solution for 30 minutes in a darkroom; after 50-fold dilution with water the solution was nebulized into an air-hydrogen flame for AAS at 328·1 nm. Standard solutions were prepared to contain similar amounts of thiosulphate to the samples.

References

1. B. M. Gatehouse and J. B. Willis, *Spectrochim. Acta,* 1961, **17,** 710.
2. W. Slavin, "Atomic Absorption Spectroscopy", Interscience, New York, 1968.
3. K. E. Zacha, M. P. Bratzel, J. D. Winefordner and J. M. Mansfield, *Anal. Chem.,* 1968, **40,** 1733.
4. T. S. West and X. K. Williams, *Anal. Chem.,* 1968, **40,** 335.
5. D. C. Manning and P. Heneage, *Atomic Absorption Newsletter,* 1968, **7,** 80.
6. D. G. Mitchell and A. Johansson, *Spectrochim. Acta,* 1970, **25B,** 175.
7. R. Lockyer and G. E. Hames, *Analyst,* 1959, **84,** 385.
8. L. Wilson, *Anal. Chim. Acta,* 1964, **30,** 377.
9. R. Belcher, R. M. Dagnall and T. S. West, *Talanta,* 1964, **11,** 1257.
10. T. S. West and X. K. Williams, *Anal. Chim. Acta,* 1968, **40,** 335.
11. M. Fixman and L. Boughton, *Atomic Absorption Newsletter,* 1966, **5,** 33.
12. B. S. Rawling, M. C. Greaves and M. D. Amos, *Nature,* 1960, **188,** 137.
13. I. Rubeska, Z. Sulcek and B. Moldan, *Anal. Chim. Acta,* 1967, **37,** 27.
14. M. C. Greaves, *Nature,* 1963, **199,** 552.
15. T. T. Chao, J. W. Ball and J. M. Nakagawa, *Anal. Chim. Acta,* 1971, **54,** 77.
16. M. P. Bratzel and C. L. Chakrabarti, *18th Spect. Symp. Canada,* Toronto, 1971.

17. P. E. Moloughney and J. A. Graham, *Talanta*, 1971, **18**, 475.
18. V. L. Ginzburg, D. M. Livshits and G. I. Satarina, *Zh. Analit. Khim.*, 1964, **19**, 1089.
19. G. Henrich, H. Spitzer and G. Tesik, *Z. Anal. Chem.*, 1967, **226**, 124.
20. M. G. Hofton and D. P. Hubbard, *Proc. 3rd CISAFA*, Paris, 1971.
21. L. G. Hickey, *Anal. Chim. Acta*, 1968, **41**, 546.
22. J. Stupar, *Mikrochim. Acta*, 1966, 722.
23. R. Emmerman and W. Luecke, *Z. Anal. Chem.*, 1969, **248**, 325.
24. T. T. Chao, M. J. Fishman and J. W. Ball, *Anal. Chim. Acta*, 1969, **47**, 189.
25. J. A. Warburton, *J. Appl. Meteorology*, 1969, **8**, 464.
26. F. K. West, P. W. West and T. V. Ramakrishna, *Environ. Sci. Technol.*, 1967, **1**, 717.
26a. R. Woodriff, B. Culver, D. Shrader and A. B. Super, *Anal. Chem.*, 1973, **45**, 230.
27. T. T. Chao and J. W. Ball, *Anal. Chim. Acta*, 1971, **54**, 166.
28. H. W. Edwards, *Anal. Chem.*, 1969, **41**, 1172.
29. D. R. Jackson, C. Salama and R. Dunn, *Can. Spectrosc.*, 1970, **15**, 17.
30. R. Smith, C. M. Stafford and J. D. Winefordner, *Can. Spectrosc.*, 1969, **14**, 38.
31. J. F. Alder and T. S. West, *Proc. 3rd SAC Conf.*, Durham, 1971.
32. R. L. Miller, L. M. Fraser and J. D. Winefordner, *Appl. Spectrosc.*, 1971, **25**, 477.
33. W. Schmidt, *Flame Notes*, 1969, **4**, 9.

13.4.54 Sodium

Na, atomic weight 22·99

Technique	Line, nm	*f*	Source	Flame	Sensitivity, ppm/1% abs.	Detection limit, ppm	Reference (see below)
AAS	589·00, 589·59	0·7	VDL	A-coal gas	0·03	—	1
	589·00/59	0·7	HCL	A-A	0·015	0·002	2
	330·23, 330·29	0·05	VDL	A-coal gas	5	—	1
	330·23/29	0·05	HCL	A-A	2·8	—	3
	330·23/29	0·05	HCL	A-H	3·0	—	3
AFS	589·59	—	Xe arc	A-H	—	0·008	4

Ionization potential: 5·14 eV.
Atomization efficiency: A-A, 1·0; N-A, 0·33.

Notes on analytical conditions

1. The simple spectrum of Na hollow-cathode and vapour-discharge lamp sources enables a wide spectral band-pass to be employed for AAS if required. The Na 598·00 nm line, however, gives higher sensitivity than the 589·59 nm line and should be isolated from the latter using 0·2-0·3 nm bandpass when highest sensitivity is needed. Argon is preferred to neon filler gas in HCL sources as neon emits several lines in close proximity to the Na lines.

Both HCL and VDL sources should be operated at low power to minimize self-absorption broadening.

2. Sodium is most commonly determined by AAS in a lean (hot) air-acetylene flame; at low concentrations some ionization of Na is experienced in this flame and excess K or Cs should be added to samples and standards as suppressor. Interference from high concentrations of mineral acid may be observed on Na absorption. This may result partly from a change in nebulizer performance and partly from chemical effects in the flame. Samples and standards may be matched with respect to acid to compensate for these effects.

3. In the presence of high concentrations of Ca the emission and absorption produced by CaOH molecules in the flame may cause interferences and deterioration of the signal : noise ratio at the Na 589·00/589·59 nm absorption lines.

Applications notes and key references

Although many workers prefer to retain the flame emission spectrometric technique for the determination of sodium a considerable volume of literature now exists concerning its determination by AAS. Sodium and potassium are very frequently both determined in the same samples; Table 13-9 lists some applications of the determination of these two elements by AAS.

TABLE 13-9

Some applications of the determination of sodium and potassium by AAS

Sample type	Technique, sample pretreatment	Reference
Molybdenum tungsten	AAS; use Cs to suppress interferences; A-A flame	5
Aluminium alloys	AAS; dissolve in HNO_3/HF, dilute; A-A flame	5a
Non-ferrous alloys	AAS; dissolve in acid. Match major element concentration in samples and standards	6
Ferrites (Na)	AAS; dissolve in 50% HCl; A-H flame	7
Silicate rocks	AAS; dissolve in HF/H_2SO_4 or fuse with Na_2CO_3; dissolve in HCl	8
Silicate and carbonate rocks	AAS; dissolve in HF/HNO_3; use La addition	9
Rocks	AAS; cation exchange procedure for alkali metal separation	10
Silicate rocks	AAS; fuse with $LiBO_2$, dissolve in dilute HNO_3, dilute	11
Silicate rocks and minerals	AAS; fuse with $LiBO_2$, dissolve in HF, add H_3BO_3	12
Sulphide minerals	AAS; decompose in bomb with HF/HCl/HNO_3 at 110°C	13
Slags	AAS; dissolve in HNO_3/HF, evaporate, redissolve in HCl; A-A flame	14

TABLE 13-9 (*continued*)

Sample type	Technique, sample pretreatment	Reference
Silicate rocks and coal ash	AAS; fuse with $LiBO_2$, dissolve in HNO_3 + tartaric acid; A-A flame	15
Coal ash	AAS; dissolve in HF; use La to suppress interferences	16
Coal ash	AAS; dissolve in $HF/HClO_4$, volatilize fluoride, dilute	17
Cement	AAS; dissolve in HCl, evaporate, redissolve in 50% HCl; match Ca concentration in samples and standards; A-A flame	18
Cement	AAS; match Ca concentration in standards with that in sample solution	19
Cement	AAS; dissolve in HF/HCl, add La	20
Glass	AAS; HCl/HF dissolution. Use Sr to suppress interference	21
Lubricating oils (Na)	AAS; dilute with 1 : 2 with MIBK	22
Gas oils (Na)	AAS	23
Propellants	AAS; oxidize with HNO_3, then $HClO_4$, evaporate, dilute	24
Serum	AAS; dilute 1 : 50; use Na 330·2 nm line; add Na to K standards or add EDTA to K samples and standards	25
Serum	AAS; A-P flame	26
Serum	AAS; non-dispersive (resonance monochromator)	27
Serum	AAS; dilute, add strontium; A-A flame	28
Cattle plasma	AAS; effect of oral Na and K on plasma composition	29
Urine, faeces	AAS	30
Urine	AAS; 100 μl sample diluted with 9·9 ml $LaCl_3/CsCl$ soln.	31
Renal dialysates	AAS; dilute. Differential spectrometric technique	32
Cell membranes	AAS; ash at low temp. in active oxygen or with $HNO_3/HClO_4$; add Cs to suppress ionization interference	33
Biological specimens	AAS; ash at low temp., dissolve in HCl; use Cs and La to suppress interferences	34
Heart muscle, milk	AAS; dry ash, dissolve in 6 N HCl	35
Plant and animal tissues	AAS; Wet digest with $HClO_4/HNO_3$	36
Diet, faeces	AAS; wet digestion with $HClO_4$ acid mixture	37
Bone ash	AAS	38
Plant materials	AAS; digest with $HClO_4/H_2SO_4/HNO_3$. Standards contain 1% H_2SO_4	39
Plants	AAS	40
Maize, wheat	AAS; digest with HNO_3, dissolve in water, evap. with HCl, redissolve in HCl, dilute	41
Fertilizers	AAS	42
Soils	AAS; dissolve in HF/HCl mixture	43

TABLE 13-9 (*continued*)

Sample type	Technique, sample pretreatment	Reference
Soil extracts	AAS; extract with 1 M NH_4Cl	44
Feeds	AAS; dry ash at 450°C; dissolve residue in acid	45
Foods	AAS	46
Feeding stuffs	AAS	47
Beer	AAS; dilute with water	48
Oils and fats	AAS; shake with water or dilute with MIBK	49
Fruit juices	AAS; dilute, add HCl, centrifuge, add La	50
Milk	AAS; use preliminary dry-ashing proc.	51
Sow milk	AAS	52
Gelatine	AAS; wet ash with acid; carbon rod atomizer	53
Dried milk products	AAS; dissolve 1 g in water, dilute to 100 ml. Dilute 1 : 10	54
Milk	AAS; precipitate protein with TCA, filter, dilute, add La	55
White sugar	AAS	56
Waters	AAS; for K analysis add Na to standards and samples	57
Sea-water	AAS	58

References

1. B. M. Gatehouse and J. B. Willis, *Spectrochim. Acta,* 1961, **17,** 710.
2. "Analytical Methods for Atomic Absorption Spectrophotometry", Perkin-Elmer Corp., Norwalk, Conn., U.S.A., 1971.
3. J. E. Schallis and H. L. Kahn, *Atomic Absorption Newsletter,* 1968, **7,** 75.
4. A. Hell and B. Ricchio, *Pittsburgh Conf. in Anal. Chem. and Appl. Spectroscopy,* Cleveland, Ohio, U.S.A., 1970.
5. G. M. Newman, *Talanta,* 1971, **18,** 1047.

5a. I. Janousek, *Chemicke Listy,* 1973, **67,** 982.

6. T. W. Sattur, *Atomic Absorption Newsletter,* 1966, **5,** 37.
7. Y. Yamamoto, T. Kumamaru, Y. Hayashi, Y. Otani and T. Mutsunobu, *Kogyo Kagaku Zasshi,* 1971, **74,** 779.
8. D. J. Trent and W. Slavin, *Atomic Absorption Newsletter,* 1964, **3,** 17, 118.
9. G. K. Billings and J. A. S. Adams *Atomic Absorption Newsletter,* 1964, **3,** 65.
10. F. W. E. Strelow, F. S. Toerien and C. H. Weinert, *Anal. Chim. Acta,* 1970, **50,** 399.
11. J. C. Van Loon and C. M. Parissis, *Analyst,* 1969, **94,** 1057.
12. Sydney Abbey, *Geol. Surv. Pap. Can.,* 1970, No. 72-73.
13. F. J. Langmyhr and P. E. Paus, *Anal. Chim. Acta,* 1970, **50,** 515.
14. D. W. Cobb, W. W. Foster and T. S. Harrison, *Metallurgia,* 1971, **83,** 239.
15. P. L. Boar and L. K. Ingram, *Analyst,* 1970, **95,** 124.
16. C. B. Belcher and K. A. Brooks, *Anal. Chim. Acta,* 1963, **29,** 202.
17. E. L. Obermiller and R. W. Freedman, *Fuel,* 1965, **44,** 199.
18. L. Capacho-Delgado and D. C. Manning, *Analyst,* 1967, **92,** 553.
19. T. Takeuchi and M. Suzuki, *Talanta,* 1964, **11,** 1391.
20. J. T. H. Roos and W. J. Price, *Analyst,* 1969, **94,** 89.
21. A. H. Jones, *Anal. Chem.,* 1965, **37,** 1761.
22. D. R. Jackson, C. Salama and R. Dunn, *Can. Spectrosc.,* 1970, **15,** 17, 24.
23. J. W. Robinson, *Anal. Chim. Acta,* 1960, **23,** 458.

24. D. M. Knight, *Atomic Absorption Newsletter*, 1970, **9**, 108.
25. J. B. Willis, *Spectrochim. Acta*, 1960, **16**, 551.
26. R. Herrman and W. Lang, *Z. Ges. Exp. Med.*, 1961, **134**, 268.
27. J. V. Sullivan and A. Walsh, *Spectrochim. Acta*, 1966, **22**, 1843.
28. K. Paschen and C. Fuchs, *Clin. Chim. Acta*, 1971, **35**, 401.
29. K. C. Scotts, U. R. Bohman and A. L. Lesperance, *J. Amin. Sci.*, 1971, **32**, 354.
30. J. B. Coombe, K. R. Christian and M. D. Holgati, *J. Agr. Sci. Cambr.*, 1971, **77**, 159.
31. B. W. Grunbaum and N. Pace, *Microchem. J.*, 1970, **15**, 666.
32. D. R. Thomerson, *Spectrovision*, 1971, **25**, 12.
33. N. Sanui and N. Pace, *Appl. Spectrosc.*, 1966, **20**, 135.
34. N. Sanui, *Analyt. Biochem.*, 1971, **42**, 21.
35. H. Feder, *Dt. Tieraerztl, Wschr.*, 1970, **77**, 642.
36. T. Jaakola, H. Puumala and J. K. Miettinin, *Proc. Intern. Symp., Radioecol. Conc. Proc.*, Stockholm, 1966, 341.
37. E. D. Rees and R. D. Hughes, *Clin. Chim. Acta*, 1968, **21**, 515.
38. D. C. Manning, *Atomic Absorption Newsletter*, 1963, No. 11.
39. D. J. David, in "Modern Methods of Plant Analysis", Vol. 5, eds. Peach and Tracy, Springer, Berlin, 1962.
40. J. E. Allan, *Analyst*, 1961, **86**, 530.
41. W. J. Garcia, C. W. Blessin and G. E. Inglett, *Cereal Chem.*, 1972, **49**, 158.
42. C. H. McBride, *J. Ass. Off. Analyt. Chem*, 1967, **50**, 401.
43. S. Pawluk, *Atomic Absorption Newsletter*, 1967, **6**, 53.
44. D. J. David, *Analyst*, 1960, **85**, 495.
45. W. A. Dewar and W. G. Siller, *Brit. Poultry Sci.*, 1971, **12**, 535.
46. H. L. Kahn, *Food Eng.*, 1965, **37**, 105.
47. Mary Heckman, *J. Ass. Off. Analyt. Chem.*, 1970, **53**, 923.
48. S. W. Frey, *Atomic Absorption Newsletter*, 1964, **3**, 127.
49. A. Prevot, *Atomic Absorption Newsletter*, 1966, **5**, 13.
50. J. T. H. Roos and W. J. Price, *J. Sci. Food Agr.*, 1970, **21**, 51.
51. G. K. Murphy and U. Rhea, *J. Dairy Sci.*, 1967, **50**, 313.
52. R. F. Elliot, G. W. Van der Noot, R. L. Gilbreath and H. Fisher, *J. Anim. Sci.*, 1971, **32**, 1128.
53. Z. T. Gakovic and T. S. West, *Proc. 3rd CISAFA*, Paris, 1971.
54. P. Uebersax, *Mitt. Geb. Lebensmittelunters. u. Forsch.*, 1971, **62**, 80.
55. I. B. Brooks, G. A. Luster and D. G. Easterly, *Atomic Absorption Newsletter*, 1970, **9**, 93.
56. T. D. Carpenter and S. E. Bichsel, *J. Am. Soc. Sug. Beet Technol*, 1969, **15**, 369.
57. M. J. Fishman and S. C. Downs, *U.S. Geol. Surv. Water Supply Papers*, 1540-C, 1966.
58. G. K. Billings and R. C. Harriss, *Texas J. Sci.*, 1965, **17**, 129.

13.4.55 Strontium

Sr, atomic weight 87·62

Technique	Line, nm	f	Source	Flame	Sensitivity, ppm/1% abs.	Detection limit, ppm	Reference (see below)
AAS	460·73	1·54	HCL	A-A	0·15	—	1
	460·73	1·54	HCL	A-A	0·05	—	2
	407·73	0·76	HCL	A-A	3·5	—	1
	460·73	1·54	HCL	N-A	0·09	0·01	3
AFS	460·73	1·54	EDL	Ar/H_2	—	0·03	4
	460·73	1·54	N_2 pumped dye laser	A-H	—	0·01	4a

Ionization potential: 5·69 eV.
Dissociation energy, SrO: 4·85 eV.
Atomization efficiency: A-A, 0·13.

Notes on analytical conditions

1. Fuel-rich conditions provide highest atomization efficiency and therefore absorption sensitivity for Sr when the air-acetylene flame is employed. The degree of ionization as well as the atomization efficiency is also dependent upon the flame stoicheiometry and is least in the cooler fuel-rich flame. Ionization suppression should be employed even in this flame, however. The stoicheiometric nitrous oxide-acetylene flame provides somewhat higher sensitivity for AAS of Sr; again alkali metal addition is recommended for suppression of the substantial (*ca.* 80%) ionization in this flame.

2. As also observed for Ca and Ba the absorption signal obtained for Sr in the air-acetylene flame is depressed in the presence of Al, Si or PO_4 and this effect may be eliminated by the use of lanthanum or calcium as releasing agent. The use of the nitrous oxide-acetylene flame has been recommended for minimization of chemical interference effects.

3. Although no extensive studies of their analytical absorption characteristics has been made, the Sr 242·8, 256·9 and 292·2 nm lines may provide for analysis when lower sensitivity than that obtained at 460·73 nm is required.

Applications notes and key references

The AAS determination of Sr in silicate and other rock samples has been described by several workers.[4-6] Either HF dissolution or carbonate or $LiBO_2$ fusion procedures have been employed for sample decomposition. In earlier work reported with the air-acetylene flame lanthanum releasing agent was used to suppress interferences from matrix elements, and $LaCl_3$ may also be added to samples and standards when the nitrous oxide-acetylene flame is used in order to suppress ionization and residual chemical interference effects. Strontium in phosphate rock has been determined by extraction with concentrated HCl, removal of cations from the solution using an ion-exchange procedure, elution with 4 M HCl, dilution and nebulization into an air-acetylene flame.[7] Moldan and Miksovsky[8] have described the determination of Sr in rocks and minerals after acid dissolution or alkaline fusion by AAS in an air-acetylene flame; 8-hydroxyquinoline and lanthanum were added to solutions to suppress interferences. Belcher and Brookes[9] determined Sr in coal ash by AAS; samples were warmed with concentrated HF, $HClO_4$ was added, the solution was warmed to fumes, cooled and treated with dilute HCl before addition of $LaCl_3$ and aspiration into an air-acetylene flame. Strontium in cement has been determined by several AAS procedures which differ in the method of acid attack employed for sample dissolution.[10-13] Capacho-Delgado and Manning[10] used HCl and Roos and Price[12] used HCl with subsequent addition of HF and also boric acid (to prevent precipitation of LAF_3 from the La buffer added subsequently); there is general agreement that the nitrous oxide-acetylene flame is most suitable for Sr in these samples. Adams and Passmore[14] have described the determination of Sr in glass by AAS with an $HF/HClO_4$ dissolution technique using EDTA and lanthanum to suppress interferences in the air-acetylene flame.

Curnow and co-workers[15] have reported the determination by AAS of Sr in serum and urine from Sr-treated subjects. In the method described by these workers Sr was co-precipitated with Ca oxalate, the precipitate was dissolved in HCl and $LaCl_3$ was added prior to nebulization. Tompsett[16] evaporated urine samples to dryness, ashed the residue and dissolved the ash in 0·1 M HCl before passing through Dowex 50 W ion-exchange resin and determination of Sr in the eluate by AAS. Montford and Cribbs[17] and Descube and co-workers[18] have also described methods for the determination of Sr in urine. Miro and Sutton[19] have determined Sr in a variety of biological materials (see Ba). David[2] utilized an anion-exchange resin separation of phosphate from ashed biological materials before determination of Sr by AAS. Tanaka and co-workers[20] used Ca addition to suppress Si and Al interference in the determination of Sr by AAS in bone, food, water and soils. David[21] and Trent and Slavin[22] have also described AAS methods for determination of Sr in plant materials, and the former worker described its determination in NH_4Cl soil extracts. Strasheim, Strelow and Norval[23] have described the determination of Sr in maize meal and bread flour by AAS. In this procedure the ashed samples are subjected to ion-exchange chromatography to separate Sr from Ca and phosphate before its determination.

Oil-field brines have been analysed for Sr by Billings and Angino[24] and several other studies of the Sr content of saline waters have been described.[25,26,27] Strontium is included in the scheme developed by Fishman and Downs[28] for the determination of trace elements in fresh water and brines by AAS. These workers recommended that both lanthanum and potassium be added to samples and standards to suppress ionization and chemical interference in the Sr determination.

Other materials in which Sr has been determined by AAS include synthetic fibres[29] and medicinal mineral waters.[30]

References

1. B. M. Gatehouse and J. B. Willis, *Spectrochim. Acta,* 1961, **17,** 710.
2. D. J. David, *Analyst,* 1962, **87,** 576.
3. "Analytical Methods for Atomic Absorption Spectrophotometry", Perkin-Elmer Corp., Norwalk, Conn., U.S.A. 1971.
4. K. E. Zach, M. P. Bratzel Jnr., J. D. Winefordner and J. M. Mansfield Jnr., *Anal. Chem.,* 1968, **40,** 1733.
4a. L. M. Fraser and J. D. Winefordner, *Anal. Chem.,* 1972, **44,** 1444.
5. D. J. Trent and W. Slavin, *Atomic Absorption Newsletter,* 1964, **3,** 17, 118.
6. G. K. Billings and J. A. S. Adams, *Atomic Absorption Newsletter,* 1964, **3,** 65.
7. A. Hofer, *Z. Anal. Chem.,* 1970, **249,** 115.
8. B. Moldan and M. Miksovsky, *Coll. Czech. Comm.,* 1971, **36,** 1673.
9. C. B. Belcher and K. A. Brookes, *Anal. Chim. Acta,* 1963, **29,** 202.
10. L. Capacho-Delgado and D. C. Manning, *Analyst,* 1967, **92,** 553.
11. T. P. Lees, *Chemy. Ind.,* 1969, **36,** 1249.
12. J. T. H. Roos and W. J. Price, *Analyst,* 1969, **94,** 89.
13. F. J. Langmyhr and P. E. Paus, *Anal. Chim. Acta,* 1969, **44,** 445.
14. P. B. Adams and W. O. Passmore, *Anal. Chem.,* 1966, **38,** 630.
15. D. C. Curnow, D. H. Gutteride and E. D. Horgan, *Atomic Absorption Newsletter,* 1968, **7,** 45.
16. S. L. Tompsett, *Proc. Ass. Clin. Biochem.,* 1968, **5,** 125.
17. B. Montford and S. C. Cribbs, *Atomic Absorption Newsletter,* 1969, **8,** 77.
18. J. Descube, N. Roques, F. Rousselet and M. L. Girard, *Anal. Biochem. Clin.,* 1967, **25,** 1011.
19. M. Miro and D. C. Sutton, *Appl. Spectrosc.,* 1970, **24,** 200.
20. G. Tanaka, A. Tamikawa, H. Kawamura and Y. Oyagi, *J. Chem. Soc. Japan, Pure Chem. Sect.,* 1968, **89,** 175.

21. D. J. David, *Analyst*, 1969, **94**, 884.
22. D. J. Trent and W. Slavin, *Atomic Absorption Newsletter*, 1964, **3**, 53.
23. A. Strasheim, F. W. E. Strelow and E. Norval, *J. S. Afr. Chem. Inst.*, 1967, **20**, 25.
24. G. K. Billings and E. E. Angino, *Bull. Can. Petrol. Geol.*, 1965, **13**, 29.
25. G. K. Billings and R. C. Harriss, *Texas J. Sci.*, 1965, **17**, 129.
26. E. E. Angino, G. K. Billings and N. Andersen, *Chem. Geology*, 1966, **1**, 145.
27. B. P. Fabricand, E. S. Imbimbo, M. E. Brey and J. A. Weston, *J. Geophys. Res.*, 1966, **71**, 3917.
28. M. J. Fishman and S. C. Downs, *U.S. Geol. Surv. Water Supply Papers*, 1540-C, 1966.
29. W. Slavin, *Atomic Absorption Newsletter*, 1965, **4**, 192.
30. C. Pulido, M. C. M. de Almeida and A. A. G. de Almeida, *Revta. Port. Quim.*, 1969, **11**, 84.

13.4.56 Sulphur

S, atomic weight 32·06

Technique	Line, nm	*f*	Source	Flame	Sensitivity, ppm/1% abs.	Detection limit, ppm	Reference (see below)
AAS	180·7	—	EDL	L'Vov furnace	8×10^{-11} g	—	1
	182·1	—	EDL	L'Vov furnace	1×10^{-10} g	—	1
	182·6	—	EDL	L'Vov furnace	1×10^{-10} g	—	1
	180·7	—	EDL	N_2 sep. N-A	9	30	2
	180·7	—	HCL	N_2 sep. N-A	1	5	3
AFS: not reported							

Ionization potential: 10·36 eV.

Notes on analytical conditions

1. In the work of L'Vov,[1] the graphite furnace was purged with argon and a vacuum monochromator was employed.

2. The transmission of the slightly fuel-rich nitrogen-separated flame (70 mm slot) at 180·7 nm is *ca.* 64% of that of nitrogen alone at this wavelength, i.e. the flame is more transparent than the atmosphere. The light path between the EDL source and the separated flame, and between flame and monochromator must be purged with nitrogen or argon. A nitrogen-filled monochromator may be used satisfactorily.

3. The sensitivity values (1% absorption) given above for the direct determination of sulphur in the nitrous oxide-acetylene flame refer to nebulization of K_2SO_4 in water. Similar sensitivities are

obtained in aqueous solution when sulphur is introduced into the flame as thiosulphate, sulphite, thiocyanate and thiourea.[2] The exploitation of the organic solvent effect leads to 1% absorption for 4·4 ppm S as thiourea in ethanol and 2·7 ppm S as dibenzyldisulphide or t-butyl disulphide in MIBK.[4] Sulphur compounds of high volatility give rise to higher sensitivity when introduced into the flame in MIBK via an indirect nebulizer assembly, e.g. for thiophen in MIBK 1% absorption is produced by 0·6 ppm S.

4. The high atomization efficiency obtained for sulphur and other elements in the nitrous oxide-acetylene flame allows AAS measurements for sulphur with considerable freedom from chemical interferences. At 50-fold weight excess over 200 ppm S in aqueous medium Al, Cu, K, Mg, Mn, Mo, Na, Ni, Zn, fluoride, phosphate, chloride bromide and iodide have been shown not to interfere.[2]

5. Several indirect AAS methods for the determination of sulphur have been reported (see Chapter 14).

Applications notes and key references

The only application of the direct determination of sulphur by AAS in the separated nitrous oxide-acetylene flame that has been reported at the time of writing is concerned with the determination of sulphur in oils.[4] Down to 0·1% total sulphur in refined and crude oils may be determined directly after suitable dilution with MIBK and comparison with standards prepared from dibenzyl disulphide dissolved in MIBK. Any viscosity mismatch between samples and standards encountered when small dilution factors are used for oils of low sulphur content may be compensated for by addition of sulphur-free base oil to the standard solutions. Alternatively the method of standard additions may be employed.

The indirect AAS determination of sulphur (via precipitation of $BaSO_4$ and estimation of Ba) in biological materials has been reported.[5] Other materials in which sulphur has been determined by this type of method include textiles,[6] soils[7,8] and waste waters.[9] See Chapter 14 for further details of these procedures.

References

1. B. V. L'Vov, in "Atomic Absorption Spectroscopy", eds. R. M. Dagnall and G. F. Kirkbright, Butterworths, London, 1970, p. 28.
2. G. F. Kirkbright and Maurice Marshall, *Anal. Chem.*, 1972, **44**, 1288.
3. G. F. Kirkbright and P. J. Wilson, *Anal. Chem.*, 1974, in press.
4. G. F. Kirkbright, Maurice Marshall and T. S. West, *Anal. Chem.*, 1972, **44**, 2379.
5. T. Kumamaru, *Anal. Chim. Acta*, 1968, **43**, 19.

6. R. Dunk, R. A. Mostyn and H. C. Hoare, *Atomic Absorption Newsletter*, 1969, **8**, 79.
7. J. A. Varley and P. Y. Chin, *Analyst*, 1970, **95**, 592.
8. G. G. Galindo, H. Applet and E. B. Schalscha, *Soil Sci. Soc. Amer., Proc.*, 1969, **33**, 974.
9. F. Y. Borden and L. H. McCormic, *Soil Sci. Sol. Amer. Proc.*, 1970, **34**, 705.

13.4.57 Tantalum

Ta, atomic weight 180·95

Technique	Line, nm	f	Source	Flame	Sensitivity, ppm/1% abs.	Detection limit, ppm	Reference (see below)
AAS	271·40	0·05	HCL	N-A	11	—	1
	271·40	0·05	HCL	N_2-O_2-A	30	—	1
	260·86	—	HCL	N_2-O_2-A	58	—	1
	277·59	—	HCL	N_2-O_2-A	58	—	1
	255·94	—	HCL	N_2-O_2-A	70	—	1
	264·75	—	HCL	N_2-O_2-A	90	—	1
	266·13	—	HCL	N_2-O_2-A	120	—	1
	275·83	—	HCL	N_2-O_2-A	120	—	1
AFS: not reported							

Ionization potential: 7·88 eV.

Notes on analytical conditions

1. Amos and Willis[1] used a spectral band-pass of 0·3 nm in their study of Ta absorption in a 60% oxygen-nitrogen-acetylene flame. Other workers have reported lower sensitivity in AAS using the 271·40 nm line[2,3] for Ta solutions containing fluoride. Headridge and Hubbard[3] obtained a value of 17 ppm Ta/1% absorption at 271·40 nm when 25% ethanol solutions were nebulized.

2. The absorption signal obtained for Ta may be enhanced in the presence of iron and HF.[1] Bond[4] recommended the use of 0·1 M ammonium fluoride solutions in samples and standards to compensate for this effect of HF.

3. Ta may be extracted into MIBK from 10M HF/6M H_2SO_4 2·2M NH_4F medium and from other strongly acidic media containing excess fluoride.

Application notes and key references

Very few applications of AAS to the determination of Ta have been reported. Bond[4] determined Ta in Co-based alloys by a procedure in which the alloy was dissolved in HF/HCL and Ta extracted into MIBK. Schlewitz and Shields[5] determined Ta in zirconium and its alloys by a similar procedure after sample dissolution with HF/H_2O_2.

References

1. M. D. Amos and J. B. Willis, *Spectrochim. Acta,* 1966, **22,** 1325.
2. "Analytical Methods for Atomic Absorption Spectrophotometry", Perkin-Elmer Corp., Norwalk, Conn., U.S.A., 1971.
3. J. B. Headridge and D. P. Hubbard, *Anal. Chim. Acta,* 1967, **37,** 151.
4. A. M. Bond, *Anal. Chem.,* 1970, **42,** 932.
5. J. H. Schlewitz and M. G. Shields, *Atomic Absorption Newsletter,* 1971, **10,** 39.

13.4.58 Tellurium

Te, atomic weight 127·60

Technique	Line, nm	*f*	Source	Flame	Sensitivity, ppm/1% abs.	Detection limit, ppm	Reference (see below)
AAS	214·28	0·08	HCL	A-H	0·36*	0·12	1
	225·90	—	HCL	A-H	3·5*	—	1
	238·58	—	HCL	A-H	67*	—	1
	214·28	0·08	HCL	A-A	0·5	—	2
	214·28	0·08	HCL	A-A	0·5	—	3
AFS	238·58 } 238·30 }	—	EDL	A-P	—	0·12	4
	214·28	—	EDL	Ar/H_2	—	0·5	5

Ionization potential: 9·0 eV.
Dissociation energy, TeO: 2·7 eV.

Notes on analytical conditions

1. Values for 1% absorption sensitivity marked * were obtained with a triple-pass optical arrangement.

2. Aluminium or nickel cathode shells should be used for Te HCL sources; early Te HCL sources with copper cathode shells show emission from Cu lines near the Te 214·28 nm line. The use of large band-pass may result in loss of sensitivity owing to transmission of the non-absorbing Te line at 214·72 nm.

3. The fuel : oxidant flow-rate ratio is not critical in the air-acetylene flame, although a lean flame is preferred by many workers (possibly to minimize interference effects).

4. The AF signal at the 238·58 and 238·30 nm lines corresponds to direct-line fluorescence ($^3S_1^0, \rightarrow {}^3P_1$ or $^3P_0$) and the energy from

both lines may be collected using sufficiently wide spectral band-pass. The resonance fluorescence emission is that observed at 214·28 nm.

5. Chakrabarti[1] reported interference with Te absorption from high concentrations of Zn, Cu, Na and Ca. Other workers[6] have found interference from Zr and Si using the air-acetylene flame.

Applications notes and references

Most applications of the determination of Te by AAS have been made in geochemistry and metallurgy. Nakagawa and Thompson[7] have examined the determination of Te in rocks and soil utilizing extraction into MIBK from HBr medium, and Davis, Ewers and Fletcher[8] have described a method in which the ground rock is mixed with a synthetic pyrite mixture and heated to 800-850°C. The volatilized Te and sulphur are swept out and condensed. The sublimate is refluxed with aqua regia, the solution is evaporated with HBr and $HClO_4$ to remove AS and Se, and then with aqua regia and $HClO_4$. The final solution, containing only Te, HCl and $HClO_4$ is introduced into an oxy-hydrogen flame fitted with a long tube burner for Te determination. Less than 0·1 ppm Te was determined in this way. Tellurium was determined in copper matrices by Sprague, Manning and Slavin;[9] matching of the copper concentration of samples and standards was found to be necessary. Musha *et al.*[6] analysed copper slime and sulphur samples for Te by AAS. The AAS determination of between 5 ppm and 0·03% Te in steel has been reported[10] utilizing a procedure in which Te is quantitatively precipitated using $SnCl_2$ solution in HCl followed by conversion to the diethyldithiocarbamate complex and its extraction into ethyl acetate for nebulization.[10] In the direct AAS determination of Te in steel using an air-acetylene flame Barnett and Kahn recommend the use of a deuterium background corrector to compensate for the broad band absorption at 214·28 nm from the matrix material.[11] Sattur[12] has reported the determination of Te in non-ferrous alloys and Roth, Bohl and Sellers[13] have determined Te in bismuth by AAS.

Dagnall, Taylor and West[14] have described the AFS determination of Te in nickel-based alloys after its separation by solvent extraction.

Kinser[15] described the determination of Te in tissue samples after their digestion with nitric acid, and Koirtyohann and Feldman[16] have determined Te in blood and urine after extraction with sodium diethyldithiocarbamate.

References

1. C. L. Chakrabarti, *Anal. Chim. Acta,* 1967, **39,** 293.
2. "Analytical Methods for Atomic Absorption Spectrophotometry", Perkin-Elmer Corp., U.S.A., 1971.
3. J. E. Allan, *Spectrochim. Acta,* 1962, **18,** 259.
4. R. M. Dagnall, K. C. Thompson and T. S. West, *Talanta,* 1967, **14,** 557.
5. K. E. Zacha, M. P. Bratzel, J. D. Winefordner and J. M. Mansfield, *Anal. Chem.,* 1968, **40,** 1733.
6. Soichiro Musha, Makoto Munemori and Taketoshi Nakahara, *J. Chem. Soc. Japan, Pure Chem. Sect.,* 1968, **89,** 495.
7. H. M. Nakagawa and C. E. Thompson, *Prof. Paper U.S. Geol. Surv.,* 600B, 1968, B123-125.
8. C. E. S. Davis, W. E. Ewers, and A. B. Fletcher, *Proc. Australas. Inst. Min. Metall.,* 1967, 67.
9. S. Sprague, D. C. Manning and W. Slavin, *Atomic Absorption Newsletter,* 1964, 3, 27.
10. M. V. Marcec, K. Kinson and C. B. Belcher, *Anal. Chim. Acta,* 1968, **41,** 447.
11. W. B. Barnett and H. L. Kahn, *Atomic Absorption Newsletter,* 1969, **8,** 21.
12. T. W. Sattur, *Atomic Absorption Newsletter,* 1966, **5,** 37.
13. D. J. Roth, D. R. Bohl and D. E. Sellers, *Atomic Absorption Newsletter,* 1968, **7,** 87.
14. R. M. Dagnall, M. R. G. Taylor and T. S. West, *Laboratory Practice,* 1971, **20,** 209.
15. R. E. Kinser, *Am. Ind. Hyg. Assoc. J.,* 1966, **27,** 260.
16. S. R. Koirtyohann and C. Feldman, in "Developments in Applied Spectroscopy", Vol. II, eds. J. E. Forrette and E. Lanterman, Plenum Press, New York, 1964.

13.4.59 Terbium

Tb, atomic weight 158·925

Technique	Line, nm	*f*	Source	Flame	Sensitivity, ppm/1% abs.	Detection limit, ppm	Reference (see below)
AAS	432·65	–	Xe arc	O_2/A	–	30	1
	432·65	–	HCL	N-A	7	2	2
	431·88	–	HCL	N-A	9	–	2
	390·13	–	HCL	N-A	12	–	2
	406·16	–	HCL	N-A	14	–	2
	433·84	–	HCL	N-A	16	–	2
	410·54	–	HCL	N-A	28	–	2
AFS: not reported							

Ionization potential: 5·98 eV.

Notes on analytical conditions

1. Jaworowski *et al.*[3] reported a sensitivity of 25 ppm Tb/1% absorption at 435·68 nm using a nitrous oxide-acetylene flame and introducing Tb in 80% ethanolic solutions.

2. Appreciable ionization of Tb occurs in the nitrous oxide-acetylene flame. The values for sensitivity achieved by Manning[2] shown above were obtained in the presence of excess alkali metal ion to suppress ionization.

3. Little information is available concerning chemical interferences in the determination of Tb.

Applications notes and key references

See Lanthanum.

References

1. V. G. Mossotti and V. A. Fassel, *Spectrochim. Acta,* 1964, **20,** 1117.
2. D. C. Manning, *Atomic Absorption Newsletter,* 1966, **5,** 127.
3. R. J. Jaworowski, R. P. Weberling and D. J. Bracco, *Anal. Chim. Acta,* 1967, **37,** 284.

13.4.60 Thallium

Tl, atomic weight 204·37

Technique	Line, nm	*f*	Source	Flame	Sensitivity, ppm/1% abs.	Detection limit, ppm	Reference (see below)
AAS	276·78	0·27	VDL	A-coal gas	0·8		1
	276·78	..	HCL	A-A	0·5		2
	276·78	..	HCL	A-P	2·5		3
	377·57	0·13	VDL	A-coal gas	3		1
AFS	377·57	0·13	VDL	O_2/H_2	—	0·04	4
	377·57	0·13	EDL	A-H	—	0·12	5
	377·57	..	EDL	Ar/H_2	—	0·008	6
	377·57	..	HIL	A-H	—	0·1	7
	377·57	..	150 W Xe	A-H	—	0·2	7
	377·57	..	HCL	A-H	—	0·1	8

Ionization potential: 6·11 eV.
Dissociation energy; TlO: $<$ 3·9 eV.
Atomization efficiency: A-A, 0·36.

Notes on analytical conditions

1. When a vapour-discharge lamp source is used for Tl best sensitivity is obtained at low operating currents[9] (300-350 mA); the source must be operated without a glass envelope for use at the Tl 276·78 nm line.

2. With an air-propane flame Veenendaal and Polak[3,10] observed

enhanced Tl absorption in the presence of most common cations and anions; use of 0·5% Li, Na or K added as a buffer reduced the effect of these to less than 10% interference. Few interference data are available for AAS of Tl in the air-acetylene flame.

3. As it is relatively volatile and easily atomized, there has been considerable interest in the detection of Tl by AAS and AFS in non-flame atom cells (see Chapter 7 and refs. 11-14).

Applications notes and key references

The toxic nature of thallium makes its determination important in the body and the environment. Its determination in urine and blood has been undertaken directly[15] and after ashing and extraction with sodium diethyldithiocarbamate into MIBK.[16] In the determination of Tl in blood, urine, tissue or faeces, Alexander *et al.*[17] digested samples with a mixture of H_2SO_4, HNO_3 and $HClO_4$ and extracted Tl with hydrobromic acid into ether, evaporated the organic solvent and took the residue into acid before nebulization. Curry, Read and Knott[18] describe three methods for the determination of Tl in biological material by atomic emission and absorption spectrometry. In these methods the samples are either (a) digested with HNO_3 and the Tl preconcentrated by extraction with HBr or diethyldithiocarbamate into MIBK, or (b) examined directly in a tantalum boat after evaporation of a 50 to 200 μl sample (blood or urine), and, for blood, ashing with HNO_3 in the boat. Thompson, Morgan and Purdue[19] have determined Tl in atmospheric particulate matter by AAS after sample collection on a glass fibre or membrane filter, ashing of the filter in an oxygen plasma at 150°C for 1 hour, and extraction of the ash with HCl/HNO_3. Danchik, Boltz and Hargis[20] have developed an indirect method for Tl in which Tl molybdophosphate is precipitated, washed and dissolved in NaOH for determination of its Mo content at 313·2 nm in a reducing air-acetylene flame.

References

1. B. M. Gatehouse and J. B. Willis, *Spectrochim. Acta,* 1961, **17,** 710.
2. "Analytical Methods for Atomic Absorption Spectrophotometry", Perkin-Elmer Corp., U.S.A. 1971.
3. W. A. Veenendaal and H. L. Polak, *Z. Anal. Chem.*, 1966, **223**, 17.
4. J. M. Mansfield, J. D. Winefordner and C. Veillon, *Anal. Chem.*, 1965, **37**, 1049.
5. R. F. Browner, R. M. Dagnall and T. S. West, *Talanta,* 1969, **16,** 75.
6. K. E. Zacha, M. P. Bratzel, J. D. Winefordner and J. M. Mansfield, *Anal. Chem.*, 1968, **40,** 1733.

7. D. C. Manning and P. Heneage, *Atomic Absorption Newsletter,* 1967, **6,** 124.
8. G. Rossi and N. Omenetto, *Talanta,* 1969, **16,** 263.
9. J. Ramirez-Munoz, *Flame Notes,* 1966, **1,** 85.
10. W. A. Veenendaal and H. L. Polak, *Spectrosc. Letters,* 1969, **2,** 173.
11. B. V. L'Vov, *Spectrochim. Acta,* 1969, **24B,** 53.
12. H. Brandenberger, *Chimia,* 1968, **22,** 449.
13. M. P. Bratzel, R. M. Dagnall and J. D. Winefordner, *Appl. Spectrosc.,* 1970, **24,** 518.
14. R. G. Anderson, I. S. Maines and T. S. West, *Anal. Chim. Acta,* 1970, **51,** 355.
15. W. Slavin, S. Sprague, F. Reiders and V. Crodova, *Atomic Absorption Newsletter,* 1964, (17).
16. E. Berman, *Atomic Absorption Newsletter,* 1967, **6,** 57.
17. G. V. Alexander, R. E. Nusbaum and N. S. McDonald, *J. Biol. Chem.,* 1951, **192,** 489.
18. A. S. Curry, J. F. Read and A. R. Knott, *Analyst,* 1969, **94,** 744.
19. Richard J. Thompson, George B. Morgan and Larry J. Purdue, *Atomic Absorption Newsletter,* 1970, **9,** 53.
20. R. S. Danchik, D. F. Boltz and L. G. Hargis, *Analyt. Letters,* 1968, **1,** 891.

13.4.61 Thorium

Th, atomic weight 232·038

Technique	Line, nm	f	Source	Flame	Sensitivity, ppm/1% abs.	Detection limit, ppm	Reference (see below)
AAS	324·58	–	HCL	N-A	850	–	1
AFS: not reported							

Ionization potential: 6·2 eV.
Dissociation energy, ThO: 8·6 eV.

Notes on analytical conditions

1. Even in the hot fuel-rich nitrous oxide-acetylene flame the atomization efficiency for Th is very low owing to the high stability of thorium oxide.

2. The value for the 1% absorption sensitivity given above was obtained using a heated spray chamber.

3. The indirect determination of Th by AAS is described in Chapter 14.2.

Reference

1. J. B. Willis, in "Analytical Flame Spectroscopy", ed. R. Mavrodineanu, Philips, Eindhoven, 1968.

13.4.62 Thulium

Tm, atomic weight 168·93

Technique	Line, nm	f	Source	Flame	Sensitivity, ppm/1% abs.	Detection limit, ppm	Reference (see below)
AAS	409·42	–	Xe arc	O_2/A	–	0·1	1
	409·42	–	Tm in O_2/H_2 Flame	O_2/A	7·0	–	2
	410·58	–	Tm in O_2/H_2 Flame	O_2/A	7·6	–	2
	418·76	–	Tm in O_2/H_2 Flame	O_2/A	13·4	–	2
	420·37	–	Tm in O_2/H_2 Flame	O_2/A	17·0	–	2
	371·79	–	HCL	N-A	0·7	0·2	3
AFS: not reported							

Ionization potential: 6·2 eV.

Notes on analytical conditions

1. Mossotti and Fassel[1] used a triple-pass optical arrangement and three oxy-acetylene flames supported at total consumption nebulizer-burner units. Skogerboe and Woodriff[2] used an oxy-hydrogen flame supplied with Tm as a source for the determination of this element in ethanolic sample solutions nebulized into a fuel-rich oxy-acetylene flame.

2. Suppression of ionization of Tm would be expected to be required in the nitrous oxide-acetylene flame as for other rare-earths of similar ionization potential.

Applications notes and key references

See Lanthanum.

References

1. V. G. Mossotti and V. A. Fassel, *Spectrochim. Acta,* 1964, **20,** 1117.
2. R. K. Skogerboe and R. A. Woodriff, *Anal. Chem.,* 1963, **35,** 1977.
3. "Analytical Methods for Atomic Absorption Spectrophotometry", Perkin-Elmer Corp., Norwalk, Conn., U.S.A., 1971.

13.4.63 Tin

Sn, atomic weight 118·69

Technique	Line, nm	f	Source	Flame	Sensitivity, ppm/1% abs.	Detection limit, ppm	Reference (see below)
AAS	244·61	0·41	HCL	A-H	1·1	0·1	1
	286·33	0·23	HCL	A-H	1·8		1
	253·48	0·27	HCL	A-H	2·2		1
	224·61	0·41	HCL	N-A	3·0		1
	286·33	0·23	HCL	N-A	5·4		1
	235·48	0·27	HCL	N-A	3·8		1
	224·61	0·41	HCL	A-A	2·8		1
	224·61	0·41	HCL	Ar/H_2	0·15	0·02	2
AFS	303·4	—	EDL	$Ar/O_2/H_2$	—	0·1	6

Ionization potential: 7·34 eV.
Dissociation energy, SnO: 5·7 eV.
Atomization efficiency: A-A, $<10^{-4}$; N-A, 0·76.

Notes on analytical conditions

1. The fuel-rich air-hydrogen flame gives higher sensitivity than the air-acetylene flame, probably as a result of the formation of easily dissociated SnH species in the former flame rather than predominantly SnO in the latter. The hotter nitrous oxide-acetylene flame is required to dissociate SnO effectively. The Sn 235·48 nm line originates from above the ground state (1692 K) and the sensitivity at this line is enhanced relative to that at the 224·61 nm resonance line when a hotter nitrous oxide flame is used.

2. Use of the air-hydrogen or Ar/H_2-air flames may give rise to problems from chemical interferences.[2,3] H_3PO_4 and H_2SO_4 interfere and should be added to standards if present in sample solutions.[11] Ni and Cu, or combinations of these with Pb and Zn, enhance absorption in the air-hydrogen flame; no interference is encountered for these elements, however, in hotter flames—Burke and Albright[4] observed depressed absorption for Sn in the Ar/H_2-air flame in the presence of S and 10% solns. of HCl, HNO_3, H_2SO_4 and H_3PO_4, while 2% solutions of NH_4 salts enhanced Sn absorption. Harrison and Juliano[5] found that the presence of an alcohol, ketone or organic acid (even at C $\approx$ 8%) considerably depressed Sn absorption at 224·6 nm in the air-hydrogen flame; the effect increased with increasing carbon number and chain-branching. These workers proposed that the organic solvent decreased the effective H-atom population available for direct reduction of SnO or formation of easily dissociated SnH. In AFS in the Ar-O_2-H_2 flame of 34 cations

and anions examined for interference at 100-fold excess over 5 ppm of Sn, only Ca, Mg, Mo(VI), Sr and V(VI) interfere.[6]

Applications notes and key references

Most applications of the determination of Sn by AAS have been in metallurgical and geochemical analysis. Perry[7] has described the determination of 10 to 50 ppm of Sn in lead and lead alloys by AAS in A-H flame at 224·6 nm after dissolution of the sample (10 g) in warm concentrated HNO_3 to which NH_4F is added. The determination of Sn in Cu alloys,[8] tin-plate,[9] ferrosilicon[10] and U-fission products[11] by AAS has also been reported. Mastalerz and Raczynska[12] have determined Sn in slag, flue dust and Zn concentrates by AAS in the A-A flame after dissolution of samples (containing 0·2-10% Sn) in HCl or HNO_3/HF, with $Na_2CO_3/Na_2B_4O_7$ fusion of the metastannic acid obtained and HCl extraction of the melt. Moldan *et al.*[13] have described the application of AAS to the determination of Sn in geological materials, while Bowman[14] determined Sn in Sn ores and concentrates by AAS after its separation by sublimation as SnI_4. In this latter method the ground ore or concentrate (0·2 g) is heated with 1 g of NH_4I in a test-tube; SnI_4 sublimes and condenses on the walls and may be dissolved in 2 N HCl for direct aspiration into the N-A flame. Harrison and Juliano[15] have investigated the feasibility of determination of Sn by direct aspiration of aqueous suspensions of tin ore concentrates into turbulent hydrogen-air or C_2H_2/O_2 flames at a large-bore total-consumption burner and obtained reasonable results when standards and samples had a similar composition. The determination of Sn in fruit juices has been reported by several workers.[16,17] In the method recommended by Price and Roos, concentrated HCl (10 ml) is added to 20 ml of sample and diluted to 100 ml. This solution is then centrifuged before nebulization into the N-A flame for determination of Sn at 286·3 nm. For samples with low PO_4 or high Fe content the addition of 1% H_3PO_4 to the diluted samples is recommended to avoid the interference of citrate. Tin has also been determined in H_2O_2[18] and in polymers[19] by AAS.

References

1. L. Capacho-Delgado and D. C. Manning, *Spectrochim. Acta,* 1966, **22,** 1505.
2. H. L. Kahn and J. E. Schallis, *Atomic Absorption Newsletter,* 1968, **7,** 5.
3. M. D. Amos and J. B. Willis, *Spectrochim. Acta,* 1966, **22,** 1325.
4. K. E. Burke and C. H. Albright, *Develop. Appl. Spectrosc.,* 1970, **8,** 33.

5. W. W. Harrison and P. O. Juliano, *Anal. Chem.*, 1969, **41**, 1016.
6. R. F. Browner, R. M. Dagnall and T. S. West, *Anal. Chim. Acta*, 1969, **46**, 207.
7. B. Perry, *Spectrovision*, 1971, **25**, 8.
8. P. Johns and W. J. Price, *Metallurgia*, 1970, **81**, 75.
9 C. Clauss, P. Laugel and M. Hasselman, *Chim. Analyt.*, 1971, **53**, 102.
10. F. L. Langmyhr and P. E. Paus, *Anal. Chim. Acta*, 1969, **45**, 173.
11. J. Grison, *Chim. Analyt.*, 1969, **51**, 605.
12. Roman Mastalerz and Danuta Raczynska, *Chimia Analit*, 1971, **16**, 85.
13. B. Moldon, I. Rubeska, Miroslav Miksovsky and Miroslav Huka, *Anal. Chim. Acta*, 1970, **52**, 91.
14. J. A. Bowman, *Anal. Chim. Acta*, 1968, **42**, 285.
15. W. W. Harrison and P. O. Juliano, *Anal. Chem.*, 1971, **43**, 248.
16. W. J. Price and J. T. H. Roos, *J. Sci. Food. Agric.*, 1969, **20**, 437.
17. J. C. Meranger, *Bull. Envir. Contam. Toxicol.*, 1970, **5**, 271.
18. E. J. Agazzi, *Anal. Chem.*, 1965, **37**, 364.
19. M. Olivier, *Z. Analyt. Chem.*, 1969, **248**, 145.

13.4.64 Titanium

Ti, atomic weight 47·90

Technique	Line, nm	*f*	Source	Flame	Sensitivity, ppm/1% abs.	Detection limit, ppm	Reference (see below)
AAS	365·35	0·22	HCL	N-A	1·6	0·1	1
	365·25	0·22	HCL	N-A	3·5	—	2
	364·27	0·25	..	N-A	1·8	—	1
	364·27	0·25	HCL	N-A	3·5	—	2
	337·14	0·20	HCL	N-A	2·0	—	1
	319·99	—	..	..	2·0	—	1
	375·28	—	..	..	2·5	—	1
	374·10	—	..	..	2·6	—	1
	319·19	—	..	..	2·6	—	1
	335·46	—	..	..	2·9	—	1
AFS	319·99	—	HIL	N_2 sep. N-A	—	4	2a

Ionization potential: 6·82 eV.
Dissociation energy, TiO: 6·9 eV.
Atomization efficiency: N-A, 0·3.

Notes on analytical conditions

1. Willis[3] reported the first determination of Ti in the fuel-rich nitrous oxide-acetylene flame and Amos and Willis[2] reported the sensitivities obtainable in AAS for Ti at a number of lines using both nitrous oxide-acetylene and oxygen-nitrogen-acetylene flames. Cartwright, Sebens and Manning[1] have reported sensitivity and noise data in the nitrous oxide flame for absorption at 35 lines of titanium emitted from a high-brightness HCL. The determination of Ti by

AAS had earlier been reported by Fassel and Mossotti[4] and Slavin and Manning[5] using oxy-acetylene flames supported at total-consumption nebulizer-burner units and aspirating Ti in organic solvent media.

2. Ti absorption in the fuel-rich nitrous oxide-acetylene flame is enhanced by HF and by the presence of large amounts of iron or aluminium.[2] Bond[6] recommended use of Ti standards and samples containing 0·1 M NH_4F_7 to compensate for the enhancement from fluoride. Headridge and Hubbard[7] added iron (5000 ppm) to Ti solutions and found no interference from Cr, Co, Mn, Mo, Nb, W, Ta and Cu; some enhancement, however, was still observed from Al even when iron was present. Ti is *ca.* 10% ionized at the concentrations usually introduced into the nitrous oxide-acetylene flame and some elements may interfere by suppression of this ionization rather than by their effect on the TiO dissociation equilibrium. Mostyn and Cunningham[8] observed enhancement of Ti absorption by Li, Na, K and Ca as well as other elements; these workers recommended addition of excess KCl to control these effects.

3. Ti atomic fluorescence has been reported by Dinnin[9] and Zacha *et al.*[10] but it seems that the signal observed in the air-hydrogen flame was probably scattered radiation from the source.[11] Dagnall, Kirkbright, West and Wood[12] were unable to detect atomic fluorescence from Ti in an inert gas-separated nitrous oxide-acetylene flame.

4. Indirect procedures for determination of Ti by AAS are described in Chapter 14.

Applications notes and key references

The determination of Ti in ferro-alloys by AAS has been reported by Smith, Johnson and Soth;[13] these authors recommended dissolution of ferrotitanium in HF/HCl, ferrovanadium in dilute HNO_3 and ferroboron and ferrochrome by fusion with Na_2CO_3 and Na_2O_2 respectively. Welcher and Kriege[14] described the determination of Ti in Co-base alloys by AAS in the nitrous oxide-acetylene flame. The determination of Ti in iron ores has been studied by Endo *et al.*[15] and Langmyhr and Paus;[16] the latter authors dissolved samples by heating with HF in a sealed PTFE bomb. Chrome ores and chrome-magnesites have been examined by Lucas and Ruprecht[17] and Ti determined by AAS in a nitrous oxide-acetylene flame after sample dissolution by treatment with H_3PO_4 and heating to 310°C. Several workers have reported the AAS determination of Ti in minerals and in slags with HF pretreatment[16,18] or fusion with $Na_2CO_3/Na_2B_4O_7$.[19] Boar and Ingram[20] have recommended a procedure for the determination of Ti in coal ash in which samples are fused with $LiBO_2$ at 900°C and the melt is dissolved in dilute HNO_3 containing tartaric acid. A similar decomposition procedure had earlier been recommended by Van Loon and Parissis[21] ($LiBO_2$) and Omang[22] (Li_2CO_3/H_3BO_3) for fusion of silicate samples prior to determination of Ti by AAS. Other workers, however, have utilized an HF dissolution procedure for silicates and rock samples.[23] Bauxite samples were examined by both Langmyhr and Paus[24] and Omang.[22] Capacho-Delgado and Manning[25] described a sample preparation procedure suitable for the determination of Ti in cement samples; samples were heated with HCl, evaporated to dryness, redissolved in 50% HCl and filtered before dilution to volume for the Ti determination. Lees[26] has also described the determination of Ti in cement after acid dissolution.

Olivier[27] determined Ti and other elements in polymer samples. The polymers were first dissolved or wet ashed with H_2SO_4 and H_2O_2; wool cotton and synthetic fibres were examined.

References

1. J. S. Cartwright, C. Sebens and D. C. Manning, *Atomic Absorption Newsletter,* 1966, **5,** 91.
2. M. D. Amos and J. B. Willis, *Spectrochim. Acta,* 1966, **22,** 1325.
2a. P. L. Larkins and J. B. Willis, *Spectrochim. Acta,* 1971, **26B,** 491.
3. J. B. Willis, *Nature,* 1965, **207,** 715.
4. V. A. Fassel and V. G. Mossotti, *Anal. Chem.,* 1963, **35,** 252.
5. W. Slavin and D. C. Manning, *Anal. Chem.,* 1963, **35,** 253.
6. A. M. Bond, *Anal. Chem.,* 1970, **42,** 932.
7. J. B. Headridge and D. P. Hubbard, *Anal. Chim. Acta,* 1967, **37,** 151.
8. R. A. Mostyn and A. F. Cunningham, *Atomic Absorption Newsletter,* 1967, **6,** 86.
9. J. I. Dinnin, *Anal. Chem.,* 1967, **39,** 1491.
10. K. E. Zacha, M. P. Bratzel Jnr., J. D. Winefordner and J. M. Mansfield Jnr., *Anal. Chem.,* 1968, **40,** 1733.
11. J. D. Winefordner, *Anal. Chem.,* 1970, **42,** 206R.
12. R. M. Dagnall, G. F. Kirkbright, T. S. West and R. Wood, *Anal. Chem.,* 1970, **42,** 1029.
13. D. C. Smith, J. R. Johnson and G. C. Soth, *Appl. Spectrosc.,* 1970, **24,** 576.
14. G. G. Welcher and O. H. Kriege, *Atomic Absorption Newsletter,* 1970, **9,** 61.
15. Yoshihide Endo, Toshihiko Hata and Yuki Nakahara, *Japan Analyst,* 1968, **18,** 838.
16. F. J. Langmyhr and P. E. Paus, *Anal. Chim. Acta,* 1969, **45,** 157.
17. R. P. Lucas and B. C. Ruprecht, *Anal. Chem.,* 1971, **43,** 1013.
18. F. J. Langmyhr and P. E. Paus, *Anal. Chim. Acta,* 1970, **50,** 515.
19. J. L. Jimenez-Seco and Aurora Gomez Coedo, *Revta Metal,* 1970, **6,** 324.
20. P. L. Boar and L. Ingram, *Analyst,* 1970, **95,** 124.

21. J. C. Van Loon and C. M. Parissis, *Analyst*, 1969, **94**, 1057.
22. S. H. Omang, *Anal. Chim. Acta*, 1969, **46**, 225.
23. F. J. Langmyhr and P. E. Paus, *Anal. Chim. Acta*, 1969, **47**, 371.
24. *Idem*, ibid., 1968, **43**, 508.
25. L. Capacho-Delgado and C. C. Manning, *Analyst*, 1967, **92**, 553.
26. T. P. Lees, *Chemy. Ind.*, 1969, **36**, 1249.
27. M. Olivier, *Z. Anal. Chem.*, 1969, **248**, 145.

13.4.65 Tungsten

W, atomic weight 183·85

Technique	Line, nm	*f*	Source	Flame	Sensitivity, ppm/1% abs.	Detection limit, ppm	Reference (see below)
AAS	255·13	0·8	HCL	N-A	5·3	—	1
	294·44	—	HCL	N-A	12	—	1
	400·87	—	HCL	N-A	18	3	1, 2
AFS: not reported							

Ionization potential: 7·98 eV.
Dissociation energy, WO: 7·2 eV.

Notes on analytical conditions

1. Amos and Willis[1] examined the absorption at eleven W lines in both nitrous oxide-acetylene and a premixed oxygen (60%)-nitrogen(40%)-acetylene flame. The lines given above were found to be the most suitable in the former flame; a spectral band-pass of 0·08 nm was employed. Manning[3] also examined the absorption of W in this flame. Although its use gives lower sensitivity than the 255·13 or 294·44 nm lines, the W 400·87 nm line has been observed to give better signal : noise characteristics in the nitrous oxide flame.

2. Very little information is available concerned with interference in AAS for W. Large amounts of iron may depress the absorption signal when present in $HClO_4$ medium; this effect is not observed in steel analysis when $H_3PO_4/H_2SO_4/HClO_4$ is used for dissolution.

3. Wendt and Fassel[4] determined W by AAS in an induction-coupled high-frequency plasma and reported a sensitivity similar to that obtainable in the nitrous oxide-acetylene flame.

4. Tungsten AFS has not been reported; this may result from lack of sufficiently intense line sources and its low atomization efficiency in many atom cells.

Applications notes and key references

Knight and Pyzyna[5] have determined W in tool steel by AAS after sample dissolution in $H_3PO_4/H_2SO_4/HClO_4$, and other workers have also described its determination in ferrous alloys.[6] Welcher and Kriege[7] have determined W in cobalt-based alloys by AAS. Rooney and Pratt[8] have reported the determination of W in steels and Ni alloys by AAS after sample dissolution in $HNO_3/HCl/HF$. Keller and Parsons[9] determined W in silicate ores. In this method the pulverized sample is fused with $LiBO_2$ at 1000°C, the melt is extracted with hot water and hot-concentrated HCl, and the solution volume is then reduced until W precipitates. The precipitate is dissolved in hot concentrated ammonia and W determined at 400·87 nm by AAS in a nitrous oxide-acetylene flame. Tungsten in ores may also be determined by fusion with $LiBO_2$, extraction of the melt with H_3PO_4 and solvent extraction with methyltricaprylammonium chloride into di-isobutylketone for nebulization into a nitrous oxide flame.[10]

References

1. M. D. Amos and J. B. Willis, *Spectrochim. Acta,* 1966, **22**, 1325.
2. "Analytical Methods for Atomic Absorption Spectrophotometry", Perkin-Elmer Corp., Norwalk, Conn., U.S.A., 1971.
3. D. C. Manning, *Atomic Absorption Newsletter,* 1966, **5**, 127.
4. R. H. Wendt and V. A. Fassel, *Anal. Chem.,* 1966, **38**, 337.
5. D. M. Knight and M. K. Pyzyna, *Atomic Absorption Newsletter,* 1969, **8**, 129.
6. J. L. Jimenez Seco and A. Gomez Coedo, *Revta Metal,* 1970, **6**, 64.
7. G. G. Welcher and O. H. Kriege, *Atomic Absorption Newsletter,* 1970, **9**, 61.
8. R. C. Rooney and C. G. Pratt, *Analyst,* 1972, **97**, 400.
9. E. D. Keller and M. L. Parsons, *Atomic Absorption Newsletter,* 1970, **9**, 92.
10. P. Dharma Rao, *Atomic Absorption Newsletter,* 1970, **9**, 131.

13.4.66 Uranium

U, atomic weight 238·03

Technique	Line, nm	*f*	Source	Flame	Sensitivity, ppm/1% abs.	Detection limit, ppm	Reference (see below)
AAS	358·49	2·1	HCL	N-A	100	—	1, 2
	351·46	1·2	HCL	N-A	250	30	1
	356·66	1·9	HCL	N-A	130	—	1
	394·38	0·94	HCL	N-A	250	—	1
	348·94	1·2	HCL	N-A	300	—	1, 2
AFS: not reported							

Ionization potential: 6·1 eV.
Dissociation energy, UO: 7·7 eV.

Notes on analytical conditions

1. Manning[1] also reported absorption data for the U 365·92, 404·27 and 415·39 nm lines and also for the U(II) 355·08 nm ion line in the rich nitrous oxide-acetylene flame. The absorption signal at the latter line indicated appreciable ionization of uranium in this flame; the data tabulated above, however, were obtained without the addition of KCl to suppress ionization.

2. The U 358·49 nm line lies in the strong (1, 0) sequence of the CN violet band system at 359·0 nm and its use in AAS with the nitrous oxide-acetylene flame gives rise to the high noise levels (i.e. poor detection limits). The use of the U 356·66 nm line is preferred for this reason.

3. Apart from ionization interference by alkali metal ions, vapour phase interferences may be observed from other elements which form refractory oxides. Martin,[3] for example, observed enhancement of U absorption at both 358·49 and 356·66 nm in the presence of excess Al; Fe, Co, Pb and Ni were also reported to enhance absorption and the absorption is also sensitive to the presence of different mineral acids. Bond[4] has reported enhancement of U absorption by the use of NH_4F in sample solutions.

4. Zacha *et al.*[5] have reported an AFS detection limit of 5 ppm for U at 381·2 nm using an EDL source and an air-hydrogen flame. It is possible, however, that the signal observed resulted from a scattering effect rather than a fluorescence emission.[6]

Applications notes and key references

AAS has found little application to the determination of U in real samples. This almost certainly results from the low sensitivity attainable in the nitrous oxide-acetylene flame due to the high thermal stability of uranium oxides. Martin[3] has described the AAS determination of U in uranium-based uranium-nickel catalysts; in order to eliminate the interference effects observed without separation, uranium was extracted into ethyl acetate from nitrate medium and back-extracted into water for nebulization into the nitrous oxide-acetylene flame.

Goleb[7,8] has determined the U^{235} and U^{238} isotopes by AAS by a technique in which single isotope sources and a water-cooled hollow-cathode absorption cell were employed (see Chapter 14.3). Ferraris and Proksch[9] have more recently described a similar procedure using electrodeless discharge lamp sources for uranium isotope analysis.

References

1. D. C. Manning, *Atomic Absorption Newsletter*, 1966, **5**, 127.
2. M. D. Amos and J. B. Willis, *Spectrochim. Acta*, 1966, **22**, 1325.
3. M. J. Martin, *Analyst*, 1971, **96**, 843.
4. A. M. Bond, *Anal. Chem.*, 1970, **42**, 932.
5. K. E. Zacha, M. P. Bratzel Jnr., J. D. Winefordner and J. M. Mansfield Jnr., *Anal. Chem.*, 1968, **40**, 1733.
6. J. D. Winefordner, *Anal. Chem.*, 1970, **42**, 206R.
7. J. A. Goleb, *Anal. Chem.*, 1963, **35**, 1978.
8. J. A. Goleb, *Anal. Chim. Acta*, 1966, **34**, 135.
9. M. M. Ferraris and G. Proksch, *Proc. 3rd CISAFA*, Paris, 1971.

13.4.67 Vanadium

V, atomic weight 50·94

Technique	Line, nm	*f*	Source	Flame	Sensitivity, ppm/1% abs.	Detection limit, ppm	Reference (see below)
AAS	318·54	—	HCL	N-A	1·0	0·02	1
	318·34 } 318·40 }	—	. .	. .	1·3	—	1
	306·64	—	. .	. .	2·9	—	1
	306·05	—	. .	. .	2·8	—	1
	305·63	—	. .	. .	3·3	—	1
	385·54 } 385·58 }	—	. .	. .	6·2	—	1
	437·92	—	. .	. .	4·4	—	1
AFS	318·34 } 318·40 }	—	EDL	Ar sep. N-A	—	0·07	2

Ionization potential: 6·74 eV.
Dissociation energy, VO: 5·5 eV.
Atomization efficiency: N-A, 0·91.

Notes on analytical conditions

1. A hot fuel-rich nitrous oxide-acetylene or oxy-acetylene flame is required for atomization of V. Although a number of early reports of the use of the oxy-acetylene for V AAS were made,[3,4,5] the nitrous oxide-acetylene flame is now most commonly employed. Amos and Willis[6] studied the use of several V lines for AAS in this flame and obtained highest sensitivity at 318·40 nm (1·5 ppm V/1% absorption). Cartwright, Sebens and Manning[1] and Sachdev *et al.*[5] obtained similar sensitivity in the fuel-rich flame; in addition the former workers recorded the attainable sensitivity and noise at 28 other vanadium absorption lines. Weak vanadium absorption has also been observed in very rich air-acetylene[7] and oxy-cyanogen[8] flames.

2. V may be ionized to the extent of *ca.* 10% in the nitrous oxide-acetylene flame, so that unless excess of an ionization buffer is present an enhancement of V absorption by the alkali metals may result. Enhancement of V absorption is also observed in the presence of Al and Ti[5,9] (see Chapter 12.2.4) owing to their competition with V for oxygen in the flame gases; the addition of excess Al to samples and standards has been recommended to control this effect.[10] Similar, though less pronounced, enhancement may be observed in the presence of high concentrations of Fe and other elements. Capacho-Delgado and Manning[11] observed enhancement of V absorption in the presence of phosphoric acid. Sachdev *et al.*[12] were able to increase sensitivity for V by 50% by addition of *ca.* 8% diethylene glycol to aqueous sample solutions. Of 21 ions examined for their effect at 100-fold weight excess over 25 ppm of V in aqueous solution on the atomic fluorescence signal at 318·4 nm, only Al and W interfered by enhancement of the signal, and Co produced a slight depression.[2]

3. Vanadium may be extracted into MIBK as its cupferron or APDC complex. Additional sensitivity enhancement has been shown to result when a mixture of MIBK and oleic acid (78 : 22 v/v) is employed for extraction of vanadium cupferrate.[5] V(V) has also been extracted into dichloromethane with triphenylarsine from HCl medium prior to determination by AAS.

Applications notes and key references

AAS provides a rapid and sensitive technique for the determination of V; most applications have been for analysis of metallurgical and geological samples. Capacho-Delgado and Manning[11] reported the determination of V in steels by AAS in a nitrous oxide-acetylene flame without interference from other elements present; these workers dissolved samples in a mixture of H_3PO_4 and H_2SO_4. Knight and Pyzyna[14] employed a mixture of H_2SO_4, H_3PO_4 and $HClO_4$ for dissolution of tool steels prior to determination of V by AAS. Smith, Johnson and Soth[15] determined V in ferrovanadium by AAS after dissolving samples in warm HNO_3, and Goecke[10] and Endo *et al.*[16] have described the determination of V in iron ores. Goecke[10] has also reported the determination of V in Cu ores; the samples were dissolved in H_2SO_4 or HCl. In a comprehensive scheme for analysis of silicate materials by AAS, Bernas[17] determined V after sample decomposition in HF/H_3BO_3 in a sealed vessel at 110°C. Langmyhr and Paus[18] also employed an HF decompositoin technique in the determination of V in iron ores and slags by AAS. Pearton *et al.*[19] determined V in silicates in a procedure in which samples were dissolved in $HF/HNO_3/HClO_4$, and V was oxidized to V(V) with Mo(VI) in H_2SO_4 and extracted as vanadium cupferrate into butyl acetate for nebulization into a nitrous oxide-acetylene flame.

Vanadium has been determined in gas oils by Capacho-Delgado and Manning[11] using AAS in a nitrous oxide flame; these workers diluted samples with *p*-xylene (1 + 2) before nebulization. Standard V solutions were prepared from bis-(1-phenyl-1,3-butanediono)oxovanadium(IV) in oil and diluting with *p*-xylene. Omang[20] has employed a heated graphite furnace as atom cell for the AAS determination of V in oil residues, fuel and mineral oils. Amos *et al.*[21] have described the use of a heated carbon rod atomizer for the direct determination of V in lubricating oils. Vanadium in brines has been determined by AAS after its concentration by extraction with cupferron in MIBK at pH 1.[22] After examination of the suitability of 13 chelating agents for extraction of V, Chau and Lum Shue Chan[23] chose 5,7-dichloro-8-hydroxyquinoline for extraction of V from lake waters for determinations by AAS. Other materials in which V has been determined by AAS include $TiCl_4$[13] and atmospheric particulate matter.[24]

References

1. J. S. Cartwright, C. Sebens and D. C. Manning, *Atomic Absorption Newsletter*, 1966, **5**, 91.
2. R. M. Dagnall, G. F. Kirkbright, T. S. West and R. Wood, *Anal. Chem.*, 1970, **42**, 1029.
3. V. A. Fassel and V. G. Mossotti, *Anal. Chem.*, 1963, **35**, 252.
4. W. Slavin and D. C. Manning, *Anal. Chem.*, 1963, **35**, 253.
5. S. L. Sachdev, J. W. Robinson and P. W. West, *Anal. Chim. Acta*, 1967, **37**, 12.
6. M. D. Amos and J. B. Willis, *Spectrochim. Acta*, 1966, **22**, 1325.
7. B. M. Gatehouse and J. B. Willis, *Spectrochim. Acta*, 1961, **17**, 710.
8. J. W. Robinson, *Anal. Chem.*, 1961, **33**, 1067.
9. T. V. Ramakrishna, P. W. West and J. W. Robinson, *Anal. Chim. Acta*, 1967, **39**, 81.
10. R. Goecke, *Talanta*, 1968, **15**, 871.
11. L. Capacho-Delgado and D. C. Manning, *Atomic Absorption Newsletter*, 1966, **5**, 1.
12. S. L. Sachdev, J. W. Robinson and P. W. West, *Anal. Chim. Acta*, 1967, **37**, 156.
13. R. Bock and B. Jost, *Z. Anal. Chem.*, 1970, **250**, 358, 366.
14. D. M. Knight and M. K. Pyzyna, *Atomic Absorption Newsletter*, 1969, **8**, 129.
15. D. C. Smith, J. R. Johnson and G. C. Soth, *Appl. Spectrosc.*, 1970, **24**, 576.
16. Y. Endo, T. Hata and Y. Nakahara, *Japan Analyst*, 1969, **18**, 833.

17. B. Bernas, *Anal. Chem.*, 1968, **40**, 1682.
18. F. J. Langmyhr and P. E. Paus, *Anal. Chim. Acta*, 1969, **45**, 173, 157.
19. D. G. Pearton, J. D. Taylor, P. K. Faure and T. W. Steele, *Anal. Chim. Acta*, 1969, **44**, 353.
20. S. H. Omang, *Anal. Chim. Acta*, 1971, **56**, 470.
21. M. D. Amos, K. G. Brodie and J. P. Matousek, *Pittsburgh Conf. Anal. Chem. and Appl. Spectrosc.*, Cleveland, Ohio, 1971.
22. H. J. Crump-Wiesner, H. R. Feltz and W. C. Purdy, *Anal. Chim. Acta*, 1971, **55**, 29.
23. Y. K. Chau and K. Lum Shue Chan, *Anal. Chim. Acta*, 1970, **50**, 201.
24. R. J. Thompson, G. B. Morgan and L. J. Purdue, *Atomic Absorption Newsletter*, 1970, **9**, 53.

13.4.68 Ytterbium

Yb, atomic weight 173·04

Technique	Line, nm	*f*	Source	Flame	Sensitivity, ppm/1% abs.	Detection limit, ppm	Reference (see below)
AAS	398·80	—	Xe arc	O_2/A	—	0·2	1
	398·80	—	HCL	N-A	0·25	—	2
	346·44	—	HCL	N-A	0·8	—	2
	246·45	—	HCL	N-A	1·6	—	2
	398·80	—	Yb in O_2/H_2 flame	O_2/A	2·0	—	3
AFS: not reported							

Ionization potential: 6·22 eV.

Notes on analytical conditions

1. Skogerboe and Woodriff[3] used an oxy-hydrogen flame supplied with Yb as a source for the AAS determination of this element in ethanolic sample solutions nebulized into a fuel-rich oxy-acetylene flame.

2. Yb ionization in the nitrous oxide-acetylene flame is suppressed by the addition of excess potassium (as KCl) to sample and standard solutions.

3. Little information is available concerning chemical interferences in the determination of Yb by AAS.

Applications notes and key references

See Lanthanum.

References

1. V. G. Mossotti and V. A. Fassel, *Spectrochim. Acta*, 1964, **20**, 1117.
2. M. N. Amos and J. B. Willis, *Spectrochim. Acta*, 1966, **22**, 1325.
3. R. K. Skogerboe and R. A. Woodriff, *Anal. Chem.*, 1963, **35**, 1977.

13.4.69 *Yttrium*

Y, atomic weight 88·906

Technique	Line, nm	f	Source	Flame	Sensitivity, ppm/1% abs.	Detection limit, ppm	Reference (see below)
AAS	410·24	0·21	HCL	N-A	2	0·3	1
	410·24	0·21	HCL	N-A	5·0	—	2
	412·83	0·18	HCL	N-A	5·4	—	2
	407·74	0·27	HCL	N-A	5·7	—	2
	414·29	0·20	HCL	N-A	11	—	2
AFS: not reported							

Ionization potential: 6·51 eV.
Dissociation energy, YO: 7·0 eV.

Notes on analytical conditions

1. A hot fuel-rich flame is required for AAS of Y. Mossotti and Fassel[3] in early work described Y absorption characteristics at the four lines listed above using a fuel-rich oxy-acetylene flame supported at total consumption nebulizer-burner units. With a xenon arc continuum source these workers reported a sensitivity of 10 ppm Y at 407·74 nm.

2. Y is appreciably ionized in the nitrous oxide-acetylene flame; Amos and Willis[2] added K or Cs to sample and standard solutions to suppress this effect. Interference with Y absorption at 410·24 nm was reported by Kinnunen and Lindsjo[4] in the presence of HNO_3, H_2SO_4, $HClO_4$, H_3PO_4 and fluoride; these workers also observed interference from Al, Ce and Si.

3. Wendt and Fassel[5] obtained a sensitivity of 30 ppm Y by using the Y ion line at 371·03 nm in an induction-coupled plasma.[5]

Applications notes and key references

Kinnunen and Lindsjo[4] applied AAS to the determination of Y in phosphate rock, and Van Loon and co-workers[6] determined Y in minerals after its separation with the rare-earths by precipitation as hydroxides. Muntz[7] has described the determination of Y in stabilized hafnia and zirconia; samples were dissolved in HF/H_2SO_4 and nebulized into a nitrous oxide-acetylene flame. Y has been determined in yttrium oxide by AAS in a nitrous oxide-acetylene flame after sample dissolution in warm dilute HCl.[8] Riley and Taylor[9] utilized preconcentration of Y with a chelating ion-exchange resin before its determination by AAS in sea-water samples.

References

1. "Analytical Methods for Atomic Absorption Spectrophotometry", Perkin-Elmer Corp., Norwalk, Conn., U.S.A., 1971.
2. M. D. Amos and J. B. Willis, *Spectrochim. Acta*, 1966, **22**, 1325.
3. V. G. Mossotti and V. A. Fassel, *Spectrochim. Acta*, 1964, **20**, 1117.
4. J. Kinnunen and O. Lindsjo, *Chemist-Analyst*, 1967, **56**, 25.
5. R. H. Wendt and V. A. Fassel, *Anal. Chem.*, 1966, **38**, 337.
6. J. C. Van Loon, J. H. Galbraith and H. M. Aarden, *Analyst*, 1971, **96**, 47.
7. J. H. Muntz, *Atomic Absorption Newsletter*, 1971, **10**, 9.
8. R. W. Cattrall and S. J. E. Slater, *Anal. Chim. Acta*, 1971, **56**, 143.
9. J. P. Riley and D. Taylor, *Anal. Chim. Acta*, 1968, **4**, 479.

13.4.70 Zinc

Zn, atomic weight 65·38

Technique	Line, nm	*f*	Source	Flame	Sensitivity, ppm/1% abs.	Detection limit, ppm	Reference (see below)
AAS	213·86	1·2	HCL	A-A	0·02	—	1
	213·86	1·2	HCL	A-A	0·015	0·002	2
	307·59	<0·001	HCL	A-A	150	—	—
AFS	213·86	1·2	EDL	Air/H_2	—	0·00004	3
	213·86	1·2	VDL	A-H	—	0·0005	4
	213·86	1·2	VDL	N_2 sep. A-A	—	0·0002	5
ND	213·86	1·2	VDL	Air/H_2	—	0·0002	6
ND	213·86	1·2	VDL	A-town gas	—	0·00001	7
ND	213·86	1·2	HIL	N_2 sep. A-A	—	0·0003	8
	213·86	1·2	VDL	O_2/H_2	—	0·0001	9
ND	213·86	1·2	Pulsed HCL	N_2 sep. A-A	—	0·001	10

Ionization potential: 9·39 eV.
Dissociation energy, ZnO: 4·0 eV.
Atomization efficiency: A-A, 0·45; N-A, 0·91.

Notes on analytical conditions

If the HCL source has a brass cathode the Zn determination by AAS requires use of a half-band pass of 2·0 nm or less to avoid transmission of Cu lines from the lamp.

2. The air-acetylene flame exhibits some absorption at 213·86 nm; this does not lead to unacceptable noise levels in AAS for Zn, however, although the fuel flow-rate should be closely regulated. The background absorption and noise levels may be decreased by inert gas shielding of the flame.[11]

3. T-piece adaptors[12] and long path absorption tube devices[13,14] have been shown to yield considerable increase in sensitivity in the determination of Zn by AAS.

4. Few interference effects are observed in the AAS determination of Zn in an air-acetylene flame. In early work Gidley and Jones[15] observed no effect on Zn absorbance in an air-propane flame in the presence of 1000 ppm of many elements; only silicon was found to depress the absorbance significantly. Allan[1] reported no interference in air-acetylene, air-propane or air-coal gas from 0·5 N solutions of $HClO_4$, H_2SO_4, HNO_3 and HCl. In the determination of Zn in soil extracts by AFS no interference was seen from 100 ppm of Al, Ba, Bi, Ca, Cd, Co, Cr, Fe, Hg, K, Mg, Mn, Na, Ni, Pb, Th, Ti, V, chloride, fluoride, nitrate and sulphate, using 1 ppm Zn solutions.[16] In this work the Zn fluorescence was enhanced by copper and depressed by phosphate; these effects were removed by using strontium as releasing agent. Cresser and West[17] obtained a detection limit of 0·0001 ppm for Zn by AFS in a turbulent oxy-hydrogen flame but recommended use of the premixed air-acetylene flame for the determination of Zn in the presence of foreign ions.

Applications notes and key references

AAS is the technique of choice for the determination of zinc in a wide variety of sample types; AAS methods for Zn are sensitive and relatively free from interelement effects.

The assay of ores for zinc was described by Farrar[18] after extraction with a mixture of HCl and HNO_3, and Slavin[19] has reported a method for Zn in mine and mill samples in which the extraction is effected with $HClO_4$ and HNO_3. Vlastnik[20] determined Zn in Fe ores and blast furnace materials after dissolution of samples in HCl, filtration of the residual silica, and nebulization of the solution into an air-propane flame. Endo and co-workers[21] assayed Fe ores for Zn by a method in which after HCl dissolution the residual silica was filtered off, fused with $K_2S_2O_7$ and the solution of the melt combined with the HCl fraction before nebulization into an air-acetylene flame. In the determination of Zn in aluminous materials Calkins[22] employed sodium carbonate sample fusion and subsequent separation of the zinc by anion exchange from HCl medium. Several workers have examined the determination of Zn in silicates;[23,24,25] sample decomposition has been effected by dissolution in acid or by fusion with Li metaborate. Bond and

Canterford[26] have discussed the interference observed from high concentrations of Li when LiF/H_3BO_3 mixture is used for fusion of rock samples; these workers recommend addition of greater than 1000 ppm of Li to standards and samples to compensate for this enhancement. The AAS determination of Zn in sulphide minerals after decomposition in a bomb at 110°C for 10 minutes with a mixture of HF, HCl and HNO_3 has been described,[27] and Roos and Price[28] determined Zn in cement in an air-acetylene flame, by dissolution in HCl, addition of HF to dissolve SiO_2, boric acid to prevent LaF_3 precipitation, and $LaCl_3$ to compensate for the signal enhancement caused by some elements.

The determination of zinc by AAS in metals and alloys is usually without complication after acid dissolution and dilution and direct nebulization or after solvent extraction. In early work Gidley and Jones[29] determined Zn in metallurgical materials, and Wallace[30] also reported its determination in Cu and Al alloys. Other workers have since determined Zn in these alloys,[31,32] lead and tin alloys,[18,33] tin-lead solders[34] and ferrous alloys.[35] The determination of Zn in Al alloys by AFS using an air-acetylene flame and direct aspiration of dilute sample solutions has been described.[36] Atsuya[37] determined zinc in high-purity lead, aluminium or electrolytic iron by AAS using solvent extraction of zinc into MIBK with 8-hydroxyquinoline or dithizone. Donaldson, Charette and Rolko[38] have described the AAS determination of Zn in high-purity Nb, Ta, Mo and W by extraction of Zn as its complex with diantipyrinylmethane and thiocyanate into chloroform, back extraction into ammonia buffer, evaporation to dryness and dissolution of the residue in dilute HCl for nebulization into an air-acetylene flame. Walker and Vita[39] determined Zn by AAS in uranium compounds after separation of the uranium by extraction with tributylphosphate and other workers have recently described the determination of traces of Zn in uranium and other metals using an air-hydrogen flame after solvent extraction.[40] The determination of traces of Zn in copper has been undertaken by AAS[41] and by AFS using a non-dispersive system[7] to provide a detection limit corresponding to 0·1 ppm Zn in copper. Shafto[42] investigated the determination of Zn by AAS in nickel-plating solutions, and Parker[43] has also determined Zn in these and chromium plating solutions.

A large volume of literature exists concerned with the determination of Zn in biological materials. The recommended procedures for Zn in serum, blood and urine are similar to those proposed for cadmium (q.v.); tissue samples have been analysed after both wet or dry ashing procedures. Table 13-10 lists some applications of AAS and AFS to Zn determinations in these matrices.

Riley and Topping[44] used 5,7-dibromo-8-hydroxy-quinoline for cocrystallization of zinc from sea-water between pH 4·5 and 8·0 prior to its determination by AAS, while Burrell and Guerner Wood[45] determined Zn in sea-water directly using the boat-in-flame technique without pretreatment. Marshall and Smith[46] have determined Zn in high-purity waters by AFS in an air-hydrogen flame using a non-dispersive system, and other workers have used AFS to determine Zn in sea-water samples after preconcentration by solvent extraction with NaDDTC.[47] Zinc may be determined in lubricating oils after dilution with MIBK and nebulization into air-propane[48] or air-acetylene[49] flames. Musha, Munemori and Nakanishi[50] have reported the determination of Zn in PVC by AAS after sample dissolution in 0·5% dimethylacetamide, and Toma and Crisan[51] determined Zn in polystyrene after dry-ashing at 400°C and wet-ashing with a mixture of HNO_3 and $HClO_4$.

TABLE 13-10

Some applications of the determination of zinc by AAS and AFS to biological materials

Sample type	Technique, pre-treatment	Reference
Plasma, red blood cells, urine	AAS; precipitate protein + trichloroacetic acid (TCA)	52
Serum	AAS; either dilute with water and nebulize with 3-slot burner or add TCA, centrifuge, nebulize supernate	53
Serum, urine	AAS; H_2/air flame + absorption tube	54
Serum	AAS; 2 ml sample + TCA, heat, cool centrifuge, spray supernate	55
Serum	AAS; TCA does not affect results but deproteinization not essential. Dilute sample 5 x with water, nebulize A-A flame	56
Serum	AAS; study of effects of nebulizer capillary diameter and aspiration rate. Dilute 1 : 2 with water, use 0·58 nm capillary	57
Serum, red blood cells	AAS; serum deproteinized + TCA; red cells heated + 1·5 N HCl at 90°C then add TCA	58
Plasma	AAS; citrated plasma diluted 1 : 1 with water; standards prepared to contain 3% dextran. A-A flame, 3-slot burner	59

TABLE 13-10 (*continued*)

Sample type	Technique, pre-treatment	Reference
Plasma, blood, urine	AAS; whole blood diluted with 0·1 N HCl, 1 : 100; plasma diluted 1 : 20, urine 1 : 10 with 0·1 N HCl. Prepare standards of similar composition. A-A flame, 3-slot burner	60
Serum, urine	AFS; VDL source	61
Urine, organs, tissue	AAS; organs and tissue digested + $HNO_3/HClO_4$; urine acidified and examined directly. A-P flame	62
Brain tissue	AAS; 5 g sample digested with HNO_3 and $HClO_4$, diluted for nebulization	63
Bones, tissue	AAS; dry ash at 600°C	64
Brain, hair	AAS; digest + $KClO_3/HClO_4$	65
Human hair	AAS; wet digest + $HNO_3/HClO_4$. 3-slot burner, A-A flame	66
Insulin and insulin preps.	AAS; dissolve 5 mg cryst. insulin or dilute 1 ml protamine insulin solution with water; dilute to 50 ml; nebulize	67
Agricultural materials	AAS; extract Zn with APDC into MIBK	68
Leaf analysis	AAS; dry ash at 400°, digest + $HNO_3/HClO_4$	69
Chloroplasts	AAS; ash at 500°C, dissolve in 2 N HCl, evaporate to dryness, dissolve in water	70
Plants	AAS; dry-ash, dissolve in 0·1 N HCl	71
Tree leaves	AAS; dry-ash at 550°C	72
Soil	AAS; heat in Ta boat, AAS in supernatant vapour detection limit 0·05 ppb Zn	73
Fertilizers	AAS; official methods	74
Soil extracts	AAS; extract + EDTA; mechanically separated A-A flame	75
Soil extracts	AFS; extract with EDTA, ammonium acetate or acetic acid. A-A flame	16
Stock and mixed feeds	AAS; extract ground sample + EDTA for 1 hour, centrifuge, dilute supernate, nebulize; A-A flame	76
Foods	AAS; wet-ashing procedure	77
Sea foods	AAS	78
Shrimps	AAS; dry-ash at 550°C, dissolve in HNO_3, dilute and nebulize	79
Wines	AAS; ethanol removed by evaporation	80
Beers and wine	AAS; no special sample preparation	81
Fruit juice, soft drinks	AAS; no special sample preparation	82

References

1. J. E. Allan, *Analyst,* 1961, **86,** 530.
2. "Analytical Methods for Atomic Absorption Spectrophotometry", Perkin-Elmer Corp., Conn., U.S.A., 1971.
3. K. E. Zacha, M. P. Bratzel, J. D. Winefordner and J. M. Mansfield Jnr., *Anal. Chem.,* 1968, **40,** 1733.

5. R. S. Hobbs, G. F. Kirkbright, M. Sargent and T. S. West, *Talanta*, 1968, **15**, 997.
6. T. J. Vickers and R. M. Vaught, *Anal. Chem.*, 1969, **41**, 1477.
7. P. D. Warr, *Talanta*, 1970, **17**, 543.
8. P. L. Larkins, *Spectrochim. Acta*, 1971, **26B**, 477.
9. T. L. Martin, F. M. Hamm and P. B. Zeeman, *Anal. Chim. Acta*, 1971, **53**, 437.
10. D. Mitchell and D. Demers, *Proc. Technicon Intern. Conf.*, New York, 1970.
11. L. Ranson, Ph.D. Thesis, University of London, 1972.
12. T. V. Ramakrishna, J. W. Robinson and P. W. West, *Anal. Chim. Acta*, 1967, **37**, 20.
13. K. Fuwa, P. Pulido, R. McKay and B. L. Vallee, *Anal. Chem.*, 1965, **36**, 2407.
14. K. Fuwa and B. L. Vallee, *Anal. Chem.*, 1963, **35**, 942.
15. J. A. F. Gidley and J. T. Jones, *Analyst*, 1960, **85**, 249; 1971, **86**, 271.
16. R. M. Dagnall, G. F. Kirkbright, T. S. West and R. Wood, *Anal. Chem.*, 1971, **43**, 1765.
17. M. S. Cresser and T. S. West, *Anal. Chim. Acta*, 1970, **50**, 517.
18. B. Farrar, *Atomic Absorption Newsletter*, 1965, **4**, 325.
19. W. Slavin, *Atomic Absorption Newsletter*, 1965, **4**, 243.
20. J. Vlastnik, *Hutn. Listy*, 1966, **21**, 730.
21. Y. Endo, T. Hata and Y. Nakahara, *Japan Analyst*, 1968, **17**, 679.
22. R. C. Calkins, *Appl. Spectrosc.*, 1965, **45**, 21.
23. D. C. Burrell, *Norsk. Geol. Tidsskr.*, 1965, **45**, 21.
24. A. J. Thompson, *Spectrovision*, 1968, (20), 7.
25. M. Miksovsky and B. Moldan, *Chemicke Zvesti*, 1970, **24**, 128.
26. A. M. Bond and D. R. Canterford, *Anal. Chem.*, 1971, **43**, 134.
27. F. J. Langmyhr and P. E. Paus, *Anal. Chim. Acta*, 1970, **50**, 515.
28. J. T. H. Roos and W. J. Price, *Analyst*, 1969, **94**, 89.
29. J. A. F. Gidley and J. T. Jones, *Analyst*, 1961, **86**, 271.
30. F. J. Wallace, *Hilger Journal*, 1962, **7**, 39.
31. J. L. Jimenez Seco and A. Gomez Coedo, *Revta Metal*, 1969, **5**, 33, 290.
32. P. Johns and W. J. Price, *Metallurgia*, 1970, **81**, 75.
33. J. L. Jimenez Seco, Aurora Gomez Coedo and Maria T. Dorado, *Revta Metal*, 1969, **5**, 602.
34. J. Y. Hwang and Linda M. Sandonato, *Anal. Chem.*, 1970, **42**, 744.
35. J. L. Jimenez Seco and Aurora Gomez Coedo, *Revta Metal*, 1970, **6**, 64.
36. R. M. Dagnall, G. F. Kirkbright, T. S. West and R. Wood, *Analyst*, 1972, **97**, 245.
37. I. Atsuya, *J. Chem. Soc. Japan, Pure Chem. Sect.*, 1967, **88**, 179.
38. E. M. Donaldson, D. J. Charette and V. H. E. Rolko, *Talanta*, 1969, **16**, 1305.
39. C. R. Walker and O. A. Vita, *Anal. Chim. Acta*, 1968, **43**, 27.
40. T. Kuroha, I. Tsukahara and S. Shibuya, *Japan Analyst*, 1971, **20**, 1137.
41. Bruce Gandrud and John C. Marshall, *Appl. Spectrosc.*, 1970, **24**, 367.
42. R. G. Shafto, *Atomic Absorption Newsletter*, 1964, **3**, 115.
43. J. E. Parker, *Spectrovision*, 1970, (24), 10.
44. J. P. Riley and G. Topping, *Anal. Chim. Acta*, 1969, **44**, 234.
45. D. C. Burrell and G. Guerner Wood, *Anal. Chim. Acta*, 1969, **48**, 45.
46. G. B. Marshall and A. C. Smith, *Proc. 3rd CISAFA*, Paris, 1971.
47. M. Jones, G. F. Kirkbright, L. Ranson and T. S. West. *Anal. Chim. Acta*, 1973, **63**, 210.
48. Takeshi Iishi, Seichiro Musha and Makoto Munemori, *Japan Analyst*, 1968, **17**, 27.

49. D. R. Jackson, C. Salama and R. Dunn, *Canad. Spectrosc.*, 1970, **15**, 17.
50. S. Musha, M. Munemori and Y. Nakanishi, *Japan Analyst*, 1964, **13**, 330.
51. O. Toma and T. Crisan, *Mater. Plast.*, 1970, **7**, 83.
52. A. S. Prasad, D. Oberlens and J. A. Halsted, *J. Lab. Clin. Med.*, 1965, **66**, 508.
53. M. M. Parker, F. L. Humoller and D. J. Mahler, *Clin. Chem.*, 1967, **13**, 40.
54. K. W. Fuwa, P. Pulido, R. McKay and B. L. Vallee, *Anal. Chem.*, 1964, **36**, 2407.
55. A. D. Olson and W. B. Hamlin, *Atomic Absorption Newsletter*, 1968, **7**, 69.
56. H. Matsumota, K. Tsunematsu and T. Shiraishi, *Japan Analyst*, 1968, **17**, 703.
57. K. Oiwa, T. Kimura, H. Makino and M. Okuda, *Japan Analyst*, 1968, **17**, 810.
59. B. M. Hackley, J. C. Smith and J. A. Halsted, *Clin. Chem.*, 1968, **14**, 1.
60. J. B. Dawson and B. E. Walker, *Clin. Chim. Acta*, 1969, **26**, 465.
61. R. Klaus, *Z. Klin. Chem.*, 1966, **4**, 299.
62. M. L. Girard and F. Rousselet, *Ann. Pharm. Fr.*, 1967, **25**, 353.
63. T. L. Chang, T. A. Gover and W. W. Harrison, *Anal. Chim. Acta*, 1966, **34**, 17.
64. F. H. Nielson, M. L. Sunde and W. G. Hoekstra, *J. Nutr.*, 1966, **89**, 24.
65. E. T. Backer, *Clin. Chim. Acta*, 1969, **24**, 233.
66. W. W. Harrison, J. P. Yurachek and C. A. Benson, *Clin. Chim. Acta*, 1964, **23**, 83.
67. G. I. Spielholtz and Gloria C. Toralballa, *Analyst*, 1969, **94**, 1072.
68. J. E. Allan, *Analyst*, 1961, **86**, 530.
69. E. E. Bradfield, *J. Sci. Fd. Agric.*, 1964, **15**, 469.
70. R. Herrman and W. Neu, *Experientia*, 1965, **21**, 436.
71. K. Trier, P. Neubert and R. Ebeting, *J. Prakt. Chem.*, 1968, **37**, 234.
72. J. R. Buchanan and T. T. Muraoka, *Atomic Absorption Newsletter*, 1964, (24), 1.
73. R. Emmermann and W. Luecke, *Z. Anal. Chem.*, 1969, **248**, 325.
74. C. H. McBride, *J. Ass. Off. Analyt. Chem.*, 1965, **48**, 406.
75. A. M. Ure and M. L. Berrow, *Anal. Chim. Acta*, 1970, **52**, 247.
76. Jan. Tusl, *J. Ass. Off. Analyt. Chem.*, 1970, **53**, 1190.
77. G. R. Rogers, *J. Ass. Off. Analyt. Chem.*, 1968, **51**, 1042.
78. J. C. Meranger and E. Somers, *Bull. Envir. Contam. Toxicol.*, 1968, **3**, 360.
79. G. A. Knauer, *Analyst*, 1970, **95**, 476.
80. J. C. Meranger and E. Somers, *J. Ass. Off. Analyt. Chem.*, 1968, **51**, 922.
81. J. P. Weiner and L. Taylor, *J. Inst. Brew.*, 1969, **75**, 195.
82. J. C. Meranger, *Bull. Envir. Contam. Toxicol.*, 1970, **5**, 271.

13.4.71 Zirconium

Zr, atomic weight 91·22

Technique	Line, nm	*f*	Source	Flame	Sensitivity, ppm/1% abs.	Detection limit, ppm	Reference (see below)
AAS	360·12	–	HCL	N-A	15	–	1
	360·12	–	HCL	N-A	20	5	2
	360·12	–	HCL	N-A	6	5	3
	351·96	–	HCL	O_2-N_2-A	22	–	1
	301·18	–	HCL	O_2-N_2-A	34	–	1
	301·18	–	HCL	N-A	34	–	2
	354·77	–	HCL	O_2-N_2-A	50	–	1
	354·77	–	HCL	N-A	29	–	2
AFS: not reported							

Ionization potential: 6·84 eV.
Dissociation energy, ZrO: 7·8 eV.

Notes on analytical conditions

1. Other less sensitive Zr absorption lines examined by Amos and Willis[1] and Manning[2] were 362·39, 302·95, 298·54, 389·03 and 350·93 nm. Headridge and Hubbard[4] obtained a value of 13 ppm Zr for 1% absorption at 360·12 nm in the fuel-rich nitrous oxide-acetylene flame nebulizing 90% ethanolic zirconium solutions containing HF. Zr is ionized to the extent of *ca.* 10% at analytical concentrations in the nitrous oxide-acetylene flame; this may be suppressed by addition of potassium chloride to sample solutions.

2. Zr absorption is enhanced by the presence of HF[1] or HCl[5] and by iron, and has been reported to be depressed by sulphate, nitrate and nickel (when present as $NiBr_2$).[3]

3. Although Zr atomic fluorescence has been reported by Dinnin[6] and by Zacha *et al.*[7] it seems probable that the signals observed in an air-hydrogen flame for this element resulted from the scatter of incident radiation.[8]

Applications notes and key references

Like hafnium, although Zr absorption has been discussed in several papers concerned with interference effects in the nitrous oxide-acetylene flame,[9,10,11] little information is available concerning the determination of Zr in real samples. Bond and Canterford[12] described the effect of Li on Zr and Hf absorption and recommended the use of Li in standards and samples when determining Zr and Hf in rock samples. Foster[13] determined Zr in the presence of lanthanum by extraction from 4-8 M HCl or HNO_3 with a 10% solution of thenoyltrifluoroacetone and nebulization of the organic phase into a nitrous oxide-acetylene flame. Grossman and Inglis[14] have described a method for the determination of Zr in hides in which the sample was dissolved in HF, ferric chloride was added to minimize interferences, and the solution was nebulized directly into a nitrous oxide-acetylene flame.

References

1. M. D. Amos and J. B. Willis, *Spectrochim. Acta,* 1966, **22,** 1325.
2. D. C. Manning, *Atomic Absorption Newsletter,* 1966, **5**, 127.
3. "Analytical Methods for Atomic Absorption Spectrophotometry", Parkin-Elmer Corp., Norwalk, Conn., U.S.A., 1971.
4. J. B. Headridge and D. P. Hubbard, *Anal. Chim. Acta,* 1967, **37,** 151.
5. W. Slavin, A. Venghiattis and D. C. Manning, *Atomic Absorption Newsletter,* 1966, **5,** 84.
6. J. I. Dinnin, *Anal. Chem.,* 1967, **39,** 1491.
7. K. E. Zacha, M. P. Bratzel, Jnr., J. D. Winefordner and J. M. Mansfield Jnr., *Anal. Chem.,* 1968, **40,** 1733.
8. J. D. Winefordner, *Anal. Chem.,* 1970, **42,** 206R.
9. V. S. Sastri, C. L. Chakrabarti and D. E. Willis, *Can. J. Chem.,* 1969, **47,** 587.
10. *Idem, Talanta,* 1969, **16,** 1093.
11. A. M. Bond, *Anal. Chem.,* 1970, **42,** 932.
12. A. M. Bond and D. R. Canterford, *Anal. Chem.,* 1971, **43,** 134.
13. R. R. Foster, *Atomic Absorption Newsletter,* 1968, **7,** 110.
14. V. B. Grossman and A. S. Inglis, *Anal. Chem.,* 1971, **43,** 1903.

CHAPTER 14

Special Techniques in AAS and AFS

14.1 INTRODUCTION

This chapter is concerned with the description of various techniques and methods in which either AAS or AFS is applied for indirect determinations or for special purposes. Thus the application of AAS for isotopic analysis and of AAS and AFS for the determination of temperatures in flame and non-flame atom cells is reviewed. The present status of the application of laser sources in AFS is also described.

14.2 APPLICATION OF INDIRECT METHODS IN ANALYSIS BY ATOMIC ABSORPTION SPECTROSCOPY*

A considerable number of elements may not be determined directly by AAS with conventional instrumentation, or may be

* Taken from a review by G. F. Kirkbright and H. N. Johnson, *Talanta,* 20, 433.

determined only with low sensitivity (>5 ppm for 1% absorption) so that they are not readily determined at trace levels in solution. Thus, for example, the elements F, Cl, Br, I, S, P, C, O and N exhibit their principal resonance lines in the vacuum ultraviolet (see Table 14-1) and are not directly determinable by AAS without modified equipment (see I, S and P in Chapter 13). A second large group of elements for which direct AAS is difficult includes those elements for which, although their resonance lines lie at accessible wavelengths, it is extremely difficult to form and maintain a significant population of atoms in flame and non-flame cells. This group includes Th and Ce, for which no report of direct determination by AAS with appreciable sensitivity has been made, and in decreasing order of difficulty the elements U, B, W, Ta, La, Nb, Zr, Re, Lu, Hf Hg, Gd, Ir, Pr, Nd, Sm and Tb. In addition, the attainable sensitivity

TABLE 14-1

Elements with principal resonance lines below 200 nm

Element	Line (nm)	Element	Line (nm)
F	95·5, 95·2	P	178·8, 178·3, 177·5
Cl	138·0, 134·7	S	182·6, 182·1, 180·7
Br	157·6, 148·9	As	197·3, 193·8, 189·0
I	183·0	Se	196·1
C	165·7, 156·1	Hg	184·9

for several other elements that are frequently to be determined may be inadequate for their determination at the trace levels commonly encountered in samples to be analysed; these include As, Ge, Ti, V, Se and Te.

In order to extend the range of application of AAS, and to increase the attainable sensitivity for those elements where it is inadequate for trace analysis, much attention has been devoted to the development of indirect methods of analysis for many of the elements listed above. In addition, some work has been reported for the application of indirect methods to permit the determination of organic species by AAS.

14.2.1 General Considerations

One of the principal advantages of direct methods of analysis by AAS is the high selectivity for particular elements in dilute solution. The success of indirect methods of analysis by AAS, and their applicability to practical analyses, depends on the degree to which the selectivity for the determination of the species concerned can be retained via the chemical reaction utilized before the final flame absorbance measurement. Even in those methods where some loss of selectivity is experienced owing to the nature of the initial chemical reaction employed, the ability to increase the sensitivity, or to achieve the AAS determination at all, is frequently of considerable value.

Several types of procedure may be employed in indirect determinations by AAS. In each of these the species to be determined is made to undergo a reaction with one or more other ions, and the concentration of the reaction product or of the unreacted excess of added reagent (which are directly and inversely proportional, respectively, to the concentration of the analyte species in the sample) is measured. The following specific types of procedure may be employed.

(1) Measurement of the suppression or enhancement of the atomic-absorption signal for a given element by its reaction in the flame with the species to be determined.

(2) Precipitation of the ion or compound to be determined by a second ion which is determined by AAS in the precipitate or filtrate.

(3) Formation of an ion-association or chelate complex between an organic compound or an anion to be determined and a metal or metal complex; the metal is then determined by AAS after extraction with an organic solvent.

(4) Formation of a binary or ternary heteropoly acid species between the analyte ion and other ions (one of which is usually molybdate), and determination of its concentration by AAS after separation of the complex by solvent extraction.

(5) Application of the reducing (or oxidizing) properties of the analyte species to yield an equivalent concentration of the lower (or higher) oxidation state of a metallic ion, only one oxidation state of which forms an extractable chelate complex, which may be determined by AAS.

Indirect AAS could be considered under these five headings, and this approach has been adopted elsewhere in reviews of indirect AAS methods.[1,2] However, as the analytical chemist is most frequently interested in the determination of a particular element or group of elements, indirect methods are classified here by the element determined.

14.2.2 Metals

Only rarely are the metallic elements not directly determinable by AAS; the indirect methods reported have therefore been concerned principally to provide increased sensitivity in their determination.

Aluminium

An indirect AAS method for aluminium has recently been reported.[3] The sensitivity obtained in the determination of iron by AAS is low in the fuel-rich air-acetylene flame. An enhancement of the absorbance for iron is observed in the presence of small amounts of aluminium when this fuel-rich flame is employed. The enhancement by aluminium of the absorbance produced on nebulization of a 25 ppm iron solution is linear with respect to the aluminium content of the solution in the range 0·25-1·2 ppm.

The sensitivity obtained in the determination of aluminium by this method may be varied by the nebulization of solutions containing different concentrations of iron; in this way with a 25 ppm iron solution the most sensitive working range for aluminium was found to be 0·01-0·09 ppm. Aluminium was also observed to enhance the absorbance of the elements cobalt, nickel and chromium in the flame employed. Titanium, calcium and zirconium at the same concentration levels as aluminium interfere with its determination. At concentrations equivalent to the iron concentration nickel and chromium interfere. The method is useful only for relatively pure aluminium solutions or analyses in which a separation procedure is included.

Germanium

The indirect determination of germanium by AAS, based on the formation of molybdogermanic acid (MGA) in which the molybdenum : germanium combining ratio is 12 : 1, has been described by Jakubiec and Boltz.[4] After its formation in hydrochloric acid solution at pH 1·5, the MGA complex is extracted with a mixture of 1-pentanol and diethyl ether (1 : 4 v/v). The complex is then back-extracted and decomposed by shaking the organic layer with an ammoniacal buffer at pH 9·3. The resulting molybdate may be measured by AAS at 313·2 nm in an air-acetylene flame. A sensitivity for germanium of 0·05 ppm was reported with linear calibrations up to 1·6 ppm. No interference was found for 500 ppm of Ca, Co, K, Mg, Mn, Na, Ni, Zn, acetate, chloride, perchlorate, nitrate or sulphate ions in the determination of 1·5 ppm of germanium. The maximum permissible concentrations of foreign ions for non-interference in the determination of 1·5 ppm of germanium were established as: Al, Cd, Cr(III), Cu, Mo(VI) 500 ppm; Pb 50 ppm; Fe(II), Fe(III), Ti(IV), Cr(VI), W(VI) 10 ppm; As(III), As(V), P(V) and Si interfered at any concentration.

Mercury

Mercury(II) may be converted into its anionic bromo-complex and extracted into 1,2-dichloroethane as an ion-association complex with the zinc 2,2-bipyridyl chelate cation. Yamamoto *et al.*[5] have reported the determination of the zinc content of the extract, which is equivalent to the mercury present in the sample, by AAS in an air-acetylene flame. The extraction was found to be quantitative between pH 3·5 and 5·5 and a linear calibration graph was obtained for 0·8 to 4·0 ppm of mercury in the initial aqueous phase with a sensitivity (for 1% Zn absorption) of 0·04 ppm of mercury. The following maximum permissible concentrations of foreign ions for non-interference with the determination of 2 ppm of mercury were established: Cu, 0·1 ppm; Co and I^-, 5 ppm; Cd, 15 ppm; Fe, 20 ppm; Pb, 200 ppm; Cl^-, 400 ppm.

Niobium

Kirkbright *et al.*[6] have utilized the formation of the ternary complex between niobium, phosphate and molybdate as the basis of an indirect method for the determination of niobium. The method is of the general type illustrated schematically in Fig. 14-1. As the

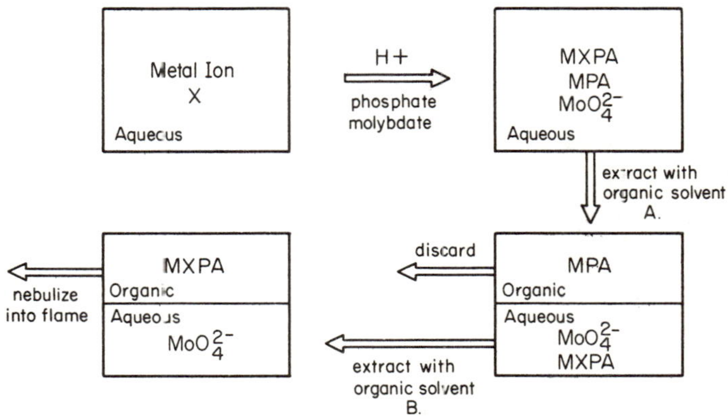

Fig. 14-1. General scheme for indirect determination of elements by AAS via formation of ternary heteropoly acids (MXPA) with phosphate and molybdate. MPA represents molybdophosphoric acid formed simultaneously with MXPA.

molybdoniobophosphoric acid (MNPA) cannot be formed in aqueous solution without the simultaneous formation of molybdophosphoric acid (MPA) a prior selective extraction of MPA is required. Both MPA and MNPA are formed in 0·5 M hydrochloric acid solution. The MPA may then be extracted with isobutyl acetate before extraction of MNPA into a single portion of n-butanol. The molybdenum associated with the niobium in the MNPA is then determined by direct AAS in the n-butanol phase which has been washed with dilute hydrochloric acid. A sensitivity of 0·015 ppm of niobium was reported, compared to a sensitivity by conventional direct AAS determination of 20 ppm for 1% absorption at the 334·4 nm line (Table 14-3); the calibration curve obtained was linear in the range 0·22-2·2 ppm of niobium. Out of twenty-seven ions studied for their interference with the determination of 1·3 ppm of niobium at the 30 to 700-fold weight ratio level, only As(V), Ge, Si, Ti, V(V) and Cr(VI) interfered. The method is sensitive because of the amplification accruing via the 11 : 1 combining ratio of molybdenum to niobium in MNPA and the "organic solvent effect" in the nebulizer of the flame spectrometer.

Thallium

Molybdophosphoric acid (MPA) has been used as a precipitant for thallium; the molybdate obtained by dissolution of the precipitate may be determined by AAS at 313·2 nm in an air-acetylene flame.[7] Thallium(I) is precipitated with MPA as $Tl_2HPMo_{12}O_{40}$ from 0·5 M perchloric acid medium. The precipitate is centrifuged, washed with perchloric acid and dissolved in 0·1 M sodium hydroxide solution. The molybdate equivalent to a definite amount of thallium is then determined. The optimum concentration range was found to be 2–20 ppm of thallium. The effect of other ions was not examined but as the method may also be used for the determination of ammonium by co-precipitation, this ion may thus interfere, as would other heavy metals which precipitate as molybdophosphates. As thallium can be determined by AAS in an air-acetylene flame with a sensitivity for 1% absorption of *ca.* 0·5 ppm at the 276·8 nm line it is doubtful whether, in view of the decreased selectivity, this method represents any advance on the direct method.

Thorium

Thorium may be determined indirectly by a method similar to that used for niobium.[8] Molybdothorophosphoric acid (MThPA) is formed in 0·15 M hydrochloric acid in the presence of excess MPA. The latter is then selectively extracted into isobutyl acetate and discarded. MThPA is extracted with n-butanol which is washed with butanol-saturated dilute hydrochloric acid. Although Kirkbright *et al.* claim to extract MThPA as such it is probable that only the MPA associated with the ternary complex was actually extracted; treatment with n-butanol has been shown to break down MThPA into thorium and MPA.[9,10] Eleven molybdenum atoms are associated with one thorium atom in the complex and the molybdenum is measured by AAS at 313·2 nm in a nitrous oxide-acetylene flame. A linear calibration curve is obtained for 1–6 ppm of thorium in the initial aqueous solution with a sensitivity of 0·063 ppm of thorium for 1% absorption. Only Zr, Fe(III), Si, As(V), Ti and Nb interfere seriously and required removal or masking.

Titanium

Titanium may be determined via its ternary heteropoly complex, molybdotitanophosphoric acid (MTPA).[11] The dissolution of the sample in hydrofluoric acid before the addition of phosphate, molybdate and acid requires the addition of aluminium to mask the excess of fluoride ion, which otherwise interferes in the determination. The excess of MPA is selectively extracted with chloroform : n-butanol (4 : 1 v/v) after formation of MTPA in 0·5 M hydrochloric acid. The MTPA is extracted with n-butanol which is then washed with butanol-saturated dilute hydrochloric acid before nebulization into a nitrous oxide-acetylene flame. Eleven molybdenum atoms are associated with two titanium atoms in the complex and the molybdenum is measured by AAS at 313·2 nm. A linear calibration curve is obtained for 0·3 to 3 ppm of titanium in the initial aqueous solution, corresponding to a sensitivity (for 1% Mo absorption) of 0·013 ppm of titanium. Interference is observed from As(V), Ge, Zr, Ta and Nb.

Ottaway *et al.*[12] have used the enhancement by titanium of the atomic absorption signal of iron in a fuel-rich air-acetylene flame to determine titanium in the range 0·01–10 ppm. The iron is nebulized as ferric sulphate in 10^{-3} M sulphuric acid. Chloride, calcium, aluminium and zirconium also give enhancements at similar concentrations. A method based on the suppression of the absorbance of strontium by titanium in the presence of oxalate ions when nebulized into a stoicheiometric air-acetylene flame has been used to determine titanium in the range 1 to 10 ppm.[13] The suppression is thought to be due to compound formation by strontium with titanium and oxalate. A solution 2×10^{-4} M in strontium and containing 100 ppm of oxalate, 10% by volume of hydrochloric acid, and the titanium, is nebulized into an air-acetylene flame. If the strontium concentration is increased higher ranges may be covered; these can, however, equally well be determined directly with a nitrous oxide-acetylene flame. A sensitivity of 0·15 ppm of titanium was recorded. Slight negative interference is experienced with EDTA, whereas Mn(II), Cu, Mg, Ca, Sn(IV) and Ni present at 200 ppm produce a small positive interference with 2 ppm of titanium. Al, Zr, H^+, Cr(III), PO_4^{3-} and F^- all interfere, and therefore the method has limited applicability but does allow determination of titanium in the presence of high hydrochloric acid concentrations. Titanium causes similar depressions in the absorbances of calcium and barium in the presence of oxalate but methods based on these effects are less sensitive.

Vanadium

Vanadium(V) has been determined indirectly by AAS via the formation of molybdovanadophosphoric acid (MVPA).[14,15] In the method of Jakubiec and Boltz[14] a mixed phosphate and molybdate reagent is added to an aliquot of the sample at pH 1·5–2·0. The excess of MPA is extracted with diethyl ether, and the MVPA complex is then extracted with an n-pentanol : diethyl ether mixture (1 : 4 v/v). The organic layer is washed with dilute hydrochloric acid, and the MVPA is decomposed by back-extraction into an ammonia-ammonium chloride buffer solution. The molybdate content of this solution, which is equivalent to the vanadium, is then determined in a reducing air-acetylene flame with a sensitivity for vanadium of 0·05 ppm for 1% absorption. The following ions caused interference in the determination of 2·17 ppm of vanadium at the concentrations indicated: Cr(III), NO_3^-, SO_4^{2-} 250 ppm, Cd, $C_2H_3O_2^-$, Cr(VI) 100 ppm; As(V), Sn(IV) 50 ppm; Ti, Si, W(VI) 10 ppm; Fe(III) 5 ppm; Fe(II) interfered at any concentration.

A method has been described which does not require back-extraction of MVPA from the organic solvent.[15] Phosphate and molybdate are added separately to the vanadium(V) solution to form MVPA in a 0·4 M nitric acid medium. An n-butanol : chloroform mixture (1 : 4 v/v) is used to extract the MPA selectively and has the advantage of forming the denser phase so that it is readily separated and discarded without the need for transfer of the sample solution. MVPA is extracted with n-butanol, which is then washed with butanol-saturated dilute acid and nebulized into a fuel-rich nitrous oxide-acetylene flame. Eleven molybdenum atoms are associated with each vanadium atom in the complex and the molybdenum is measured at 313·2 nm. A sensitivity (for 1% Mo absorption) of 0·011 ppm of vanadium was obtained with a linear calibration from 0·2 to 2·0 ppm. Interference with the determination of 0·8 ppm of vanadium was experienced from F^- (16 ppm), As(V), Sn(IV) (8 ppm); Cr(VI) (2·4 ppm); Zr (0·8 ppm); Ti and Si interfere at any concentration. A scheme has been recommended to improve the selectivity of the procedure in the presence of ions which form binary or ternary heteropolymolybdates. These ions are removed as their corresponding heteropoly acids before formation and extraction of MVPA. This is possible as vanadium exists both as V(IV) and V(V) and only V(V) apears to form the ternary heteropoly acid. Vanadium in aluminium has been determined in this manner.

14.2.3 Non-metals

Most of the published work that deals with indirect methods of analysis by AAS has been concerned with the determination of the non-metallic elements and other anionic species.

Arsenic

The reported indirect AAS methods for arsenic are all based on the formation of the binary heteropoly acid, molybdoarsenic acid (MAA). Danchik and Boltz[16] used bromine water to ensure that arsenic was present as As(V) before its reaction with molybdate at pH 1·8. The MAA was extracted with methyl isobutyl ketone (MIBK) which, after being washed with dilute hydrochloric acid, was shaken with a basic buffer solution to decompose and back-extract the MAA. The molybdenum associated with the arsenic was measured by AAS at 313·2 nm in an air-acetylene flame in the basic aqueous phase. A sensitivity of 0·07 ppm was achieved. Ramakrishna, Robinson and West[17] have described a method in which MAA formed at pH 0·9 is selectively extracted into an ethylacetate : butanol : isopentylacetate mixture (1 : 1 : 2 v/v) in the absence of MPA. A sensitivity of 0·025 ppm was reported for measurement of the associated molybdenum in a nitrous oxide-acetylene flame. The effect of 40-fold weight ratios of 40 ions was studied, none of which was found to interfere. Arsenic has been determined with a sensitivity of 1 ppm in urine.[18] In this procedure the MAA was formed at pH 1, MPA was extracted with isobutyl acetate and the pH was then increased to 2 before extraction of MAA with cyclohexane.

By judicious choice of the extraction system arsenic may be determined with high sensitivity and good selectivity via its heteropoly acid species with molybdate.

Fluorine

The presence of fluoride ion depresses the absorbance produced at 285·2 nm for magnesium in an air-coal gas flame; this effect may be utilized as the basis of an indirect AAS method.[19] The change in absorbance of the magnesium at 285·2 nm has been reported to be proportional to the fluoride concentration over the range 0·2-20 ppm of fluoride. When the concentration ratio of magnesium to fluoride in the sample solution reached 1 : 2 a constant absorbance was obtained. The method is subject to some interference, principally from phosphate and sulphate. Aluminium, oxalate and acetate also interfere at greater than 1000-fold molar ratio to fluoride.

An alternative, but less sensitive, technique based on the enhancement of zirconium absorbance by fluoride in a nitrous oxide-acetylene flame at 360·1 nm has been reported to allow the determination of fluoride in aqueous solution with a high degree of freedom from interference over the range 5–200 ppm.[19] Phosphate again interferes, but the similar enhancement observed for titanium absorption at 364·3 nm can be used for fluoride determinations in the range 40-400 ppm in the presence of phosphate.[19]

These indirect AAS methods for fluoride are considerably more rapid than the classical methods.

Chlorine

All the indirect AAS methods which have been described for the determination of chloride ion have been based on its reaction with silver nitrate. Either the excess of silver in the supernatant liquid is measured by AAS at 328 nm after removal of the silver chloride precipitate,[20-24] or the precipitate is redissolved in ammonia solution and the silver content of the ammoniacal solution is measured.[25,26,27] Chloride has been determined in serum,[20] plant liquors,[21] wine,[22] and polyvinyl chloride[23] by the first technique. A variation of this method involves the addition of a range of silver ion concentrations to aliquots of the sample containing chloride. After filtration the excess of silver in each filtrate is measured by AAS; a graph of absorbance *vs* silver concentration should then show a point of inflection at the concentration of silver equivalent to the chloride content of the original solution.

The first reported determination of chloride by AAS was based on precipitation of silver chloride, dissolution of the precipitate in ammonia solution and the determination of the silver in the resulting solution.[27] This method has been applied to the determination of chloride in titanium metal with an air-hydrogen flame[25] and to the determination of chlorine in selenium.[26]

Perchlorate

Perchlorate forms a complex with copper(I) and neocuproin (2,9-dimethyl-1,10-phenanthroline) corresponding to the composition $Cu(neocuproin)_2ClO_4$; this species is extractable into ethyl acetate. When the copper is determined by AAS at 324·7 nm the method may be applied to the determination of 12·5-125 μg of perchlorate with a sensitivity of 0·025 ppm.[28] In the determination of 3 ppm of perchlorate by this method Collinson and Boltz[28] observed interference when nitrate, ammonia, magnesium or aluminium was present at a concentration greater than 10 ppm.

Iodine

Iodine has been determined by AAS by use of a cadmium hollow-cathode lamp.[29,30] The method is based on solvent extraction into nitrobenzene of the ion-association complex formed between tris(1,10-phenanthroline)Cd(II) and iodide at pH 5; subsequent AAS determination of the cadmium content of the extract may then be made at 228·8 nm in an air-acetylene flame. The effect of a number of diverse anions upon the determination of 2×10^{-5} M iodide has been studied. Interference was experienced at concentrations greater than the molar ratios relative to cadmium indicated for the following ions: IO_4^- (0·02); ClO_3^-, ClO_4^- (0·03); NO_2^-, NO_3^- (0·1); Br^- (0·2); Cl^-, BrO_3^- (2); CN^- (10); IO_3^- (100); F^-, CO_3^{2-}, $B_4O_7^{2-}$ (1000). Christian and Feldman[31] have described the determination of iodide by utilization of its reducing action on a metal ion, coupled with extraction of the remaining oxidized form into an organic solvent and its estimation by AAS. Thus, for example, iodide reduces chromium(VI) in acid medium and the excess of chromium (VI) may be extracted from 3 M hydrochloric acid into MIBK. The absorbance of chromium(III) in the aqueous layer increases linearly and the absorbance of the unreacted chromium(VI) in the MIBK decreases linearly with increasing amounts of iodide. The reaction is slow and does not go to completion in dilute solutions. Interferences have not been studied but most reducing species, e.g. Fe^{2+}, would probably have a similar effect and thus interfere.

Selenium(IV) is reduced by iodide to elemental selenium in acid solution. When iodide is allowed to react with excess of selenium(IV) and the solution is filtered through a Millipore filter, the decrease in absorbance of the filtrate is proportional to the iodide ion concentration of the original sample solution.[31] Again, reducing species would be expected to interfere, though no results were reported for the effects of other ions.

Iodate

Iodate oxidizes iron(II) to iron(III) in acid medium; the iron(III) may be extracted into diethyl ether from 9 M hydrochloric acid solution. The absorbance of iron in the ether phase is then proportional to the iodate concentration. The solutions must be deaerated to prevent aerial oxidation. As with the reduction method for iodine, this method lacks specificity but may be useful when no other oxidizing species are present.

Nitrogen compounds

Ammonia. Bond and Willis[32] have used the enhancement of zirconium absorbance in a nitrous oxide-acetylene flame at 360·1 nm by nitrogen-containing compounds for the determination of ammonia. For ammonia, the magnitude of this enhancement was found to be proportional to the concentration of the base in the range 1×10^{-4} to 5×10^{-3} M and was used for its determination in the absence of phosphate and certain nitrogen-containing compounds. In the presence of interfering species, prior separation of the ammonia by distillation is required. The method has been applied to the determination of ammonia in biological systems.[32]

Danchik, Boltz and Hargis[7] have shown the co-precipitation of ammonium molybdophosphate with a known amount of thallium molybdophosphate and subsequent dissolution of the precipitate to provide the basis for an indirect AAS method for the determination of ammonia. The molybdate obtained by dissolution of the precipitate, which is equivalent to a known amount of thallium and ammonia, is determined at 313·2 nm in an air-acetylene flame. Molybdophosphoric acid is used as precipitant in 0·5 M perchloric acid. The method is unselective, as many heavy metals are precipitated as phosphomolybdates.

Cyanide. Cyanide may be determined via the equivalent silver associated with it in the $Ag(CN)_2^-$ complex.[33] Pieces of metallic silver are added to the solution containing the cyanide to be determined and the mixture is then shaken for one hour. The dissolved silver is determined by AAS to produce a rectilinear calibration graph for up to 3 ppm of cyanide ion. No interference is experienced from Br^-, I^-, SCN^- or SO_3^{2-} ions but $S_2O_3^{2-}$ enhances the signal. Danchik and Boltz[34] report two methods for the indirect determination of cyanide. The first is based on the formation and extraction of the dicyano-bis(1,10-phenanthroline)Fe(II) complex. The complex is extracted into chloroform, which is then evaporated, and the residue is dissolved in ethanol. The associated iron is determined by AAS in an air-acetylene flame at 248·3 nm. A sensitivity of 0·06 ppm of cyanide has been obtained by this method; the calibration graph is linear up to 5 ppm. Precipitation of silver cyanide and measurement of the excess of silver in the supernatant solution by AAS in an air-acetylene flame at 328·1 nm forms the basis of a second method. A working range of 0·3 to 2·5 ppm of cyanide has been reported with a sensitivity of 0·03 ppm.

Thiocyanate. Thiocyanate forms a dithiocyanatodipyridine copper(II) complex, $Cu_2(Py)_2(SCN)_2$, which is extractable with chloroform. The copper equivalent to the thiocyanate is determined at 324·7 nm. A sensitivity of 0·2 ppm of thiocyanate has been reported with an optimum concentration range of 7 to 18 ppm.[35] The sensitivity may be improved to 0·05 ppm by evaporating the chloroform solution almost to dryness and diluting with ethyl acetate. The optimum concentration range is then between 0·5 to 2·0 ppm.

Nitrate. A small amount of nitrate ion is extractable into MIBK when neocuproin (2,9-dimethyl-1,10-phenanthroline) is contained in the organic phase and copper(I) ions are present in the aqueous phase. The copper associated with nitrate ions in the complex, $Cu(I)(neocuproin)_2NO_3$, is measured in an air-acetylene flame at 324·7 nm.[36,37] Beer's Law is obeyed in the range 1 to 7×10^{-5} M nitrate in the aqueous phase. Nitrite and nitro-groups in organic compounds may also be determined via formation of the ion-pair with the Cu(I) neocuproin complex after their oxidation to nitrate with ceric sulphate or potassium permanganate.[38] The complex may be extracted into MIBK before measurement of the associated copper. Prior reduction of nitrite and nitro groups to amino groups by sulphamic acid permits determination of nitrate in the presence of such groups.

Phosphorus

Orthophosphate has been determined indirectly, mainly via the formation of molybdophosphoric acid (MPA) in the presence of molybdate in acid solution. The general procedure most frequently employed is illustrated in Fig. 14-2. Molybdate (ammonium or sodium salt) is added to the sample solution containing orthophosphate and the acidity is adjusted to its optimum value. The solution is mixed and allowed to stand for 5-10 minutes to ensure complete formation of MPA. The MPA is then extracted into an

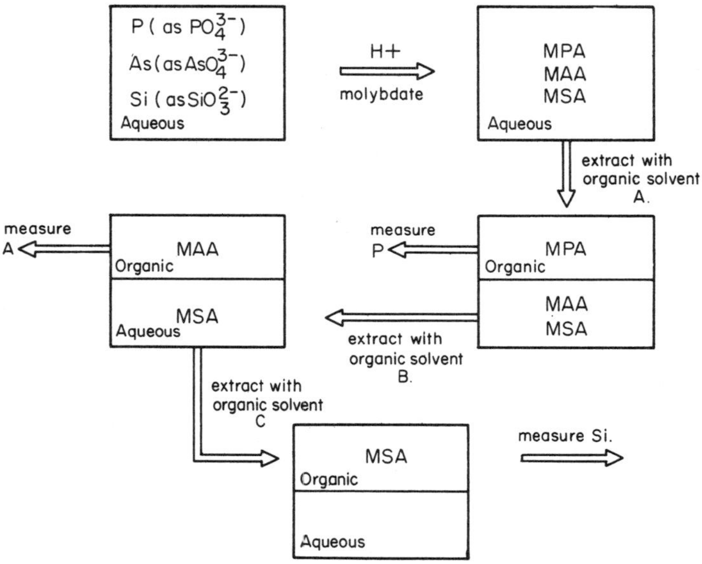

Fig. 14-2. General scheme for indirect determination of elements by AAS via formation of heteropoly acid species with molybdate.

organic solvent which is washed either with dilute acid[39,40] to remove molybdate transferred to the organic layer, or with citrate solution to complex the excess of molybdate.[43,44] The molybdate associated with the phosphate in the complex is measured by AAS at 313·2 nm in an air-acetylene[40,46] or a nitrous oxide-acetylene flame.[17,39,47] Kumamaru *et al.*[47] employed 0·5 N hydrochloric acid to form MPA and performed the extraction with n-butyl acetate to obtain a linear, calibration graph up to 0·35 ppm of phosphorus. A general method for the determination of inorganic phosphate using 2-octanol as a solvent[44] has been applied to the determination of phosphate in fresh-water and sea-water and in blood, serum, bone, biological tissues and enzyme-containing reaction mixtures.[45] 2-Octanol was chosen as solvent after examination of n-butanol, isobutyl alcohol, 1-hexanol, 1-heptanol, 1-octanol and 2-octanol.[45] Phosphate has been determined in urine, the MPA being formed in 0·37 M hydrochloric acid and extracted into isobutyl acetate.[41] Linden *et al.*[42] have determined phosphorus in milk products and blood serum via MPA, using perchloric acid medium and MIBK as solvent, for the range 100-200 μg of phosphorus. The results obtained agreed with those established colorimetrically. Phosphate has been determined in small (0·2 ml) samples of plasma.[43] The MPA is formed at pH 1·9 in a DL-α-alanine-hydrochloric acid buffer and is extracted with MIBK. The authors claim to extract MPA with an Mo : P ratio of 15 : 1. No evidence concerning the establishment of this apparently anomalous combining ratio is given. The procedure is applicable to the determination of between 5 and 200 ppm of phosphorus.

Several schemes have been devised for the sequential determination of phosphorus, arsenic and silicon,[17] phosphorus and arsenic,[18] and phosphorus and silicon[39,40] (Fig. 14-2). All are based on the ability to extract MPA selectively. Thus MPA has been selectively extracted with isobutyl acetate after formation at pH 0·7 in nitric acid medium, to result in a final sensitivity of 0·01 ppm of phosphorus.[17] Devoto[18] used isobutyl acetate to extract MPA formed at pH 1 in an ammonium nitrate-nitric acid medium. Isobutyl acetate has also been used by Kirkbright *et al.*[39] to selectively extract MPA formed at an acidity of 0·96 M in hydrochloric acid in a single portion of solution containing phosphate and silicate, before formation of molybdosilicic acid (MSA). A sensitivity of 0·007 ppm of phosphorus and rectilinear calibration from 0·08 to 1 ppm of phosphorus were obtained by measuring the associated molybdate in the organic phase at 313·2 nm in a nitrous oxide-acetylene flame. Only W(VI) is reported to interfere out of 26 ions examined. MPA formed at pH 1·3 has also been extracted from 1·2 M hydrochloric acid medium with diethyl ether.[40] After being washed with dilute hydrochloric acid, the organic layer may be shaken with an ammoniacal buffer and the molybdate released measured in the aqueous phase by AAS in an air-acetylene flame. The optimum concentration range reported is 0·05 to 1·3 ppm of phosphorus. The use of isobutyl acetate coupled with the formation of MPA produces an extremely selective method for the determination of phosphorus by indirect AAS.

Phosphorus has been determined by several workers using the direct chemical interference of phosphate on a metal species in a flame to produce a depression of the absorption, which is a function of the phosphate concentration.[31,48,49] Orthophosphate up to $2{\cdot}5 \times 10^{-4}$ M added to 5×10^{-4} M calcium solution produces a marked depression in the absorbance obtained for calcium at 427·7 nm with a total consumption burner.[31] The limiting concentration of phosphate may be increased by increasing the calcium concentration. Singhal *et al.* have determined phosphorus in rock phosphate by depression of both calcium[49] and strontium[48] absorbance. In both cases the powdered rock phosphate is treated with a cation-exchange resin (H^+-form) before addition of calcium or strontium solution. A modification of the quinolinium molybdophosphate method has been proposed whereby total phosphorus is determined indirectly by the AAS measurement of molybdenum.[50] The familiar and well-tried gravimetric method is followed to the stage where the quinolinium molybdophosphate is formed. The precipitate is then dissolved in ammonia solution and the molybdenum determined by AAS at 313·2 nm. The method has been applied to the phosphorus analysis of Magruder collaborative fertilizer samples.[50]

Selenium

Selenium may be difficult to determine directly with high sensitivity by AAS in some flames because of strong absorption of the resonance radiation at 196 nm by the flame and optical components of the spectrometer. An indirect method has been described based on two selective reactions of selenite with 2,3-diaminonaphthalene (DAN) and palladium(II) to form the complex $Pd(DANSe)_2Cl_2$.[51] When excess of DAN is added to a solution of selenite at pH 1·5-2·5 the following reaction occurs:

$$\text{(I)}\quad C_{10}H_6(NH_2)_2 + SeO_3^{2-} + 2H_3O^+ \rightarrow C_{10}H_6N_2Se + 5H_2O$$

In the presence of palladium chloride three further reactions are possible:

$$\text{(II)}\quad 2\,\text{DANSe} + PdCl_2 \xrightarrow{pH \leqslant 6} Pd(DANSe)_2Cl_2\,\text{(purple)}$$

$$\text{(III)}\quad 2\,\text{DANSe} + 2PdCl_2 \xrightarrow{pH \leqslant 6} Pd_2(DANSe)_2Cl_4\,\text{(green)}$$

$$\text{(IV)}\quad \text{DAN} + PdCl_2 \xrightarrow{pH\ 0\text{–}12} \text{white polymeric complex.}$$

Whether it is reaction II or III that occurs depends on the pH, relative concentration of reagents and also on the solvent used for the DANSe. Reaction IV always occurs and thus DANSe is removed from excess of DAN by extraction with chloroform. With chloroform as solvent the complex from reaction III is formed. The optimum calibration is from 0·05 to 1 ppm of selenium. A sensitivity of 0·017 ppm has been obtained. A study of the effect of 60-fold molar ratios of 50 ions on the determination of 0·5 ppm of selenium showed only Fe(III), Sn(IV) and Ce(IV) to interfere.

Silicon

Most indirect methods for silicon are based on the formation of molybdosilicic acid (MSA).[17,39,40,52] In one case MSA is formed at pH 1·3, extracted with a 1-pentanol : diethyl ether mixture (1 : 5 v/v) from 2M hydrochloric acid medium, and decomposed by back-extraction into a 0·1 M ammonia solution.[52] A known excess of lead(II) ion is then added to the aqueous solution to precipitate the molybdate as lead molybdate in amount equivalent to the amount of silicon present. The lead remaining in solution may be measured by AAS at 217·1 nm in an air-acetylene flame. An optimum concentration range of 0·005 to 0·07 ppm of silicon has been reported.[52] No interference was experienced from 500 ppm of Al^{3+}, NH_4^+, Fe^{3+}, Mg^{2+}, Cl^-, NO_3^-, ClO_4^- or SO_4^{2-}.

Silicon has also been determined sequentially with phosphorus[39,40] and arsenic[17] by using their respective molybdoheteropoly acids (Fig. 14-2). Kirkbright *et al.*[39] first investigated the determination of silicon by this means. They determined phosphorus as phosphate by forming MPA in 0·96 M hydrochloric acid medium and selectively extracting it into isobutyl acetate. The silicon in the same aliquot of solution is determined after the MPA extraction by adjusting the acidity of the solution of 0·15 M hydrochloric acid with 4 M ammonia solution to allow efficient formation of MSA. The MSA may then be extracted into n-butanol and the twelve molybdenum atoms associated with each silicon atom determined by AAS at 313·2 nm in a nitrous oxide-acetylene flame. A sensitivity of 0·008 ppm of silicon and an optimum concentration of 0·08 to 1 ppm of silicon are obtainable. The presence of a 100-fold weight ratio of 24 ions causes no interference in the determination of 12 μg of silicon; only arsenic(V) and germanium(IV) interfere by formation of their respective heteropoly acids. Hurford and Boltz[40] have described a similar sequential procedure in which the MSA is formed at pH 1·3 after the MPA has been extracted and is then extracted from 2 M hydrochloric acid medium with a 1-pentanol : diethyl ether mixture (1 : 5 v/v). After being washed with dilute hydrochloric acid the organic layer is shaken with basic buffer solution which extracts and decomposes the MSA. In this method the molybdenum is then measured in an air-acetylene flame. A calibration range between 0·05 and 1·1 ppm is obtained. Ramakrishna *et al.*[17] have separated MSA formed at pH 0·7 from large amounts of MPA and molybdoarsenic acid by extraction into MIBK in the presence of citrate. A sensitivity of 0·01 ppm of silicon is then obtainable, the optimum concentration range being from 0·1 to 1 ppm. Of 40 ions tested only germanium(IV) interferes with the determination of silicon.

Silicate has been determined in waste waters by atomic-absorption inhibition titration.[53] The method is based on the strong inhibition by silicate of the magnesium absorbance produced in an air-hydrogen flame. The absorbance is measured continuously while the nebulized solution is titrated; this procedure contrasts with the usual direct suppression methods when discrete sample solutions are nebulized. The sample is adjusted to pH 3-4 and passed through a cation-exchange resin (H^+-form) to remove magnesium and other interfering cations. The eluate from the column is titrated with standard magnesium solution and the titrated solution is simultaneously nebulized into an air-hydrogen flame. The end-point is detected by a sharp increase in the magnesium absorption signal at 285·2 nm. High accuracy and precision are obtained with a limit of detection of 0·1 ppm of silica. The removal of cations by ion-exchange ensures the selectivity of the method.

Sulphur

The first reported indirect AAS method for sulphur is based on the precipitation of barium sulphate on addition of excess of barium chloride to the sample solution. In this method the precipitate is dissolved in a solution of disodium EDTA and the resulting barium concentration is determined at 553·5 nm in an air-acetylene flame.[54] Lanthanum is added to prevent phosphate interference. The method has been used to estimate sulphur in urine, faeces and biological tissues after oxidation by one of several techniques.

In a procedure that avoids the necessity to dissolve the barium sulphate precipitate, a known excess of barium is added to the sample solution, and the excess is determined after the precipitation of barium sulphate.[55,56] A 0·02M hydrochloric acid medium is used to minimize loss of selectivity through co-precipitation of other ions. The error caused by the partial solubility of the precipitated barium sulphate and the incomplete reaction is overcome by calibration against standard sulphate solutions treated in a similar manner. The method has been applied to the determination of sulphate in textiles.[56] A similar method has been applied to the determination of water-soluble sulphate in acidic sulphate soils.[57] Calcium and phosphate above 30 and 10 ppm respectively interfere with the determination of 50 ppm of sulphate. A sensitivity of 1·5 ppm of sulphur has been reported for a method that involves precipitation of sulphate with barium chloride, centrifugation and measurement of the excess of barium.[58] An improvement in sensitivity is claimed to be achieved by reduction of the concentration or volume of the barium chloride solution added. The method has been applied to determination of sulphur in samples from deep mine waste water which have a high cation and sulphate concentration, and in the percolate from pyrite-bearing coalmine spoils. Good correlation has been reported between the results of the indirect AAS method and standard methods.

Sulphur in soils derived from volcanic ash is extracted with potassium dihydrogen phosphate solution; the extracted sulphur is converted into sulphate, and the organic matter is destroyed with 30% hydrogen peroxide solution and sulphate is precipitated with 15% barium chloride solution. The barium sulphate may be dissolved in a solution of EDTA (diammonium salt), and the barium measured by AAS in the presence of 1500 ppm of sodium ions to suppress barium ionization, to give a sensitivity (1% absorption) of 0·8 ppm of sulphur.[59]

Sulphur dioxide, after its preliminary conversion into sulphate with hydrogen peroxide, has been determined after precipitation and centrifugation of lead sulphate.[60] The excess of lead remaining in the supernatant liquid may be measured at 283·3 nm in an air-acetylene flame. The solution is made 25% with respect to ethanol to aid the precipitation of lead sulphate. A concentration range of 2 to 20 ppm of sulphur dioxide was recommended. Interference was experienced above the concentrations shown for the determination of 5 ppm of sulphur dioxide with the following ions Fe(III), PO_4^{3-} (0 ppm); Ba, Al (10 ppm); EDTA (20 ppm), $CH_3CO_2^-$ (250 ppm).

Sulphite or sulphur dioxide have been measured by AAS, with utilization of the high stability of the $Hg(SO_3)_2^{2-}$ complex.[61] Addition of sulphite ions to a suspension of mercury(II) oxide leads to dissolution of the oxide and formation of the soluble complex.

The amount of oxide dissolved is proportional to the increase in absorbance at 253·7 nm observed for mercury in the supernatant fluid, and is then indicative of sulphite concentration. An optimum concentration range of 12 to 84 ppm of sulphite ions (10 to 77 ppm of sulphur dioxide) with a sensitivity of 13 ppm of sulphite has been reported. The reaction is not specific and iodide, thiosulphate and thiocyanate interfere.

Sulphate has been determined by atomic-absorption inhibition titration.[62] The test solution containing 0·05-1 mg of sulphate is titrated at a constant delivery rate with standard magnesium chloride solution. The titrated solution is simultaneously nebulized into an air-hydrogen flame. The magnesium absorbance is inhibited until the concentration ratio of magnesium to sulphate approaches unity. Thereafter the graph representing increasing magnesium content becomes rectilinear and extrapolation to the base-line is used to locate the end-point. Interfering cations such as those of the alkaline earths metals are removed by passing the solution at pH 3-4 through a cation-exchanger (H^+-form). Silicate and phosphate interfere.

14.2.4 Organic Compounds

No systematic study of the determination of organic compounds by AAS appears to have been undertaken. Of the methods reported some are extremely selective and rely on the formation of a particular complex and its subsequent extraction,[63-66] while others utilize a more general reaction which can be applied for the determination of a range of compounds.[17,67]

2-Hydroxynaphthoic acid has been determined indirectly through the selective extraction of tris(1,10-phenanthroline)Ni(II)-2-hydroxy-naphthoate into nitrobenzene.[66] Measurement of the nickel concentration by AAS in an air-acetylene flame as a function of the 2-hydroxynaphthoic acid concentration initially present in the aqueous phase leads to a calibration that is linear in the range from 8×10^{-5} to 4×10^{-4} M 2-hydroxynaphthoic acid. No interference is experienced from 2-naphthol.

Phthalic acid is extractable into MIBK as the ion-pair formed between bis(neocuproin) Cu(I) and the univalent anion of phthalic acid. The copper extracted is determined by AAS in an air-acetylene flame at 324·7 nm. The absorbance of the extract gives a linear relationship with the phthalic acid concentration initially present in the aqueous phase in the range 4×10^{-6} to 4×10^{-5} M.[63,64] Isophthalic and terephthalic acid do not interfere if present in amounts equimolar with the phthalic acid.

A similar method has been employed for the determination of pentachlorophenol.[65] The basis of the method is the solvent extraction into nitrobenzene of the ion-pair formed between tris(1,10-phenanthroline)Fe(III) and the pentachlorophenol anion. The iron is subsequently determined in the extract by AAS at 248·3 nm in an air-acetylene flame. A concentration range up to 3×10^{-4} M in pentachlorophenol was studied.

Vitamin B_{12} contains one atom of cobalt per molecule. On dissolution of the vitamin the cobalt may be determined by AAS at 242·5 nm. The method has formed the basis for the determination of vitamin B_{12} in pharmaceutical dosage forms.[68,69]

Christian and Feldman[31] have determined several organic compounds by a variety of methods. Glucose oxidase and ribonuclease may be determined by their depressive effect on the absorbance of calcium. The upper concentration can be increased by raising the calcium concentration. Glucose below 10^{-6} M is found to produce a marked decrease in the absorbance obtained for calcium, but above that concentration it produces an increase, up to a limiting concentration of 10^{-5} M. Glucose in the range 10^{-6} to 5×10^{-5} M was also found to give a linear enhancement in the calcium absorbance which had been depressed by the addition of a constant amount of sulphate to the solution. Ammonium pyrrolidine dithiocarbamate (APDC) has been determined by measurement of the equivalent amount of copper or cobalt with which it complexes; the complex is extracted into MIBK at pH 3. With 1 ml of 1% metal ion solution a concentration of 10^{-5} M APDC may be determined. Similarly oxine is determinable via extraction of its copper complex into MIBK or ethyl acetate from an ammonium acetate buffer at pH 6·5 and subsequent AAS measurement of the copper. With 1 ml of 10^{-4} M copper solution a concentration of 2×10^{-5} M oxine is determinable. EDTA may be determined through its action as a masking agent for the copper oxinate chelate. The decrease in absorbance of the extracted copper is proportional to the concentration of EDTA in the initial aqueous phase.

Sugar in plant materials has been determined by measurement of unreduced copper.[70] Copper is reduced in alkaline solution by the reducing sugars to form insoluble copper(I) oxide. The copper(I) oxide is centrifuged from solution and the unreduced copper(II) in solution is measured by AAS. The concentration of sugar in test solutions is determined from standard graphs prepared from known amounts of sugar.

Several organic compounds have been determined by precipitation of their silver compounds and measurement by AAS of the concentration of the excess of silver left in solution.[67] Thus iodoform, theobromine, sodium salicylate, mercaptobenzothiazole and xanthate have all been determined in this way.[67]

Non-ionic surfactants may be determined indirectly by AAS.[71,72] In one method the surfactant is precipitated as the MPA-Ba complex and the molybdenum in the supernatant liquid is assayed by AAS.[71] The molybdenum concentration is related to the quantity of surfactant by an empirical factor which must be determined for each batch and multicomponent system owing to the non-stoicheiometric nature of the precipitate. A somewhat better method, which does not require the introduction of an empirical factor, has been developed.[72] The basis is the extraction of a cobaltithiocyanate-surfactant adduct from aqueous solution into benzene, and measurement of the associated cobalt in the benzene by AAS. Pollution by non-ionic surfactants in the Atlantic has been estimated in this way with a precision better than 10%.

14.2.5 Conclusion

As shown in Table 14-2, indirect AAS methods have been reported for the determination of inorganic cations and anions in solution as well as for the organic species described above. Tables 14-3 and 14-4

TABLE 14-2

Summary of available indirect AAS methods for elements

H																	He
Li	Be											B	C	N	O	F	Ne
Na	Mg											Al	Si	P	S	Cl	Ar
K	Ca	Sc	Ti	V	Cr	Mn	Fe	Co	Ni	Cu	Zn	Ga	Ge	As	Se	Br	Kr
Rb	Sr	Y	Zr	Nb	Mo	Tc	Ru	Rh	Pd	Ag	Cd	In	Sn	Sb	Te	I	Xe
Cs	Ba	La	Hf	Ta	W	Re	Os	Ir	Pt	Au	Hg	Tl	Pb	Bi	Po	At	Rn
Fr	Ra	Ac															
			Ce	Pr	Nd	Pm	Sm	Eu	Gd	Tb	Dy	Ho	Er	Tm	Yb	Lu	
			Th	Pa	U	Np	Pu	Am	Cm	Bk	Cf	Es	Fm	Md	No	Lw	

Not determined by AAS with conventional equipment.
Determinable by indirect AAS procedures.

illustrate that the sensitivities obtainable are usually clearly superior to those in the corresponding direct AAS determinations, where these are possible; with careful choice of experimental conditions many of the indirect methods reported exhibit sufficient selectivity

TABLE 14-3

Indirect AAS methods based on formation of molybdo-heteropoly compounds

Element	Sensitivity[a] Indirect	Direct[b]	Range, indirect, ppm	Ref.
P	0·003	250	0·04-2	44
As	0·025	0·25	0·2-2	17
Ge	0·05	2·5	0-1·6	4
Si	0·008	2·0	0·08-1·2	39
V	0·011	1·5	0·2-2	15
Nb	0·015	20	0·22-2·2	6
Th	0·063	—	1-6	8
Ti	0·013	2·0	0·3-3	11

[a] Concentration in ppm for 1% absorption

[b] From Perkin-Elmer Methods Book.

TABLE 14-4

Indirect methods based on miscellaneous procedures

Element	Sensitivity[a] Indirect	Direct	Range, indirect, ppm	Ref.
Al		1·3	0·01-10	3
Hg	0·04	11	0·8-4	5
Tl		0·45	2-20	7
Se	0·017	2	0·05-1	51
F			0·2-20	19
Cl			0-0·4	25
ClO_4	0·025		0·5-5	28
I			0·5-5	30
CN	0·03		0·3-2·5	34
SCN	0·2		7-18	35

[a] Concentration in ppm to give 1% absorption.

for the analyte species to be applicable for practical analyses. Although several workers have described the direct determination of elements such as carbon,[73] phosphorus,[73,74] sulphur and iodine[74] in non-flame cells, and sulphur[75] and iodine[76] in the inert gas-separated nitrous oxide-acetylene flame, using the resonance lines of these elements in the vacuum ultraviolet, these procedures require modified or specialized instrumentation and have not yet found widespread application.

14.3 ISOTOPE ANALYSIS BY ATOMIC ABSORPTION SPECTROSCOPY

Atomic absorption spectrometry may be utilized to advantage in the study of the isotopic composition of various elements. In order to view the isotopic splitting and hyperfine structure components of atomic lines by emission spectroscopy instrumentation of high resolving power is required and the broadening processes occurring in the source must be minimized. While for many elements the latter requirement also holds for AAS measurements, by this technique isotopic composition may be studied with apparatus of modest resolving power similar to that employed in many commercially available atomic absorption spectrophotometers.

The general approach involves the use of a light source emitting radiation of a single isotope of the element studied and measurement of the absorption of this radiation at one of the ground state lines of this isotope by an atomic vapour of the sample isotope mixture produced in a flame or non-flame cell. Alternatively the sample isotope mixture itself may be used in a light source such as a sputtering cell or high-frequency electrodeless discharge lamp. The absorption of the radiation emitted by one of the isotopes in the sample is then observed for the atomic vapour produced from a single isotope of the element (usually in a non-flame absorption cell).

Whichever method is employed the wavelength separation between the lines for the two or more isotopes of the element studied must be sufficient to preclude significant "spectral overlap" of the emission and absorption profiles. In order to achieve this it may be necessary to operate the radiation source and absorber cell at low pressure, temperature and optical thickness to achieve narrow line profiles. L'Vov[77] has concluded that isotopic analysis is only possible for twelve elements; this was deduced from a consideration of the isotope shift of the resonance lines of various elements and an estimate of the degree of overlap between the isotope components assuming minimal Doppler broadening in a hollow-cathode source cooled with liquid nitrogen. The elements for which isotopic analysis is possible under these conditions are Ar, B, He, Hg, Li, Ne, Os, Pb, Pt, Re, Tl and U. While some of these elements (e.g. the rare gases,

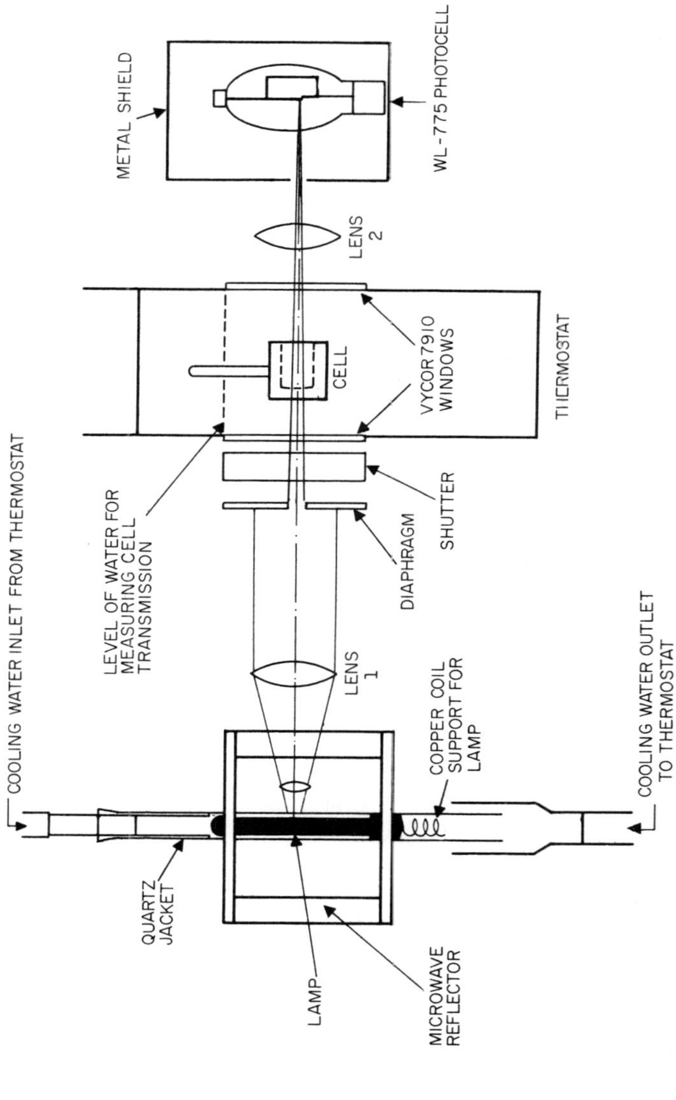

Fig. 14-3. Optical system for mercury isotope AAS used by Osborn and Gunning.[78]

Hg and Li) are either already monatomic gases or easily converted into atomic vapour in low temperature emission or absorption cells for isotopic measurements, others (e.g. B, Os, Pt) are less readily atomized and few observations of their isotopic composition have been made by AAS. The following section reviews briefly the application of AAS to the study of isotopic composition of the more easily atomized elements.

Mercury

AAS was applied to the determination of the Hg^{202} isotope in mercury by Osborn and Gunning.[78] As shown in Fig. 14-3 these workers employed as the light source a water-cooled microwave-excited electrodeless-discharge lamp containing the Hg^{202} isotope. Radiation at the Hg 253·7 nm line was passed through the vapour of the mercury sample contained in a silica cell maintained at a constant temperature which could be varied between 12 and 40°C. Under these conditions there was little overlap between the Hg^{202} emission line and the absorption lines of the other mercury isotopes; the results obtained for Hg samples containing widely differing concentrations of the Hg^{202} isotope were in good agreement with those obtained by mass spectrometry.

Lead

Kirchof[79] has described a method for the determination of the stable isotopes of lead by using the absorption of the hyperfine components of the Pb 283·3 and 405·8 nm resonance lines. Samples and standards containing known isotopic mixtures were incorporated into hollow cathodes used as primary radiation sources. The Pb^{208} (or Pb^{206}) content in the sample could be determined by measuring the absorbance of this radiation by lead vapour known to contain only Pb^{208} (or Pb^{206}). The absorbing vapour was produced by cathodic sputtering in a second hollow-cathode discharge. Figure 14-4 shows the absorption calibration curves obtained for Pb^{208}; the Pb 283·3 nm resonance line from the source hollow-cathode containing a Pb^{206}/Pb^{208} mixture is here absorbed by atomic vapour of the Pb^{208} isotope in the absorption cell. The absorption sensitivity is seen to be proportional to the concentration of the Pb^{208} isotope and to the operating current in the absorption cell. The examination by this technique of various geological samples containing varying Pb^{208} and Pb^{206} concentrations gave results which were in good agreement with those obtained by mass spectrometry.

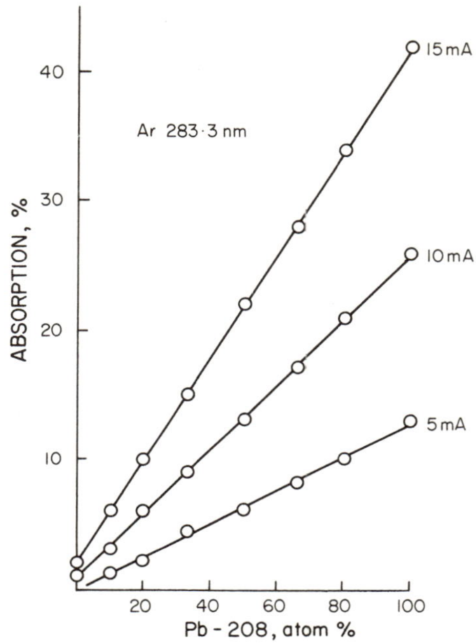

Fig. 14-4. Variations of absorption with Pb^{208} content of the Pb 283·3 nm resonance line from a Pb^{206}/Pb^{208} hollow-cathode source by Pb^{208} in a hollow-cathode absorption cell. Current values are those for the absorption cell. Argon carrier gas pressure 1·5 torr.

Brimhall[80] has demonstrated that it is possible to utilize a conventional long path air-acetylene flame for examination of the isotopic composition of lead sample solutions using a commercially available flame spectrophotometer and three different hollow-cathode lamps each enriched with a particular isotope. Three aqueous standard solutions, each likewise enriched in one of the three lead isotopes and containing a total of 10 ppm Pb were also employed. Figure 14-5 shows the instrument response obtained for each standard solution using each of the three sources. The absorption signal is dependent on the isotopic composition of the solution examined although overlapping absorption occurs. From the absorption signals measured using each of the three sources, it is possible to calculate the isotopic concentrations in a sample solution by solving a set of three simple simultaneous equations.

Lithium

The lithium 670·8 nm line is a doublet ($^2P_{3/2} \rightarrow {}^2S_{1/2}$, $^2P_{1/2} \rightarrow {}^2S_{1/2}$) in which the two components have a separation of approximately 350 mK (0·015 nm). The isotope shift between Li^6 and Li^7 is

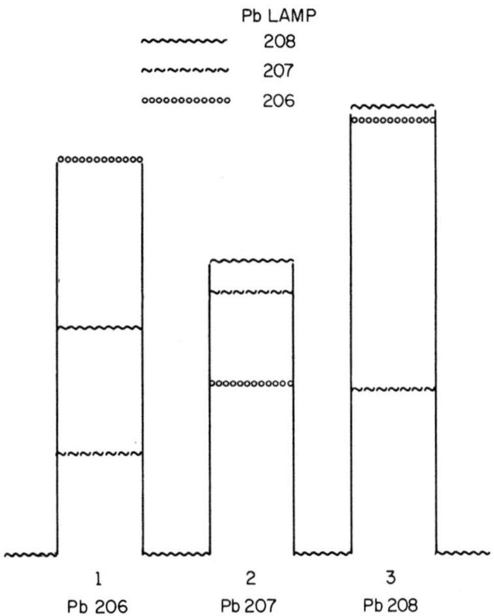

Standard	Concentration of isotope, ppm		
	Pb^{206}	Pb^{207}	Pb^{208}
1.	9·74	0·069	0·192
2.	0·216	9·240	0·548
3.	0·012	0·041	9·947

Fig. 14-5. Absorption response obtained for standard solutions 1, 2 and 3 for hollow-cathode lamps enriched in Pb^{206}, Pb^{207} and Pb^{208} respectively.

also near to this value; one of the Li^6 hyperfine components is therefore almost exactly coincident with one of the Li^7 lines (see Fig. 14-6). Even with a sharp line source of a single lithium isotope and a narrow absorption line profile in a suitable absorbing cell, spectral overlap interference results in the absorption of Li^6 radiation by Li^7 and vice versa. In early work Zaidel and Korennoi[81] described a method for lithium isotope analysis in which the radiation from Li^6 or Li^7 enriched hollow-cathode lamp sources absorbed by the lithium sample nebulized into an air-acetylene flame was measured. The difficulty of the overlap of the Li^6 and Li^7 lines was overcome by making simultaneous measurement of the

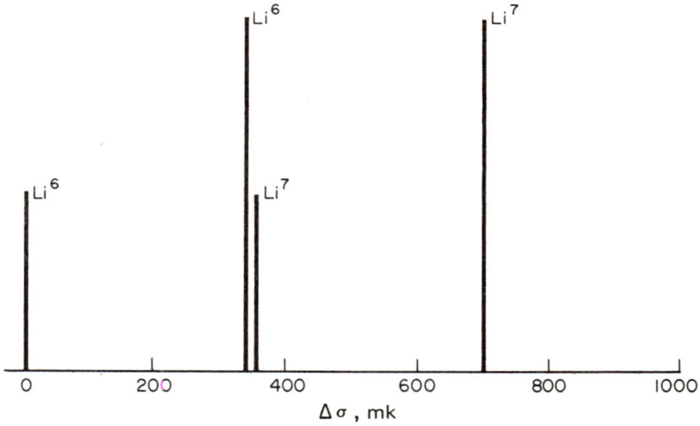

Fig. 14-6. Fine structure of the lithium 670·8 nm doublet.

total lithium content of the flame by flame emission spectroscopy; the isotope composition of samples was determined using standards of known composition to prepare calibration graphs on which Li^6 or Li^7 absorbance was plotted versus the relative concentration of Li^6 (i.e. $[Li^6]/[Li^6]+[Li^7]$).

In another study of lithium isotope composition Zaidel and Korennoi[82] utilized the technique of "isotope filtration". In this method the lithium sample was introduced into the flame and the total line emission at 670·8 nm from both isotopes was passed through Li^7 vapour contained in a heated tube. The Li^7 vapour absorbs the radiation emitted by the Li^7 isotope in the flame and also part of that from the Li^6 isotope owing to the absorption overlap. The Li^6 isotope is then determined from a calibration graph of the intensity of the transmitted (Li^6) radiation versus Li^6

concentration; linear calibration graphs were obtained over the range 2 to 91% Li^6.

Manning and Slavin[83] have described the determination of Li^6 and Li^7 by AAS using an air-acetylene flame as the absorption cell and an oxy-hydrogen source flame into which Li^6 and Li^7 isotopes were introduced separately and alternately. These workers observed that small variations in the total lithium content of sample solutions did not significantly affect the accuracy of the determination. Manning and Slavin have also reported work in which excess Li^7 isotope was added to the sampling flame to absorb the strong Li^6 component (at the weaker overlapping Li^7 absorption line) from the Li^6 emission in the source flame. The measured and known absorption by the excess Li^7 was subtracted from the absorption produced by the sample containing excess Li^7; the net absorbance with a Li^6 source flame after the subtraction is attributable to Li^6 in the sample. Between 0·0001 and 0·1% Li^6 may be determined in this way.

Goleb and Yokoyama[84] have determined the natural abundance ratio Li^7/Li^6 by AAS using Li^6 and Li^7 in separate water-cooled hollow cathodes as the light sources, and a Schüler-Gollnow water-cooled hollow-cathode sputtering chamber to atomize samples. Li^6 and Li^7 absorption readings were obtained by passing radiation from each source independently through the sputtering chamber containing vapour of either samples or standards. The analytical calibration graph was established by plotting the ratio of the absorbances for Li^6 and Li^7 against Li^6 concentration; over the narrow range of concentrations studied (2·5 to 10% Li^6) a straight line calibration graph was obtained. The result obtained for the Li^7/Li^6 natural abundance ratio (12·1 ± 3·3%) compared favourably with the mass spectrometric value of 12·3.

Uranium

The determination of U^{238} and U^{235} isotopes by AAS has been described by Goleb.[85,86] In an early paper the U 591·54 and 502·74 nm lines, for which the isotope shifts are 0·007 nm and 0·01 nm respectively, were employed. With these large isotope shifts no absorption overlap problems were encountered when using a water-cooled, single-isotope, hollow-cathode lamp as light source and a water-cooled, open-ended cylindrical hollow-cathode sputtering cell for samples. Table 14-5 shows the absorption results for standards at the 591·54 and 502·74 nm lines. It was suggested that samples could

TABLE 14-5

Incident uranium light absorbed by a uranium hollow-cathode discharge tube used as an absorption source[a]

Composition of uranium		Incident light absorbed, %	Uranium line, nm
Emission tube	Absorption tube		
99·3% U^{238} 0·7% U^{235}	99·3% U^{238} 0·7% U^{235}	60	591·540
7·0% U^{238} 93·0% U^{235}	99·3% U^{238} 0·7% U^{235}	5	591·540
99·3% U^{238} 0·7% U^{235}	99·3% U^{238} 0·7% U^{235}	42	502·738
7·0% U^{238} 93·0% U^{235}	99·3% U^{238} 0·7% U^{235}	3	502·738

[a] From ref. 85.

be examined directly as solids inserted into the absorption tube. In the report of a later, more detailed study Goleb[86] described a similar procedure using Schüler-Gollnow water-cooled hollow-cathode lamp light sources for each isotope and a water-cooled hollow-cathode absorption cell. An isotope filtration method was finally adopted,

TABLE 14-6

Comparison of atomic absorption and mass spectrographic results for uranium isotope analysis[a]

(Samples sputtered in emission tube)

Sample no.	Uranium-238 (at-%) Atomic absorption	Mass. spectr.	Dev. (at-%)	Rel. dev. (%)
1. 50 (mg) Metal	78·8	79·7	0·90	1·1
2. 50 (mg) Metal	60·3	59·9	0·40	0·7
3. 50 (mg) Metal	51·5	50·4	1·10	2·2
4. 50 (mg) Metal	40·6	39·5	1·10	2·8
5. 50 (mg) Metal	19·6	19·1	0·50	2·6
6. 50 (mg) Metal	5·60	5·59	0·01	0·2
7. 50 (mg) U_3O_8	49·8	50·4	0·60	1·2
8. 36 (mg) U_3O_8 (92 at-% U-236)	2·41	2·38	0·03	1·3
9. 1 (mg) Uranyl nitrate (98 at-% U-234)	0·58	0·60	0·02	3·3

[a] From ref. 86.

however, in which radiation emitted from samples and standards in the hollow-cathode discharge was absorbed selectively by the pure uranium isotope contained in the absorption cell. The percentage transmission was then related to the isotope concentration of the sample placed in the source tube. The calibration graph obtained using this method for the determination of U^{238} shows that good correlation between the experimental data and the theoretical relationship was observed; results for the analysis of samples containing between 0·58 and 78·8% of U^{238} agreed well with mass spectrometric values. Table 14-6 shows a comparison of AAS and mass spectrometric results. An advantage of the isotope filtration technique is that samples can be used in the emission tube in metallic, oxide or compound form.

Boron

Goleb[86] has described an attempt to determine the boron natural abundance ratio, B^{11}/B^{10}, by AAS using similar instrumentation and procedures to those employed for his work with the uranium isotopes. With single isotope sources no difference in absorption at the B 249·6 and 249·7 nm lines was observed when different isotopes were used in the absorption discharge cell. Goleb attributed this to the fact that the half-widths of the lines were greater than the isotopic displacements. L'Vov has commented that the comparatively high temperature of the hollow cathodes and the use of xenon as filler gas may be responsible for the negative results obtained. Because of the high current density and the poor thermal conductivity of boron the gas temperature within the hollow cathodes could be 400-500°C, and the line broadening by xenon would be large compared with that obtained if neon had been employed; both factors would result in overlapping absorption owing to increases in the half-widths of the isotopic components relative to the isotopic displacement.

Helium

L'Vov and Mosichev[87] examined the isotopic composition of mixtures of He^3 and He^4 using a gas discharge absorption cell providing sufficient excitation to obtain an atomic population in the ^3S level. The absorption at 1083·0 nm due to the transition $2^3S - 2^3P_{0,1,2}$ may then be observed using a He^3 or He^4 high-frequency gas discharge lamp as light source. This technique is necessary to avoid the use of the resonance lines of helium which lie in the vacuum ultraviolet. The isotope shift of each of the three lines of the multiplet is 1180 mK and is similar to the separation of the lines of the multiplet (1070 mK); therefore some overlap of the isotope components occurs. Self-absorption of the emission line from the source was found to lead to deviation from linearity in the calibration graph for He^4 but this did not prevent successful analysis under carefully controlled experimental conditions.

14.4 LASER-EXCITED ATOMIC FLUORESCENCE SPECTROSCOPY

The source power delivered to the atom cell in AFS is relatively low when vapour-discharge lamps, hollow-cathode lamps, electrodeless-discharge tubes or continuum sources are employed. As the magnitude of the fluorescence signal is proportional to the source power transferred to the sample atoms in the flame or non-flame cell employed, any method whereby source power delivery may be increased would have a beneficial effect on sensitivity in AFS. As pointed out by Fraser and Winefordner,[88] the use of a stable, repetitively pulsed source of excitation with a small duty cycle, i.e. small ratio of on-to-off time, might permit an increase in the fluorescence signal-to-noise ratio owing primarily to decreased noise; during the short on-time the signal may be of the same magnitude as the average signal obtained using a continuously operated source, whereas the noise would be small because of the small number of random photo-detector pulses from dark current, flame background emission, etc.[89] Thus the high peak power output from a repetitively pulsed, tunable laser should provide for high power delivery to an analytical atom cell and permit the attainment of high detection sensitivity. This type of system should in most respects be the ideal source for analytical AFS; the general characteristics of the most suitable lasers available at present have been discussed in Section 5.4.3.

Denton and Malmstadt[90] have described the use of a ruby laser-pumped tunable organic dye laser for excitation of atomic fluorescence of barium in a flame. The experimental arrangement employed is shown in Fig. 14-7. The ruby laser was fitted with a cryptocyanine saturable Q-switch cell; second harmonic generation was achieved in a 10 mm KDP crystal. A 5×10^{-4} M solution of 2′,7′-dichlorofluorescein at pH 11 was used in the 10 mm dye laser cell. Beam splitter and photodiode assemblies were employed to monitor the output power of both the ruby and dye laser. A spectrograph was employed to assist in setting up the system. The ruby rod was allowed to equilibrate thermally for one minute between each firing. The problem of scatter of the intense source radiation by water droplets or salt particles in the flame was overcome by use of an ultrasonic nebulizer-burner system which produced a very fine mist to ensure complete desolvation and conversion of the sample into atomic vapour. A linear atomic

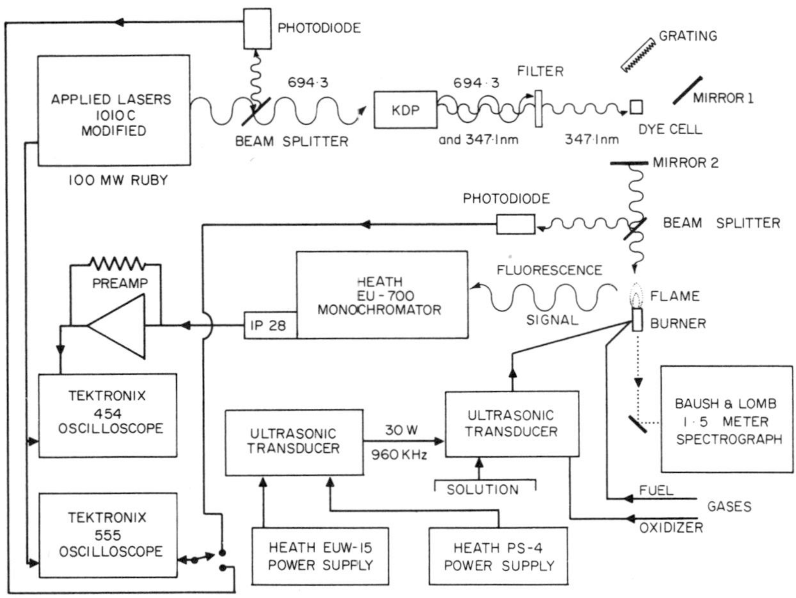

Fig. 14-7. Diagram of experimental arrangement for laser excited AFS used by Denton and Malmstadt.[90]

fluorescence calibration graph was constructed in the concentration range 1 to 10 ppm of barium at the Ba 553·5 nm resonance line.

Fraser and Winefordner[88,91] have reported the use of a faster repetition rate pulsed system utilizing a nitrogen laser-pumped tunable dye laser for excitation of atomic fluorescence from a wide range of elements in air-hydrogen, air-acetylene and nitrous oxide-acetylene flames. A block diagram of the experimental system used is shown in Fig. 14-8. The peak power output of the dye laser over a spectral band-width of 0·1–1 nm is *ca.* 10 kW, the average power output of the dye laser is *ca.* 0·001 W, the pulse half-width is 2–8 ns, and the repetition rate is 1–30 Hz depending on the dye used. Ten dyestuffs were used, each covering a spectral range of about 30 nm in order to provide a system tunable over the range 360 to 650nm. Figure 14-9 shows the output radiant flux and wavelength range covered by each of the ten laser dyes used. As the output of the tunable dye laser can be varied over a range of 10-30 nm, depending on the dye, it is possible to wavelength scan the output of the dye laser. This provides a convenient means of correcting for any background scatter signals and obtaining spectral information by the

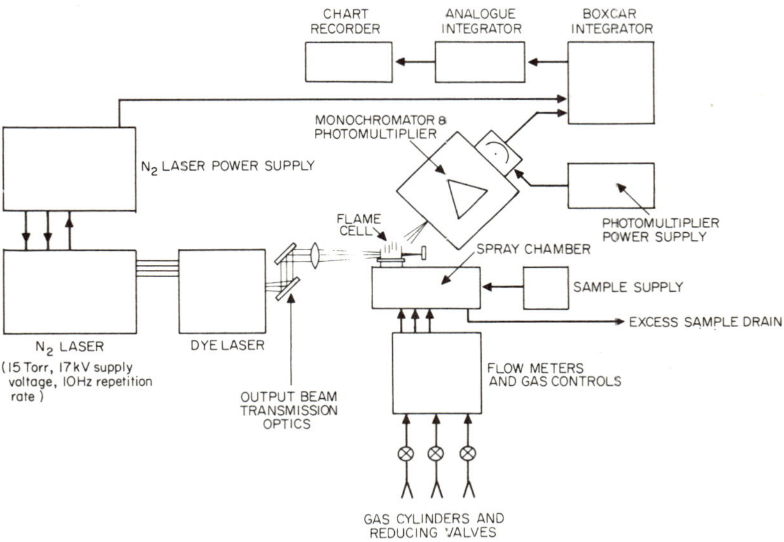

Fig. 14-8. Block diagram of experimental system used for laser-excited flame AFS by Fraser and Winefordner.[91]

use of a fast response photomultiplier tube and a boxcar integrator capable of aperture gate widths of the order of 10 ns. Figure 14-10 demonstrates the wavelength scanning facility and shows the spectral distribution (obtained by varying the dye laser wavelength) over a wavelength range of *ca.* 1 nm centred on 422·7 nm and the relative signal magnitude (mV or V) obtained by deliberate scattering of the incident radiation or nebulizing water and Ca solution into an air-hydrogen flame. With this laser technique, therefore, atomic fluorescence is effectively excited with a wide line source with a power output of 10 kW, i.e. during the laser "on-time" fluorescence is excited with a source having an effective black body temperature of 75,000 K at 400 nm and during the "off-time" no signal or noise

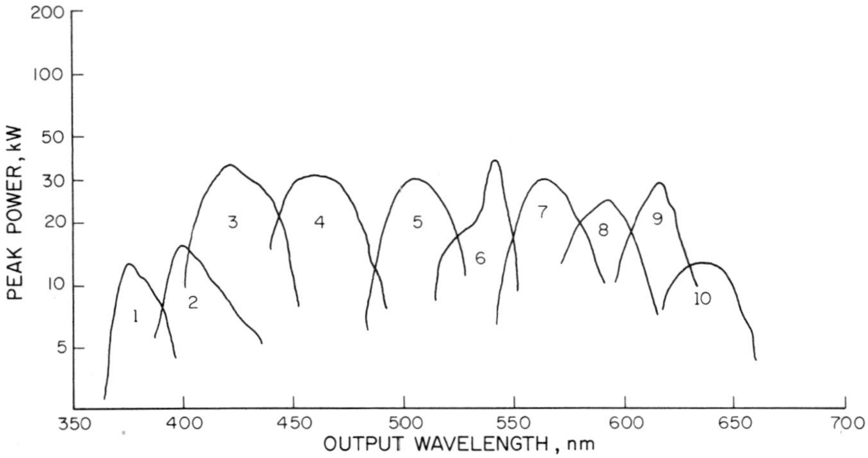

Fig. 14-9. Output radiant flux and wavelength range of laser dyes used by Fraser and Winefordner.[91]

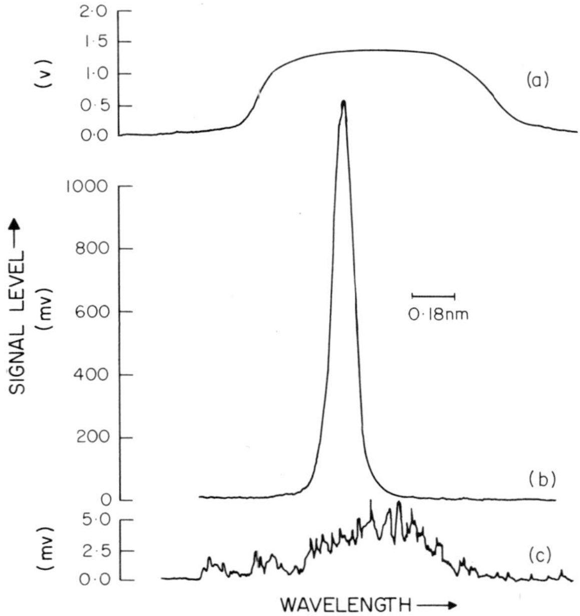

Fig. 14-10. (a) Spectrometer slit function obtained by wave-length scanning the dye laser radiation scattered from a Teflon sheet used in place of the flames (monochromator wavelength set at 422·7 nm). (b) Spectral distribution of fluorescence radiation of Ca (1000 μg/ml) obtained by varying the dye laser wavelength over about 1 nm surrounding 422·7 nm (monochromator wavelength set at 422·7 nm). Solution aspiration rate is 4 ml/min. (c) Spectral distribution of scattered radiation by the flame gases obtained by varying the dye laser wavelength over about 1 nm surrounding 422·7 nm (monochromator wavelength set at 422·7 nm).

is being measured. With a duty cycle of *ca.* 10^{-7}, i.e. with the laser "on" for only 1 part in 10^{7} parts, random noise effects from flame flicker, electronic measurement noise and fluctuations in sample introduction are negligible at most concentrations. Fraser and Winefordner have pointed out that the resonance fluorescence signal: noise ratio at high concentrations is dependent on shot noise in the signal owing to the randomness in the emission of photons, pulse-to-pulse amplitude variations in the dye laser output and also scatter noise. At low concentrations, i.e. near the limit of detection, the resonance fluorescence signal:noise ratio is primarily dependent on scatter noise from particles and optical inhomogeneities in the flame gases.

Linear analytical fluorescence calibration graphs were constructed by Fraser and Winefordner for fluorescence of Al, Ca, Co, Cr, Ga, In, Fe, Mn, Mo, Ni, Sr, Ti and Tl over three or four orders of magnitude concentration range. Resonance and non-resonance fluorescence effects were utilized. The corresponding detection limits obtained in this work are shown in Table 14-7. The values of the detection limits obtained for Al, Ga, In and Ti are equal to, or better than, any reported values by any other flame spectroscopic method, and the values for Cr, Mn, Ni, Mo and Tl are within a factor of 10 of those obtained by any other flame method; of the remaining values only those for Ca, Co, Sr and Fe are not within a factor of ten of other flame spectroscopic methods. Although it might be expected from the use of a "wide-line" source (0·1–1·0 nm) for excitation of fluorescence, for the elements investigated no appreciable spectral interference resulting from simultaneous excitation of atomic fluorescence from other elements was observed in this work. The limits of detection obtained may be substantially improved by reducing the laser beam scatter noise to the extent where the principal noise limitation arises from photon noise due to the atomizer background and from detector noise. A limitation of the present dye laser excitation system is the lower useful wavelength limit of 360 nm which precludes the possibility of excitation of intense fluorescence for many elements.

With the laser excitation system two-photon excitation fluorescence was observed for Cd and Zn.[91] In this process two photons are absorbed simultaneously of combined energy equal to the difference in energy between the two energy levels involved in the absorption process. Thus the two-photon excitation fluorescence of Cd involves excitation of analyte atoms with intense radiation at 457·6 nm (or of

TABLE 14-7

Limits of detection by laser excited atomic fluorescence[91]

Element	Type of fluorescence[a]	Type of flame Ox(1/min)/F(1/min)[b]	Mono-chromator slit, μm	Dye type[c]	Analytical line, nm[d]	Energy levels, cm^{-1} [e]	*gf* Value[f]	Limit of detection, ppm[g]	Upper limit, ppm[h]
Al	RF	N_2O(14·5)/C_2H_2(6·5)	150	17/66B	394·4	0-25348	0·15	1×10^{-1}	2×10^{2}
	E-RF	N_2O(14·5)/C_2H_2(6·5)	150	17/66B	396·1	112-25348	0·31	3×10^{-2}	1×10^{2}
	N-DLF	N_2O(14·5)/C_2H_2(6·5)	150	17/66B	394·4/396·1	0-25348/112-25348	0·15/0·31	5×10^{-3}	2×10^{2}
Ca	RF	Air(7·5)/H_2(15·2)	250	17/52	422·7	0-23652	0·28	5×10^{-3}	5×10^{1}
Cr	RF	Air(7·5)/H_2(11·8)	150	17/66D	357·8	0-27935	1·60	2×10^{-2}	2×10^{1}
	RF	Air(7·5)/H_2(11·8)	150	17/66D	359·3	0-27820	1·40	2×10^{-2}	2×10^{1}
Co	E-RF	N_2O(12·0)/C_2H_2(6·4)	100	17/66D	357·5	816-28777	0·22	5×10^{1}	ND[i]
	AS-DLF	N_2O(12·0)/C_2H_2(6·4)	100	17/66D	347·4/357·5	0-28777/816-28777	0·64/0·22	2×10^{-1}	2×10^{2}
Ga	RF	Air(7·5)/H_2(20·0)	150	17/66B	403·2	0-24788	0·24	1×10^{-1}	5×10^{2}
	N-DLF	Air(7·5)/H_2(20·0)	150	17/66B	403·2/417·2	0-24788/826-24788	0·24/0·53	2×10^{-2}	2×10^{1}
In	RF	Air(7·5)/H_2(18·3)	250	17/66B	410·4	0-24373	0·47	1×10^{-2}	5×10^{1}
	N-DLF	Air(7·5)/H_2(18·3)	250	17/66B	410·5/451·1	0-24373/2213-24373	0·47/0·66	2×10^{-3}	5×10^{1}
Fe	RF	Air(7·5)/H_2(11·8)	250	17/66D	372·0	0-26875	0·52	3×10^{-1}	2×10^{3}
Mn	RF	Air(7·5)/H_2(20·0)	150	17/66B	403·0	0-24802	0·33	1×10^{-2}	2×10^{1}
Mo	RF	N_2O(12·0)/C_2H_2(6·4)	100	17/66D	379·8	0-26321	1·30	1×10^{1}	1×10^{2}
	RF	N_2O(12·0)/C_2H_2(6·4)	100	17/66D	390·3	0-25614	0·47	3×10^{-1}	1×10^{2}
Ni	E-RF	Air(14·3)/C_2H_2(2·6)	200	17/66D	361·0	880-28569	0·15	2×10^{0}	5×10^{2}
	E-AS-DLF	Air(14·3)/C_2H_2(2·6)	200	17/66D	361·0/352·4	880-28569/205-28569	0·15/0·85	5×10^{-2}	2×10^{2}
Sr	RF	Air(7·5)/H_2(15·2)	150	17/53	460·7	0-21698	0·27	1×10^{-2}	2×10^{1}
Ti	E-RF	N_2O(14·5)/C_2H_2(6·5)	150	17/66B	399·8	387-25388	1·50	1×10^{-1}	2×10^{2}
Tl	RF	Air(14·3)/C_2H_2(2·3)	200	17/66D	377·6	0-26478	0·22	1×10^{0}	5×10^{1}
	N-DLF	Air(14·3)/C_2H_2(2·3)	200	17/66D	377·6/535·0	0-26478/7793-26478	0·22/0·92	2×10^{-2}	1×10^{2}

[a] RF = resonance fluorescence involving ground state. E-RF = thermally assisted resonance fluorescence involving excited lower state. N-DLF = Stokes direct line fluorescence involving ground lower state in absorption process. AS-DLF = anti-Stokes direct line fluorescence involving ground lower state in fluorescence process. E-AS-DLF = excited anti-Stokes direct line fluorescence involving excited lower states for both absorption and fluorescence processes.

[b] Ox(1/min)/F(1/min) = oxidant (flow rate)/fuel (flow rate).

[c] Dye types as designated by manufacturer of dye laser (AVCO Everett Research Laboratories).

[d] If only one line is listed, the absorption and fluorescence lines are identical.

[e] Taken from C. H. Corliss and W. R. Bozman, "Experimental Transition Probabilities for Spectral Lines of Seventy Elements", NBS Monograph 53, U.S. Government Printing Office, Washington, D.C., 1962 (and from W. F. Meggers, C. H. Corliss and B. F. Scribner, "Tables of Spectral-Line Intensities", NBS Monograph 32, Part I, U.S. Government Printing Office, Washington, D.C., 1961).

[f] Taken from C. H. Corliss and W. R. Bozman (see footnote *e*).

[g] Concentration resulting in a signal-to-rms noise of 2.

[h] Highest concentration of analyte resulting in a signal deviating from linearity by no more than 5%.

[i] Not determined.

two different lines of comparable total energy) and observation of the cadmium resonance line at 228·8 nm.

The nitrogen pumped dye laser system has also been used by Winefordner and co-workers in a study of the laser excited atomic

TABLE 14-8

Detection limits obtained for the rare-earth elements by laser-excited atomic and ionic fluorescence spectrometry (from ref. 92)

Element	Wavelength, nm[a]	Detection limit, ppm AFS[b,c]	Detection limit, ppm IFS[c]
Ce	569·92	n.d.	
	394·21 (371·64)		0·5
Dy	418·68 (418·68)	0·6	
	353·60 (364·54)		0·3
Er	400·80 (400·80)	0·5	
	337·27 (369·26)		2·5
Eu	462·72 (459·40)	0·02	
	393·05 (420·50)		0·2
Gd	368·41 (368·41)	5	
	336·22 (376·84)		0·8
Ho	410·38 (405·39)	0·15	
	—— ——		n.d.
Lu	513·51 (465·80)	3	
	—— ——		n.d.
Nd	489·69 (463·42)	2	
	424·74 (430·36)		40
Pr	495·14 (495·14)	10	
	430·58 (427·23)		1
Sm	429·67 (375·64)	0·6	
	373·91 (366·14)		0·15
Tb	433·84 (432·65)	1·5	
	350·92 (370·28)		0·5
Tm	409·42 (371·79)	0·1	
	379·58 (376·13)		5
Yb	346·44 (398·80)	0·01	
	328·94 (369·42)		0·03

[a] Laser excitation wavelength given in parentheses.
[b] 1000 ppm of potassium as KCl added to suppress ionization.
[c] n.d. = not detectable at 500 ppm.

fluorescence determination of the rare-earth elements in a premixed nitrous oxide-acetylene flame.[92] Both resonance and non-resonance fluorescence was studied; the appreciable ionization of many of the rare-earth elements in the nitrous oxide-acetylene was also found to permit observation of resonance and non-resonance fluorescence for the ionic species. The detection limits obtained on nebulization of aqueous solutions of the rare-earths are shown in Table 14-8. These values, except for erbium, are within a factor of three of the best values reported by AAS. The authors point out, however, that because in the majority of cases non-resonance fluorescence was utilized, AFS should be superior to AAS for the practical determination of rare-earth mixtures, i.e. when spectral interferences may possibly be encountered. The laser-excited AFS detection limits reported to date are inferior to those attainable by flame emission spectroscopy for six elements (Dy, Er, Eu, Ho, Nd and Pr) and are similar to the FES values for the other elements. As few spectral interferences occur in the determination of the rare-earths by flame emission spectroscopy in high-temperature flames (owing to the simplicity of the emission spectra obtained[93,94,95]) at the time of writing FES is probably preferable to AAS or AFS for the determination of these elements.

14.4.1 Saturation Effects with Laser Excitation

In flame AAS and AFS with continuum sources and with dilute atomic vapours the integral of the atomic absorption coefficient, A_T, is generally dependent only on the wavelength and the optical density of the absorber (equation 3-17). This is strictly correct, however, only in the limit of zero incident light flux, i.e. when the incident radiation does not cause significant population of the excited state relative to the ground state. This is a good approximation only for low incident light fluxes available from conventional sources such as hollow-cathode or electrodeless-discharge lamps. When the source flux is very high, however, as in laser excitation, this expression does not provide an accurate description of the attenuation of the light beam. Under these conditions *the absorption coefficient becomes also a function of the incident radiant flux density* (non-linear absorption).[96-103] The intense radiation may then induce in the sample a state of near-saturation of the energy levels in which the excited state population becomes substantially equal to that of the ground state. Winefordner and co-workers,[102,103] and Piepmeier[104,105] have described the theoretical and practical consequences of the occurrence of saturation in AFS for a broad-band laser excitation source and a monochromatic laser excitation source respectively. With a high incident radiation flux, where the integrated absorption coefficient becomes a function of the source irradiance, a more correct expression for the integrated absorption coefficient at low concentrations is given by[97,102]

$$A_T = \int (1 - \exp(-k_v \,.\, E^s_v \,.\, l)\, dv = c_1 N_T \left[\frac{(E^s_{vmn})_{\text{saturation}}}{E^s_{vmn} + (E^s_{vmn})_{\text{saturation}}} \right] \tag{14-1}$$

where (E^s_{vmn}) saturation is the source spectral irradiance at v_{mn}, the frequency of the transition, for which the steady state integrated absorption coefficient is reduced by a factor of two over the low light flux absorption coefficient,[97] E^s_{vmn} is the spectral irradiance of the excitation source, N_T is the total number of atoms per unit volume of flame gases, and c_1 is a factor including fundamental constants and atomic parameters.

If $E^s_{vmn} \ll (E^s_{vmn})_{\text{saturation}}$ the equations given in Chapter 4 relating fluorescence signal strength to source irradiance apply. However, when the available source irradiance E^s_{vmn} is much greater than that required for saturation, $(E^s_{vmn})_{\text{saturation}}$, the expression for the fluorescence flux, F_{nm}, is given for a broad-band laser source by

$$F_{nm} = A_{mn} \,.\, N^* = c_2 Y_{mn} N_T \left[\frac{(E^s_{vmn})_{\text{saturation}}}{hv_{mn}} \right] \tag{14-2}$$

Which is independent of E^s_{vmn}, the source irradiance. A_{mn} is the Einstein spontaneous transition probability, N^* is the number of atoms per unit volume in the excited state (in excess of the thermal population of the level), Y_{mn} is the quantum yield of the fluorescence process and c_2 is a proportionality constant.

Thus above a certain value of the source irradiance there is a *non-linear* dependence between the fluorescence flux and the spectral irradiance of the source; the condition is approached in which the fluorescence flux becomes independent of the source spectral irradiance. Effectively the vapour becomes transparent to a further increase in E^s_{vmn}. There is still a linear dependence for a dilute vapour, however, on the concentration of absorbing atoms. This effect of working close to saturation conditions, and the observation by Piepmeier[104,105] and Winefordner *et al.*[102-103] that under these conditions the fluorescence flux is influenced very little by collisional quenching, have several consequences of great importance from an analytical point of view. These are:

(1) The fluorescence signal is not greatly influenced by the source stability. The saturation effect above a certain value of E^s_{vmn}

ensures that any pulse-to-pulse variation in the source irradiance does not affect the stability of the fluorescence signal.

(2) The linearity of the fluorescence working curve is extended to higher concentrations. At saturation the medium becomes transparent at v_{mn} because the excited and ground state populations are similar and further absorption in the irradiated volume cannot occur. If the irradiance of the source is such that the atomic system can be kept at saturation for any value of N, the fluorescence flux will be linearly related to N, and the calibration curves will have a slope of unity even at high optical densities. If the medium is transparent self-absorption cannot take place, also the source irradiance is no longer a function of the length along the absorption path, L (see Chapter 4 and Fig. 4-10). Obviously this consideration of transparency under saturation conditions applies only for the irradiated volume; the factors of self-absorption and self-reversal must still be considered for the post-filter effect produced in any unilluminated volume between the fluorescing volume and the detector. Winefordner *et al.*[106] have demonstrated this effect of increased linear range by varying the spectral irradiance of the laser source with neutral density filters and/or by defocusing the laser spot on the flame. Figure 14-11 shows the resonance fluorescence growth curves for indium and strontium obtained at high and low spectral irradiance obtained by defocusing the source.

(3) The proportional dependence of fluorescence signal on the quantum efficiency which is observed at low irradiance is removed under saturation conditions. Thus, provided that the atomization efficiency does not change, the magnitude of the saturated fluorescence signal would not be expected to be greater in the oxygen-argon-hydrogen flame than in a hydrocarbon flame containing nitrogen in spite of the large difference in quantum yield for these flames. Thus the nitrous oxide-acetylene flame is just as effective as other flames from this viewpoint and can be used for AFS to maximize atomization efficiency and minimize chemical interferences without the penalty of lower quantum yield experienced under low irradiance conditions.

Piepmeier[104] has pointed out that the use of high source irradiance with a laser may result in deterioration in detection limits

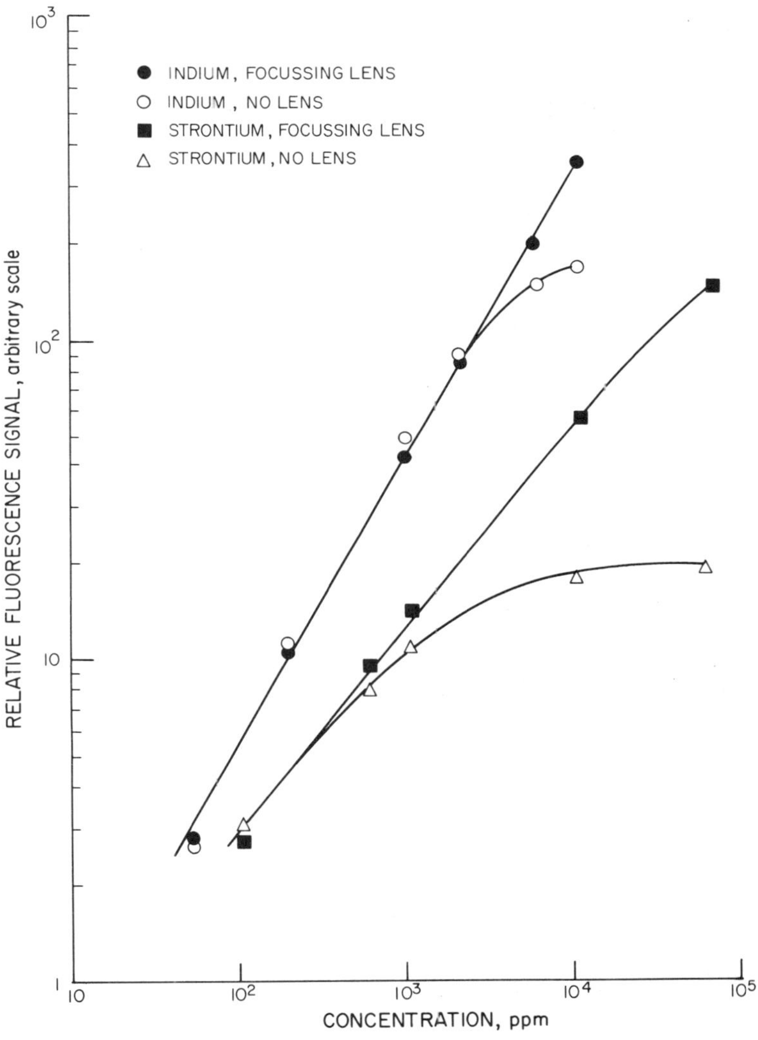

Fig. 14-11. Resonance fluorescence calibration curves for indium and strontium. The fluorescence signals for the different curves are not reported on the same scale.

owing to the fact that scattered light is proportional to the source irradiance. The greater possibilities for employment of non-resonance transitions when a laser source is used for excitation, however, should minimize this possible disadvantage. In addition it is possible to improve the fluorescence-to-scattering intensity ratio by enlarging the laser beam in the flame.[102] It is, however, necessary to reach a compromise between scattering and signal stability. As there is a limiting value of the fluorescence radiance, even if the source irradiance is increased indefinitely, it is apparent that it is the signal : noise ratio rather than the signal itself which should be maximized.

Current developments in AFS using laser excitation lend credence to earlier claims that this type of source is probably ideal in many respects; the fundamental advantages of using high source irradiance described above should certainly lead to the more widespread adoption of these sources for practical AFS measurements.

14.5 MEASUREMENT OF FLAME TEMPERATURES BY AAS AND AFS

In the study of atomization processes and interference phenomena encountered in analytical atomic spectroscopy it is often necessary to have a knowledge of the temperature within the flame or non-flame atom cell employed. It is frequently not satisfactory to assume the temperature in the atom cell under investigation from literature values for similar cells. In flames, for example, the measured temperature will depend on the values of the gas flow-rate and the design of burner head used and considerable misinterpretation may occur if literature values for a particular flame temperature are used to treat experimental data obtained with different apparatus. Except for some thermocouple methods of measurement, which may often be unsatisfactory owing to the influence of the sensing element on the atom cell environment and measured temperature, the two most widely used spectroscopic techniques of temperature measurement are the line reversal and two-line emission procedures. Both of these methods require special apparatus and are frequently time-consuming.[107,108,109] Additionally both methods require that atomic emission can be stimulated from a species (often Na in the line-reversal technique and Fe for two-line emission) added to the atom cell. Although little work has been reported to date concerned with the temperature of atomic vapours produced in non-flame cells, this latter requirement of these methods renders them of little value for these purposes. There exists, therefore, a need for simple spectroscopic techniques of temperature measurement which are applicable to flame and non-flame atom cells and which require mainly absorption and fluorescence instrumentation similar to that used by many workers undertaking studies of atomization processes and interference effects. Two such techniques have been reported; these depend on the measurement of relative absorption and fluorescence respectively at two atomic lines of the analyte element introduced into the atom cell.

14.5.1 Two-line Atomic Absorption Method

Browner and Winefordner[110] have described a straightforward atomic absorption technique for temperature measurement using a simple uncalibrated continuum light source (quartz iodine lamp). The method is applicable directly, with no modification required, to any atomic absorption spectrometer possessing a monochromator that will allow the use of a spectral band-pass of *ca.* 0·04 nm.

The two-line absorption method depends on the measurement of the absorption at each of two lines for an element introduced into the sample flame. One of these lines originates from the ground state and the other is chosen to originate from a thermally populated level less than 1 eV above the ground state. The ratio of these absorption signals may then be applied to calculate the effective electronic excitation temperature of the atoms within the flame. At low optical density the relative absorption, α, of the radiation from a continuum light source by atoms in the light path is given by

$$\alpha = \frac{\phi_A}{\phi_0} = \frac{\pi e^2 l \lambda_0^2 n f}{m s c^2} \tag{14-3}$$

where e and m are the electronic charge and mass, c is the velocity of light, l is the absorption path length, λ_0 is the peak absorption wavelength, n is the concentration of atoms at the lower energy level involved in the transition, f is the absorption oscillator strength, ϕ_0 is the radiant flux of the continuum source transmitted with only a blank introduced into the flame, ϕ_A is the radiant flux from the source absorbed by the analyte atoms in the flame, and s is the spectral band-width of the monochromator. Equation (14-3) only holds provided that s is greater than, or equal to, $4\Delta\lambda_a$, where $\Delta\lambda_a$ is the total half-width of the absorption line.

At high optical densities, equation (14-3) becomes

$$\alpha = \frac{1}{s}\left[\frac{2\pi e^2 l \lambda_0^2 n f a\, \Delta\lambda_D}{m c^2 (\ln 2)^{1/2}}\right]^{1/2} \tag{14-4}$$

where $\Delta\lambda_D$ is the Doppler half-width of the line, and a is the damping constant (see Chapter 3.2). By substitution for a and neglecting the natural line width, equation (14-4) becomes

$$\alpha = \frac{1}{s}\left[\frac{2\pi e^2 l\lambda_0^2 n f \Delta\lambda_L}{mc^2}\right]^{1/2} \tag{14-5}$$

where $\Delta\lambda_L$ is the Lorentz (collisional) half-width.

Equations (14-4) and (14-5) are not valid at very high optical densities (α greater than *ca.* 0·3) because at these values parts of the absorption line wings may extend beyond the spectral band-width of the monochromator.

When thermal equilibrium exists in the atomic vapour the relative populations n_0 and n_1 of two electronic levels 0 and 1 are given by

$$\frac{n_1}{n_0} = \frac{g_1}{g_0}\exp\left[\frac{E_0 - E_1}{kT}\right] \tag{14-6}$$

where E_0 and E_1 are the energy levels of states 0 and 1 and g_0 and g_1 their statistical weights, k is the Boltzmann constant and T is the absolute temperature. Atoms with degenerate ground states have one or more electronic levels close to the ground state. For a particular atom with a degenerate ground state it is possible to relate accurately the relative absorption at the resonance line and at a non-ground-state line to the relative populations of these levels by combining equations (14-4) and (14-6). At low optical densities the temperature is then given by

$$T = \frac{E_1 - E_0}{k\ln\left[\left(\frac{g_1 f_1}{g_0 f_0}\right)\left(\frac{\lambda_1}{\lambda_0}\right)^2\left(\frac{\alpha_0}{\alpha_1}\right)\right]} \tag{14-7}$$

and at high optical densities by

$$T = \frac{E_1 - E_0}{k\ln\left[\left(\frac{g_1 f_1}{g_0 f_0}\right)\left(\frac{\lambda_1}{\lambda_0}\right)^2\left(\frac{\alpha_0}{\alpha_1}\right)^2\left(\frac{\Delta\lambda_{L_1}}{\Delta\lambda_{L_0}}\right)\right]} \tag{14-8}$$

where α_0 and α_1 refer to the relative absorption signals at the lines of wavelengths λ_0 and λ_1, whose lower states have energies E_0 and E_1 and whose collisional half-widths are $\Delta\lambda_{L_0}$ and $\Delta\lambda_{L_1}$. To achieve optimum signal levels for the measurement, the energy separation of states 0 and 1 should not be too great (less than 1 eV for temperatures less than 4000 K) and the absorption oscillator strengths must be comparable. Only relative values of the f values are required in order to obtain an absolute electronic excitation temperature at low optical densities. If the low optical density region is used, however, this limits both α_0 and α_1 to *ca.* $< 0{\cdot}03$; at these low values of α unacceptable errors in the temperature measurement may result. If the high optical density region is used it is apparently necessary to know the ratio of the collisional half-widths. Browner and Winefordner[110] were able to show from curve of growth measurements with In in a number of flames of different temperatures and composition that the ratio of the collisional half-widths, $\Delta\lambda_{L_1}/\Delta\lambda_{L_0}$, given by the intersection point of the low and high

TABLE 14-9

Experimental values of Lorentz half-width ratios from growth curves for indium (from ref. 110)

Flame	Temperature, K	$\Delta\lambda_{L_{451\cdot1}}/\Delta\lambda_{L_{410\cdot2}}$ $(=\alpha_1'/\alpha_0')^a$	α_1/α_0 Low O.D.^{a}	α_1/α_0 High O.D.^{a}
$Ar/O_2/H_2$	2098 (R)^b	0·56	0·56	0·58
	2250 (R)	0·69	0·71	0·66
Air/C_2H_2	2430 (R)	0·76	0·78	0·73
	2470 (R)	0·78	0·77	0·74
N_2O/C_2H_2	2995 (Fe)^c	0·90	0·85	0·91

^{a} Values of $\Delta\lambda_{L_{451\cdot1}}/\Delta\lambda_{L_{410\cdot2}}$ from the growth curves and of α_1/α_0 measured at low optical densities (O.D.) have an error of approximately 10%. The error in the ratio of α_1/α_0 at high optical densities (O.D.) is about 1%.

^{b} R = reversal temperature.

^{c} Fe = iron two-line temperature.

optical density asymptotes for each curve, was identical with α_1/α_0 within the limits of experimental accuracy. Thus

$$\frac{\Delta\lambda_{L_1}}{\Delta\lambda_{L_0}} = \frac{\alpha_1'}{\alpha_0'} = \frac{\alpha_1}{\alpha_0} \tag{14-9}$$

where the values of α' refer to the intersection points of the high- and low-density asymptotes. The experimental data obtained in different flames to show that α_1'/α_0' is equal to α_1/α_0 for indium are shown in Table 14-9. On substitution for $\Delta\lambda_{L_1}/\Delta\lambda_{L_0}$, equation (14-6) for high optical densities becomes identical to equation (14-5) for low optical densities. This equation may then be used up to α values of *ca.* 0·3 without the need to evaluate $\Delta\lambda_{L_1}/\Delta\lambda_{L_0}$. In order to obtain high sensitivity $dT/d(\alpha_0/\alpha_1)$ should be small and thus widely different α values should be measured. This leads to poor precision in the ratio α_0/α_1, however, and to obtain good precision the approximate range $1{\cdot}1 \gtrsim \alpha_0/\alpha_1, \gtrsim 5$ was recommended. When the two-line absorption method is applied with Ga, In and Tl this results in restriction of the working ranges of temperature to Ga, 650-2200 K; In, 1200-3500 K and Tl, 3650-7750 K. Table 14-10 lists some temperatures obtained by Browner and Winefordner for different flames using the two-line absorption method and values obtained by the same workers for the same flames using other spectroscopic methods. Systematic and random errors of about 1% were claimed for the technique.

TABLE 14-10

Temperature of flames measured by two-line atomic absorption method (from ref. 110)

Flame	Gas flow rates ($1\ min^{-1}$)		Temperature (K)	
	Fuel	Oxidant	Absorption method	Other methods[a]
$Ar/O_2/H_2$	6·2	O_2-2·1[b] Ar-13·2	2078	2098 (R)
	6·2	O_2-2·6[b] Ar-13·2	2260	2250 (R)
Air/C_2H_2	1·3	9·5[b]	2452	2430 (R)
	1·3	9·5[c]	2473	2470 (R)
N_2O/C_2H_2	4·5	8·2[d]	2961	2995 (Fe)

[a] R = reversal temperature, Fe = iron two-line temperature.
[b] Flame shielded Meker burner. Same gas flows through outer and inner flames.
[c] 50-mm slot burner.
[d] 10-mm diameter capillary burner.

The two-line absorption technique should prove to be of great value in the measurement of electronic excitation temperatures in non-emissive atom cells such as graphite rod and tube devices.

14.5.2 The Two-line Atomic Fluorescence Method

Omenetto, Benetti and Rossi[111] have described four possible methods whereby flame electronic excitation temperature measurements may be made by atomic fluorescence spectroscopy. These methods may be related to the processes shown schematically in Fig. 14-12 for thallium. When the flame is irradiated by a continuum source (compact arc or broad line vapour-discharge lamp) from which suitable excitation wavelengths are selected with filters, the following possibilities present themselves:

(a) It may be shown[112] that the ratio of the anti-Stokes fluorescence (F_{20}) to direct line fluorescence (F_{21}) at low atomic concentrations (no self-absorption) is given by

$$\frac{F_{20}}{F_{21}} = \frac{E(\lambda_{12})f_{12}}{E(\lambda_{02})f_{02}} \cdot \left(\frac{\lambda_{12}}{\lambda_{02}}\right)^3 \cdot \frac{n_1}{n_0} \cdot \frac{\phi_{20}}{\phi_{21}} \qquad (14\text{-}10)$$

where the $E(\lambda)$ values represent the spectral irradiance of the excitation source at the wavelengths concerned, the f values are the

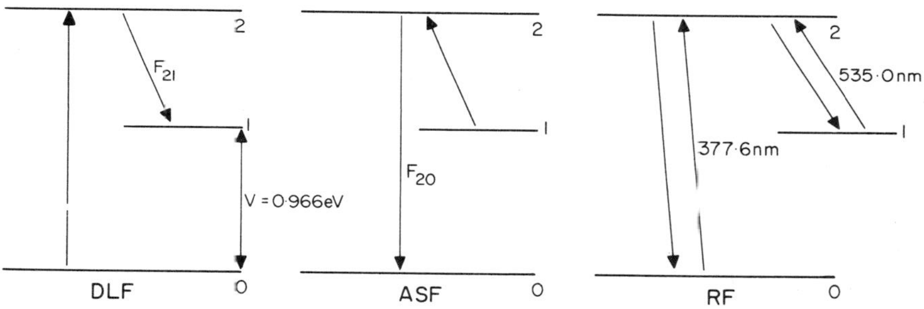

Fig. 14-12. Schematic diagram of the fluorescence process for thallium in flames. DLF = direct-line fluorescence; ASF = anti-Stokes fluorescence; RF = resonance fluorescence.

corresponding oscillator strengths of the lines, λ_{12} and λ_{02} are the central wavelengths of the observed transitions, n_1 and n_0 are the atomic concentrations in the corresponding states and ϕ_{20} and ϕ_{21} are the quantum efficiencies of the fluorescence processes considered. Alkemade[112] has also demonstrated that $\phi_{20}/\phi_{21} = A_{20}/A_{21}$ where A_{20} and A_{21} are the Einstein transition probabilities. Furthermore, for thermal equilibrium conditions n_1 and n_0 are related via the Boltzmann distribution. Thus equation (14-10) becomes

$$\frac{F_{20}}{F_{21}} = \frac{E(\lambda_{12})}{E(\lambda_{02})} \cdot \left[\frac{\lambda_{12}}{\lambda_{02}}\right]^5 . 10 \exp\left[\frac{-5040V_1}{T_f}\right] \qquad (14\text{-}11)$$

where V_1 is the excitation energy of level 1 above the ground state. Hence, solving for T_f the flame temperature,

$$T_f = \frac{5040V_1}{\left[\ln\frac{E(\lambda_{12})}{E(\lambda_{02})} + 5\ln\frac{\lambda_{12}}{\lambda_{02}} - \ln\frac{F_{20}}{F_{21}}\right]} \qquad (14\text{-}12)$$

The experimental data required to calculate the flame temperature are the spectral irradiances of the source at the excitation wavelengths and the intensities of the anti-Stokes and direct-line fluorescence signals.

(b) The thermodynamic radiation law implies that the ratio $E(\lambda_{12})\backslash E(\lambda_{02})$ corresponds to the temperature T of a black body. A temperature T_c, often called the colour temperature, can therefore be formally assigned to the continuum source. Using Wien's law it can be written

$$\frac{E(\lambda_{12})}{E(\lambda_{02})} = \frac{\lambda_{12}^{-5} \exp\left(\dfrac{-h\nu_{12}}{kT_c}\right)}{\lambda_{02}^{-5} \exp\left(\dfrac{-h\nu_{02}}{kT_c}\right)}$$

$$= \lambda_{12}^{-5} \lambda_{02}^{5} \exp\left(\frac{V_1}{kT_c}\right) \qquad (14\text{-}13)$$

where ν_{12} and ν_{02} are the central frequencies of the transitions and k is the Boltzmann constant. By substitution into equation (14-11),

$$\frac{F_{20}}{F_{21}} = 10 \exp\left[5040 V_1 \left(\frac{1}{T_c} - \frac{1}{T_f}\right)\right] \qquad (14\text{-}14)$$

Thus when the intensity of the anti-Stokes fluorescence equals that of the direct-line fluorescence the colour temperature of the source will be equal to the flame temperature. If the colour temperature of the source could be varied continuously a value of unity for F_{20}/F_{21} could be achieved. In practice several values of F_{20}/F_{21} can be obtained by changing the filter transmission characteristics and these values may be used to construct a graph of F_{20}/F_{21} against $1/T$; the straight line obtained may then be extrapolated to $\ln(F_{20}/F_{21}) = 0$.

(c) If the temperature of one flame is known by independent measurement, the measurement of the ratio F_{20}/F_{21} provides a means of calculating the temperature of several other flames. If equation (14-11) is written for two temperatures T_{f_1} (known) and T_{f_2} (unknown), we may simplify to

$$\log\left(\frac{F_{20}}{F_{21}}\right)_{T_{f_1}} \Big\backslash \log\left(\frac{F_{20}}{F_{21}}\right)_{T_{f_2}} = \frac{5040 V_1}{T_{f_2}} - \frac{5040 V_1}{T_{f_1}} \qquad (14\text{-}15)$$

The use of this relationship then permits the calculation of T_{f_2}.

Methods of flame temperature measurement based on (a) and (b) require the accurate calibration of the relative efficiency of the monochromator and detector system at the wavelengths λ_{12} and λ_{02}, and the ratio of the spectral irradiances of the source must be measured experimentally. It should be pointed out, however, that a knowledge of the transition probabilities for the lines used is not required; this represents an advantage over many other methods. Omenetto and co-workers[111] described a fourth method of flame

TABLE 14-11

Values for electronic excitation temperatures of various premixed flames obtained by two-line atomic fluorescence methods (from ref. 111)

		Flame temperatures (K)			
		Fluorescence (thallium)			
Gas mixture	Method	(a)	(d)	Reversal (thallium)	Two-line (iron)[a]
Air Hydrogen	9·8 l/min 11·6 l/min	2070	2085	2090	2160
Air Acetylene (nitrogen-separated)	9·8 l/min 1·5 l/min	2435	(2440)[b]	2440	ND[c]
Air Acetylene	9·8 l/min 1·5 l/min	2540	2530	2540	2480
Nitrous oxide Acetylene (nitrogen-separated)	7·2 l/min 4·7 l/min	2860[d]	2870[d]	2850	2820
Nitrous oxide Acetylene	7·2 l/min 4·7 l/min	3085[d]	3090[d]	3070	3000

[a] Iron line pairs used: 3733·32/3734·87; 3820·43/3895·66 (air-hydrogen). 3722·56/3734·87 (air-acetylene). 3722·56/3749·49 (nitrous oxide-acetylene) after Kirkbright *et al.* (ref. 109).

[b] Temperature obtained by the "line reversal" method and used for the calculation of the other flame temperatures.

[c] ND = not determined. Consistent data could not be obtained.

[d] Philips metal vapour discharge lamp used as excitation source.

temperature measurement by AFS based on determination of the ratio F^*_{21}/F_{21}, where F^*_{21} is the number of $2 \rightarrow 1$ transition/s due to excitation involving thermal assistance between levels 0 and 1, and F_{21} is the number of direct $2 \rightarrow 1$ fluorescence transition/s due to excitation via $0 \rightarrow 2$. This results in an equation similar to that used in method (a) but requiring a knowledge of the *gf* values of the transitions studied. Method (c) does not require the separate calibration of the monochromator and detector assembly or the measurement of the source irradiance.

Omenetto and co-workers used concentrations of between 50 and 200 ppm of thallium for temperature measurements and did not find significant dependence of the measured temperatures on thallium concentration. Table 14-11 shows values of the thallium electronic excitation temperature obtained in various flames by the AFS methods (a) and (c) described above. Good agreement was obtained between these values and those from the line-reversal method with thallium and the two-line emission method with iron. The temperature values in Table 14-11 obtained by method (c) were calculated taking the line-reversal temperature of the nitrogen-separated air-acetylene flame as the known temperature. From the coefficient of variation of fluorescence measurements for air-acetylene and air-hydrogen flames (1·6 and 3% for F_{21} and 3% and 2·2% for F_{20} respectively) and the estimated precision in the instrumental calibration (*ca.* 5%) a possible overall precision of about 1-2% was claimed by the authors. The AFS methods probably require greater care in the instrument calibration and fluorescence measurements using the continuum source than does the AAS method described in Section 14.5.1.

REFERENCES

1. D. F. Boltz, in E. L. Grove and A. J. Perkins, "Developments in Applied Spectroscopy", Vol. 7A, p. 222, Plenum Press, New York, 1969.
2. M. Pinta, *Méthodes Phys. Anal.*, 1970, **6**, 268.
3. J. M. Ottaway, D. T. Coker and B. Singleton, *Talanta*, 1972, **19**, 787.
4. R. J. Jakubiec and D. F. Boltz, *Anal. Chem.*, 1969, **41**, 78.
5. Y. Yamamoto, T. Kumamaru, Y. Hayashi and Y. Otani, *Anal. Letters*, 1968, **1**, 955.
6. G. F. Kirkbright, A. M. Smith and T. S. West, *Analyst*, 1968, **93**, 292.
7. R. S. Danchik, D. F. Boltz and L. G. Hargis, *Anal. Letters*, 1968, **1**, 891.
8. G. F. Kirkbright, A. P. Rao and T. S. West, *Spec. Letters*, 1969, **2**, 69.
9. V. F. Barkovskii and T. V. Velikanova, *Russ. J. Inorg. Chem.* (English Transl.), 1970, **15**, 829.
10. *Idem*, ibid., 1966, **11**, 1230.
11. G. F. Kirkbright, A. M. Smith, T. S. West and R. Wood, *Analyst*, 1969, **94**, 754.
12. J. M. Ottaway, D. T. Coker and J. A. Davies, *Anal. Letters*, 1970, **3**, 385.
13. C. L. Chakrabarti and M. Katyal, *Anal. Chem.*, 1971, **43**, 1302.
14. R. J. Jakubiec and D. F. Boltz, *Anal. Letters*, 1968, **1**, 347.
15. H. N. Johnson, G. F. Kirkbright and T. S. West, *Analyst*, 1972, **97**, 696.
16. R. S. Danchik and D. F. Boltz, *Anal. Letters*, 1968, **1**, 901.
17. T. V. Ramakrishna, J. W. Robinson and P. W. West, *Anal. Chim. Acta*, 1969, **45**, 43.
18. G. Devoto, *Boll. Soc. Ital. Biol. Sper.*, 1968, **44**, 425.
19. A. M. Bond and T. A. O'Donnell, *Anal. Chem.*, 1968, **40**, 560.
20. H. Bartels, *At. Absorption Newsletter*, 1967, **6**, 132.
21. J. B. Ezell, Jr., *ibid.*, 1967, **6**, 84.
22. M. D. Garrido, C. Llagund and J. Carrindo, *Am. J. Enol. Viticult.*, 1971, **22**, 44.

23. E. D. Truscott, *Anal. Chem.*, 1970, **42**, 1657.
24. Y. Tofuku and S. Hirano, *Bunseki Kagaku,* 1971, **20**, 142.
25. H. Fujinuma, K. Kasama, K. Takeuchi and S. Hirano, ibid., 1970, **19**, 1487.
26. W. Reichel and L. Acs, *Anal. Chem.*, 1969, **41**, 1886.
27. U. Westerlund-Helmerson, *At. Absorption Newsletter,* 1966, **5**, 97.
28. W. J. Collinson and D. F. Boltz, *Anal. Chem.*, 1968, **40**, 1896.
29. Y. Yamamoto, T. Kumamaru, Y. Hayashi and Y. Otani, *Bunseki Kagaku,* 1968, **17**, 92.
30. T. Kumamaru, *Bull. Chem. Soc. Japan,* 1969, **42**, 956.
31. G. D. Christian and F. J. Feldman, *Anal. Chim. Acta,* 1968, **40**, 173.
32. A. M. Bond and J. B. Willis, *Anal. Chem.*, 1968, **40**, 2087.
33. E. Jungreis, *Israel J. Chem.*, 1969, **7**, 583.
34. R. S. Danchik and D. F. Boltz, *Anal. Chim. Acta,* 1970, **49**, 567.
35. *Idem, Anal. Chem.*, 1968, **40**, 2215.
36. T. Kumamaru, E. Tao, N. Okamoto and Y. Yamamoto, *Bull. Chem. Soc. Japan,* 1965, **38**, 2204.
37. Y. Yamamoto, T. Kumamaru, Y. Hayashi and Y. Otani, *Bunseki Kagaku,* 1969, **18**, 359.
38. M. E. Houser and M. I. Fauth, *Microchem. J.*, 1970, **15**, 399.
39. G. F. Kirkbright, A. M. Smith and T. S. West, *Analyst,* 1967, **92**, 411.
40. T. R. Hurford and D. F. Boltz, *Anal. Chem.*, 1968, **40**, 379.
41. G. Devoto, *Boll. Soc. Ital. Biol. Sper.*, 1968, **44**, 424.
42. G. Linden, S. Turk and B. Tarodo de la Fuente, *Chem. Anal.*, 1971, **53**, 244.
43. J. A. Parsons, B. Dawson, E. Callahan and J. T. Potts, Jr., *Biochem. J.*, 1970, **119**, 791.
44. W. S. Zaugg and R. J. Knox, *Anal. Chem.*, 1966, **38**, 1759.
45. *Idem, Anal. Biochem.*, 1967, **20**, 282.
46. W. S. Zaugg, *At. Absorption Newsletter,* 1967, **6**, 63.
47. T. Kumamaru, Y. Otani and Y. Yamamoto, *Bull. Chem. Soc. Japan,* 1967, **40**, 429.
48. K. C. Singhal, A. C. Banerji and B. K. Banerjee, *Technology* (*Sindri*), 1968, **5**, 117.
49. K. C. Singhal and B. K. Banerjee, *Technology* (*Sindri*), 1968, **5**, 239.
50. J. R. Melton, W. L. Hoover, P. A. Howard and V. S. Green, *J. Assoc. Offic. Anal. Chemists,* 1971, **54**, 373.
51. H. K. Y. Lau and P. F. Lott, *Talanta,* 1971, **18**, 303.
52. L. A. Trudell and D. F. Boltz, *Mikrochim. Acta,* 1970, 1220.
53. R. W. Looyenga and C. O. Huber, *Anal. Chem.*, 1971, **43**, 498.
54. D. A. Roe, D. S. Miller and L. Lutwak, *Anal. Biochem.*, 1966, **15**, 313.
55. B. Magyar and F. Santos, *Helv. Chim. Acta,* 1969, **52**, 820.
56. R. Dunk, R. A. Mostyn and H. C. Hoare, *At. Absorption Newsletter,* 1969, **8**, 79.
57. J. A. Varley and P. Y. Chin, *Analyst,* 1970, **95**, 592.
58. F. Y. Borden and L. H. McCormic, *Soil. Soc. Am. Proc.*, 1970, **34**, 705.
59. G. G. Galindo, H. Applet and E. B. Schalscha, ibid., 1969, 33, 974.
60. S. A. Rose and D. F. Boltz, *Anal. Chim. Acta,* 1969, **44**, 239.
61. E. Jungreis and Z. Anavi, ibid., 1969, **45**, 190.
62. R. W. Looyenga and C. O. Huber, ibid., 1971, **55**, 179.
63. T. Kumamaru, Y. Hayashi, N. Okamoto, E. Tao and Y. Yamamoto, ibid., 1966, **35**, 524.
64. T. Kumamaru, ibid., 1968, 43, 19.

65. Y. Yamamoto, T. Kumamaru and Y. Hayashi, *Talanta*, 1967, **14**, 611.
66. Y. Yamamoto, T. Kumamaru, Y. Hayashi and Y. Otani, *Bull. Chem. Soc. Japan*, 1969, **42**, 1774.
67. H. K. L. Gupta and D. F. Boltz, *Microchem. J.*, 1971, **16**, 571.
68. D. G. Berge, R. T. Pflaum, D. A. Lehman, and C. W. Franck, *Anal. Letters*, 1968, **1**, 613.
69. F. J. Diaz, *Anal. Chim. Acta*, 1972, **58**, 455.
70. A. L. Potter, E. D. Ducay and R. M. Macready, *J. Assoc. Offic. Anal. Chemists*, 1968, **51**, 748.
71. J. C. Sheridan, E. P. K. Lau and B. Z. Senkowski, *Anal. Chem.*, 1969, **41**, 247.
72. J. Courtot-Coupez and A. Le Bihan, *Anal. Letters*, 1969, **2**, 567.
73. A. Walsh, *Proc. Xth Colloq. Spectrosc. Internationale*, p. 127, Hilger, 1962.
74. B. V. L'Vov and A. D. Khartsyzov, *Zh. Prikl. Spektr.*, 1969, **11**, 413.
75. G. F. Kirkbright and M. Marshall, *Anal. Chem.*, 1972, **44**, 1228.
76. G. F. Kirkbright, T. S. West and P. J. Wilson, *At. Absorption Newsletter*, 1972, **11**, 53.
77. B. V. L'Vov, "Atomic Absorption Spectrochemical Analysis", Hilger, London, 1970, p. 260.
78. K. R. Osborn and H. E. Gunning, *J. Opt. Soc. Am.*, 1955, **45**, 552.
79. H. Kirchof, *Spectrochim. Acta*, 1969, **24B**, 235.
80. W. H. Brimhall, *Anal. Chem.*, 1969, **41**, 1349.
81. A. N. Zaidel and E. P. Korennoi, *Opt. i. Spektrosk.*, 1961, **10**, 570.
82. A. N. Zaidel and E. P. Korennoi, *Zav. Lab.*, 1963, **29**, 1449.
83. D. C. Manning and W. Slavin, *At. Absorption Newsletter*, 1962, **1**, 39.
84. J. A. Goleb and Y. Yokoyama, *Anal. Chim. Acta*, 1964, **30**, 213.
85. J. A. Goleb, *Anal. Chem.*, 1963, **35**, 1978.
86. J. A. Goleb, *Anal. Chim. Acta*, 1966, **34**, 135.
87. B. V. L'Vov and V. I. Mosichev, *Zh. Prikl. Spektrosk.*, 1966, **4**, 491.
88. L. M. Fraser and J. D. Winefordner, *Anal. Chem.*, 1971, **43**, 1693.
89. J. D. Winefordner, *Accounts Chem. Res.*, 1969, **2**, 361.
90. M. B. Denton and H. V. Malmstadt, *Appl. Phys. Letters*, 1971, **18**, 485.
91. L. M. Fraser and J. D. Winefordner, *Anal. Chem.*, 1972, **44**, 1444.
92. N. Omenetto, N. N. Hatch, L. M. Fraser and J. D. Winefordner, *Anal. Chem.*, 1973, **45**, 195.
93. A. P. D'Silva, R. N. Kniseley, V. A. Fassel, R. H. Curry and R. B. Myers, *Anal. Chem.*, 1964, **36**, 532.
94. D. N. Hingle, G. F. Kirkbright and T. S. West, *Analyst*, 1969, **94**, 864.
95. R. N. Kniseley, C. C. Butler and V. A. Fassel, *Anal Chem.*, 1969, **41**, 1494.
96. T. J. McDrath, in "Modern Optical Methods in Gas Dynamic Research", ed. D. S. Dosanjh, Plenum Press, New York, 1971.
97. M. Hercher, *Appl. Optics*, 1967, **6**, 947.
98. M. Hercher, W. Chu and D. L. Stockman, *I.E.E.E. J. Quantum Elect.*, 1968, **QE4**, 954.
99. L. Huff and L. G. DeShafer, *J.O.S.A.*, 1970, **60**, 157.
100. F. Gires, *I.E.E.E. J. Quantum Elect.*, 1966, **QE2**, 624.
101. C. R. Guiliano and L. D. Hess, *I.E.E.E. J. Quantum Elect.*, 1967, **QE3**, 358.
102. N. Omenetto, P. Benetti, L. P. Hart, J. D. Winefordner and C. Th. J. Alkemade, *Spectrochim. Acta*, 1973, **28B**, 289.
103. J. D. Winefordner, *Proc. 4th International Conf. on Atomic Spectroscopy*, Toronto, 1973.

104. E. H. Piepmeier, *Spectrochim. Acta,* 1972, **27B**, 431.
105. E. H. Piepmeier, *Spectrochim. Acta,* 1972, **27B**, 445.
106. N. Omenetto, L. P. Hart, P. Benetti and J. D. Winefordner, *Spectrochim. Acta,* 1973, **28B**, 301.
107. A. G. Gaydon and H. G. Wolfhard, "Flames, Their Structure, Radiation and Temperature", Chapman and Hall, London, 3rd ed., 1970.
108. R. H. Tourin, "Spectroscopic Gas Temperature Measurements", Elsevier, Amsterdam, 1966.
109. G. F. Kirkbright, M. Sargent and S. Vetter, *Spectrochim. Acta,* 1970, **25B**, 465.
110. R. F. Browner and J. D. Winefordner, *Anal. Chem.,* 1972, **44,** 247.
111. N. Omenetto, P. Benetti and G. Rossi, *Spectrochim. Acta,* 1972, **27B**, 453.
112. C. T. J. Alkemade, *Pure Appl. Chem.,* 1970, **23**, 73.

Subject Index

A

B

D

E

G

H

I

J

K

L

M

Q

R

S

U

V

W